Living Dangerously

Living Dangerously

THE EARTH, ITS RESOURCES, AND THE ENVIRONMENT

Heinrich D. Holland and Ulrich Petersen

PRINCETON UNIVERSITY PRESS · PRINCETON, NEW JERSEY

Published by Princeton University Press, 41 William Street,
Princeton, New Jersey 08540
In the United Kingdom: Princeton University Press,
Chichester, West Sussex

Library of Congress Cataloging-in-Publication Data

Holland, Heinrich D.
Living dangerously : the earth, its resources, and the
environment / Heinrich D. Holland and Ulrich Petersen.
p. cm.
Includes bibliographical references and index.
ISBN 0-691-03266-1 (pbk : alk. paper)
1. Earth sciences. 2. Natural resources. 3. Human ecology.
I. Petersen, Ulrich, 1927– . II. Title.
QE28.H74 1995 333.7—dc20 95-12429

This book has been composed in Times Roman

Princeton University Press books are printed on acid-free paper and meet the
guidelines for permanence and durability of the Committee on Production Guidelines
for Book Longevity of the Council on Library Resources

Printed in the United States of America

10 9 8 7 6 5 4 3 2 1

We dedicate this book to our grandchildren with much love and great expectations

---

Benedict Merwyn Holland • Carl Simon Bourque

Esther Ruth Holland • Christine Eileen Bourque

Alexander Michael Petersen • Stephanie Anh Bourque

Mark Edmund Petersen • Alexander Minh Bourque

Thomas Marten Zu-Chuan Loh • Andrew Tien Bourque

Sebastian Peter Chien-Chuan Loh • Ryan Frederic Lahey

Francis Erik Fu-Chuan Loh • Dylan Patrick Lahey

Contents

Preface

Some years ago, during the discussion of the proposed Core Curriculum at Harvard, one of our colleagues rose to inquire about the purpose of the new program. Was it to reduce the level of ignorance? If so, he ventured, the new program was not worthwhile, because, as he looked around the faculty room, he could plainly see that ignorance was no impediment to success. In spite of this indictment, the faculty approved the Core Curriculum. Our course, Science B-34, "The Earth, Its Resources, and the Environment," has been part of its offerings since 1986. The course was designed for students who were not intending to concentrate in the Earth Sciences. It seems to us that these students, like all educated persons, should understand how the Earth works, how renewable and nonrenewable natural resources are formed and distributed, and how humanity is transforming the Earth. We wished our students to be able to judge whether we are apt to run out of natural resources during the next century, and whether we are apt to "drown in our own garbage."

A textbook is a useful adjunct for such a course. There are a number of excellent introductory geology textbooks; there are also some fine introductions to natural resources, and there is no dearth of introductions to environmental science. However, we know of no book that combines all three subjects. For some years we have asked our students to read relevant portions of several textbooks, but this method did not prove very successful; we therefore made the fateful decision to write our own text. By the time this volume is published, four years will have been spent in its preparation. The effort has been considerable, but so have the rewards. We have been able to assemble materials that we deem important, to raise questions that we consider critical, to juxtapose the need for natural resources with the need for a livable environment, and to venture an assessment for the likely effects of natural resource depletion and environmental change on humanity during the next century.

And so we are launching this book. Its sea trials in manuscript form have been auspicious. One of our students even went so far as to say that the book is not as dry as most textbooks, and our colleagues have been generally enthusiastic. We are therefore encouraged to believe that the book in its final form will have a successful voyage. It is a rather full book. In a sense, we have compressed three books into one, and we have included enough data to support most of our major contentions. Some readers may feel that there are too many facts, somewhat like the Emperor Joseph II in *Amadeus*, who felt that there were too many notes in *The Abduction from the Seraglio*. Although, like Mozart, we believe in the usefulness of the included material, the book can surely be used in abbreviated form, tailored to the needs of particular courses. We have assumed a knowledge of mathematics and chemistry at the high school level because their use, together with that of minor doses of physics and biology, is necessary for the understanding of many of the subjects we cover and the issues we raise.

The first chapters were written during the first part of H.D.H.'s sabbatical leave at The Pennsylvania State University. We owe the title of our book to Judy

Kiusalas of Penn State, who felt that what has become the subtitle was entirely too stodgy for a main title. James F. Kasting and Lee R. Kump reviewed the first chapters and supplied much needed encouragement. Deb Detweiler shepherded the Penn State portion of the manuscript through its first typing.

The next group of chapters was written during the second part of H.D.H.'s sabbatical at the University of Hawaii in Manoa Valley, Honolulu. Chapter 6 in particular was heavily influenced by this beautiful locale. The chapters written in Hawaii benefited greatly from critical reviews by Ralph Moberly, Fred T. Mackenzie, and George P. Walker, and by the secretarial help of Alice Daniels and Leona Anthony.

The last six chapters were written at Harvard and have been seen through many alterations and revisions by the devoted attention of Pauline Solomon and Robert Quirk. Of the students who have taken the time to think critically about the manuscript, the contributions of Beth McPherson, Albert Colman, and Phillip Kuo have been particularly helpful.

Among outside reviewers the thoughtful comments of Kenneth S. Deffeyes, Yuk L. Yung, Michel Meybeck, Brian J. Skinner, and W. Gary Ernst have been taken much to heart. The most detailed, critically sharp, and immeasurably helpful reviews were prepared by two former students: Mark Logsdon and Mark A. McCaffrey. Without their hours of careful reading and commenting, many passages would still be obscure, incomplete, or just plain wrong. We owe them an enormous debt of thanks.

Previously published figures and plates are reprinted with the permission of the respective publishers, authors, and photographers as cited in the captions and references (for figures) and list of credits (for plates).

Finally, we wish to thank our friends at Princeton University Press. Alice Calaprice edited the manuscript with care and cheer, Sara Van Rheenen bent many of the rules charmingly at the Press to speed publication, Christopher Brest improved the figures immeasurably, and the production department converted the mass of manuscript into an elegant book.

Cambridge, Massachusetts, February 1995

Living Dangerously

1 Introduction

We have known the shape and the size of the Earth for more than 2,000 years, and we have known our place in the solar system for more than 300 years. But not until the astronauts viewed the Earth during the lunar expeditions were these facts engraved on our consciousness. We finally saw the Earth as a whole, and we saw that its individual parts are inextricably connected. Political and economic developments during the past 20 years have reinforced this view. Many of our natural resources are now transported across the globe, and political developments in countries where natural resources are produced can have serious repercussions the world over. Pollution also knows no political boundaries. The burning of fossil fuels and deforestation are increasing the carbon dioxide content of the atmosphere everywhere. The release of chlorofluorocarbons into the atmosphere, largely by the industrialized nations of the Northern Hemisphere, is responsible for the ozone hole in the Antarctic. Nuclear contamination from the reactor accident at Chernobyl in the USSR in 1986 blanketed much of Europe with radioactive fallout and increased levels of radioactivity over the entire planet.

These events, together with the rapid rise in population, have increased the uneasiness of the world's people regarding the future availability of natural resources and the effects of pollution on the habitability of our planet. There are many disparate voices in the land, some very strident. *The Population Bomb* by Paul Ehrlich (1971), *Limits to Growth* by Meadows et al. (1972), *Global 2000* by a committee commissioned by the Carter administration (1980), and *Earth in the Balance* by Al Gore (1992) are rather pessimistic. Their views have been countered by a group of much more optimistic books that include Simon and Kahn's *The Resourceful Earth* (1984), Dixy Lee Ray's *Environmental Overkill* (1993), and Ronald Bailey's *Eco-Scam* (1993).

This book contains the authors' views of these contentious matters. It has grown out of a course that was designed to introduce undergraduate students to those parts of the Earth Sciences that are, or should be, of interest to everyone. It is not a traditional geology text, because we have left out much that is generally covered in such books. Neither is it in the tradition of books devoted to natural resources, nor in the tradition of books dedicated to environmental science, because we believe that the availability of natural resources and the effects of their use on the environment are both matters of great importance.

1.1 Our Home Planet as a System

The abundance and the distribution of renewable and nonrenewable natural resources can best be understood in terms of the operation and the history of the entire Earth system. The availability and the distribution of renewable resources such as food and water largely reflect the present state of the system. Nonrenewable resources have accumulated during much of Earth history; their distribution therefore reflects the evolution of the Earth system during its entire history. The same strong link exists between environmental problems and the operation of the Earth system. Humans are disturbing the Earth and will probably disturb it even more in the future. The response of the Earth to all of this can be understood only if we know how the Earth system works.

The Earth is not a system in the sense of a political, philosophical, or mathematical system. It is much closer in kind to digestive systems, heating systems, and plumbing systems. These are all complex arrangements of many component parts, whose interactions involve a wide range of physical, chemical, and biological processes. The components of the Earth system range in size from subatomic particles to air and ocean currents that traverse much of the globe. Among other things, the system contains some 5 billion-plus people—each one of them a complex world of their own—and millions of other kinds of living organisms, including several hundred thousand species of beetles.

The Earth system is powered by external and internal sources of energy. The Sun is by far the most intense of these energy sources. Solar energy drives most of the processes at and above the surface of the Earth. The flux of solar energy is some five thousand times more intense than the flux of energy from the interior of the Earth, which is derived from the cooling of the initially hot planet and from the radioactive decay of naturally occurring uranium, thorium, potassium, and other radionuclides that are present in trace quantities in the Earth's interior. Although the supply of internal energy is so much smaller than the supply of external energy, the Earth's internal energy drives all of the interior motions of the planet. These give rise to volcanoes, earthquakes, mountain building, the movements of the continents, and the Earth's magnetic field.

Attempts to predict the behavior of this complicated system require a certain amount of optimism and gall. Fortunately, some parts of the Earth system are now quite well understood; but others are not. Weather forecasts for more than a few days, for instance, are still virtually impossible, because very small disturbances during a given day can grow enormously in a matter of four or five days. Similar degrees of uncertainty beset forecasts for the climatic effects of the large quantities of carbon dioxide that are being added to the atmosphere by fossil fuel burning and deforestation.

Many of the current formulations of the Earth system involve the use of box models. The Earth is divided into reservoirs, "boxes," that are taken to be chemically homogeneous and can be treated as units responding fairly simply to a variety of inputs. Boxes in models are connected by lines that represent inputs and outputs (see fig. 1.1 and later chapters). The response of the whole system depends on the nature of these connections. If they are well defined, we can predict whether a system will be stable or unstable, whether it will oscillate or vary randomly, whether it will respond quickly or slowly to disturbances. One of the challenges for scientists concerned with the Earth system is to define these connections well enough, so that we can predict the response of the Earth as a whole to the massive changes that are being imposed on it by humanity.

1.2 Natural Resources

The energy stored in fossil fuels is a minuscule fraction of the total quantity of solar energy that has bathed our planet during the last 600 million years. We are releasing this stored energy at a prodigious rate, far in excess of the rate at which energy is being stored by the formation of new fuels today. Fossil fuels are there-

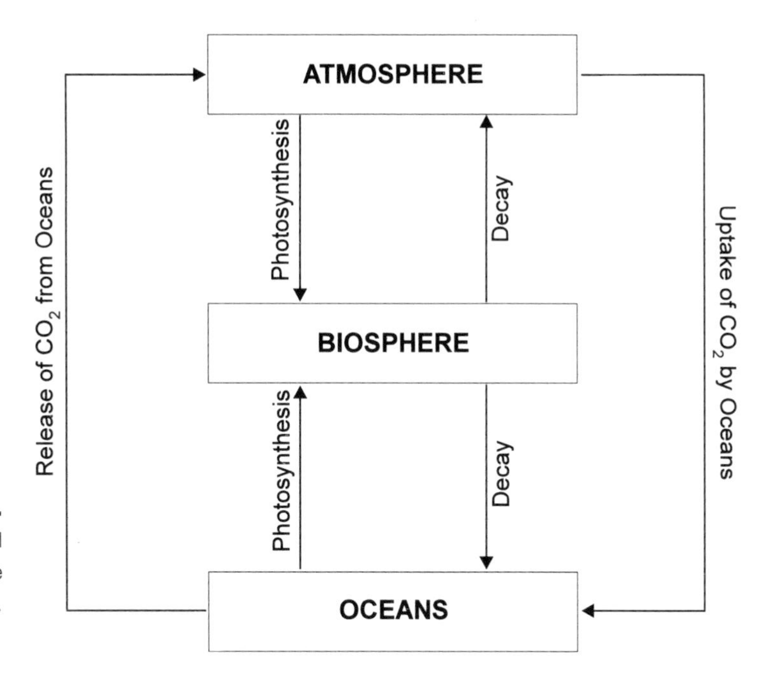

Figure 1.1. A simplified box model of a part of the carbon cycle.

fore being depleted. Once they have been used, they are no longer available. They belong to a large group of nonrenewable natural resources whose quantity and distribution are critical for the economy of the world and for the course of international politics. The magnitude of the world inventory of fossil fuels and their rate of use will determine how long humanity can count on this source of energy. In the course of the book we will try to estimate how long it will be before we "run out," and whether we will then have to "freeze in the dark."

Fossil fuels belong to the group of nonmetallic natural resources that also includes building and construction materials, fertilizers, and salt. Some nonmetallic natural resources have a property they share with metallic natural resources but not with fossil fuels: they can be recycled. But to what extent will recycling be able to slow the depletion of these nonrenewable resources? Will we run out of them? Will the grade of their mined ores become so low that the resources will become inordinately expensive, or will improvements in technology offset the foreseeable decreases in ore grades?

Renewable resources are the second great class of natural resources. They differ from nonrenewable resources in that they are replenished on a timescale comparable to the timescale of their use. Crops are planted and harvested each year. Water for hydroelectric power flows continuously and can be reused by power plants farther downstream. The oxygen that we breathe is regenerated continuously by plant photosynthesis. The distinction between renewable and nonrenewable resources is not completely sharp. Forests can be replanted and reharvested after they are cut, but old-growth forests require more time and patience to regrow than modern society is willing to muster. Overgrazing can destroy pastureland, and it may take a long time and a good deal of effort to restore land to productivity where erosion and nutrient loss have been severe. Even water is sometimes nonrenewable. In arid areas where groundwater is pumped extensively, the rate of groundwater renewal is frequently much slower than the rate of groundwater extraction. In such areas groundwater is effectively being mined much like oil and natural gas, and must be classed as a nonrenewable resource. During the next century, the most seriously limiting renewable resources will almost certainly be food and water. It is fair to ask, and it is important to know, whether there will be enough food and water to support the world's burgeoning population at a humane level. Will the world be more hungry or less in the year 2100 than it is now?

1.3 The Environment

The welfare of the human family also depends on the state of its environment. Humanity has literally changed the face of the Earth. Forests, grasslands, and deserts have been replaced by farms, villages, towns, cities, and highways. The composition of the atmosphere, of rivers, and even of the oceans has been affected. The explosive growth of world population during the twentieth century and the rapid increase in economic well-being since World War II have increased the rate of waste production enormously, and have added to their variety and toxicity.

Environmental problems are legion, and they are worth classifying. Perhaps it is most useful to classify them by their extent, their intensity, and their duration. Many environmental problems are local: small, toxic garbage dumps and localized sources of water pollution come to mind. Some environmental problems are regional; the fallout of acid rain from power plants that burn high-sulfur coal can affect areas 1,000 km from their source. Still other problems are global; CO_2 added to the atmosphere by fossil fuel burning and chlorofluorocarbons from refrigeration units mix through the entire atmosphere of the planet on a timescale of a few years. The policy implications of environmental problems at these disparate length scales are obviously quite different.

Environmental problems also vary greatly in intensity. Water pollution in the developed countries can create serious problems, but they are rarely fatal. Water pollution in developing countries is arguably the single most important environmental problem facing the world today. Waterborne diseases probably account for the death of more than 5 million children annually, and are responsible for a world of misery.

Environmental problems also vary greatly in duration. The flow of rivers tends to flush out contaminants. Rivers that have become devoid of fish due to industrial contamination can become repopulated within a few years after contamination has stopped. On the other hand, the flow of most groundwater is extremely slow. Many areas of groundwater contamination are not self-cleaning on a timescale of a few years. The technical and policy approaches to the remediation of heavily contaminated groundwater systems are therefore quite different from the approaches to decontaminating heavily contaminated rivers.

We are faced with so many environmental problems that it is surely worth inquiring to what extent we are fouling our nest, and whether we are apt to drown in our own garbage. Will we destroy the diversity of the biosphere? Will the remedies for healing the ills of large, technologically advanced populations be benign, or will they generate even greater evils? The answers to many of these questions are not clear, but we can make a start. The glass is dark, but it is not opaque.

References

Bailey, R. 1993. *Eco-scam: The False Prophets of Ecological Apocalypse*. St. Martin's Press, New York.

Barney, G. O., ed. 1980. *The Global 2000 Report to the President*: vol. 1, *Entering the Twenty-first Century*; vol. 2, *The Technical Report*; vol. 3, *Documentation on the Government's Global Sectoral Models: The Government's Global Model*. Penguin Books, New York.

Ehrlich, P. R. 1969. *The Population Bomb*. Ballantine Books, New York.

Gore, A. 1992. *Earth in the Balance: Ecology and the Human Spirit*. Houghton Mifflin, Boston.

Meadows, D. H., Meadows, D. L., Randers, J., and Behrens, W. W. III. 1972. *The Limits to Growth: A Report for the Club of Rome's Project on the Predicament of Mankind*. Universe Books, New York.

Ray, D. L., and Guzzo, L. 1993. *Environmental Overkill: Whatever Happened to Common Sense?* Regnery Gateway, Washington, D.C.

Simon, J. L., and Kahn, H., eds. 1984. *The Resourceful Earth: A Response to Global 2000*. Basil Blackwell, New York.

2 Solar Energy

2.1 Introduction: From Nuclear Reactions to Sunlight

The Sun supplies nearly all of the Earth's external energy and is responsible for much that goes on in the atmosphere, at the Earth's surface, and in the oceans. As the major external driver for the Earth system, the Sun deserves a chapter in this book; unfortunately, descriptions of solar energy sources and of the solar spectrum involve a good deal of physics. Some readers may find this somewhat unsettling, perhaps even downright unpalatable. They may feel that, like the multiplication tables, such material is to be committed to memory and left there. The authors feel otherwise. Nuclear reactions are the source of the Sun's energy; they are also the source of much of the Earth's internal energy. They supply a significant percentage of the world's electrical energy today, and they may become the dominant source of electrical energy during the next 200 years. Some of the solar physics described in this chapter will therefore be handy when reading chapters 11 and 12.

A similar argument can be made for the description of the solar spectrum, i.e., the manner in which energy is radiated from the Sun. The Earth intercepts a minuscule part of that energy. Different parts of the solar spectrum interact differently with the Earth's atmosphere. One part gives rise to the photochemistry of the atmosphere, a second is responsible for the greenhouse effect. Photosynthesis, the basis of nearly all life, depends on a third part of the solar spectrum. None of these important aspects of the Earth system can be understood without delving into the mysteries of the nature of light. So, please bear with us, or—failing the necessary patience—continue on, and return when you meet material that builds on the contents of this chapter.

2.2 The Source of Solar Energy

The discovery that the Earth and the Sun are several billion years old (see chapter 5) proved that chemical reactions and gravitational collapse cannot possibly be responsible for the Sun's energy output. Only the enormous energy of nuclear reactions can account for the continuous luminosity of the Sun, i.e., the output of solar energy. During the 1930s, Weizsäcker, Bethe, and Critchfield proposed most of the nuclear reactions that are now considered to be responsible for the Sun's energy output. Since then, laboratory measurements have defined the major parameters of these reactions. By far the largest part of solar energy is derived from the conversion of four hydrogen nuclei into a helium nucleus. At Earth surface temperatures, the element hydrogen occurs as molecules consisting of two hydrogen atoms. As shown in figure 2.1, hydrogen atoms can be represented by a nucleus and a single electron. In hydrogen molecules the electrons are shared by the two nuclei. This sharing creates the forces that bind the atoms together.

Most of the mass of atoms is concentrated in their nuclei. The nucleus of the simplest atom, ^{1}H, contains a single proton, which is a stable elementary particle

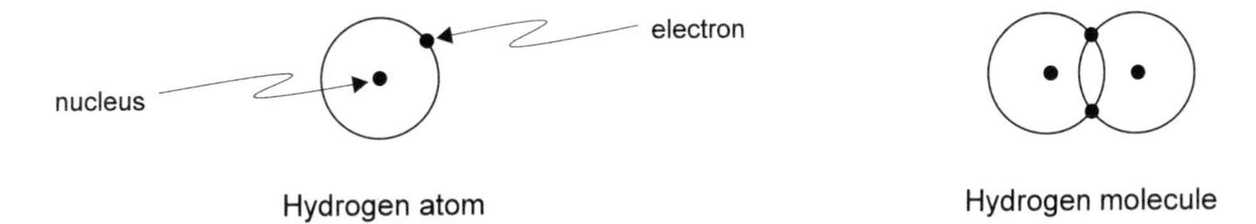

Figure 2.1. Highly schematic representation of a hydrogen atom and a hydrogen molecule.

of charge +1 and a mass of 1 on the atomic scale. The mass of a proton on the metric scale is 1.66×10^{-24} grams. The number of protons in one gram of hydrogen, 6.02×10^{23}, is usually referred to as *Avogadro's number.*

A small fraction of hydrogen on Earth has nuclei that contain one neutron in addition to one proton. Neutrons are electrically neutral particles and have a mass that is slightly greater than that of protons. The mass of nuclei containing one proton and one neutron is therefore close to 2 on the atomic scale. This type of hydrogen is usually represented either by the symbol ^2H or by the letter D, short for deuterium. The nuclei of an extremely small fraction of terrestrial hydrogen contain one proton and two neutrons. This type of hydrogen is usually represented by the symbol ^3H or as T, short for tritium. ^1H and ^2H are stable nuclei in Earth surface environments, tritium is not. It is radioactive, has a half-life of 12.5 years, and decays to an isotope of helium, ^3He. This means that 100 grams of ^3H are reduced to 50 grams in 12.5 years. After another 12.5 years, only 25 grams of the original 100 grams of ^3H remain. After another 12.5-year period, 12.5 grams of ^3H remain, and during each successive 12.5-year interval the quantity of ^3H decreases by another factor of two.

Tritium decays by the ejection of an electron from its nucleus. Electrons have a charge of −1, i.e., their charge is of the same magnitude but of opposite sign to that of protons. Their mass is only 1/1,850 that of protons. Neutrons can be regarded formally as particles consisting of a proton plus an electron. The charges of the two particles cancel, and the mass of the two together is barely larger than that of a proton. Free neutrons decay into a proton and electron with a half-life of about 15 minutes. The decay is rather complicated. A neutrino is also emitted during neutron decay. This is an energetic particle without charge and a mass that is either zero or very small, even compared to the mass of an electron. We shall return to the role of neutrinos later.

The three types of hydrogen nuclei, ^1H, ^2H, and ^3H, are called *isotopes* of hydrogen, i.e., they are nuclei with the same number of protons but with different numbers of neutrons. In hydrogen atoms the positive charge of the proton is balanced by the negative charge of one orbiting electron. Since the chemical properties of atoms depend almost entirely on the number of electrons associated with a nucleus, the chemistry of all of the isotopes of hydrogen is very similar.

If we add a second proton to a hydrogen nucleus, we must also add a second orbiting electron to balance the extra charge. The chemical properties of nuclei with two protons are therefore different from those of hydrogen. Atoms with nuclei containing two protons belong to the element helium, He. Only nuclei of helium that contain one or two neutrons in addition to two protons are stable, i.e., nonradioactive. These nuclides have masses of 3 and 4 and are designated ^3He and ^4He, respectively. The addition of a third proton and a third electron yields

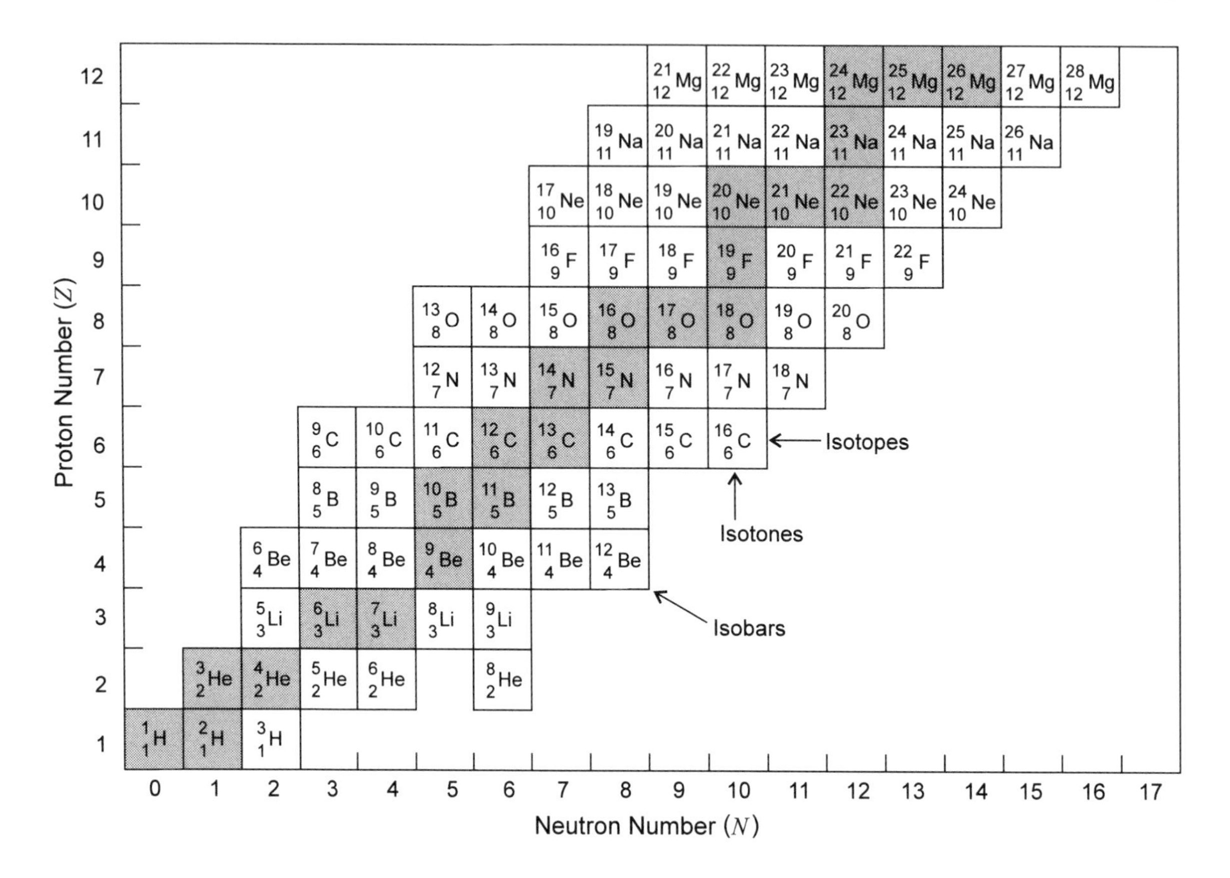

Figure 2.2. Chart of the light nuclides. Each square represents a particular nuclide, which is defined in terms of the number of protons (Z) and neutrons (N) that make up its nucleus. The shaded squares represent stable atoms; the white squares represent unstable, radioactive nuclides. Isotopes are atoms having the same Z but different values of N. Isotones have the same N but different values of Z. Isobars have the same value of (Z+N) but different values of Z and N. Isotopes are atoms of the same element and have nearly identical chemical properties.

the element lithium, Li, an element with two stable isotopes, ^{6}Li and ^{7}Li. Figure 2.2 shows the stable and some of the radioactive isotopes of elements 1 to 12.

Nearly all nuclei with proton numbers less than 83 have at least one stable isotope. As shown in figure 2.3 the entire set of stable nuclides falls along a single, thin band. In stable nuclei of the lighter elements, the ratio of neutrons to protons is approximately one. In stable nuclei with a large number of protons the ratio of neutrons to protons is greater than one and approaches 1.6 for the heaviest elements. None of the nuclides containing more than 83 protons have stable isotopes. Beyond uranium, the element with a proton number of 92, the half-life of all nuclides is short compared to the age of the Earth. Nuclides with proton numbers greater than 92 therefore occur only where they are being generated or have been generated recently. Large quantities of the transuranic elements are produced in nuclear power reactors. The half-life of some of the isotopes of these elements is long on the human timescale, and questions surrounding their safe

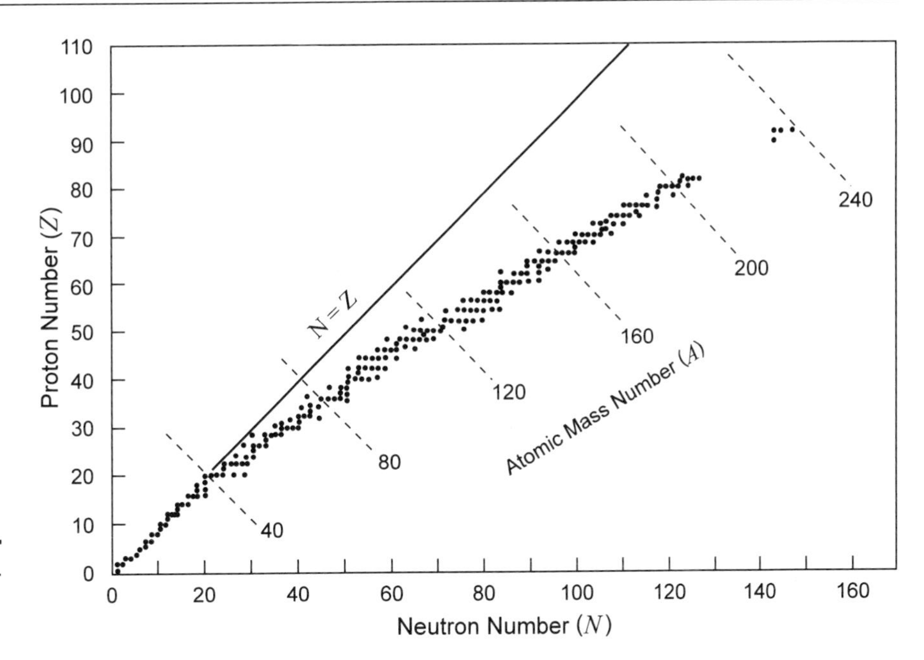

Figure 2.3.
Chart of the nuclides.

long-term disposal are an important part of the controversy regarding the use of nuclear energy (see chapter 11).

At high temperatures hydrogen tends to be present in atomic rather than in molecular form. As the temperature rises, the proportion of hydrogen atoms increases relative to that of hydrogen molecules. At very high temperatures, such as those that exist in the Sun, electrons tend to become separated from nuclei. Under these conditions hydrogen gas is a plasma consisting of a mixture of electrons and hydrogen nuclei.

The velocity of molecules, atoms, nuclei, and electrons in hydrogen gas increases with increasing temperature. At temperatures in the millions of degrees, hydrogen nuclei can collide with such force that they can overcome the repulsive force between particles of like charge, and react with each other. In the course of these reactions hydrogen is converted into helium. The overall reaction can be written

$$4\ {}^{1}H^{+} \rightarrow {}^{4}He^{2+} + 2e^{+} + 2\upsilon, \tag{2.1}$$

where e^{+} represents a positive electron (a positron), and υ represents a neutrino. Some of the reactions by which hydrogen is converted to helium in the Sun involve isotopes of lithium, Li; beryllium, Be; and boron, B. Others involve carbon, C; nitrogen, N; and oxygen, O, which act as catalysts in the conversion of hydrogen to helium (Parker 1986). The energy released during the conversion of hydrogen to helium is many orders of magnitude greater than the energy released by burning an equal mass of fossil fuels. Hydrogen-helium conversion is responsible not only for solar energy but for the destructive power of the hydrogen bomb. Controlled in a sustained fashion, this reaction could supply a virtually inexhaustible supply of fusion energy for human society (see chapter 11).

Although most of the nuclear reactions in the Sun seem to be well understood, a major puzzle has persisted since the 1970s. Neutrinos are produced during several of the likely nuclear reactions. Large experiments designed to detect these elusive particles have found a neutrino flux that is smaller than the predicted flux by a factor of two or more (McDonald 1992). The puzzle is potentially important for understanding the evolution of the Earth's atmosphere. Current models of solar evolution indicate that during the early history of the solar system the luminosity of the Sun was less than it is today, perhaps by as much as 30%. There is convincing evidence that the Earth's climate 3,800 million years ago was not very different from present-day climates (see, for instance, Holland 1984). If at that time the solar luminosity was only 70% of its present value, the composition of the atmosphere must have been rather different from today's atmosphere to prevent a general freeze-up of the earth. For people interested in the chemical evolution of the atmosphere, the outcome of the current neutrino flux experiments is therefore of more than passing interest.

2.3 The Solar Spectrum

The energy released by nuclear reactions in the Sun's interior is radiated into space. A very small fraction of the Sun's energy is intercepted by the Earth. Most of this energy arrives as electromagnetic radiation, i.e., light in the broadest sense of the term. Its interaction with the atmosphere, the biosphere, and the oceans drives most of the processes that take place at and close to the Earth's surface.

Light possesses properties of both waves and particles. Like ripples in a pond, light can be described in terms of its wavelength, frequency, and velocity. As shown in figure 2.4, the wavelength, λ, of a wave is the distance between successive crests or troughs. Waves in a pond tend to have wavelengths of a few tens of

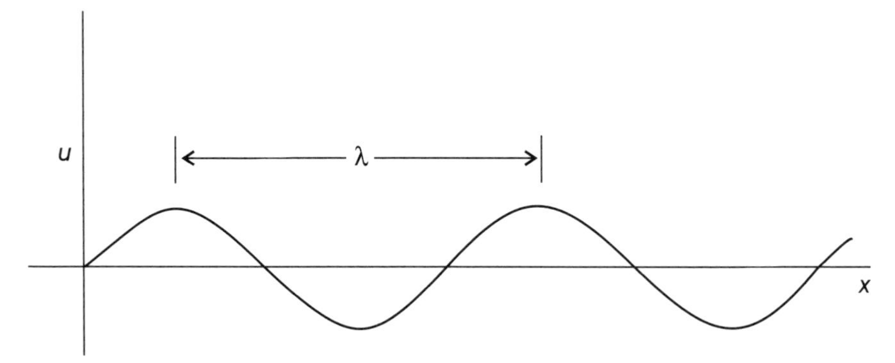

Figure 2.4. A symmetrical wave of wavelength λ.

centimeters. The frequency, f, of a wave is the number of wave crests that pass an observer in a unit of time. If you stand on a bridge over a pond into which a pebble has been dropped, you can determine the frequency of a passing wave by counting the number of crests that pass below you every minute. The velocity, c, of the wave is the product of its wavelength and its frequency:

$$c = \lambda \cdot f. \tag{2.2}$$

The range of wavelengths represented in solar light is enormous. The wavelength of radio waves emitted by the Sun is about 10^{12} (a million million) times greater than the wavelength of the soft X rays of the solar spectrum. The extreme

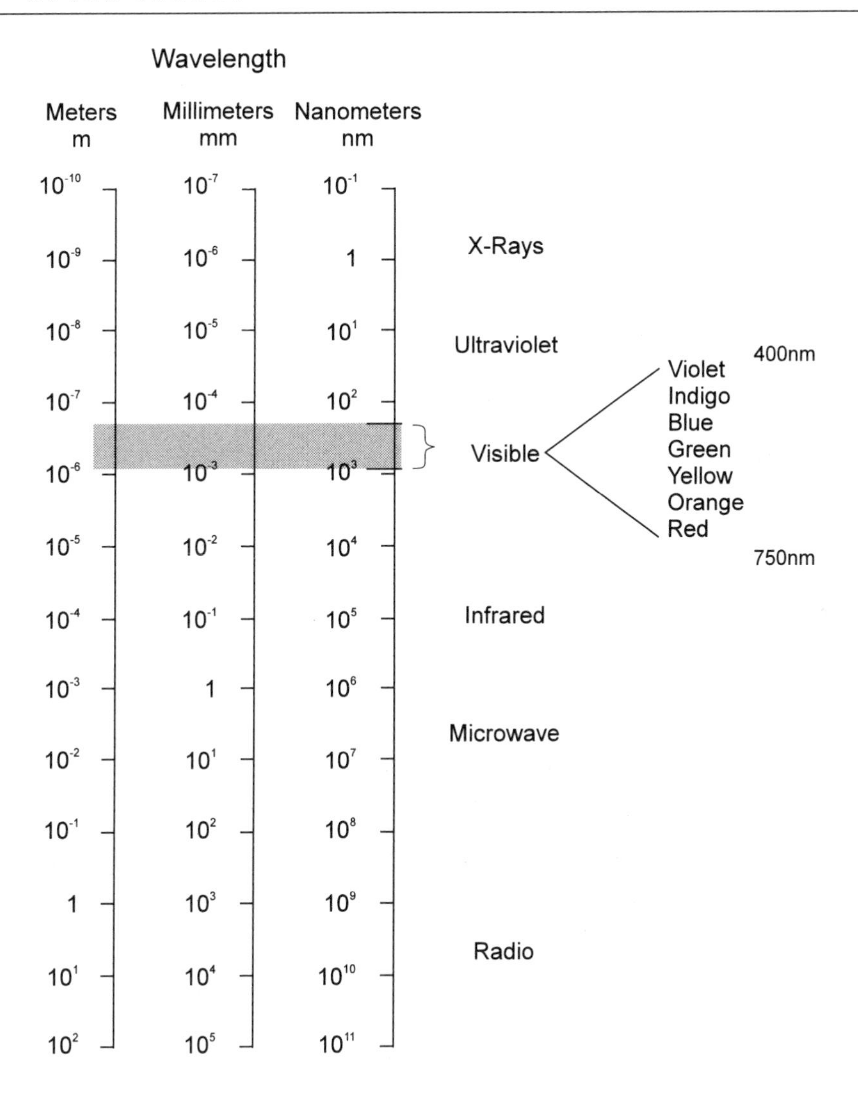

Figure 2.5. The electromagnetic spectrum.

parts of the solar spectrum and the intermediate parts are identified in figure 2.5. Three wavelength scales are shown along the left-hand margin of the figure: wavelengths in meters, in millimeters (10^{-3} m), and in nanometers (10^{-9} m). The wavelength of X rays and of ultraviolet, visible, and infrared light is usually expressed in nanometers (nm). The wavelength of radio waves is usually given in meters or millimeters. Normally, radio waves are described in terms of their frequency rather than their wavelength. Since the velocity of electromagnetic waves in a vacuum is 3.0×10^8 m/sec, the frequency of a radio wave with a wavelength of 10 meters is

$$f = \frac{3.0 \times 10^8 \text{ m/sec}}{10 \text{ m}} = 30 \times 10^6 \text{ sec}^{-1}.$$

A frequency of 1 sec^{-1} is usually called 1 hertz. A 10-meter radio wave therefore has a frequency of 30×10^6 hertz, or 30 megahertz (Mhz). As a point of refer-

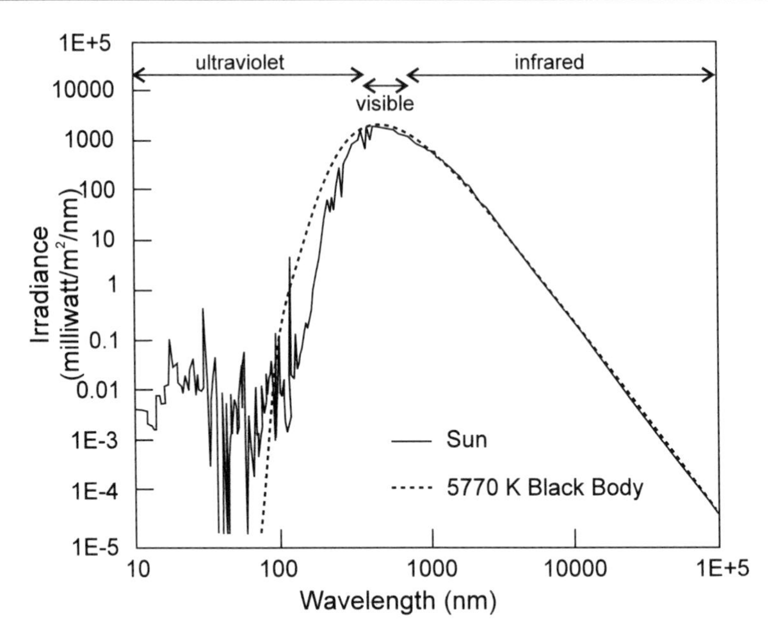

Figure 2.6. The Sun's spectral irradiance typical of solar minimum conditions compared with the spectrum of a blackbody radiator at 5,770 K (Lean 1991). UV = ultraviolet; VIS = visible; IR = infrared.

ence, FM radio station WGBH in Boston broadcasts on a frequency of 89.7 Mhz. Light that can be seen by the human eye occupies a very small part of the solar wavelength range. Violet light, at the extreme lower wavelength limit of visible light, has a wavelength of about 400 nm. Red light, at the extreme upper wavelength limit of visible light, has a wavelength of about 750 nm.

The intensity of sunlight varies greatly with wavelength. Most of the energy (intensity) of sunlight is concentrated in the visible part of the spectrum. Our eyes are therefore well adapted to detect sunlight. The intensity of the solar spectrum between 10 nm and 100,000 nm is shown in figure 2.6. The energy per nanometer of the wavelength spectrum reaches a maximum near 450 nm. The dashed curve in figure 2.7 represents the light spectrum emitted by a "black body" (i.e., a perfect radiator) at a temperature of 5,500°C. The two curves agree quite well, except in the UV part of the spectrum.

The shape of the solar spectrum at wavelengths both shorter and longer than the visible range is complicated and—in part—highly variable on a variety of timescales. Even the energy in the visible part of the spectrum is not completely constant. It is difficult to make precise measurements of the amount of solar

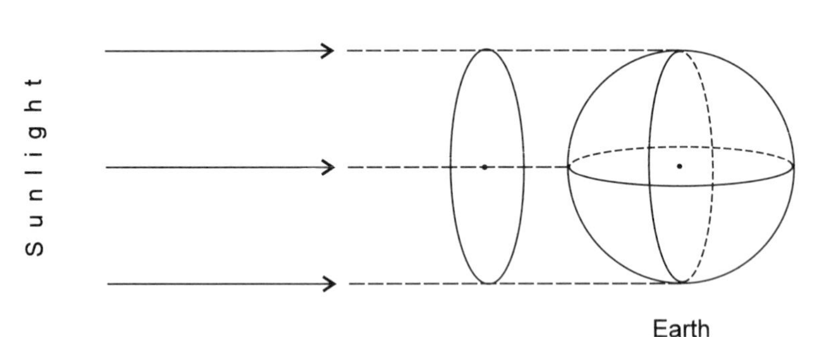

Figure 2.7. The intensity of sunlight falling on the Earth is 1.95 cal/cm^2 min of surface area perpendicular to the Sun-Earth direction. One calorie is the energy required to heat one gram of water 1°C.

energy received by the Earth at ground level, because sunlight is partially reflected and absorbed during passage through the atmosphere (see chapter 3). Measurements from space have now solved this problem. They have shown that solar luminosity varied by about 0.1% during the 1980s. Luminosity was highest when the number of sunspots was greatest, and lowest when the number of sunspots was smallest (Lean 1991). Variations of 0.1% in solar luminosity probably influence climate very little. There are, however, hints of larger variations in solar luminosity in the past. Stellar observations have shown that the luminosity of solar-type stars can vary by considerably more than 0.1%, and the record of ^{14}C production in the atmosphere together with historical records suggest that when the Sun gets quiet, the Earth gets colder (Wigley and Kelly 1990). This suggestion raises all sorts of interesting questions about the influence of variations in solar luminosity on world climate. How much of the warming trend during the past century has been due to changes in solar luminosity? Was the "little ice age" between about A.D. 1500 and 1800 caused by abnormally low solar luminosity? How about the major ice ages that have visited the Earth about once every hundred thousand years during the past 2 million years? And how is solar luminosity apt to vary in the future? At this point no one knows. We are very much in need of good techniques for reconstructing the history of solar luminosity and for predicting its future.

The energy in sunlight incident on the Earth today amounts to 1.95 cal/min per square centimeter of surface area measured perpendicularly to the line connecting the Sun and the Earth, as shown in figure 2.7. The total amount of sunlight received by the Earth is therefore 1.95 cal/min per square centimeter times the area of the disk that is intercepted by sunlight. Since the radius of the Earth is 6,371 km, i.e., 6.371×10^8 cm, the area A of the disk is

$$A = \pi r^2 = 3.14 \times (6.371 \times 10^8)^2 \text{ cm}^2 = 1.27 \times 10^{18} \text{cm}^2. \tag{2.3}$$

The total amount of solar energy, E, received by the Earth in one year is therefore

$$\begin{aligned} E &= 1.27 \times 10^{18} \text{cm}^2 \times 1.95 \frac{\text{cal}}{\text{cm}^2/\text{min}} \times 5.26 \times 10^5 \frac{\text{min}}{\text{yr}} \\ &= 1.30 \times 10^{24} \text{ cal/yr}. \end{aligned} \tag{2.4}$$

The energy received annually by each square centimeter of Earth surface is the total energy divided by the area $A_\oplus$ of the Earth. Since the area of the Earth is

$$A_\oplus = 4\,\pi r^2 = 5.08 \times 10^{18} \text{ cm}^2, \tag{2.5}$$

the energy received per square centimeter of the Earth's surface is

$$\frac{1.30 \times 10^{24} \text{ cal/yr}}{5.08 \times 10^{18} \text{ cm}^2} = 256{,}000 \text{ cal/cm}^2\text{/yr}.$$

This energy flux is about five thousand times greater than the flux of energy from the interior of the Earth. The large difference between the two fluxes explains why the Earth's internal energy plays only a small role in surface and near-surface processes, even though the Earth's internal energy is the dominant driver for the processes within the Earth that build mountains, cause earthquakes, fuel volcanoes, and move continents.

One calorie can heat one gram of water 1°C (but note that 1 dietary calorie equals 1,000 gram calories). About 100 calories are needed to heat one gram of water from its freezing point at 0°C to its boiling point at 100°C. Five hundred and fifty calories are required to convert one gram of water into steam. About one-quarter of the solar energy that reaches the Earth is used to evaporate water (see chapter 3). The average evaporation rate of water is therefore approximately

$$\frac{1}{4} \times \frac{256{,}000 \text{ cal/cm}^2\text{/yr}}{550 \text{ cal/gm } H_2O} = 116 \text{ gm } H_2O\text{/cm}^2 \text{ yr.}$$

Since the density of water is 1 gm/cm^3, 116 gm H_2O/cm^2 represents a column 116 cm, or a little less than 4 ft, in height. This is the water that is responsible for the hydrologic cycle of the Earth (see chapter 4).

2.4 The Flux of Particles from the Sun

The Sun consists largely of hydrogen and helium. Temperatures in the Sun's corona are so high that both of these elements can escape from the Sun's gravity field. Their outward flow is hindered by the Sun's strong magnetic field, but flow can take place from coronal holes where the magnetic lines of force are open (Stix 1989). The flux of particles has been dubbed the solar wind, but it is a strange wind indeed. Its velocity is much greater than winds on Earth, and its density is much smaller. In hurricanes on Earth, wind velocities can exceed 100 miles per hour. The solar wind has velocities between 100,000 and 1,000,000 miles per hour. On the order of 3×10^{19} molecules are present in each cubic centimeter of hurricane air, whereas there are only one to ten particles in one cubic centimeter of the solar wind. It might be preferable to dub the particle emissions from the Sun extremely high speed breezes.

Since the particles of solar wind are ejected from a very high temperature part of the Sun, the atoms of both elements are ionized. Their electrons are stripped off and accompany the nuclei as separate particles. The solar wind therefore consists largely of a stream of protons (nuclei of ^1H), alpha particles (nuclei of ^4He), and electrons. Most of the solar wind that is directed toward the Earth does not reach the surface of our planet. The Earth has its own magnetic field, and this deflects much of the stream of charged particles from the Sun. The solar wind can penetrate into the Earth's atmosphere only near the North and South poles. Their penetration there results in spectacular auroral displays. The solar wind may be adding hydrogen and helium to the Earth's atmosphere, but today the quantities of both elements added in this fashion are much smaller than the loss of these gases from the Earth's atmosphere by escape into interplanetary space. Early in Earth history the roles may well have been reversed; if so, the addition of hydrogen and helium from the solar wind may have had a significant effect on the chemistry of the early atmosphere.

References

Holland, H. D. 1984. *The Chemical Evolution of the Atmosphere and Oceans*. Princeton University Press, Princeton, N.J.

Lean, J. 1991. Variations in the Sun's Radiative Output. *Rev. of Geophys*. 29:505–35.

McDonald, A. B. 1992. Neutrino Astrophysics Experiments. *Physics in Canada* 48:120–26.

Mewaldt, R. A., Cummings, A. C., and Stone, E. C. 1994. Anomalous Cosmic Rays: Interstellar Interlopers in the Heliosphere and Magnetosphere. *EOS Transactions* 75:185, 193.

Parker, P. D. MacD. 1986. Thermonuclear Reactions in the Solar Interior. In *Physics of the Sun*, ed. P. A. Sturroch et al., chap. 2, vol. 1. D. Reidel, Dordrecht, Netherlands.

Stix, M. 1989. *The Sun: An Introduction*. Springer-Verlag, New York.

Wigley, T.M.L., and Kelly, P. M. 1990. In *The Earth's Climate and Variability of the Sun over Recent Millennia*, ed. J. C. Pecker and S. K. Runcorn. The Royal Society, London.

3 Sunlight and the Earth's Atmosphere

3.1 Introduction: Sunburn, Heat Blankets, and Ice Ages

The Sun is by far the largest energy source for the Earth. The previous chapter described the nuclear reactions that produce the Sun's energy and the transfer of a small fraction of this energy to the Earth as sunlight. This chapter deals with the interaction of sunlight with the atmosphere (plate 1). The many complex effects of sunlight on the Earth's atmosphere are subjects of intense research that is driven by much more than scientific curiosity. People are altering the composition of the Earth's atmosphere in ways that are affecting the level of ultraviolet radiation at the Earth's surface (see chapter 12), and that may have a significant effect on surface temperatures and on the global distribution of rainfall (see chapter 11). We need to know the consequences of the continued use of chlorofluorocarbons, the effects of fossil fuel burning on the Earth's climate, and the potential impacts of new and still untried chemicals on the environment. This chapter serves as an introduction to these problems. We will first take a look at the chemical composition and the temperature structure of the atmosphere, and then at the interaction of sunlight in different parts of the wavelength spectrum with the atmosphere.

3.2 The Composition of the Atmosphere

Our atmosphere consists of a few major components, several minor components, and a host of trace components. Many of these are listed in table 3.1. Nitrogen, N_2, and oxygen, O_2, are by far the most abundant constituents of the atmosphere. Together they account for 99% of the mass of the atmosphere. The high oxygen content of the atmosphere is unique in the solar system. Atmospheric oxygen is clearly the product of green-plant photosynthesis (see chapter 5), but the mechanisms that control the quantity of oxygen in the atmosphere are still poorly understood. Argon, one of the rare gases, is a distant third in atmospheric abundance. Argon owes its relatively high atmospheric concentration to the radioactive decay of a potassium isotope, ^{40}K, to an isotope of Argon, ^{40}Ar, in the solid Earth, and to the transport of ^{40}Ar from the Earth's interior into the atmosphere.

Among the minor components of the atmosphere, water vapor is the most abundant and the most variable. Its concentration in cold, dry regions can be as low as 40 ppmv (parts per million by volume). Its concentration in hot, humid regions can be as high as 40,000 ppmv, i.e., 4% by volume. This means that one million liters of very dry air contain about 40 liters of water vapor, and that one million liters of very damp air contain about 40,000 liters of water vapor. The great variability of the water content of air is, of course, related to the evaporation of water, largely from the oceans, and to its removal from the atmosphere as rain, snow, and ice (see chapter 4).

The concentration of carbon dioxide, CO_2, in the atmosphere is currently about 360 ppmv. It is the source of carbon in green-plant photosynthesis (see chapter 5) and is regenerated by the oxidation of organic matter. Its concentration

Table 3.1. The Composition of the Atmosphere

Component	Concentration
Major Components (concentration in percent by volume in dry air)	
Nitrogen, N_2	78.08
Oxygen, O_2	20.95
Argon, Ar	0.93
Minor Components (concentration in parts per million by volume, ppmv)	
Water vapor, H_2O	40–40,000
Carbon dioxide, CO_2	360
Neon, Ne	18.2
Helium, He	5.24
Methane, CH_4	1.7
Krypton, Kr	1.1
Trace Components (incomplete list; concentration in parts per billion by volume, ppbv)	
Hydrogen, H_2	550
Nitrous oxide, N_2O	330
Xenon, Xe	87
Carbon monoxide, CO	60–200
Ozone, O_3	10–30
Ammonia, NH_3	4–20
Formaldehyde, CH_2O	0–10
Nitric oxide, NO	1
Nitrogen dioxide, NO_2	1
Sulfur dioxide, SO_2	1–4
Chlorofluorocarbons	
F11 ($CFCl_3$)	0.18
F12 (CF_2Cl_2)	0.38
Carbon tetrachloride, CCl_4	0.13
Methyl chloride, CH_3Cl	0.6

Sources: Holland 1978; Warneck 1988; Rowland and Isaksen 1987.

in the atmosphere has been rising steadily during the past 200 years, partly due to deforestation and land clearance, but mostly due to the burning of fossil fuels (see chapter 11).

Methane, CH_4, is the only other carbon-containing gas among the minor components. It is the major constituent of natural gas, and is produced during the decomposition of organic matter in oxygen-free environments such as swamps and marshes (hence the name *marsh gas*). The concentration of methane, like that of CO_2, is increasing in the atmosphere. The increase is probably due to agriculture and to the release of natural gas. Higher methane concentrations could have a significant effect on the Earth's climate. The geochemistry of methane is therefore a topic of considerable interest.

The other three gases classified as "minor" in table 3.1—Ne, He, and Kr—are members of the rare-gas family. The abundance of these gases decreases in a general way with increasing atomic weight. The two exceptions are helium, the lightest of the rare gases, and argon, the third lightest. Argon is more abundant than expected, because ^{40}Ar has been produced by the decay of ^{40}K and degassed continuously during Earth history. Helium has been generated in comparable quantities by the radioactive decay of two uranium and one thorium isotope. The rarity of helium in the atmosphere compared to ^{40}Ar is due to the escape of He from the top of the atmosphere into interplanetary space, while the more massive atoms of ^{40}Ar are almost entirely retained in the atmosphere. Because of its escape, ^4He has been called the ultimate nonrenewable, nonrecyclable resource.

The concentration of the atmospheric trace components in table 3.1 is listed in parts per billion by volume. At a concentration of 1 ppbv, 1 liter of a gas is present in 1 billion liters of air, which is a volume roughly equal to that of an enclosed football stadium. The concentrations of many of the trace constituents in the atmosphere are highly variable, because the anthropogenic and natural processes that produce trace gases vary from place to place. The concentrations of carbon monoxide, CO, and ammonia, NH_3, for instance, are heavily influenced by local pollution. The concentration of sulfur dioxide, SO_2, is also strongly enhanced by human activities, but its concentration is also high in volcanic gases and hence in the vicinity of volcanoes. The level of ozone, O_3, in the upper atmosphere depends strongly on the interplay between natural photochemical processes and the effects of synthetic chlorofluorocarbon compounds (CFCs). In the lower atmosphere, ozone levels are largely determined by anthropogenic effects. Human disturbances are more noticeable in the concentration of the atmospheric trace constituents than in the concentration of the major and minor constituents, because even rather modest anthropogenic emissions can exert a significant effect on the concentration of trace components in the atmosphere (see chapter 12).

The atmosphere is reasonably well mixed up to a height of about 100 km, but there are changes in the concentration of a few trace components that are important for the absorption of solar ultraviolet radiation in the atmosphere. Above

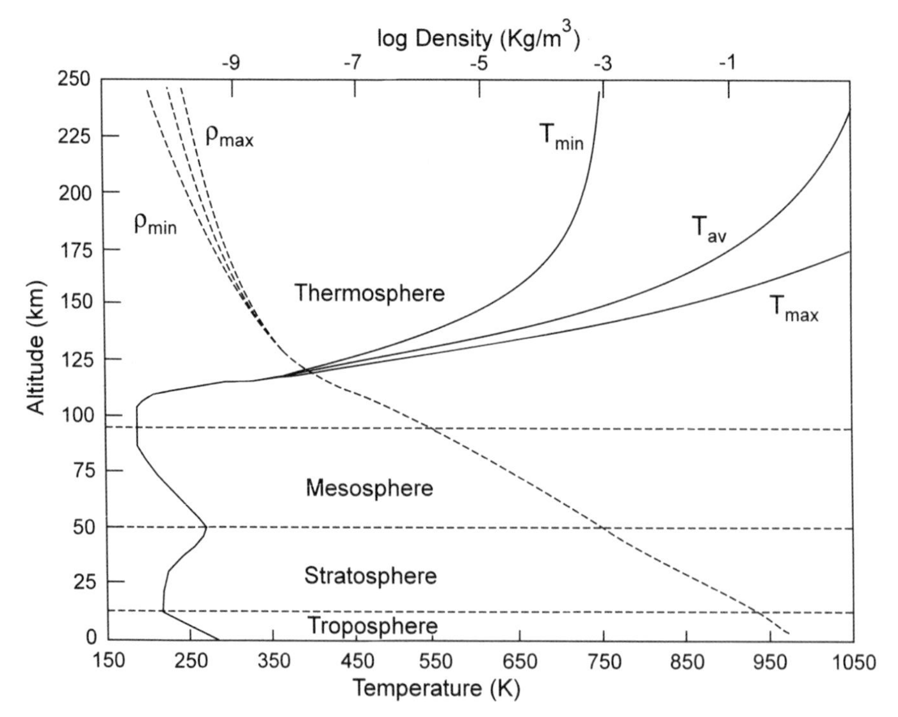

Figure 3.1. Temperature (T) and density (ρ) of the Earth's atmosphere from the surface to 250 km. Boundaries of the atmospheric regions, called "spheres," are based on inflections in the temperature profile, which is determined largely by solar radiative heating. Temperature and density in the thermosphere vary significantly according to the level of solar activity because of changes in the UV radiation absorbed at these altitudes; profiles are shown for maximum, average, and minimum solar activity. (Lean 1991)

100 km the atmosphere is extremely thin (see fig. 3.1) and chemically somewhat stratified. In this region the concentrations of the lightest gases, hydrogen and helium, gradually increase upwards at the expense of the concentration of the heavier gases, nitrogen, oxygen, and argon.

3.3 The Temperature Structure of the Atmosphere

The temperature structure of the lower atmosphere presents something of a paradox: the higher we rise in a balloon or the higher we climb on a mountain, the closer we are to the Sun; yet the temperature decreases during our ascent. Most long-distance jetliners fly at elevations between 30,000 and 40,000 ft, i.e., between 9 and 12 km. As passengers, we are not normally aware that the outside temperature during flight is close to a frigid –40°C, which happens to be equivalent to –40°F. The progressive decrease in temperature with elevation in the lower atmosphere is illustrated in figure 3.1. The temperature scale may require a bit of explanation. A temperature rise of 1°C is exactly equal to a temperature rise of 1 K. However, the zero point on the Centigrade scale is the freezing point of water, while the zero point on the Kelvin scale is absolute zero. Absolute zero corresponds to –273.16° on the Centigrade scale. The freezing point of water at 0°C is therefore +273.16 on the Kelvin scale. To convert a temperature in °C to K, simply add 273.16.

Figure 3.1 holds more surprises than a decrease in temperature with increasing elevation in the troposphere, the lowest 10 km of the atmosphere. At about 10 km—at the tropopause—the temperature ceases to decrease with increasing height. It rises again in the next region, the stratosphere, becomes constant in the stratopause close to a height of 50 km, then decreases again in the mesosphere up to about 80 km, and finally rises again in the thermosphere. At a height of 250 km, temperatures are on the order of a thousand degrees Kelvin and vary widely with the intensity of solar activity. The temperature at all levels of the atmosphere varies with latitude, with the time of day, and with the time of year.

3.4 The Absorption of Sunlight by the Atmosphere

The curiously complicated temperature structure of the atmosphere is largely due to heating, cooling, and convection of the atmosphere and of the Earth's surface by the absorption of sunlight. The interaction of light with atoms and molecules depends on its frequency. Many of the properties of light are best explained by considering it to consist of a stream of particles which carry energy but that have no mass when they are stopped. The energy of these particles, called *photons*, is proportional to the frequency (i.e., inversely proportional to the wavelength) of the associated light wave. A photon of ultraviolet light carries more energy than a photon of visible or infrared light. As might be expected, collisions of high-energy photons with atoms and molecules produce more violent effects than collisions of low-energy photons.

Sunlight contains a small proportion of photons that are sufficiently energetic to knock electrons out of atoms with which they collide. The absorption of the energy of these solar photons in the outermost reaches of the atmosphere is largely responsible for the very high temperature of the thermosphere. It also accounts for the presence of electrons in the thermosphere and for the positively charged atoms (ions) that are produced when electrons are removed from neutral atoms (see fig. 3.2). The thermosphere is almost coextensive with the ionosphere,

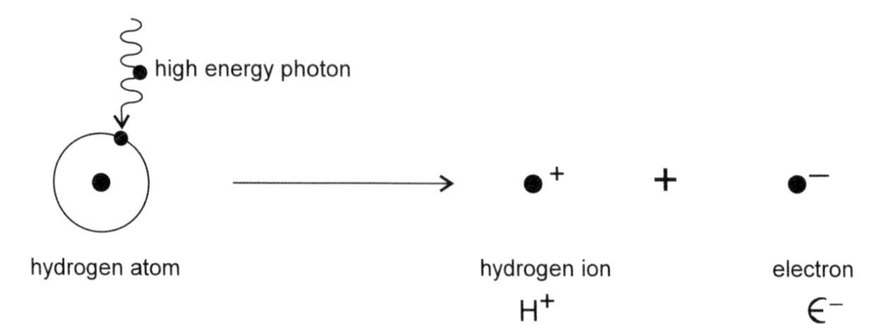

Figure 3.2. The conversion of a neutral hydrogen atom into a hydrogen ion and an electron by a high energy photon.

the region in which electrons and ions are abundant. Radio waves are reflected by ions in the ionosphere, and long-distance radio transmissions are made possible by these reflections. The variability of long-range radio transmissions is related to temporal changes in the distribution of electrons and ions in the ionosphere. Their distribution depends on the rate at which atoms are dissociated into electrons and ions, and on the rate at which these particles recombine to form neutral atoms. Above 60 km the energy available for ionization increases with elevation. However, the atmosphere also becomes progressively more rarefied, and the number of neutral atoms and molecules decreases rapidly, as shown in figure 3.3. This means that the number of collisions of high-energy photons with atoms and molecules decreases upward. It also means that electrons and ions collide and recombine less frequently at progressively greater altitudes. Electrons and ions are produced in greatest abundance near 180 km. However, the highest concentration of ions and electrons is found much higher, close to 300 km. There the ionization rate of atoms is lower, but the concentration of ions and electrons is higher, because the rate of recombination of electrons with ions is very much lower still.

Solar UV photons associated with light of wavelength less than about 120 nm are largely absorbed in the atmosphere at altitudes above 100 km (see, for instance, Banks and Kockarts 1973). Photons of solar radiation in the wavelength region 120–300 nm are not sufficiently energetic to ionize atoms and molecules in the atmosphere and are absorbed mainly in the mesosphere and stratosphere at altitudes between 20 and 80 km. The major absorption mechanisms involve the excitation of electrons within atoms and molecules and the dissociation of molecules. At low temperatures and in the absence of other excitements, elec-

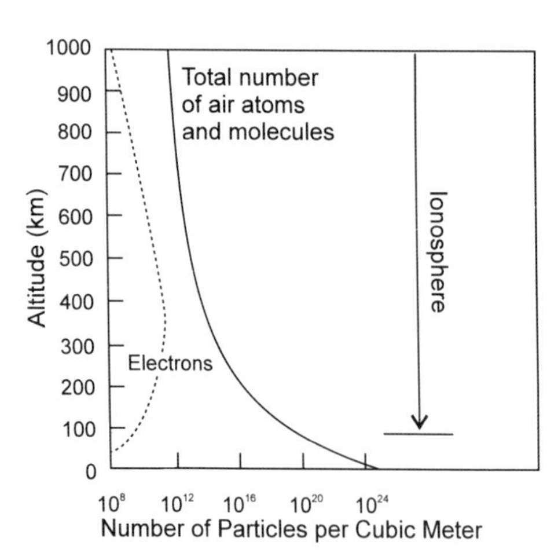

Figure 3.3. The total number of atoms and molecules in a cubic meter of atmosphere decreases exponentially with altitude. The number of electrons is extremely small below 60 km, rises to a maximum at about 300 km, and then decreases rapidly toward the outer reaches of the atmosphere. (Ahrens 1991)

trons in atoms tend to be in their "ground state," i.e., in orbits of lowest energy. Photons can be absorbed by such atoms, and can boost electrons into higher-energy orbits.

Nitrogen, the most abundant constituent of the atmosphere, is not an efficient absorber of photons between 120 and 300 nm. However, oxygen molecules are excellent absorbers of photons with wavelengths below 200 nm; such photons are effective in dissociating oxygen molecules into oxygen atoms by the reaction

$$O_2 + \text{photon} \rightarrow O + O. \tag{3.1}$$

Oxygen atoms produced by this reaction can attach themselves to oxygen molecules to form ozone, O_3:

$$O_2 + O + M \rightarrow O_3 + M. \tag{3.2}$$

In this equation M denotes an inert atom or molecule, usually N_2 or O_2, that acts as a catalyst to carry off the energy of the collision; without M the O_2 molecule and the O atom would fly apart again. Ozone is a minor constituent of the stratosphere, but it is extremely important, because it is the only efficient atmospheric absorber of ultraviolet radiation in the wavelength range between 200 and 300 nm. Stratospheric ozone prevents potentially lethal doses of ultraviolet radiation from reaching the Earth's surface. Without such an ultraviolet screen, life on the continents as we know it today would not be possible.

Stratospheric ozone is destroyed by a number of mechanisms. One of the most important is a catalytic cycle involving NO and NO_2

$$NO + O_3 \rightarrow NO_2 + O_2 \tag{3.3}$$

$$NO_2 + O \rightarrow NO + O_2 \tag{3.4}$$

$$O_3 + \text{photon} \rightarrow O_2 + O \tag{3.5}$$

$$\text{Net: } 2\,O_3 \rightarrow 3\,O_2$$

In this cycle NO is converted to NO_2 by reaction (3.3) and is returned to NO by reaction (3.4). In the process, two O_3 molecules are converted into three O_2 molecules.

Chlorofluorocarbons, fine refrigerant gases that they are, have turned out to contribute to the destruction of O_3 in the stratosphere (see chapter 12). This has given rise to the famous Antarctic ozone hole and to the potentially more destructive reduction of the stratospheric UV screen at lower latitudes, where most of us live (see chapter 12).

3.5 The Surface Temperature of the Earth

The quantity of sunlight that is absorbed per square centimeter of Earth surface and that is reemitted from the Earth's surface is very much less than the quantity of energy emitted per square centimeter of the Sun's surface. The surface temperature of the Earth is therefore much lower than the temperature of the Sun, and the spectrum of the light emitted by the Earth peaks at much longer wavelengths than the solar spectrum (see fig. 3.5). If you turn on an electric stove or a toaster, the color of the heating elements gradually changes from dull red to orange as the

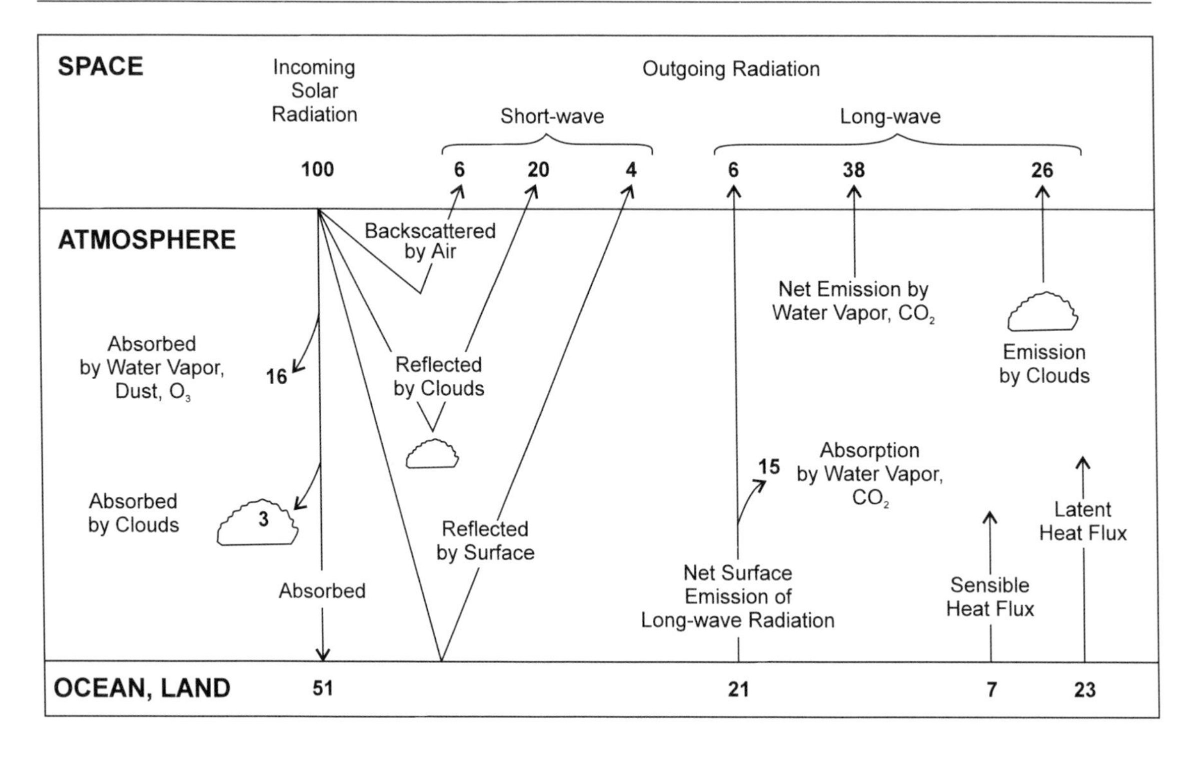

Figure 3.4. Schematic representation of the relative amounts of the various energy inputs and outputs in the earth-atmosphere system. (Crowley and North 1991)

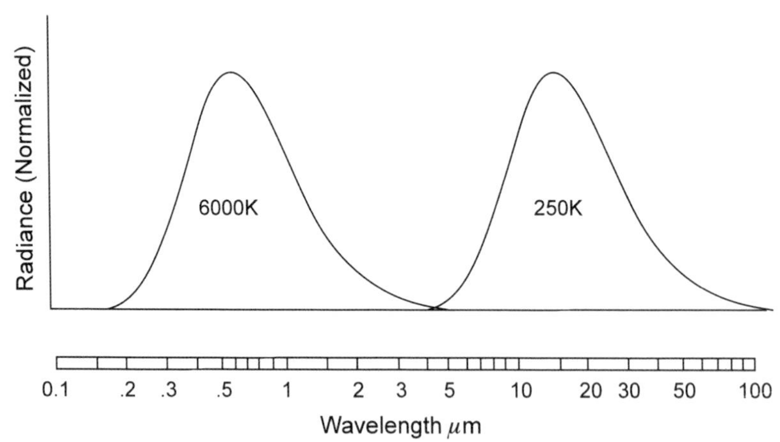

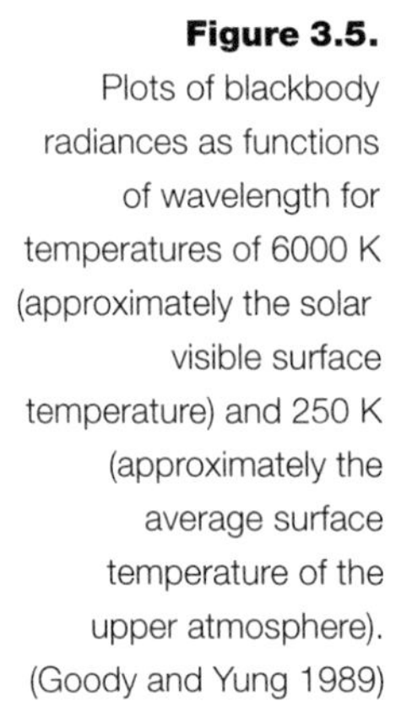

Figure 3.5. Plots of blackbody radiances as functions of wavelength for temperatures of 6000 K (approximately the solar visible surface temperature) and 250 K (approximately the average surface temperature of the upper atmosphere). (Goody and Yung 1989)

temperature rises; but invisible radiant energy is emitted from the heating elements well before they turn dull red. The light emitted at these lower temperatures has wavelengths greater than red light, in a part of the spectrum called the infrared. The spectrum of light emitted by the Earth is in this range. Astronauts looking back at the Earth see the splendor of reflected sunlight. They cannot see the infrared glow of the Earth itself.

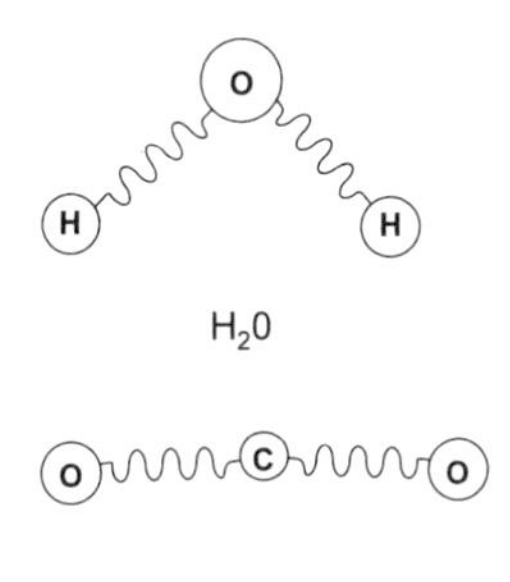

CO_2

Figure 3.6. Molecules of water and carbon dioxide can be regarded as atoms connected by springs.

The atmosphere acts as a heat blanket for the Earth, because it is an excellent absorber of infrared radiation. Photons of infrared light do not possess enough energy either to ionize atoms or to dissociate molecules. They do, however, have enough energy to impart rotational and vibrational energy to molecules. Molecules such as water and carbon dioxide can be regarded as groups of atoms held together by springs, somewhat as illustrated in figure 3.6. These molecules can absorb light energy that gives them rotational motions and that causes the individual atoms to vibrate. Quite a few components of the atmosphere are infrared absorbers. Water vapor is by far the most important of these. Carbon dioxide is a rather distant second. Such molecules enter excited states by absorbing infrared radiation in sunlight entering the atmosphere from above and from light emitted by the Earth, that enters the atmosphere from below. They reradiate the absorbed infrared energy when they return to their unexcited states. The radiated energy is emitted equally in all directions, i.e., the energy emitted upward is equal to the energy emitted downward. The proportion of infrared light in the solar spectrum is quite small. This means that most of the infrared energy enters the atmosphere from below. Infrared absorbers in the atmosphere redirect a good deal of this Earth light back toward the Earth. The total flux of energy that must be emitted by the Earth to maintain an energy balance is therefore considerably greater than it would be in the absence of infrared absorbers in the atmosphere. The mean surface temperature of the Earth therefore has to be higher than it would be in the absence of infrared absorbers. In the absence of an atmosphere, the mean surface temperature of the Earth would be 255 K, i.e., −18°C. The actual mean surface temperature is 288 K, i.e., +15°C. The difference of 33° has enormous consequences for life on Earth. At −18°C water is solid, at +15°C it is liquid. At +15°C our environment is reasonably comfortable on land as well as in the oceans.

3.6 Climate Changes and Climate Controls

The climate of the Earth has varied considerably but not drastically during its long history. Figure 3.7a contains an estimate of temperature fluctuations during the last million years. Temperatures were repeatedly about 4°C lower than the present mean temperature of the Earth. These temperature minima occurred during the ice ages, when ice sheets covered significant parts of the Earth and left their mark on the landscape of the continents at intermediate and high latitudes (see chapter 6).

Figure 3.7b is an expanded view of the extreme right-hand part of figure 3.7a. Temperatures rose rapidly at the end of the last ice age, about 10,000 years ago. They reached a maximum about 5,000 years ago, and have fluctuated somewhat since then. Figure 3.7c is an expanded view of the Earth's surface temperatures during the last 1,000 years. Historical records are reasonably abundant and reliable for this time period. A warm era during the Middle Ages was followed by the "little ice age" and by a rise in temperature that began during the nineteenth century and is still continuing today.

During the last 2 million years the Earth has alternated between glacial and interglacial periods, and there have been a series of unusual but probably not unique cold snaps. During much of Earth history prior to 2 million years ago (m.y.a.), temperatures were higher, glaciers were absent or confined to very small areas at high latitudes, and flora and fauna that are now confined to the tropics ranged much closer to the poles (fig. 3.8). During the Cretaceous period, some 140–65 m.y.a. (see fig. 3.9), the ancestors of alligators lived as far north as

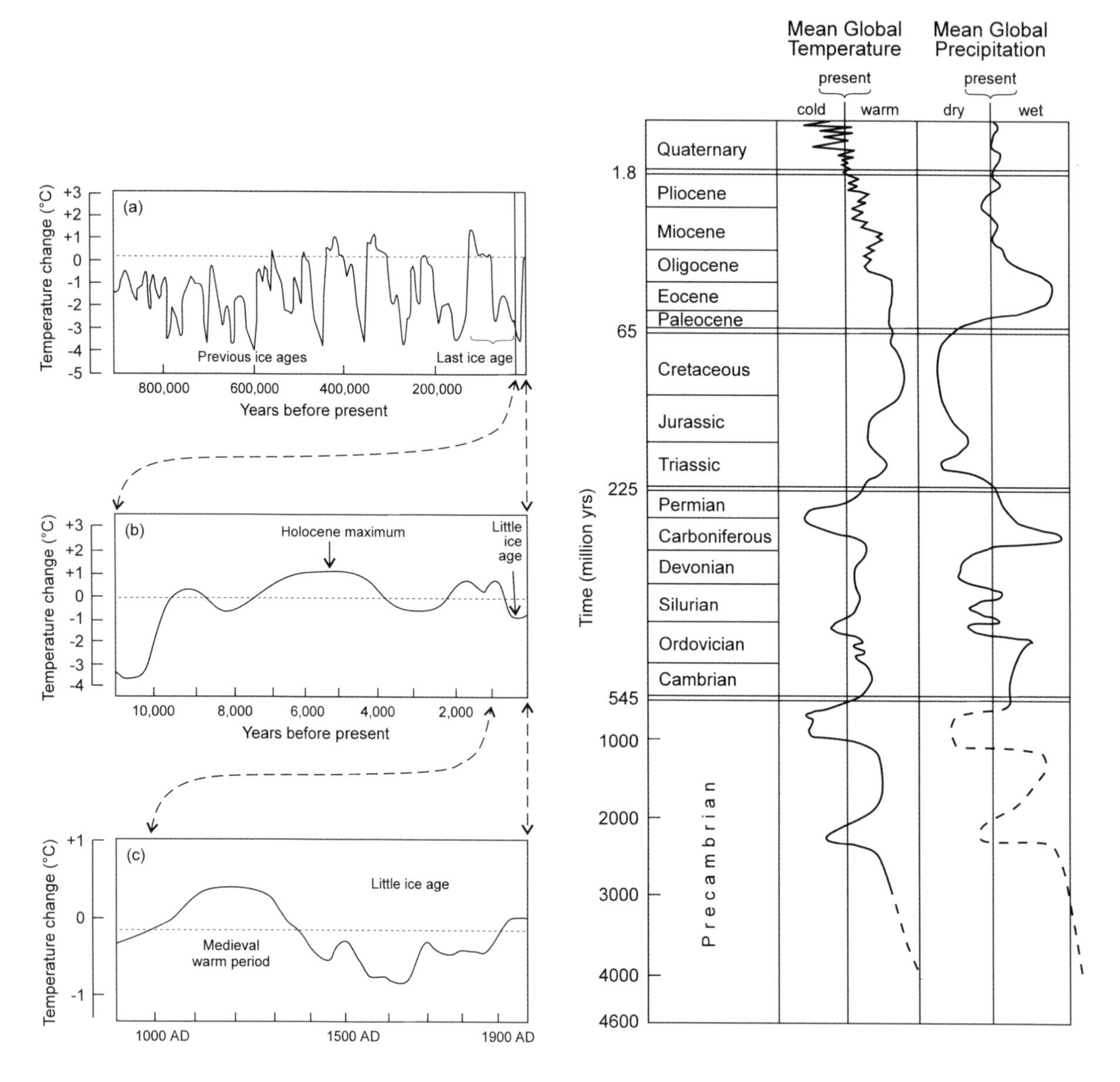

Figure 3.7 (*left*). Temperature variations during the past million years. The dotted line represents conditions near the beginning of the twentieth century (Folland et al. 1990). (a) the last million years; (b) the last ten thousand years; (c) the last thousand years.

Figure 3.8 (*right*). Generalized temperature and precipitation history of the Earth. Trends are dashed where data are very sparse. The curves are drawn to represent departures from present global means, but only relative values are indicated. Note that the timescale is progressively expanded in younger time units. (Frakes 1979)

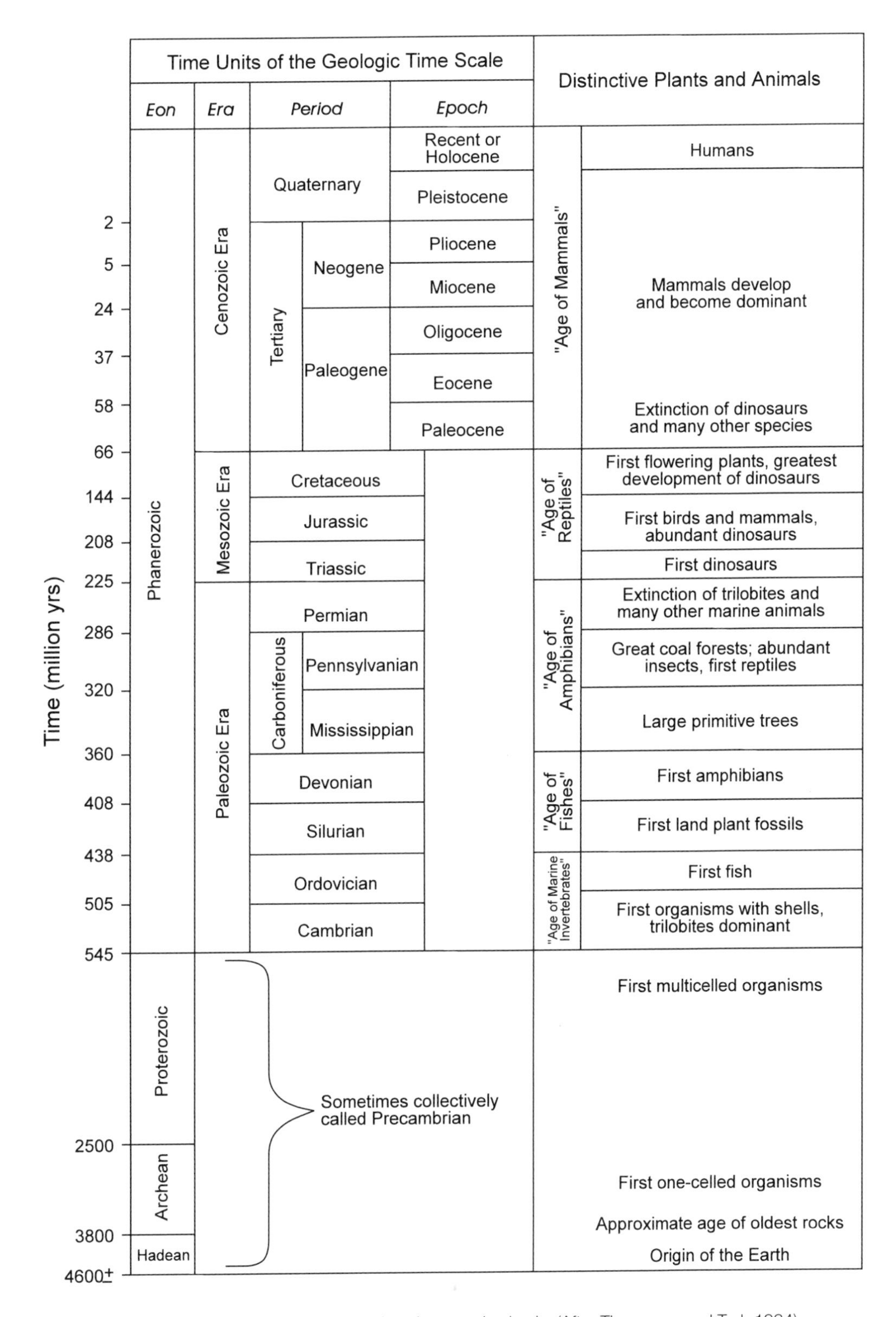

Figure 3.9. The geological timescale together with distinctive plants and animals. (After Thompson and Turk 1994)

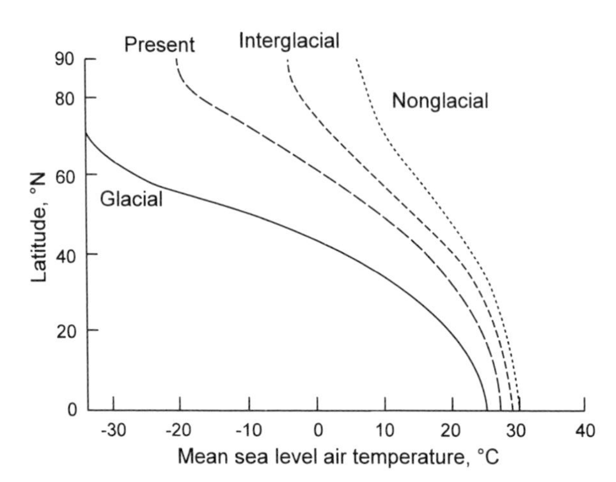

Figure 3.10. Approximate latitudinal variation of air temperature at mean sea level during glacial, present, interglacial, and nonglacial time. (Fairbridge 1964)

latitude 85°. Currently available data (see fig. 3.10) indicate that during these warm, nonglacial periods the tropics were probably not much warmer than at present, but polar temperatures were very much higher than they are now.

Glacial periods have visited the Earth at intervals of a few hundred million years. The Permo-Carboniferous glaciation about 300 m.y.a. and the late Precambrian glaciations about 600 m.y.a. were clearly extensive. The geological record becomes more and more fragmentary as we penetrate more deeply into the earliest parts of Earth history. There is good evidence for what is called the Gowganda glaciation about 2,300 m.y.a., and there are indications that one or more glaciations occurred between 2,500 and 3,000 m.y.a.

Life began more than 3,500 m.y.a., and there is a continuous record of living organisms since then. Until about 400 m.y.a., life existed mainly, if not exclusively, in water. The fossil record demonstrates that open water existed in some parts of the globe during all of the past 3,500 million years (m.y.). Ice could not have covered the entire globe. The nature of sedimentary rocks of all ages corroborates this conclusion. Their chemical compositions can be understood only if the hydrologic cycle has been operating efficiently for the past 3,800 m.y. (see chapter 4), and if chemical weathering was about as intense during this long period of time as it is today (see chapter 6, and Holland 1984, chapter 5). The continuity of life on Earth also sets an upper limit on temperatures in the past. Primitive organisms can live quite happily at temperatures close to the boiling point of water: witness the abundance of algae in many hot springs. However, life is difficult above 100°C, and may turn out to be impossible above about 120°C, because organic compounds that are necessary for life decompose too rapidly at higher temperatures.

The climate of the Earth is and has been determined by a complex interplay of numerous processes and circumstances. These can be divided into outside effects, surface and near surface effects, and interior effects. Outside effects determine the irradiance of the Earth and the distribution of this irradiance between continents and oceans. The irradiance depends on the luminosity of the Sun and on the manner in which solar energy is radiated into space. There is reason to believe that the luminosity of the Sun has increased gradually during the last

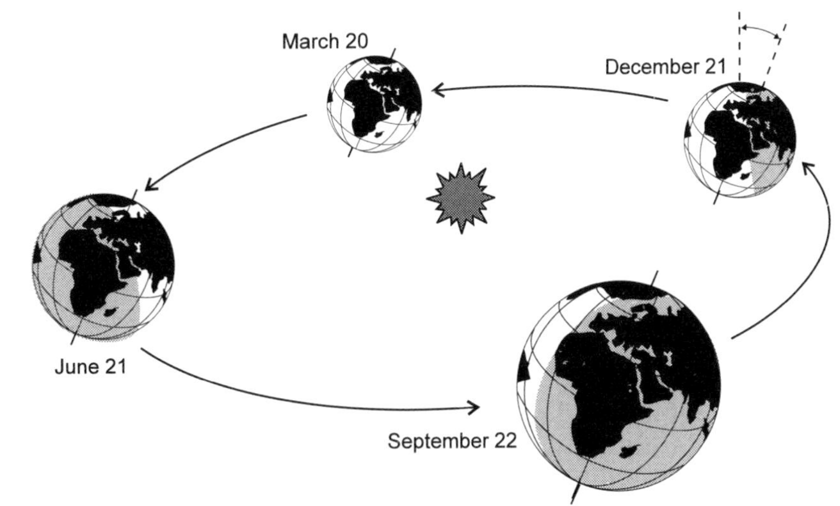

Figure 3.11. March of the seasons; as the tilted Earth revolves around the Sun, changes in the distribution of sunlight cause the succession of the seasons. (Imbrie and Imbrie 1986)

4,400 m.y., but the magnitude of this increase is still somewhat uncertain (see chapter 2). Small (0.08%) variations in solar luminosity have been detected during the last sunspot cycle, and it will be of great interest to see whether larger variations, perhaps related to the length of the sunspot cycle, will be observed during the coming decades.

The distribution of sunlight on the Earth's surface varies considerably on a daily and on a seasonal basis. The differences between daytime and nighttime radiation levels are large, and the difference between radiation levels during summer and winter months is large at high latitudes (see fig. 3.11). There are also smaller fluctuations that occur on a longer timescale. These involve changes in the shape of the Earth's orbit around the Sun, changes in the tilt angle between the rotation axis of the Earth and the plane of the ecliptic, which is defined by the Earth's path around the Sun, and changes in the Earth-Sun distance at the time of the equinoxes, the times of the year when the length of the day is equal to the length of the night. Early in this century, Milutin Milankovitch proposed that cyclical changes in irradiance due to these orbital changes might be responsible for the alternation of glacial and interglacial periods (see, for instance, Imbrie and Imbrie 1986). Intense research on the timing of the ice ages during the last million years has lent support to this theory, but a good many difficult questions remain to be answered before the Milankovitch hypothesis can be considered proven.

The surface and near-surface processes that affect the Earth's climate are more complex. The albedo of the Earth, i.e., the fraction of the Sun's energy that is reflected back into space, determines the quantity of solar energy that is absorbed by the Earth. The extent and nature of the cloud cover and the nature of the clouds, the degree of ice cover, the relative proportion of the Earth covered by land and sea, and the reflectivity of the land surface all influence the Earth's albedo. Some of these parameters serve as components of feedback systems. Consider, for instance, the role of ice and snow cover on climate. If, for some reason, the area covered by ice and snow increases, the albedo of the Earth increases. This means that more of the Sun's energy is reflected back into space; hence less of the Sun's energy is absorbed by the Earth, and the temperature continues to fall. This in turn tends to increase the area covered by snow and ice, producing a further decrease in temperature. Ultimately the entire Earth would be covered by snow and ice; the Earth would then have entered the "golfball" mode. Fortunately this has never happened, although it may have been a close-run thing during the late Precambrian glaciations. Something has always reversed the cooling process. It is not yet clear whether this "something" is an increase in solar irradiance, or terrestrial processes, or a combination of external and internal climatic forces (see chapter 11).

The composition of the atmosphere also affects climate. Without atmospheric oxygen there would be no ozone screen, and much less solar ultraviolet radiation would be absorbed in the atmosphere by photochemical reactions unless other shields were present. The greenhouse effect in the troposphere depends critically on the concentration of water vapor and carbon dioxide, and here we have the basis for another positive feedback. As the temperature of the Earth rises, more water vapor is likely to be present in the troposphere. This increases the greenhouse warming of the troposphere. The water vapor content is therefore apt to increase even more, and one could imagine a scenario in which the oceans evaporate and the Earth is turned into an extremely hot, dense steam bath in which life is impossible. Fortunately this—the runaway greenhouse—has not happened on Earth, because the luminosity of the Sun has been too low (Kasting 1988). Clouds probably reduce the intensity of the positive feedback between temperature and atmospheric water vapor. An increase in atmospheric water vapor is probably accompanied by an increase in cloud cover; this tends to increase the Earth's albedo, and hence moderates the warming. Clouds, however, are very complex, and too little is known about cloud physics to test this notion rigorously.

The distribution of temperature and rainfall on Earth depends on the transport of heat and moisture from the equatorial regions toward the poles. Roughly half of the heat is transported by the atmosphere, the rest by ocean currents. The motions of air and water are complicated, difficult to model, and difficult to predict. The development of high-speed computers has led to the evolution of complex general circulation models of the atmosphere and oceans, but the use of these models still has severe limitations for predicting the climatic effects of anthropogenic changes in the composition of the atmosphere. At least another decade of research and computer development will almost certainly be required before models of the atmosphere-ocean system are sufficiently reliable, so that predictions based on their outputs can be used confidently for assessing the effects of fossil fuel burning and for formulating energy policy (see chapter 11).

World climate is also influenced by processes in the Earth's interior. Violent volcanic eruptions can affect climate on timescales of days to years, largely due to the sudden injection of volcanic dust and gases into the stratosphere. The emission of volcanic gases also has important consequences for the long-term chemistry of the atmosphere. Until recently the CO_2 content of the atmosphere has been determined by the balance between the injection of CO_2 into the atmosphere by volcanoes and hot springs and by the removal of CO_2 with sediments (see chapter 7). This removal process is influenced by the presence of living organisms. Life on Earth has played a significant role in the evolution of the atmosphere. In later chapters we shall examine the interplay between the atmosphere and the other parts of the Earth system to see how well the biosphere has served as a regulator of climate, and to determine whether and to what extent the biosphere will moderate climate changes in the future.

References

Ahrens, C. D. 1991. *Meteorology Today: An Introduction to Weather, Climate and the Environment*. 4th ed. West Publishing, St. Paul, Minn.

Banks, P. M., and Kockarts, G. 1973. *Aeronomy*, part A. Academic Press, New York.

Crowley, T. J., and North, G. R. 1991. *Paleoclimatology*. Oxford Univeristy Press, New York.

Fairbridge, R. W. 1964. The Importance of Limestone and Its Ca/Mg Content to Paleoclimatology. In *Problems in Paleoclimatology*, ed. A.E.M. Nairn, pp. 431–477. Interscience, New York.

Folland, C. K., Karl, T. R., and Vinnikov, K. Ya. 1990. Observed Climate Variations and Change. In *Climate Change: The IPCC Scientific Assessment*, ed. J. T. Houghton, G. J. Jenkins, and J. J. Ephraums, chap. 7. Cambridge University Press, Cambridge, U.K.

Frakes, L. A. 1979. *Climates throughout Geologic Time*. Elsevier, Amsterdam.

Goody, R. M., and Yung, Y. L. 1989. *Atmospheric Radiation: Theoretical Basis*. 2d ed. Oxford University Press, New York.

Holland, H. D. 1978. *The Chemistry of the Atmosphere and Oceans*. Wiley-Interscience, New York.

Holland, H. D. 1984. *The Chemical Evolution of the Atmosphere and Oceans*. Princeton University Press, Princeton, N.J.

Imbrie, J., and Imbrie, K. P. 1986. *Ice Ages: Solving the Mysteries*. Harvard University Press, Cambridge, Mass.

Kasting, J. F. 1988. Runaway and Moist Greenhouse Atmospheres and the Evolution of Earth and Venus. *Icarus* 74:472–94.

Kelley, K. W., ed. 1988. *The Home Planet*. Addison-Wesley and MIR Publishing, Reading, Mass.

Lean, J. 1991. Variations in the Sun's Radiative Output. *Rev. of Geophys.* 29:505–35.

Rowland, F. S. and Isaksen, I.S.A. eds. 1987. *The Changing Atmosphere*. Wiley-Interscience, New York.

Thompson, G. R., and Turk, J. 1994. *Essentials of Modern Geology: An Environmental Approach*. Saunders College Publishing, Orlando, Fla.

Warneck, P. 1988. *Chemistry of the Natural Atmosphere*. Vol. 41 of the International Geophysics Series. Academic Press, Orlando, Fla.

4 The Hydrologic Cycle

4.1 Introduction: In Praise of Water

Water was the cradle of life, and life as we know it is still impossible without water. In the American West, where water is scarce, it is the subject of myths and legends. In another desert, the children of Israel were saved in their wanderings by Moses' discovery of water in an unlikely spot. Sometimes, of course, there is too much water: witness the story of Noah's flood and the great inundations before and since. Hydrogeologists and engineers continually face twin challenges: providing water where it is scarce and controlling the destructive force of water where it is in excess.

Water is abundant on our planet. It covers 70% of the Earth's surface to an average depth of nearly 4,000 meters. There is so much water that we should, perhaps, consider calling our planet Oceanus rather than Earth. Water in the oceans is not static. Evaporation rates are high, and seawater is cycled so rapidly through the atmosphere that the biography of a water molecule would include a tiresomely long list of round-trips from the oceans through the atmosphere and the biosphere, back to the oceans.

Water has unusual physical properties. It is one of the very few liquids that expands on freezing and is therefore denser than ice. This creates problems, such as cracked water pipes in winter; but it also makes ice skating possible, and—of more importance—prevents bottom-dwelling organisms in lakes from being crushed by ice during winter freezes.

Water also has unusual chemical properties. It is, for instance, an excellent solvent for many salts and for some groups of organic compounds. Sea water is salty in part because water is such a good solvent, and the salinity of our blood is probably a memento of our distant marine ancestry. The saltiness of the sea has its drawbacks; witness the anguished cry of the Ancient Mariner: "Water, water everywhere, nor any drop to drink." Today this complaint is not confined to sailors. It includes members of many communities whose water supply has been contaminated by salts, organic compounds, or pathogens. Hence the challenge in many parts of the world is and will continue to be not only to supply adequate quantities of water but to furnish water that nourishes and sustains human life.

4.2 The Evaporation of Water from the Oceans

If we observe a jar half-filled with water, its lid tightly closed, sitting on a table in a quiet room, we see no motion. The water level is constant, and nothing betrays the frantic movement of water molecules in the gas space, nor the more sedate movement of water molecules in the liquid. These movements can, however, be measured, and it can be shown that water molecules continually travel from the liquid into the gas phase and from the gas phase into the liquid. The situation is somewhat like that at the entrance of a hive, where bees are continually leaving and arriving. If we raise the temperature of the jar, water molecules move more rapidly in both the liquid and in the gas phase; the traffic at the water

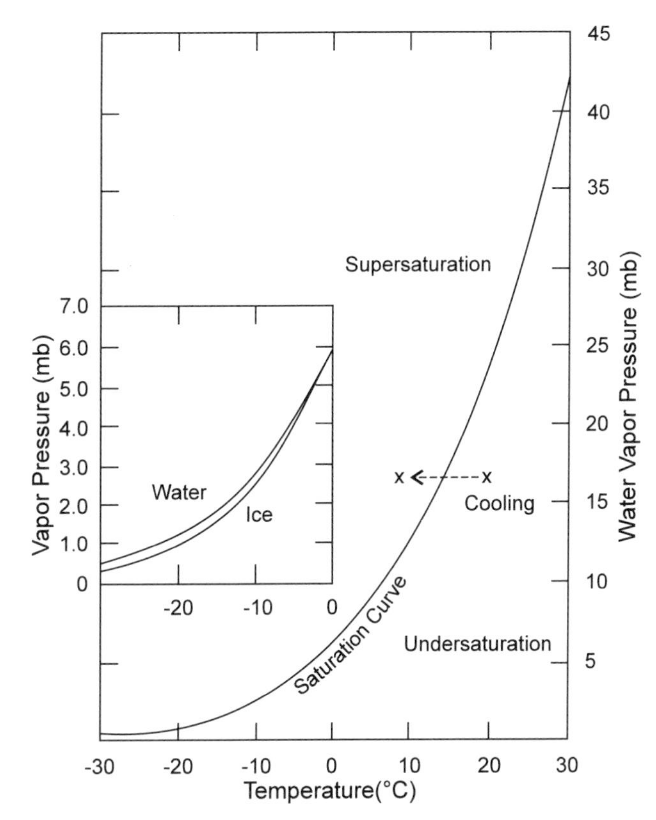

Figure 4.1. The vapor pressure of ice and water between –30° and 30°C (mb = millibar) (Berner and Berner 1987)

surface increases, as does the number of water molecules in the gas phase. The frequency and the energy with which water molecules collide with the walls and with the lid also increase. Hence the pressure exerted by water vapor in the jar increases (see fig. 4.1). In fact, if the temperature of the water in the jar increases sufficiently, the pressure on the lid and on the walls can cause the jar to explode. The analogy with the beehive wears thin at this point, although the effect of an increase in water temperature could be compared to the effect of an increased level of excitement in the bee community (see, for instance, Milne 1926). If the temperature of the jar is lowered, the water molecules slow down. Quite suddenly, at 0°C their motion in the liquid becomes literally frozen. Water turns into ice, and the rather random arrangement of water molecules in liquid water is replaced by the highly ordered arrangement of water molecules in ice crystals. This arrangement, shown in figure 4.2, is responsible for the marvelous symmetry of the snowflakes in figure 4.3.

In some ways the oceans can be compared to a huge, open-topped jar in which the temperature is rather variable. Mean surface temperatures on the Earth for June 1988 are shown in plate 2. The high temperatures in the equatorial regions are shown in red, progressively colder temperatures in yellow, green, and blue. The South polar regions were the coldest parts of the Earth during this month, since June is the depth of winter in the Southern Hemisphere.

In all parts of the globe the temperature decreases with increasing height in the troposphere (see chapter 3). The amount of water in the atmosphere therefore tends to decrease both poleward from the equator and upward from the Earth's

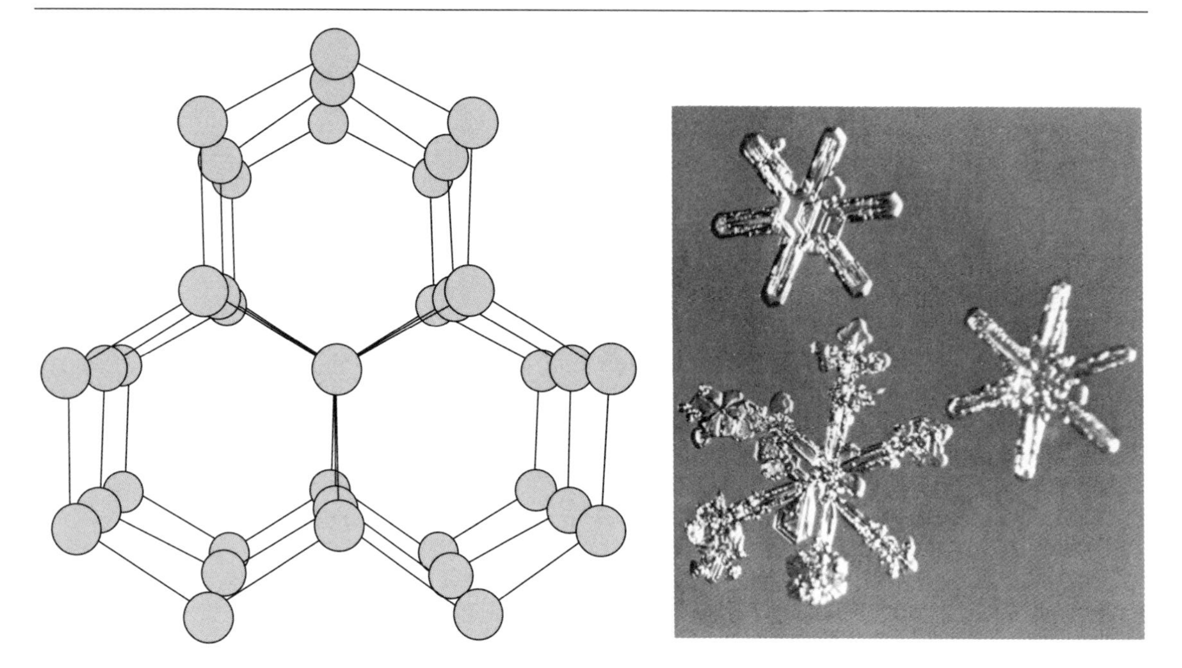

Figure 4.2 (*left*). The arrangement of oxygen atoms in ice. The rods joining the oxygen atoms represent bonds. The hydrogen atoms are not shown. (Hobbs 1974)

Figure 4.3 (*right*). Snow crystals. (La Chapelle 1969)

surface. The Earth is somewhat like a large room with a humidifier that puts water vapor into the air in one part, and a dehumidifier that removes water vapor from the air in another part. As shown in figure 4.4, evaporation of seawater is therefore intense in the lower latitudes and minor at high latitudes. The dip in the evaporation rate close to the equator is due to the low average wind velocity in this part of the world.

The average annual evaporation rate from the world's oceans is about 125 cubic centimeters of water per square centimeter of ocean surface (Sellers 1965). This represents the annual evaporation of a water column 125 centimeters, i.e., 1.25 meters or a shade more than 4 feet, in height. Since the mean depth of the oceans is 3,900 meters, the time required to evaporate the entire ocean is

$$\frac{3{,}900 \text{ m}}{1.25 \text{ m/yr}} = 3{,}120 \text{ years.} \tag{4.1}$$

This does not mean that the entire ocean will be in the atmosphere as water vapor in 3,120 years. Most of the water that evaporates from the oceans returns to the oceans in about 10 days, largely as rainfall at sea; most of the remainder returns to the oceans as river runoff from the continents. The number 3,120 years can therefore be thought of as the length of time required for the entire ocean to pass once through the atmosphere. From the point of view of an individual water molecule, this means that on average it will make one round trip from the ocean into the atmosphere and back every 3,120 years. Strictly speaking this is true only if the oceans are well mixed on this timescale. If the oceans are stratified,

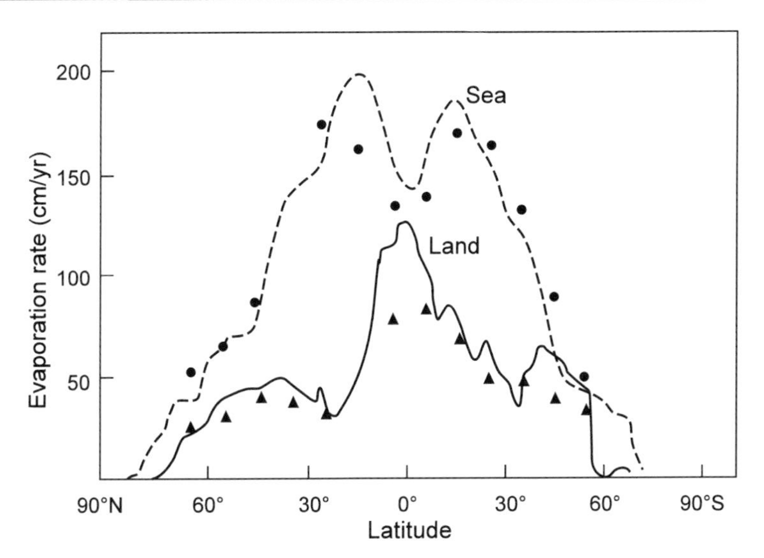

Figure 4.4. The latitudinal distribution of the simulated annual evaporation rate broken down into means over land (solid line) and over sea (dashed line). For comparison, values for evaporation from land and sea areas derived from observed data are indicated by triangles and dots, respectively. (Holland 1978)

then water molecules in surface waters are apt to make many more round-trips than water molecules in the deep oceans. It turns out that the oceans today mix on a timescale of 1,000–2,000 years.

The Earth is some 4,500 million years old (Dalrymple 1991), and the oceans have almost certainly existed during at least the last 3,800 million years of Earth history (Holland 1984, chap. 5). A water molecule that has been part of the oceans since then has made approximately

$$\frac{3{,}800 \times 10^6 \text{ years}}{3{,}100 \text{ years per trip}} = 1{,}230{,}000 \text{ trips} \tag{4.2}$$

through the atmosphere, if the average rate of evaporation from the oceans has been equal to the present-day evaporation rate. But we shall see that the tedium of shuffling between the atmosphere and the oceans has been relieved for most water molecules by many side trips through the biosphere, through sediments and rocks, and even through volcanoes.

4.3 Rain and Other Precipitation

Some 88% of all evaporation takes place from the oceans. This is due in part to their large extent—they cover 70% of the area of the globe—and in part to their composition—they consist largely of water. It is not surprising, then, that most of the precipitation, some 79% of it, falls on the oceans. The remaining 21% falls on continental areas. Weather patterns can move across the United States in a matter of a few days. Since the average residence time of water molecules in the atmosphere is about 10 days, water molecules can, but rarely do, travel across entire continents without being rained out. The distribution of rainfall on the globe is therefore a complex function of the evaporation of water, the circulation patterns of the atmosphere, and the condensation of water vapor.

Figure 4.5 is a highly simplified view of the general circulation of the atmosphere. In the equatorial regions warm air rises in what are called Hadley cells.

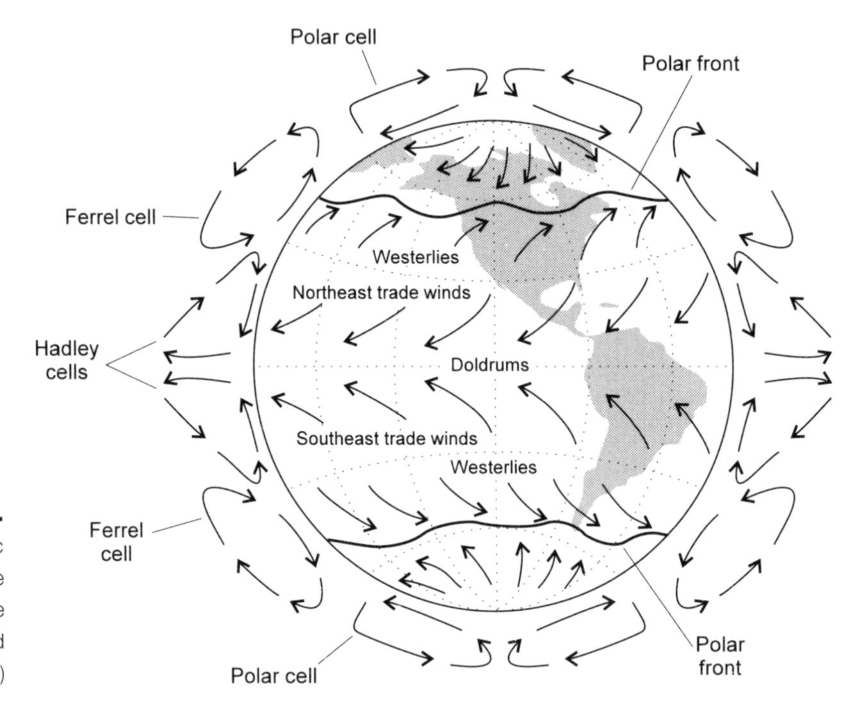

Figure 4.5. Schematic representation of the general circulation of the atmosphere. (Modified from Miller et al. 1983)

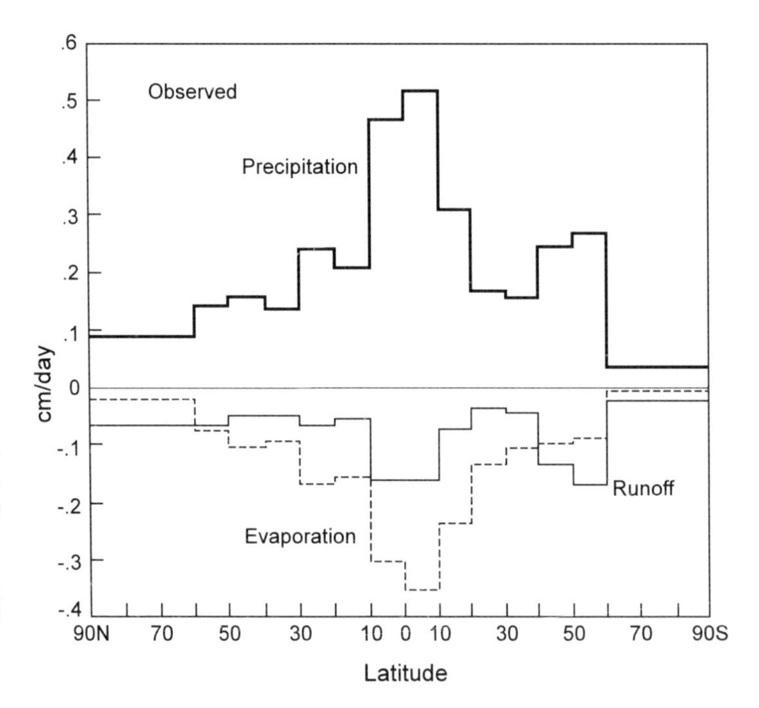

Figure 4.6. The latitude dependence of continental precipitation, evaporation, and runoff. (Holland 1978)

North of the equator the rising air moves northward at considerable heights, descends, and then returns on a southwestward course as the northeast trade winds. South of the equator the rising air moves south, descends, and then returns northward as the southeast trade winds. In the midlatitudes of the Northern Hemisphere, airflow is to the northeast near the surface. These air masses rise near the polar front and return southward at high altitudes. Airflow in the midlatitudes of the Southern Hemisphere has a similar pattern. Air moving south from the North Pole meets the midlatitude winds at the polar front near latitude 50°N. The polar air rises there and returns to the pole, as shown in figure 4.5. A similar pattern prevails in the Southern Hemisphere. The boundaries between the several cells are not constant. To the people living along the East Coast of the United States, the Westerlies that come up the coast are known as Bermuda Highs. They bring hot, muggy, and very unwelcome air during the summer, and pleasantly warm air during the winter. When air from the north invades the East Coast during the summer, it brings the beautiful cool, clear air that makes many New England summers so pleasant. In winter the arrival of these winds is welcomed largely by ski enthusiasts. Known as the Montreal or Calgary Express, they are accompanied by frigid temperatures and generate high heating bills.

In Hawaii the trade winds from the northeast bring the cool, dry weather that makes even the summers pleasant there. When the weather shifts and comes from the southwest, the Kona wind brings with it moisture-laden air that dampens the enthusiasm of even the most devoted admirers of the islands, but improves the waves that are prized by surfers.

The dependence of continental rainfall on latitude is dramatic. As shown in figure 4.6, the highest precipitation rate is near the equator, the lowest in the polar regions. In a rough way, the pattern of precipitation tends to parallel the pattern of evaporation. However, a look at figure 4.7 shows that the distribution of rainfall on the continents and in the oceans is highly complex. Most of the tropics are wet, but near latitude 20°N rainfall is quite variable. Africa at this latitude is a desert, and southeast Asia is well watered. At 30°S the interior of Australia is extremely dry, but the eastern part of South America is well supplied with rainwater. The major arid lands are associated with cold oceanic currents offshore.

Rainfall in any given area depends on the moisture content of the air masses that pass over it and on the local inducement to lose this moisture. Cooling by movement toward higher latitudes and to higher elevations is the most important of these inducements. Figure 4.8 shows a particularly striking example of the effect of elevation on rainfall. Moisture-laden air rises on the flanks of Mount Waialeale on the island of Kauai in Hawaii. In the process, it cools and drops some 11.8 meters (470 in) of rain annually on this, one of the rainiest spots on Earth. The west side of the mountain is very dry, because the air passing over this area has been relieved of so much of its moisture, and because it is warmed as it moves toward lower elevations. Kauai is not the only example of this kind in Hawaii. Downtown Honolulu, on the south shore of the island of Oahu, is quite dry, but just 10–15 km north of the city, at the head of Manoa Valley, rainfall is frequently intense. Dense, dark clouds often shroud these hills while Honolulu is bathed in sunshine. Visitors to Honolulu can watch with satisfaction as these clouds gradually thin into wisps and finally disappear altogether on their way southward toward the city. Residents of Honolulu may worry about their water supply, but more on this later.

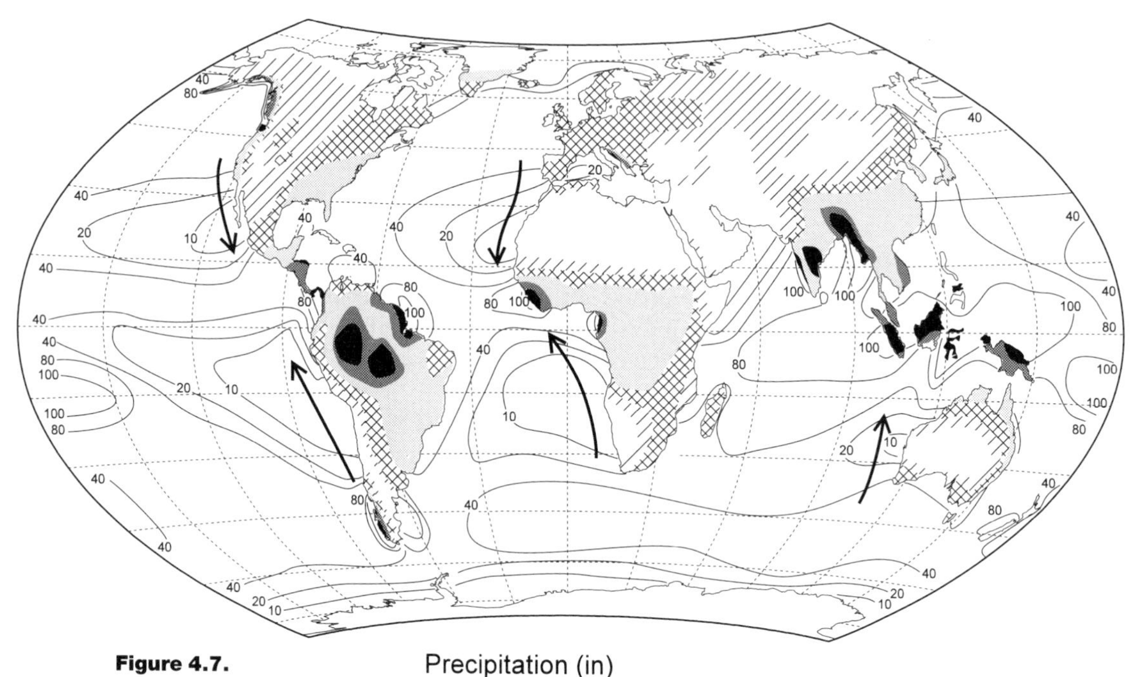

Figure 4.7. World distribution of average annual precipitation. The arrows indicate the position of cold oceanic currents. (From ESSA, *Climates of the World*, 1969)

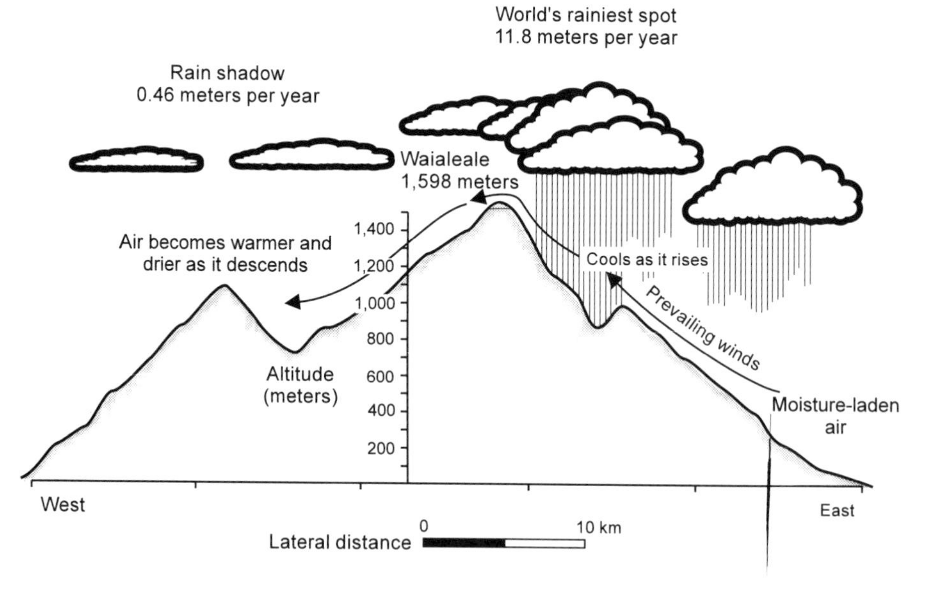

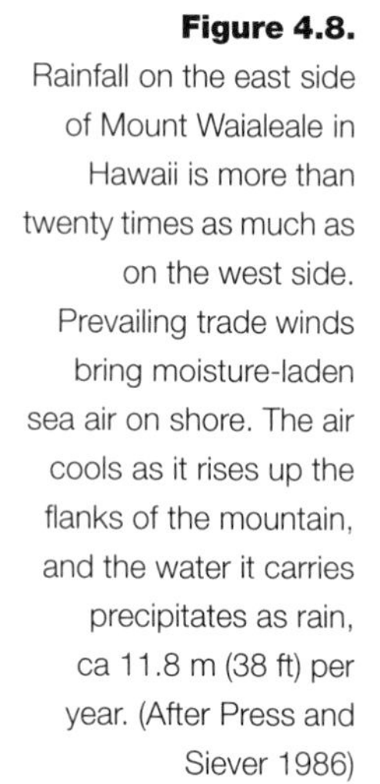

Figure 4.8. Rainfall on the east side of Mount Waialeale in Hawaii is more than twenty times as much as on the west side. Prevailing trade winds bring moisture-laden sea air on shore. The air cools as it rises up the flanks of the mountain, and the water it carries precipitates as rain, ca 11.8 m (38 ft) per year. (After Press and Siever 1986)

4.4 Evaporation and Runoff from the Continents

In all but very high latitudes, precipitation on land either evaporates or flows back to the sea. The magnitudes of the important fluxes are summarized in figure 4.9. Of the approximately 456,000 km^3 (4.56×10^{20} cm^3) of seawater that evaporate annually from the oceans, some 410,000 km^3 (4.10×10^{20} cm^3) rain out over the oceans. The remaining 46,000 km^3 of water are transported to the continents. There they are joined by 62,000 km^3 of water that evaporate from the continents. Together, they account for the 108,000 km^3 of water that precipitate annually on the continents. The difference between rainfall and evaporation on land, 46,000 km^3 of water, returns to the oceans, largely as river water. This flow restores the 46,000 km^3 of water that are transported annually as water vapor from the oceans to the continents. The fraction of water contributed to the hydrologic cycle by volcanoes, hot springs, comets, and all other sources of water is very small.

Figure 4.9 shows that about 60% of all the rain that falls on the continents reevaporates. Most of the remaining 40% is restored to the sea by river flow; the remainder returns to the oceans as groundwater, i.e., as water that flows below the land surface and enters the ocean below sea level. It is easy to see that river flow is a small fraction of the water that participates in the hydrologic cycle. River runoff only amounts to about 10% of the water evaporated from the oceans and to only about 8% of the water that evaporates from the Earth as a whole. This modest fraction of the hydrologic cycle together with groundwater sustains life on the continents and is one of the most important resources for humanity.

The quantity of runoff, frequently designated by the symbol Δf, varies a great deal from area to area on the continents. In well-watered terrain, runoff is extensive, while in many deserts there is virtually no runoff. The ratio of the runoff, Δf, to the rainfall, r, is defined as the runoff ratio, $\Delta f/r$. The runoff ratio is a complicated function of many geographic and biologic parameters; but two of these, the rainfall and the mean annual temperature, account for most of the variability in the runoff ratio. The higher the annual rainfall in a watershed,

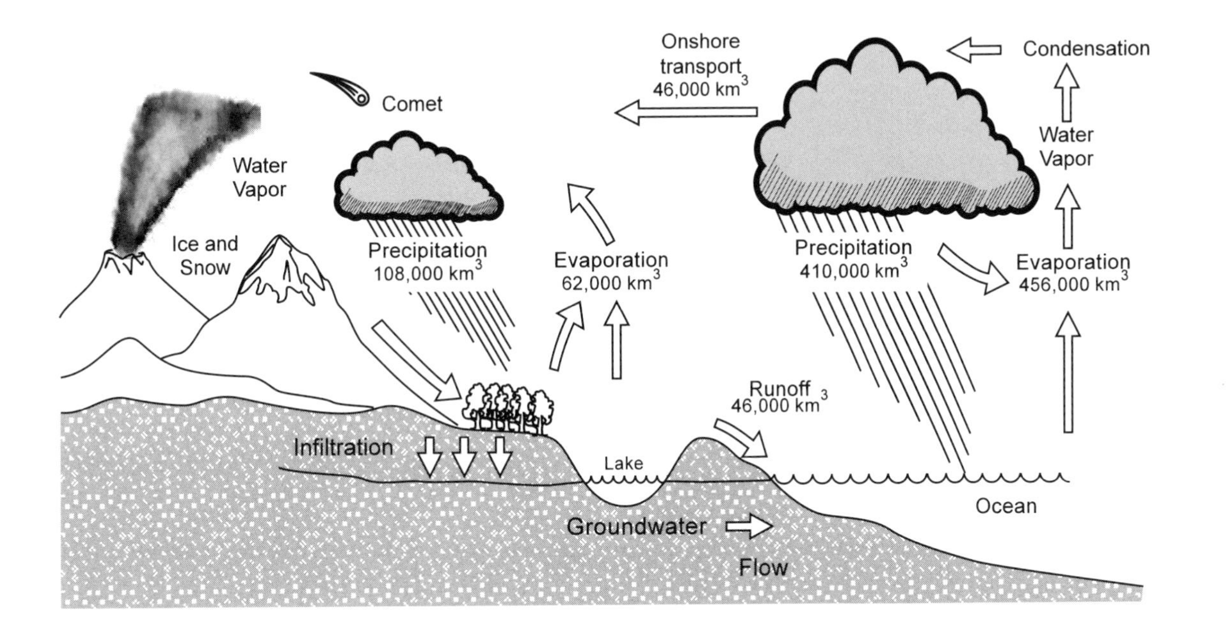

Figure 4.9. The hydrologic cycle moves water constantly between aquatic, atmospheric, and terrestrial compartments driven by solar energy and gravity. Total annual flows shown here are in cubic kilometers. (After Press and Siever 1986)

the greater the fraction of rainwater that flows out as river water. The lower the annual rainfall in a watershed, the greater the fraction of rainwater that reevaporates.

A rather simple equation describes the relation between rainfall and the runoff ratio in numerous watersheds in central Europe:

$$\frac{\Delta f}{r} = e^{-\alpha/r}. \tag{4.3}$$

In this equation e is 2.71, the base of natural logarithms, and α is nearly constant for all of central Europe. In well-watered river basins r is large, the value of α/r is small, and the value of $\Delta f/r$ is close to 1. In watersheds with a small rainfall α/r is large, and the value of $\Delta f/r$ is much smaller than 1. Interestingly, equation (4.3) can be used to describe the runoff ratio from watersheds not only in central Europe but in all parts of the world (Holland 1978, chap. 3). The parameter α does, however, vary from region to region. It turns out that α is equal to E_o, the maximum amount of annual evaporation in a given watershed. This is the quantity of water that evaporates annually from the surface of lakes in a watershed and from seawater held at the same mean annual temperature. As might be expected, the evaporation rate E_o increases with the mean annual temperature, T. The expression that relates E_o to T on a worldwide basis is

$$E_o = 1.2 \times 10^9 \, e^{-4.6\times10^3/T} \, \frac{\text{cm}}{\text{yr}}, \tag{4.4}$$

where the temperature is in degrees Kelvin. The value of the coefficient 4.6×10^3 is related quite simply to the energy that is required to convert liquid water into water vapor. Figure 4.10 is a graphical representation of the relationship between annual runoff and annual precipitation in watersheds of different mean annual

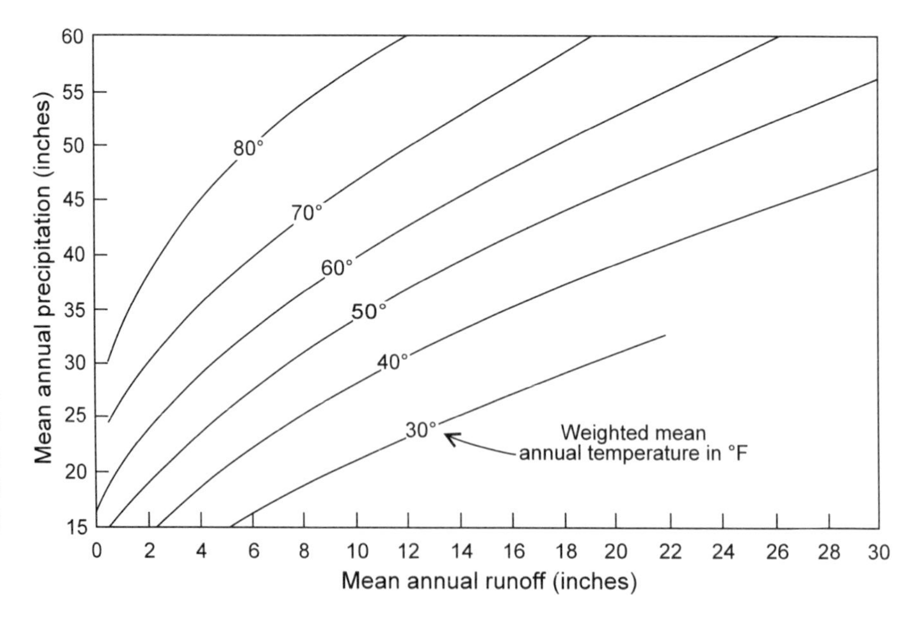

Figure 4.10.
Relation of annual runoff to precipitation and temperature. (Moss and Lins 1989)

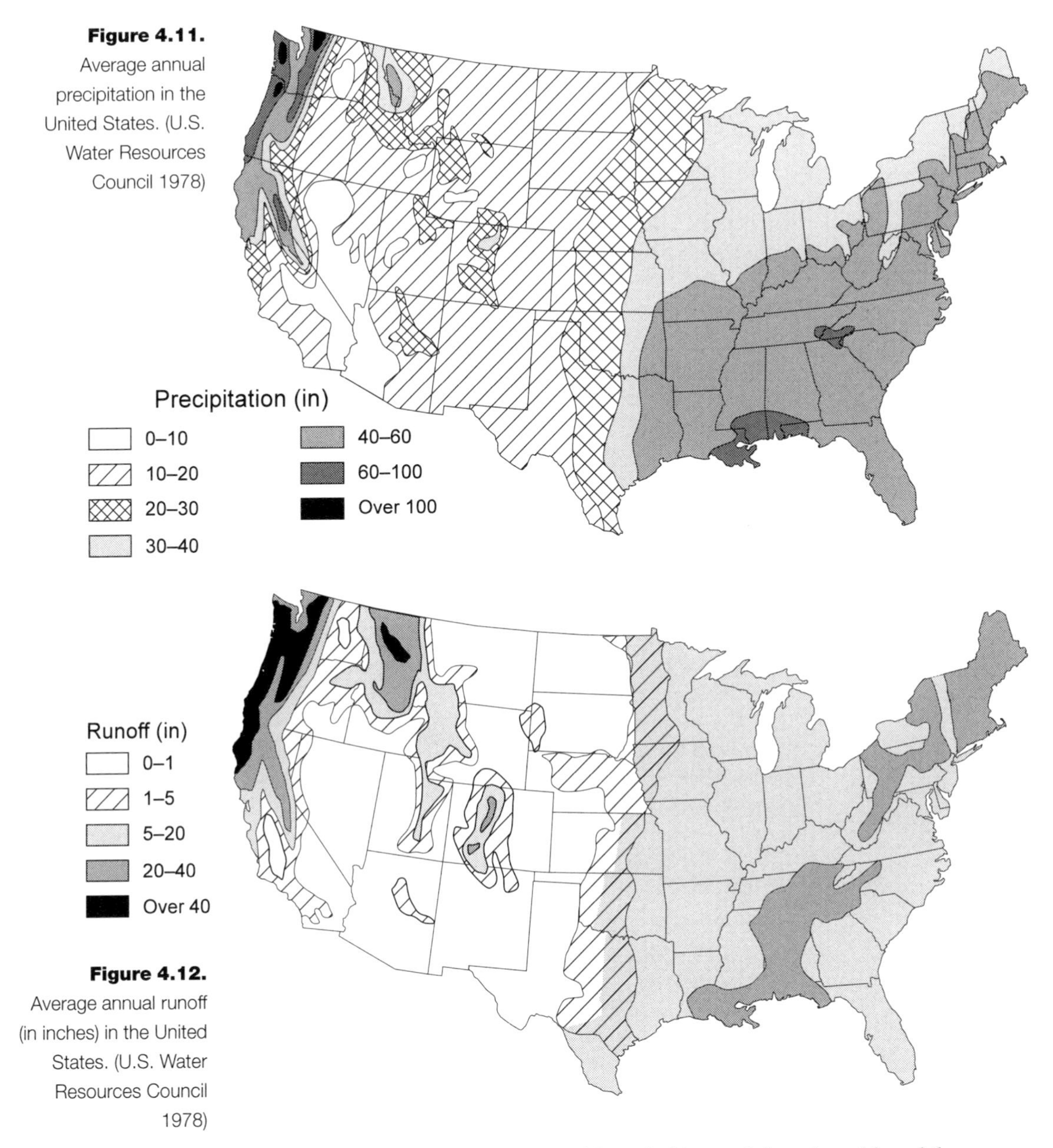

Figure 4.11. Average annual precipitation in the United States. (U.S. Water Resources Council 1978)

Figure 4.12. Average annual runoff (in inches) in the United States. (U.S. Water Resources Council 1978)

temperature in the United States. Although this graph is rather old, and the curves were drawn only for watersheds in the United States, the data can be applied to watersheds the world over. The position and the shape of the curves agree quite satisfactorily with predictions based on equations (4.3) and (4.4), and could be used to help forecast changes in worldwide precipitation and runoff as a result of greenhouse warming (see chapter 11).

Rainfall and temperature vary a great deal within the United States. As shown in figure 4.11, rainfall is highest along the coast in the northwestern states, intermediate in the east, and lowest in the southwestern desert regions. Runoff is even more variable than rainfall. As shown in figure 4.12, runoff is abundant along the northwestern coastal areas and minimal in the southwestern states. The runoff ratio in the well-watered northwest is about 0.4. In the desert regions it is less than 0.1. Most of the western states therefore face water shortages. The low rainfall in this region leads to high evaporation rates and leaves little in the way of usable freshwater.

4.5 Rivers

Rivers have played an important role in the history of many great civilizations. It is difficult to imagine Egypt without the Nile; the Assyrians and the Babylonians without the Tigris and the Euphrates; the civilizations of India without the Indus, the Brahmaputra, and the Ganges; or China without the Yellow and the Yangtze Rivers. These and countless other rivers have served as sources of water and food and as means of transportation. Many have been revered for their healing powers and have shaped and been shaped by local mythology. Rivers flowed through Eden; the Nile turned red before the Children of Israel were allowed to leave Egypt; they wept by the waters of Babylon, and they were purified in the waters of the Jordan. In the German sagas, the Rhine was immortalized by the Rhine maidens, the Rheingold, and the Lorelei. For Americans, the Mississippi will always be Ole Man River, Tom Sawyer, Huckleberry Finn, and the people in *Life on the Mississippi*. Even lakes and ponds have become important as symbols and as treasured memories. The life of Christ is closely connected with the Sea of Galilee, the history of the Aztecs with the lake that is now Mexico City, the memories of Scotland with Loch Lomond, and the intellectual heritage of New England with Walden Pond.

Rivers come in all sizes. The twenty largest rivers are listed in table 4.1 together with the size of their drainage basins, their annual discharge, and the runoff, Δf, in their drainage. Among the smallest rivers there is the Piankatank in Virginia, whose limited width is celebrated by the toast,

> Here's to the River Piankatank,
> Where the frogs can jump from bank to bank.

The Amazon is by far the largest of the world's rivers. Its discharge is some 200×10^3 m^3/sec, about ten times the discharge of the Mississippi River. The total runoff of water from the continents is about 46,000 km^3/yr (see fig. 4.11). If we convert cubic kilometers to cubic meters and years to seconds, this becomes

$$\frac{46{,}000 \text{ km}^3 \times 10^9 \text{ m}^3/\text{km}^3}{1 \text{ yr} \times 3.14 \times 10^7 \text{ sec/yr}} = 1{,}465 \times 10^3 \frac{\text{m}^3}{\text{sec}}. \tag{4.5}$$

The twenty largest rivers together account for about 30% of the entire flow of freshwater to the oceans. The Amazon alone contributes about 14% of the total continental water flows. The Amazon and the Orinoco together drain much of the northern part of South America, a very wet area of enormous rain forests that are an important natural resource and the center of much controversy (see chapters 5 and 12).

North America is drained by four of the twenty largest rivers: the Mississippi (no. 6 in the table), the St. Lawrence (no. 10), the Columbia (no. 15), and the Mackenzie (no. 17). Figure 4.13 shows the major rivers of what are frequently

Table 4.1. The Twenty Largest Rivers Arranged in Order of Their Water Discharge

	Drainage Area ($10^3 km^2$)	Discharge ($10^3 m^3/sec$)	Runoff Δf (cm/yr)	*Suspended Sediment* Yield ($kg/km^2 yr \times 10^3$)	*Suspended Sediment* Concentration (mg/kg)	*Dissolved Solids* Yield ($kg/km^2 yr \times 10^3$)	*Dissolved Solids* Concentration (mg/kg)
1. Amazon	5,930	~200	92	55.3	66		53
2. Congo	4,000	40	31	14.5	51		
3. Orinoco	950	23	72	82.5	77		
4. Yangtze	1,030	22	67	444.0	700		
5. Bramaputra	560	19.9	110	1,179.0	1,070		
6. Mississippi	3,268	18.4	18	82.5	510	43.7	245
7. Yenisei	2,480	17.5	22	3.8	190		
8. Mekong	390	15.0	120	395.0	365		
9. Parana	2,300	14.9	20	31.7	175		
10. St. Lawrence	1,300	14.2	34	2.5	8		
11. Ganges	1,060	14.2	42	1,270.0	3,400		
12. Irrawaddy	370	13.6	115	744.0	710		
13. Ob	2,440	12.5	16	5.4	37		
14. Volga	1,350	8.0	19	12.7	73		
15. Columbia	669	8.0	38	19.0	56	29.0	85
16. Pearl-West	310	7.9	80	79.8	110		
17. Mackenzie	1,700	7.4	14	2.7	21		
18. Indus	1,050	6.8	20	408.0	2,200		
19. Danube	810	6.2	24	21.8	100		
20. Niger	1,100	6.1	17	3.8	25		

Source: Updated from Holland 1978.

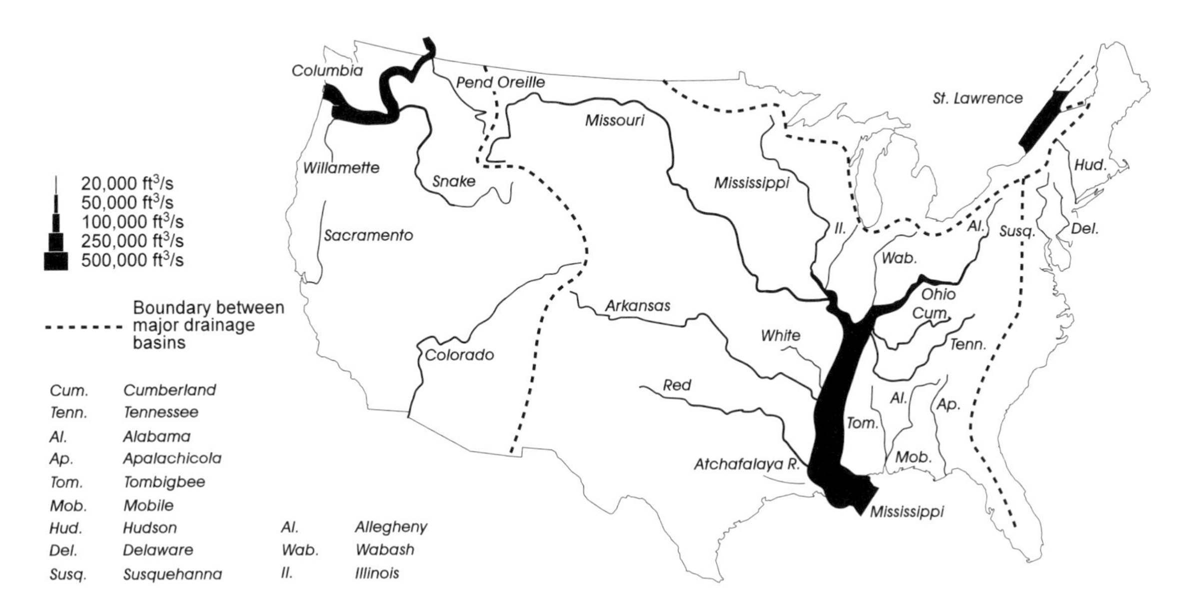

Figure 4.13. Major river drainages in the United States.

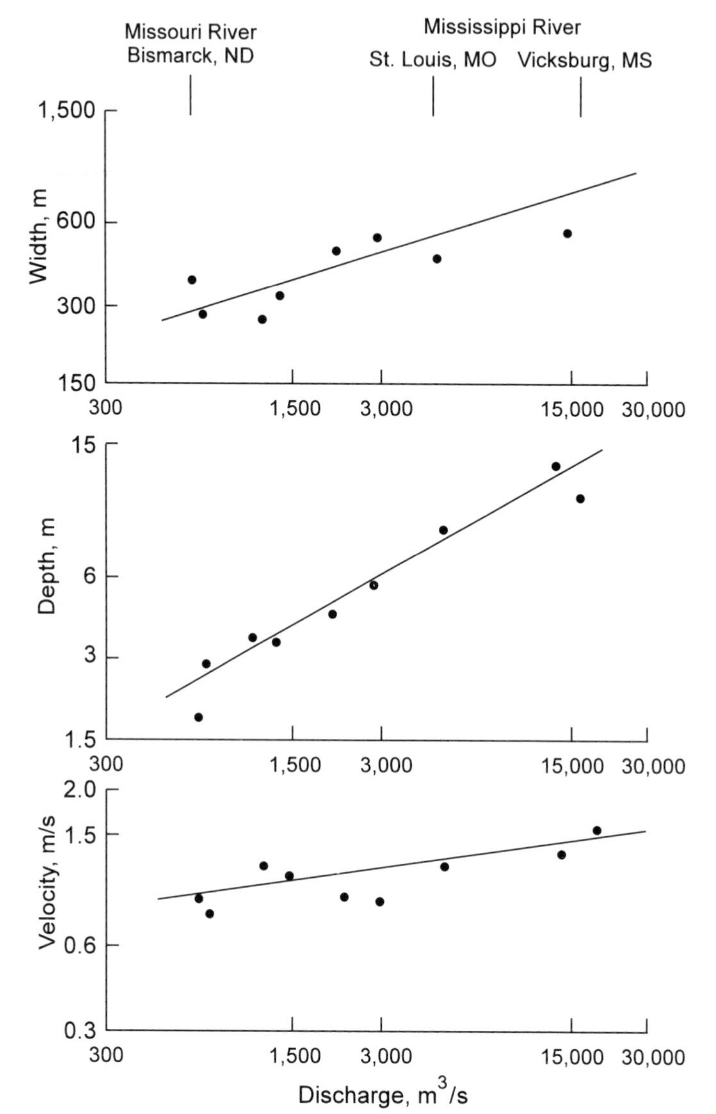

Figure 4.14. The stream velocity, depth, and width of a channel increase as the discharge of a stream increases downstream. Measurements in this example were taken at mean annual discharge along a section of the Mississippi-Missouri river system. Note that the scales are logarithmic. (Leopold and Maddock 1953)

called the "Lower 48" states. The Mississippi River dominates the drainage pattern of this part of the world. Its discharge of roughly 18,400 m^3/sec is equivalent to 650,000 ft^3/sec. It combines rivers from the well-watered eastern and the dry western parts of the United States, and funnels their flow into the Gulf of Mexico. The Mississippi drainage is bounded on the east and on the north by a divide that separates it from the watersheds of rivers that drain into the Atlantic Ocean. On the west the Mississippi drainage is bounded by the continental divide; the rivers on the west side of this divide drain into the Pacific Ocean.

As rivers pick up tributaries, their width, depth, and velocity increase. Figure 4.14 shows the increase in these parameters in the Missouri-Mississippi river system between Bismarck, North Dakota, and Vicksburg, Mississippi. The depth of this river system increases by a much greater factor than the width and the

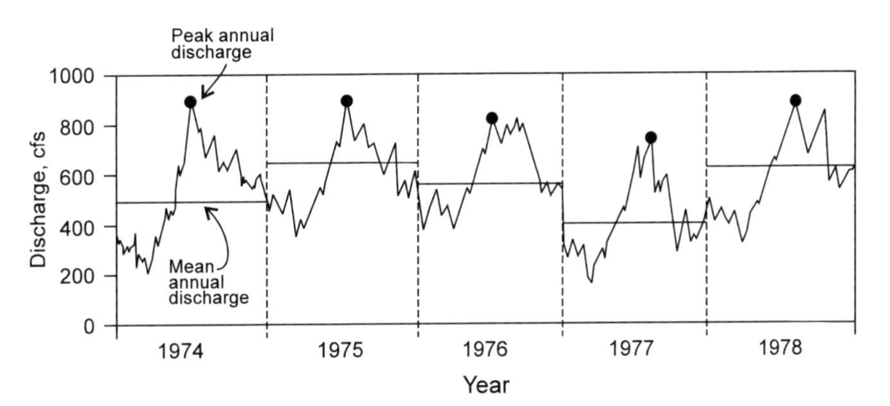

Figure 4.15. Stream hydrograph showing the fluctuation of river discharge with time (heavy line) and the mean annual discharge for each of the years shown. (Freeze 1982)

velocity of the river. All three of these measures vary seasonally and from year to year. Figure 4.15 shows how the discharge of a typical river varied between 1974 and 1978. Each year the flow rate peaked during the summer months and bottomed during the winter months, but the details of the discharge pattern varied from year to year. The maximum and minimum discharges did not always occur at the same time, and the mean annual discharge varied considerably. These differences are due to weather changes in the drainage basin. Years of heavy rainfall can be followed by years of light rainfall or drought. The timing of rains and the melting of snow accumulated during the previous winters tend to vary from year to year, from decade to decade, and from century to century.

The discharge of a river at a particular point along its course reflects the addition and the loss of water everywhere upstream. Heavy rains are apt to produce flooding and the host of problems that accompany abnormally high river stands. There is always some time delay between heavy rains and the peak of river drainage. Figure 4.16 shows the relationship between the timing of a rainfall and the response of the discharge of a river disposing of the resulting runoff. The delay between rainfall and the peak of river discharge is determined by the time

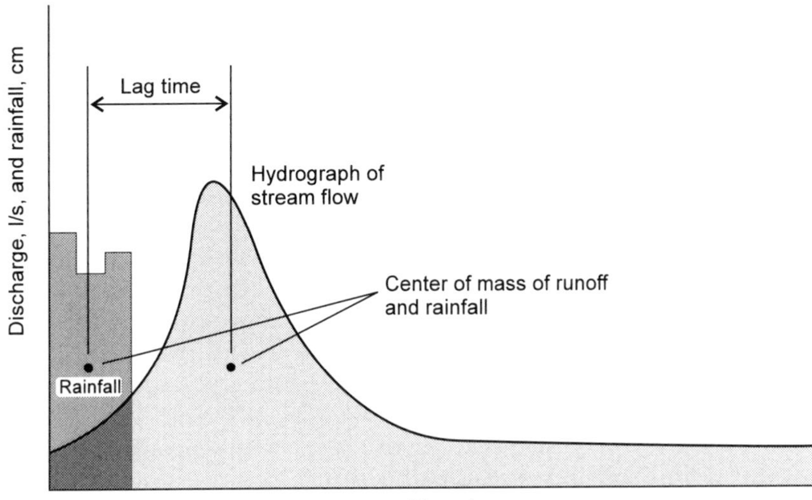

Figure 4.16. A hypothetical river hydrograph and its relationship to rainfall. (Leopold 1968)

that it takes rainwater to make its way to the river as surface runoff and via slower routes, which take it through forest litter, field vegetation, soil, and deeper subsurface parts of the local landscape.

4.6 Floods

The data in figure 4.15 are for a river whose maximum stream discharge did not vary wildly between the years 1974 and 1978. If the record had been longer, much larger fluctuations would probably have been observed. Flooding is endemic in river systems. Great floods are remembered with pain and awe. As the rate of discharge increases, the water level rises until rivers fill their channels. This condition is called the *bankfull stage*. A further increase in the river discharge leads to flooding. Water spills over the river banks and begins to cover the floodplains, which are areas that have always been attractive for human habitation. Water coursing over floodplains can therefore create enormous amounts of property damage and has frequently led to serious loss of life.

In 1539 Garcilaso de la Vega, a member of De Soto's ill-fated expedition to the New World, wrote that while they were building boats to take them down the Mississippi River to the Gulf of Mexico, "God, our Lord, hindered the work with a mighty flood, which began to come down with an enormous increase of water." The rising torrent "overflowed the wide level ground between the river and the cliffs that loomed some distance away; then, little by little, it rose to the top of the cliffs. Soon it began to overflow the meadows in an immense flood" (Clark 1982, chap. 3). The Mississippi was a law unto itself for more then three centuries after the flooding in 1539. Even a century ago Mark Twain, who was an experienced paddle-wheel pilot on the Mississippi, commented that men "cannot tame that lawless stream, cannot curb it or confine it, cannot say to it, 'go here or go there,' and make it obey."

Fortunately, Mark Twain was only partly right. Flood control measures implemented since the nineteenth century have managed to make the river obey most of the time. Levees along the banks of the river managed to keep it within its channel during many floods but have proved incapable of containing the river during periods of major flooding. Matters came to a particularly serious pass in the spring of 1927. During August of the previous year, unusually heavy rains began to fall over much of the Mississippi drainage basin. By September tributaries in eastern Kansas, northwestern Iowa, and parts of Illinois lapped over their banks. The rains continued throughout the autumn and winter. In April 1927 the Mississippi broke loose. Levees failed in many places along the lower part of the river. Water, 18 feet deep and 80 miles across in some places, spread over more than 16.5 million acres (67,000 km^2) in seven states. The official death toll was reported at 246 but may have been as high as 500. Nearly 650,000 people were driven from their homes. Officials put the damage at more than $230 million (Clark 1982, chap. 3). As William Faulkner wrote, the river "was now doing what it liked to do, had waited patiently the ten years in order to do, as a mule will work for you ten years for the privilege of kicking you once."

Clearly, levees alone had not been sufficient to control the worst of the Mississippi River floods (see fig. 4.17). Even the Army Corps of Engineers, a staunch advocate of levees for many decades, agreed that levees do not give sufficient protection. Since then, dams, protective coverings to stabilize the river banks,

Figure 4.17. Old Man River lies abed with a bad case of the floods in this cartoon from the Philadelphia *Ledger* of May 3, 1927. Secretary of Commerce Herbert Hoover and the commander of the U.S. Army Corps of Engineers, Major Gund Jadwin, press upon him the age-old remedy for high water—higher levees—while more modern (and, as it turned out, more effective) remedies lie ignored on the bedside table. (Clark 1982)

diversionary floodways to lessen the volume of water in the main stream, and an additional outlet to the Gulf of Mexico have been built, and the course of the river has been shortened by cutting channels across some of the great loops, the meanders, of the Mississippi. These improvements undoubtedly lessened the impact of the great floods in 1937, 1973, and 1993. The area that was flooded in 1973 was about the same as that in 1927; without the new flood control works it would probably have been nearly twice as great. Property losses in 1973 were estimated at $1 billion. An additional $15 billion in damages were probably avoided in areas protected by flood control measures. Only twenty-three deaths were ascribed to this flood.

However, the damage from the 1993 flood was much greater and is now estimated to be more than $12 billion. Flooding was major in the nine-state flood region, and new records were set along many stretches of the river system (see plate 3). The Missouri River at Kansas City, Missouri, and the Mississippi River at St. Louis, Missouri, crested nearly 50 feet (16 m) above their normal levels (see fig. 4.18 and plate 4). More than eight hundred of the fourteen hundred levees along the banks of the overflowing rivers either crumbled or were overrun by water. Levees turned out to be a mixed blessing: while they contained some cresting waters, they increased the chances of greater flooding downstream by drying up wetlands, changing the river channel, and increasing the force of the river. Thought is therefore being given to converting entire towns and some large tracts of farmlands into wetlands rather than rebuilding levees to protect them. The cost of rebuilding the levees relative to the cost of buying land and turning it into wetlands will largely determine the fate of the conversion proposal.

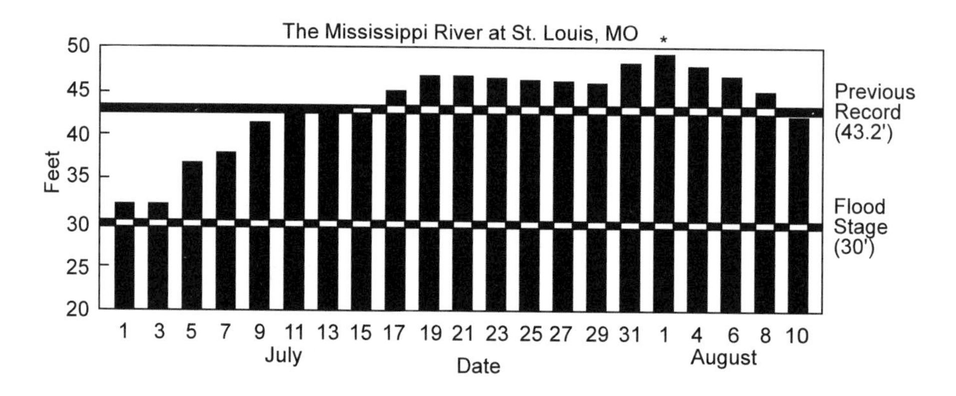

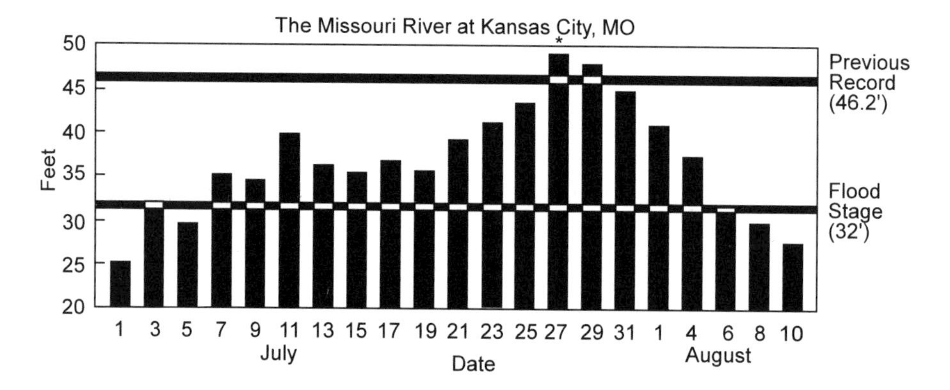

Figure 4.18. Histograms showing river levels at two locations along the Mississippi River (top and bottom) and on the Missouri River (middle) during the Midwest flood of 1993. (Halpert et al. 1994)

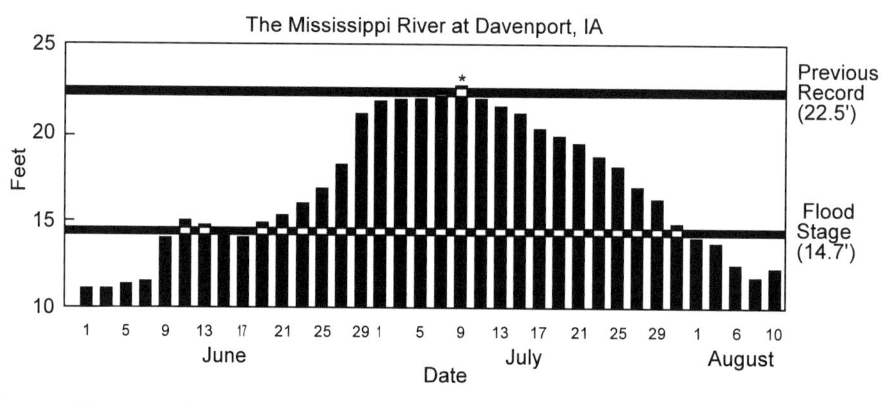

Many of the measures adopted to control flooding of the Mississippi have been known for a long time. Florence, that magnificent city of northern Italy, is bisected by the Arno River, which has flooded roughly once every 26 years. Between 1500 and 1510 Leonardo da Vinci submitted plans for a complex diversion of the Arno involving a large retention basin, a canal, a tunnel, and floodgates. His plans were ignored. In 1547 the river again flooded the low-lying parts of the city, devastated the Santa Croce district, and claimed more than 100 lives. After that disaster Bernardo Segni commented: "Because very great numbers of trees had been cut down for timber in the Faltevona and other mountains, the soil was more easily loosened by water and carried down to silt up the beds of rivers. In these ways man had contributed to the disaster." In spite of its long history of flooding, Florence was singularly unprepared for the great flood of November 1966, which drenched a great number of priceless books and art treasures. Fortunately, this disaster produced an outpouring of labor and money from many parts of the world, which prevented the damage from assuming even more catastrophic proportions.

The death tolls of the Mississippi floods and those of the Arno have been very minor compared to those in several parts of Asia. The floods of the Yellow River have been an important part of the history of China for thousands of years. The flood of 1887 alone was responsible for at least 900,000—and possibly as many as 2.5 million—deaths by drowning, starvation, and disease. In 1931 the Yangtze River burst its banks, and more than 3.7 million Chinese died, mainly of famine. Flood control measures taken since then have been largely successful. In 1981, when the Yangtze again rose to levels close to those of 1931, the floods were so well contained that fewer than 1,000 people died. Flood controls are still sadly lacking in some parts of the world. Bangladesh, a country of some 50 million people built largely on the low-lying flood plain of the Ganges-Brahmaputra, is particularly susceptible to flooding. Rains during 1990 were unusually heavy; this led to very extensive flooding, which claimed the lives of some 100,000 people.

Engineers and hydrologists responsible for flood control measures in river basins need to know the frequency and the intensity of past flooding events to predict the likelihood of future floods of a given magnitude. Diagrams such as the one in figure 4.19 have turned out to be very useful, though not foolproof, tools in making such predictions. The data in this figure are based on a 93-year record of floodings along the Cumberland River at Nashville, Tennessee. The maximum flood discharge for each of the 93 years has been listed and then rank

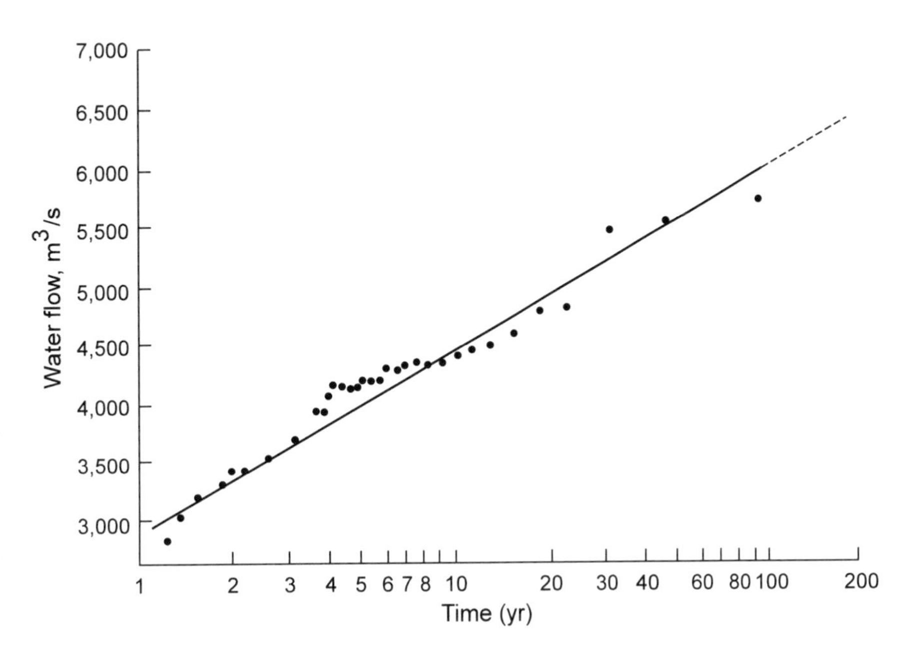

Figure 4.19. Average interval between floods of various sizes during 93 years on the Cumberland River at Nashville, Tennessee. The vertical scale is linear while the horizontal scale is logarithmic. (From Jarvis et al. 1936, and Judson and Kauffman 1990)

ordered by size from the largest to the smallest. The largest discharge has been assigned the number 1, the smallest the number 93. The flood recurrence, R, in years has then been calculated using the formula

$$R = \frac{N+1}{M}, \tag{4.6}$$

where

N = number of years of the record, and
M = the rank of each maximum annual discharge.

For the stream record in figure 4.19, the number N is 93 years. For the largest maximum river discharge during this period,

$$R = \frac{93+1}{1} = 94 \text{ years.} \tag{4.7}$$

The river discharge during this event was 5,750 m^3/sec. This flood therefore plots at an R value of 94 and a water flow rate of 5,750 m^3/sec. The second ranked annual discharge of 5,570 m^3/sec has an R value of 94/2 = 47 years, the third ranked annual discharge an R value of 94/3 = 31 years. As for the Cumberland River, a plot of the maximum water flow of many rivers during any given year versus the logarithm of their rank R approximates a straight line. Such graphs are useful for describing the behavior of rivers during the period of the flood record and, to a certain extent, for predicting the flooding behavior of the river. Figure 4.19 shows, for instance, that a water discharge of 5,750 m^3/sec occurred once in 93 years. A water discharge of 5,570 m^3/sec or more occurred twice during this time interval, or, on average, once in 46.5 years. A discharge of 5,570 m^3/sec is therefore called a 47-year flood. The lowest maximum river discharge in the record represents the one-year flood, since water discharges of at least this magnitude were recorded every year.

During the 93 years following this record, the frequency of flooding of a given intensity will probably be similar to that in figure 4.19, unless there are significant climatic changes within the river basin. It is likely, however, that at least one larger flood event will occur. This prophesy can be justified by dividing the 93-year record into two parts. The flood with an R value of 94 years will appear in only one of these 46.5-year records. A 186-year record is therefore apt to contain a flood event that is larger than the 94-year event in figure 4.19. The magnitude of this event can be estimated by extending the line in figure 4.19 to 187 years and reading off the corresponding water flow on the y-axis. This turns out to be 6,500 m^3/sec. It is reasonable to expect a discharge of this magnitude, but the estimate is not really firm. The magnitude of very large flood events is difficult to predict, because the statistics of rare events are uncertain. However, in designing flood control measures, modest extrapolations of a record as long as that in figure 4.19 are usually justified.

Floods are not always unmitigated disasters. It is hardly coincidental that the ancient Egyptians called their country "Black Land" in honor of the rich mud that is carried down from the Abyssinian Plateau during the annual floods of the Nile.

For the farmers of the Nile Valley, the greatest problems have occurred not when the river flooded, but when it failed to flood, denying them the life-giving waters necessary to nourish their crops and the mud to fertilize their fields. Food storage programs to fend off famine in the Nile Valley during periods of drought go back at least as far as the biblical story of Joseph in the Book of Genesis.

The construction of the high dam at Aswan on the Nile now prevents silt from reaching the Nile Delta. Many of the effects of this change on the ecosystem of the river were unforeseen and—in some ways—unfortunate (see chapter 11). The land in the Nile Delta is subsiding and fisheries along the coast have declined. The Aswan Dam is not alone in this respect. All large dams have a significant impact on the life of their river, and various objections have always been raised against their construction. Some are based on economic grounds, some on environmental grounds, some on sociological grounds. In his essay, "A River," John McPhee (1971) describes the encounter between David Brower, an ardent conservationist, and Floyd Dominy, a fervent advocate of dams. The essay is an eloquent account of the clash between two men with very different values that set them far apart in their opinions on the desirability of dams. Large dams are major human interventions in the workings of nature and cannot be regarded simply as flood control measures, as ways to make water more accessible for human use, or as sources of hydroelectric power (see chapter 11).

4.7 Groundwater

In most parts of the globe, at least some rainwater penetrates into the ground. This water makes its way to a river, to a lake, or to the ocean by paths that are often long and circuitous. Its flow is driven by gravity and is determined by the physical properties of the materials that it encounters. Soil, the first material usually encountered by rain, is the epidermis of loose particles and fragments that supports the terrestrial biosphere. Except in swamps and tundras, soils are not usually water saturated—i.e., completely filled with water—right to the surface. Unless they are quite deep, trenches dug in soil may not fill with water, except, of course, after heavy rains. The level below which a soil is completely water saturated is called the *water table*. Soil and rock above the water table are said to be in the *unsaturated zone*. The depth from the surface to the water table depends on a host of variables. Among these rainfall, temperature, topography, and subsurface geology are particularly important. In the cool, damp English climate the depth to the water table is rarely more than 1 or 2 meters. In winter, groundwater tends to creep up into the walls of houses, much like it creeps into a piece of blotting paper. In houses that are poorly heated this gives rise to an unpleasant phenomenon, appropriately called "the damp." At the other extreme, in desert areas, the water table may be hundreds of meters below ground. From the point of view of water availability, this is a distinct problem. On the other hand, it is a very real asset for siting repositories of nuclear and other wastes, since the access of groundwater to stored wastes can be very dangerous (see chapter 11).

In unpolluted areas of the world, CO_2 is usually the most reactive component of rainwater. Within soils, atmospheric CO_2 is joined by CO_2 released by root respiration and decay (see chapter 6) and combines with water to form carbonic acid, H_2CO_3. Together with organic acids these compounds react with near-surface materials to generate new soil. In the process, soil water gradually picks up small quantities of dissolved salts. Most of these are carried along with groundwater and ultimately flow into the oceans as constituents of river water (see chapter 7). In seasonally hot and dry climates, groundwater can move many meters back up to the surface by capillary action, evaporate there, and deposit its dissolved salts. Among these deposits, $CaCO_3$ and SiO_2 are the most common. The precipitation of $CaCO_3$ gives rise to hard crusts called *calcrete*; the deposition of SiO_2 leads to the formation of hard crusts called *silcrete*.

Groundwater flow below the water table depends on the local topography and on the subsurface geology. The simplest type of flow occurs in areas where the subsurface material has a constant permeability, i.e., where the transmission of groundwater is equally easy throughout. A solid block of steel has an extremely low permeability; beach sand has a very high permeability. The permeability of steel is low, because the amount of interconnected space between adjacent grains of metal is exceedingly small. The permeability of beach sand is high, because there is a good deal of interconnected space between adjacent grains. Materials that are highly porous may not be particularly permeable. Swiss cheese comes to mind. Its numerous large holes give this material a high degree of porosity, but its permeability is low, because few of the holes are connected. Some geologic materials have similar properties. Certain types of lava, for instance, are so porous that they float on water; yet the pores tend not to fill up with water, because few of them are connected.

The rate of groundwater flow through a body of soil, sediment, or rock depends on the viscosity of water, the acceleration due to gravity, the permeability of the material through which the water is flowing, and the gradient in the flow direction. The quantity, q, of water flowing per unit area in unit time through the circular cylinder in figure 4.20 is given by the expression

$$q = K\left(\frac{h_1 - h_2}{l}\right), \tag{4.8}$$

where h_1 = fluid level in the upper tube, h_2 = fluid level in the lower tube, l = distance between the two tubes, K = hydraulic conductivity, a parameter that is proportional to the permeability of the packing material in the cylinder. Equation (4.8) is known as *Darcy's law* (see, for instance, Hubbert 1956). The elevation

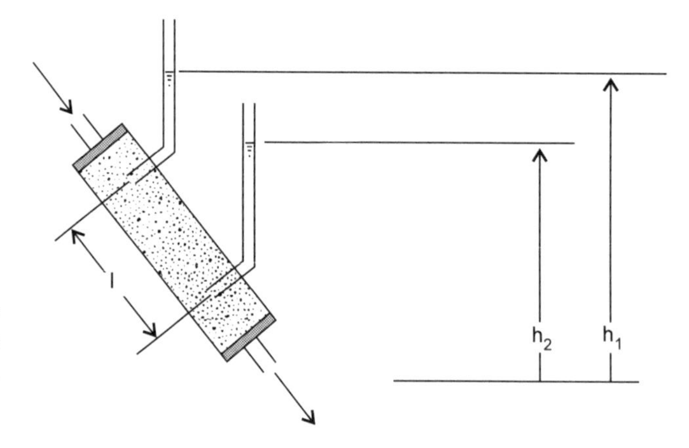

Figure 4.20. Experimental apparatus to illustrate Darcy's law.

Table 4.2.
The Hydraulic Conductivity of Geological Materials

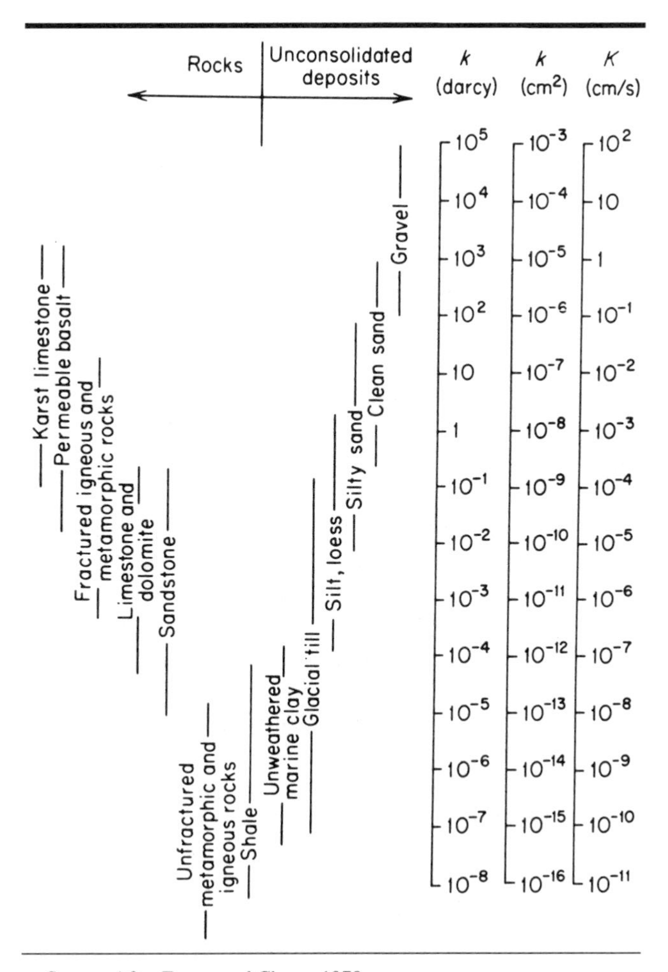

Source: After Freeze and Cherry 1979.

difference $(h_1 - h_2)/l$ is the hydraulic gradient. For example, if the difference in head between the two tubes is 1 meter and the distance between them is 100 meters, the hydraulic gradient is 0.01. The significance of the hydraulic gradient is similar to but not identical with that of a highway gradient.

The value of K is highly variable. For essentially impermeable rocks, its value is less than 10^{-8} cm/sec. For highly permeable material such as beach sand, K is about 10^{-2} cm/sec. Table 4.2 lists the values of K for a wide range of geologic materials. In ordinary aquifers—i.e., rock units that act as water reservoirs—groundwater typically flows at rates greater than 1.5 m/yr and less than 1.5 m/day. Since these flow rates are very much less than the flow decay rate of water in rivers, the travel times of water in aquifers are much longer than in rivers. If the distance between a recharge area of groundwater and its discharge into a river is 15 km, and if the mean velocity of the groundwater en route is 1.5 m/day, the travel time of the groundwater between intake and discharge is about 30 years. If the mean velocity en route is 1.5 m/yr, the travel time is 10,000 years. Aquifers can therefore contain quite "old" water, and this has implications

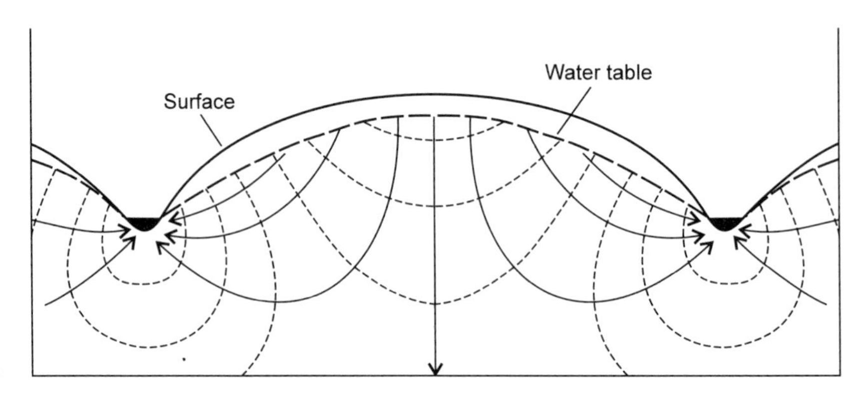

Figure 4.21. Diagram of flow lines for groundwater moving from recharge areas through a homogeneous, permeable unit to low-lying discharge areas. (Based on Hubbert 1940)

both for groundwater use (see next section) and for the cleanup of contaminated groundwaters (see chapter 12).

Groundwater paths tend to be complex, even in homogeneous ground. Figure 4.21 shows a simple hill between two rivers. In the valleys, the water table lies at stream level; the depth to water table is greatest at the top of the hill. Groundwater does not flow in straight lines from the recharge area to the discharge area. Rather, the flow lines are curved, so much so that rainwater falling close to the top of the hill actually enters the discharge area from below. The physics of groundwater flow in such settings is now quite well understood (see, for instance, Freeze and Cherry 1979). However, the quantitative modeling of groundwater flow in geologically complex areas—and most areas are geologically complex—is impossible without the use of high-speed computers. Even with such computers, models are often no more than rough approximations, because the physical properties of real geologic settings cannot be known at all points and for all times of interest.

Qualitatively, the flow of groundwater even in geologically inhomogeneous areas can be understood without recourse to computers. If the permeable aquifer at the top of the hill in figure 4.21 were underlain by an impermeable rock unit as shown in figure 4.22, rainwater entering the water table in the recharge area would be deflected and would tend to move along the boundary between the permeable and the impermeable units until it reaches the hillside. There it would issue as a series of springs.

The contact between permeable and impermeable rock units is rarely flat. Figure 4.23 shows a common geometry, where the aquifer is a stratum of permeable

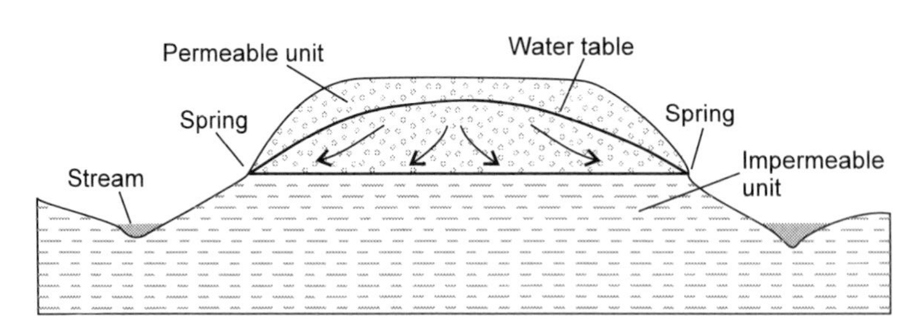

Figure 4.22. Flow of groundwater through a hill capped by a permeable unit that is underlain by an impermeable unit. Groundwater issues as springs at the contact between the two units. (Judson and Kauffman 1990)

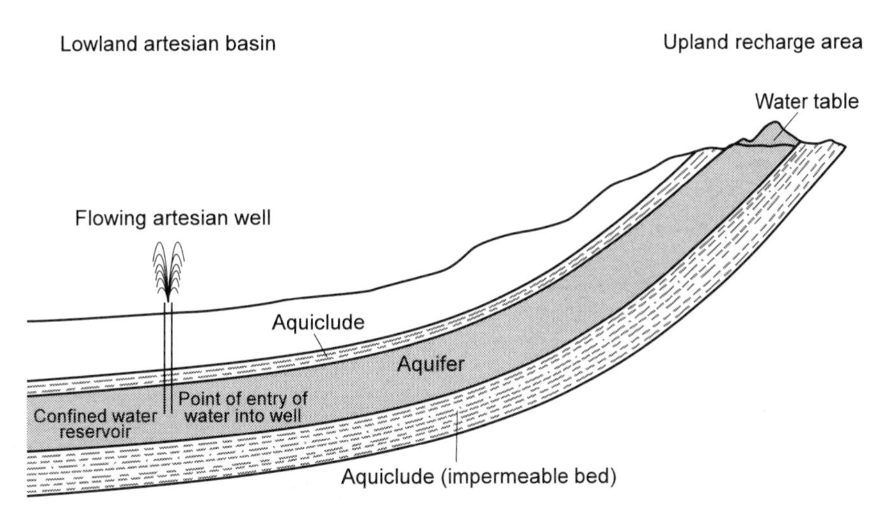

Figure 4.23. A confined water reservoir is created where water enters an aquifer sandwiched between two confining aquicludes. The artesian well flows in response to the pressure difference between the height of the water table in the recharge area and the bottom of the well before the well was drilled. (Press and Siever 1986)

rock that is sandwiched between layers of impermeable rock. Rainwater that falls in the upland recharge area is constrained to flow within the aquifer. If a well is drilled at low elevations, where ground level is below the level of the water table in the recharge area of the aquifer, water will flow from the well without pumping. Such artesian wells are obviously highly desirable.

In many parts of the world, groundwater is held not in homogeneous permeable aquifers but in fractures. Rocks such as granites are essentially impermeable unless they are fractured. Many limestones consist of tightly cemented grains of calcite (see chapter 7). The porosity and permeability of limestones are generally small, and groundwater can flow through them only where they are fractured. Since calcite is slightly soluble in most groundwaters, fractures in many limestones are gradually enlarged into connected channel ways that can become accessible to caving enthusiasts (see chapter 6). Two students wanted to test this interconnectedness by adding fluorescein dye as a tracer to groundwater in the limestone aquifers of their home area. They did not know quite how much fluorescein to add, so, to be on the safe side, they added a whole bucketful. The consequences were spectacular. For several weeks all of the surrounding households that were drawing water from wells in this limestone aquifer were treated to brightly fluorescing water. The experiment was clearly a success; needless to say, the students failed to claim credit for it.

If the solution of limestones by groundwaters goes far enough, remnants of the original rock units that have escaped dissolution can create memorable landscapes. One of the most spectacular examples of what is called "Karst topography" is the scenery along the river Li near Guilin in China (see fig. 4.24), which has been the subject of so many incredible Chinese landscape paintings.

The flow of groundwater does not need to be purely through pores or purely through fractures. If, for instance, a porous aquifer is cut by fractures that also pass through impermeable rock layers above or below the aquifer, groundwater will move from the aquifer into these fractures and is apt to travel through them until it reaches the surface or until the fractures intersect another permeable aquifer.

Figure 4.24. Karst topography along the River Li near Guilin, China. (Photo: Hans J. Holland)

In coastal areas and on oceanic islands, the flow of rain-related, low-salinity groundwater is complicated by the presence of seawater. The transition zone from freshwater to seawater can have a complex geometry, and submarine freshwater springs are common features along some coasts. Thick lenses of freshwater can form below oceanic islands. The flow of freshwater in such a lens is shown in figure 4.25. The downward water pressure at its base is equal to the upward water pressure of seawater just below the base of the lens. This condition is satisfied if the depth to the base of the freshwater lens below sea level is about forty times the height of the groundwater table above sea level. On an island in which the crest of the water table is 2 meters above sea level, the base of the freshwater lens is apt to be about 80 meters below sea level. Groundwater flow within "real" islands is much more complex than in the island of figure 4.25, and the transition from fresh- to saltwater is much more gradual, because the permeability of sediments and rocks in most oceanic islands is not uniform.

The major driving force in all of the examples of groundwater flow discussed above is simply gravity; but this is not entirely true where there are large temperature differences within a groundwater area. In volcanic areas such as Yellowstone Park, the geysers in California, and on the island of Hawaii, temperatures 100 meters below the surface can be well above the boiling point of water. Rainwater that penetrates into such areas is heated and tends to reappear at the surface as hot springs and geysers. Such areas are important not only as tourist attractions but as sources of geothermal power (see chapter 11) and as the locus of ore deposits (see chapters 8 and 9).

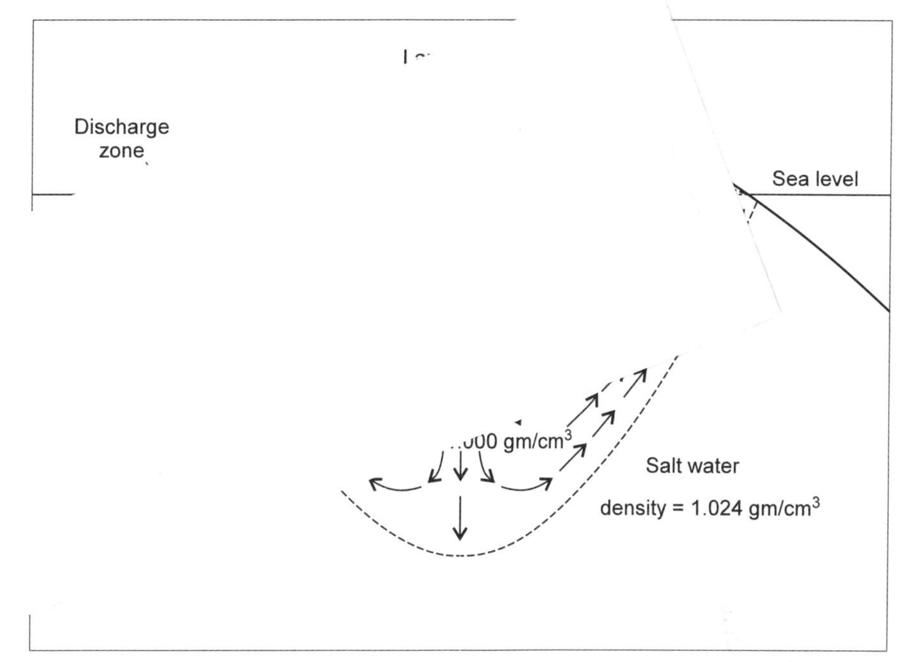

Figure 4.25. A freshwater lens below an oceanic island that consists of homogeneous, permeable material.

4.8 Water Use

Water is something we tend to take for granted. When we turn on a tap, we expect water to come pouring out. We also expect the water to be drinkable, to have a reasonably pleasant taste, and to be cheap. Water bills are usually much smaller than bills for property taxes, mortgage payments, and car repairs. Where they are not, water rates tend to be hot political issues. The supply of water in rainfall is very uneven in the United States (see fig. 4.13). Differences in runoff from the several parts of the United States are even more pronounced (see fig. 4.14). Water is much more plentiful in the well-watered eastern parts of the country and in the American Northwest than in the dry Southwest. The range of water supply issues that faces these different regions is large and includes many of the same issues that face other parts of the world.

An enormous amount of rain falls on the United States as a whole. The average rainfall is ca. 75 cm/yr. Since the area of the United States is 7.8 million km^2, the total rainfall is

$$75 \frac{\text{cm}}{\text{yr}} \times 7.8 \times 10^{16}\ \text{cm}^2 = 585 \times 10^{16} \frac{\text{cm}^3}{\text{yr}} = 585 \times 10^{10} \frac{\text{m}^3}{\text{yr}}.$$

The average runoff from the United States is 22.5 cm/yr. The total runoff is therefore

$$22.5 \frac{\text{cm}}{\text{yr}} \times 7.8 \times 10^{16}\ \text{cm}^2 = 175 \times 10^{16} \frac{\text{cc}}{\text{yr}} = 175 \times 10^{10} \frac{\text{m}^3}{\text{yr}}.$$

This is equivalent to 1,320 billion gallons per day, or about 5,100 gallons per person per day, a figure that is huge compared to our daily consumption of drink-

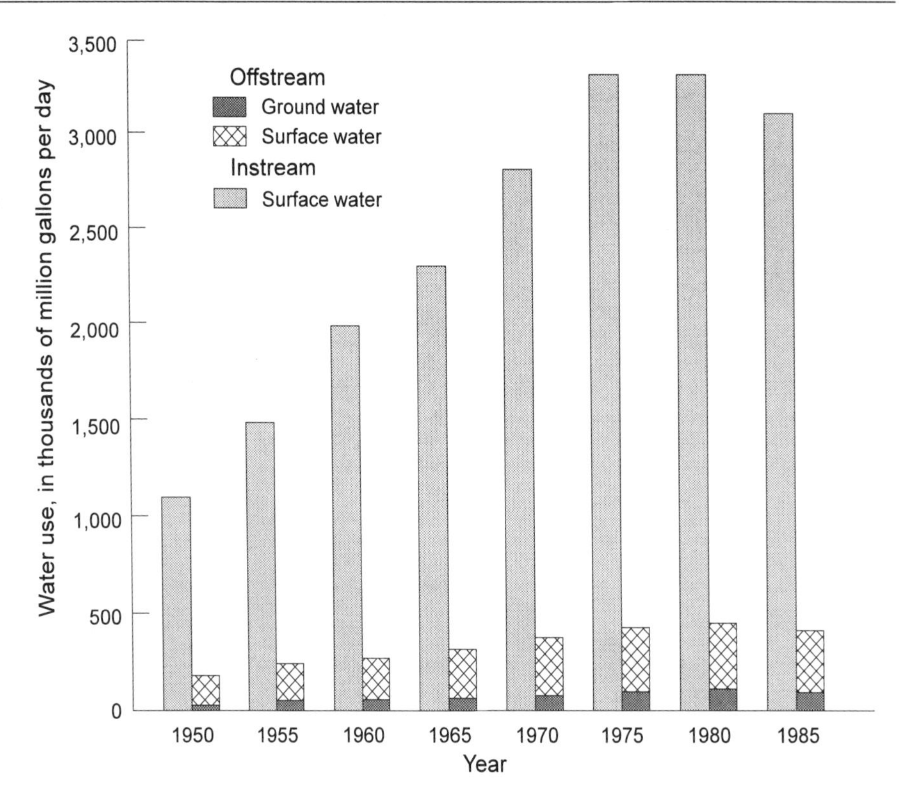

Figure 4.26. Trends in offstream and instream water uses in the United States, 1950–85. (Solley et al. 1988)

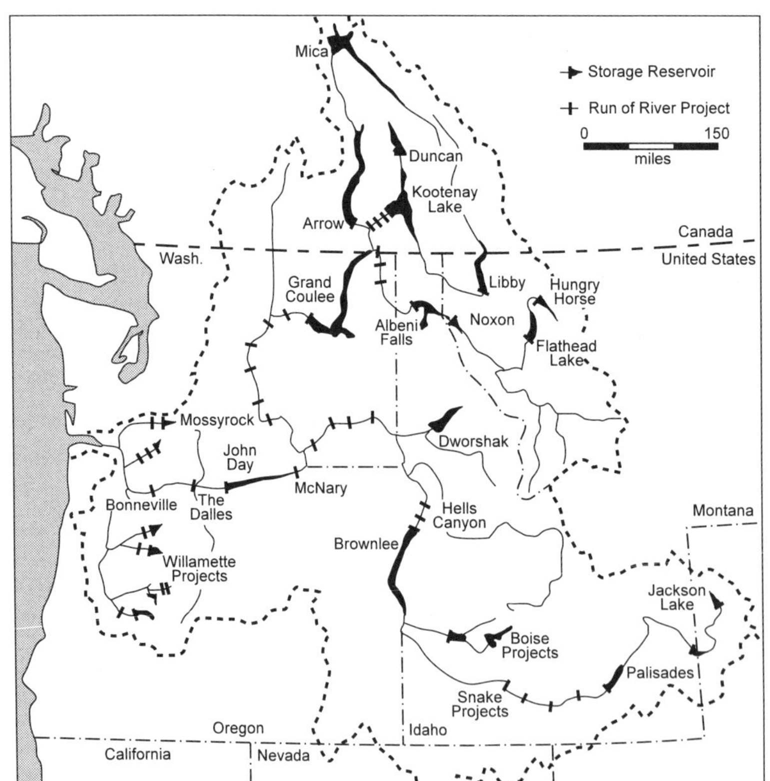

Figure 4.27. Map of the Columbia River basin showing the principal dams. (U.S. Army Corps of Engineers, Portland, Oregon)

ing water. It is even very large compared to our total home consumption, which includes showers, baths, the washing of dishes and clothes, and the flushing of toilets. It is not large, however, if we include water that is used for industrial and agricultural purposes.

Figure 4.26 shows the trends in water use for the United States between 1950 and 1985. Two types of use are distinguished: instream and offstream. By far the most important instream use was for the generation of hydroelectric power. In 1985 approximately 11.4 billion m^3 (3,000 billion gal/day) were used for this purpose (Solley et al. 1988). This is equivalent to $410 \times 10^{16} cm^3/yr$. At first glance, such a rate looks preposterously high, since it exceeds the entire runoff from the United States. However, river water flowing through mountainous terrain can be used repeatedly along its course to generate hydroelectric power. Figure 4.27 shows the location of the principal dams in the Columbia River basin. The amount of instream water use in this basin is many times the flow of the river. The loss of water due to instream use is small but not negligible. However, the evaporation of water from large reservoir lakes behind dams in hot, dry parts of the country is significant and has been used as an argument against building dams in such areas.

The offstream use of water between 1950 and 1985 was much smaller than the instream use. This is shown graphically in figure 4.26 and numerically in table 4.3. For the country as a whole, the offstream use in 1985 was some 1.5 billion m^3/day (400 billion gal/day), i.e., about 13% of the instream use rate and about one-third of the total daily runoff of 5 billion m^3/day (1,320 billion gal/day). Approximately 80% of the offstream use consisted of surface water; the remaining 20% consisted of groundwater. The sketch in figure 4.28 shows some of the main features of offstream water use. Water is withdrawn from a river and from wells into the public water supply. From there it is distributed to water users. Some of these users are partially self-supplied by wells and/or by direct water withdrawals from rivers. Wastewater can be discharged either directly into the river or after undergoing wastewater treatment. The sketch in figure 4.28 is obviously simplified; for some cities it is incomplete. Boston, for instance, draws most of its water from distant reservoirs and has, until recently, discharged most of its wastewater into Boston harbor. A concerted and expensive tunneling effort is underway to discharge Boston wastewaters several miles out at sea (see chapter 12). This diversion should rescue Boston from the onus of having one of the most polluted harbors in the United States.

In figure 4.29 and in table 4.3, the offstream uses of water in the United States are divided into five categories. Thermoelectric power generation accounts for nearly half of all offstream water use. Water in this category is used largely for condenser and reactor cooling in power plants. Since this use is nonpolluting, cooling water can be reused either directly or after cooling. A small fraction of the water used for thermoelectric cooling is lost by evaporation.

Irrigation accounts for about 35% of all offstream withdrawals. A large part, approximately 70%, of this water is lost, largely by evaporation; the remainder is reusable as return flow to surface and groundwater supplies. The high loss rate of irrigation makes this use a major target for water conservation. However, irrigation has played a major role in the increase of agricultural productivity of the United States, particularly in the rain-poor states. The western states accounted for some 90% of the total water withdrawn for irrigation between 1980 and 1985;

Table 4.3. Summary of Estimated Water Use in the United States at 5-Year Intervals, 1950–85 (in thousands of millions of gallons per day)

	Year								Percentage Change
	1950[a]	*1955*[a]	*1960*[b]	*1965*[b]	*1970*[c]	*1975*[d]	*1980*[d]	*1985*[d]	*1980–85*
Population, in millions	150.7	164.0	179.3	193.8	205.9	216.4	229.6	242.4	+6
Offstream use									
Total withdrawals	180	240	270	310	370	420	440[e]	400	−10
Public supply	14	17	21	24	27	29	34	37	+7
Rural domestic and livestock	3.6	3.6	3.6	4.0	4.5	4.9	5.6	7.8	+39
Irrigation	89	110	110	120	130	140	150	140	−6
Industrial use									
Thermoelectric power	40	72	100	130	170	200	210	190	−13
Other	37	39	38	46	47	45	45	31	−33
Source of water									
Ground									
Fresh	34	47	50	60	68	82	83[e]	73	−12
Saline	( )[f]	.6	.4	.5	1	1	.9	.7	−29
Surface									
Fresh	140	180	190	210	250	260	290	260	−8
Saline	10	18	31	43	53	69	71	60	−16
Reclaimed sewage	( )[f]	.2	.6	.7	.5	.5	.5	.6	+22
Consumptive use	( )[f]	( )[f]	61	77	87[g]	96[g]	100[g]	92[g]	−9
Instream use									
Hydroelectric power	1,100	1,500	2,000	2,300	2,800	3,300	3,300	3,100	−7

Source: Solley 1988.
Note: The data generally are rounded to two significant figures; percentage changes are calculated from unrounded numbers.
[a] 48 States and District of Columbia.
[b] 50 States and District of Columbia.
[c] 50 States, District of Columbia, and Puerto Rico.
[d] 50 States, District of Columbia, Puerto Rico, and Virgin Islands.
[e] Revised.
[f] Data not available.
[g] Freshwater only.

California and Idaho alone accounted for some 37% of this water. These regions contribute a sizable fraction of the entire agricultural yield of the country (see chapter 5). The growth of agriculture in the western United States led to the abandonment of many farms in the East. In many parts of New England this has been followed by the regrowth of forests on land that was farmed during the eighteenth and nineteenth centuries.

In dry areas, the availability of water can make the difference between wealth and poverty, even between life and death. Sugar, a mainstay of the Hawaiian Islands economy, would be impossible to grow on a large-scale without irrigation. On the island of Maui, rainfall is heavy on the slopes of Mount Haleakala, but the best areas for sugar growing are in the dry plains. These plains started to

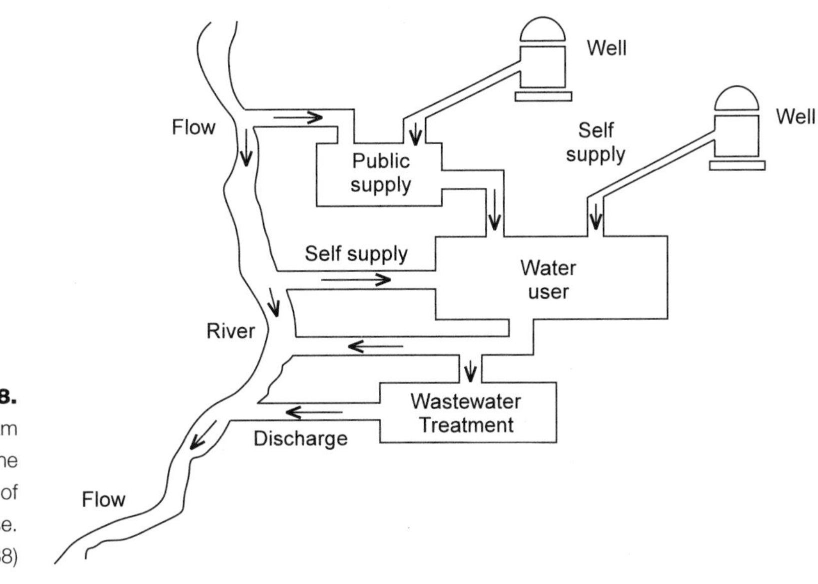

Figure 4.28. Schematic diagram showing some of the main features of offstream water use. (Solley et al. 1988)

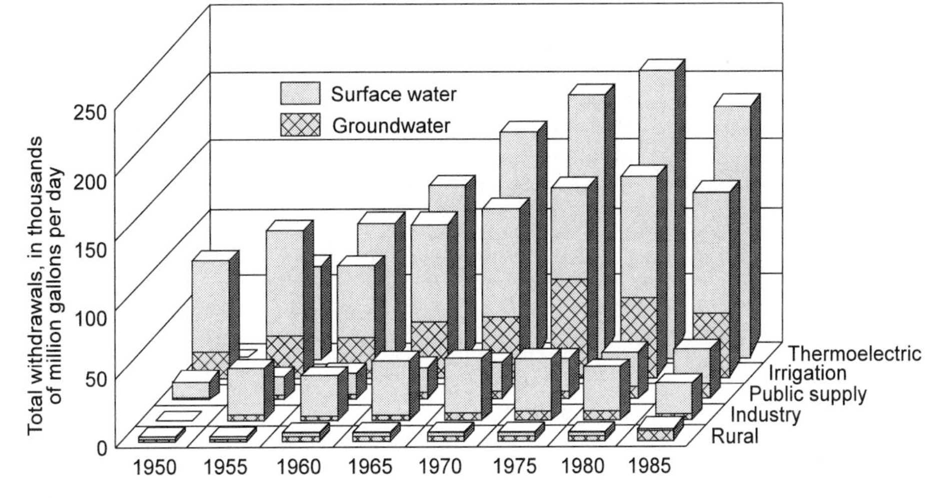

Figure 4.29. Trends in water withdrawals by water-use category and source of supply for rural, industrial, public supply, irrigation, and thermoelectric power uses, 1950–85. (Solley et al. 1988)

become productive only after major irrigation projects brought water down into the fields in the late 1870s. The water rights for these projects were frequently obtained by bribery and political chicanery (see, for instance, Adler 1966). Hawaii was not alone in this. The bitter feuds over water rights in the Western states are part of American history. The U.S. courts enforce a "first-use doctrine." This rankles particularly in states like Wyoming, because early irrigators in Nebraska established rights to major amounts of water in the Platte River. Struggles for water will undoubtedly intensify as the population of the Sunbelt increases and as the demands for water escalate.

The withdrawal of water by public and private water suppliers for domestic, industrial, and commercial uses is third in importance among offstream uses.

About half of the 140 million cubic meters (37×10^9 gal) per day supplied in 1985 were delivered to domestic users. This amounted to some 400 liters (105 gal) per day for each person served. Of this, about 23% represented consumptive use. The remaining 77% were reusable as return flow. The two smallest categories of offstream water withdrawal are for industrial uses other than thermoelectric power and for rural domestic and livestock use, i.e., for nonirrigation farm consumption. Together, these uses accounted for about 10% of the total offstream water withdrawals in 1985.

4.9 Water Supplies in the Future

Though the population of the world is increasing rapidly, the economic welfare of people in most parts of the globe is improving. The need for water is increasing and will probably continue to increase during the coming century. Will we be able to meet this demand, and if so, at what financial and environmental costs?

More dams will undoubtedly be built, particularly on rivers with slightly tapped hydroelectric potential. The instream use of water will therefore rise. Much of the power required by countries that are mountainous and well supplied with water will continue to be supplied by hydroelectric power. In countries like the United States, where hydroelectric power is already well developed but accounts for only a modest fraction of the total energy budget (see chapter 11), increases in instream water use are unlikely to supply more than a small fraction of the additional energy required during the next century. Countries like Canada could expand their hydroelectric power production considerably, but the large proposed Hydro-Québec project in eastern Canada has run into formidable opposition for environmental and social reasons; it remains to be seen how these controversies will be resolved and to what extent similar concerns will limit the exploitation of hydroelectric power in other parts of the globe.

The offstream demands for water will probably be difficult to meet in some parts of the world. In the well-watered regions, which include the eastern United States, water is not and probably will not become seriously limiting in the foreseeable future, although drought cycles will continue to occur, as they always have, and may seriously affect farmers in some areas. In the drier regions, which include the western United States, water will surely become a progressively scarcer commodity. It is already transported hundreds of miles via the California aqueduct to supply the offstream needs of southern California. The well-watered American and Canadian Northwest could deliver considerable volumes of water to the dry Southwest, but large-scale water transport from these areas may not be economically or politically feasible in the twenty-first century.

One of the largest public works projects ever conceived has been proposed by China's Communist party leadership to save its capital, Beijing, from perpetual drought in the next century. The plan calls for the diversion of water from the huge reservoir at Danjianku to Beijing and Tianjin some 860 miles to the northeast at a cost of at least $5 billion. A huge aqueduct will carry some 15 billion cubic meters (4,000 trillion gal) per year of water, more than seven times the quantity used by New York City and twice as much as the amount of water delivered by California's largest water diversion system. The proposed South-North Water Diversion project would be second in magnitude only to China's Three Gorges project (see chapter 11), which was approved after four years of contentious debate.

Opponents of the aqueduct have pointed out that Chinese cities waste billions of gallons of water each year through poor conservation practices and leaky toilets. President Jiang Zemin complained that "if a country can send satellites and missiles into space, it should be able to dry up its toilets." His remarks sent Communist party cadres skittering to mount a nationwide campaign against makers of shoddy toilets. Their success would surely help the water shortage but is unlikely to reduce the shortfall of water in the Beijing region significantly.

As the price of water increases, water will surely be used more efficiently. New irrigation techniques are replacing old standbys, and less water-intensive crops will probably replace water-intensive varieties. In the meantime, a good deal of the groundwater that is used in dry parts of the world is literally being mined out. This practice is no more reprehensible than the mining of gold and of other nonrenewable commodities. It means, however, that aquifers are being depleted, and that other sources of water must be found when they are exhausted. Perhaps the most important example of water drawdown in a U.S. aquifer is the rapid depletion of the Ogallala Aquifer, which supplies groundwater to much of the Great Plains. This aquifer extends about 400 miles north-south, and has a maximum width of about 80 miles east-west (see, for instance, Peterson and Keller 1990).

On Oahu in the Hawaiian Islands, abundant artesian water was discovered during the late 1800s in volcanic rocks west of Pearl Harbor. After about 30 years of exuberant use, the artesian fountains that had initially arched to a height of 12 meters (40 ft) were reduced to mere seeps (Culliney 1988). Nearby Honolulu has grown so large that water supply problems are imminent. It is virtually certain that water for large new apartment buildings will have to be supplied by desalination plants.

Desalination plants will also become an important source of water in southern California and in other dry regions of the world close to seacoasts (Petersen 1994). Desalinated seawater is already a major source of freshwater on the Arabian Peninsula and will probably become a staple in Israel and among its neighbors. An intriguing alternative to desalinated water in Arabia was proposed some years ago: icebergs. The idea was promoted at cocktail parties where the drinks were, of course, cooled with bits of iceberg. Unfortunately, the economics of transporting icebergs from the Arctic to the Persian Gulf have proved daunting.

Desalinated water is more expensive than freshwater obtained locally. In most parts of the United States, domestic water costs between forty cents and a dollar per cubic meter ($1.60–$4.00/1,000 gal). At this rate the price of water is about 1/1,000 the price of milk and gasoline, and about 1/10,000 the price of a modest wine. The data in figure 4.30 show that the price of water in the United States is lower than in many European countries. Australia has very expensive water. The price of water in Germany may shortly exceed the price in Australia, because huge sums will be required to upgrade the former East Germany's water systems.

A doubling of the price of domestic water would not be a major calamity. Even at a price of $2.65 per cubic meter (about $10.00/1,000 gal.) the cost of water for an average U.S. customer using 375 liters (100 gal) per day would only be a dollar per day. Water costs in this range would, however, have a serious impact on agriculture. About two tons of water are required to produce one kilogram of

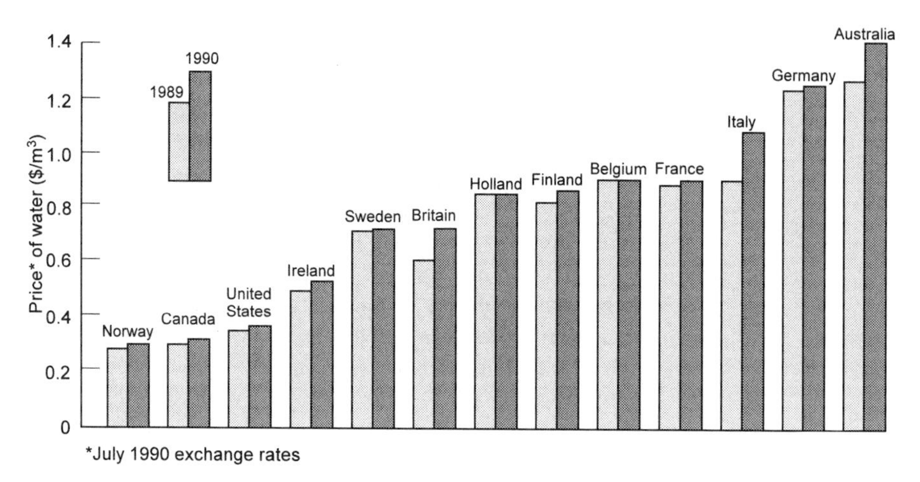

Figure 4.30. The price of water in the United States, Australia, and a number of European countries (*The Economist*, 1990). Note that a price of $1.00 per cubic meter is the equivalent of $3.80 per 1,000 gallons.

sugar with traditional irrigation techniques. Even at a cost of only fifty cents per cubic meter ($2.00/1,000 gal), two tons of water would cost a dollar. The cost of sugar is currently about sixty cents per kilogram ($0.30/lb). Water used in sugar production is currently very cheap, and sugar growers who are forced to switch to desalinated water would no longer be able to compete in the open market. Significant increases in the price of irrigation water would almost certainly shift agriculture from the dry, western parts of the United States to the wetter eastern parts. Similar trends would appear in other parts of the world. However, in the long run, the increasing demand for food to feed the burgeoning population of the world (see chapter 5) may make farming profitable even where the cost of irrigation water adds significantly to the price of agricultural products.

The environmental consequences of farming in dry areas can become very severe. The precipitation of salts in soils due to the evaporation of irrigation water can affect soil productivity, and the buildup of toxic metals in soils can put an end to agriculture completely (see chapter 12). The depletion of underground aquifers can lead to land subsidence, and excessive pumping of water in near-shore areas such as Florida and Long Island, New York, can lead to the intrusion of seawater into freshwater supplies.

Predictions of water availability generally assume that the effects of any changes in climate will be minimal. This assumption may not be correct. Memories of the dust bowl years of the 1930s are still vivid in the minds of older Americans and of readers and viewers of John Steinbeck's *The Grapes of Wrath.* Pueblo Indians may have been forced to abandon their original homes in the West as a result of intense drought early in the present millennium. Neither of these fluctuations was caused by people. In addition to natural variations, climate during the next century may well be affected by anthropogenic changes in the chemistry of the atmosphere (see chapter 11). Little is presently known about the effects of changes in the composition of the atmosphere on the hydrologic cycle (see, for instance, Waggoner 1990). The total amount of rainfall may increase, and the distribution of rainfall over the globe may change considerably. The effect of major shifts in rainfall distribution on farming areas would be dramatic, and could be catastrophic. We are indeed living dangerously.

References

Adler, J. 1966. *Claus Spreckels: The Sugar King of Hawaii.* University of Hawaii Press, Honolulu.

Berner, E. K. and Berner, R. A. 1987. *The Global Water Cycle: Geochemistry and Environment.* Prentice Hall, Englewood Cliffs, N.J.

Byers, H. R. 1965. *Elements of Cloud Physics.* University of Chicago Press, Chicago.

Chang, A.T.C., Chiv, L. S., and Wilheit, T. T. 1993. Oceanic Monthly Rainfall Derived from SSM/I. *EOS Transactions* 74:505, 513.

Clark, C. 1982. *Planet Earth: Flood.* Time-Life Books, Alexandria, Va.

Culliney, J. L. 1988. *Islands in a Far Sea: Nature and Man in Hawaii.* Sierra Club Books, San Francisco.

Curtis, W. F., Culbertson, J. K., and Chase, E. B. 1973. Fluvial Sediment Discharge to the Oceans from the Conterminous United States. U.S. Geol. Survey Circ. 670.

Dalrymple, G. B. 1991. *The Age of the Earth.* Stanford University Press, Stanford, Calif.

EOS Program Office. 1993. *EOS: A Mission to Planet Earth.* NASA, Washington, D.C.

ESSA Environmental Data Service. 1969. *Climates of the World.* Ashville, N.C.

Freeze, R. A. 1982. Hydrogeological Concepts in Stochastic and Deterministic Rainfall-Runoff Predictions. In *Recent Trends in Hydrogeology*, pp. 63–79. Special Paper 189, Geological Society of America.

Freeze, R. A., and Cherry, J. A. 1979. *Groundwater.* Prentice Hall, Englewood Cliffs, N.J.

Gibbs, R. J. 1972. Water Chemistry of the Amazon River. *Geochim. Cosmochim. Acta* 36:1061–66.

Halpert, M. S., Bell, G. D., Kousky, V. E., and Ropelewski, C.F., eds. 1994. *Fifth Annual Climate Assessment, 1993.* Climate Analysis Center, National Oceanic and Atmospheric Administration.

Hobbs, P. V. 1974. *Ice Physics.* Oxford University Press, Oxford.

Holeman, J. N. 1968. The Sediment Yield of Major Rivers of the World. *Water Resour. Res.* 4:737–47.

Holland, H. D. 1978. *The Chemistry of the Atmosphere and Oceans.* Wiley-Interscience, New York.

Holland, H. D. 1984. *The Chemical Evolution of the Atmosphere and Oceans.* Princeton University Press, Princeton, N.J.

Hubbert, M. K. 1940. The Theory of Groundwater Motion. *Jour. Geol.* 48:785–994.

Hubbert, M. K. 1956. Darcy's Law and the Field Equations of the Flow of Underground Fluids. *Trans. Amer. Inst. Mining Metall. Engineers* 207:222–39.

Jarvis, C. S., et al. 1936. Floods in the United States—Magnitude and Frequency. U.S. Geol. Survey Water Supply Paper 771.

Judson, S., and Kauffman, M. E. 1990. *Physical Geology.* 8th ed. Prentice Hall, Englewood Cliffs, N.J.

Kozoun, V. I., et al. 1974. *World Water Balance and Water Resources of the Earth.* Prepared by the USSR National Committee for the International Hydrological Decade, V. I. Kozoun, ed. in chief. Leningrad.

La Chapelle, E. R. 1969. *Field Guide to Snow Crystals.* University of Washington Press, Seattle.

Langbein, W. B., et al. 1949. Annual Runoff in the United States. U.S. Geol. Survey Circ. 52.

Leifeste, D. K. 1974. Dissolved Solids Discharge to the Oceans from the Conterminous United States. U.S. Geol. Survey Circ. 685.

Leopold, L. B. 1968. Hydrology for Urban Land Planning: A Guidebook on the Hydrologic Effects of Urban Land Use. U.S. Geol. Survey Circ. 554.

Leopold, L. B., and Maddock, T., Jr. 1953. The Hydraulic Geometry of Stream Channels and Some Physiographic Implications. U.S. Geol. Survey Prof. Paper 252.

Livingstone, D. A. 1963. Chemical Composition of Rivers and Lakes. U.S. Geol. Survey Prof. Paper 440G.

MacKichan, K. A. 1951. Estimated Water Use in the United States, 1950. U.S. Geol. Survey Circ. 115.
MacKichan, K. A. 1957. Estimated Water Use in the United States, 1955. U.S. Geol. Survey Circ. 398.
MacKichan, K. A., and Kammerer, J. C. 1961. Estimated Use of Water in the United States, 1960. U.S. Geol. Survey Circ. 456.
McPhee, J. 1971. *Encounters with the Archdruid.* Farrar, Straus and Giroux, New York.
Miller, A. and Thompson, J. C. 1983. *Elements of Meteorology.* 4th ed. Charles E. Merrill, Columbus, Ohio.
Milne, A. A. 1926. *Winnie the Pooh.* Methuen, London.
Moss, M. E., and Lins, H. F. 1989. Water Resources in the Twenty-first Century—A Study of the Implications of Climate Uncertainty. U.S. Geol. Survey Circ. 1030.
Murray, C. R. 1968. Estimated Use of Water in the United States, 1965. U.S. Geol. Survey Circ. 556.
Murray, C. R., and Reeves, E. B. 1972. Estimated Use of Water in the United States, 1970. U.S. Geol. Survey Circ. 676.
Murray, C. R., and Reeves, E. B. 1977. Estimated Use of Water in the United States, 1975. U.S. Geol. Survey Circ. 765.
Petersen, U. 1994. Mining the Hydrosphere. *Geochim. Cosmochim. Acta* 58:2387–2403.
Peterson, D. F., and Keller, A. A. 1990. Irrigation. In *Climate Change and U.S. Water Resources*, ed. P. E. Waggoner, chap. 12. John Wiley, New York.
Press, F., and Siever, R. 1986. *Earth.* 4th ed. W. H. Freeman, San Francisco.
Richey, J. E., and Victoria, R. L. 1990. Fluvial Sources of Carbon, Nutrients, and Sediments for the Amazon Estuary (Abstr.). *EOS Transactions* 71:1365.
Sarin, M. M., Krishnaswami, S., Dilli, K., Somayajulu, B.L.K., and Moore, W. S. 1989. Major Ion Chemistry of the Ganga-Brahmaputra River System: Weathering Processes and Fluxes to the Bay of Bengal. *Geochim. Cosmochim. Acta* 53:997–1009.
Sarin, M. M., Krishnaswami, S., Somayajulu, B.L.K., and Moore, W. S. 1990. Chemistry of Uranium, Thorium, and Radium Isotopes in the Ganga-Brahmaputra River System: Weathering Processes and Fluxes to the Bay of Bengal. *Geochim. Cosmochim. Acta* 54:1387–96.
Sellers, W. D. 1965. *Physical Climatology.* University of Chicago Press, Chicago.
Solley, W. B., Chase, E. B., and Mann, W. B. IV. 1983. Estimated Use of Water in the United States, 1980. U.S. Geol. Survey Circ. 1001.
Solley, W. B., Merk, C. F., and Pierce, R. R. 1988. Estimated Use of Water in the United States, 1985. U.S. Geol. Survey Circ. 1004.
U.S. Water Resources Council. 1978. *Summary.* Vol. 1 of *The Nation's Water Resources, 1975–2000.* U.S. Government Printing Office, Washington, D.C.
Van der Leeden, F., Troise, F. L., and Todd, D. K. 1990. *The Water Encyclopedia.* 2d ed. Lewis Publishers, Chelsea, Mich.
Waggoner, P. E., ed. 1990. *Climate Change and U.S. Water Resources.* John Wiley, New York.

5 The Biosphere

5.1 Introduction: "All Creatures Great and Small"

On approaching the Earth, a space traveler from another part of the galaxy was puzzled by some very odd, long-wavelength electromagnetic radiation. He was also puzzled by the rather curious composition of the Earth's atmosphere: there was much more oxygen than could reasonably be explained by "natural" processes; it was an atmosphere quite different from that of the other planets he had encountered in this solar system. As an intelligent space traveler, he concluded that life must exist on the Earth, that the oxygen content of the atmosphere was somehow produced by organisms on the planet, and that the odd electromagnetic radiations were probably radio and television signals.

At closer range, his intuition was confirmed. He found that parts of the planet were covered by a green mat. This mat, as well as parts of the oceans, gave a significant signal in the visible parts of the light spectrum that suggested the presence of chlorophyll, a sign of active photosynthesis. All this the space traveler found inviting. He realized, of course, that intelligent life is apt to be unpleasant, but he decided to take his chances and to land on the planet. As trekkies well know, all of the inhabitants of the galaxy speak fluent English. He was therefore able to confirm his conjectures and to communicate freely and happily with the originators of the radio and television signals, especially with those in the English-speaking parts of the Earth. Some of his observations are contained in this chapter.

5.2 The Size of the Biosphere

The biosphere is the sum total of all living beings on Earth. The population of the biosphere changes very rapidly. Some microorganisms live less than a day; even the oldest living organisms, the bristle cone pines, live less than 10,000 years, an age that is minuscule compared to the age of the Earth. In spite of this constant change of occupants, the biosphere has been a very active part of the Earth for much of the history of the planet.

The mass of the biosphere is not known very precisely, even though nearly all of it is easily accessible at the surface of the Earth. Even determining the mass of living matter in a single forest is no easy task. One can determine the mass of a single tree quite accurately by chopping it down, digging up all the roots, and weighing the parts. However, extrapolating this information to a whole forest yields results that are imprecise. Measuring the mass of the marine biosphere is even more difficult, and we have to content ourselves with fairly rough estimates of biospheric mass. For most purposes these rough estimates serve quite well; for others, as we will see, they do not.

Recent estimates of the mass of the terrestrial (land) biosphere range from about $1{,}000 \times 10^{15}$ to $2{,}000 \times 10^{15}$ gm, i.e., 1,000 to 2,000 billion tons, and estimates of the marine (oceanic) biosphere range from 2×10^{15} to 4×10^{15} gm (see, for instance, Sundquist 1985). The striking difference between the mass of the terrestrial biosphere and the mass of the marine biosphere is due in large part to the absence of marine equivalents of trees.

The mass of the biosphere is minuscule compared to the mass of the Earth:

$$\frac{\text{mass of biosphere}}{\text{mass of the earth}} = \frac{1500 \times 10^{15}\ \text{gm}}{6.0 \times 10^{27}\ \text{gm}} \approx 0.25 \times 10^{-9}.$$

Since the biosphere amounts to less than one billionth of the mass of the Earth, the biosphere can be safely neglected in discussing the deeper parts of the planet. However, a certain amount of humility is inspired even if we confine ourselves to comparing the size of the biosphere with that of other near-surface features. Let us imagine for a moment a very large blender, which can convert the entire biosphere into a green paste, and a very large spatula with which to spread this paste uniformly over the continents. The thickness of the layer of green paste would then be about 1 cm, comparable in thickness to a peanut butter and jelly sandwich, but not nearly as tasty.

Humanity is a small part of the biosphere, and would contribute only a very small part to this paste. The human population of the Earth is now between 5 and 6 billion (see sec. 5.5). The average weight of human beings is about 70 kilograms (150 lb), and their density is about that of water, i.e., 1.0 gm/cc. It is easy to show that all of us would fit quite comfortably into a cube that is 700 meters long, a little less than half a mile, on each side. This is quite a large box, but it would take up only a very small part of the Grand Canyon of Arizona and would hardly be noticed there, except possibly by the local animal population. Too much humility in the face of these facts would, however, be misplaced. The effect of the biosphere on the atmosphere and on many near-surface processes is out of all proportion to its mass. The effect of humanity on the Earth system has been even more disproportionately large. We are enormously active, and this has been our bane and our glory.

5.3 The History of the Biosphere

The two most important concepts that the science of geology has contributed to humankind are probably the length of Earth history and the history of life. The Earth is much older than the estimate of some 6,000 years based on literal interpretations of biblical chronology. The Earth is not, however, infinitely old. The evolution of data and opinions regarding the age of the Earth makes for fascinating reading (see, for instance Gould 1987 and Dalrymple 1991). The age of the Earth is now quite well established, give or take 50 million years or so. That is, of course, a large uncertainty in human terms, but it is small compared to the approximately 4,500 million years that have elapsed since the formation of the planet.

The best evidence for events during the earliest history of the Earth comes from studies of meteorites and the moon. Meteorites are the only pieces of extraterrestrial matter that come to us free of charge. Most are fragments of larger, very ancient bodies. Many meteorites have been altered fairly little since they were formed early in the history of the solar system; they therefore retain a good deal of information about events during that distant period. The moon also formed early in the history of the solar system, probably by the accumulation of material splashed out of the Earth by the impact of a planet-sized object.

Since the Moon is rather small by planetary standards, with a mass only a little more than 1% of the mass of the Earth, it cooled rapidly, and has been a more or less inert lump of rock for most of its history. The only major excitement in its history has been provided by the infall of large meteorites. These visitations have become progressively less frequent, and the material at the surface of the moon still retains quite a good record of ancient events, even of those that occurred during the first 500 million years of its history. On Earth, virtually all of the history of this period has been obliterated by later events, in much the same way that the early history of many ancient towns has been obliterated by repeated sackings and burnings or by later burial by sand, mud, and new dwellings.

Fifty years ago the oldest known rocks were about 2,600 million years old. Since then much older rocks have been found. We now know of several areas in Greenland where many of the rocks are 3,800 million years old, and one area has been found in Canada where some rocks are 4,000 million years old. There is still, however, very little direct geological evidence concerning the momentous events during the first 500 million years of Earth history, when our planet was formed and when it grew to its present size.

The oldest rocks that have not been heated and deformed extensively are the 3,500-million-year-old Warrawoona sedimentary rocks in Western Australia. These rocks contain what are almost certainly the remains of microorganisms (Schopf 1992). It is hard to know what these minute organisms were like when they were alive, but they were probably blue-green algae (cyanobacteria). We do not know when these organisms evolved. The origin of life is still poorly understood and hotly debated.

The record of life becomes progressively more vivid and complete in younger rocks. Superbly preserved microorganisms were discovered nearly 40 years ago in sedimentary rocks that are about 2,000 million years old. Some of these are shown in figure 5.1. All of the fossils discovered in rocks from this period of Earth history are the remains of small, unicellular (single-celled) organisms that did not contain nuclei. The development of nucleated cells probably occurred between about 2,000 and 1,750 million years ago (m.y.a.) (Knoll, 1992) (see

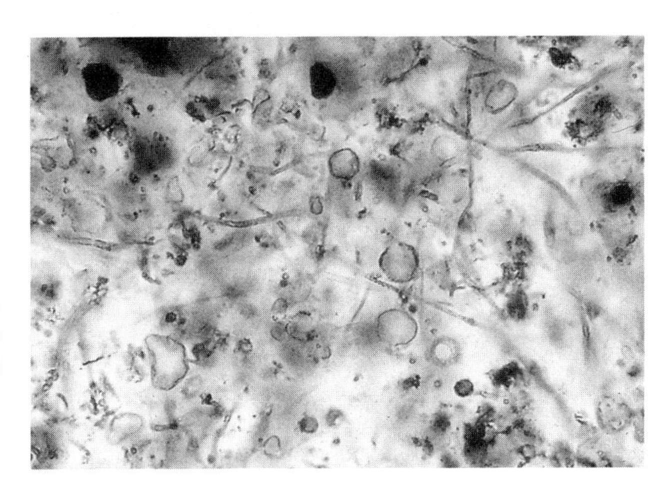

Figure 5.1. Cyanobacteria from the ca. 2,000 m.y. old Gunflint Formation, Ontario, Canada. (Courtesy of A. H. Knoll)

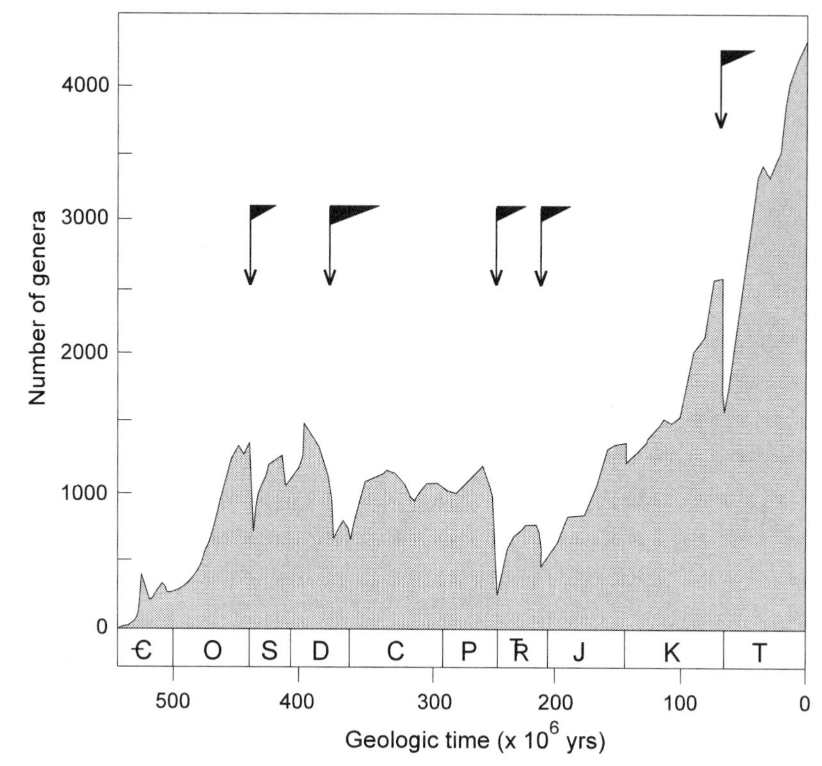

Figure 5.2. The diversity of animal genera in the oceans through time. The five biggest mass extinctions are identified by arrows with flags spanning the subsequent faunal recovery times. (Sepkoski 1994)

fig. 3.9), but another 1,000 million years had to elapse before the first animals made their appearance. Animals with shells and/or bones developed quite suddenly (on a geological timescale) about 540 m.y.a. Their evolution and that of the higher plants occupy only the last 13% of Earth history. Land plants and fish developed about 450 m.y.a., amphibians about 400, reptiles about 300, and flowering plants about 100 m.y.a. Mammals burgeoned during the past 65 million years. Humankind developed during the past 2 million years, and has become the cultural mammal par excellence. We have learned to speak about our history, and—quite recently—we have even learned to write about it.

The number of genera of marine animals has increased dramatically but rather jerkily during the last 540 million years of Earth history (see fig. 5.2). Periods of growth in the number of genera have been followed by periods of shrinkage. Most of the latter have been quite sharp and rapid. Just how sharp and rapid is shown in figure 5.3, which indicates the percentage of genera that became extinct during eighty-four stratigraphic intervals, each about 7 million years in duration. The largest marine extinction occurred 240 m.y.a., at the end of the Permian. Nearly 70% of all marine animal genera disappeared at that time. At the end of the Cretaceous, 65 m.y.a., the dinosaurs became extinct together with nearly 50% of the genera of marine animals. This extinction was probably due to the impact of one or more large meteorites. It is tempting to infer that all extinctions are extraterrestrial in origin; but it now looks as if that was not the case. Some decreases in the total number of genera seem to be due to a lack of development of new genera rather than to the rapid destruction of existing genera. During

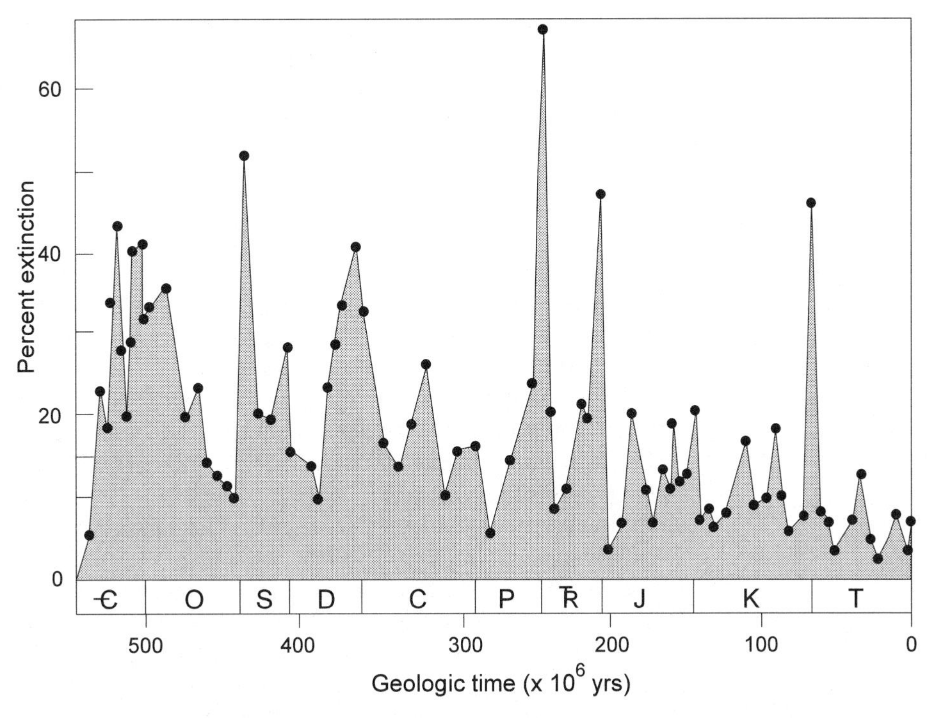

Figure 5.3. Intensity of extinctions of well-skeletonized marine animal genera during the Phanerozoic. There have been five major extinctions during the Phanerozoic: at the end of the Ordovician (O), the Devonian (D), the Permian (P), the Triassic (Tr), and the Cretaceous (K). The extinctions define the boundary between these and the following geological periods. At present there is strong evidence for an extraterrestrial cause only for the extinction at the end of the Cretaceous (K). (Sepkoski 1994)

several extinctions the tell-tale marks of large meteorite impacts are missing. The periodicity of the extinction record in figure 5.3 is striking, but there is no consensus regarding its cause and meaning.

The great length of Earth history can be illustrated by two analogies. Let all of Earth history be compressed into a single year. Since a year contains about 31 million seconds, each second represents about 143 years of Earth history. If the Earth was formed on January 1 of this year, the last second of December 31 will contain the history of the past 143 years. The last minute of that day will contain all of recorded human history.

The history of the Earth can also be represented in terms of a journey from Los Angeles to Boston, a distance of about 5,000 kilometers. Each kilometer along this route can therefore be considered to represent a little under a million years of Earth history, and one meter corresponds to a little less than 1,000 years of Earth history. On this scale, the length of a desk contains world history since the fall of the Roman Empire, and the length of a pen contains everything that has happened since the beginning of the Civil War. In this context Alexander Pope's famous dictum, "The proper study of mankind is man," looks somewhat cramped. If the study of our planet during the 260 years since Pope's time has taught us anything, it is surely that we must study humans not in isolation but embedded in the large context of geologic time and biological evolution.

5.4 The Dynamics of the Biosphere

The biosphere can be divided into two groups of organisms: the producers of organic matter and the users of organic matter. Plants are the most important producers. Among the plants, those that generate O_2 as a by-product of photosynthesis are the most abundant. In an oversimplified manner, the reaction by which organic matter is produced by green plants can be written as follows:

$$6CO_2 + 6H_2O \overset{\text{sunlight}}{\rightarrow} C_6H_{12}O_6 + 6O_2. \tag{5.1}$$

This implies that CO_2 from the atmosphere is combined with water to yield organic compounds and O_2. The energy required for this process is supplied by light in the visible part of the solar spectrum.

$C_6H_{12}O_6$ is the composition of certain sugars; these are carbohydrates. The average composition of the carbohydrate group (see table 5.1) is fairly close to $C_6H_{12}O_6$. To this extent at least, equation (5.1) is not unreasonable. However, the other groups of organic compounds that are important constituents of plants have somewhat different bulk chemical compositions (see table 5.1). Fats, on average, contain more carbon; they also tend to contain phosphorus, nitrogen, and sulfur. Proteins, on average, contain much less combined oxygen than carbohydrates, but they contain quite large amounts of nitrogen as well as some phosphorus, sulfur, and a small amount of iron. The carbon, oxygen, and hydrogen content of dried alfalfa (see table 5.2), or for that matter of dehydrated humans, is not very different from that of the average carbohydrate, but other elements are also important constituents of alfalfa, *Homo sapiens*, and the remainder of the biosphere (see table 5.3).

If equation (5.1) were strictly correct as a description of green-plant photosynthesis, the only requirements for plant growth would be the availability of CO_2, H_2O, and sunlight. CO_2 is rarely limiting in nature, since it is reasonably abundant in the atmosphere. Water is limiting in deserts, and sunlight can be limiting on the floor of dense forests, where little light penetrates through the canopy down to ground level. Even where all three of these requirements are met, plant growth may be minimal, because one or more of the other essential plant nutrients is in short supply. The addition of fertilizers containing combined nitrogen, phosphorus, and potassium usually increases the agricultural productivity of soils considerably. The Green Revolution of the last 40 years owes much of its success in developing countries to the increased use of fertilizers. Although intense fertilization can produce significant environmental problems (see chapter 12), the continued application of fertilizers will be absolutely essential if the world's population is to be fed.

The biosphere is distributed very unevenly over the globe. In Africa the belt of desert between latitudes 10° and 30°N grades southward into equatorial jungles. In South America the western Amazonian rain forests grade southward into Andean deserts. In all but the high northern and southern latitudes, where temperature is a controlling factor, the density of vegetation depends primarily on the availability of water. People have been able to make deserts bloom, but only by importing large quantities of water.

In spite of the large extent of deserts, the biosphere as a whole is enormously active. Estimates of the rate of photosynthesis on land vary a good deal. The best figures for the rate of fixation of carbon into terrestrial biomass are in the range

Table 5.1. Average Chemical Composition of Organic Matter

	Percentage Composition by Weight		
Element	*Carbohydrates*	*Fats*	*Proteins*
O	49.38	17.90	22.4
C	44.44	69.05	51.3
H	6.18	10.00	6.9
P		2.13	0.7
N		0.61	17.8
S		0.31	0.8
Fe			0.1
Total	100.00	100.00	100.00

Source: Rankama and Sahama 1950.

Table 5.2. Average Total Composition of Dehydrated Living Matter

	Percent of Dry Weight	
Element	*Adult (Homo sapiens)*	*Alfalfa (Medicago sativa)*
C	48.43	45.37
O	23.70	41.04
N	12.85	3.30
H	6.60	5.54
Ca	3.45	2.31
S	1.60	0.44
P	1.58	0.28
Na	0.65	0.16
K	0.55	0.91
Cl	0.45	0.28
Mg	0.10	0.33
Total	99.96	99.96

Source: Rankama and Sahama 1950.

Table 5.3. Distribution of Elements in Organisms as Percentage of Body Weight

Invariable			*Variable*		
Primary 60–1	*Secondary 1–0.05*	*Micro-constituents <0.05*	*Secondary*	*Micro-constituents*	*Contaminants among Others*
H	Na	B	Ti	Li	He
C	Mg	F	V	Be	A
N	S	Si	Zn	Al	Se
O	Cl	Mn	Br	Cr	Au
P	K	Cu		Co	Hg
	Ca	I		Ni	Bi
	Fe			Ge	Tl
				As	
				Rb	
				Sr	
				Mo	
				Ag	
				Cd	
				Sn	
				Cs	
				Ba	
				Pb	
				Ra	

Source: Rankama and Sahama 1950.

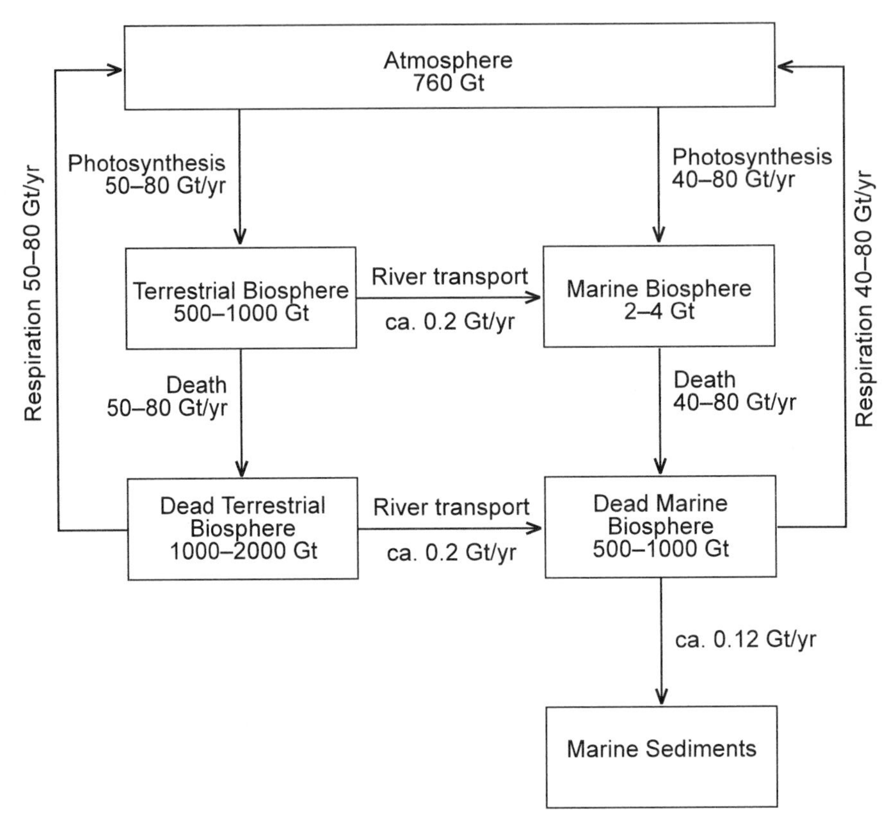

Figure 5.4. The biological parts of the carbon cycle. The carbon content of the several reservoirs is in Gt carbon (1 Gt = 10^{15} gm C). (Data from the compilation of Sundquist 1985)

of 50 to 80 × 10^{15} gm/yr (Sundquist 1985). 10^{15} grams are equivalent to 10^9 tons, a quantity that is frequently called a *gigaton*, or Gt. We will use this shorthand notation for the amount of carbon in the various near-surface reservoirs and for the transfer rate of carbon between these reservoirs. In figure 5.4 the rate of carbon transfer via photosynthesis from the atmosphere to the terrestrial biosphere is therefore shown as 50–80 Gt/yr.

The total amount of carbon in the atmosphere is now 760 Gt; nearly all of this is present as a constituent of CO_2 (see table 3.1). Carbon in methane, CH_4, is a distant second as a carrier of atmospheric carbon. The CO_2 content of the atmosphere is increasing at a rate of about 2 Gt/yr, largely due to the addition of CO_2 by fossil fuel burning (see chapter 11).

Photosynthesis on land is so rapid that all of the CO_2 in the atmosphere is converted into organic matter in about a decade. If there were no return flow of CO_2, photosynthesis would deplete the CO_2 content of the atmosphere very quickly. It is now known that the CO_2 content of the atmosphere has been remarkably stable at about 270 ppmv for some 10,000 years between the end of the last ice age and the beginning of the industrial age. During this period of time, the removal of CO_2 from the atmosphere must therefore have been balanced quite precisely by an equivalent return flow of CO_2. Nearly all of the organic matter synthesized by plants must have been reoxidized by the reverse of reactions such as equation (5.1):

$$C_6H_{12}O_6 + 6O_2 \rightarrow 6CO_2 + 6H_2O + \text{energy}. \tag{5.2}$$

The energy used in creating organic matter via reaction (5.1) is released during the oxidation of organic matter by reactions such as (5.2). The users of organic matter in the biosphere therefore serve the important function of restoring CO_2 to the atmosphere, so that it again becomes available for photosynthesis. In the process, the users release the energy required to sustain their own life. Animals, it has been said, are organisms that have forgotten how to photosynthesize; but this comment may reflect a rather too botanical view of the biosphere. The size range of users of organic matter is truly enormous. Bacteria are among the most important users, but the efficiency with which larger users such as goats, sheep, and a variety of other grazers convert plant matter into CO_2, CH_4, and H_2O is also truly impressive.

Eating and reproduction are surely among the most important functions of animals. The biosphere can be described in terms of long food chains and complex food webs that connect its inhabitants and that define the pathways by which organic matter is oxidized and returned to the atmosphere. Humankind is part of this process. After dispatching Polonius, Hamlet was accosted by his uncle, the king:

KING: Now, Hamlet, where's Polonius?
HAMLET: At supper.
KING: "At supper"? Where?
HAMLET: Not where he eats, but where he is eaten; a certain convocation of politic worms are e'en at him. (act 4, scene 3)

Much of the oxidation of plant material takes place in forest litter and in soils (see chapter 6). The quantity of dead terrestrial biosphere in this transitional state is probably about twice the quantity of the living biosphere and amounts to some 1000–2000 Gt. Nearly all of it is returned to the atmosphere on a timescale of a few to about 1,000 years. Rivers carry a very small fraction of dead terrestrial biosphere to the oceans, together with a small fraction of the living organic matter (see fig. 5.4).

The rate of photosynthesis in the oceans is probably 20–40 Gt of carbon per year, i.e., somewhat smaller than terrestrial photosynthesis. Like photosynthesis on land, the distribution of photosynthesis in the oceans is very heterogeneous (see plate 5). Unlike photosynthesis on land, this heterogeneity is not due to geographic differences in the availability of water but due to geographic differences in the availability of nutrients, especially phosphate, PO_4^{3-}, and nitrate, NO_3^-. Coastal areas where rivers bring nutrients to the oceans tend to be highly productive, as are some coastal areas that border deserts. The west coast of South America is a good example. Very few rivers drain the western slopes of the Andes, yet the coastal areas off Ecuador and Peru fairly teem with life. Marine productivity in these areas reaches values as high as 10 g $C/m^2/day$, exceeding the productivity of the open ocean by as much as a factor of one thousand. They are prime fishing grounds, because nutrients are brought into these areas from the deeper parts of the ocean by upwelling currents. Other areas of upwelling-induced productivity include north and southwest Africa, coastal Alaska, and coastal Oman. The nutrients are used there by minute photosynthesizing organisms, the phytoplankton. These form the base of complex food chains and webs.

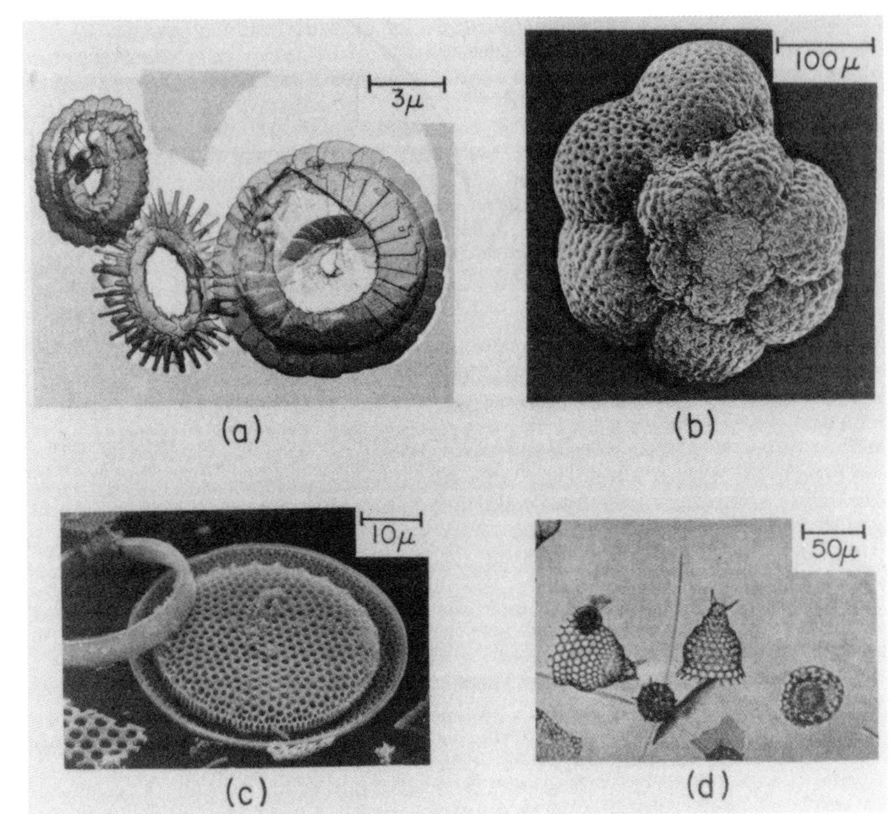

Figure 5.5. Photographs of $CaCO_3$ hard parts produced by (a) coccolithophorida and (b) foraminifera, and of SiO_2 hard parts produced by (c) diatoms and (d) radiolarians. (Broecker and Peng 1982)

Some of the smallest of the phytoplankton and zooplankton construct the most beautiful of all skeletons (see fig. 5.5). These can become quantitatively important components of marine sediments. Of the hard-part producing plankton, coccoliths and diatoms are among the more important plants, while foraminifera and radiolarians are among the more important animals.

Nearly all marine organic matter is produced in the upper 100 meters of the oceans. Below that depth, light levels are too low to sustain extensive photosynthesis. Most of the organic matter synthesized in the oceans is eaten in the upper 200 meters. The nutrients contained in this organic matter are released to seawater and are reused in successive rounds of photosynthesis. Only about 7%–10% of the organic matter synthesized in the upper 100 meters escapes destruction and sinks into the deeper parts of the oceans. Most of this residue is eaten on the way to the ocean floor, and most of the remainder is eaten by bottom-dwelling organisms. Only a tiny fraction of the organic matter that enters the oceans from land and is generated in the oceans escapes being eaten and is buried with marine sediments at a rate of about 0.1 Gt carbon/yr (see fig. 5.4). Although this is a minuscule fraction of the total annual production of organic matter, we shall return to it in chapter 7, because this material has been the source of most of our oil and natural gas.

The organic matter eaten in the lower parts of the ocean releases its nutrients there. The oceans mix slowly today. It takes on the order of 1,000–2,000 years for surface water to sink into the deeper parts of the North Atlantic Ocean, to travel southward at depth into the Antarctic Ocean, and ultimately to reappear at the sea surface in the Indian or Pacific Ocean. In the course of its journey, seawater accumulates phosphate, PO_4^{3-}, and nitrate, NO_3^-, released from organic matter that is eaten in the deeper parts of the oceans as well as SiO_2, Ca^{2+}, and HCO_3^- from the dissolution of the hard parts of marine organisms. Upwelling seawater is therefore enriched in nutrients and is highly fertile when it is bathed in sunlight on returning to the ocean surface. Upwelling is particularly intense along the western coasts of continents. This explains the great intensity of photosynthetic activity off the northwestern coast of South America.

Plate 5 shows that many parts of the open oceans are essentially biological deserts. Nevertheless, photosynthesis in these oceans accounts for a large fraction of total marine photosynthesis. Per unit area, the coastal regions are much more productive than the open oceans, but the total area of open ocean is so large compared to that of the coastal areas that total primary production in the open oceans is much greater than total primary production in the coastal regions. Although the annual rate of photosynthesis in the oceans is not much less than the annual rate of photosynthesis on land, the mass of the marine biosphere is very small compared to the mass of the terrestrial biosphere (see fig. 5.4). The difference is largely due to the presence of trees on land. Photosynthesis in the surface oceans does not require large support structures, whereas photosynthesis on land is helped a great deal by the support supplied by tree trunks and tree limbs.

The size of the dead marine biosphere is comparable to that of the dead terrestrial biosphere. There is, of course, no direct marine equivalent of forest litter and of humus layers. However, organic molecules are present throughout the oceans. The total amount of carbon contained in this material is very large, even though the concentration of dissolved organic compounds in seawater is very small.

Figure 5.4 summarizes much of the information that we have discussed in this section. Perhaps the most salient feature of the biological part of the carbon cycle is the rapidity with which carbon cycles from the atmosphere through the biosphere and back again. We are part of an enormously dynamic system, driven by solar energy and kept in balance by the interlocking processes of organic matter production and organic matter use on land and in the oceans. This system has nourished and sustained us. Will it be able to do so during the next century? The answer to this depends on the size of the human population and on the ability of the biosphere to produce the required quantity of food.

5.5 Population

Imagine a large island that is well endowed with grass and is largely bare of animal life. A passing ship releases a pair of rabbits on this island. These find the island to be a congenial home, and begin to multiply, as rabbits do. The number of rabbits, N, increases exponentially with time, t:

$$N = 2e^{bt}, \tag{5.3}$$

where b is a constant, and the number 2 is the rabbit population at time $t = 0$, i.e., when the initial rabbit pair lands on the island. If time is expressed in years, the value of b is approximately 0.16. The rabbit population increases very rapidly

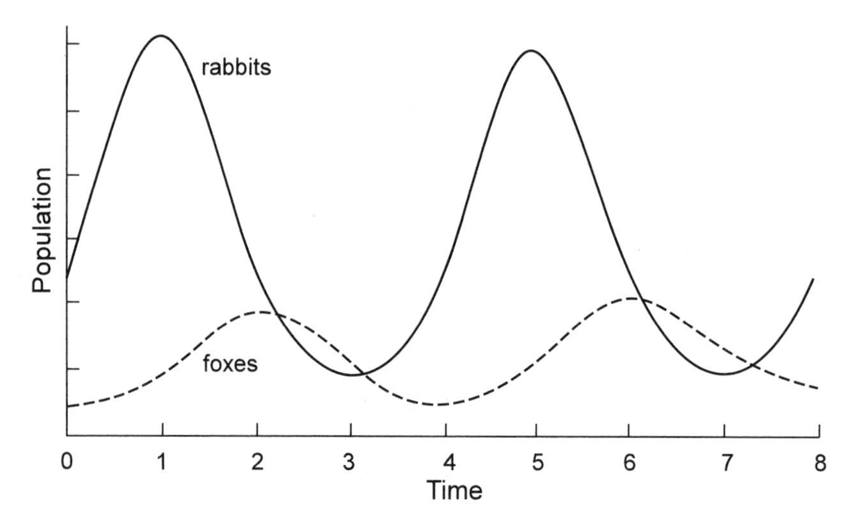

Figure 5.6. Fluctuations in the rabbit and fox populations on an island. (After Hedrick 1984)

indeed. In 1859 a "small shipment" of rabbits was introduced into Australia near Geelong in southern Victoria. By 1949 the rabbit population in southeastern Australia was estimated to have reached 40 to 50 million. Myxomatosis, a virus lethal to rabbits, was introduced in 1950 to control what had clearly become an infestation (Andrewartha 1970).

On our island the increase in the rabbit population is ultimately limited by the availability of grass. The island becomes overgrazed, rabbits die of starvation, and the rabbit population decreases precipitously. This allows the grass to recover. In the long run the rabbit population either stabilizes, or—more likely—fluctuates, somewhat as shown in figure 5.6. These fluctuations are due to periodic overshoots in the rabbit population followed by overgrazing, famine, rabbit die-off, and recovery, as grass again becomes more available. The magnitude and the period of these fluctuations depend on the reproduction rate of the rabbits and on the recovery rate of the grass cover on the island.

On a return visit some years later, the ship that deposited the pair of rabbits on the island leaves a pair of foxes. These are happy to discover the large rabbit population, and multiply rapidly at the expense of the rabbits. As the rabbit population dwindles, the grass cover increases, and the fox population dwindles as well. Ultimately, the rabbit and the fox populations begin to oscillate as shown in figure 5.6. The peaks in the fox population are offset from the peaks in the rabbit population, because it takes some time for the fox population to respond to changes in the rabbit population. Obviously, the simple pattern in figure 5.6 can be disturbed by many environmental changes. Droughts can diminish the grass cover; harsh winters or illness can reduce the rabbit and fox populations; the ship can return and introduce additional herbivores (plant eaters) or carnivores (meat eaters), or perhaps even a group of people (omnivores), who proceed to cover the island with tarmac and with the other blessings of civilization. The simple population fluctuations in figure 5.6 will then tend to become very complex (see, for instance, Pielou 1974 and May 1979), unless, of course, the rabbits and the foxes are both driven to extinction.

The Earth as a whole is an island in the planetary system, and *Homo sapiens* is one species on this island. It is fair, then, to ask whether the tale of grass, rabbits, and foxes applies to us as well. Thomas Robert Malthus answered this question with a resounding "yes." In the first edition of his *First Essay on Population* (Malthus 1798), he puts his case as follows:

> I think I may fairly make two postulata.
>
> First, That food is necessary to the existence of man. Secondly, That the passion between the sexes is necessary, and will remain nearly in its present state. . . . Assuming then, my postulata as granted, I say, that the power of population is indefinitely greater than the power in the earth to produce subsistence for man. Population, when unchecked, increases in a geometrical ratio. Subsistence increases only in an arithmetical ratio. A slight acquaintance with numbers will show the immensity of the first power in comparison of the second. By the law of our nature which makes food necessary to the life of man, the effects of these two unequal powers must be kept equal. This implies a strong and constantly operating check on population from the difficulty of subsistence. This difficulty must fall some where; and must necessarily be severely felt by a large portion of mankind.
>
> Through the animal and vegetable kingdoms, nature has scattered the seeds of life abroad with the most profuse and liberal hand. She has been comparatively sparing in the room, and the nourishment necessary to rear them. The germs of existence contained in this spot of earth, with ample food, and ample room to expand in, would fill millions of worlds in the course of a few thousand years. Necessity, that imperious all pervading law of nature, restrains them within the prescribed bounds. The race of plants, and the race of animals shrink under this great restrictive law. And the race of man cannot, by any efforts of reason, escape from it. Among plants and animals its effects are waste of seed, sickness, and premature death. Among mankind, misery and vice. The former, misery, is an absolutely necessary consequence of it. Vice is a highly probable consequence, and we therefore see it abundantly prevail; but it ought not, perhaps, to be called an absolutely necessary consequence. The ordeal of virtue is to resist all temptation to evil.
>
> This natural inequality of the two powers of population, and of production in the earth, and that great law of our nature which must constantly keep their effects equal, form the great difficulty that to me appears insurmountable in the way to the perfectibility of society. All other arguments are of slight and subordinate consideration in comparison of this. I see no way by which man can escape from the weight of this law which pervades all animated nature. No fancied equality, no agrarian regulations in their utmost extent, could remove the pressure of it even for a single century. And it appears, therefore, to be decisive against the possible existence of a society, all the members of which should live in ease, happiness, and comparative leisure; and feel no anxiety about providing the means of subsistence for themselves and families.
>
> Consequently, if the premises are just, the argument is conclusive against the perfectibility of the mass of mankind.

The essay raised a storm of criticism, and in the second edition, entitled *An Essay on the Principle of Population; or A View of its Past and Present Effects on Human Happiness*, Malthus (1803) modified his position somewhat: "Throughout the whole of the present work, I have so far differed in principle from the former, as to suppose another check to population possible, which does not strictly come under the head either of vice or misery; and, in the latter part, I have endeavored to soften some of the harshest conclusions of the first essay."

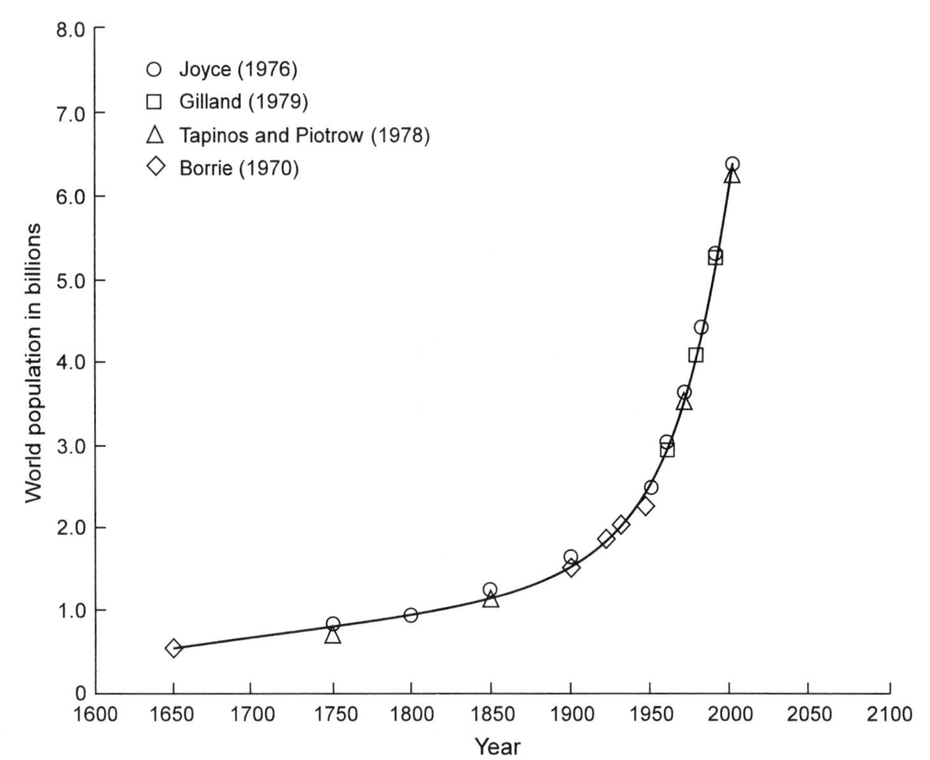

Figure 5.7. Estimates of world population, 1650–2000.

Nevertheless, Malthusianism has always implied a pessimistic view of the future of humanity on the grounds that the growth of populations will always outstrip the means of subsistence. How well did Malthus predict the future? In the nearly 200 years since the *First Essay*, Malthus's two postulates have fared badly in Europe and in North America, but have done rather well in the developing countries. In Europe they fared badly in part because North America was able to absorb the European overflow. In North America they fared badly in part because there was a virgin western frontier into which the population could expand. One of the most important tasks facing humanity today is to prove Malthus wrong in the long run and for the world as a whole.

At the beginning of the nineteenth century, birthrates were high in Europe, but the death rate among children was also high. In *The Wealth of Nations*, Adam Smith remarked that "in some places one-half of the children die before they are four years of age; in many places before they are seven; and in almost all places before they are nine or ten." Wolfgang Amadeus and Constanze Mozart had six children; only two of these survived to maturity. Between 1748 and 1782 Thomas Jefferson and Martha Wayles also had six children; only two lived to maturity, and only one past age 26. This mortality rate was by no means unusual. The population of the world therefore rose rather slowly. The data in figure 5.7 for the population of the world prior to 1900 are not particularly accurate; they are certainly not based on carefully designed censuses; but the figures are generally accepted and are probably not grossly in error (Joyce 1976; Gilland 1979; Tapinos and Piotrow 1978; Borrie 1970; Repetto 1987). They indicate that some 150 years elapsed after 1650 before the population of the world doubled. Changes in the doubling time of populations are best shown on a graph such as

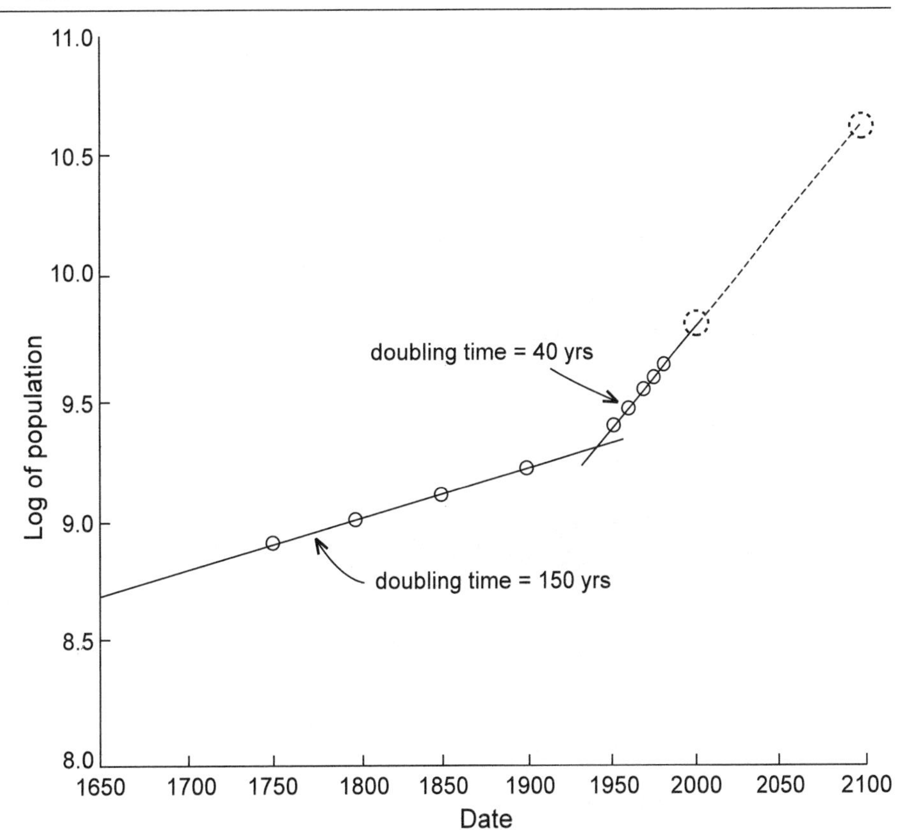

Figure 5.8. Logarithm of the estimated and projected world population, 1650–2100.

figure 5.8, where the logarithm of the world population is plotted against time. In such a graph, a straight line represents a historical period during which the doubling time remains constant. The data between 1650 and 1900 fall along a straight line whose slope is defined by a doubling time of 150 years. In contrast, the data for the period between 1950 and 1990 fall along a much steeper line, whose slope indicates a doubling time of only 40 years. World population has not only increased rapidly during the post-World War II era, but its rate of increase has been much greater than during the 1650–1900 period. The population trend of the past 40 years projected linearly in figure 5.8 to the year 2100 forecasts a world population of 40 billion, some seven times the present population. It is not at all clear that the Earth can sustain such a large population, at least not one that in Malthus's (1798) words would "live in ease, happiness and comparative leisure, and feel no anxiety about providing the means of subsistence for themselves and families."

Fortunately, there are reasons to believe that the population of the world will be a good deal less than 40 billion in the year 2100. In the developed nations, population increased rapidly in the nineteenth century, in large part due to a drastic decline in infant mortality. The death rate among children in Europe and North America is only about 1%, and it no longer plays an important role in limiting the population of these parts of the world. In spite of the increase in population, the standard of living in Europe and North America has risen and the birthrate has declined. During the 1980s, birthrates in several countries were below the average level of 2.1 children per woman that is required to maintain a steady-state population. The fertility (birth) rate in West Germany was 1.4, in

Table 5.4
Fertility Rates around the World: The Number of Children a Woman of Childbearing Age in Each Country Is Expected to Have in Her Lifetime, Estimated for the Period between 1990 and 1995

Lowest		*Highest*	
Country	*No. Children*	*Country*	*No. Children*
Italy	1.3	Rwanda	8.5
Hong Kong, Singapore	1.4	Malawi	7.6
Greece, Portugal, Austria, Germany	1.5	Ivory Coast	7.4
Japan, Denmark, Belgium, Holland, Switzerland	1.7	Uganda	7.3
		Angola, Yemen	7.2
Korea, Bulgaria, Hungary, Finland, France, Canada	1.8	Benin, Mali, Niger	7.1
		Ethiopia, Somalia	7.0
Britain, former Yugoslavia, Cuba, Australia	1.9	Afghanistan	6.9
Norway, Lithuania, Latvia, Estonia	2.0	Burundi, Tanzania	6.8
United States, New Zealand, Ireland, Poland, Romania	2.1		
China, Thailand, Puerto Rico	2.2		
Averages:			
Industrialized countries	1.9		
Developing countries	3.6		
World	3.3		

Source: United Nations Population Fund.

Italy 1.6, in Hungary 1.9, and in the United States 1.8 (Jarmul 1984). Estimates of fertility rates in these countries for the years 1990–1995 are shown in table 5.4. These countries have reached "zero population growth" (ZPG) and will see a gradual decline in their population during the twenty-first century unless birthrates increase somewhat, or unless immigration supplies the population deficit.

The evolution of populations in the developing countries is similar to that of the developed countries, but it is offset in time by roughly 150 years. High birthrates were matched by high rates of infant mortality until this century. Improved sanitation and health care have dropped the infant mortality rate, although it is still unacceptably high (see chapter 12). Unfortunately, birthrates have not dropped commensurately, the average life span has increased, and populations in these regions have increased dramatically. Most of the world's people live in the developing countries. The population of the world during the twentieth century therefore reflects this increase much more than the simultaneous stabilization of populations in the developed regions of the world (see fig. 5.9).

The trend toward lower birthrates that served to stabilize the population of the developed regions is finally underway in most of the world's developing regions (see, for instance, Coale 1983). Figure 5.10 shows the decline in birthrates since 1950 in the major regions of the world. Birthrates declined dramatically except in sub-Saharan Africa (Sai 1984; Caldwell and Caldwell 1990), but even in Africa there have recently been significant decreases (Cohen 1993). The decline in the birthrate for the world as a whole has been significant. This does not mean that world population has stabilized; far from it. The birthrate for the world as a whole is still much greater than the death rate. World population is still increasing rapidly (see fig. 5.11) and will continue to do so well into the next century,

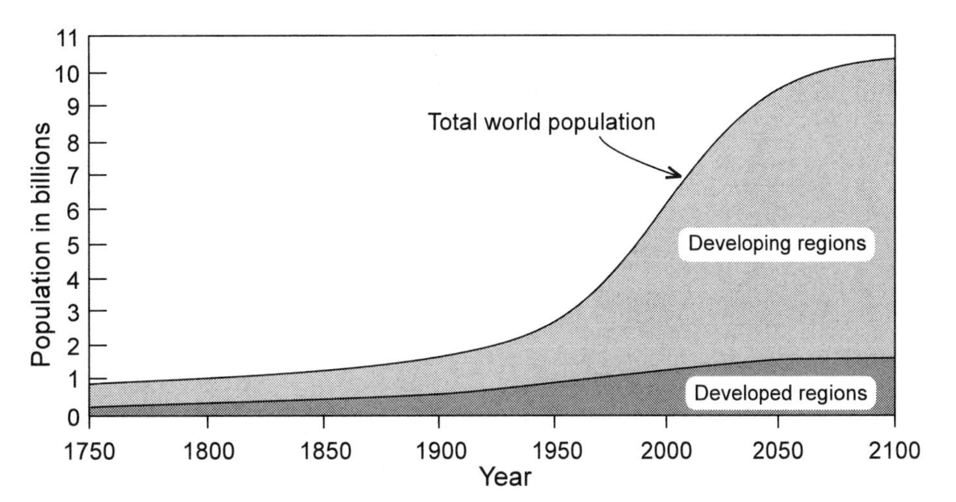

Figure 5.9. Population growth from 1750 to 1980 projected to 2100 in the world, in developing, and in developed regions. (Repetto 1987)

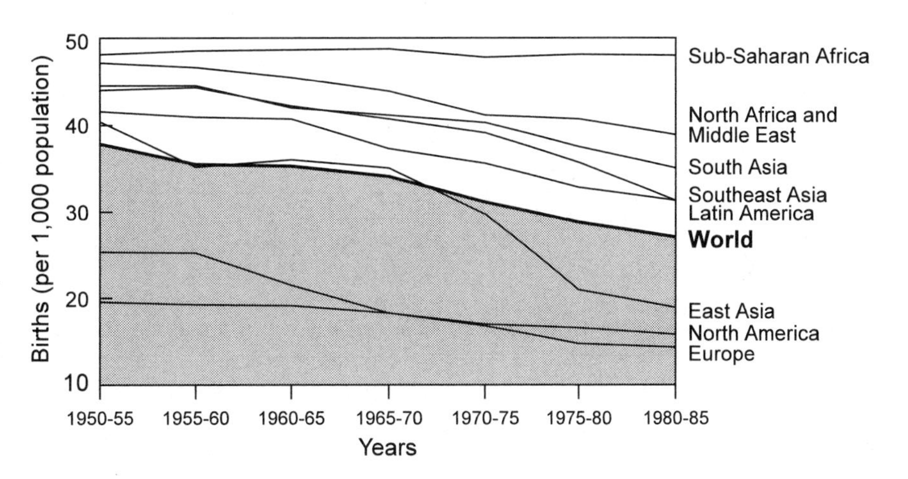

Figure 5.10. Birthrates in regions throughout the world have declined since the end of World War II. The only exceptions to this trend are birthrates in sub-Saharan Africa. Africa could account for nearly a quarter of the world's population by the late twenty-first century. (Caldwell and Caldwell 1990)

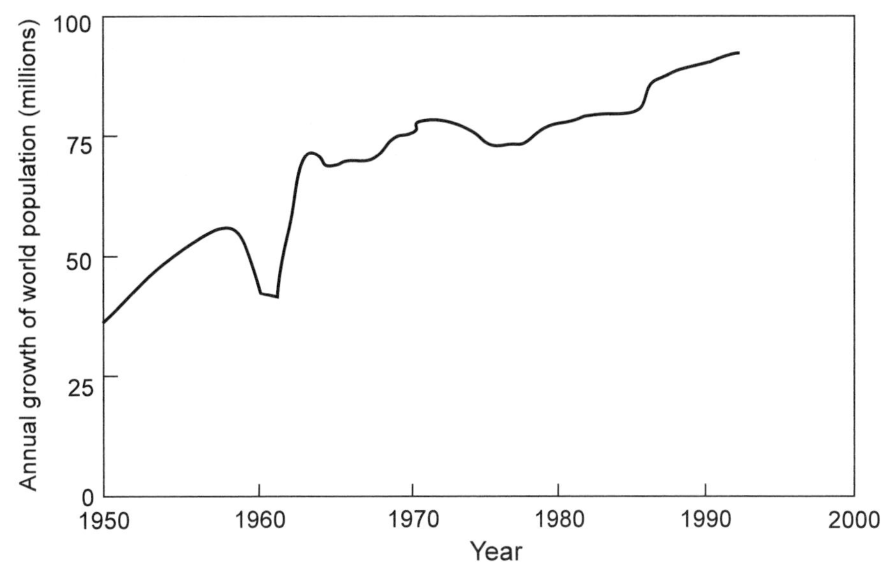

Figure 5.11. Annual additions to world population between 1950 and 1992. (Brown, Kane, and Ayres 1993)

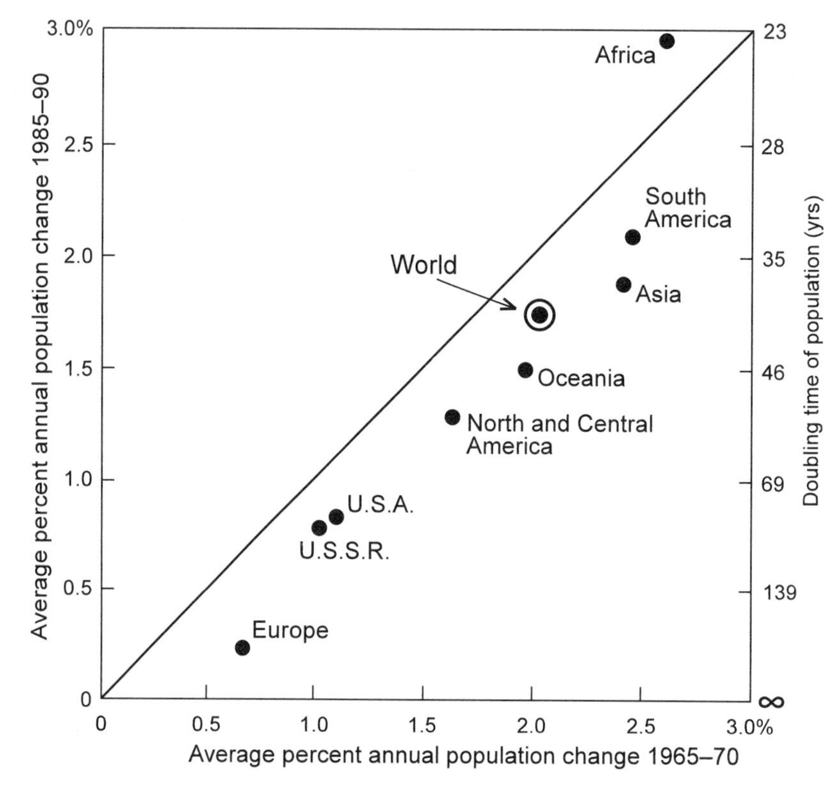

Figure 5.12. Comparison of annual population changes between 1985 and 1990 with those between 1965 and 1970 in several parts of the world.

even if the decrease in birthrates continues on its present course. Figure 5.12 compares the average annual population change for the major regions of the world during the period 1985–90 with the rate of 1965–70. In all of the regions except Africa, the population increase between 1985 and 1990 was significantly less than the population increase between 1965 and 1970. Nevertheless, populations increased everywhere. For the world as a whole, the doubling time of the population increased from about 35 to about 40 years. This increase is significant even though it is too small to be apparent in figure 5.8.

The adage that it is difficult to make predictions, particularly about the future, is most apt for estimating future world populations. In 1823 William Ellis, an English missionary, undertook to explore the island of Hawaii in the company of three American missionaries for the purpose of learning more about the country and its people, with a view of establishing missions there. Upon the return of the party to Honolulu, a joint journal of their journey and observations was prepared. A copy of this, together with a report by the American missionaries, was printed in Boston in 1825. Ellis (1825, p. 4) reported that "the population [of the island of Hawaii] at present is about 85,000, and will most probably be greatly increased by the establishment of Christianity, whose mild influence, it may be reasonably expected, will effect a cessation of war, an abolition of infanticide, and a diminution of those vices, principally of foreign origin, which have hitherto so materially contributed to the depopulation of the islands." Not so. Between 1800 and 1870 the native population of the Hawaiian Islands as

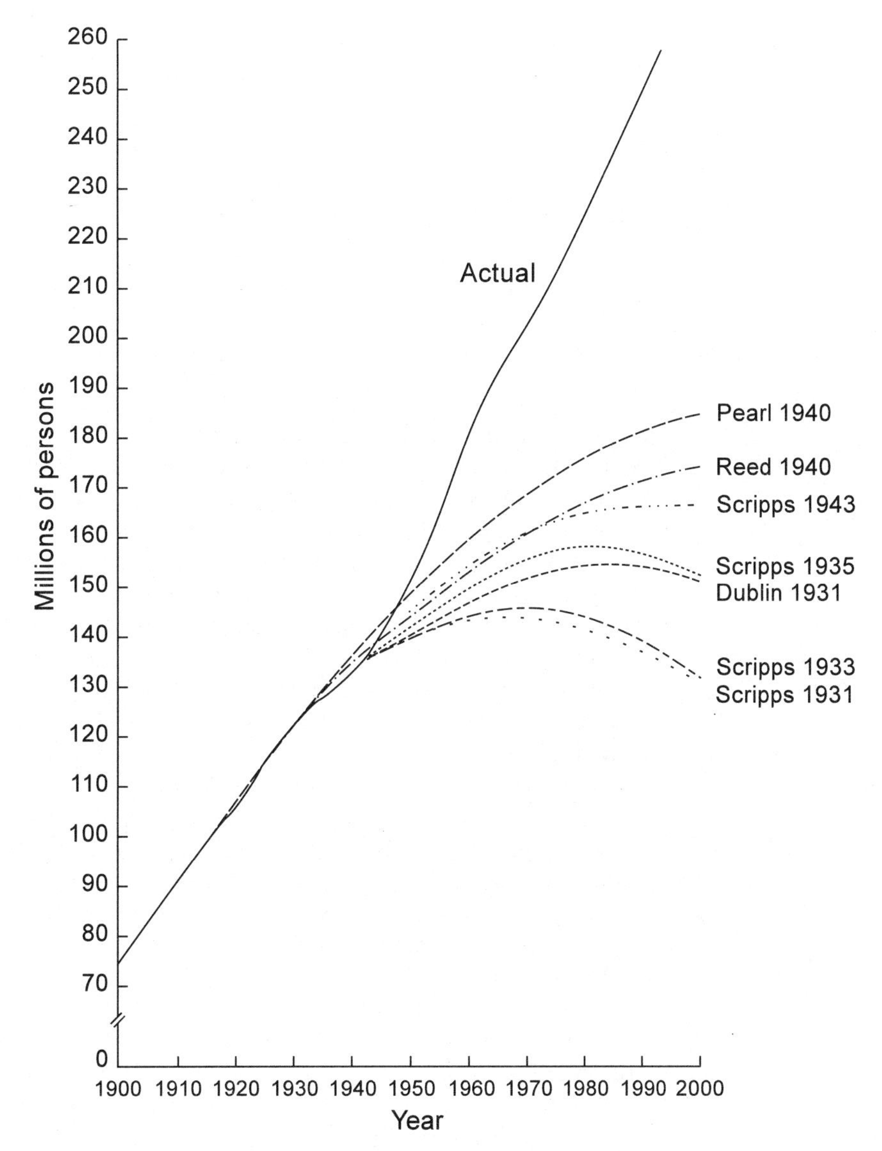

Figure 5.13. U.S. population forecasts made between 1931 and 1943, and the actual population increase during the twentieth century.

a whole decreased from about 300,000 to 60,000 (Adler 1966), largely due to disease.

It is also instructive to compare population forecasts that were prepared for the United States between 1931 and 1943 with the actual population trend since 1940 (Perlman 1984). All of the forecasts (see fig. 5.13) predicted that the population would begin to level off before the end of the twentieth century. Several predicted a population decline. None of the predictions was unreasonable. As shown in figure 5.14, the birthrate in the United States had been dropping consistently since the beginning of the century. During the years of the Great Depression, there was no reason to expect a return to higher birthrates. And yet that is what happened. After World War II the size of families increased dramatically.

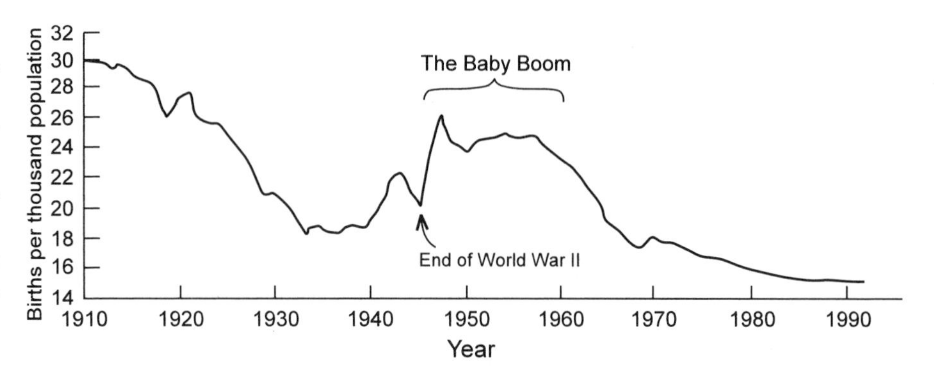

Figure 5.14. The birthrate in the United States between 1910 and 1992. (After Population Profile, Population Reference Bureau, March 1967; recent data from U.S. Census Bureau)

Families with four and five children were common. It was the time of the "baby boom," whose ripples and repercussions have been felt ever since. Birthrates have now dropped back to historic lows. The forecasts of the 1930s are useful reminders of the uncertainties inherent in forecasting populations.

We have belabored this point to emphasize that it is unwise to base national and global policies too firmly on long-range population forecasts. Nevertheless, it has been clear for several decades that humanity is poised between overpopulation and nuclear annihilation. The threat of nuclear war has receded, and the threat of overpopulation is diminishing, but both dangers are still very real and require careful watching and planning. At this time it is unlikely that world population will reach 40 billion by the end of the next century: the trend toward lower birthrates is very strong, and the historical precedent of decreasing birthrates in the developed regions of the world is compelling. It is also unlikely that world population will be less than 6 billion at the end of the twentieth century. Even if birthrates were to drop instantaneously, the children of the large number of young people who are now living in developing regions will bring the population of the world close to 6 billion. This still leaves a very wide range for the likely size of the world population in the year 2100. The forecast in figure 5.9 suggests that by the end of the next century the world will reach steady state at a population of about 10 billion. Gilland (1979) has proposed that between 10 and 16 billion people will inhabit the Earth in 2100. If nuclear war and other man-made disasters are avoided, if the AIDS epidemic is brought under control, and if other disasters akin to AIDS do not strike us, the Earth will become more populous during the next century, but we will have gone through the demographic transition.

In 1994 the 10th International Conference on AIDS estimated that, worldwide, at least 4 million people have developed AIDS and that, cumulatively, at least 17 million people have been infected by the human immunodeficiency virus (HIV). By the year 2000, some 30 to 40 million people will probably have been infected by HIV since the start of the epidemic. At present, there is neither an effective vaccine nor a definitive cure. Although only a small fraction of the world population has suffered from AIDS, it is possible that the disease could ultimately have a large demographic impact (Stoto 1993), especially in Africa (see fig. 5.15), where the majority of HIV infections have been recorded.

Most would agree that a major increase in population is not desirable, and believe that the population of the world should be stabilized as quickly as possi-

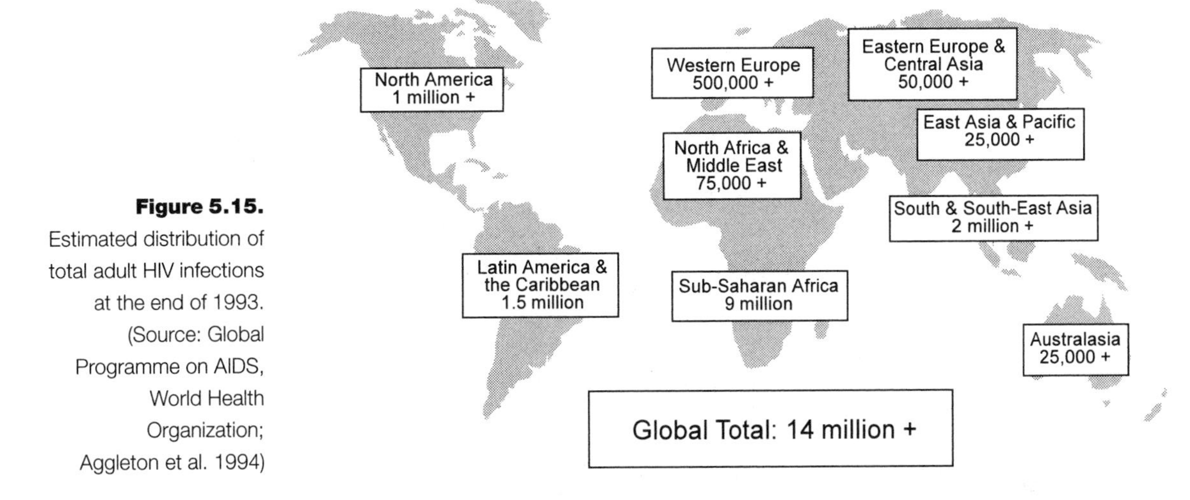

Figure 5.15. Estimated distribution of total adult HIV infections at the end of 1993. (Source: Global Programme on AIDS, World Health Organization; Aggleton et al. 1994)

ble at the lowest feasible level, and we believe that those who advocate the opposite are acting irresponsibly. The reduction in the birthrate in the developed regions of the world followed an increase in the standard of living, an increase in the availability of birth control measures, and public policies that favored lower birthrates. All three factors together will probably be needed to achieve the required decrease in the birthrate in developing regions. If these measures fail, the population of the developing regions may well be limited by the availability of food, and Malthus will have been proven right.

5.6 Agriculture and Fisheries

Most of our food is produced on land, but a significant fraction grows in water. Table 5.5 lists the land areas that are used for forests and woodlands, for grassland and pasture, and for cropland in ten regions of the world and for the sum of these regions. The areas in table 5.5 are given in millions of hectares, a unit that may need a little explanation, at least in the United States. One hectare is an area 100 meters by 100 meters, roughly one football field long and equally as wide. One hectare is 1/100 of a km^2, 10^8 cm^2, or 2.25 acres. Hence the world's cropland area in 1980 was 0.15×10^{18} cm^2 (0.15×10^{10} ha). Since the total area of the continents is 1.6×10^{18} cm^2, croplands covered a little less than 10% of the continents; forests and woodlands occupied about 30%, and grasslands and pasture about 40%. The remaining 20% was occupied by deserts, water, ice, and human settlements.

Since 1850, the cropland areas of the world increased by nearly a factor of three (see table 5.5), largely at the expense of forests and woodlands. During this period the population of the world increased by a little more than a factor of five. The feeding of this multitude has been accomplished by expanding the cropland area and increasing the agricultural yield.

Figure 5.16 shows the history of the average yield of wheat in the United Kingdom and in the United States between 1880 and 1980. The average wheat yield in the two countries increased very gradually at a rate of 3–4 kg/ha/yr, i.e., about 0.3%/yr, up to about 1950. Since then the average wheat yield has

Table 5.5. World Land Use, by Region, 1850–1980

Region	*Land Type*	*Area (million hectares)* 1850	1900	1950	1980	*Percent Change 1850–1980*
Ten regions, total	Forests and woodlands	5,919	5,749	5,345	5,007	–15
	Grasslands and pasture	6,350	6,284	6,293	6,299	–1
	Cropland	538	773	1,169	1,501	179
Tropical Africa	Forests and woodlands	1,336	1,306	1,188	1,074	–20
	Grassland and pasture	1,061	1,075	1,130	1,158	9
	Cropland	57	73	136	222	288
North Africa and Middle East	Forests and woodlands	34	30	18	14	–60
	Grassland and pasture	1,119	1,115	1,097	1,060	–5
	Cropland	27	37	66	107	294
North America	Forests and woodlands	971	954	939	942	–3
	Grassland and pasture	571	504	446	447	–22
	Cropland	50	133	206	203	309
Latin America	Forest and woodlands	1,420	1,394	1,273	1,151	–19
	Grassland and pasture	621	634	700	767	23
	Cropland	18	33	87	142	677
China	Forests and woodlands	96	84	69	58	–39
	Grassland and pasture	799	797	793	778	–3
	Cropland	75	89	108	134	79
South Asia	Forests and woodlands	317	299	251	180	–43
	Grassland and pasture	189	189	190	187	–1
	Cropland	71	89	136	210	196
Southeast Asia	Forests and woodlands	252	249	242	235	–7
	Grassland and pasture	123	118	105	92	–25
	Cropland	7	15	35	55	670
Europe	Forests and woodlands	160	156	154	167	4
	Grasslands and pasture	150	142	136	138	8
	Cropland	132	145	152	137	–4
USSR (former)	Forests and woodlands	1,067	1,014	952	941	–12
	Grassland and pasture	1,078	1,078	1,070	1,065	–1
	Cropland	94	147	216	233	147
Pacific developed countries	Forests and woodlands	267	263	258	246	–8
	Grassland and pasture	638	634	625	608	–5
	Cropland	6	14	28	58	841

Source: Repetto 1987.

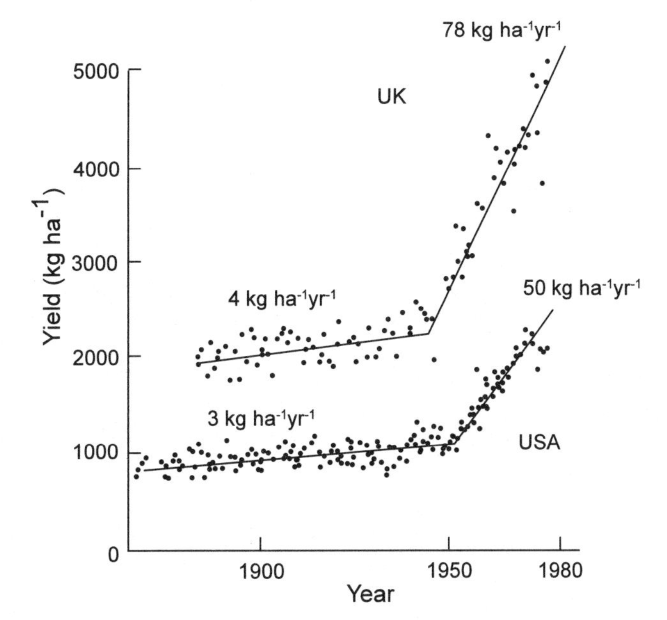

Figure 5.16. Average wheat yields in the United Kingdom during the course of the past century. (Van Keulen and Wolf 1986)

increased at about 78 kg/ha/yr in the U.K. and at a rate of about 50 kg/ha/yr in the United States. This means that annual yields have grown nearly twenty times more rapidly since 1950 than prior to 1950. Productivity since 1950 has increased at a rate of about 2%–3% per year. This is much faster than the rate of population growth in the industrialized nations.

The increase in the average grain yields between 1954 and 1980 in Africa, Asia (excluding China), and South America is shown in figure 5.17. The annual improvement is considerably less than in the industrialized part of the world. In Africa the average grain yield is about 1,000 kg/ha/yr, and this is increasing at a rate of about 10 kg/ha/yr, i.e., at about 1% per year. However, the growth rate of the population in Africa is 2%–3% per year, and the increase in cropland area has not been sufficient to make up the difference.

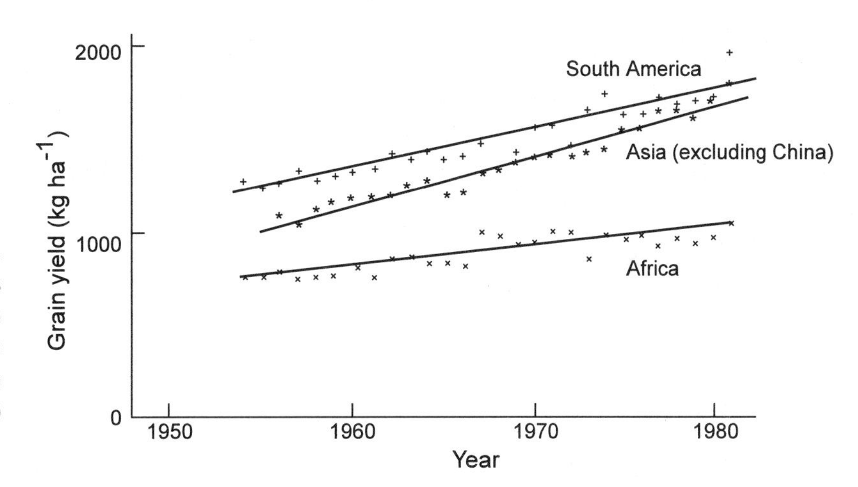

Figure 5.17. Average grain yields from 1954 to 1980 in Africa, Asia, and South America. (Van Keulen and Wolf 1986)

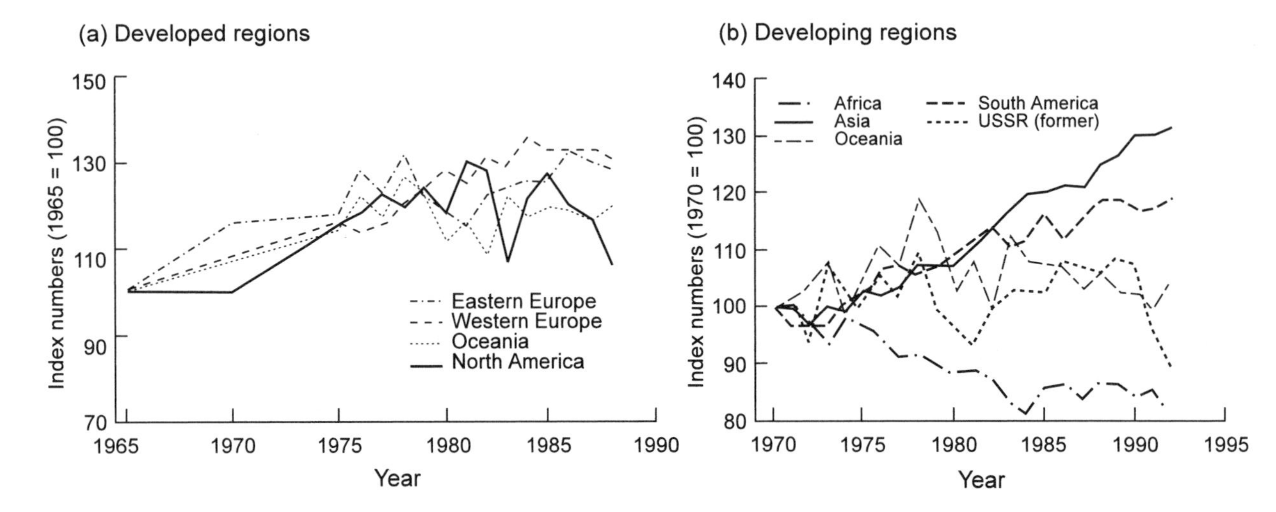

Figure 5.18. Index of per capita food production in the developed and developing regions. (*World Resources 1990–91*, 1990; and *World Resources 1994–95*, 1994)

Figure 5.18 shows that the per capita food production in the developed regions of the world has increased about 20% in the last 25 years. Among the developing regions, Asia has done very well, South America has shown a gain, Oceania and the former USSR have more or less held their own, but in Africa per capita food production has declined considerably. Sub-Saharan Africa has therefore become the region with the world's most serious food problem. Although overall food production is growing, per capita production has been falling ever since 1970. Africa was a food exporter in the early 1960s, but in the 1970s Africa became a food importer. By the 1980s cereal imports had quadrupled, and the region had become heavily dependent on food aid.

Table 5.6 shows the estimated number of undernourished people in four regions of the world in 1969–71, 1979–81, and 1988–90. People are defined as undernourished when their daily caloric intake is less than 1.4 times the estimated energy requirements in a state of fasting and complete rest, approximately

Table 5.6. Prevalence of Chronic Undernutrition in Developing Regions

	1969–71		*1979–81*		*1988–90*	
Region	*Millions of Under-nourished*	*Percent of Total Population*	*Millions of Under-nourished*	*Percent of Total Population*	*Millions of Under-nourished*	*Percent of Total Population*
Africa	101	35	128	33	168	33
Asia	751	40	645	28	528	19
Latin America	54	19	47	13	59	13
Middle East	35	22	24	12	31	12
Total developing regions	941	36	844	26	786	20

Source: *World Resources*, 1994–95.

Table 5.7.
Food Available for Direct Human Consumption (nutritional calories per capita per day)

	1961–63	*1969–71*	*1979–81*	*1984–86*
World total	2,300	2,440	2,600	2,690
Developing countries				
Africa (sub-Saharan)	2,040	2,100	2,140	2,060
Near East/North Africa	2,240	2,390	2,870	3,050
Asia	1,830	2,030	2,260	2,430
Asia[a]	1,970	2,070	2,200	2,280
Latin America	2,380	2,520	2,670	2,700
Low-income countries	1,850	2,020	2,200	2,360
Middle-income countries	2,160	2,340	2,620	2,680
Developed countries	3,060	3,230	3,340	3,380
North America	3,180	3,380	3,510	3,620
Western Europe	3,090	3,230	3,370	3,380
Other developed market economies	2,590	2,810	2,900	2,930
European centrally planned economies	3,140	3,330	3,390	3,410

Source: *World Resources, 1990–91*.
[a] Does not include China.

1,600 to 1,700 kilocalories per day (one "nutritional" calorie is the same as one kilocalorie as used in physics and chemistry). The percentage of undernourished people has decreased in all of the regions, and the actual number of undernourished people has also decreased.

The caloric food intake that has been available for direct consumption since 1961–63 in different parts of the world is shown in table 5.7. Available food in the developed regions of the world amounts to about 3,000–3,600 kcal (kilocalories) per day. In the developing countries, the number of kilocalories per person per day is much lower; in sub-Saharan Africa it is barely above 2,000 kcal/day. During the past 25 years the production of cereals, a major component of world food, has increased more rapidly than world population (see fig. 5.19). These gains have not, however, been distributed uniformly. While North America and Europe have done well, the food situation in sub-Saharan Africa has deteriorated.

Adequate nourishment of the world's population does not require that all parts of the world grow an adequate food supply to feed their own populations. Food can be, and is, traded extensively. During the 1980s, the United States and Canada exported very large amounts of grain to many parts of the world. This is shown graphically in figure 5.20 and in more detail for 1988 in table 5.8. We tend to think of the United States as basically an industrial nation, and we tend to forget that agricultural exports play a major and positive role in our balance of trade with the rest of the world. The increases in U.S. wheat production during the last 70 years have been brought about largely by an increase in productivity. American agriculture has shifted somewhat from the Northeast and Southeast to the West (see fig. 5.21a). The total area of cropland has, if anything, decreased slightly since the mid-1920s, as has the total number of farms (see fig. 5.21b). In spite of these changes there has been a constant danger of food gluts rather than of food shortages, and it is clear that much more food could be produced in the United States, in Canada, and in several other parts of the world.

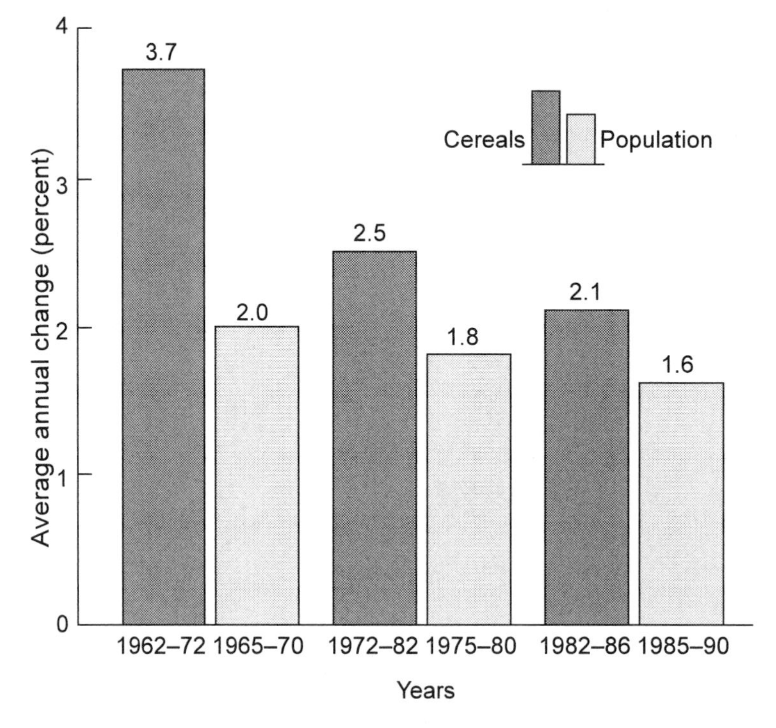

Figure 5.19. World food production is growing faster than world population. The figure shows the annual increase in total production of cereals (darker) and in the world's population (lighter). (Crosson and Rosenberg 1989)

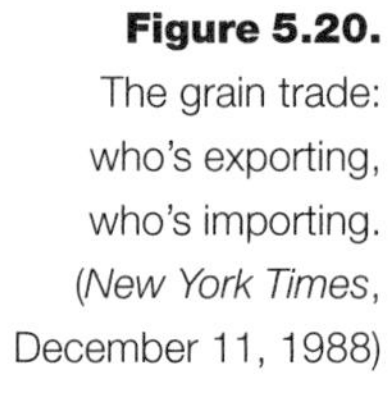

Figure 5.20. The grain trade: who's exporting, who's importing. (*New York Times*, December 11, 1988)

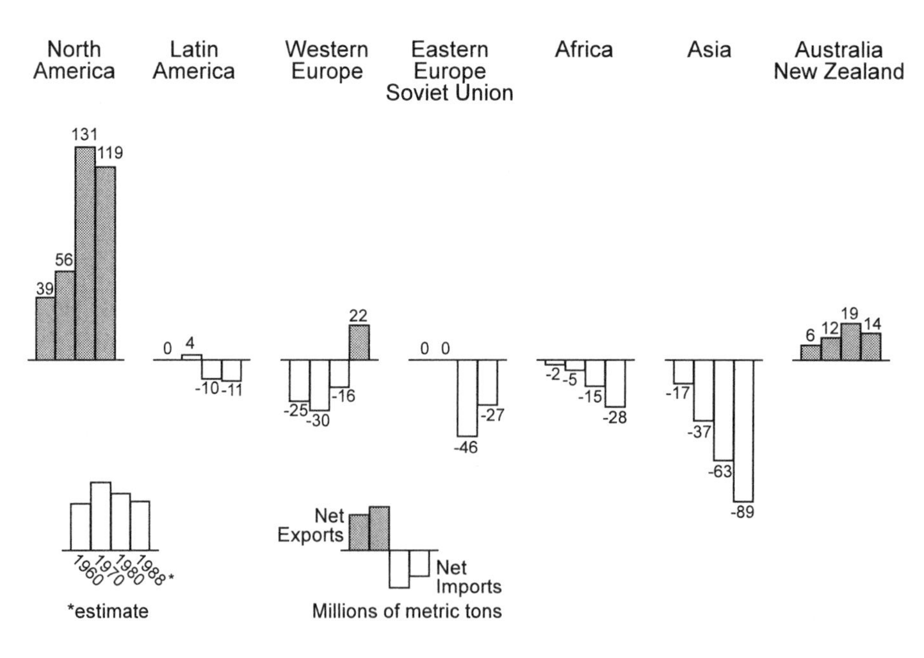

Table 5.8. Major Net Cereal Importers and Exporters, 1987

Importers		*Exporters*	
Country	*Million Metric Tons*	*Country*	*Million Metric Tons*
USSR (former)	29	United States	83
Japan	27	Canada	28
China	16	France	26
Egypt	9	Australia	18
Korea	9	Argentina	9
Saudi Arabia	8	Thailand	6
Iran	6	United Kingdom	4
Italy	5	South Africa	2
Mexico	5	Denmark	1
Iraq	4	New Zealand	0.2

Source: *World Resources*, 1990–91.

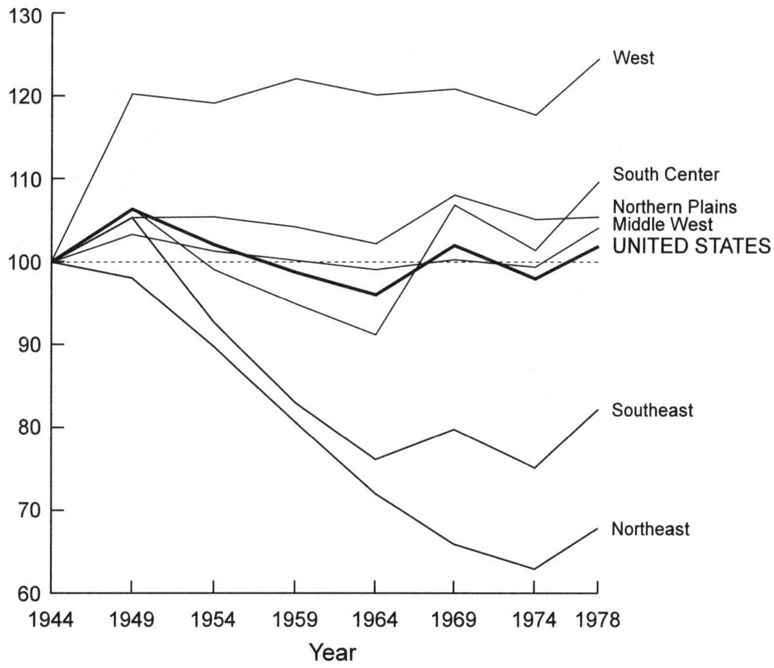

Figure 5.21a. Changes in total cropland area in the major regions of the United States between 1944 and 1978 (1944 = 100). (Hart 1984)

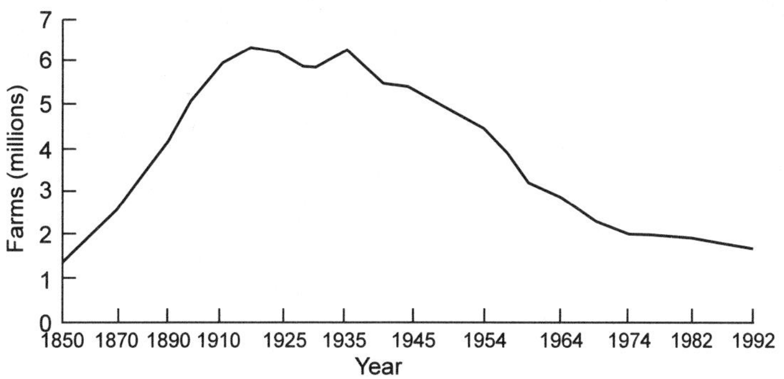

Figure 5.21b. The number of farms (in millions) in the United States from 1850 to 1992. (*New York Times*, November 10, 1994)

Figure 5.22. (a) Wheat prices received by farmers. (b) Milk prices received by farmers. (c) Chicken prices received by farmers. (Deflated by 1967 = 100 W.P.I.; Johnson 1984)

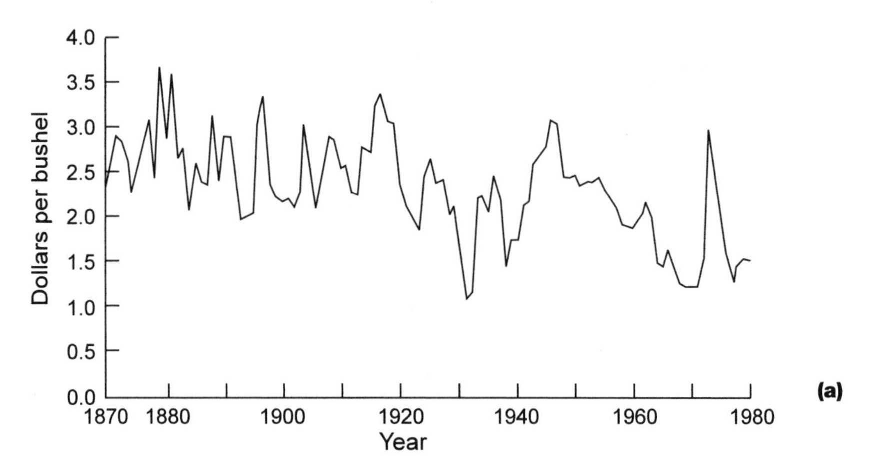

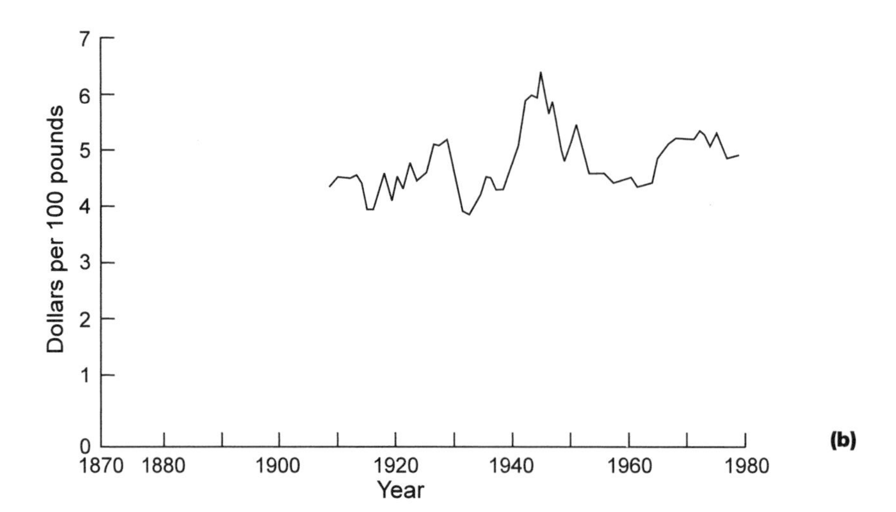

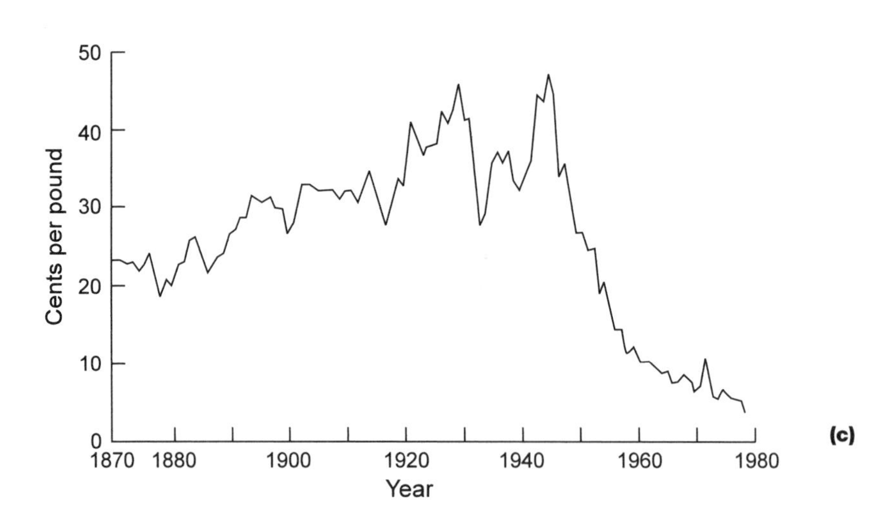

In a free market economy the price of a commodity is a good indicator of its scarcity, and price trends tend to reflect changes in the balance between supply and demand. For agricultural products this connection is weakened but not obliterated by the effects of government subsidies and controls. The data in figure 5.22 show that wheat prices have decreased in constant dollars since the last century, and that milk prices have increased, but only slightly since 1910. The price of chicken rose between 1870 and about 1945 and then dropped rapidly as a result of more efficient production techniques. In the process, chicken dinners were demoted from "something special on Sundays" to a cheap, everyday dinner and fast-food commodity. The price histories of other food categories show similar trends. Clearly, the availability of food today is limited neither by physical nor biological boundaries but by fiscal policies, poverty, and political instability.

Annually, nearly 4,000 million tons, or 4 Gt, of food are grown on land. The total catch of fish is quite small in comparison. Although the annual catch has grown rapidly, from about 40 million tons in 1960 to nearly 100 million tons in 1988 (see fig. 5.23), fish protein remains a minor part of the total protein consumption in most parts of the world (see fig. 5.24).

Before we ask how much food the Earth can reasonably be expected to produce in the future, we should consider present-day food production within the context of the carbon cycle as a whole. Earlier in this chapter we noted that approximately 50–80 Gt of carbon are fixed annually by the terrestrial biosphere. Food production now amounts to about 4 Gt/yr. This weight of food requires the growth of roughly 12 Gt of plants. The edible part of corn plants, for instance, amounts to about one-third the weight of the entire plant. Carbon accounts for about 50% of the 12 Gt of plants. Thus, approximately 6 Gt of carbon are fixed annually to supply human nutrition. This is about 10% of the estimated total carbon fixation of the terrestrial biosphere. The estimate is rough, but it is consistent with the fact that croplands cover approximately 10% of the land area. The annual fish catch of 0.1 Gt accounts for less than 1% of the carbon that is fixed photosynthetically in the oceans.

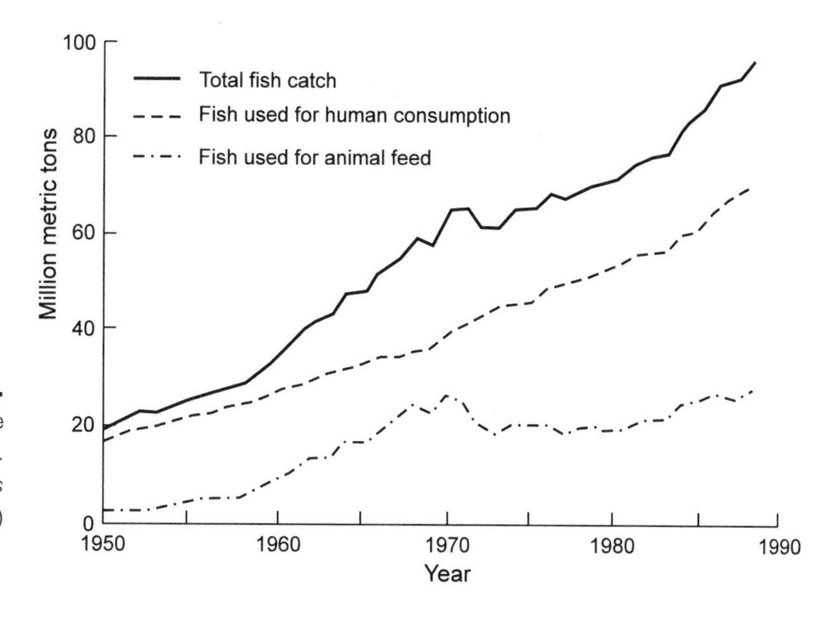

Figure 5.23. World fish catch by use from 1950 to 1988. (*World Resources 1990–91*, 1990)

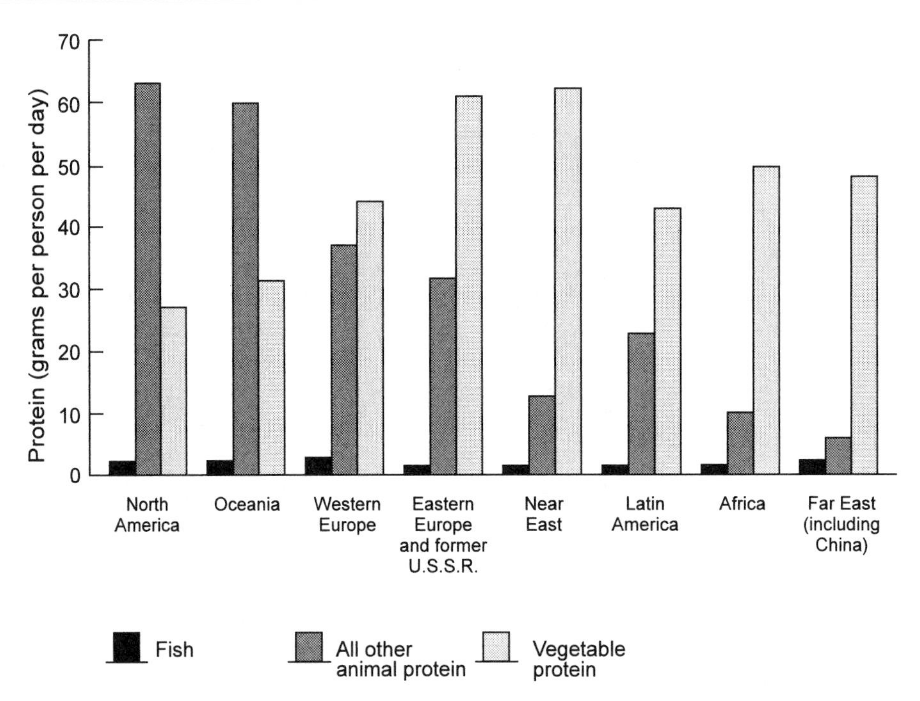

Figure 5.24. The relatively minor role played by fish in the world's total consumption of protein is apparent when the grams of fish eaten per person per day in various parts of the world (left column in each group) are compared with the consumption of other animal protein (middle column) and vegetable protein (right column). (Holt 1969)

These figures suggest that food production for human consumption can be raised considerably above present levels. It could be argued, however, that food production might be limited by the availability of solar energy. The conversion of one gram of carbon in atmospheric CO_2 to carbon in plant material requires about 10 kilocalories. The annual conversion of 50 Gt of carbon in CO_2 to organic matter therefore requires

$$50 \times 10^{15} \frac{\text{gmC}}{\text{yr}} \times 10 \times 10^3 \frac{\text{cal}}{\text{gm}} = 5.0 \times 10^{20} \text{ cal/yr}.$$

The total amount of solar energy received annually by each square centimeter of the Earth is 2.55×10^5 cal (see chapter 2). The solar energy input to the continents is therefore approximately

$$2.55 \times 10^5 \frac{\text{cal}}{\text{cm}^2\text{yr}} \times 1.6 \times 10^{18} \text{ cm}^2 = 4.0 \times 10^{23} \text{ cal}.$$

Thus, only about 0.1% of the solar energy reaching the Earth is used to produce biomass. This low efficiency is due to a number of factors. Only solar energy in the wavelength range of 400–700 nm is used in photosynthesis, and this wavelength range encompasses only half of the energy in the solar spectrum. Not all of the sunlight that reaches ground level is intercepted by plants. Some of the sunlight that does reach plants is reflected rather than used in photosynthesis. Finally, the process of photosynthesis itself is not very efficient. Table 5.9 shows that crops are not much more efficient than the biosphere as a whole. Under ideal conditions, efficiencies as high as 5% can be achieved (Good and Bell 1980). The

Table 5.9
Efficiency of Conservation of Light Energy in Various Crops

Crop	*Efficiency of Use of Sunlight (%)*	*Crop*	*Efficiency of Use of Sunlight (%)*
Wheat (Netherlands)	0.35	Soybeans (Canada)	0.18
Wheat (world average)	0.10	Soybeans (world average)	0.10
Corn (United States)	0.35	Sugar cane (Hawaii)	0.95
Corn (world average)	0.17	Sugar cane (Cuba)	0.30
Rice (Japan)	0.42	Sugar beets (Netherlands)	0.56
Rice (world average)	0.18		
Potatoes (United States)	0.31		
Potatoes (world average)	0.17		

Source: Good and Bell 1980.

increase in agricultural productivity during the course of this century is due in considerable part to improving growth conditions in croplands. Further increases in productivity are clearly possible, perhaps aided by genetic engineering. It is unlikely, however, that conversion efficiencies for sunlight will reach the theoretical limit of about 10%. Artificial light can, of course, be used to grow some crops.

Figure 5.25 shows a recent, reasonably optimistic forecast for grain production expected between 1976 and 2176. A likely increase in grain production of about a factor of 3.5 is proposed if only conventional agriculture is used. An increase by a factor of about 11 is deemed possible, if a variety of new techniques, including genetic engineering, are used to increase food production. How these increases would translate into per capita food consumption depends on the growth of the human population and on the proportion of meat to grain in our

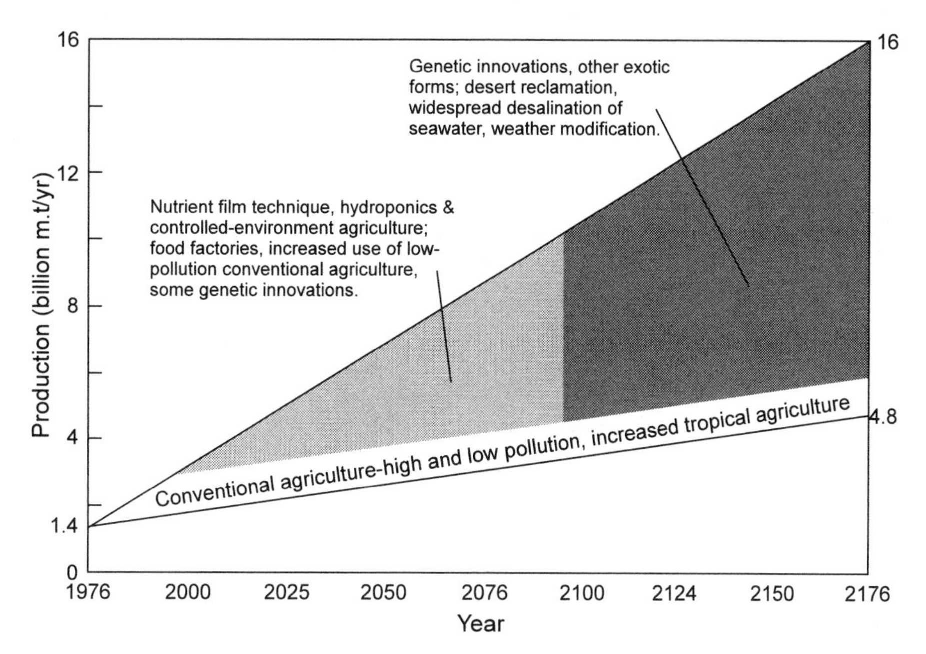

Figure 5.25. A reasonably optimistic forecast for the increase in equivalent grain production between 1976 and 2176. (Kahn et al. 1976)

food. If world population stabilizes at two to three times the present population, even the low projection implies a potential increase in food availability. If world population approaches 40 billion people in the year 2176, the low projection clearly predicts starvation; along the high projection there would still be adequate amounts of food.

The projections in figure 5.25 do not include fisheries, but this may not be a serious omission. Fish are currently such a minor part of the world's caloric intake that a very great increase in the worldwide catch would be required to increase the total quantity of the world food supply. This is impossible with present-day fishing methods. Overfishing along the Atlantic Coast in Canada and in the United States, and along the West Coast, except for Alaska, has already led to a serious depletion of available fish. Aquaculture is growing in importance, but it is currently only a source of rather expensive fish. Theoretically, marine aquaculture could supply a good deal more food, but it remains to be seen whether large-scale fishing, or for that matter, seaweed farming, will become practicable. In the distant future one can foresee the development of synthetic foods. Coal, oil, and natural gas would be the logical starting materials, but limestones would also do, if sufficient energy is available to reduce the carbon in $CaCO_3$ to its valence state in food. In most science fiction dealing with such matters, synthetic food is not considered particularly palatable; but it is surely not beyond our ingenuity to solve this problem, and if it becomes a matter of survival, taste will hardly be a deterrent to the consumption of such foodstuffs.

5.7 The Endangered Biosphere

During the last 40 years the well-being of the human family has increased at an unprecedented rate. Tables 5.10 and 5.11 show that in nearly all countries per capita income and life expectancy have risen significantly. Figure 5.26 shows that even in countries with very modest per capita incomes the average life expectancy is now more than 60 years. All this has happened while the population of the world has doubled.

But, as the First Law of Thermodynamics states, "There is no free lunch"; nor are there free breakfasts, dinners, or midnight snacks. The ecological costs of the expansion and material improvement of humanity have already been considerable, and they promise to be severe (see chapter 12). During the nineteenth century much of the overflow of humanity in Europe moved into North America. The forests that covered much of the eastern half of America were destroyed and turned into farmland. The feats of the early lumberjacks were celebrated in the Paul Bunyan stories. The conversion of forests into farms was celebrated in books such as Laura Ingalls Wilder's *Little House in the Big Woods*. The conversion of prairie into cropland was celebrated in books such as Willa Cather's *O, Pioneers!* Now, less than a century later, there is enough cleared land for all Americans, and the remaining wilderness is becoming a cherished treasure. The "tamers of the wilderness" have become the "despoilers of nature." The taming was as necessary then as the conserving is now.

The processes that transformed North America are now at work in the developing nations. Croplands are increasing at the expense of forests and grasslands. Native habitats are disappearing, and with them the way of life of indigenous peoples and a large number of animal and plant species. The destruction of the rain forests in South America, Africa, and Asia could bring about an enormous amount of biological extinction (see chapter 12). One can argue about the impor-

Table 5.10. Socioeconomic Indicators, Selected Countries, about 1960 and 1984

Country	*Population (millions) mid-1984*	*Average Annual Growth of GNP per Capita (percent) 1965–84*	*Average Index of Food Production per Capita (1965–67 = 100) 1982–84*	*Life Expectancy at Birth (years)*	
				1960	*1984*
Low-income	2,389.5	2.8	111.4	36	60
China	1,029.2	4.5	138.2	51	69
India	749.2	1.6	117.7	42	56
Kenya	19.6	2.1	72.2	43	54
Middle-income	1,187.6	3.0	108.2	49	61
Brazil	132.6	4.6	131.1	56	64
Egypt	45.9	4.3	94.6	45	60
Indonesia	158.9	4.9	140.4	40	55
Mexico	76.8	2.9	101.9	56	66
Nigeria	96.5	2.8	85.4	34	50
Industrial market	733.4	2.4	117.9	70	76
Japan	120.0	4.7	97.4	67	77
United States	237.0	1.7	119.7	70	76
East European nonmarket	389.3	—	117.4	66	68
USSR (former)	275.0	—	114.1	68	67

Source: Repetto 1987.

Table 5.11. Life Expectancy at Birth, African and Other Developing Countries, 1960 and 1979

Country Group	*Life Expectancy*			*1979 GNP per Capita (U.S. Dollars)*	*Index of per Capita Food Production 1977–79 (1969–71 = 100)*	*Daily Calorie Supply per Capita, 1977*	
	1960	*1979*	*Increase*			*Calories*	*Percentage of Requirement*
Africa*							
Low-income	38	46	8	239	91	2,040	91
Low-income, semi-arid	37	43	6	187	88	1,992	89
Low-income, other	39	47	8	247	91	2,086	93
Middle-income oil importers	41	50	9	532	95	2,180	97
Middle-income oil exporters	39	48	9	669	86	1,970	89
Sub-Saharan Africa	—	47	—	411	91	—	—
Selected low-income countries	42	50	8	200	97	2,052	91
India	43	51	8	180	100	2,021	91
Bangladesh	40	47	7	90	90	1,812	78
Developing countries by per capita income							
Less than $390	42	50	8				
$390–1,050	46	55	9				
$1,060–2,000	47	64	7				
$2,040–3,500	65	71	6				

Source: Johnson 1984. * Excludes South Africa.

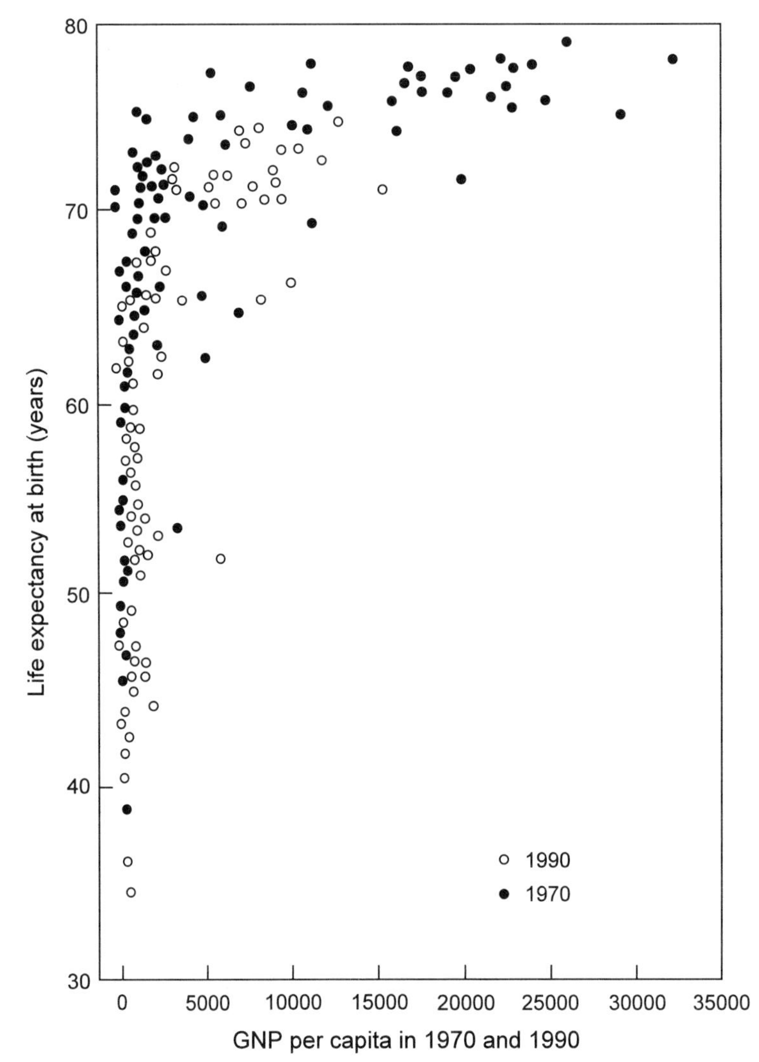

Figure 5.26. Life expectancy at birth in relation to income in countries for which data were available in 1970 and in 1990. Income is computed as gross national product per capita (GNPPC) and is plotted in 1990 U.S. dollars. (Wilkinson 1994)

tance of maintaining the present diversity of mosquitoes, fleas, and even ants, but it is difficult to believe that the extinction of the large and magnificent range of birds, reptiles, and mammals in the rain forests would be anything but an ecological disaster (see, for instance, Wilson and Peter 1988; Wilson 1989; Nisbet 1991). A balance will have to be found between humanity's need for food and the need to conserve sufficient wilderness to preserve a large fraction of the plants and animals that inhabit the Earth. Good stewardship and the reverence for life that was championed so eloquently by Albert Schweitzer demand that we consider both needs.

The destruction of the remaining wilderness can be slowed by changing land and fiscal policies, by trading wilderness for national debt, by reducing the rate of population growth, and by increasing the productivity of the land that is already under cultivation. More conservative fishing limits will be required to preserve fish populations. The productivity of the land has been increased successfully during the twentieth century, but there have been environmental costs. Fertilizers, fungicides, and pesticides have polluted surface and groundwaters (see chapter 12), and intense farming has increased rates of erosion and soil loss (see chapter 6). Continued increases in soil productivity during the next several centuries should therefore not be taken for granted. The increasing CO_2 content of the atmosphere will probably improve photosynthetic efficiency somewhat (see chapter 11), but it may also affect the global pattern of rainfall distribution and hence the agricultural productivity of the world's great "bread baskets."

References

Adler, J. 1966. *Claus Spreckels: The Sugar King of Hawaii*. University of Hawaii Press, Honolulu.

Aggleton, P., O'Reilly, K., Slutkin, G., and Davies, P. 1994. Risking Everything? Risk Behavior, Behavior Change and AIDS. *Science* 265:341–45.

Andrewartha, H. G. 1970. Population Growth and Control: Animal Populations. In *Population Control*, ed. A. Allison, pp. 45–69. Penguin Books, New York.

Barghoorn, E. S., and Tyler, S. A. 1965. Microorganisms from the Gunflint Chert. *Science* 147:563–77.

Borrie, W. D. 1970. *The Growth and Control of World Population*. Weidenfeld and Nicolson, London.

Broecker, W. S., and Peng, T.-H. 1982. *Tracers in the Sea*. Eldigo Press, Palisades, N.Y.

Brown, L. R., Kane, H., and Ayres, E. 1993. *Vital Signs 1993: The Trends That Are Shaping Our Future*. W. W. Norton, New York.

Caldwell, J. C., and Caldwell, P. 1990. High Fertility in Sub-Saharan Africa. *Scientific American* 262 (May):118–25.

Carlson, P. S., ed. 1980. *The Biology of Crop Productivity*. Academic Press, New York.

Coale, A. J. 1983. Recent Trends in Fertility in Less Developed Countries. *Science* 221:828–32.

Cohen, B. 1993. Fertility Levels, Differentials, and Trends. In *Demographic Change in Sub-Saharan Africa*, ed. K. A. Foote, K. H. Hill, and L. G. Martin, chap. 2. National Academy Press, Washington, D.C.

Crosson, P. R., and Rosenberg, N. J. 1989. Strategies for Agriculture. *Scientific American* 261 (Sept.):128–35.

Dalrymple, G. B. 1991. *The Age of the Earth*. Stanford University Press, Stanford, Calif.

Eden, H. F., Elero, B. P., and Perkins, J. N. 1993. Nimbus Satellites: Setting the Stage for Mission to Planet Earth. *EOS Transactions* 74:281.

Ellis, W. 1825. *Journal of William Ellis*. Reprinted 1963. Advertiser Publishing, Honolulu.

Feeney, G. 1994. Fertility Decline in East Asia. *Science* 266:1518–23.

Gilland, B. 1979. *The Next Seventy Years: Population, Food and Resources*. Abacus Press, Tunbridge Wells, U.K.

Good, N. E., and Bell, D. H. 1980. Photosynthesis, Plant Productivity, and Crop Yield. In *The Biology of Crop Productivity*, ed. P. S. Carlson, pp. 3–50. Academic Press, New York.

Gould, S. J. 1987. *Time's Arrow, Time's Cycle: Myth and Metaphor in the Discovery of Geological Time*. Harvard University Press, Cambridge, Mass.

Hammond, A. L., ed. 1990, *World Resources 1990–91*. World Resources Institute. Oxford University Press, New York.

Hammond, A. L., ed. 1994. *World Resources 1994–95*. World Resources Institute. Oxford University Press, New York.

Hart, J. F. 1984. Cropland Change in the United States, 1944–78. In *The Resourceful Earth*, ed. J. L. Simon and H. Kahn, pp. 224–49. Basil Blackwell, Oxford.

Hedrick, P. W. 1984. *Population Biology: The Evolution and Ecology of Populations.* Jones and Bartlett, Boston.

Holt, S. J. 1969. The Food Resources of the Ocean. In *The Ocean*, pp. 99–106. W. H. Freeman, San Francisco.

Jarmul, D. 1984. World Population Is Still Growing, but Rates Vary Greatly by Country. News Report, National Academy of Sciences, no. 4, 4–9.

Johnson, D. G. 1984. World Food and Agriculture. In *The Resourceful Earth*, ed. J. L. Simon and H. Kahn, pp. 67–112. Basil Blackwell, Oxford.

Joyce, J. A. 1976. *World Population Year*. Vol. 4 of *World Population: Basic Documents.* Oceana Publications, Dobbs Ferry, N.Y.

Kahn, H., Brown, W., and Martel, L. 1976. *The Next 200 Years: A Scenario for America and the World.* Quill, New York.

Knoll, A. H. 1992. The Early Evolution of Eukaryotes: A Geological Perspective. *Science* 256:622–27.

Malthus, T. R. 1798. *An Essay on the Principle of Population.* J. Johnson, London.

Malthus, T. R. 1803. *An Essay on the Principle of Population: or a View of Its Past and Present Effects on Human Happiness, with an Inquiry into our Prospects Respecting the Future Removal or Mitigation of the Evils which It Occasions.* J. Johnson, London.

May, R.M. 1979. The Structure and Dynamics of Ecological Communities. In *Population Dynamics*, ed. R. M. Anderson, B. D. Turner, and L. R. Taylor, chap. 18. Blackwell Scientific, Oxford.

Nisbet, E. G. 1991. *Leaving Eden: To Protect and Manage the Earth.* Cambridge University Press, New York.

Perlman, M. 1984. The Role of Population Projections for the Year 2000. In *The Resourceful Earth*, ed. J. L. Simon and H. Kahn, pp. 52–66. Basil Blackwell, Oxford.

Petersen, W. 1979. *Malthus.* Harvard University Press, Cambridge, Mass.

Pielou, E. C. 1974. *Population and Community Ecology: Principles and Methods.* Gordon and Breach, New York.

Rankama, K., and Sahama, Th. G. 1950. *Geochemistry.* University of Chicago Press, Chicago.

Repetto, R. 1987. Population, Resources, Environment: An Uncertain Future. *Population Bulletin* 42 (no. 2):1–44.

Sai, F. T. 1984. The Population Factor in Africa's Development Dilemma. *Science* 226:801–5.

Schopf, J. W. ed. 1992. *Major Events in the History of Life.* Jones and Bartlett, Boston.

Sepkoski, J. J., Jr. 1994. Extinction and the Fossil Record. *Geotimes* 39 (March):15–17.

Stanley, S. M. 1986. *Earth and Life through Time.* W. H. Freeman, New York.

Stoto, M. A. 1993. Models of the Demographic Effect of AIDS. In *Demographic Change in Sub-Saharan Africa*, ed. K. A. Foote, K. H. Hill, and L. G. Martin, chap. 9. National Academy Press, Washington, D.C.

Sunquist, E. T. 1985. Geological Perspective on Carbon Dioxide and the Carbon Cycle. In *The Carbon Cycle and Atmospheric CO_2: Natural Variations, Archean to Present*, ed. E. T. Sunquist and W. S. Broecker, pp. 5–59. Geophysical Monograph 32. American Geophysical Union, Washington, D.C.

Tapinos, G. B., and Piotrow, P. T. 1978. *Six Billion People: Demographic Dilemmas and World Politics.* McGraw-Hill, New York.

van Keulen, H., and Wolf, J. 1986. *Modelling of Agricultural Production: Weather, Soils and Crops.* Pudoc, Wageningen, Netherlands.

Wilkinson, R. G. 1994. The Epidemiological Transition: From Material Society to Social Disadvantage? *Daedalus* 123 (no. 4):61–78.

Wilson, E. O. 1989. Threats to Biodiversity. *Scientific American* 261 (Sept.):108–16.

Wilson, E. O., and Peter, F. M., eds. 1988. *Biodiversity.* Nat. Acad. Press, Wash., D.C.

6 Weathering and Erosion

6.1 Introduction: Weaning Away the Land

"There was once a very high mountain. Every year a bird came, pecked at the mountain, and carried away what it could hold in its beak. When the entire mountain had been carried away, one second of eternity had passed."

This is a fine parable for the immensity of time. However, for geologists the story is somewhat unsatisfactory. They are apt to ask a number of embarrassing questions. What, for instance, happened to the mountain between visits by the bird? Did rainstorms alter and then wash away material from the hill slopes? Did heavy winds lash the summit? Were there snowstorms and avalanches? Were there perhaps even glaciers that scoured out deep valleys and spread mountain debris over surrounding plains? Was the mountain near an ocean, and did storm waves break against its flanks, weaken its fabric, and carry its substance off to sea? Might there have been earthquakes to move the mountain and even perhaps to have it sink into the Earth's crust? In this chapter we will show that these destructive processes are much more efficient than the avian visits of the parable. Mountains have short life spans compared to the age of the Earth. Some, like the Appalachians, that are modest today were once as high as the Andes or even the Himalayas; in a few tens of millions of years the Andes and the Himalayas will probably be no more than modest hills.

Although many parts of the Earth would serve the purpose, we have chosen the Hawaiian Islands to illustrate the building and the wearing away of the land. The geology of the Islands is relatively simple and well understood (see, for instance, Macdonald et al. 1983; Walker 1990). Many of the questions raised by the parable of the bird and the mountain can be answered in Hawaii, and those that cannot be answered there require only occasional trips to other parts of the world. The human history of the Islands has been superbly described by Daws (1968); the recent struggles for land and power have been documented incisively by Cooper and Daws (1985).

6.2 The Building of an Island Chain

The Hawaiian Islands consist of eight major islands that lie along a northwesterly trending belt some 650 km in length (see fig. 6.1). The southernmost member, the island of Hawaii, is the home of Kilauea, the most active volcano in the world, and of the two giant volcanoes, Mauna Loa and Mauna Kea (see plate 6). Lava flows from Kilauea and Mauna Loa frequently reach the sea (see plate 7). The views of lava flowing into the ocean are spectacular, and the addition of new land to the island has given Hawaii an excellent claim to being "the fastest growing state in the Union." The island of Hawaii is known as the Big Island; it is by far the largest in the chain. Among volcanic islands it is second in size only to Iceland. Among the fifty states, it is three times the size of Rhode Island. Mauna Kea and Mauna Loa rise to slightly more than 4,100 meters above sea level. That elevation does not assure them of any records. However, if one reckons their

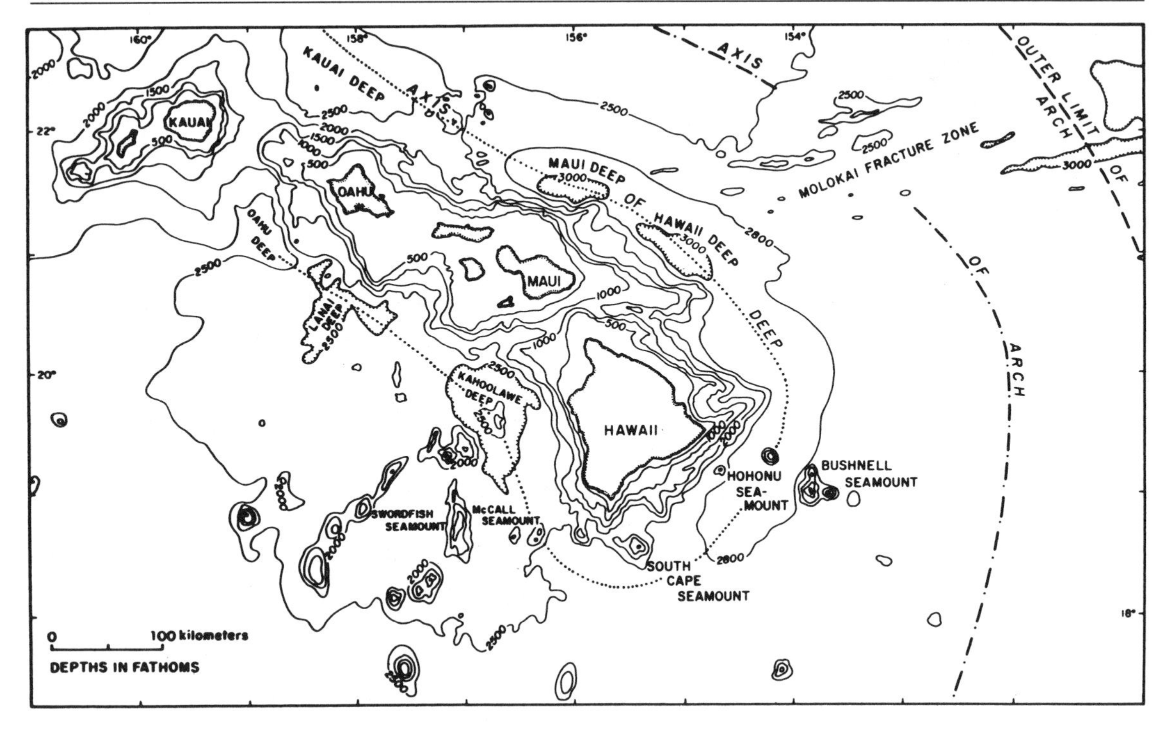

Figure 6.1. Geography of the Hawaiian Islands.

height from the ocean floor in the vicinity of the Islands, the height of both volcanoes exceeds that of Mount Everest.

Maui is the next island to the northwest. Its major volcano, Haleakala, rises to just over 3,000 meters above sea level; it last erupted some 200 years ago. None of the volcanoes on the other islands has erupted in historic times. As shown in figure 6.2, their age increases in a very regular fashion from island to island. The major volcanoes that built the island of Oahu, the site of Honolulu and Pearl Harbor, were active between 2 and 4 million years ago. There has, however, been some minor volcanic activity during the past several hundred thousand years. Diamond Head, the most famous landmark of Honolulu (fig. 6.3), is one of these geologically recent afterthoughts.

The island of Kauai was built between about 3.5 and 6 million years ago. Beyond Kauai the chain barely extends above sea level, but it continues northwestward beyond Midway Island (see fig. 6.4). The chain then turns more to the north and becomes a series of seamounts, the Emperor Sea Mounts, which extend all the way to the Aleutians. The oldest of these seamounts is about 70 million years old. There are other island and seamount chains in the Pacific Ocean, but none can match the Hawaiian chain's superb 2,400-kilometer string of progressively older beads. Mark Twain rightly called them "the loveliest fleet of islands that lies anchored in any ocean." They owe their origin to a large, nearly stationary source of magma that has brought basaltic lava to the surface for at least the past 70 million years. The island chain was formed by the movement of the Earth's crust under the Pacific Ocean. One can compare each island to a stitch in a piece of cloth that has been moved through a stationary sewing machine (see chapter 7).

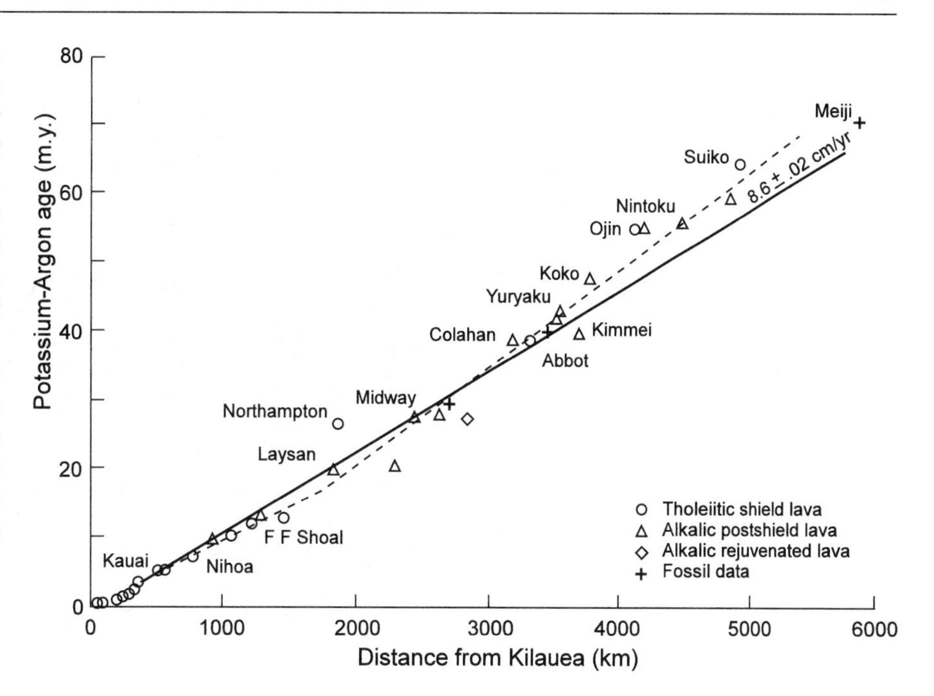

Figure 6.2a. Age of volcanoes in the Hawaiian-Emperor chain as a function of distance from Kilauea. The solid line represents an average rate of propagation of volcanism of 8.6 ± 0.2 cm/yr. The dashed line is a two-segment fit using the data from Kilauea to Gardner and from Laysan to Suiko. (Clague and Dalrymple 1989)

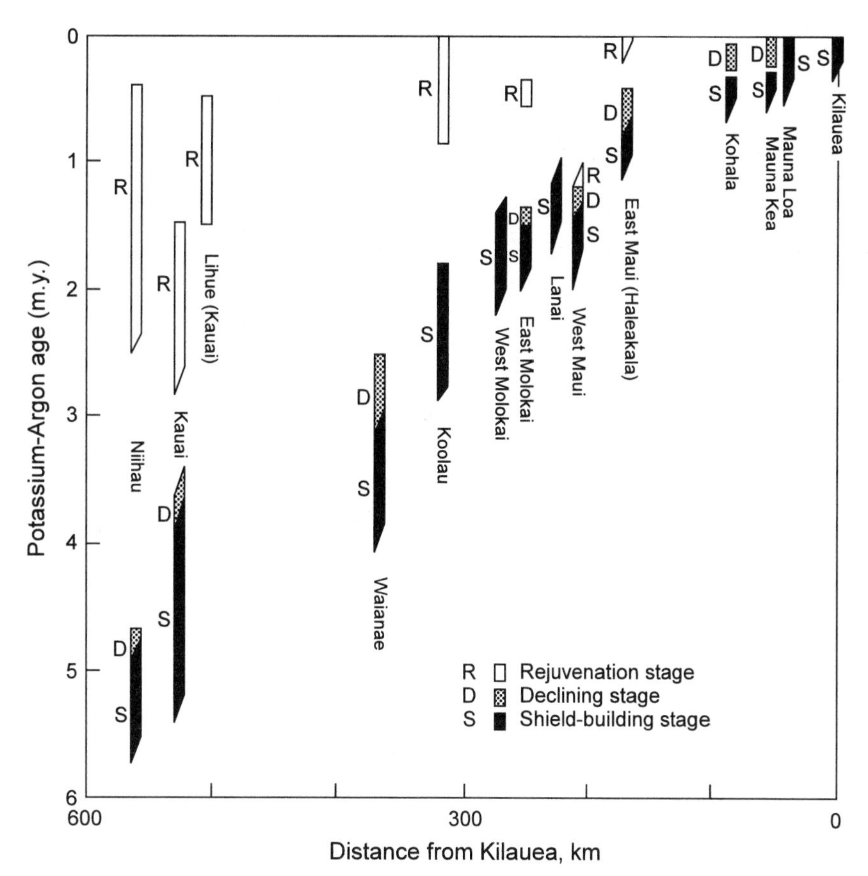

Figure 6.2b. Potassium-argon ages of rocks from the Hawaiian Islands in millions of years plotted against distance of the volcano centers from Kilauea (from data referenced and collated by Clague and Dalrymple 1987, 1989). Note that the dated rocks were collected from the island tips of largely submerged volcanic edifices and do not give the time of inception of each edifice (Walker 1990). S: The main stage of volcano construction; D: the waning of the construction stage; R: late-stage volcanism after the end of the major volcanic episode.

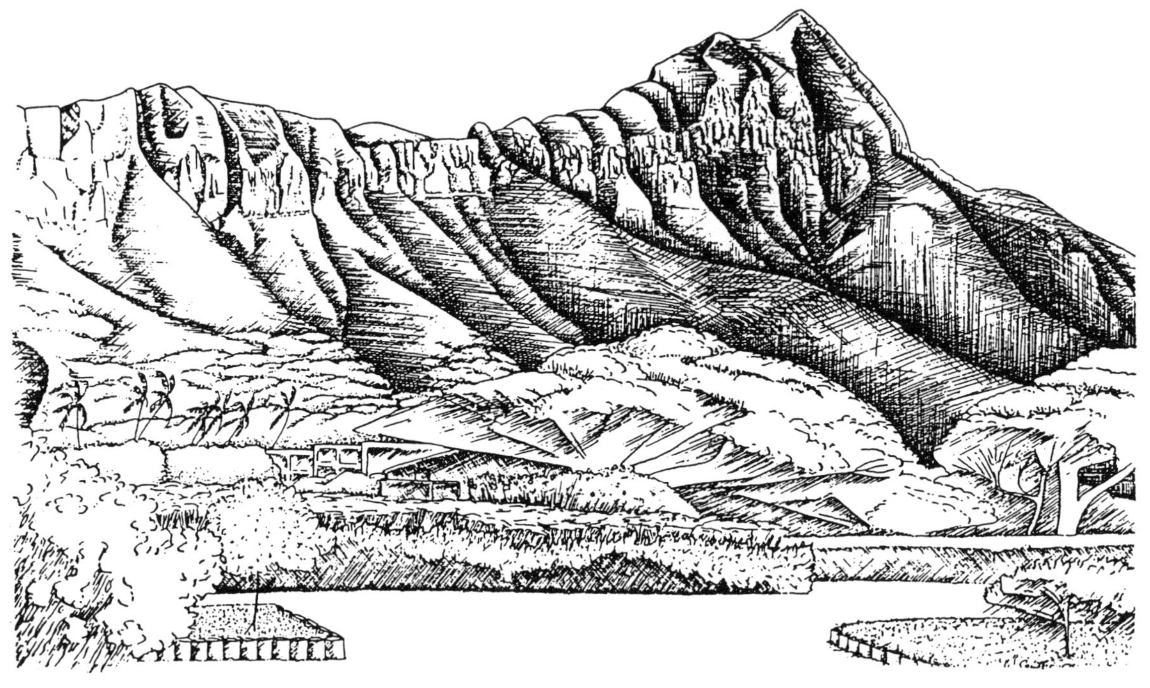

Figure 6.3. Diamond Head, the most famous landmark of Honolulu, Hawaii. (Pen-and-ink sketch by William Coulbourn)

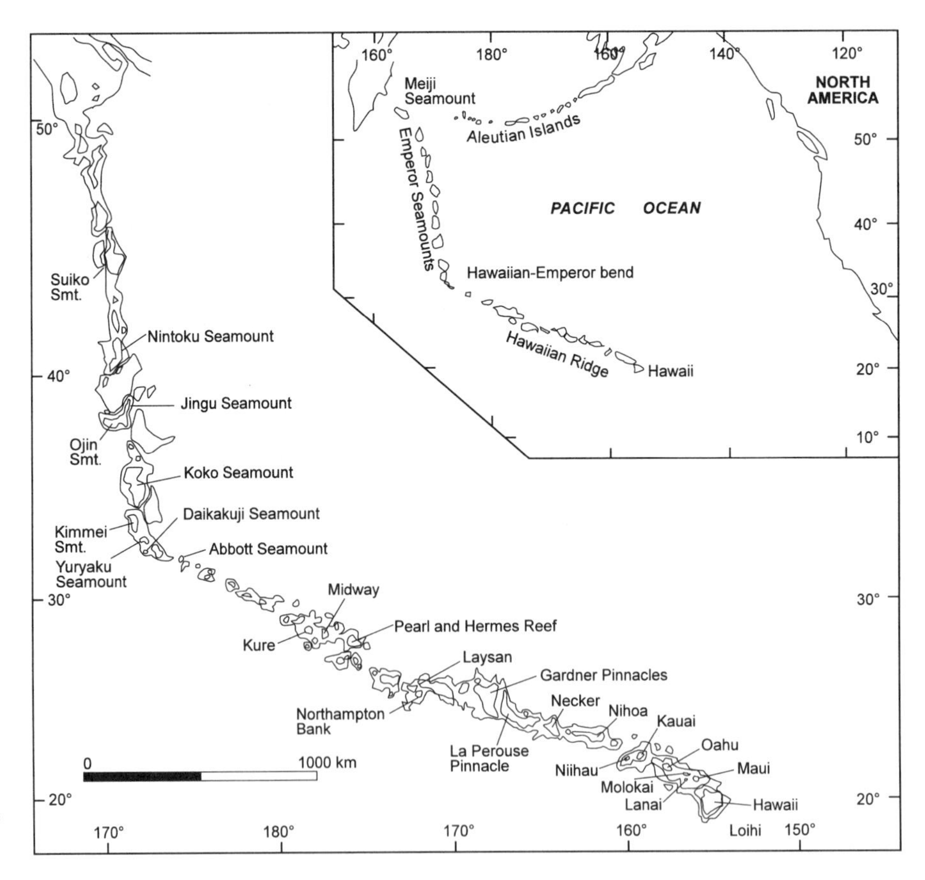

Figure 6.4. Bathymetry of the Hawaiian-Emperor volcanic chain (modified from Chase and others, 1970). Contours at 1 km and 2 km depths are shown in the area of the chain only. The inset shows the location of the chain (outlined by 2 km depth contour) in the central North Pacific. (Clague and Dalrymple 1989)

6.3 Hawaiian Lavas

The Islands' volcanoes are impressive. Mark Twain visited Kilauea in 1866. His description of the volcano in *Roughing It* carries much the same information and feeling as the descriptions of other visitors, but it is considerably more quotable than most:

> Shortly the crater came into view. I have seen Vesuvius since, but it was a mere toy, a child's volcano, a soup-kettle, compared to this. Mount Vesuvius is a shapely cone thirty-six hundred feet high; its crater an inverted cone only three hundred feet deep, and not more than a thousand feet in diameter, if as much as that; its rims meager, modest, and docile—But here was a vast perpendicular, walled cellar, nine hundred feet deep in some places, thirteen hundred in others, level-floored, and ten miles in circumference! Here was a yawning pit upon whose floor the armies of Russia could camp, and have room to spare.

The frequent lava flows from the active volcanoes on the island of Hawaii do not all have the same chemical composition. Most, however, are tholeiitic or—more rarely—alkalic basalts. Their average chemical compositions (table 6.1) are similar to those of basalts elsewhere in the world. They contain the same components in approximately the same proportions, but alkali basalts contain somewhat more Na_2O and K_2O than tholeiitic basalts. The compositions of both types of basalt are expressed in terms of oxide components, because O^{2-} is their dominant anion (negative ion). The negative charge of O^{2-} in the lavas is largely balanced by eleven positive ions, of which Si^{4+}, Al^{3+}, Fe^{2+}, Ca^{2+}, and Mg^{2+} are quantitatively the most important.

Basaltic lavas are quite fluid, somewhat like noncrystallized honey. They flow rapidly down steep slopes and congeal into elegant forms that are mementos of their last motions (fig. 6.5). Lava that moves after its crust solidifies tends to break and to be torn apart into pieces of clinker. Since much of the solidified lava

Table 6.1a.
Average Chemical Compositions of Hawaiian Rock Types (in wt %)

Constituent	*Tholeiitic Basalt*	*Alkalic Basalt*
SiO_2	49.4	46.5
TiO_2	2.5	3.0
Al_2O_3	13.9	14.6
Fe_2O_3	3.0	3.3
FeO	8.5	9.1
MnO	0.2	0.1
MgO	8.4	8.2
CaO	10.3	10.3
Na_2O	2.1	2.9
K_2O	0.4	0.8
P_2O_5	0.3	0.4

Source: Macdonald et al. 1983.

Table 6.1b.
Average Composition of Five Classses of Volcanic Rocks (in wt %)

Constituent	*Rhyolite*	*Dacite*	*Andesite*	*Basalt*	*Phonolite*
SiO_2	73.66	63.58	54.20	50.83	56.90
TiO_2	0.22	0.64	1.31	2.03	0.59
Al_2O_3	13.45	16.67	17.17	14.07	20.17
Fe_2O_3	1.25	2.24	3.48	2.88	2.26
FeO	0.75	3.00	5.49	9.05	1.85
MnO	0.03	0.11	0.15	0.18	0.19
MgO	0.32	2.12	4.36	6.34	0.58
CaO	1.13	5.53	7.92	10.42	1.88
Na_2O	2.99	3.98	3.67	2.23	8.72
K_2O	5.35	1.40	1.11	0.82	5.42
P_2O_5	0.07	0.17	1.28	0.23	0.17
H_2O	0.78	0.56	0.86	0.91	0.96
Total	100.0	100.0	100.0	100.0	100.0

Source: Carmichael, Turner, and Verhoogen 1974.

Figure 6.5. Elegant flow features in lavas from Kilauea. (Photo: H. D. Holland)

is highly porous, lava clinkers are usually jagged, and crossing clinker fields without appropriate footwear can be quite an ordeal.

6.4 Minerals in Hawaiian Lavas

Basaltic lava that is chilled in a matter of minutes cools to a dark, almost opaque glass. As in window glass, the arrangement of atoms in quickly chilled lava is almost completely random. In lava that cools slowly the chemical components have time to crystallize. Groups of atoms arrange themselves in patterns that are repeated many times in all three dimensions. Bathroom tile and wallpaper patterns are two-dimensional analogues of crystals (see plate 8). The patterns in crystals are frequently imperfect, but the difference between crystals and glasses is dramatic. The monotonous repetition of basic patterns in crystals (see fig. 6.6) is completely absent in glasses.

The geometry of these patterns is highly variable. The arrangement of water molecules in ice crystals (see chapter 4) is determined by the geometry of H_2O molecules and by the nature of the bonding between neighboring H_2O molecules. The patterns of the atoms in crystals that form when basaltic lavas cool slowly depend in large part on the relative sizes of their constituents and on the bonding between them.

O^{2-} is by far the most abundant of the anions in basaltic lavas, and—for that matter—in virtually all lavas. It can be regarded as an O atom to which two electrons have been added. The addition of these electrons increases the radius of O significantly. Atomic and ionic radii are usually expressed in Ångstroms (Å). One Å is equal to 1×10^{-8} cm or 0.1 nm. The radii of common atoms and ions are listed in table 6.2 and are illustrated in figure 6.7. Most positive ions (cations) are smaller than most anions, because cations are created by removing electrons from neutral atoms. The greater the number of electrons that are removed from an atom, the smaller the radius of the resulting cation. For example, the radius

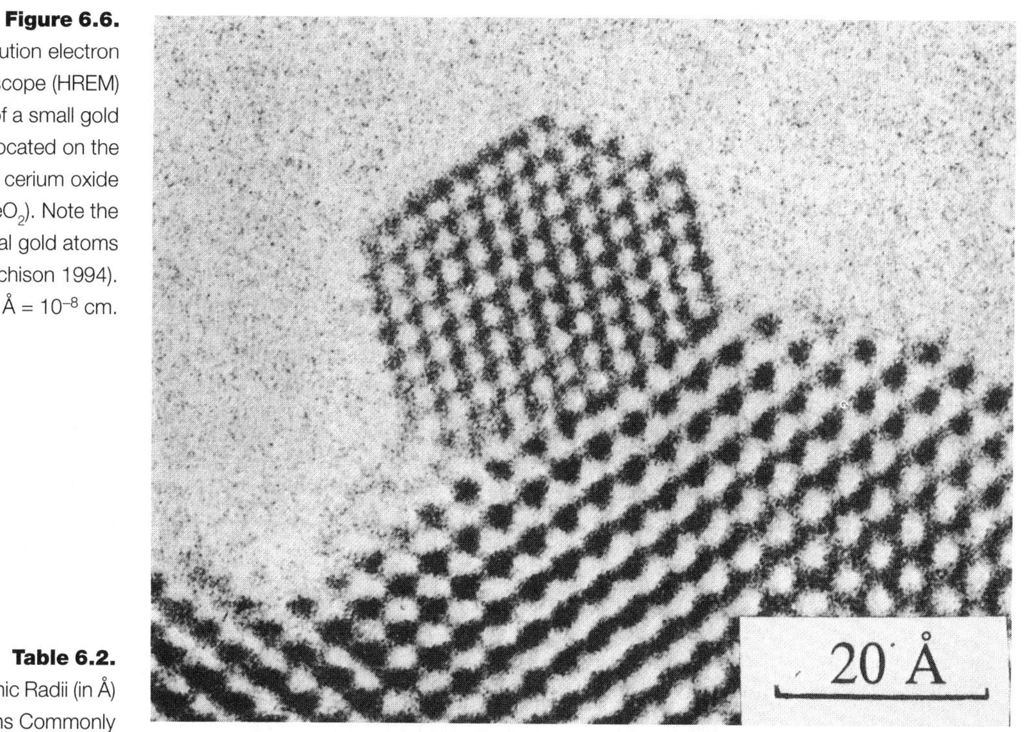

Figure 6.6. High resolution electron microscope (HREM) image of a small gold crystallite located on the surface of cerium oxide (CeO_2). Note the individual gold atoms (Hutchison 1994). 1 Å = 10^{-8} cm.

Table 6.2. Effective Ionic Radii (in Å) of Ions Commonly Found in Minerals

I	II	Transition elements										III	IV	V	VI	VII ←Column	Row
Li^{+} 0.59 (4) 0.74 (6) 0.92 (8)	Be^{2+} 0.16 (3) 0.27 (4) 0.45 (6)											B^{3+} 0.11 (4) 0.27 (6)	C^{4+} −0.08 (3) 0.15 (4) 0.16 (6)	N^{5+} −0.10 (3) 0.13 (6)	O^{2-} 1.36 (3) 1.38 (4) 1.40 (6) 1.42 (8)	F^{-} 1.31 (4) 1.33 (6)	2
Na^{+} 0.99 (4) 1.02 (6) 1.18 (8) 1.24 (9) 1.39 (12)	Mg^{2+} 0.57 (4) 0.72 (6) 0.89 (8)											Al^{3+} 0.39 (4) 0.48 (5) 0.54 (6)	Si^{4+} 0.26 (4) 0.40 (6)	P^{5+} 0.17 (4) 0.29 (5) 0.38 (6)	S^{2-} *1.84 (4)* S^{6+} 0.12 (4) 0.29 (6)	Cl^{-} *1.81 (6)*	3
K^{+} 1.38 (6) 1.51 (8) 1.55 (9) 1.59 (10) 1.64 (12)	Ca^{2+} 1.00 (6) 1.12 (8) 1.18 (9) 1.23 (10) 1.34 (12)	Sc^{3+} 0.75 (6) 0.87 (8)	Ti^{4+} 0.42 (4) 0.61 (6) 0.74 (8)	V^{5+} 0.36 (4) 0.46 (5) 0.54 (6)	Cr^{3+} 0.62 (6) Cr^{4+} 0.41 (4) 0.55 (6) Cr^{6+} 0.26 (4)	Mn^{2+} 0.83 (6) 0.96 (8) Mn^{3+} 0.65 (6) Mn^{4+} 0.53 (6)	Fe^{2+} 0.63 (4) 0.78 (6) 0.92 (8) Fe^{3+} 0.65 (6) 0.78 (8)	Co^{2+} 0.74 (6) 0.90 (8)	Ni^{2+} 0.55 (4) 0.69 (6)	Cu^{+} 0.46 (2) 0.77 (6) Cu^{2+} 0.57 (4) 0.65 (5) 0.73 (6)	Zn^{2+} 0.60 (4) 0.74 (6) 0.90 (8)	Ga^{3+} 0.47 (4) 0.55 (5) 0.62 (6)	Ge^{4+} 0.39 (4) 0.53 (6)	As^{3+} 0.58 (6) As^{5+} 0.34 (4) 0.46 (6)	Se^{2-} *1.98 (6)*	Br^{-} *1.96 (6)*	4
Rb^{+} 1.52 (6) 1.61 (8) 1.66 (10) 1.72 (12)	Sr^{2+} 1.18 (6) 1.26 (8) 1.36 (10) 1.44 (12)	Y^{3+} 0.90 (6) 1.02 (8)	Zr^{4+} 0.72 (6) 0.78 (7) 0.84 (8) 0.89 (9)	Nb^{5+} 0.64 (6) 0.69 (7) 0.74 (8)	Mo^{4+} 0.65 (6) Mo^{6+} 0.41 (4) 0.59 (6)			Rh^{4+} 0.60 (6)	Pd^{2+} 0.64 (4) 0.86 (6)	Ag^{+} 1.15 (6) 1.28 (8)	Cd^{2+} 0.58 (4) 0.74 (6) 0.90 (8)	In^{3+} 0.62 (4) 0.80 (6) 0.92 (8)	Sn^{4+} 0.69 (6) 0.81 (8)	Sb^{3+} *0.76 (6)* Sb^{5+} 0.60 (6)	Te^{2-} *2.21 (6)*	I^{-} *2.20 (6)*	5
Cs^{+} 1.67 (6) 1.74 (8) 1.81 (10) 1.85 (11) 1.88 (12)	Ba^{2+} 1.35 (6) 1.42 (8) 1.47 (9) 1.52 (10) 1.61 (12)	La^{3+} 1.03 (6) 1.16 (8) 1.22 (9) 1.27 (10)	Hf^{4+} 0.71 (6) 0.76 (7) 0.83 (8)	Ta^{5+} 0.64 (6) 0.69 (7) 0.74 (8)	W^{6+} 0.42 (4) 0.51 (5) 0.60 (6)	Re^{4+} 0.63 (6) Re^{7+} 0.38 (4) 0.53 (6)			Pt^{2+} 0.80 (6)		Hg^{2+} 0.96 (4) 1.02 (6) 1.14 (8)		Pb^{2+} 1.19 (6) 1.29 (8) 1.35 (9) 1.40 (10)	Bi^{3+} 0.96 (5) 1.03 (6) 1.17 (8)			6
			Th^{4+} 0.94 (6) 1.05 (8) 1.09 (9) 1.13 (10)		U^{4+} 0.89 (6) 1.00 (8) U^{6+} 0.52 (4) 0.73 (6)												

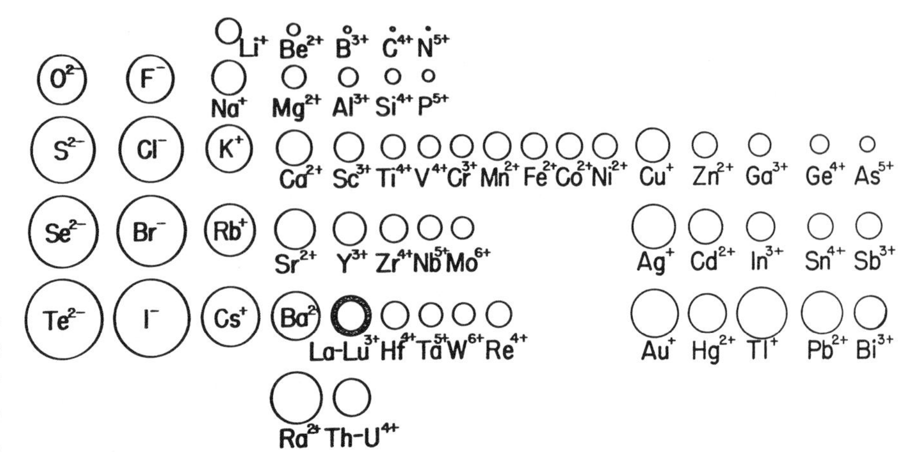

Figure 6.7. Relative size of some mineralogically important ions. One inch represents 5 Å.

of Fe^{3+} is smaller than that of Fe^{2+}, which in turn is smaller than the radius of neutral iron atoms. The radius of atoms also depends on their atomic number. However, the effect of increasing atomic number on the radius of neutral atoms is rather modest, because the radius increase that accompanies an increase in the number of electrons is nearly balanced by the decrease in size caused by the increased attraction of the orbital electrons by the larger number of protons in the nucleus.

Since particles of like charge repel each other, and particles of opposite charge attract each other, cations and anions in crystals tend to arrange themselves so as to maximize attractive interactions while minimizing the effect of the repulsive interactions. In some instances, this principle leads to rather simple arrangements of cations and anions. In the mineral periclase, MgO, the O^{2-} ions are arranged in a close-packed array, much like ping-pong balls stacked to leave as little empty space as possible. In two dimensions, this arrangement looks like figure 6.8a. Each ball in such a layer is surrounded by six others. A second layer can be fitted above the first layer, so that the balls in the second layer fit into depressions between balls in the first layer. A third layer can be added in the same fashion. Close inspection shows that the third layer can be arranged in two ways: in one, as in figure 6.8b, the center of each ball is directly above the center of a ball in the first layer; in the other, as in figure 6.8c, none of the balls in the third layer is directly above balls in the first layer. If a fourth layer is added, its balls are directly above either first-layer balls or second-layer balls.

Close-packed layers of balls in which third-layer balls are directly above first-layer balls have 12121212 . . . or ABABABAB stacking, as in figure 6.8d. In contrast, close-packed layers in which the fourth-layer balls are directly above first-layer balls are called 123123123 . . . or ABCABCABC stacking, as in figure 6.8e. Not surprisingly, 121212 . . . stacking leads to crystals with hexagonal symmetry. In the mineral corundum, Al_2O_3, the O^{2-} ions have this arrangement. Corundum frequently occurs as hexagonal prisms such as those shown in plate 9. Clear, dark red crystals of corundum are highly valued as rubies.

Rather surprisingly, 123123123 . . . stacking gives rise to crystals that have cubic symmetry. To convince yourself that this is so, you might want to build a ping-pong ball model with this type of stacking. You will find that the stacking axis of the close-packed layers is the body diagonal of a cube. The arrangement

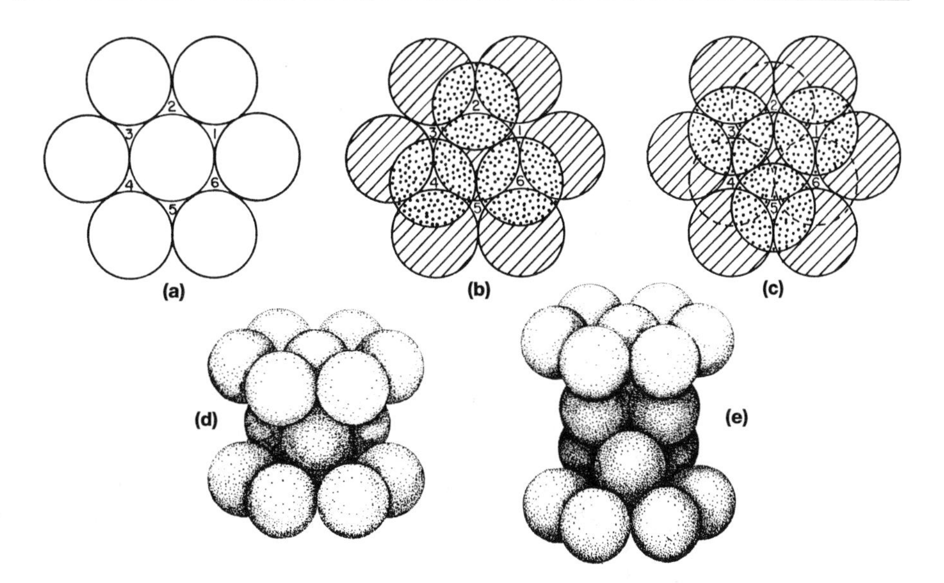

Figure 6.8. (a) Six atoms surrounding a central atom of the same size. (b,c) Two types of packing of 12 spheres around a central sphere of the same size. (b) Spheres in the upper and lower layers are centered on holes 2, 4, and 6. (c) Spheres in the upper layer are centered on holes 1, 3, and 5; those in the lower layer are centered on holes 2, 4, and 6. (d) Hexagonally close-packed spheres. Note that the spheres in the third layer are directly above the spheres in the first layer. (e) Cubic close-packed spheres. Note that the spheres in the fourth layer are directly above the spheres in the first layer.

of O^{2-} ions in the mineral periclase, MgO, is of this type. It is also the arrangement of Cl^{-} ions in table salt, the mineral halite, NaCl.

A close-packed array of O^{2-} ions alone would fly apart, since the negative ions would simply repel each other. Stable oxide crystals require the presence of cations to hold the crystals together by balancing the negative charge of the O^{2-} ions. Since most cations are smaller than O^{2-} ions, many can fit into the holes in close-packed O^{2-} structures. These holes are of three kinds: holes surrounded by six, four, and three O^{2-} ions, respectively. Figure 6.9 shows the geometry of these holes and the size of the cations that can just fit into them. It is easy to show that when the ratio of the radius of the cation to that of the six surrounding anions in figure 6.9a is 0.414, the cation touches all of the surrounding anions, and the

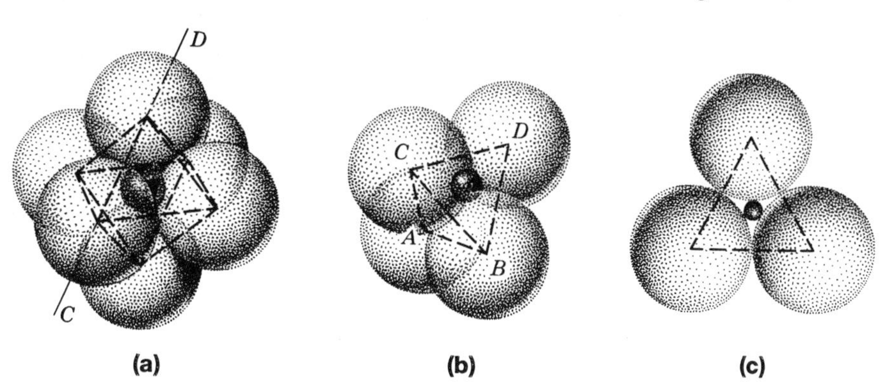

Figure 6.9. (a) Octahedral, or six-coordination of anions about a cation. (b) Tetrahedral, or four-coordination of anions about a cation. (c) Triangular, or three-coordination of anions about a cation. (Klein and Hurlbut 1985)

anions all touch each other. For four-coordinated holes the radius ratio at which all of the ions just touch is 0.225 (see fig. 6.9b), and for three-coordinated holes this radius ratio is 0.155 (see fig. 6.9c).

The ratio of the radius of Mg^{2+} ions to that of O^{2-} ions (see table 6.2) is

$$\frac{0.66\ \text{Å}}{1.40\ \text{Å}} = 0.47.$$

If a Mg^{2+} ion is placed in a six-coordinated hole in a close packed O^{2-} array, it will touch its six neighbors. Since it is slightly too large for the hole, these O^{2-} ions will not quite touch, but will be spread apart slightly. This is energetically desirable, since the spreading apart of the O^{2-} ions decreases the repulsion between them. As expected, this is the basic structure of periclase, MgO. A careful count shows that the ratio of six-coordinated holes to O^{2-} ions in close-packed structures is exactly one. Since the ratio of Mg^{2+} to O^{2-} in MgO is also one, the Mg^{2+} ions in periclase fill all of the six-coordinated holes in this structure.

In the mineral corundum, Al_2O_3, the Al^{3+} ions also occupy six-coordinated holes in the O^{2-} array. However, there are only two-thirds as many Al^{3+} ions as O^{2-} ions. Hence only two-thirds of the six-coordinated holes in Al_2O_3 is occupied by Al^{3+} ions. The remaining third of the six-coordinated holes is empty. This raises the question whether the Al^{3+} ions are arranged randomly among the six-coordinated holes or whether they are arranged in an ordered fashion. It turns out that they are ordered, as shown in figure 6.10, but the cations in many such structures are disordered, especially at temperatures close to their melting points.

The mineral halite, NaCl, has the same structure as that of periclase, MgO. The Cl^- ions are cubic close packed, and the Na^+ ions fit into the six-coordinated holes between Cl^- ions. Since the ratio of Na^+ to Cl^- ions in NaCl is 1, all of the six-coordinated holes in this mineral structure are filled.

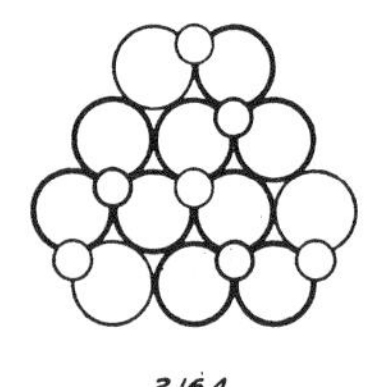

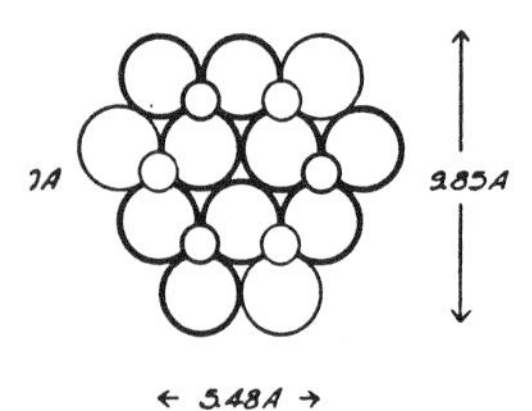

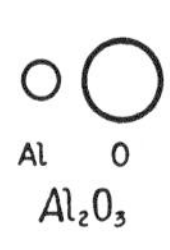

Figure 6.10. The corundum structure (Bragg 1937). The O^{2-}–Al^{3+} layer (top) fits on top of the O^{2-}–Al^{3+} layer (bottom), so that the O^{2-} ions are close packed.

Basaltic lavas contain a rather large amount of Mg^{2+} (see table 6.1). One might therefore expect that periclase would be a common mineral in Hawaiian lavas that crystallize slowly. However, no periclase has been found. Instead, one finds minerals that contain Si^{4+} as well as Mg^{2+} and O^{2-}. One of these, the mineral olivine, frequently occurs as crystals several millimeters to a centimeter in diameter, embedded in glassy lava or in lava that cooled to extremely small crystals. The occurrence of such large crystals (phenocrysts) together with glass is evidence for a "two-stage" cooling history. The large olivine crystals form during slow cooling of the lava at depth before eruption of the lava-crystal mix and rapid cooling at the surface. Carried along with the molten lava, the crystals give the cooled product somewhat the appearance of a plum pudding. Olivine in Hawaiian lavas has a composition close to Mg_2SiO_4. However, some Fe^{2+} is always present in these olivines, and a careful study shows that it proxies for Mg^{2+} in the olivine structure. In many basaltic olivines about 10% of the Mg^{2+} ions are replaced by Fe^{2+}. The formula for olivine is therefore best written as $(Mg^{2+}, Fe^{2+})_2SiO_4$, indicating that for every four O^{2-} ions there is one Si^{4+}, and a total of two Mg^{2+} and Fe^{2+} ions. Olivine has the composition Mg_2SiO_4 and is called *forsterite*; olivine which contains no Mg^{2+} has the composition Fe_2SiO_4 and is called *fayalite*. The complete interchangeability of Mg^{2+} and Fe^{2+} in olivine is largely due to the similarity of their ionic radii and is referred to as *solid solution*. The ratio of Mg^{2+} to Fe^{2+} in olivine crystals depends largely on the relative pro-

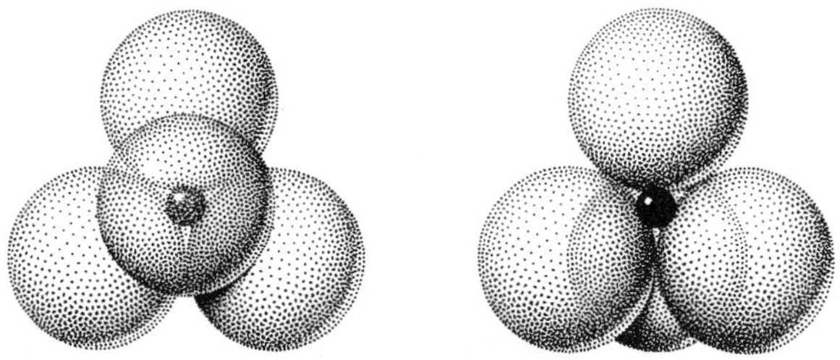

Figure 6.11a. Close packing representations of a $[SiO_4]^{4-}$ tetrahedron. (Klein and Hurlbut 1985)

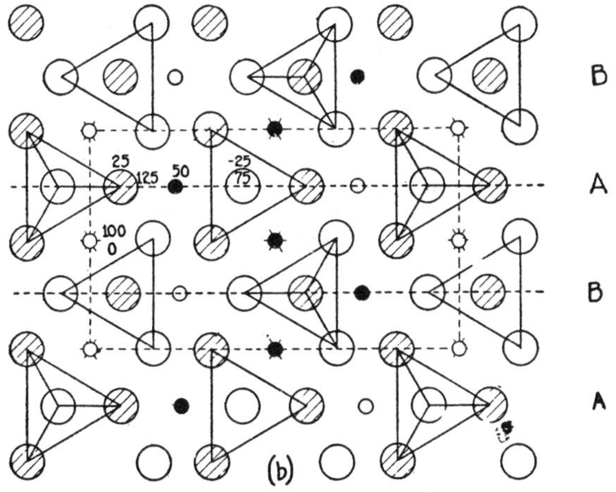

Figure 6.11b. The structure of olivine, somewhat idealized. Si^{4+} ions are located at the centers of the tetrahedra and are not shown. Large circles: O^{2-} ions; small circles: Mg^{2+} ions. (Bragg 1937)

portion of the two cations in the silicate melt from which they crystallize and on the temperature of crystallization.

At Earth surface pressures, Si^{4+} ions are too small to fill six-coordinated holes in close-packed O^{2-} arrays, but they fill four-coordinated holes very nicely. There are two such holes for each O^{2-} ion in close-packed O^{2-} arrays. In olivine crystals there is only one Si^{4+} ion per four O^{2-} ions. There are therefore eight times as many four-coordinated holes as there are Si^{4+} ions to fill them. In a rough way we can describe the structure of olivine as an array of O^{2-} ions in which one-eighth of the four-coordinated holes is filled with Si^{4+} ions, and where one half of the six-coordinated holes is filled with Mg^{2+} and Fe^{2+} ions. The O^{2-} array in olivine is not, however, precisely close packed. The Si^{4+}-O^{2-} bonds are so strong that they distort the structure slightly. In fact, the structure can be described more accurately as consisting of $[SiO_4]^{4-}$ tetrahedra that are bound together by Mg^{2+} and Fe^{2+} ions (see fig. 6.11b). This turns out to be a useful way of looking at the structure of silicate minerals in general. In all but very high-pressure silicates formed in the Earth's interior and during meteorite impacts on the Earth, Si^{4+} is surrounded by four O^{2-} ions. $[SiO_4]^{4-}$ tetrahedra are therefore the building blocks of nearly all silicate minerals at and near the surface of the Earth.

Basaltic lavas contain so much Si^{4+} that only a small fraction can be accommodated in olivine, and it is easy to show that other silicate minerals must also form during the crystallization of these magmas. The most important are the pyroxenes and the feldspars. The simplest pyroxenes have the composition $(Mg^{2+}, Fe^{2+})SiO_3$. As in the olivines, Mg^{2+} and Fe^{2+} ions occupy six-coordinated holes surrounded by O^{2-} ions, and the Si^{4+} ions occupy four-coordinated holes sur-

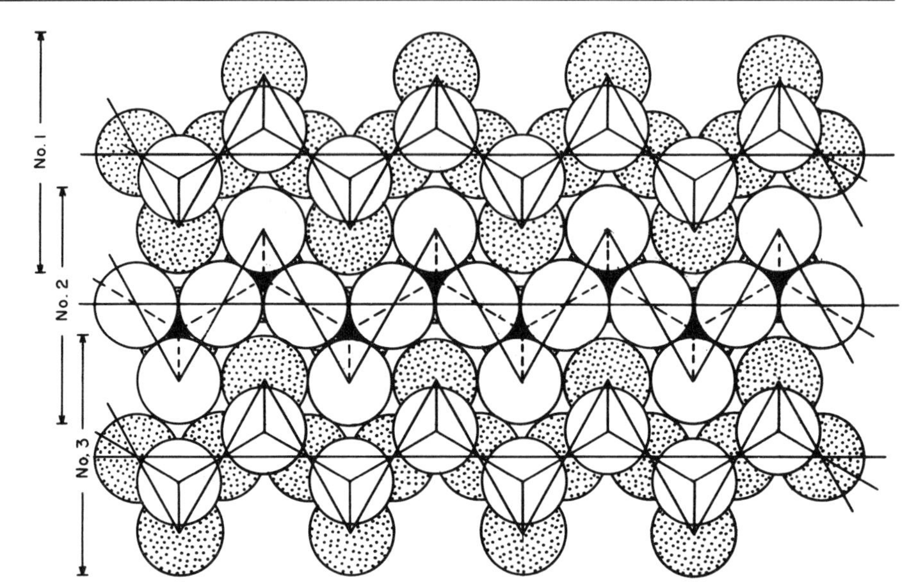

Figure 6.12a (*left*). Pyroxene single chains of $[SiO_4]^{4-}$ tetrahedra viewed from above. Each silicon ion has two of the oxygen ions of its tetrahedron bonded exclusively to itself and shares the other two with the neighboring tetrahedra fore and aft. The individual pyroxene chains are bonded to one another by cations such as Mg^{2+} and Fe^{2+}. These bonds are weaker than the silicon-oxygen bonds within the chains; cleavages therefore develop parallel to the length of the chains. (Klein and Hurlbut 1985)

Figure 6.12b (*above*). Schematic drawing of the orthopyroxene structure. Of the three parallel chains of $[SiO_4]^{4-}$ tetrahedra shown, No. 1 and No. 3 have their apex ions pointing upward, No. 2 downward. The large, unshaded circles represent O^{2-} ions in the upper plane; the large, shaded circles represent O^{2-} ions in the lower plane. The heavily shaded circles represent Si^{4+} ions.

rounded by O^{2-} ions. However, the O^{2-} ions are not quite close packed, and the structure is best understood by looking at the arrangement of the $[SiO_4]^{4-}$ building blocks. These form single strands (see fig. 6.12a and plate 10a). Each $[SiO_4]^{4-}$ tetrahedron shares an O^{2-} ion with each of its two neighbors in the chain. In orthopyroxene crystals these chains are arranged parallel to each other, so that Mg^{2+} and Fe^{2+} ions in six-coordinated holes bind the chains together (fig. 6.12b).

Pyroxene chains can also be arranged so that some of the holes between them are surrounded by eight O^{2-} ions. Such holes are large enough to accommodate Ca^{2+} ions, which have a radius considerably larger than that of Mg^{2+} and Fe^{2+} (see fig. 6.7 and table 6.2). The composition of minerals in this group, the clinopyroxenes, can be written $Ca_{1/2}^{2+}(Mg^{2+}, Fe^{2+})_{1/2}SiO_3$, or more commonly $Ca^{2+}(Mg^{2+}, Fe^{2+})Si_2O_6$. Most of the Mg^{2+} and Fe^{2+} in basaltic lavas that is not accommodated in olivine finds a home in the pyroxenes.

When lavas crystallize underground and in the presence of water, some of the Mg^{2+} and Fe^{2+} can crystallize as components of yet another group of silicates, the amphiboles. In these minerals, the $[SiO_4]^{4-}$ tetrahedra are arranged in double chains (fig. 6.13 and plate 10b). These chains consist of links of six-membered $[SiO_4]^{4-}$ rings. Each link has the formula $[Si_4O_{11}]^{6-}$. One OH^- is associated with each link, giving a total link composition of $[Si_4O_{11}(OH)]^{7-}$. The double chains

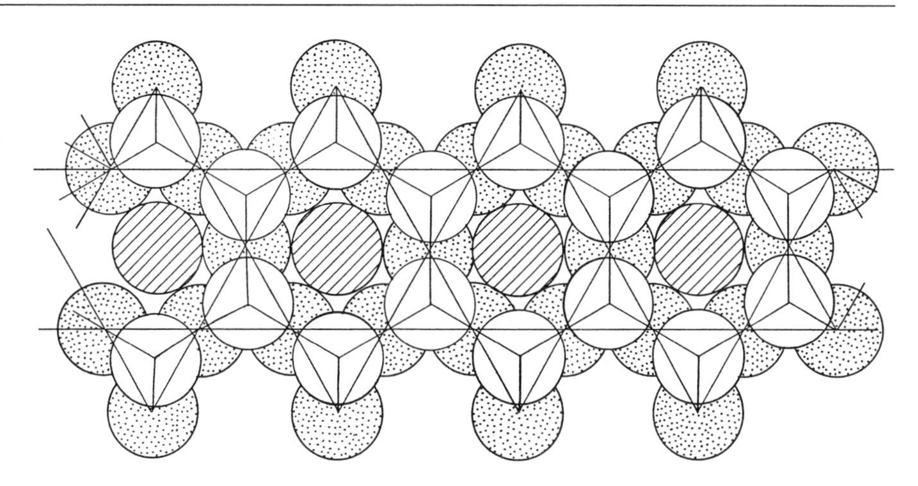

Figure 6.13. Double chain characteristic of the amphiboles, formed by joining two pyroxene chains. The stippled O^{2-} ions form the base of $[SiO_4]^{4-}$ tetrahedra. The unshaded O^{2-} ions form the apexes of these tetrahedra. The circles with diagonal lines represent OH^- groups at the center of the rings of tetrahedron apexes. The Si^{4+} ions are located directly beneath the tetrahedron apexes.

can either be linked by Mg^{2+} and Fe^{2+}, or by these cations and Ca^{2+}. The formula of the orthoamphiboles can be written as $(Mg^{2+}, Fe^{2+})_7Si_8O_{22}(OH)_2$; the formula of the clinoamphiboles can be written as $Ca_2^{2+}(Mg^{2+}, Fe^{2+})_5Si_8O_{22}(OH)_2$.

Olivines, pyroxenes, and amphiboles can accommodate all or nearly all of the Mg^{2+} and Fe^{2+} in basaltic lavas. However, much of the Ca^{2+} and nearly all of the Al^{3+}, Na^+, and K^+ still remain unaccommodated. They find a ready home in the feldspar group of minerals. These are three-dimensional in their structures; the $[SiO_4]^{4-}$ tetrahedra form "cages" that can accommodate large, singly and doubly charged cations. The cages are surrounded by double chains of $[SiO_4]^{4-}$ tetrahedra (fig. 6.14). The links of these chains consist of four rather than six $[SiO_4]^{4-}$ tetrahedra, and the chains are kinked rather than flat.

Basalts contain a good deal of Ca^{2+}, not very much Na^+, and only small amounts of K^+. The feldspars that crystallize from these lavas are therefore Ca^{2+}-rich. Pure Ca^{2+} feldspar has the composition $CaAl_2Si_2O_8$. The Al^{3+} ions in this mineral replace Si^{4+} ions in $[SiO_4]^{4-}$ tetrahedra, which are linked as described above. The charge on the chains is balanced by Ca^{2+} ions. The Al^{3+} and the Si^{4+} ions are randomly distributed in the $(Al^{3+}, Si^{4+})O_4^{2-}$ tetrahedra.

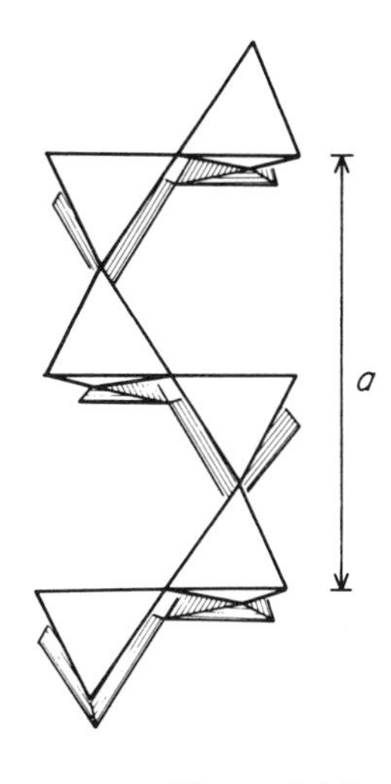

Figure 6.14. The two chains in this figure are linked to form a feldspar double chain consisting of four-membered $[SiO_4]^{4-}$ rings. (Papike and Cameron 1976)

Most basaltic lavas contain sufficient Na^+, so that the compositions of their feldspars differ significantly from anorthite, the pure calcium feldspar. Na^+ substitutes freely for Ca^{2+} in feldspars, but since its charge is only +1, the substitution of one Na^+ for one Ca^{2+} must be accompanied by the substitution of one Si^{4+} for one Al^{3+}. Albite, the pure Na^+ feldspar, has the composition $NaAlSi_3O_8$, and the composition of feldspars intermediate between anorthite and albite can be written as $Ca_xNa_{(1-x)}Al_{(1+x)}Si_{(3-x)}O_8$. The entire range of solid solution compositions from pure anorthite to pure albite has been found. The group as a whole is called the *plagioclase feldspars*.

Of the major constituents of basalts in table 6.1, only TiO_2 and Fe_2O_3 now remain unaccommodated. Both are common constituents of oxide minerals: TiO_2 as a constituent of the mineral ilmenite, $FeTiO_3$, and Fe_2O_3 as a constituent of the mineral magnetite, Fe_3O_4. In both minerals the O^{2-} ions are close packed. In ilmenite, Fe^{2+} and Ti^{4+} occupy six-coordinated holes; in magnetite Fe^{2+} and Fe^{3+} occupy some of the four-coordinated and some of the six-coordinated holes in the O^{2-} framework.

6.5 The Weathering of Hawaiian Lavas

Lava that has cooled very recently tends to be black and lustrous. Before long, hardy and beautiful Ohia plants take root in such lavas and blossom there (see plate 11). In a few years the lustrous black veneer is replaced by a rusty-colored surface: weathering has started. In reasonably unpolluted areas, such as most of the island of Hawaii, the three major weathering agents are H_2O, CO_2, and O_2. In the absence of water, weathering is extremely slow. In areas of heavy rainfall, weathering is rapid, because water is both a catalyst and an active agent in the weathering process. CO_2 is present in rainwater, combined with H_2O to form carbonic acid, H_2CO_3, which makes rainwater slightly acid even in such unpolluted areas. O_2 is also dissolved in rainwater and acts as a strong oxidant during weathering.

Olivine dissolves very slightly in rainwater. The reaction of forsterite, i.e., pure Mg_2SiO_4, with H_2O and CO_2 can be written as

$$\underset{\text{forsterite}}{Mg_2SiO_4} + 4H_2O + 4CO_2 \rightarrow 2Mg^{2+} + 4HCO_3^- + H_4SiO_4. \quad (6.1)$$

This is a shorthand notation for the series of complex acid-base surface reactions by which forsterite dissolves. H_2O and CO_2 attack the surfaces of forsterite crystals, detach Mg^{2+} and $[SiO_4]^{4-}$ groups, and donate $4H^+$ ions to each $[SiO_4]^{4-}$ group to convert it to dissolved silica, H_4SiO_4. In the process, HCO_3^- ions are generated. If rainwater and forsterite remain in contact for a period of years or decades, new Mg^{2+} minerals precipitate from solution. However, rainwater moves downward through most soils so rapidly that the products of reaction (6.1) are normally flushed out of the soil zone and into the zone of groundwater. Reaction (6.1) therefore depletes soils of magnesium and silica.

The initial reaction of rainwater with pure fayalite, Fe_2SiO_4, is quite similar to its reaction with pure forsterite:

$$\underset{\text{fayalite}}{Fe_2SiO_4} + 4H_2O + 4CO_2 \rightarrow 2Fe^{2+} + 4HCO_3^- + H_4SiO_4. \quad (6.2)$$

However, this reaction is followed very quickly by the oxidation of Fe^{2+} to Fe^{3+} and by the precipitation of $Fe(OH)_3$:

$$4Fe^{2+} + O_2 + 8HCO_3^- + 2H_2O \rightarrow 4Fe(OH)_3 + 8CO_2. \quad (6.3)$$

The overall reaction can be obtained by multiplying equation (6.2) by two and adding it to equation (6.3).

Equation (6.4) shows that the weathering of fayalite leads to the precipitation of $Fe(OH)_3$ and to silica loss.

$$2Fe_2SiO_4 + 8H_2O + 8CO_2 \rightarrow 4Fe^{2+} + 8HCO_3^- + 2H_4SiO_4$$
$$4Fe^{2+} + O_2 + 8HCO_3^- + 2H_2O \rightarrow 4Fe(OH)_3 + 8CO_2$$

$$\underset{\text{fayalite}}{2Fe_2SiO_4} + 10H_2O + O_2 \rightarrow 4Fe(OH)_3 + 2H_4SiO_4. \quad (6.4)$$

In many soils, $Fe(OH)_3$ gradually loses water. Partial water loss leads to the formation of FeO(OH):

$$Fe(OH)_3 \rightarrow \underset{\text{goethite}}{FeO(OH)} + H_2O. \quad (6.5)$$

In the FeO(OH) group of minerals, Fe^{3+}, O^{2-}, and OH^- ions can be arranged in several ways. The most common gives rise to the mineral goethite. Complete loss of water from $Fe(OH)_3$ leads to the formation of the mineral hematite. The common brown crust on recent basaltic lavas and the rusty red color of well-developed soils (see plate 12) are largely due to $Fe(OH)_3$, FeO(OH), and Fe_2O_3.

$$\underset{\text{goethite}}{2FeO(OH)} \rightarrow \underset{\text{hematite}}{Fe_2O_3} + H_2O. \quad (6.6)$$

Olivine usually has a composition intermediate between that of pure forsterite and pure fayalite. Its weathering can be thought of as the weathering of a "forsterite component" and a "fayalite component." The Mg^{2+} and H_4SiO_4 released during the weathering of the forsterite component are normally flushed out of weathering zones; the Fe^{2+} is oxidized to Fe^{3+} and is largely or entirely retained in weathering zones as $Fe(OH)_3$, FeO(OH), or Fe_2O_3. Pyroxenes and amphiboles weather in much the same fashion as olivine. Mg^{2+} and Ca^{2+} are released into solution, their positive charge balanced by HCO_3^-. Fe^{2+} is oxidized to Fe^{3+} and precipitates as one or more of the Fe^{3+}-oxides and -hydroxides.

The chemical weathering of feldspar is quite different from that of olivine, pyroxenes, and amphiboles. Na^+ and Ca^{2+} in feldspars are removed readily in solution, but Al^{3+} is largely retained as a constituent of one or more clay minerals. These are all very fine grained, and nearly all have sheetlike atomic arrangements. In the initial stages of basalt weathering, the concentration of H_4SiO_4 in soil water is relatively high, and the clay minerals are all silicates. They consist of $[SiO_4]^{4-}$ tetrahedra linked in sheets as shown in figure 6.15. The sheets can be regarded as an infinite number of linked $[SiO_4]^{4-}$ chains. Sheets of $[SiO_4]^{4-}$ tetrahedra are commonly linked in one of two ways. Figure 6.16 shows a single ring of $[SiO_4{}^{4-}$ tetrahedra, apex down. The unsatisfied negative charge of the O^{2-} ions is balanced by Al^{3+} ions, which are located in six-coordinated holes. Their three upper anion neighbors are two O^{2-} ions that are linked to Si^{4+} ions in the centers of $[SiO_4]^{4-}$ tetrahedra and one OH^- ion. The three lower neighbors of Al^{3+} are all OH^- ions, which form part of a nearly close-packed layer.

Clays of this type are given the name 1:1 clays, because they consist of one $[(SiO_4)^{4-}, OH^-]$ layer and one $[Al^{3+}, OH^-]$ layer (fig. 6.14). They can be regarded as open-face sandwiches, in which the $[(SiO_4)^{4-}, OH^-]$ layer is the bread and the $[Al^{3+},OH^-]$ layer is whatever you like to put on bread. Crystals of 1:1 clays consist of stacks of these packets, which are not strongly bound to each other. This explains why clay minerals are soft, and why crystals of the clay minerals tend to be very small. The composition of the 1:1 Al-Si clay mineral shown in figure 6.14 is $Al_2Si_2O_5(OH)_4$. Open-face clay sandwiches of this composition can be stacked in different ways. The most common stacking is that of the mineral kaolinite.

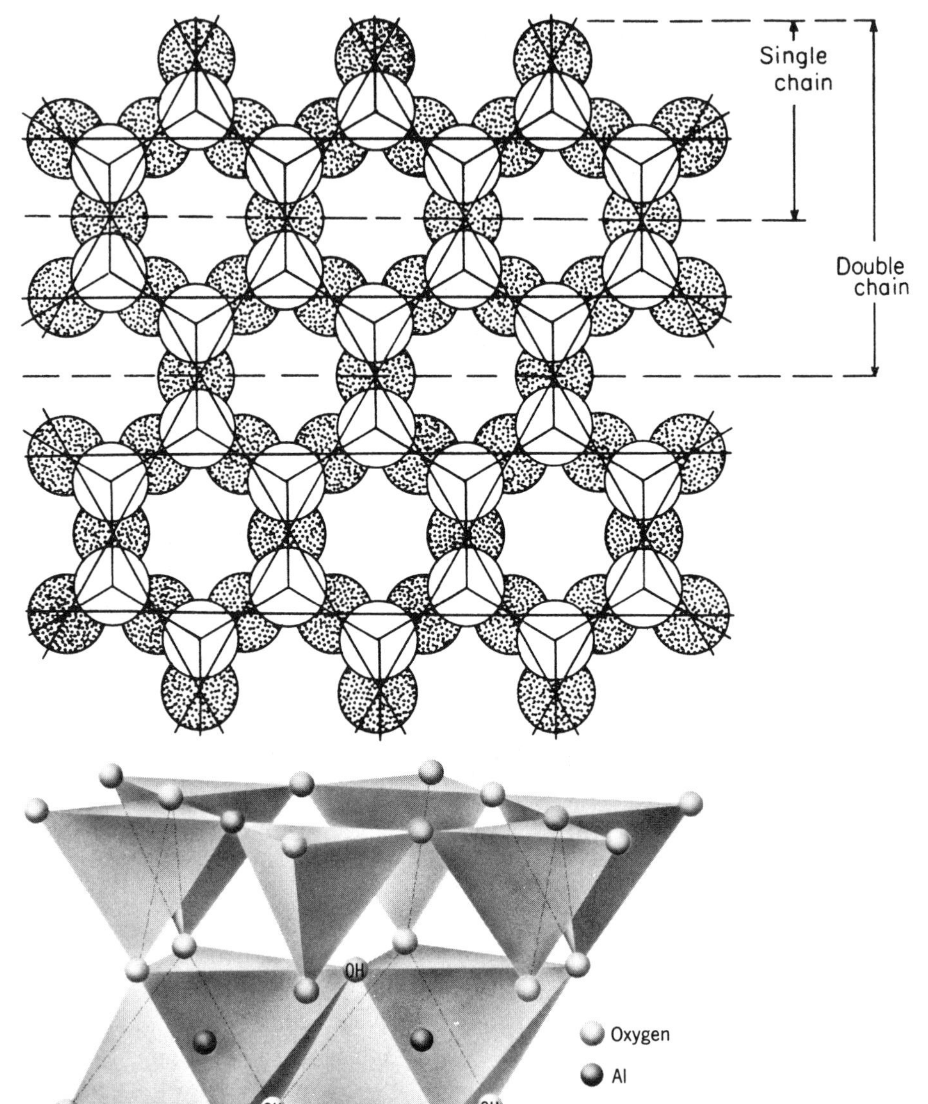

Figure 6.15. Part of a silicon-oxygen sheet produced by linking an effectively infinite number of parallel pyroxene chains. The shaded circles represent O^{2-} ions in the base of tetrahedra. The unshaded circles represent the apex O^{2-} ions. Si^{4+} ions lie directly beneath the apex O^{2-} ions and are not shown.

Figure 6.16. The structure of kaolinite: a tetrahedral Si^{4+}–O^{2-} sheet is bonded on one side by an octahedral Al^{3+}–O^{2-}–OH^- layer. (Klein and Hurlbut 1985)

The second major type of clay minerals is based on a three-layer structure: two $[(SiO_4)^{4-}, OH^-]$ sheets, one with the apexes of the tetrahedra pointing up, the other with the apexes of the tetrahedra pointing down. The two sheets are held together by a layer of cations that are six-coordinated by four O^{2-} ions and by two OH^- ions. Pyrophyllite, the aluminum silicate clay mineral in this group, has the composition $Al_2Si_4O_{10}(OH)_2$.

When pure anorthite weathers at temperatures near 25°C, it normally turns into amorphous material, or into kaolinite, or into halloysite. The formation of kaolinite or halloysite from anorthite proceeds via the reaction

$$\underset{\text{anorthite}}{CaAl_2Si_2O_8} + 3H_2O + 2CO_2 \rightarrow Ca^{2+} + 2HCO_3^- + \underset{\text{kaolinite, halloysite}}{Al_2Si_2O_5(OH)_4}. \quad (6.7)$$

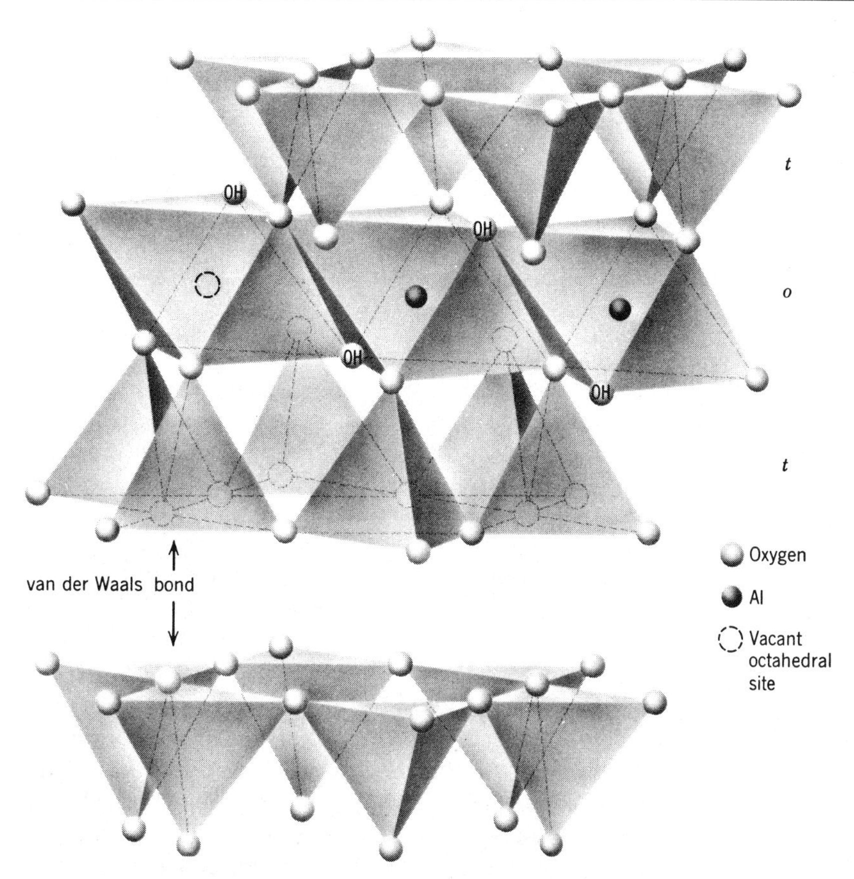

Figure 6.17. The structure of pyrophyllite. (Klein and Hurlbut 1985)

The Ca^{2+} and HCO_3^- released during this reaction are usually flushed out of the soils.

The weathering of pure anorthite and albite can also yield 1:1 clays via the reaction

$$\underset{\text{albite}}{2NaAlSi_3O_8} + 11H_2O + 2CO_2 \rightarrow 2Na^+ + 2HCO_3^- + \underset{\text{kaolinite}}{Al_2Si_2O_5(OH)_4} + 4H_4SiO_4. \quad (6.8)$$

Since the Si/Al ratio in kaolinite is 3/1, the weathering of albite releases more H_4SiO_4 than is needed to form kaolinite. The excess H_4SiO_4 tends to be flushed out of soils together with Na^+ and HCO_3^-. Intermediate feldspars, such as those common in basaltic rocks, tend to weather to kaolinite or to halloysite and to release Na^+, Ca^{2+}, HCO_3^-, and H_4SiO_4, which are usually flushed out of the soil zones in groundwater.

The chemistry of clays in soils developed on basalts is usually more complex than that of clays developed by the weathering of pure plagioclase feldspar. Mg^{2+} and Fe^{2+} can substitute for Al^{3+} in octahedral sites, and Al^{3+} can substitute somewhat for Si^{4+} in tetrahedral sites. The composition of the 2:1 (fig. 6.17) clays can be particularly complicated. Clays of this type are usually called *smectites*. The overall pattern of the changes in the chemical weathering of basaltic rocks is not,

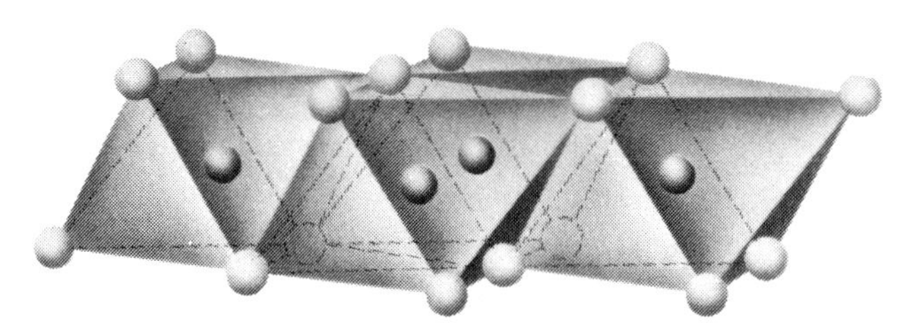

Figure 6.18. The structure of gibbsite consists of infinitely extended sheets of linked $[Al(OH)_6]^{3-}$ octahedra. The large spheres represent OH^- ions and the small spheres represent Al^{3+} ions. (Klein and Hurlbut 1985)

however, altered by these complexities. Na^+, Ca^{2+}, and Mg^{2+} are removed almost completely. Al^{3+} and iron are retained almost completely, the former as a constituent of aluminosilicate clay minerals, the latter as a constituent of these clays and of one or more of the Fe^{3+}–O^{2-}–OH^- minerals.

After all of the primary minerals in a basalt have disappeared, further weathering leads to the decomposition of the 1:1 and 2:1 clay minerals. Kaolinite and halloysite are destroyed by the reaction

$$\underset{\text{kaolinite}}{Al_2Si_2O_5(OH)_4} + 5H_2O \rightarrow \underset{\text{gibbsite}}{2Al(OH)_3} + 2H_4SiO_4. \tag{6.9}$$

The H_4SiO_4 is carried away by groundwater. Most of the Al^{3+} remains in soils as a constituent of $Al(OH)_3$, usually the mineral gibbsite. The structure of this mineral is shown in figure 6.18. All of the Al^{3+} ions are six-coordinated by OH^- ions and form OH^--Al^{3+}-OH^- layers. These layers are stacked into gibbsite crystals. One can imagine producing kaolinite from gibbsite by replacing two of the three OH^- ions below the Al^{3+} layer by an $[SiO_4]^{4-}$ sheet, and one can imagine producing pyrophyllite from kaolinite by replacing two of the three OH^- ions above the Al^{3+} layer by a second $[SiO_4]^{4-}$ sheet.

6.6 Soils and Soil Profiles

Hawaiian rocks are born of lavas at temperatures in excess of 1,000°C and are seriously out of equilibrium with water and air at Earth surface temperatures. Hawaiian soils are the reaction zones between Hawaiian rocks and the hydrous, oxygenated surface environments of the Earth. Soils in other parts of the world are reaction zones between a wide variety of rock types and Earth surface environments. The physical and chemical properties of the world's soils vary a great deal, and very elaborate schemes have been devised to classify them. Fortunately, most soils share a good many properties, and this section is devoted to these commonalities.

Soils in Hawaii range from zero to over 20 meters in thickness. As expected, the thinnest soils are the youngest, and these developed in the driest areas. The thickest soils are the oldest and developed in the wettest areas (plate 12); most of these are called *laterites*. Kauai is the oldest island in the Hawaiian chain and boasts one of the wettest spots on Earth (see chapter 4). In the wetter parts of this island soils are thick and particularly well developed. Figure 6.19 illustrates some of the important features of these soils. In well-drained areas, hard, fresh basalt passes upward into soft, weathered basalt. Weathering is rarely uniform. The lower parts of soils usually contain residual boulders of unweathered basalt surrounded by deeply weathered material. Toward the upper parts of soils these

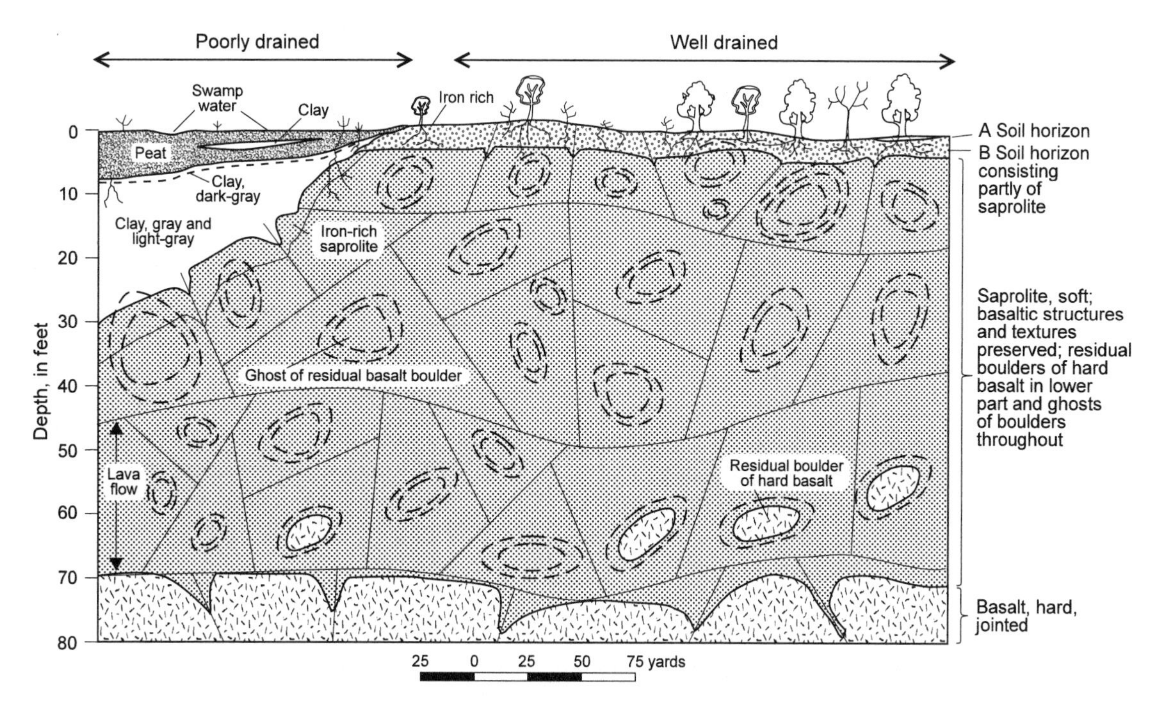

Figure 6.19. Diagrammatic geologic cross section through a deep soil in a wet area on Kauai. (Patterson 1971)

residual boulders gradually become mere ghosts of their former selves, somewhat like the Cheshire Cat in *Alice in Wonderland*. In soft, highly weathered basalt only the textures of these boulders are preserved.

The chemical composition and the mineralogy of basalts change dramatically during weathering. The data in plate 13 show that nearly all of the Ca^{2+}, Mg^{2+}, Na^{+}, and K^{+}, and a good deal of the SiO_2 in the basalts are lost in the lowest parts of the soil profiles. Fe^{2+} is converted to Fe^{3+}; very little Fe^{3+}, Al_2O_3, and TiO_2 is removed, and the H_2O content of the soil increases rapidly.

These changes reflect the mineralogical changes in the profile. The primary minerals of the basalt disappear almost completely in the lowest parts of the soil profile and are replaced largely by halloysite (a member of the kaolinite group of minerals), goethite, hematite, and amorphous material. Ilmenite ($FeTiO_3$) and magnetite (Fe_3O_4) resist oxidative decomposition most successfully. The destruction of halloysite and its replacement by gibbsite accounts for the lower SiO_2 content in the upper few meters of the soil profile in plate 18.

All of these changes are easily understood in terms of the processes described in the previous section. The loss of Mg^{2+}, Ca^{2+}, Na^{+}, and K^{+} is due to the reaction of CO_2 and H_2O with olivine, the pyroxenes, and feldspar in the parent basalt. The cations are flushed out of the weathering zone and join the groundwater reservoir. Fe^{2+} in olivine and pyroxenes is oxidized to Fe^{3+} by O_2 dissolved in soil water, and precipitates as a constituent of goethite and/or hematite. Al^{3+} from the dissolution of feldspars reprecipitates as halloysite. Some of the SiO_2 released during weathering reactions is retained in halloysite; most of the remainder is flushed out of the soil. Toward the top of the soil zone, halloysite is decomposed by SiO_2-poor soil water. Most of the Al^{3+} reprecipitates as a constituent of gibbsite, and the SiO_2 released during this reaction is flushed out of the soil zone.

Below poorly drained, swampy areas of Kauai (see fig. 6.19), soils contain much less iron, usually only 2%–3%. In some regions these soils grade downward into soils such as those described above. In other regions the iron-poor clays have developed directly on basalt. The low iron content of these soils is due to poor drainage and to the large quantity of organic matter in the overlying swamps. O_2 dissolved in rainwater is used up completely by the oxidation of organic matter in the swamps. Since the soils are waterlogged, O_2 is not replenished from the atmosphere. Fe^{2+} released from olivine and the pyroxenes is therefore not oxidized to Fe^{3+}. Rather, it is removed from the soil zone together with the other cations. This process is helped by the high acidity of the soil water caused by the generation of organic acids during organic matter decay. Intense, prolonged weathering under these conditions can produce very Al_2O_3-rich soils that are very poor in everything else except H_2O. Such weathering products are called *bauxites*. They are the most important ores of aluminum. Their abundance and distribution are discussed later in this chapter.

Deep soil profiles that were developed on basalts in other parts of the world during the past 2,000 million years (m.y.) tend to look very much like typical modern soils in Hawaii. However, the upper parts of deep soil profiles that developed on basalts more than 2,200 million years ago (m.y.a.) contain very little iron. Since higher land plants did not evolve until about 450 m.y.a., swamps as we know them today did not exist 2,200 m.y.a. The difference between soil profiles older than 2,200 m.y. and those younger than 2,000 m.y. is almost certainly due to a very considerable rise in the O_2 content of the atmosphere between 2,200 and 2,000 m.y.a. This change is also responsible for the difference between the nature of uranium ore deposits that are older than 2,200 m.y. and those that are younger than 2,000 m.y. (see chapter 7).

Soils tend to be vertically zoned and are normally subdivided into a number of horizons (e.g., Birkeland 1984). In very simple classifications, soils are divided into three horizons, quite reasonably labeled A, B, and C. "A" horizons consist largely of decomposed and decomposing organic matter mixed with a mineral fraction. "B" horizons underlie "A" horizons. They contain less organic matter and retain little of the original rock structures. The even deeper "C" horizons have properties that are transitional between those of the parent rock and the material in "B" horizons. "C" horizons tend to grade downward into the parent material of the soil.

In soil profiles such as the one in figure 6.19 the changes from rock to soil are easily documented. This is not the case where soils have developed from heterogeneous parent rocks and where soils did not develop in place. Soils in the floodplains of rivers often consist of material that was carried along in suspension during river flooding and that was deposited during the waning stages of floods (see chapter 4). In some floodplains, even soil experts cannot agree whether particular soils were developed in place or whether they were transported mechanically to their present site.

The preceding discussion focused mostly on the basaltic lavas and rocks of Hawaii, and on the soils derived from them. Magmas of different compositions (see table 6.1) form andesitic and rhyolitic lavas that solidify on the Earth's surface to andesites and rhyolites. These rocks are collectively called "extrusive" or "volcanic" igneous rocks, because they formed from magmas that were extruded from the interior of the Earth, largely via volcanoes. Below the Earth's surface these magmas may crystallize to form "intrusive" or "plutonic" igneous rocks, because the magma intruded the Earth's crust (Pluto was the Greek god of the lower world). Gabbro, diorite, granodiorite, and granite are the intrusive equivalents of basalt, andesite, dacite, and rhyolite. Peridotite is a plutonic igneous rock consisting mostly of olivine. "Alkalic" igneous rocks, such as syenite or nepheline syenite, contain larger amounts of Na and K (see chap. 9).

As a consequence of their varying chemical compositions, igneous rocks have a range of mineralogical compositions. As the Si, Na, and K contents increase and the proportions of Mg, Fe, and Ca decrease from gabbro/basalt to granite/rhyolite (table 6.1), olivine disappears, pyroxenes and amphiboles diminish, feldspars increase, micas appear, and quartz becomes the dominant mineral. Two types of micas are most common: biotite, $K(Mg,Fe)_3(Si_3Al)O_{10}(OH,F)_2$, which is black, and muscovite, $KAl_2(Si_3Al)O_{10}(OH,F)_2$, which is white or greenish. Since olivine, the pyroxenes, and the amphiboles are dark minerals, while the feldspars, muscovite, and quartz are essentially white, the color of igneous rocks changes from dark to light as one goes from gabbro/basalt to granite/rhyolite. Rocks at the ends of this compositional spectrum are often referred to as *mafic* and *felsic*. In alkalic igneous rocks the large amounts of Na and K combine with SiO_2 to form mostly feldspars (albite and orthoclase) and nepheline, $(Na,K)AlSiO_4$, leaving not enough silica to form quartz.

The weathering of all igneous rocks is similar to that described for the Hawaiian basalts, except that in the more felsic types the micas also transform into kaolinite and the quartz basically remains stable. However, because felsic rocks contain much less Fe than mafic rocks, they generate a smaller amount of iron oxides/hydroxides. The weathering reactions are all similar, in that they involve an original mineral plus H_2O, CO_2, and/or O_2 as reactants, and a solid weathering product (typically kaolinite and an iron oxide/hydroxide) plus dissolved components (typically Mg^{2+}, Ca^{2+}, Na^+, K^+, HCO_3^- and H_4SiO_4) as reaction products. The more felsic rocks yield higher proportions of Na^+, K^+, and H_4SiO_4 in solution, because they are more abundant in these rock types. All the dissolved species end up either in the groundwater or in surface drainage. In most cases, Al_2O_3 and TiO_2 are retained completely, or nearly so, in the soil.

6.7 Soils and Agriculture

A relative who is an avid gardener lives on a rather recent lava flow. Since very little soil has developed on this flow, he digs out blocks of lava and fills the holes with imported soil. In this way he cultivates a marvelous garden. His technique is highly successful, but it is not a recipe for agriculture on a large scale, although it has found application in Tenerife, in the Canary Islands. Successful agriculture requires a great deal of soil. Furthermore, it requires soils with the proper physical properties and with the proper chemical composition. Sustained agriculture depends on the continued availability of such soils. For successful agriculture, the physical properties of soils must be such that water can drain freely through them. Gravel, cobbles, and stones must not be so common as to interfere with the operation of modern farm equipment, and the depth to bedrock must be great enough so that water storage is adequate, and that roots can attain their normal extension (see, for instance, Troeh et al. 1980). The chemical composition of soil must be able to sustain plant growth. Soil acidity or alkalinity must be suitable for the crops to be grown, and the salt content of soils should not be excessive. Adequate quantities of plant nutrients must be present. These nutrients include phosphorus as soluble phosphates, nitrogen in forms that are accessible to plants, potassium, and frequently a number of "micronutrients," elements that are essential for plant growth but are needed only in very small quantities. Since plants remove nutrients from soils, soil productivity can only be maintained if nutrients are continually restored.

Traditional farming systems included a fallow period in the cropping sequence to restore nutrients to the top soil. From at least the time of Cato the Censor (234–139 B.C.), the Romans were aware of the need to increase soil fertility by means of fallowing, as well as by crop rotation, the liming of acid soils, and the addition of manure (Hillel 1991). In more recent times, the inclusion of nitrogen-fixing legumes such as clover, beans, and peas on a rotational basis was found to improve soil fertility. These plants have a symbiotic association with specialized bacteria, organisms that attach themselves to plant roots and convert atmospheric nitrogen (N_2) into nitrogen compounds which can fertilize other, less gifted plants.

As agricultural production intensified, these techniques proved incapable of sustaining agricultural productivity. Artificial fertilizers were developed in the nineteenth century. These have done much to fuel the Green Revolution of the least 50 years, which has helped to feed the world's rapidly increasing population (see chapter 5). Very large quantities of fertilizers will be required during the next 100 years, but current phosphate and potash (a potassium salt) resources are probably sufficient to meet this demand (see chapter 7). The production of nitrogen fertilizers draws on the essentially infinite supply of atmospheric nitrogen. The supply of nitrogen fertilizers is therefore assured. Their cost will depend largely on the cost of energy, because the production of nitrogen fertilizers is quite energy intensive (see chapter 11).

Soil loss due to erosion is a more serious concern. George Washington noted that "our lands were originally very good; but use and abuse have made them quite otherwise (Wilstach 1916). In *Malabar Farm*, Lewis Bromfield's (1948) record of restoring the productivity of a farm in Ohio, the author thunders against the farmer who claimed that he "wore out three farms and was still young enough to wear out a fourth." He calls this "a whorish agriculture," and points out that "such an agriculture has been until recently as prevalent in this country as prostitution itself. It is difficult to say which is more devastating to the welfare, morality, health, and security of the individual and the nation."

The profligate use of farmland was economically defensible until the early 1900s when the western frontier in the United States reached the Pacific Ocean. There is probably no other country in the world where farming has caused such tremendous soil wastage in such a short period of time. In 1930 only some 18% of the 170 million hectares under cultivation in the United States were unaffected by erosion (Troeh et al. 1980). The erosion survey of the 1930s indicated that 3.6 billion metric tons of soil were washed annually off the fields, pastures, and forests of the United States. About three quarters of this was from cropland. The mean rate of soil loss translates into an annual loss rate of about 16 tons per hectare.

Severe dust storms during the 1930s also removed vast quantities of soil. A single storm on May 12, 1934, in the panhandles of Texas and Oklahoma, northeastern Colorado, and southwestern Kansas carried off an estimated 18 million metric tons of soil. Steinbeck (1939) described such a storm in the first chapter of *The Grapes of Wrath*:

> The dust lifted up out of the fields and drove gray plumes into the air like sluggish smoke. . . . The finest dust did not settle back to earth, but disappeared into the darkening sky. . . . The air and sky darkened and through them the sun shone redly, and there was a raw sting in the air . . . and the corn fought the wind with its weakened leaves until the roots were freed by the prying wind and then each stalk settled wearily sideways toward the Earth, . . . and the wind cried and whimpered over the fallen corn.

Soil loss at these rates is unsustainable in the long run. Fortunately, a good deal of effort since the 1930s has gone into understanding and reducing soil losses due to farming. Soil loss due to water erosion depends largely on rainfall and runoff, on the erodibility of the soil, on the steepness of fields, on the nature of cropping, and on the measures used to control erosion. Rates of soil loss vary widely today. They are slowest in areas with extensive ground cover and most rapid in areas of bare soil. In well-covered terrain, soil loss is typically less than 1 ton/ha/yr. In bare areas it can be in excess of 100 tons/ha/yr. The extreme range of erodibility of land in the Hawaiian Islands is illustrated in table 6.3.

Table 6.3. Actively Eroding Areas and Erosion Rates in the Hawaiian Islands

Island	*Area (ha)*	*Actively Eroding Area (ha)*	*Rate of Erosion (tons/ha/yr)*	*Island*	*Area (ha)*	*Actively Eroding Area (ha)*	*Rate of Erosion (tons/ha/yr)*
Kauai	143,300	7,892		Lanai	36,463	10,361	
Hanalei	38,000	607	5.5–27	Maunalei	19,429	5,673	5.5–136
Lihue	33,500	971	5.5–82	Kaumalapau	17,034	4,688	5.5–136
Koloa	21,000	243	5.5–82				
Waimea	31,040	3,238	5.5–136	Maui	188,548		
Kekaha	19,870	2,833	5.5–136	Lahaina	24,808	545	0.27–79
				Wailuku	21,085	498	0.27–57
Oahu	156,415	3,116		Makawao	70,700	—*	0.27–79
Kakuku	16,350	243	5.5–55	Nahiku	28,774	—*	0.54–95
Kaneohe	21,056	364	5.5–55	Kaupo	43,181	660	0.27–106
Honolulu	25,739	405	5.5–27				
Waianae	19,709	445	5.5–54	Hawaii	1,045,852		
Ewa	36,896	850	5.5–82	Hamakua	103,731	3,642	0.27–218
Waialua	36,665	809	5.5–136	Hilo	261,068	3,238	0–82
				Ka'u	252,478	81	0–163
Molokai	67,381	11,659		Kona	200,832	minimal	0.27–27
Wailau	14,103	251	5.5–57	Kohala	227,743	1,214	0–136
Kamalo	13,335	2,671	5.5–136				
Hoolehua	24,261	3,221	5.5–136				
Maunaloa	15,682	5,516	5.5–163				

Source: El-Swaify et al. 1982.

* Soil movement may be disproportionately large compared to the negligible acreage of eroding areas accounted for in this inventory.

In fields where the erosion rate is 100 tons/ha/yr, the rate of soil loss is

$$\frac{100 \times 10^6 \text{ gm/yr}}{1 \times 10^8 \text{ cm}^2} = 1.0 \text{ gm/cm}^2\text{/yr.}$$

The density of average soil is approximately 1.3 gm/cm^3. Soil loss from such a field therefore removes soil at a rate of

$$\frac{1.0 \text{ gm/cm}^2\text{/yr}}{1.3 \text{ gm/cm}^3} = 0.7 \text{ cm/yr.}$$

This is equivalent to an erosion rate of 70 cm per century. At this rate, topsoil in most parts of the world is removed in less than a century, and most soils are completely removed in a few hundred years.

Soils are, of course, continually being generated by the processes described earlier, but the rate of soil generation by weathering is much slower than 1 $gm/cm^2/yr$. Take as an example the soils of Kauai. Groundwater on this island contains some 30–40 mg SiO_2/liter. The infiltration rate of groundwater in the soil subject to the erosion described earlier is approximately 100 $cm^3/cm^2/yr$. This implies a SiO_2 loss rate of some

$$35 \times 10^{-6} \frac{\text{gm SiO}_2}{\text{cm}^3 \text{ water}} \times \frac{100 \text{ cm}^3 \text{ water}}{\text{cm}^2\text{/yr}} = 35 \times 10^{-4} \frac{\text{gm SiO}_2}{\text{cm}^2\text{/yr}}.$$

During the conversion of one gram of basalt to soil, roughly 0.38 gm of SiO_2 is lost. Thus approximately

$$\frac{35 \times 10^{-4} \text{ gm SiO}_2\text{/cm}^2\text{/yr}}{0.38 \text{ gm SiO}_2\text{/gm basalt}} \sim 0.01 \text{ gm/cm}^2\text{/yr}$$

of basalt is converted annually to soil. This is equivalent to a conversion rate of 1 ton/ha/yr.

Another type of calculation shows that this rate is not unusual. Globally, the annual river transport of soil is some 2×10^{16} gm (Milliman and Meade 1983; Milliman and Syvitski 1992). Since the area of the continents is 1.6×10^{18} cm^2, the average annual rate of soil removal is approximately

$$\frac{2.0 \times 10^{16} \text{ gm/yr}}{1.6 \times 10^{18} \text{ cm}^2} = 1.2 \times 10^{-2} \frac{\text{gm}}{\text{cm}^2\text{/yr}}$$
$$= 1.2 \text{ ton/ha/yr.}$$

In areas where the rate of soil removal is much greater, soils are almost certainly being eroded more rapidly than they are being formed. Today this is not at all unusual. Brown and Wolf's (1984) description of soil erosion as "a quiet crisis in the world economy" is apt. It may be possible to restore soil by dredging sediment from lakes behind dams and spreading it on fields. This would serve the additional purpose of lengthening the life of dams, but it would be expensive. Measures to reduce the rate of soil erosion are ecologically and almost certainly economically preferable. Soil erosion can be reduced by slowing wind and water flow. Wind speed can be reduced by planting rows of trees (windbreaks) and hedges, and by building fences at right angles to the main wind direction. The velocity of water can be reduced by contour plowing, terracing (as practiced by the ancient Incas in the Andes, plate 14) and by planting cover crops.

6.8 Soils as Ores

Ores are usually defined as rocks that can be mined at a profit. The better the ore, the greater the profit. A more considered definition broadens the term "rock" to include all of the materials of the Earth's crust, and the term "profit" should be broadened to include the requirement that the profit is obtained by legal means (see, for instance, Craig, Vaughan, and Skinner 1988). Within this expanded definition some soils are ores. Highly aluminous soils—the bauxites mentioned in the previous section—are a case in point. Bauxites are presently the only important ores of aluminum. The minimum Al_2O_3 content of economically viable bauxites varies from country to country. The value of these ores depends not only on their Al_2O_3 content but on the presence of undesirable impurities, on local energy costs, and on their proximity to markets. Typical bauxites from Jamaica contain 49% Al_2O_3 on a dry basis; high-grade ores from South America, Guinea, and Australia contain 50%–60% Al_2O_3, minable ores in Arkansas and Western Australia 40%, and most European bauxites between 45% and 65%.

Equation (6.9) shows that gibbsite, one of the main components of bauxite, forms as the result of the extreme weathering of kaolinite. Such weathering is enhanced by high temperatures and abundant rainfall, both common in tropical rain forests. Abundant precipitation, even in temperate climates such as those in the Pacific Northwest of the United States, can be very effective. In addition, the process is more efficient if the parent rocks contain a relatively high percentage of aluminum (such as anorthosites, which consist mostly of feldspars). The tendency for reaction (6.9) to run toward the right (i.e., to produce bauxite) is favored if the waters accomplishing the weathering have a low silica content. This occurs when the parent rocks contain no quartz, as is the case for basalts, anorthosites, syenites, and limestones. However, limestones generally contain only a very small percentage of silicates that can be converted into bauxite. Why then do limestones so often contain rich bauxite deposits, as in Jamaica? One possibility is that in the central part of this island silicates from igneous, sedimentary, and metamorphic rocks are being weathered to clay minerals. During torrential rains these clays are washed onto a plateau of white limestone that is riddled with sinkholes and settle in the surface depressions. Rainwater falling on the plateau is devoid of silica and does not pick up silica from the silica-poor limestone. Hence, upon percolating through the locally accumulated clays, this water promotes their weathering to bauxite. The silica released by this process drains off through the bottoms of the sinkholes (which in practice act like sieves), leaving behind the valuable bauxite. The underground water issues from caves along the coast.

Most of the bauxitic soils in the Hawaiian Islands contain 25% to 35% Al_2O_3 (Patterson 1971). They are fairly widely distributed in three of the islands (see fig. 6.20), but their low grade and relatively small extent, as well as the demand for agricultural soils in the Hawaiian Islands, make these deposits currently unattractive as aluminum ores. Figure 6.21 and table 6.4 show the distribution of the world's reserves of bauxite and the production figures of this commodity for

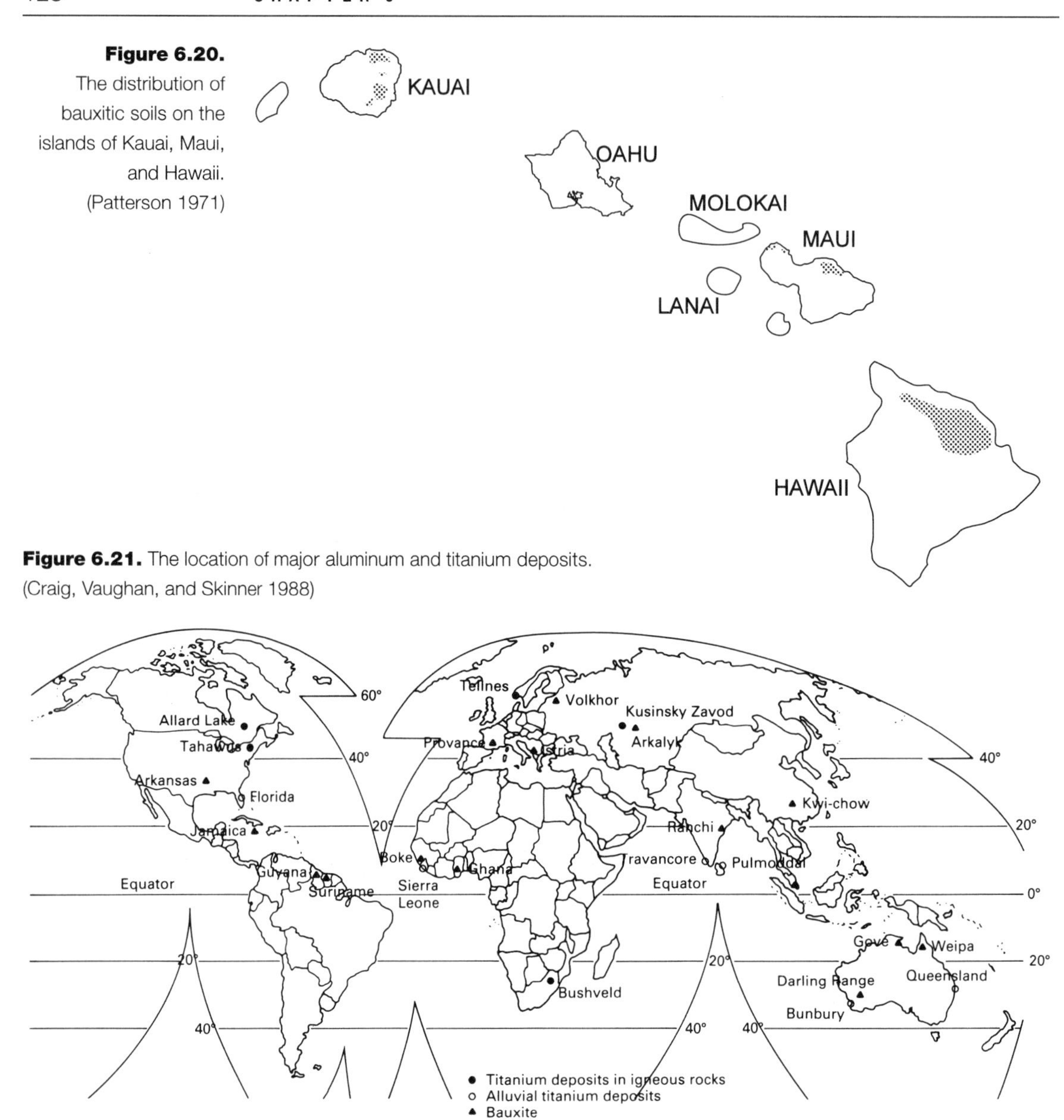

Figure 6.20. The distribution of bauxitic soils on the islands of Kauai, Maui, and Hawaii. (Patterson 1971)

Figure 6.21. The location of major aluminum and titanium deposits. (Craig, Vaughan, and Skinner 1988)

1992. Nearly all of the major bauxite reserves are in tropical countries and are geologically young. Their location in the tropics follows from the discussion in the preceding paragraphs. Their geologic youth is due to two factors: (1) unconsolidated surface materials are cheaper to mine than hard rocks at depth; highly aluminous soils are therefore prized as aluminum ores; (2) "fossil" bauxites are relatively rare. Soils are infrequently preserved in the geologic record. They are usually washed away before the deposition of the next geological layers (see chapter 7), much to the chagrin of geologists who are particularly interested in studying geologically ancient soils.

Figure 6.22 shows that the world production of bauxite has increased dramatically in the twentieth century. Aluminum was a very precious metal until the

1880s. In the early 1880s the price of aluminum ranged from fifty cents to a dollar per troy ounce. When the tip of the Washington Monument was cast by Colonel Frishmuth in Philadelphia, the 8½ pound pyramidal casting 10 inches high and 6 inches on a side of its base was one of the largest single castings of aluminum ever made (Richards 1987). Its cost in 1994 dollars was approxi-

Table 6.4. World Bauxite Resources and Production, 1992 (in thousands of metric tons)

Country	*Mine Production*	*Reserves*	*Reserve Base*
Australia	40,400	5,620,000	7,860,000
Guinea	17,000	5,600,000	5,900,000
Jamaica	11,100	2,000,000	2,000,000
Brazil	10,400	2,800,000	2,900,000
India	4,800	1,000,000	1,200,000
Subtotal	83,700	17,020,000	19,860,000
Other countries	21,300	5,980,000	8,140,000
Total	105,000	23,000,000	28,000,000
"Life"		219 yr	267 yr

Source: Mineral Commodity Summaries, 1993.

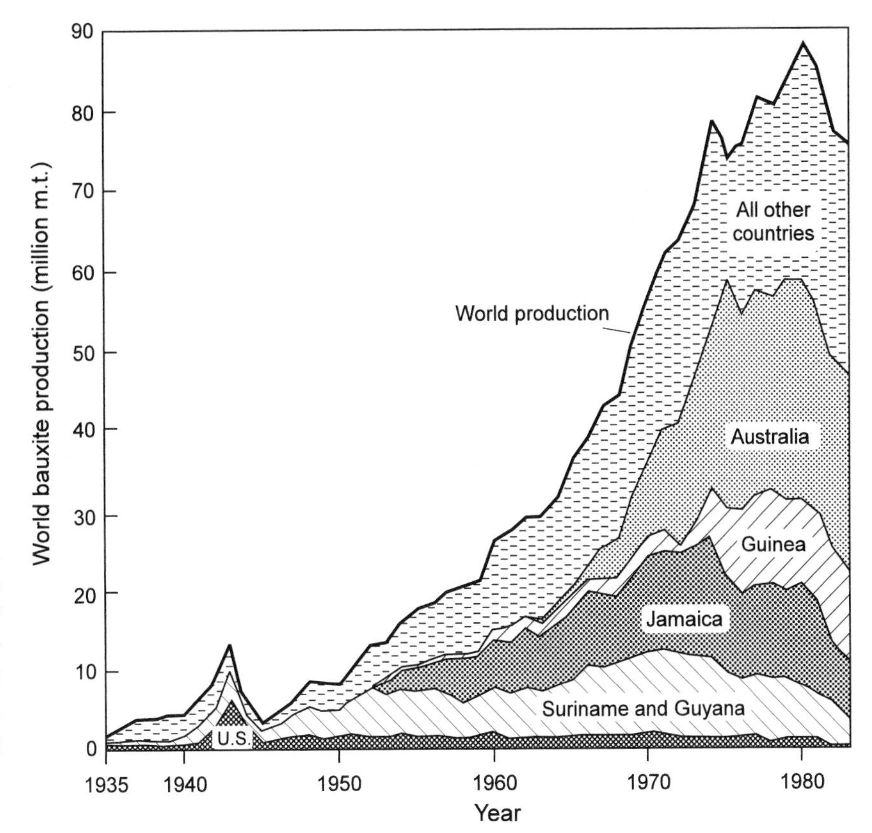

Figure 6.22. World bauxite production in millions of metric tons per year between 1935 and 1983. (Patterson et al. 1986)

mately $30,000. During the 1880s Karl Bayer, an Austrian chemist, developed a chemical process to produce pure Al_2O_3 cheaply from bauxite; almost simultaneously, Charles Martin Hall in the United States and Paul Heroult in France developed an efficient electrolytic process for converting Al_2O_3 into aluminum metal. Aluminum is therefore a "young" metal par excellence (see chapter 10). Its price in real terms decreased rapidly during the twentieth century, and it has captured many markets that formerly belonged to other metals.

Table 6.4 shows that the reserves of bauxite are approximately two hundred times its 1992 production. This observation can be read in two quite different ways. Since bauxite production will probably increase significantly during the next century, it could be argued that the world's bauxite reserves will be exhausted by the year 2100 and that the world will then face a serious shortage of aluminum ore. However, as the current bauxite reserves are depleted, aluminous soils, such as those in the Hawaiian Islands, will become economically attractive. The quantity of such soils is very large compared to the present reserves of minable bauxite. It is quite certain, therefore, that we will not run out of aluminum ore. The cost of bauxite in real terms will probably increase somewhat as lower-grade ores are mined. The total cost of producing aluminum metal will be affected only slightly by this increase, because so much of the cost of producing aluminum stems from the large amount of energy that is required to reduce Al_2O_3 to aluminum. This is an important conclusion. Aluminum is one of the three major industrial metals; iron and copper are the other two. The world's bill for metals depends largely on expenditures for these three metals. It is therefore comforting to know that the cost of aluminum ores is not apt to increase greatly during the twenty-first century. This will be true even in the very long run, well beyond the year 2100, when other aluminum-rich rocks such as syenites, anorthosites, clays, coal wastes, and oil shales are used as aluminum ores. The Al_2O_3 content of fresh Hawaiian basalt is about 14% (see table 6.2), close to the Al_2O_3 content of average crustal rock and about one-fourth the Al_2O_3 content of high-grade aluminum ores. This means that even if we were to mine average crustal rocks for aluminum, their grade would not be drastically lower than that of currently mined ores. In real terms, the cost per kilogram of contained Al_2O_3 will probably never be much more than four times the present cost (see chapter 10).

Ores of nickel, one of the other important industrial metals, are also produced by intense weathering, in this case by the weathering of olivine-rich igneous rocks. Nickel is a minor constituent of these rocks; its concentration is usually in the range of 0.1%–0.2%. Ni^{2+} substitutes readily for Mg^{2+} in olivine and in the pyroxenes. During the weathering of these minerals, Ni^{2+} is released into soil waters together with Mg^{2+} and moves downward toward the water table. In some parts of the world, a good deal of this Ni^{2+} precipitates in the lower soil horizons as a constituent of one or more hydrous silicate minerals (see fig. 6.23). Nickel concentrations in this material can be as high as 4%–5%, which is sufficient to qualify these lower soil horizons as nickel ores. Table 6.5 shows that a considerable percentage of the world's nickel production comes from mining such ores, which may contain about 80% of the world's present and potential nickel reserves.

Sulfide ores are the other important source of nickel today. These ores also occur with highly magnesian igneous rocks, but they are formed by quite

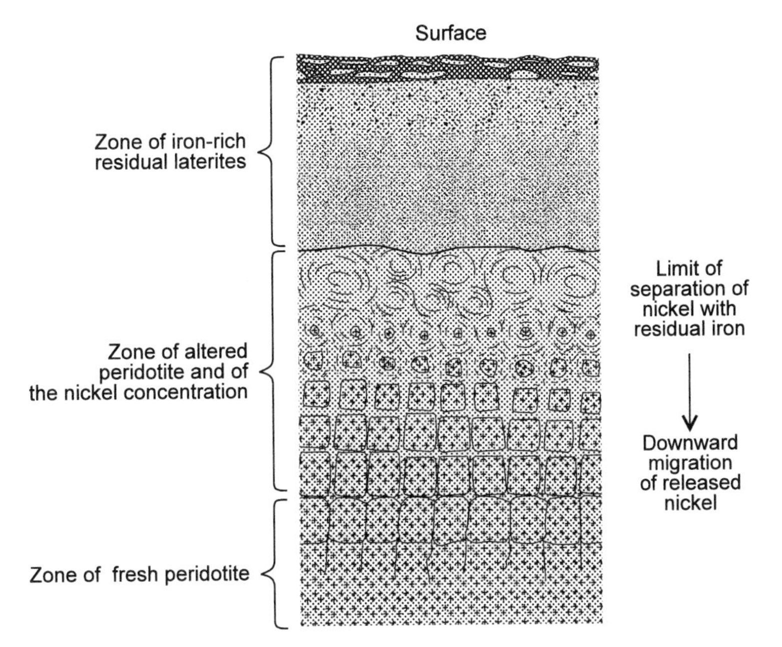

Figure 6.23. Chemical weathering of nickeliferous rocks such as peridotite releases nickel contained in olivine; the nickel is redeposited as a constituent of minerals such as the nickel silicate garnierite. Ores of this kind are worked in New Caledonia and Cuba. (Skinner 1986)

Table 6.5. Production and Estimated Resources of Nickel, 1992 (in thousands of metric tons)

		Mine Production	*Reserves*	*Reserve Base*
Mostly sulfide ores	Russia	225	6,600	7,300
	Canada	214	6,200	14,000
	Australia	70	2,200	6,800
	South Africa	30	2,500	2,600
	Botswana	20	475	900
	Zimbabwe	8	77	100
	Subtotal	567	18,052	31,700
Mostly laterite ores	New Caledonia	95	4,500	15,000
	Indonesia	70	3,200	13,000
	Cuba	35	18,000	23,000
	Brazil	24	666	4,300
	Dominican Republic	24	450	680
	Philippines	12	410	11,000
	United States	6	23	2,500
	Subtotal	266	27,249	69,480
Subtotal (sulfide and laterite)		833	45,301	101,180
	Other countries	83	1,699	8,820
World total		916	47,000	110,000
"Life"			51 yr	120 yr

Source: *Mineral Commodity Summaries*, 1993.

different processes than nickel laterites (see chapter 9). Large quantities of potential nickel ores occur on the ocean floor, particularly under the Pacific. Their origin is described in chapter 7.

6.9 Erosion

"Mending Wall," one of Robert Frost's best-loved poems, begins with the line: "Something there is that doesn't love a wall." To those even slightly familiar with physics, that "something" is clearly gravity. It operates effectively on everything above sea level, and even quite spectacularly on some sediments and rocks below sea level. We have already encountered some of its agents in this chapter and in chapter 4. Water runs toward the oceans carrying with it in solution the dissolved products of chemical weathering. The remainder of the weathering products and simply disaggregated rocks are carried in suspension and as the bed load of rivers. On a clear day, when one can see forever, rivers seem to be rather innocuous parts of the landscape. During floods, when most of the sediment transport takes place, they can be awesome. The Grand Canyon, that big ditch in Arizona (plate 15), bears silent testimony to the enormous work of the Colorado River, a thin strand of water that seems hardly capable of such a major excavation. Figure 6.24 shows the effects of erosion in the Kalalau Valley on the island of Kauai.

In this chapter, we have also encountered wind as an agent of erosion. Wind removed a great deal of soil from the Great Plains during the dust bowl years of the 1930s, and movies about the doings of the French Foreign Legion in North Africa are incomplete without at least one blinding sandstorm. A good deal of the dust bowl soil was redeposited on land, but sand from the Sahara desert is regularly blown far into the Atlantic and becomes part of the sediments accumulating in that ocean (see chapter 7). Dust from the deserts of Mongolia has been identified in Hawaiian soils.

In land areas where slopes are steep, the land may slide downhill under its own weight with or without the help of water and wind. Landslides come in all sizes. We are used to watching out for small landslides when we drive along mountain roads during and after major storms, and we worry about sliding when our house is built on a steep, unstable hillside or at the base of a hill that is frequently visited by avalanches. We rarely think of really major landslides, in large part because they tend to occur infrequently. Recent surveys of the ocean floor within a few hundred kilometers of the Hawaiian Islands have found the debris of huge land-

Figure 6.24. View of the Kalalau Valley, Kauai, Hawaii. (Pen-and-ink sketch by William Coulbourn)

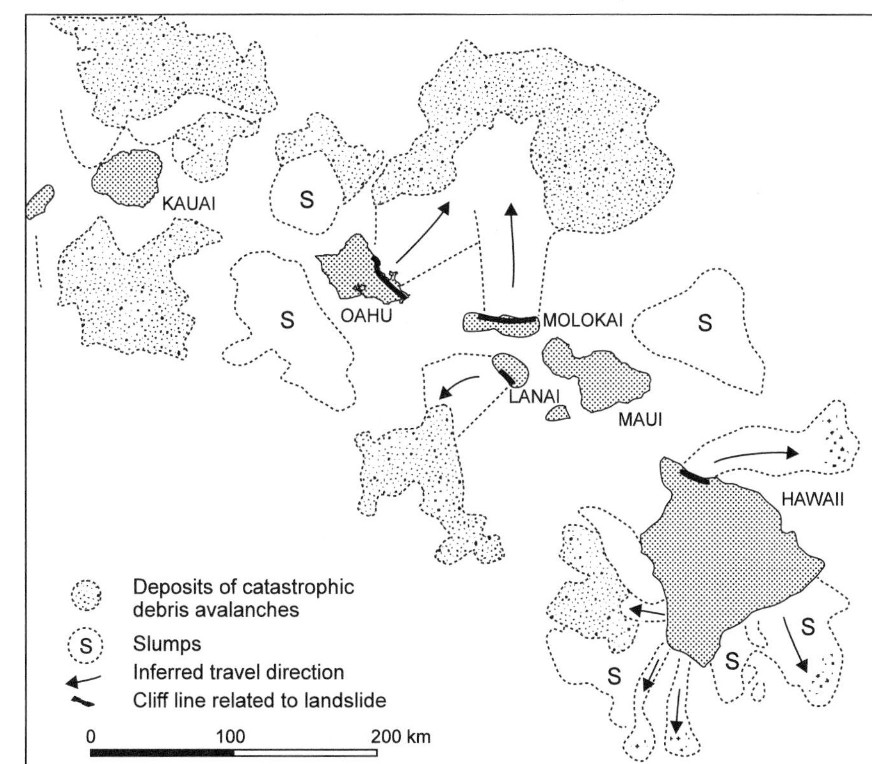

Figure 6.25. The southeastern Hawaiian Ridge, showing the location of slides and areas of hummocky bottom topography (stippled) identified as debris accumulations. (Moore et al. 1989, and Garcia and Hull 1994)

slides (see fig. 6.25). These may have removed as much as half of the present volume of the Islands. Before the volcanic mounds had grown enough to extend above sea level, they probably consisted in large part of dikes, pillow lavas, and volcanic sand and gravel. Like all very large sand piles, the islands tend to break, and the pieces tend to be scattered over the surrounding ocean floor. Such slides produce very large tidal waves (tsunamis). One of these may account for the pieces of coral that have been found 300 meters above sea level on the island of Lanai. It is conceivable that during the next major slide on the Big Island the south flank of Kilauea will founder. The effects of such an event would probably be disastrous for all of the Hawaiian Islands.

Even in the absence of such dramatic events, the oceans manage to wear away the land. Plate 16 shows some of the sea arches along the southeastern coast of the Big Island, where recent lava flows are continually pounded by the surf. The shoreline there advances when a new lava flow reaches the sea and retreats as the lava is broken up by wave action and is carried into deeper waters by ocean currents. Many of the spectacular seacliffs of the Hawaiian Islands (plate 17) and along seashores in many parts of the world owe their scenic grandeur to the erosional power of the sea.

Unlikely as it may seem, one can ski in Hawaii. In winter, snow falls on the peaks of Mauna Loa and Mauna Kea. There are no glaciers, but there is evidence of past glaciation on Mauna Loa. As befits glaciers on a tropical island, they were never large. This was not so in countries at higher latitudes. Evidence for the former presence of extensive ice sheets in northern Europe was first mustered during the early part of the last century. Since then, similar evidence has turned up in many mid- and high-latitude parts of the world. The repeated advances and retreats of ice sheets during the past 2 million years are now well documented and partially understood (see chapter 3). Ice is a fine agent of erosion: it scrapes clean the surfaces that it overrides, and it deposits the debris at its terminus, usually as a wild assortment of sand, pebbles, and boulders. In areas like northern Labrador, which was last glaciated about 10,000 years ago, the soils are very thin. Soils that had developed there before glaciation were removed very efficiently, and too little time has elapsed since the last glaciation to replenish the soil in that cold and forbidding land.

Glaciers not only scour, they gouge. Figure 6.26 shows typical U-shaped valleys produced by glaciers, as well as the deposits of glacial materials (moraines) along their sides and at their terminus. The head walls at their upper end form *cirques*. The depressions at the foot of the cirques are filled with *tarn lakes*. In the eastern part of the United States the glaciers advanced as far south as New Jersey. In Vermont, farmers are plagued by glacially deposited boulders in their fields. A tourist once asked a Vermont farmer what he was doing. "Taking out boulders," the farmer replied. Trying to make bright conversation, the tourist persisted: "Where did the boulders come from?" "Glacier brought 'em," the farmer replied. Looking surprised, the tourist inquired: "Where's the glacier now?" The farmer thought for a while and then replied: "Don't rightly know, but I think it's gone back for more boulders." There may be a considerable amount of truth in that, but the return of glaciers to Vermont is not expected for many thousands of years.

A large number of processes have now been described that conspire to wear down the land, and it is fair to ask how long it will be before all of the land that is currently above sea level will have been transported to the oceans. The total transport of land to the oceans probably amounts to about 2×10^{16}gm/yr. The mean elevation of the continents above sea level is close to 800 meters. The volume of the continents above the oceans is therefore about

$$800 \times 10^2 \text{ cm} \times 1.6 \times 10^{18} \text{ cm}^2 = 1.3 \times 10^{23} \text{ cm}^3.$$

Since the density of average crustal rocks is approximately 2.8gm/cm^3, the mass of the continents above sea level is approximately

$$1.3 \times 10^{23} \text{ cm}^3 \times 2.8 \frac{\text{gm}}{\text{cm}^3} = 3.6 \times 10^{23} \text{ gm}.$$

At the present rate of sediment transport to the oceans, it would therefore take about

$$\frac{3.6 \times 10^{23} \text{ gm}}{2 \times 10^{16} \text{ gm/yr}} = 18 \text{ million years}$$

to reduce the continents to sea level. This figure errs both on the high and on the low side. It errs on the high side because sea level would rise as continental debris are transferred to the oceans. Since the ratio of land surface to sea surface is 30/70 = 0.42, a lowering of the continents by 1 meter would raise sea level by 0.42 meter. The continents would therefore be just awash with seawater after

Figure 6.26. Glaciers and glacial deposits in East Greenland. (Photo: John Haller)

their level had dropped by 570 meters, because sea level would have risen by 230 meters. At the present rate of erosion, a drop of 570 meters in the mean level of the continents would require about 13 million years. This is, however, an underestimate, because the rate of erosion of the continents decreases rapidly with decreasing elevation above sea level. But mountain ranges behave somewhat like icebergs. As they are eroded, they tend to rise and to ride higher on the underlying crust and mantle of the Earth. This process increases the time required for eroding the continents. Nevertheless, even if this is taken into account, the time required to reduce the continents to sea level or nearly so is clearly much less than the age of the Earth.

This calculation shows that mountain building must be going on today, or must at least have occurred in the not-so-distant geologic past; otherwise the presence of high mountain ranges would be difficult to explain. The effect of the destructive processes that are removing mountains must somehow be balanced by constructive processes driven by the internal dynamics of the Earth. This idea is explored in the next two chapters.

What of the parable of the bird and the mountain that began this chapter? If the mountain was the size of Mount Fuji or Mount Shasta, its volume was about 420 km^3 and its mass about 1.0×10^{18} gm. The mass of Mauna Loa reckoned from the ocean floor is about one hundred times greater. If the bird was large enough to carry off a 1 kg (2.2 lb) chunk of rock on each visit, it would have needed one thousand trillion years to carry away Mount Fuji or Mount Shasta. That is about 200,000 times longer than the age of the Earth. The data amassed in this chapter show that the lifetime of mountains is short compared to the age of the Earth. Geologic processes clearly operate much faster than the hypothetical bird, and the questions that were raised by the geologist on hearing the parable of the bird and the mountain were surely justified.

References

Birkeland, P. W. 1984. *Soils and Geomorphology.* Oxford University Press, Oxford.

Bragg, W. L. 1937. *Atomic Structure of Minerals.* Cornell University Press, Ithaca, N.Y.

Bromfield, L. 1948. *Malabar Farm.* Harper and Brothers, New York.

Brown, L. R., and Wolf, E. C. 1984. *Soil Erosion: Quiet Crisis in the World Economy.* Worldwatch Paper 60. Worldwatch Institute, Washington, D.C.

Carmichael, I.S.E., Turner, F. J., and Verhoogen, J. 1974. *Igneous Petrology.* McGraw-Hill, New York.

Chase, T. E., Menard, H. W., and Mammerickx, J. 1970. *Bathymetry of the North Pacific.* Charts 2, 7, 8, Scale: 1:800,000. Scripps Institute of Oceanography, La Jolla, Calif.

Clague, D. A., and Dalrymple, G. B. 1987. Hawaiian Alkaline Volcanism. In *Alkaline Igneous Rocks*, ed. J. G. Fritton and B.G.J. Lipton, pp. 227–52. Special Publication 30, Geological Society of America.

Clague, D. A., and Dalrymple, G. B. 1989. Tectonics, Geochronology, and Origin of the Hawaiian-Emperor Volcanic Chain. In *The Eastern Pacific Ocean and Hawaii*, vol. N, pp. 188–217, of *The Geology of North America*, ed. E. L. Winterer, D. M. Hussong, and R. W. Decker. Geological Society of America.

Cooper, G., and Daws, G. 1985. *Land and Power in Hawaii: The Democratic Years.* Benchmark Books, Honolulu.

Craig, J. R., Vaughan, D. J., and Skinner, B. J. 1988. *Resources of the Earth.* Prentice Hall, Englewood Cliffs, N.J.

Daws, G. 1968. *Shoal of Time: A History of the Hawaiian Islands.* Macmillan, New York.

de Chételat, E. 1947. La genèse et l'évolution des gisements de nickel de la Nouvelle-Calédonie. *Bull. Soc. Géol. Fr.*, ser. 5, 17:105–60.

El-Swaify, S. A., Dangler, E. W., and Armstrong, C. L. 1982. *Soil Erosion by Water in the Tropics.* College of Tropical Agriculture and Human Resources, University of Hawaii, Honolulu.

Garcia, M. O., and Hull, D. M. 1994. Turbidites from Giant Hawaiian Landslides: Results from Ocean Drilling Program Site 842. *Geology* 22:159–62.

Grim, R. E. 1968. *Clay Mineralogy.* 2d ed. McGraw-Hill, New York.

Hillel, D. J. 1991. *Out of the Earth: Civilization and the Life of the Soil.* Free Press, New York.

Hutchison, J. L. 1994. Seeing Atoms in and around Crystals. *U.S.A. Microscopy and Analysis* (March):9–12.

Kaeppler, A. L. 1980. *Kapa, Hawaiian Bark Cloth.* Boom Books, Hilo Bay, Hawaii.

Klein, C., and Hurlbut, C. S., Jr. 1985, 1993. *Manual of Mineralogy*. 20th and 21st eds. John Wiley, New York.

MacDonald, G. A., Abbott, A. T., and Peterson, F. L. 1983. *Volcanoes in the Sea: The Geology of Hawaii*. 2d ed. University of Hawaii Press, Honolulu.

Milliman, J. D., and Meade, R. H. 1983. Worldwide Delivery of River Sediment to the Oceans. *Jour. of Geology* 91:1–21.

Milliman, J. D., and Syvitzki, J.P.M. 1992. Geomorphic/Tectonic Control of Sediment Discharge to the Ocean: The Importance of Small Mountainous Rivers. *Jour. of Geology* 100:525–44.

Moore, H. J. 1987. Preliminary Estimates of the Rheological Properties of 1984 Mauna Loa Lava. In *Volcanism in Hawaii*, ed. R. W. Decker et al., chap. 58. U.S. Geol. Survey Prof. Paper 1350.

Moore, J. G., Clague, D. A., Holcomb, R. T., Lipman, P. W., Normark, W. R., and Torresan, M. E. 1989. Prodigious Submarine Landslides on the Hawaiian Ridge. *Jour. Geophys. Res.* 94:17465–84.

Papike, J. J., and Cameron, M. 1976. Crystal Chemistry of Silicate Minerals of Geophysical Interest. *Rev. Geophys. and Space Phys.* 14:37–80.

Patterson, S. H. 1962. Investigation of Ferruginous Bauxite and Plastic Clay Deposits on Kauai and a Reconnaissance of Ferruginous Bauxite Deposits on Maui. U.S. Geol. Survey Open File Report.

Patterson, S. H. 1971. Investigation of Ferruginous Bauxite and Other Mineral Resources on Kauai and a Reconnaissance of Ferruginous Bauxite Deposits on Maui, Hawaii. U.S. Geol. Survey Prof. Paper 656.

Patterson, S. H., Kurtz, H. F., Olson, J. C., and Neeley, C. L. 1986. World Bauxite Resources. U.S. Geol. Survey Prof. Paper 1076B.

Richards, J. W. 1987. *Aluminum: Its History, Occurrence, Properties, Metallurgy and Applications, Including Its Alloys*. Henry Carey Baird and Co., Philadelphia.

Skinner, B. J. 1986. *Earth Resources*. 3d ed. Prentice Hall, Englewood Cliffs, N.J.

Steinbeck, J. 1939. *The Grapes of Wrath*. Viking Press, New York.

Troeh, F. R., Hobbs, J. A., and Donahue, R. L. 1980. *Soil and Water Conservation for Productivity and Environmental Protection*. Prentice Hall, Englewood Cliffs, N.J.

U.S. Bureau of Mines. 1992. *Mineral Commodity Summaries 1992*.

Walker, G.P.L. 1990. Geology and Volcanology of the Hawaiian Islands. *Pacific Science* 44:315–47.

Wilstach, P. 1916. *Mount Vernon: Washington's Home and the Nation's Shrine*. Doubleday, Page and Co., Garden City, N.Y.

7 The Oceans

7.1 Introduction: Down to the Sea in Ships

Humanity has had a long and complex relationship with the oceans. Ulysses, that most famous of ancient mariners, probably sailed the eastern Mediterranean about 1225 B.C. (Wood 1985), played an important role in the siege and the sack of Troy, and spent ten years trying to get home to Greece. For Ulysses, as for so many travelers before and after him, the ocean was a transportation route, sometimes safer and shorter than land routes, frequently otherwise. But for those captivated by Homer, the Odyssey is much more than a catalog of misfortunes and delays. In his old age, Tennyson's Ulysses looks back longingly to the voyage:

> Come, my friends,
> 'Tis not too late to seek a newer world.
> Push off, and sitting well in order smite
> The sounding furrows; for my purpose holds
> To sail beyond the sunset, and the baths
> Of all the western stars, until I die.

The search for adventure and for great deeds was always part of going down to the sea in ships. There also were much more mundane reasons for going to sea: food, for instance. Fishing must have started close to shore and the range of fishing boats must have been extended gradually as fish became scarce near shore and to catch more desirable, deeper-water varieties. Larger ships and improved navigational techniques finally opened all of the world's oceans to fishing. The New England whalers first depleted the local waters and ultimately helped to do much the same to the oceans as a whole. The history of the enterprise is well chronicled in the Whaling Museum at New Bedford, Massachusetts. The museum is full of ingenious devices for hunting, killing, and dismembering whales, while the gift shop sells beautiful T-shirts emblazoned with the slogan "Save the Whales." In the nearby Mariner's Chapel, plaques commemorate the loss of New Bedford men at sea and identify the pew that was Herman Melville's. The ghost of Moby Dick permeates the building. It seems a great distance from there to Titusville, Pennsylvania, where the first successful oil wells in 1859 spelled the end of the American whaling industry and ultimately the salvation of the whales.

Killing at sea has not been confined to fish and whales. Men have long been engaged in piracy and in more legal acts of homicide during wartime. *Treasure Island* has sent delicious shivers down the spines of generations of readers, as have *Captain Blood* and the many movies that documented the brutality of naval warfare and death at sea during World War II. The desire for trade routes and distant conquests were at the root of most of this violence. The explorers of the Renaissance roamed the sea largely in search of gold. They were successful, and the later Spanish bullion fleets were considered fair game for plunder. Frail ships like the *Mayflower* brought European immigrants to the New World. Their descendants' trade with Europe was frequently interrupted by war, as was their infamous triangular trade route in slaves, molasses, and rum, which contributed so much to the wealth of New England in the eighteenth and early nineteenth centuries.

Worldwide ocean travel requires good ships, good maps, and good navigational instruments. The great voyages of exploration from the fifteenth to the nineteenth centuries served to give the world an excellent self-portrait. Advances in clock making during the early part of the eighteenth century led to the first really usable ship's chronometers for precise determinations of longitude, and the explosive development of technological innovations in the twentieth century removed many of the remaining obstacles to safe travel at sea. Such innovations paralleled and intersected the increasing interest in understanding the oceans. Benjamin Franklin's map of the Gulf Stream was of considerable importance for science as well as for commerce. Captain Cook was accompanied by botanists and artists. Charles Darwin's insights and discoveries during his five-year voyage on the *Beagle* are the stuff of legends. A good deal of the research supported by the U.S. Office of Naval Research had significant military implications in addition to its great scientific importance. The support of expensive science is largely driven by the demands of human affairs; but we should not forget the cloudless nights at sea, when all is still, when stars move brilliantly across the jet-black sky, nights like those when Ulysses and his crew may well have recalled the exploits of heroes immortalized in the constellations of the heavens.

7.2 The Measurements of the Oceans

We have already encountered some of the oceans' dimensions in earlier chapters. Seawater covers 3.5×10^{18} cm^2, or about 70% of the Earth's surface. The volume of the oceans is 1.4×10^{24} cm^3, which amounts to about 0.023% of the total mass of the Earth. The mean depth of the oceans is close to 4,000 meters. The oceans are the largest reservoir of water on Earth, much larger than the ice caps today, and—for that matter—at the height of past ice ages.

These numbers tend to disguise the complex geometry of the ocean basins. The presence of a mountain range in the middle of the Atlantic Ocean was known in the last century, but most of the topographic features in figure 7.1 were discovered in this century after the invention of continuous depth recorders. The flat-topped seamounts that dot the floor of the Pacific Ocean were discovered by Harry H. Hess, then captain of a troop transport during World War II. The depth recorder of his ship was left running because, as a geologist, he was deeply interested in the topography of the ocean floor, and he named the seamounts guyots, in honor of Arnold Guyot, the first professor of natural history at Princeton University in New Jersey.

If we start at a harbor along the East coast of North America, say in Halifax, Nova Scotia, and set a course southeastward toward the west coast of North Africa (see fig. 7.2), we first cross an area of rather shallow water, the continental shelf. At the edge of the shelf the water depth increases as we cross the continental slope and then the continental rise, until we reach the rather featureless abyssal plain. From there we climb up the west slope of the Mid-Atlantic Ridge, a great mountain range that neatly divides the entire Atlantic Ocean. After crossing the ridge, we descend back to the abyssal plain. The ocean then gradually shallows. We cross the rise and the slope off the African continent, then the shelf

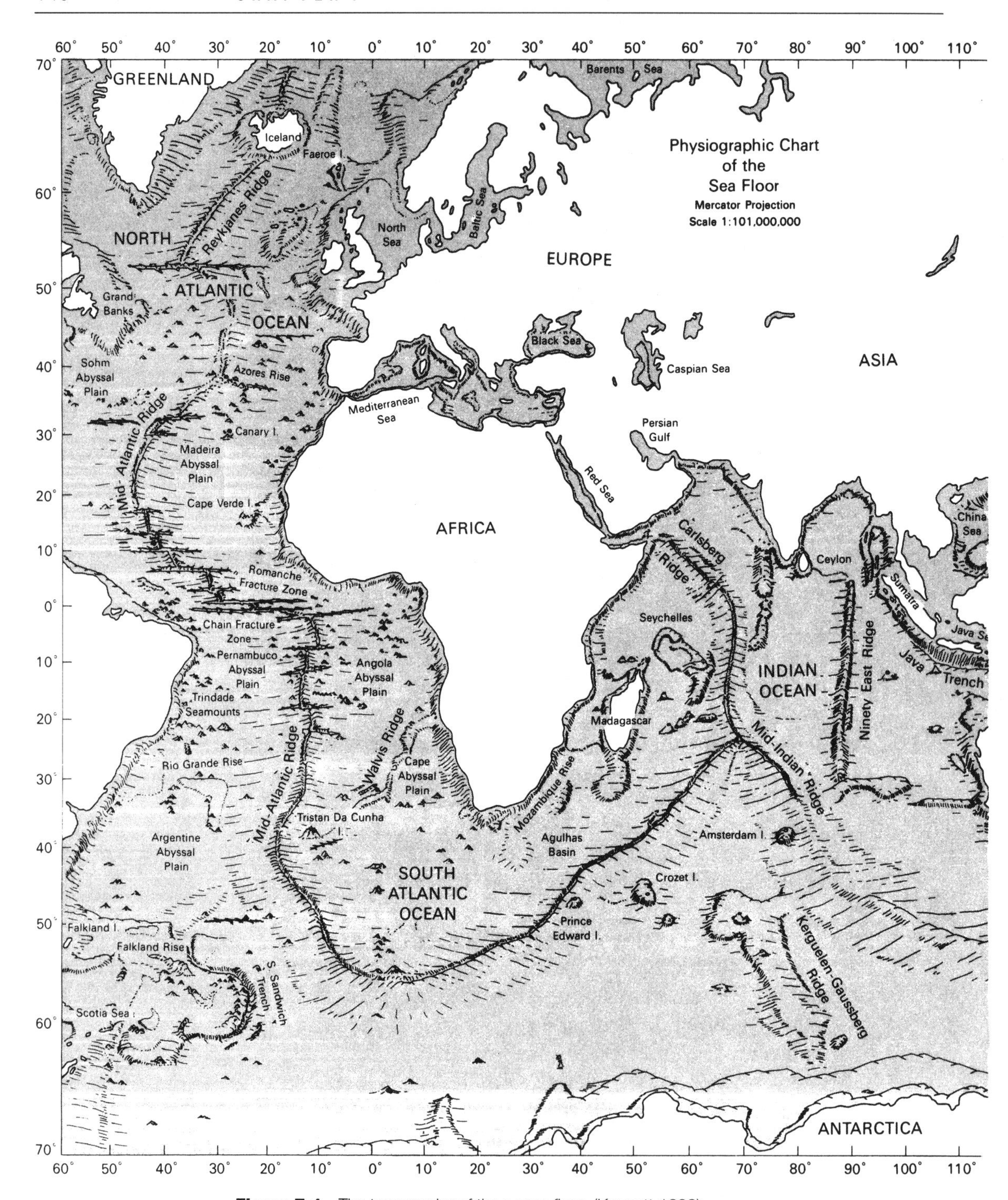

Figure 7.1. The topography of the ocean floor. (Kennett 1982)

Chukchi Sea
Bering Sea
Sea of Okhotsk
Aleutian Basin
Aleutian Trench
Aleutian Abyssal Plain
Emperor Seamount Chain
Kuril Trench
Sea of Japan
Japan
Japan Trench
Chinook Fracture Zone
Medocino Fracture Zone
Murray Fracture Zone
Molokai Fracture Zone
NORTH PACIFIC OCEAN
Bonin Trench
Mariana Trench
Philippine Sea
Philippine Trench
Mariana Basin
Mid-Pacific Mountains
Clarion Fracture Zone
Clipperton Fracture Zone
Line I.
Solomon Rise
New Guinea
Vitiaz Trench
Samoa I.
Coral Sea
Great B. Reef
Fiji I.
Tonga I.
Cook I.
Kermadec-Tonga Trench
Marquesas I.
Society I.
Austral I.
Marquesas Fracture Zone
Tuamotu Archipelago
Galapagos Fracture Zone
Galapagos I.
Cocos Ridge
East Pacific Rise
Galapago Rise
Nazca Ridge
Easter Fracture Zone
Chile Rise
Peru-Chile Trench
Middle America Trench
AUSTRALIA
Lord Howe Rise
SOUTH PACIFIC OCEAN
Chatham Rise
Campbell Plateau
Tasmania Rise
Eltanin Fracture Zone
Pacific-Antarctic Ridge
NORTH AMERICA
Hudson Bay
Hatteras Abyssal Plain
Burmuda Rise
Nares Abyssal Plain
Puerto Rico Trench
Caribbean Sea
SOUTH AMERICA

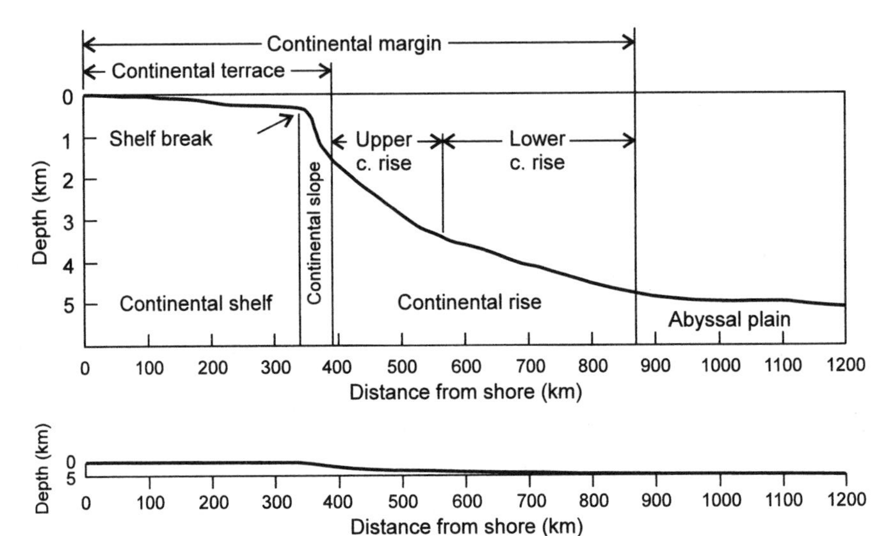

Figure 7.2a. Principal features of continental margins. (A) Vertical exaggeration 50/1; (B) without vertical exaggeration. (Kennett 1982)

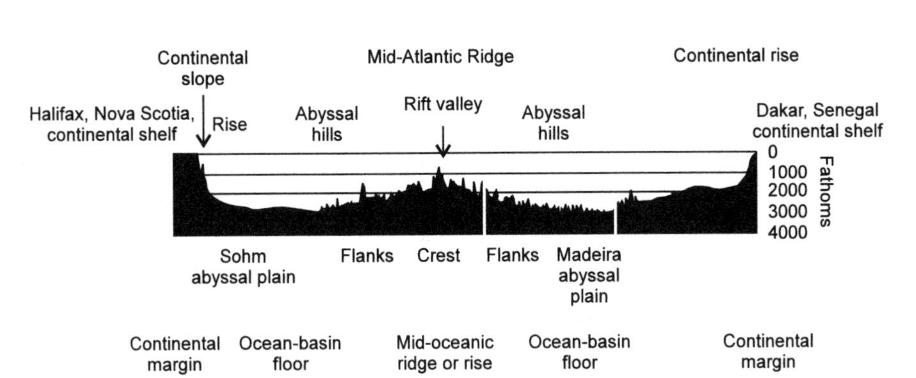

Figure 7.2b. Principal morphological features along a traverse across the North Atlantic between North America and Africa. Note vertical exaggeration. (Kennett 1982)

until we reach Dakar. This trip is typical of many Atlantic crossings, but not of all. The bottom topography of the Caribbean is very complex. The seaward side of the island arc in this sea is fringed by a deep trench, like those that are common in the Pacific Ocean.

The floor of the Pacific Ocean is quite different from that of the Atlantic. If we sail west out of Callao, the port of Lima, Peru, the ocean deepens very rapidly, and we pass over the Peru-Chile trench. Three thousand kilometers to the west we encounter a major mountain chain, the East Pacific Rise. If we follow this rise northward, we find that it heads to the North American continent and disappears into the Gulf of California. A quick look at figure 7.1 shows that the East Pacific Rise and the Mid-Atlantic Ridge are parts of an Earth-encircling range of submarine mountains that has a total length of some 80,000 kilometers, twice the circumference of the Earth. The floor of the Pacific Ocean is pimpled with seamounts. Some break the surface of the ocean and reveal themselves as volcanic islands. Their pattern is often confusing, but some, like the Hawaiian Islands and their northwestern extension, the Emperor Sea Mounts, are easily recognizable island chains (see chapter 6).

All of the oceans are cut by large fracture zones. These are clearly related to the ocean ridges and suggest that the topographic patterns of the ocean floor are products of large-scale movements. All this became clear when the first detailed maps of the ocean floor were compiled by Bruce Heezen in the 1950s.

7.3 Sediments and Sediment Inputs to the Oceans

Solid debris from the continents is transported to the oceans by water, wind, and ice. Water transport is the most important of the three, and rivers are the dominant means of transport. The products of erosion come in all sizes. Picturesque mountain streams contain large boulders. These move downstream during times of intense flooding and are gradually worn down in the process. Their disintegration products join other products of weathering and erosion in their journey downstream. The average grain size of river sediments gradually decreases, and particles larger than sand grains are uncommon in sediments at the mouths of large rivers.

En route, rivers sometimes accumulate economically important quantities of chemically and physically resistant, high-density minerals. The most famous of these minerals is gold. Grains of gold are so much denser than the common rock-forming minerals that gold tends to become separated from its less noble travel companions and accumulates in what are called *placer deposits*. These usually form where barriers allow faster-moving waters to carry away small, low-density grains while serving as traps for the denser and larger particles. Some typical sites of placer accumulation are shown in figure 7.3. Gold placers, particularly those mined in the western states during the last century, are firmly enshrined in American folklore, but they are much less important than some ancient placer gold deposits that had been buried and were recently exhumed. By far the most important fossil gold placers are those of the Witwatersrand Basin in South Africa. Gold mined from these placers accounts for about 28% of the current world production of gold (see table 7.1) and is largely responsible for the role of South Africa as the world's most significant gold-producing nation. Figure 7.4 shows the geography of the Witwatersrand Basin. Gold was discovered there in 1886 near Johannesburg on the north rim of the basin. It was found in what were once pebble beds that accumulated 2,500–3,000 m.y.a. in the deltas of rivers draining the country to the north, east, and west of the basin.

The gold is not homogeneously distributed; rather, it is concentrated in pebble layers. It follows "pay streaks," whose locations were probably determined by the mechanics of sediment distribution in the basin. During the long history of the basin since the deposition of the pebble beds, hot solutions have percolated through it and have partially redistributed the gold. The placer origin of the gold has therefore been disputed. Although no adequate primary source has been identified from which the enormous quantity of gold was eroded to form placers, the weight of geological evidence is strongly in favor of a placer origin.

The layers of gold-rich conglomerates dip toward the center of the Witwatersrand Basin. Mining follows these layers, and has now reached depths where mining is difficult due to high temperatures, and dangerous because of rock bursts. The future of South Africa as a major gold producer depends in large part on the ability to extend these mines to progressively more inhospitable depths.

The Witwatersrand conglomerates are unusual not only for their gold content, but also because they contain significant quantities of the minerals pyrite (FeS_2)

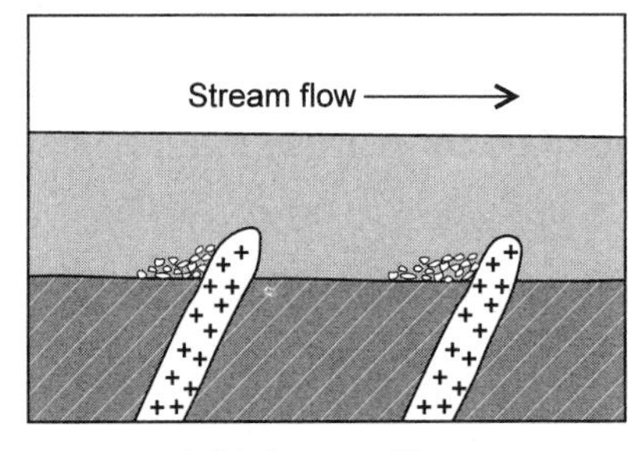

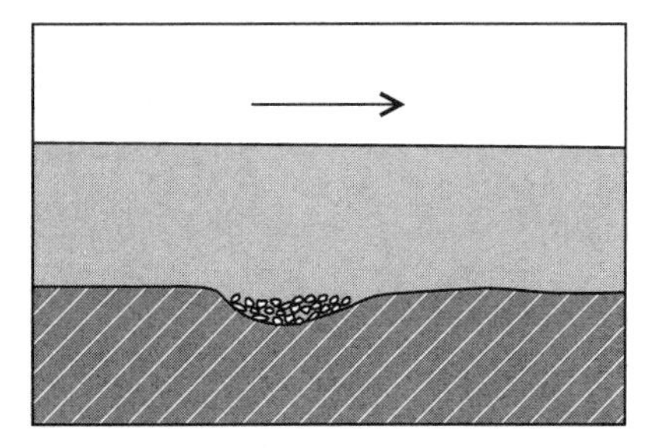

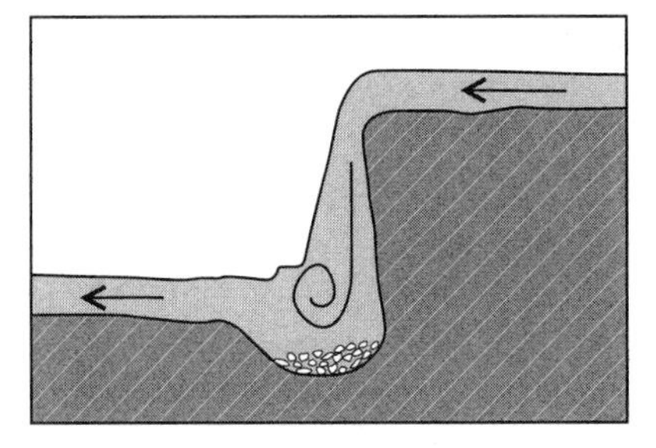

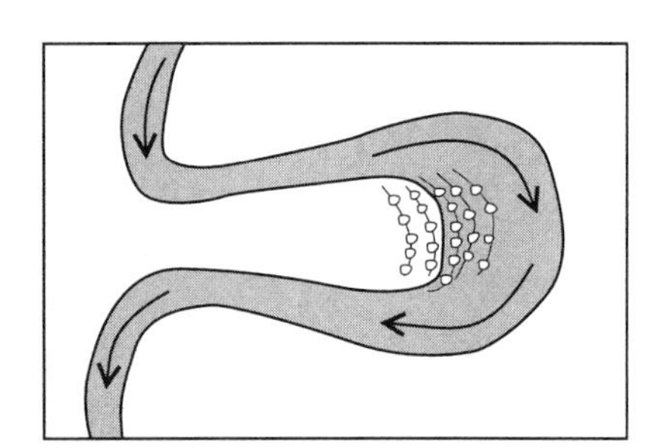

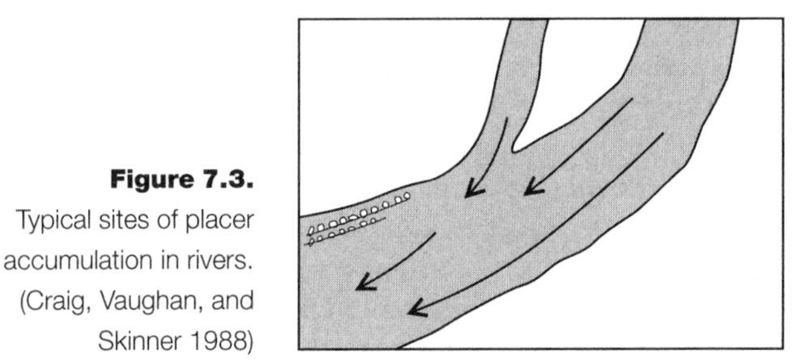

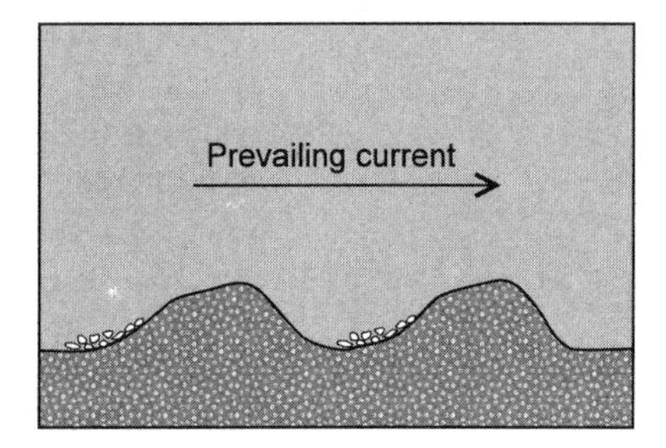

Figure 7.3. Typical sites of placer accumulation in rivers. (Craig, Vaughan, and Skinner 1988)

Table 7.1. Production and Reserves of Gold (in metric tons) (*Mineral Commodity Summaries*, 1993)

Country	*Production*	*Reserves*
South Africa	600	20,000
United States	320	4,770
Australia	240	2,150
USSR (former)	230	6,220
Canada	170	1,780
China	140	N.A.
Brazil	90	940
Subtotal	1,790	35,860
Other countries	380	8,040
World total (rounded)	2,170	43,900

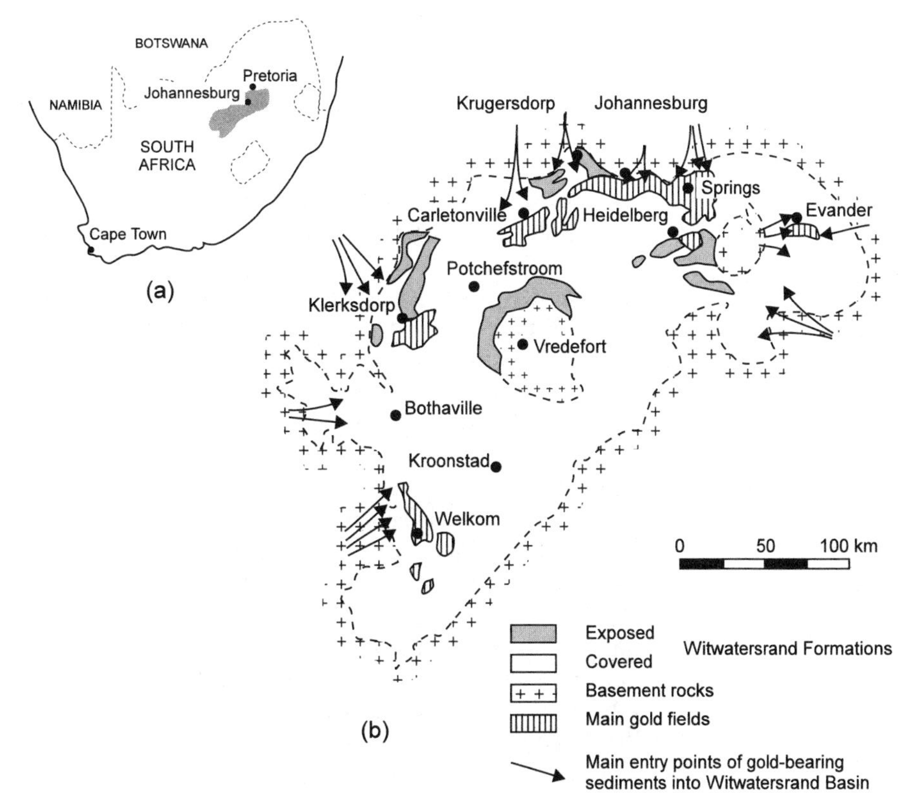

Figure 7.4. Maps of the Witwatersrand gold field: (a) location of the Witwatersrand Basin in South Africa; (b) the major gold fields, with arrows indicating the directions of transport of the conglomerates into the basin. (Craig, Vaughan, and Skinner 1988)

and uraninite, an oxide of uranium, roughly of composition $(U, Th)O_2$. Thorium, Th^{4+}, substitutes for U^{4+} in the structure of uraninite for the same reasons that Fe^{2+} substitutes for Mg^{2+} in olivine and pyroxenes (see chapter 6). The occurrence of pyrite and uraninite in these conglomerates is important both economically and scientifically. Under today's atmospheric conditions, pyrite is oxidized rapidly to ferric oxides and hydroxides (chapter 6) and uraninite is oxidized rapidly to UO_3. The oxidation products of UO_2 are sufficiently soluble so that most of the uranium transported to the oceans today is dissolved in river water. The survival of pyrite and uraninite grains during weathering, transport, and deposition of the Witwatersrand conglomerates and in the similar ancient conglomerates of the Blind River–Elliot Lake area of southern Canada is therefore rather unusual. The apparent resistance of pyrite and uraninite to oxidation in these pre-2,000-million-year-old rocks adds weight to other lines of evidence that indicate that the oxygen content of the atmosphere before 2,000 m.y.a. was very much lower than today (see chapter 6). This hypothesis is of considerable importance as a guide for uranium exploration. It suggests that economically important accumulations of uraninite in conglomerates will not be found in placers younger than about 2,100 m.y., and that other types of uranium ores, whose formation requires the presence of an O_2-rich atmosphere, will not be found in rocks older than about 2,100 m.y.

Most sediments at the mouths of large rivers are abnormally rich neither in gold nor in uranium, and are much finer grained than the Witwatersrand conglomerates. They are usually deposited in deltas, and are then redistributed by ocean currents. Along the shore, currents transport sediments parallel to shorelines and are responsible for a variety of near-shore sediment accumulations as well as for coastal erosion. Currents with a strong seaward component transport sediments to the edge of the continental shelves. From there they are moved down the continental slopes and continental rises by a variety of mechanisms. Turbidity currents are among the most spectacular and efficient of these. Submarine slumps and flows are often triggered by earthquakes and can transport marine sediments for distances in excess of 1,000 km at speeds of over 90 km/hr. If these currents begin to slow, the coarsest fraction of their sediments settles first, and the finest settles last. This gives rise to sedimentary layers that "fine upwards." Sedimentary sequences deposited from a succession of turbidity flows are called *Bouma sequences*, which consist of a succession of superimposed "fining upward" layers (see fig. 7.5).

Despite the efficient mechanisms that are available for transporting sediments to the abyssal plains, more than 90% of all river sediments today are deposited on the continental shelves, slopes and rises. Less than 10% reach the abyssal parts of the oceans. The marine sediment distribution along two sections off the eastern North American coast is shown in figure 7.6. The thickness of sediments along these sections decreases dramatically with distance from the coast. Figure 7.7 shows a similar pattern along a north-south section that starts near New Orleans at the mouth of the Mississippi River. The thickness of the sedimentary pile decreases southward into the Gulf of Mexico. The origin and significance of the evaporites in this region will be discussed later in this chapter.

Most river sediments consist largely of silicate minerals. They also contain organic matter, mainly the remnants of land plants. These include decayed and decaying wood, leaves, roots, and humus, as well as organic material derived from organisms that lived and died in the waterways themselves. On reaching the oceans, some of this organic matter is eaten by marine organisms. The unpalatable portions, joined by the remains of marine organisms, accumulate with the inorganic components of river sediments. Photosynthesis in many near-shore areas of the oceans is intense (see chapter 5). However, the consumers of organic

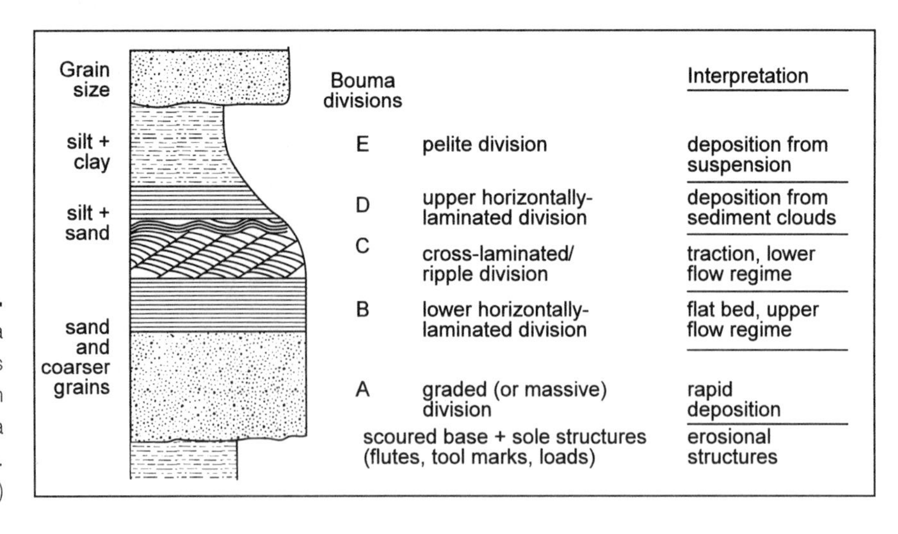

Figure 7.5. Bouma sequence of a turbidite bed. Turbidites range in thickness from several centimeters to a meter or more. (Tucker 1981)

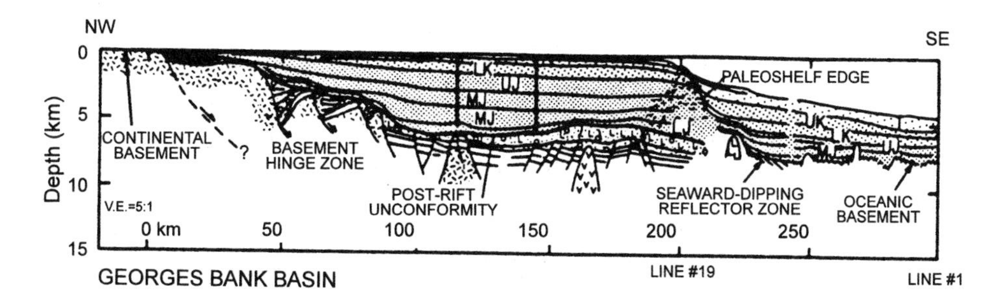

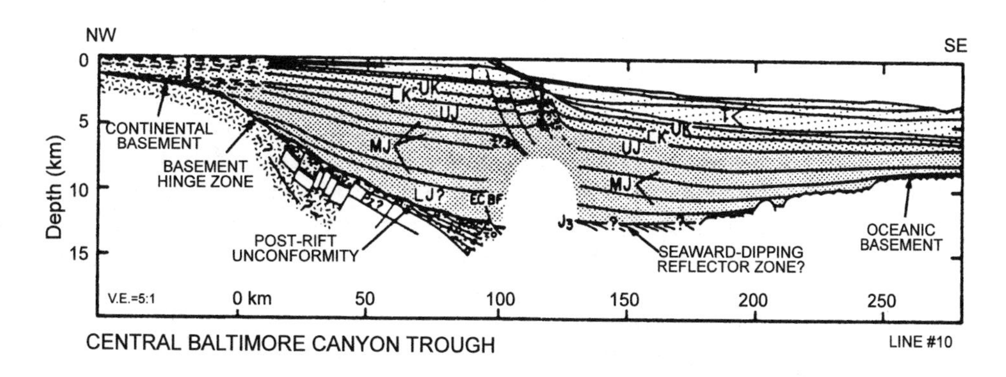

Figure 7.6. Cross sections across the Georges Bank Basin (upper) and the central Baltimore Canyon Trough (lower) (from Klitgord and Hutchinson 1985; Klitgord et al. 1988). ECBF = East Coast Boundary Fault; J_3 = deepest Atlantic continental rise reflection (Sheridan 1989). T = Tertiary; UK = Upper Cretaceous; LK = Lower Cretaceous; UJ = Upper Jurassic; MJ = Middle Jurassic; LJ = Lower Jurassic.

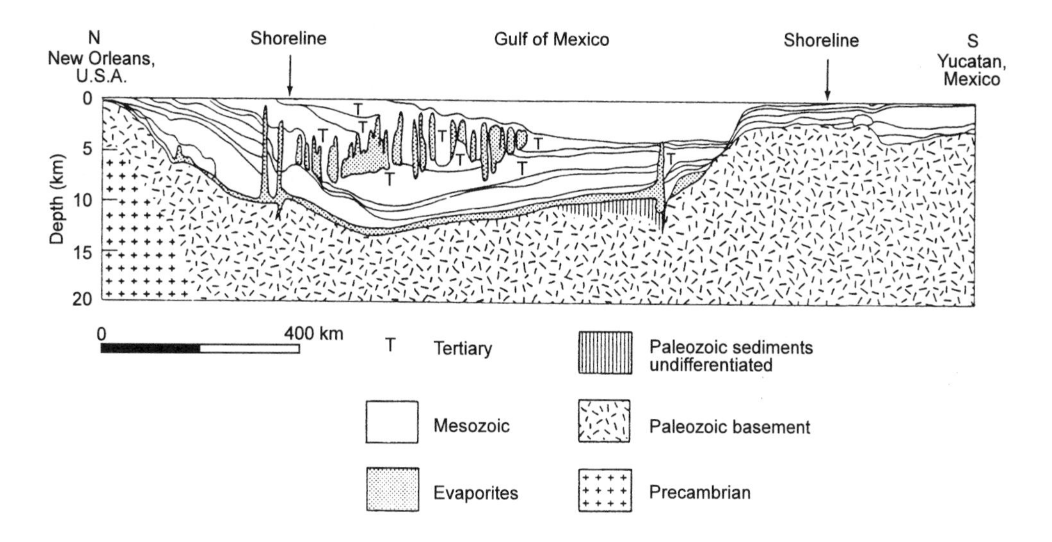

Figure 7.7. Geological cross section of the Gulf of Mexico; vertical exaggeration 20:1. (Bally 1979)

matter are very efficient, and a large fraction of the organic matter synthesized by marine phytoplankton is eaten in the water column. The small fraction that escapes this fate settles to the bottom. In shallow waters it is joined by organic matter produced by bottom-dwelling plants. In most deeper water areas, where little or no sunlight penetrates to the ocean floor, the gentle rain of organic matter from near-surface waters is the sole source of such matter. Bottom-dwelling (benthic) animals eat most of the organic matter that settles from above. Today approximately 99.7% of all the organic matter that is produced in the oceans is consumed and reoxidized. This percentage has probably not varied a great deal over time. The remnant that has been buried with sediments in the past is a minuscule fraction of the total production; but it is of more than passing interest, because it is the parent material of the world's supply of oil and natural gas.

The amount of organic matter that is buried on a particular patch of ocean floor depends on the flux of organic matter from above and on the fraction of this matter that is buried. The burial efficiency of organic matter is a function of a number of variables; by far the most important of these is the rate of sediment accumulation. The faster the sedimentation rate, the greater the chance that organic matter will be buried rather than eaten. This relation is shown in figure 7.8. Most marine sediments accumulate near-shore, where sedimentation rates are high. Marine organic matter therefore accumulates most readily in shelf, slope, and rise sediments. The best estimate for the total rate of burial of organic matter with marine sediments is about 1.5×10^{14} gm/yr (see fig. 5.4). Since rivers annually transport about 2.0×10^{16} gm of sediment to the ocean (see chapter 6), the

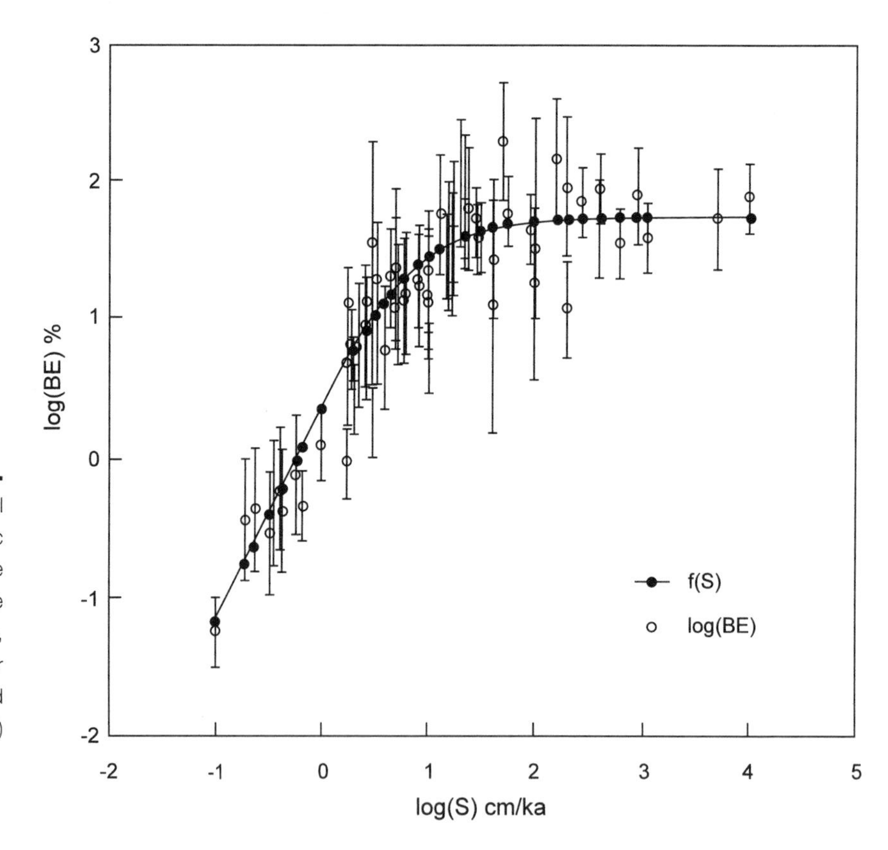

Figure 7.8. Plot of the burial efficiency, BE, of organic carbon with marine sediments vs. the sedimentation rate (S), in centimeters per 1,000 years. (Betts and Holland 1991)

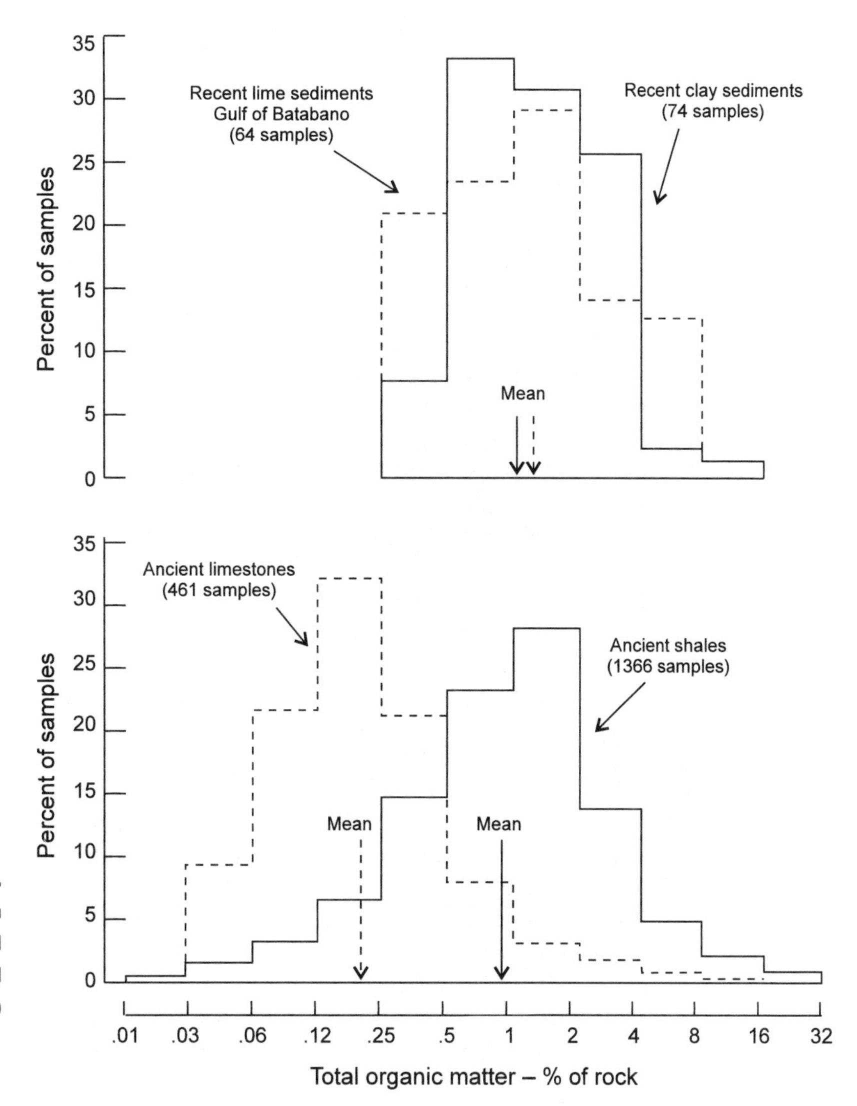

Figure 7.9.
The total organic carbon content of recent and ancient limestones and shales. (Gehman 1962)

average organic carbon content of marine sediments is about 0.7%. The observed range of the organic matter content in marine sediments is very large (see fig. 7.9). Highly oxidized, slowly accumulating, deep-water sediments in the central part of the Pacific Ocean contain virtually no organic matter. Organic-rich (> 5% organic matter) sediments tend to accumulate in near-shore settings, but carbonaceous sediments have been deposited on the ocean floor during some rather unusual episodes in Earth history.

Organic matter that is eaten in the water column and on the ocean floor can be oxidized to CO_2 and water, because small quantities of dissolved O_2 are present in seawater nearly everywhere in the oceans. However, below the sediment-water interface O_2 is rapidly exhausted by oxidation reactions. It is replenished, but very slowly, by the downward diffusion of O_2 from bottom waters. Sediments that are rich in organic matter are therefore anoxic. This discourages animal life, but a wide range of bacteria can use other sources of oxygen to decompose organic matter. Sulfate, SO_4^{2-}, the second most abundant anion in seawater, serves as a source of oxygen in bacterially mediated reactions such as

$$2CH_2O + SO_4^{2-} \xrightarrow{\text{bacteria}} 2HCO_3^- + H_2S. \tag{7.1}$$

Hydrogen sulfide generated in such reactions usually reacts with minerals containing iron in marine sediments. Ultimately, pyrite, FeS_2, is formed as a result of these reactions. Pyrite is a common minor constituent of carbonaceous marine sediments and can reach major proportions in sediments that are very rich in organic matter and in available iron.

The presence of H_2S in marine sediments serves to trap metals whose sulfides have a very low solubility. These metals include zinc, copper, lead, cobalt, nickel, silver, and molybdenum. Their concentration in seawater is very low, but slowly accumulating carbonaceous sediments can act as efficient sponges for these metals. They diffuse from bottom waters into the black, carbonaceous muds, and are precipitated there as a constituent of one or more sulfide minerals. Some elements accumulate in carbonaceous sediments simply due to their reducing nature. The most important of these elements is probably uranium. As mentioned earlier, uranium is oxidized during weathering and enters the oceans largely dissolved in river water as part of one or more "complexes" such as

$$\left[(U^{6+}O_2^{4-})(CO_3)_2^{4-}\right]^{2-}.$$

When such ions diffuse into carbonaceous sediments, they break up, the U^{6+} is reduced to U^{4+}, and UO_2, the highly insoluble oxide of U^{4+}, precipitates.

Metals in highly carbonaceous sediments can be so concentrated that the sediments can serve as the starting materials for ores. As they are gradually buried by younger sediments, they undergo a series of reactions collectively called *diagenesis*. Water is squeezed out, and, as the temperature rises, some of the original minerals recrystallize; others are replaced by new minerals, the sediments are cemented, and slowly the black muds become sedimentary rocks called *black shales*. Later uplift (see chapter 8) can bring these rocks back to the surface, where they are more easily accessible to mining. Most of these black shales are not sufficiently rich in metals to be competitive as ores at the present time. However, some that have been enriched by solutions that traveled through them after burial are currently being mined for copper, and black shales as a class will probably become minable ores well before the year 2100 (see chapters 9 and 10).

The destruction of organic matter in carbonaceous sediments is hardly ever complete. In many of these sediments, SO_4^{2-} in the trapped seawater becomes exhausted and, like O_2, is resupplied too slowly to continue serving as an oxidant of organic matter. The bacterial decomposition of organic matter after SO_4^{2-} exhaustion produces methane, CH_4, by reactions which, in simplified form, can be written as

$$2CH_2O \xrightarrow{\text{bacteria}} CH_4 + CO_2. \tag{7.2}$$

At greater depths and at temperatures between about 100°–200°C, a complex set of reactions breaks down some of the organic matter into petroleum. At temperatures in excess of 200°C, methane is produced without benefit of bacteria. Petroleum can remain where it was generated. When lifted back up to the surface, these rocks are known as *oil shales*. Since the extraction of petroleum from oil shales is relatively expensive, they are not particularly attractive commercially as long as oil can be obtained simply by drilling shallow holes and pumping the commodity in liquid form. Pilot plants to mine oil shales in Colorado and Utah have been abandoned, but oil shales will probably become attractive sources of petroleum in the next century, as the search for liquid petroleum and its recovery become more expensive. The accumulation of oil and natural gas in oil and gas fields involves their transport out of source rocks and their entrapment in reservoir rocks. These processes will be discussed in chapter 8.

Organic matter sometimes accumulates in nearly pure form in swamps, frequently in and close to gently sinking coastal areas, such as those shown in figures 7.10 and 7.11. Dissolved oxygen is consumed rapidly and efficiently by the oxidation of organic-rich sediments in such swamps. Since the concentration of SO_4^{2-} in rivers is very much lower than in seawater, H_2S generation and pyrite

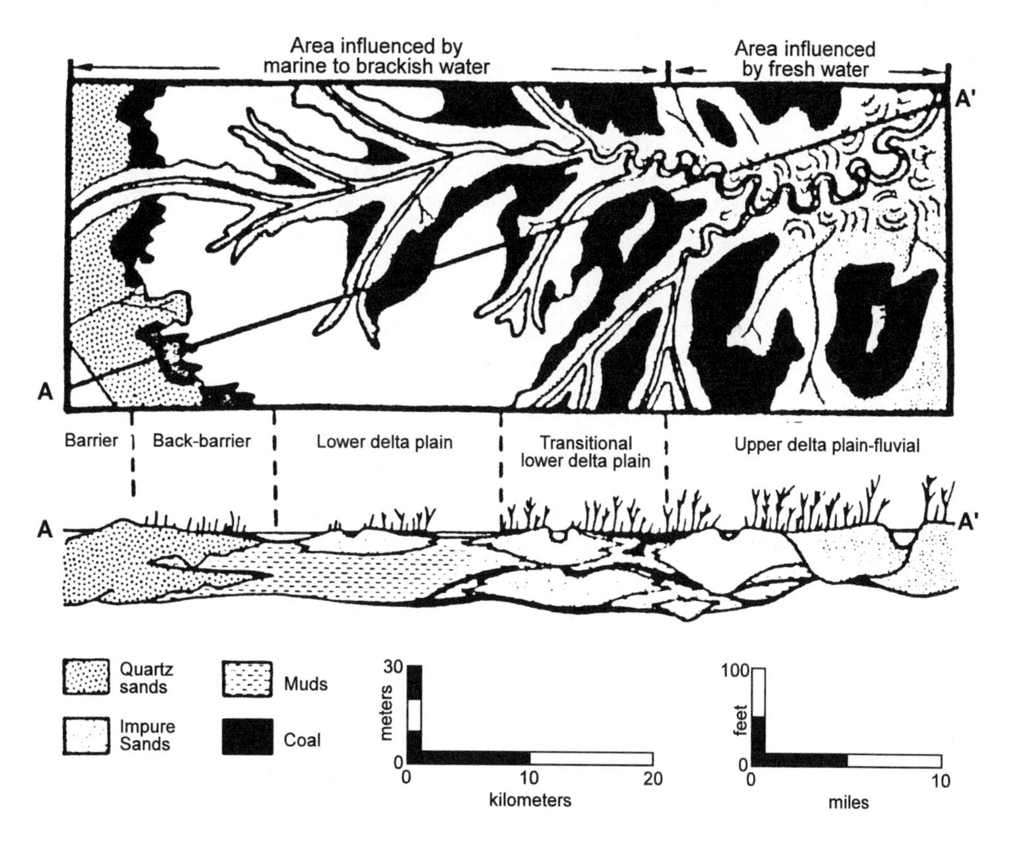

Figure 7.10. Depositional model for peat-forming environments in coastal regions. The upper part of the figure is a plan view showing sites of peat formation in modern environments; the lower part is a cross section showing the relative thickness and extent of peat beds and their relations to sands and muds along line AA′. (Modified from Ferm 1976)

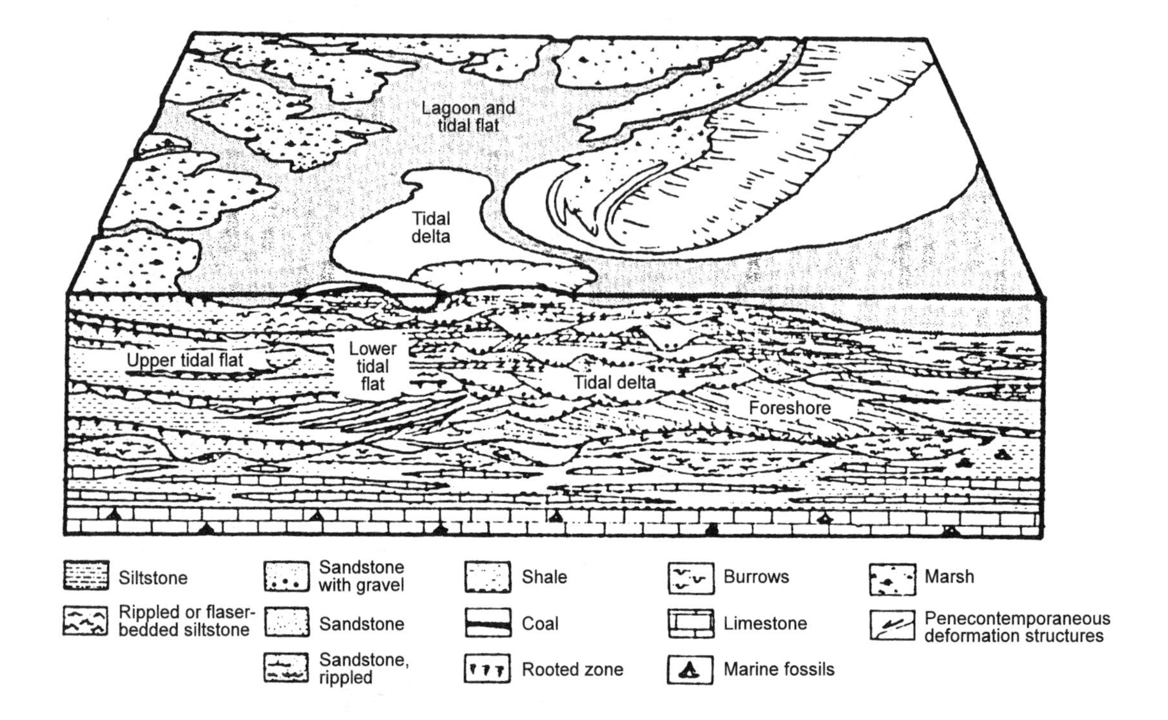

Figure 7.11. Barrier model of coal formation. Composite of exposures near Monteagle, Tennessee, showing shore, barrier, and back-barrier environments. (Ferm et al. 1972)

formation are minor in freshwater swamps, and bacterial degradation of organic matter proceeds largely via reactions such as 7.2. Methane is often produced in large quantities, hence one of its names: marsh gas. The destruction of organic matter via reaction 7.2 is usually quite slow. Where swamp residues are protected from erosion and from major admixtures with river sediments, thick layers of peat can accumulate. If these are preserved by a layer of younger sediments, the peat is gradually converted into coal. Figure 7.12 shows the major chemical changes that accompany the conversion of plant material into peat, thence to lignite, to subbituminous, bituminous, and semibituminous coal, and finally to anthracite and graphite. During this process organic matter is essentially dewatered and can ultimately be turned into the mineral graphite:

$$CH_2O \rightarrow \underset{\text{graphite}}{C^o} + H_2O. \tag{7.3}$$

High temperatures are required for organic matter to be turned into graphite. En route the carbon content of coal increases, and its oxygen and hydrogen content both decrease. A useful way to represent these changes is on a van Krevelen diagram, in which the H/C ratio of coals is plotted against their O/C ratio. In organic matter of composition CH_2O the H/C atomic ratio is 2.0, the O/C atomic ratio is 1.0. The composition of wood in figure 7.13 plots close to these ratios. During conversion to graphite, both ratios decrease, though by a rather indirect

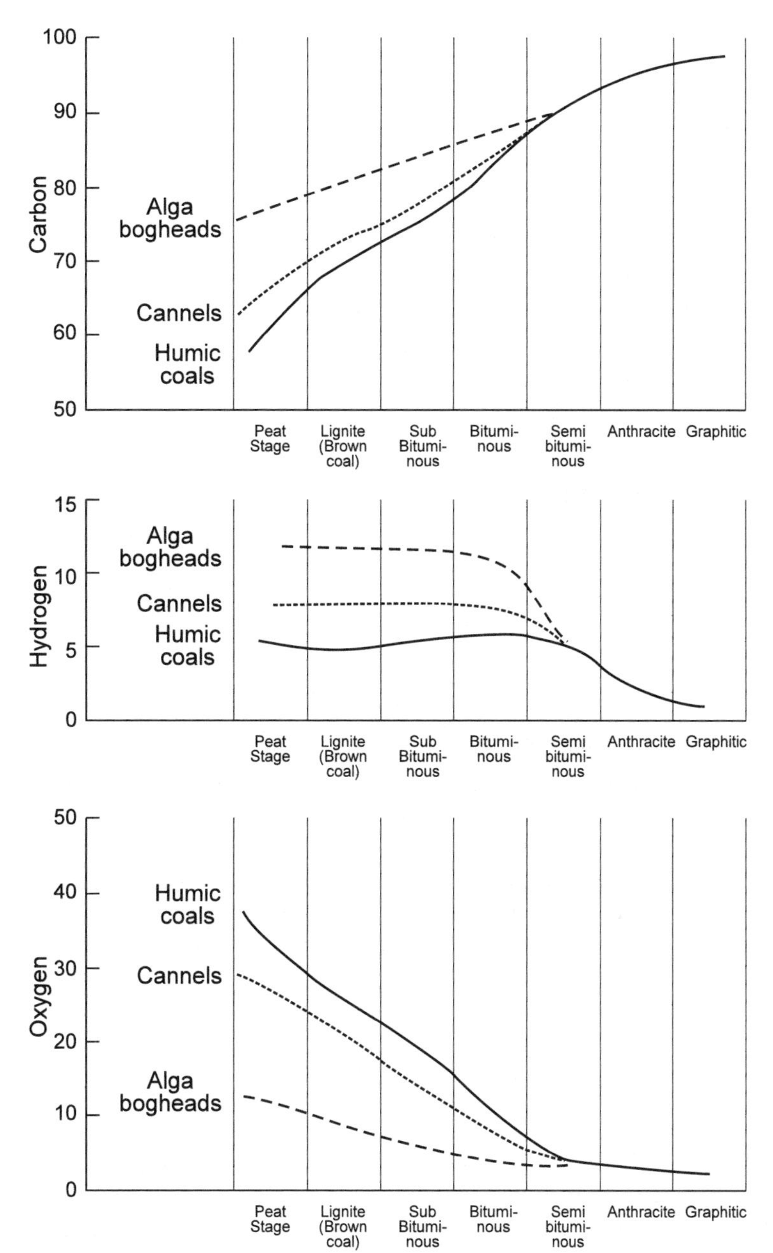

Figure 7.12. Changes in carbon, hydrogen, and oxygen content during the evolution of normal (humic) coals and algal (sapropelic) coals. (White 1925)

path that is determined by the complexities of the conversion process. Unfortunately, petroleum is not produced during the conversion of terrestrial organic matter to coal, although terrestrial organic matter disseminated in marine sediments forms a type of kerogen that can be a significant source of natural gas and a modest amount of waxy crude oil.

Swamps that develop in brackish water, i.e., in water that is a mixture of river water and seawater, are in contact with significant quantities of SO_4^{2-}. Sulfate reduction and pyrite precipitation are common in these swamps. Coal from such

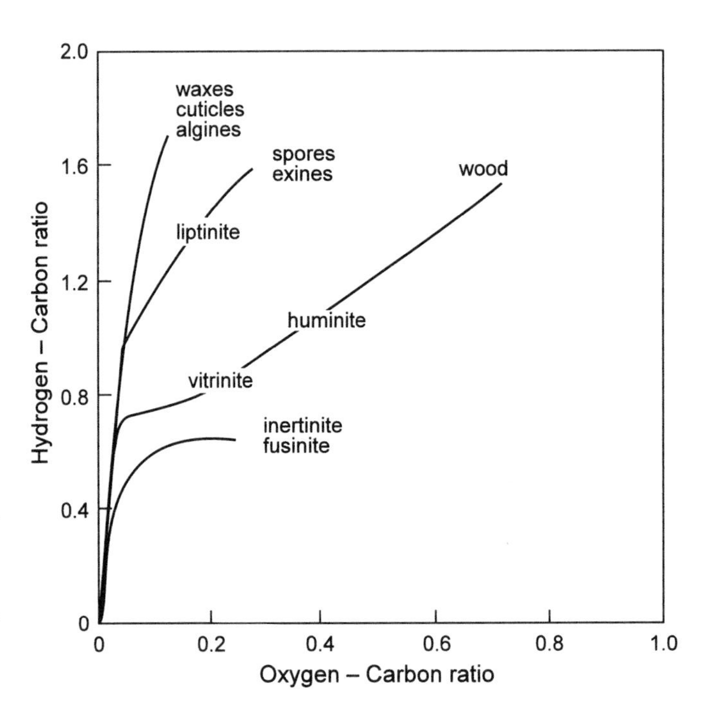

Figure 7.13. Van Krevelen diagram (H/C versus O/C atomic ratios) for the main components of coal and their predecessors with lines of dehydration, decarboxylation, demethanation, dehydrogenation, oxidation, and hydrogenation. (Modified after van Krevelen 1961; Tissot and Welte 1984; from Damberger 1991)

areas tends to be "high sulfur" coal. Unless great care is taken to remove pyrite before burning these coals, large amounts of sulfur dioxide, SO_2, are generated during combustion. Insufficient removal of this SO_2 from stack gases can lead to significant air pollution and to very acid rain downwind from their point of injection (see chapter 11). Mining "high sulfur" coal and ores that contain a great deal of pyrite and other sulfide minerals can also lead to the formation of acid mine drainage and the contamination of ground and surface waters in the vicinity of such mines (see chapter 11).

7.4 Dissolved Matter Added to the Oceans

Rivers are continuously delivering dissolved material to the oceans, and they have done so for nearly all of Earth's history. It is fair to ask if all of these dissolved salts have simply accumulated in the oceans, or if they have been removed from the oceans about as fast as they have been added. A simple calculation shows that the oceans have only served as a transitory home for dissolved salts. Table 7.2 contains much of the necessary data for the major constituents of seawater. The third column of figures lists the composition of average river water today. The numbers in this column are somewhat uncertain, because the composition of water in the world's rivers is quite variable (see, for instance, Holland 1978, and Berner and Berner 1987). Fortunately, the uncertainties are sufficiently small so that they do not affect the conclusion in any serious way. The average concentration of calcium in the world's river water is about 15 mg/kg river water. The annual flux of river water is about 4.6×10^{19} cm^3/yr, i.e., 4.6×10^{16} kg/yr (see chapter 4). The input of Ca^{2+} to the oceans is therefore approximately

$$\frac{15 \times 10^{-3} \text{ gm Ca}^{2+}}{\text{kg river water}} \times 4.6 \times 10^{16} \frac{\text{kg river water}}{\text{yr}} = 6.9 \times \frac{\text{gm Ca}^{2+}}{\text{yr}} .$$

The average concentration of Ca^{2+} in the oceans is 412 mg Ca^{2+}/kg seawater. The total quantity of Ca^{2+} in the oceans is therefore

$$\frac{412 \times 10^{-3} \text{ gm Ca}^{2+}}{\text{kg seawater}} \times 1.4 \times 10^{21} \text{ kg seawater} = 5.7 \times 10^{20} \text{ gm Ca}^{2+}.$$

At the present rate of Ca^{2+} input, rivers add an amount of Ca^{2+} equal to the entire Ca^{2+} inventory of the oceans in about

$$\frac{5.7 \times 10^{20} \text{ gm Ca}^{2+}}{6.9 \times 10^{14} \text{ gm Ca}^{2+}/\text{yr}} = 800{,}000 \text{ years.}$$

This is an interesting number. If Ca^{2+} had simply been accumulating in the oceans, only 800,000 years would have been required to add enough Ca^{2+} to account for the present inventory of this element. Eight hundred thousand years is a very small fraction of the 4,500 million years of Earth history. There is nothing to indicate that the river flux of Ca^{2+} has been so grossly abnormal during the last 800,000 years. The oceans cannot, therefore, be a simple accumulator of riverborne Ca^{2+}. Our calculation suggests very strongly that the Ca^{2+} that is brought to the oceans by rivers also leaves the oceans. Fortunately we do not have to look very far to find its exits.

The shells of marine organisms consist of $CaCO_3$, a compound that occurs with several crystal structures. The most common are those of the minerals calcite and aragonite. In both minerals, carbon is at the center of a triangle of O^{2-} ions. These CO_3^{2-} groups are linked by Ca^{2+} ions. In calcite, each Ca^{2+} ion is surrounded by six O^{2-} ions; in aragonite each Ca^{2+} ion is surrounded by nine

Table 7.2. The Concentration and Mean Residence Time of the Major Constituents of Ocean Water

Constituent	Average Concentration in Ocean Water of Salinity 35‰ (mg/kg)[b]	(mmol/kg)	Concentration in Average River Water (mg/kg)	Residence Time in Oceans (in years)
Sodium	10,760	468.0	6.9	4.8×10^7
Magnesium	1,294	53.2	3.9	1.0×10^7
Calcium	412	10.2	15.0	8.5×10^5
Potassium	399	10.2	2.1	5.9×10^6
Strontium	7.9	0.09		4.0×10^6
Chloride	19,350	545.0	8.1	7.3×10^7
Sulfate	2,712	28.2	10.6	7.9×10^6
Bicarbonate	145	2.38	55.9	8.0×10^4
Bromide	67	0.84		1.0×10^8
Boron	4.6	0.39		1.0×10^7
Fluoride	1.3	0.068		5.0×10^5

Source: Holland 1978.

somewhat irregularly arranged O^{2-} ions. Ca^{2+} is rather large for the six-coordinated holes in the CO_3^{2-} framework and props the O^{2-} ions apart. Mg^{2+} distends such holes much less. Sea urchins and some other organisms build their shells of calcite in which large amounts of Mg^{2+} occur in place of Ca^{2+} ions. These are usually called *high-magnesium calcites*. The ninefold coordination of Ca^{2+} in aragonite makes this mineral unattractive as a site for small ions like Mg^{2+} but a congenial host for large ions like Sr^{2+} (see table 6.2). This and the relatively high Sr^{2+} content of seawater explain why coral skeletons and most other marine aragonites contain significant amounts of Sr^{2+}.

Many shallow parts of the tropical oceans boast superb coral reefs. These are spectacular examples of $CaCO_3$ accumulations. Quantitatively more important accumulations of $CaCO_3$ are on the Bahama Platform, the Sunda Shelf, and north of Australia in the Gulf of Carpentaria. The carbonates in these areas are largely fine-grained carbonate muds. In the main, they are biogenic in origin, but some of their aragonite may be directly precipitated from seawater.

Far from shore in the open ocean, small organisms, the foraminifera and the coccolithophoridae (see fig. 5.6) also build their shells of $CaCO_3$. After they die, these shells settle to the ocean floor, either individually or packaged in fecal pellets or as a loose aggregate of material called *marine snow*. In some areas they accumulate as essentially pure carbonate oozes. They do not survive in the deepest parts of the oceans. At great depths, they dissolve and return their constituents to seawater. The depth in the oceans, about 4 kilometers, below which little or no $CaCO_3$ accumulates, is known as the *calcite compensation depth*.

If we total the rate of Ca^{2+} loss from the oceans as a constituent of $CaCO_3$, we find that it is roughly the same as the river input of Ca^{2+} to the oceans. The balance sheet is not perfect, but this is not surprising. We know neither the river input of Ca^{2+} nor the $CaCO_3$ precipitation rate precisely, and a close look at the Ca^{2+} metabolism of the oceans reveals several Ca^{2+} inputs other than dissolved river Ca^{2+} and several outputs other than $CaCO_3$. But these all appear to be relatively minor, and it is reasonably certain that the Ca^{2+} budget of the oceans is roughly in balance today. This casts the figure of 800,000 years to accumulate the observed Ca^{2+} in a new light. In a chemical system that is at steady state, i.e., where the concentrations of its components do not vary with time, the ratio of the total quantity of any component to its rate of input or output is called the *residence time* of that component. This means that the time that elapses between the arrival of an average Ca^{2+} ion in the oceans and its more or less permanent removal with marine sediments is approximately 800,000 years. The residence times of all the elements listed in table 7.2 are short compared to the age of the Earth. The residence times of some trace elements not listed in this table are as short as a few hundred years. The concentrations of many of these elements in seawater are not the same throughout the world ocean, because some parts of the oceans receive a larger quantity of these elements than others. The isotopic composition of some of these elements also varies from ocean to ocean, because the isotopic composition of their inputs varies geographically, and because these elements are removed from the oceans before they are completely mixed. Most, but not all, of the elements that have residence times greater than a few thousand years are homogeneously distributed throughout the oceans. This observation agrees with independent lines of evidence that indicate that the oceans mix on a timescale of 1,000–2,000 years.

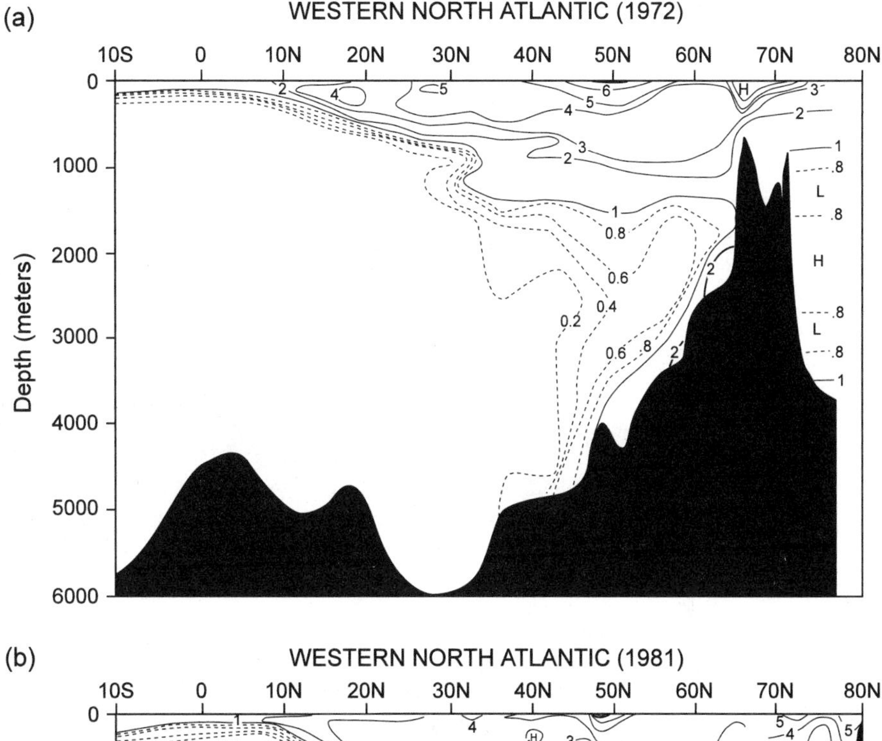

Figure 7.14. Tritium distribution in the western North Atlantic, 1972 and 1981. Note the penetration of tritium into intermediate and bottom waters. (A tritium unit is one atom of tritium in 10^{18} atoms of hydrogen). Note large vertical exaggeration. (Toggweiler 1994)

How do the oceans mix today? Surface seawater sinks where it is most dense. The density of seawater increases with salinity and with decreasing temperature. Today, surface seawater is particularly dense in the North Atlantic. Its sinking there has been well documented by the progressive downward movement of the tritium, ^3H, that was added to the atmosphere in considerable quantities during the testing of nuclear weapons in the 1950s and early 1960s before the international ban on such tests (see chapter 12). The tritium combined with O_2 in the atmosphere; the resulting water rained out into the oceans (see fig. 7.14), and can

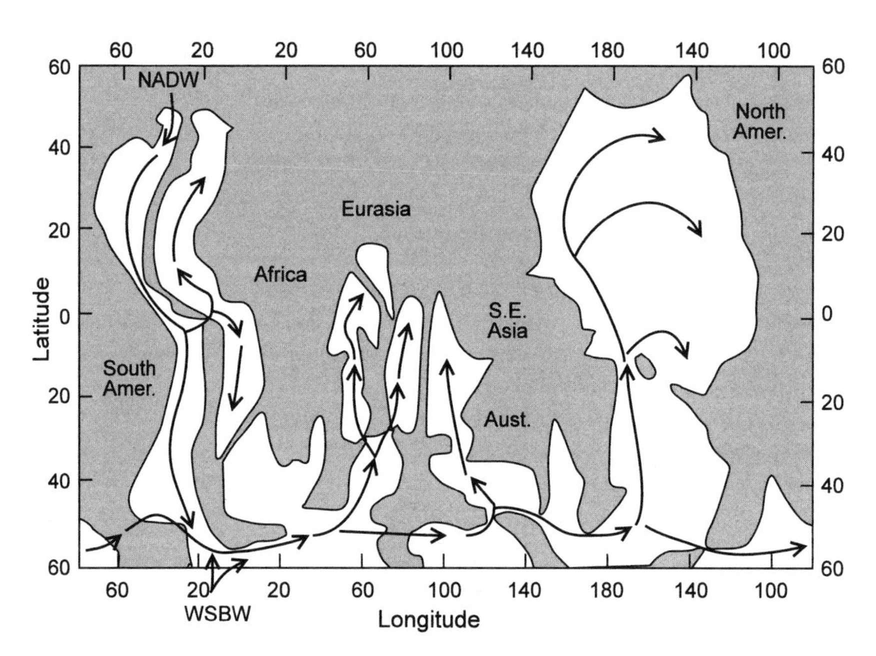

Figure 7.15. Flow pattern of ocean water at a depth of 4,000 m. The major inputs to this horizon are North Atlantic Deep Water (NADW), which enters at the northern end of the western basin of the Atlantic, and Weddell Sea Bottom Water (WSBW), which enters from the margin of the Antarctic continent adjacent to the South Atlantic. (Broecker and Peng 1982)

now be detected with very sensitive instruments down to depths of about 4 kilometers in the North Atlantic, together with freon, a chlorofluorocarbon, which also provides a fine tracer for sinking water masses.

Seawater that sinks in the North Atlantic is called North Atlantic Deep Water (NADW). This water mass gradually moves southward to the Antarctic. There it exchanges with surface seawater, which has virtually the same density. As shown in figure 7.15, it moves from the Antarctic Ocean into the Indian and Pacific Oceans. It gradually upwells there, and is carried back to the Atlantic by surface currents. The whole system acts somewhat like a conveyor belt. The movement of NADW can be monitored by a progressive decrease in its content of ^{14}C, an isotope of carbon that is produced by the interaction of cosmic rays with the Earth's atmosphere (see chapter 2). ^{14}C is oxidized to $^{14}CO_2$ in the atmosphere and is then mixed into the biosphere and the oceans. The interpretation of the ^{14}C content of seawater is complicated by the addition of this isotope during nuclear weapons testing and by the cycling of organic matter and $CaCO_3$ within the oceans; but it is unlikely that the currently accepted oceanic mixing time of 1,000–2,000 years will turn out to be seriously in error.

Given that the mixing time of the oceans is so much shorter than the residence time of Ca^{2+}, one wonders what controls the concentration of Ca^{2+} in seawater. If the oceans were a simple nonbiologic system, the answer would be fairly straightforward: the Ca^{2+} concentration would be such that inorganic $CaCO_3$ precipitation would remove Ca^{2+} at the same rate as the Ca^{2+} input to the oceans by rivers. In the real oceans the precipitation of $CaCO_3$ and the Ca^{2+} control mechanism are both strongly linked to marine biology. The precipitation of $CaCO_3$ is largely due to the shell-building activity of marine organisms, which do not know when they have exceeded their annual quota of river Ca^{2+}. An ingenious control mechanism compensates for this ignorance. A small fraction of the organic matter produced photosynthetically in the well-lit upper layer of the oceans escapes being eaten there and sinks into the lower parts of the oceans. Most of this organic matter is metabolized there via the reaction

$$CH_2O + O_2 \rightarrow CO_2 + H_2O. \quad (7.4)$$

The increase in the CO_2 content of deep ocean water leads to the dissolution of $CaCO_3$ in marine sediments below the calcite compensation depth via the reaction

$$CaCO_3 + CO_2 + H_2O \rightarrow Ca^{2+} + 2HCO_3^-. \quad (7.5)$$

This together with the lower temperature and the higher pressure accounts for the lack of $CaCO_3$ in sediments on the floor of the deepest parts of the oceans. The dissolution of $CaCO_3$ in these sediments compensates for the exuberant excesses of biologic $CaCO_3$ precipitation in the shallow parts of the oceans. As expected, the calcite compensation depth has varied during geologic time.

7.5 Phosphorites

The productivity of the marine biosphere is sustained and limited by the availability of nutrients. Phosphate and nitrate are the most important of these. Potassium, the third of the major agricultural fertilizers, is always present in abundance, because it is one of the major cations in seawater. Trace metals such as iron may be limiting photosynthesis in some parts of the oceans, but in most areas photosynthesis continues until PO_4^{3-} and NO_3^- are nearly exhausted. Phosphate enters the oceans largely via rivers. It is present there both in solution and as a constituent of river sediments. Dissolved phosphate is readily available for marine photosynthesis. Much of the phosphate that is locked in mineral grains is not available for photosynthesis, but phosphate that is loosely attached to mineral surfaces can be desorbed and used biologically. Similarly, phosphate can leave seawater by attaching itself to the surfaces of mineral grains.

The marine chemistry of nitrate is rather different from that of phosphate. NO_3^- enters the oceans via rivers in dissolved and in particulate form; but the atmosphere also supplies NO_3^- to the oceans. Oxides of nitrogen are synthesized in the atmosphere in lightning discharges and by a variety of photochemical reactions. These nitrogen oxides are ultimately rained out, largely as nitric acid, HNO_3. NO_3^- is also both created and destroyed in the oceans. Cyanobacteria, much like some land plants, can convert N_2 dissolved in seawater into biologically useful compounds. On the other hand, NO_3^- is destroyed by bacteria that use the oxygen in this ion as an oxidant of organic matter in parts of the oceans where the concentration of O_2, the preferred oxidant, is very low. Despite these complexities, the atomic ratio of carbon to nitrogen and phosphorus in many marine organisms is surprisingly constant, close to the Redfield ratio, C:N:P = 106:16:1.

Phosphate incorporated in phytophankton is passed on to organisms that occupy the next steps of the food chain. An average PO_4^{3-} ion cycles through the marine biosphere hundreds of times before it is removed more or less permanently into marine sediments. There are several important resting places for phosphate in marine sediments. In most of these, phosphate is rather dispersed and of no commercial value. In some sediments, however, phosphate is greatly concentrated. These are the sources of phosphate for agricultural fertilizers.

Table 7.3. Composition of Phosphorites from Various Submarine Localities (Cronan 1980)

	1	*2*	*3*	*4*	*5*	*6*	*7*	*8*	*9*	*10*	*11*	*12*	*13*	*14*	*15*
SiO_2	6.20	3.45	15.30	13.79		1.8[a]	2.0[a]	3.1[a]	1.6[a]	4.6[a]	22.13	8.93	0.95	0.20	
TiO_2	0.06	0.04	0.10	0.11		0.019	0.016	0.074	0.053	0.087					
Al_2O_3	1.13	0.92	2.14	2.12		0.37	0.36	1.95	1.12	2.00	5.15	1.38		0.51	
Fe_2O_3	1.40	25.80	5.58	7.34		0.54	0.77	8.6	10.7	8.7	2.85	1.00	1.83	2.80	1.14
MnO	0.01	0.16	0.01	0.02		0.01	0.09	0.017	0.025	0.016[a]					
MgO	0.97	1.49	1.34	1.45							1.07	0.71	0.50	1.02	1.83
CaO	47.02	33.42	37.04	36.54	47.4	26.2[b]	8.0[b]	9.0[b]	7.9[b]	6.7[b]	33.93	44.48	52.29	51.33	50.73
Na_2O	0.62	0.34	0.78	0.73		1.99	2.79	1.96	1.82	1.61	0.85	0.83	0.55	0.58	0.46
K_2O	0.43	0.39	1.57	1.51		0.15	0.13	2.10	0.41	1.75	1.30	0.60	0.64	0.45	0.15
P_2O_5	14.82	10.26	17.89	16.82	29.6	18.3	26.2	16.5	18.3	16.2	22.61	30.63	19.97	25.80	14.62
S	0.31	0.21	0.46	0.45		0.44[c]	0.56[c]	<0.1[c]	<0.1[c]	<0.1[c]	0.16	0.35	0.35	0.61	0.30
F	2.12	1.42	2.21	2.04	3.3						2.22	3.45	2.47	3.25	1.38
LOF	25.52	23.24	16.15	16.89							8.78[d]	9.34[d]	20.8[d]	15.2[d]	39.10[d]
Organic C						0.61	0.76	0.32	0.21	0.34					

1–4 = Agulhas Bank phosphorites, South Africa (Parker 1971, Parker and Siesser 1972, Dingle 1974).
5 = Forty Mile Bank phosphorite, Baja California (Mero 1965).
6–10 = Northwest African continental margin phosphorites (McArthur 1974).
11 = Peru-Chile continental margin, average (Burnett 1974).
12 = sea off California, average of two samples (Burnett 1974).
13 = Chatham Rise off New Zealand, average of two samples (Burnett 1974).
14 = Blake Plateau phosphorite (Burnett 1974).
15 = Necker Bank phosphorite (Burnett 1974).
[a] Quartz only.
[b] Calcite only.
[c] Pyrite sulphur only
[d] Loss on fusion (loss on ignition at 1,000° C in 11–15).

Organic matter in marine sediments retains some of the phosphate of the marine organisms from which it was derived. When sediments that are rich in organic matter are exposed to oxidation, their carbon is converted to CO_2, and some of their phosphate is precipitated as a constituent of the mineral apatite. Pure hydroxy-apatite has the composition $Ca_5(PO_4)_3(OH)$. Marine apatites usually contain some fluoride, F^-, as well as significant quantities of CO_3^{2-}. Sediments that are rich in apatite are known as *phosphorites*. Table 7.3 lists the chemical composition of phosphorites from various parts of the oceans. As expected, they are rich in CaO and P_2O_5, but they usually contain significant quantities of calcite and clay minerals.

Figure 7.16 shows the location of phosphorites on the seafloor. These are of two types: phosphorites on continental margins and phosphorites on seamounts. The economically important phosphorites are on the edges of continents, usually where upwelling of seawater is intense. Upwelling seawater is enriched in phosphate due to the release of PO_4^{3-} during the oxidation of organic matter in the deeper parts of the oceans (see fig. 7.17). Most regions of upwelling seawater are therefore well fertilized. Phosphate is used and recycled intensely there. During the production and destruction of organic matter in such areas, apatite can accumulate as a major component of sediments, provided the apatite is not exces-

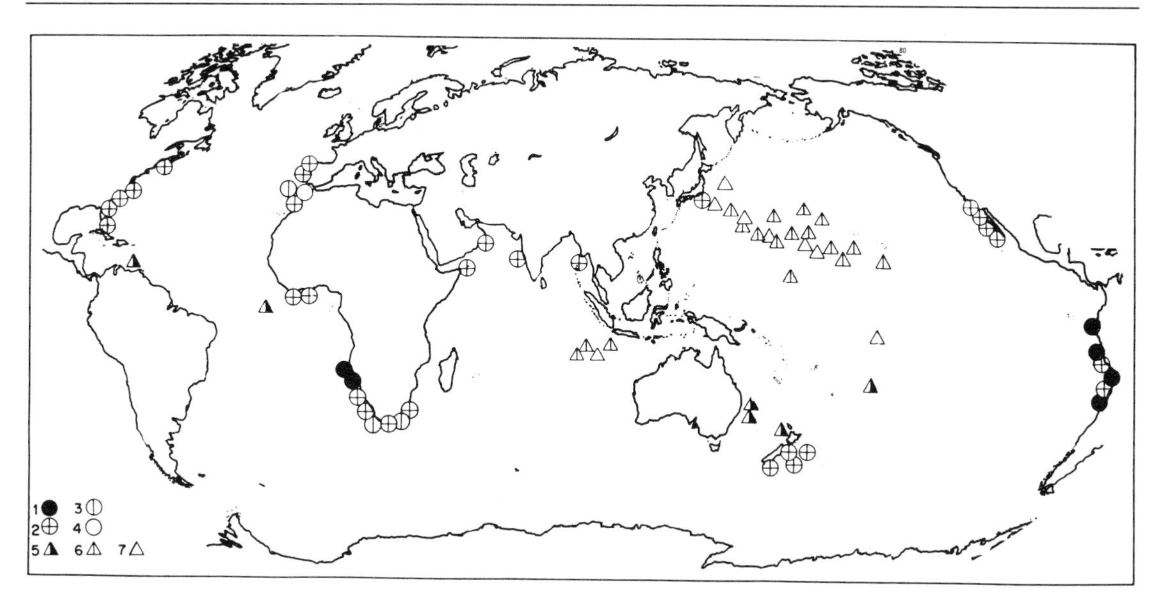

Figure 7.16. Location of phosphorites on the seafloor. 1–4, phosphorites on continental margins; 5–7, phosphorites on submerged mountains. Geological age: 1, Holocene; 2, 5, Neogene; 3, 6, Paleogene; 4, 7, Cretaceous. (From Bezrukov and Baturin 1979)

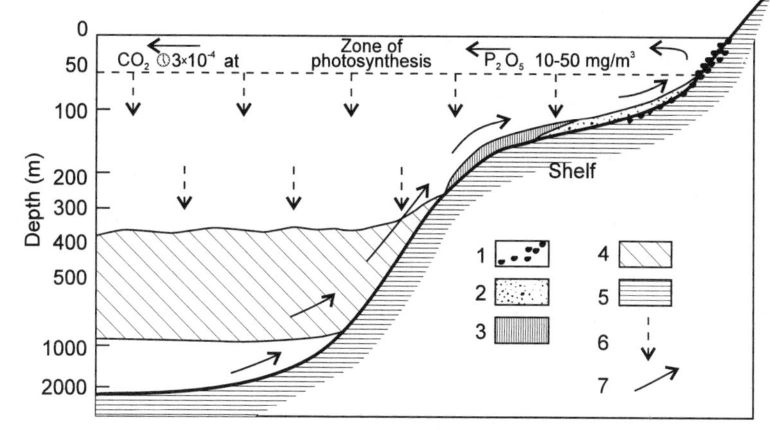

Figure 7.17. Diagram showing the formation of phosphorites (after Kazakov in Strakhov 1962). 1, facies of littoral gravel and sand; 2, phosphate facies; 3, facies of calcareous sediments; 4, zone of maximum CO_2 and organic P_2O_5 content (partial pressure of CO_2 up to 12×10^{-4} atm, P_2O_5 concentration 300 to 600 mg/m^3); 5, landmass; 6, sedimentation of plankton remains; 7, current directions.

sively diluted by terrigenous sediments. High-grade phosphate deposits tend to form along desert coasts such as in Morocco, northern Peru, and Baja California, Mexico, or where land erosion is or was minimal, as during the formation of the Florida phosphate deposits.

Figure 7.18 shows the distribution of the world's major phosphate deposits. The largest reserves are in Morocco and the western Sahara, but the United States is the largest producer (table 7.4). None of the latter are in very recent sediments. The largest phosphate mines in the United States are based on the Tertiary phosphorites of Florida. Very large phosphate reserves are also available in the

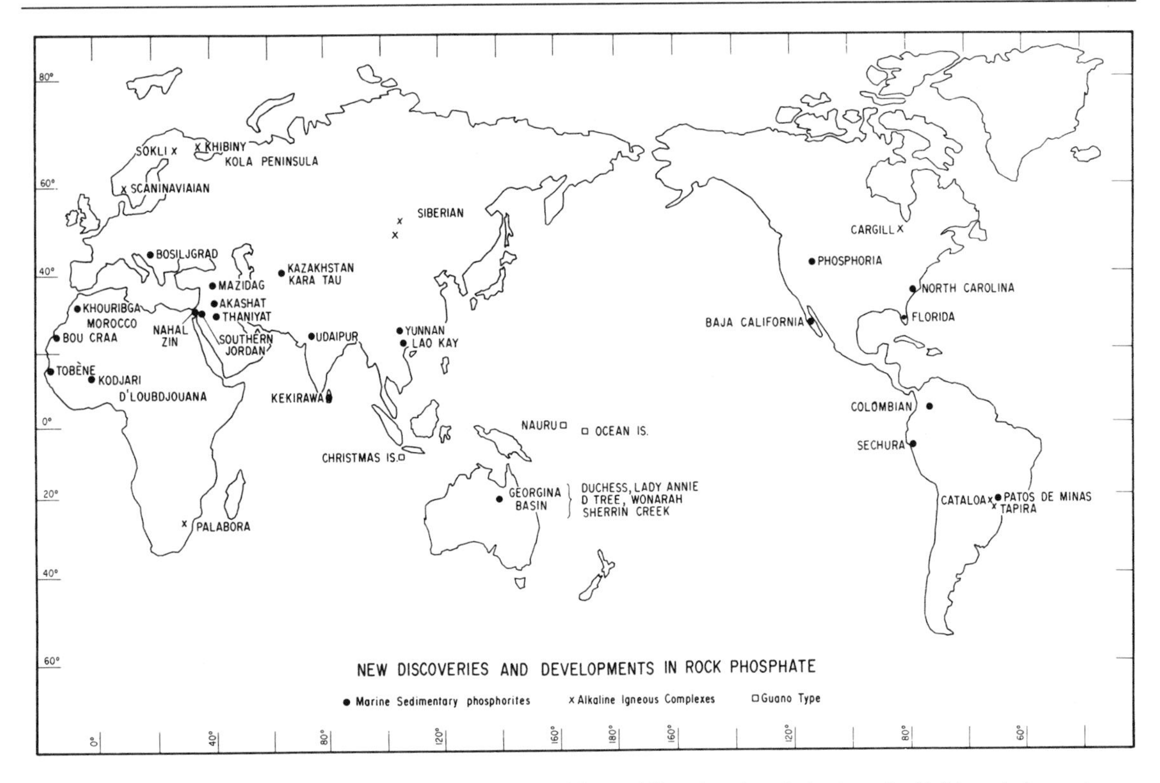

Figure 7.18. The distribution of the world's major phosphate deposits. Solid symbols are deposits from which phosphate rock is currently being produced at the rate of one million tons or more per annum. The open symbols are important deposits producing less than one million tons per annum, or deposits with potential for producing one million tons or more per annum but which are not currently realizing that potential for political, economic, or logistic reasons. (Cook 1982)

Table 7.4. Production and Reserves of Phosphate Rock, 1992 (in millions of metric tons) (*Mineral Commodity Summaries*, 1993)

Country	*Production*	*Reserves*	*Reserve Base*
United States	47	1,230	4,440
USSR (former)	24	1,330	1,330
China	20	210	210
Morocco and western Sahara	19	5,900	21,440
Subtotal	100	8,670	27,420
Other countries	31	3,330	6,580
World total	141	12,000	34,000

American West, in the Phosphoria formation, which was deposited about 250 m.y.a. The oldest known are the very large phosphorites in the Cambrian (ca. 530 m.y.) sediments of the Georgina Basin in Australia. Rather curiously, no large Precambrian (older than 545 m.y.) phosphorite deposits have been discovered. Their absence in the geological record may be due to later loss by erosion, but their rather sudden appearance could also be due to the changes in marine biology that accompanied the explosive evolution of animal life at the end of the Precambrian Eon.

The data in table 7.4 show that the identified reserves of phosphate rock are sufficient to supply the current demand for about 85 years. If one includes lower-grade resources, phosphate rock is available at present rates of mining for more than 200 years. The world is therefore assured of inexpensive sources of phosphorite at least until the end of the next century.

7.6 Evaporites

The major-element composition of seawater is reasonably constant throughout the oceans. In near-shore areas differences are frequently occasioned by freshwater inputs and evaporation. In the open oceans an excess of evaporation over precipitation or vice versa commonly produces small differences in the salinity of seawater. There are, however, a few areas where the composition of seawater differs quite considerably from that of normal seawater. In areas like the Baltic Sea, freshwater input is large relative to seawater supply and evaporation. Hence, the salinity is lower than that of normal seawater. In regions like the Bocana de Virrilá in Peru, the Persian Gulf, and the Red Sea evaporation is intense, freshwater inputs are minor and seawater access is limited. The combination of these three factors can lead to strong evaporative concentration of seawater, a large increase in salinity, and the precipitation of evaporite minerals.

Evaporation has been used for millennia to extract salt from seawater. Figure 7.19 is a map of Great Inagua Island, the southernmost island of the Bahamas. Salt has been produced there for well over a century. Seawater enters a large evaporation basin where the salinity rises from its normal value of thirty-five parts per thousand (‰), i.e., 3.5%, to nearly ten times that value. During evaporation aragonite is the first mineral to precipitate via the reaction

$$Ca^{2+} + 2HCO_3^- \rightarrow \underset{\text{aragonite}}{CaCO_3} + CO_2 + H_2O. \tag{7.6}$$

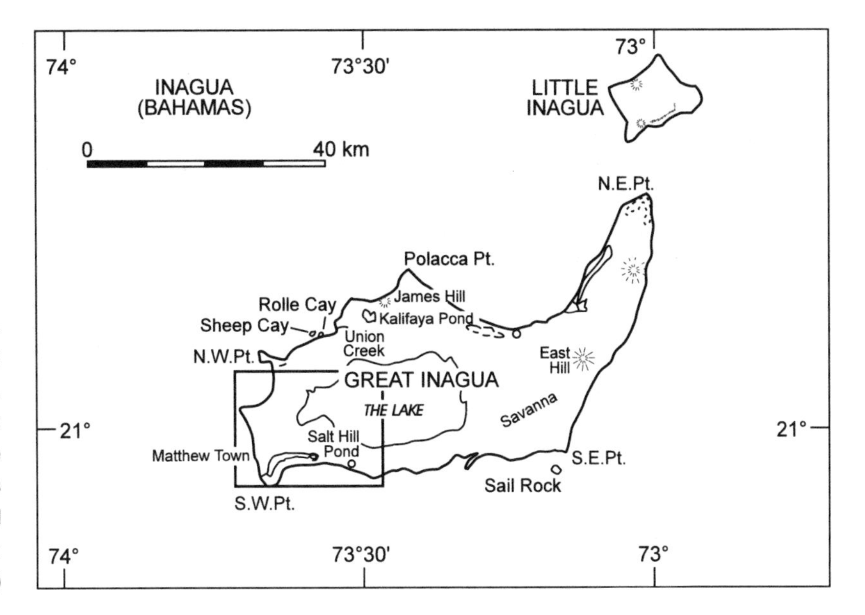

Figure 7.19. Great Inagua and Little Inagua Islands, Bahamas. The Morton Company brine reservoirs and salt pans are located in the boxed area. (McCaffrey, Lazar, and Holland 1987)

As this reaction proceeds, two moles of HCO_3^- are used for each mole of Ca^{2+}. A glance at table 7.2 shows that normal seawater contains 10.2 mmol Ca^{2+}/kg and only 2.4 mmol HCO_3^-/kg. The precipitation of aragonite therefore ends when most of the HCO_3^- has been used. At that point, only about 10% of the initial Ca^{2+} has been precipitated. Further evaporation leads to the precipitation of the mineral gypsum via the reaction

$$Ca^{2+} + SO_4^{2-} + 2H_2O \rightarrow \underset{\text{gypsum}}{CaSO_4 \cdot 2H_2O}. \tag{7.7}$$

Since seawater contains almost three times as much SO_4^{2-} as Ca^{2+} (see table 7.2), the precipitation of gypsum is limited by the Ca^{2+} content of evaporated seawater. The evolution of the composition of the brines during evaporation is shown in figure 7.20. The concentration of each of the ions is plotted against the degree of evaporation.

When seawater has evaporated to about 10% of its initial volume, the salinity has increased to about 350‰, and the brines have become saturated with respect to "table salt," the mineral halite, NaCl. The brines are then pumped into smaller and shallower ponds where NaCl precipitates via the simple reaction,

$$Na^+ + Cl^- \rightarrow \underset{\text{halite}}{NaCl}. \tag{7.8}$$

In present-day seawater the concentration of Cl^- is slightly greater than that of Na^+ (see table 7.2). Since the precipitation of NaCl requires the removal of an equal number of moles of Na^+ and Cl^-, the ratio of Na^+ to Cl^- in the residual brines drops gradually, and halite precipitation essentially stops when most of the Na^+ has been removed from the brines. At that point most of the K^+ and Mg^{2+}, about two-thirds of the initial SO_4^{2-}, and a significant amount of Cl^- remain in the brines.

Further evaporation requires an extremely low humidity in the overlying atmosphere. The air in the Bahamas is too damp for this. Residual brines from the halite ponds are therefore returned to the ocean, the salt is scraped off the floor of the halite ponds and is then readied for shipment. It does rain on Great Inagua Island. The amount of salt that can be harvested during a given year depends on the balance between rainfall and evaporation. When a hurricane passes over the island, rainfall is intense and the salt crop is usually very meager indeed.

In the past there have been vast, natural evaporite basins. In some instances, the salts that were deposited in these basins were preserved by burial under later sediments. Many of these salt deposits have since been dissolved out, but some have not. In fact, the total quantity of halite preserved in the world's sediments is roughly equal to the quantity that is now dissolved in the world's oceans. A number of evaporite deposits (see fig. 7.21) are being mined for halite. These are the chief competitors for saltworks such as those on Great Inagua Island. Table 7.5 lists the countries that are the major producers of salt. The reserves of underground salt are very large, and the supply of Na^+ and Cl^- in the oceans is essentially inexhaustible. We tend to think of salt as one of the humbler natural resources, but its roots go far back into human history. Roman soldiers were given "salarium," money for the purchase of salt, and hence their pay. The town

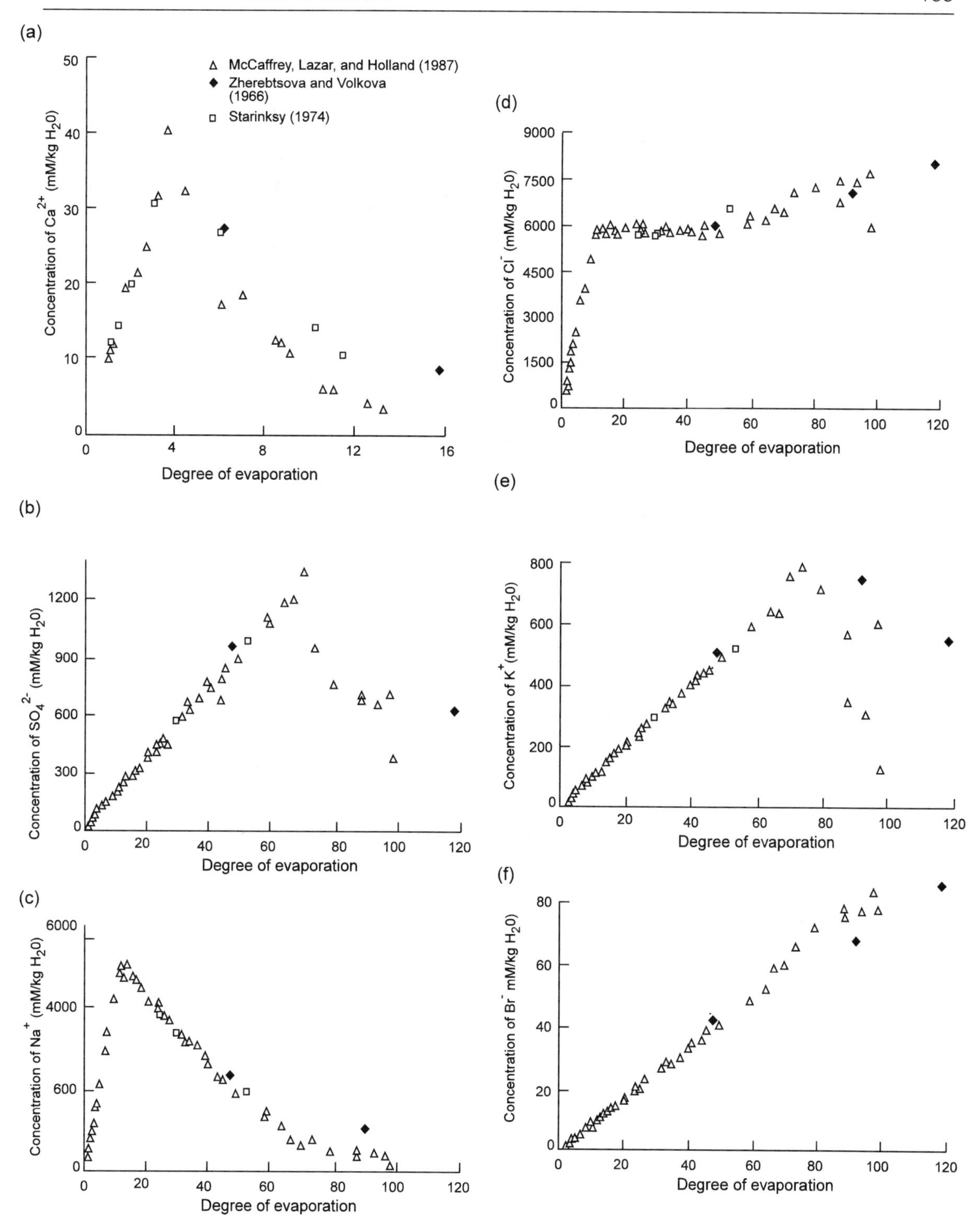

Figure 7.20. The concentration of the various ions plotted as a function of the degree of evaporation of seawater and Black Sea water. The degree of evaporation is defined as the initial volume of water divided by the volume of water after evaporation. (a) Ca^{2+}; (b) SO_4^{2-}; (c) Na^+; (d) Cl^-; (e) K^+. (McCaffrey, Lazar, and Holland 1987)

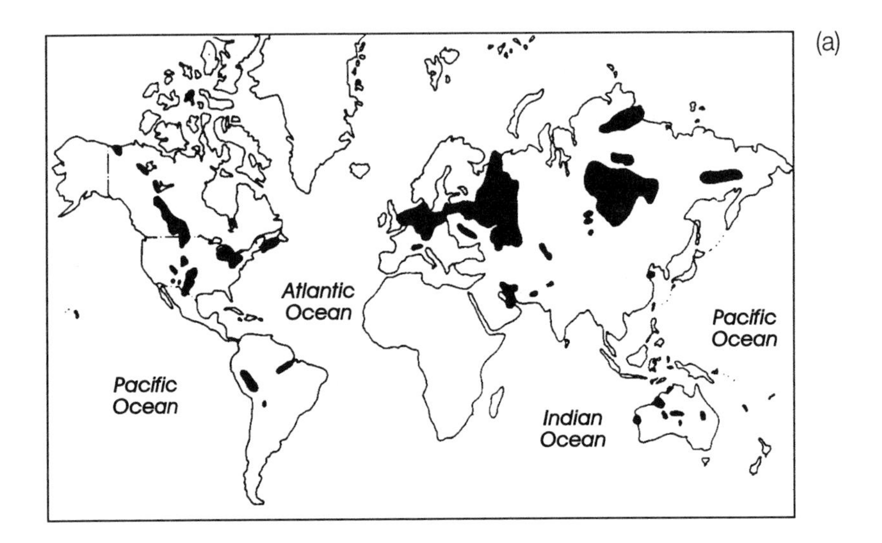

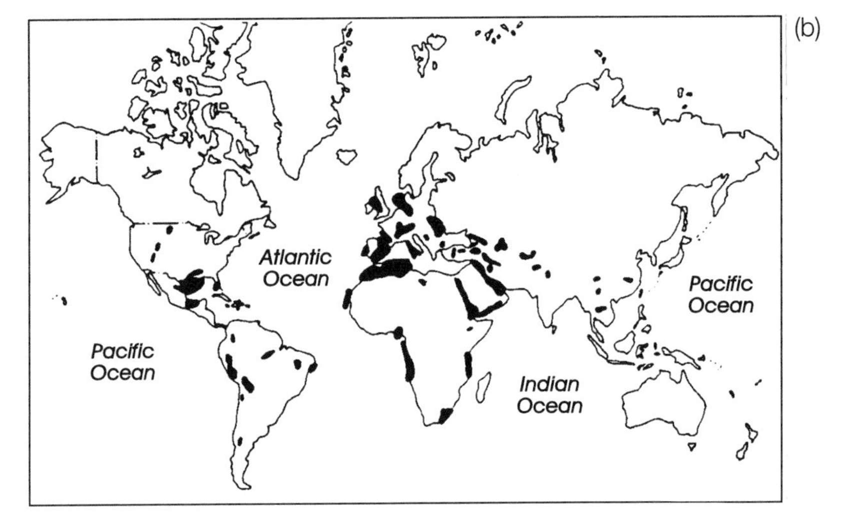

Figure 7.21. The distribution of the world's major evaporite basins. (a)Proterozoic and Paleozoic. (b) Mesozoic, Cenozoic, and Recent. (Lefond 1975)

of Salzburg in Austria may owe its fame to the music of Wolfgang Amadeus Mozart, but its name, its wealth, and its power were products of its salt mines.

The sequence of minerals in marine evaporites of all ages is monotonously the same to the end of halite precipitation (see fig. 7.22). However, the relative proportions of carbonates, gypsum, and halite differ considerably from one marine evaporite to the next. It is difficult to tell by studying the geology of the evaporites whether these differences are due to changes in the composition of seawater during the last few hundred million years or whether they are due to differences in the history of the evaporite basins. Independent lines of evidence suggest that the chemical composition of seawater has changed very little during the past 570 million years.

In some sedimentary basins, evaporation proceeded beyond the end of halite precipitation. Late-stage evaporites in such basins contain most of the world's reserves of potassium ore. The mineralogy and chemistry of late-stage evaporites

Table 7.5. Production of Salt, 1992 (in millions of short tons)

Country	*Mine Production*
United States (sold or used)	40.1
China	30.0
Germany	16.0
USSR (former)	15.4
Canada	12.1
India	10.5
Subtotal	124.1
Other countries	80.9
World total	205.0

Source: *Mineral Commodity Summaries*, 1993.

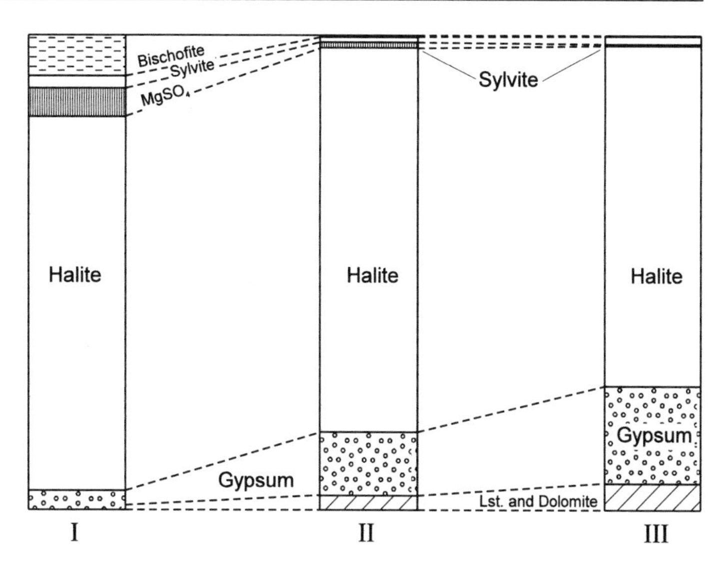

Figure 7.22. Evaporation sequences from (I) the experimental evaporation of seawater; (II) the large Zechstein evaporites in central Europe; and (III) the average of numerous other marine salt deposits. Sylvite = KCl; Bischofite = $MgCl_6 \cdot 6H_2O$. (Borchert and Muir 1964)

is quite variable, surprisingly so in view of the uniformity of the early evaporation sequences. Some of these late-stage potassium-bearing evaporites contain virtually no calcium minerals but a good many magnesium and sulfate minerals; others contain calcium and magnesium minerals but no sulfate minerals. The reasons for these differences are still a matter of debate. Admixture of brines with river salts and with hot underground solutions has been suggested, but reactions of the brines with surrounding rocks in the evaporite basins have probably played a more prominent role. For example, along the hot, dry, southern border of the Persian Gulf (see fig. 7.23) very wide stretches of carbonate muds, the sabkhas, are repeatedly flooded by seawater. These flood waters evaporate as they slowly make their way back to the gulf through the carbonate muds (see fig. 7.24). Gypsum ($CaSO_4 \cdot 2H_2O$) and anhydrite ($CaSO_4$) precipitate en route. The concentration of Mg^{2+} in the solutions rises continuously. Finally the brines react with the $CaCO_3$ sediments to produce the mineral dolomite:

$$\underset{\text{calcite and aragonite}}{2CaCO_3} + Mg^{2+} \rightharpoondown \underset{\text{dolomite}}{CaMg(CO_3)_2} + Ca^{2+}. \quad (7.9)$$

The Ca^{2+} released during this reaction then precipitates as a constituent of gypsum or anhydrite. In this way, both Mg^{2+} and SO_4^{2-} are removed from evaporite brines. If the dolomitizing reaction goes far enough, all of the SO_4^{2-} can be removed from the brines. Much of the variability in the mineralogy and chemistry of late-stage marine evaporites can be explained by differences in the extent to which dolomitization affected the chemistry of brines in particular evaporite basins.

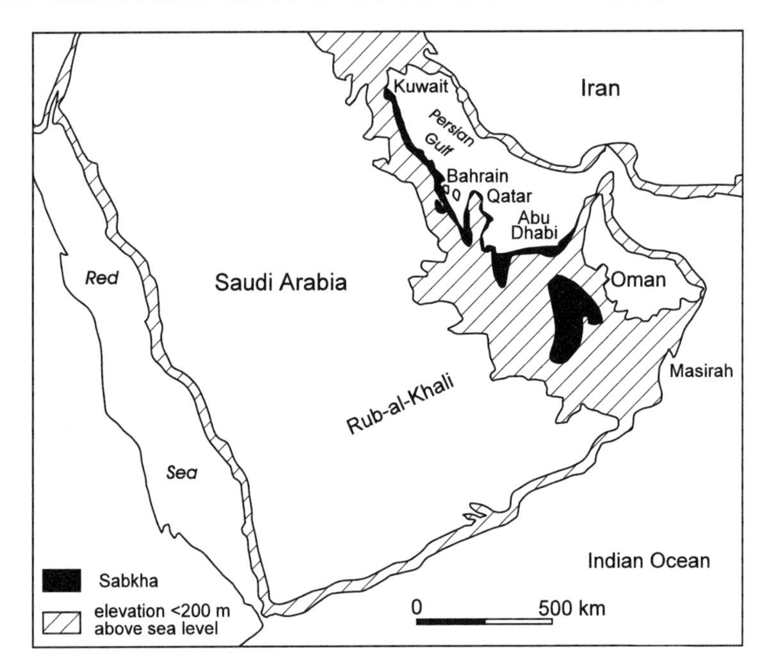

Figure 7.23. Location of sabkhas in the Persian Gulf. (Patterson and Kinsman 1977)

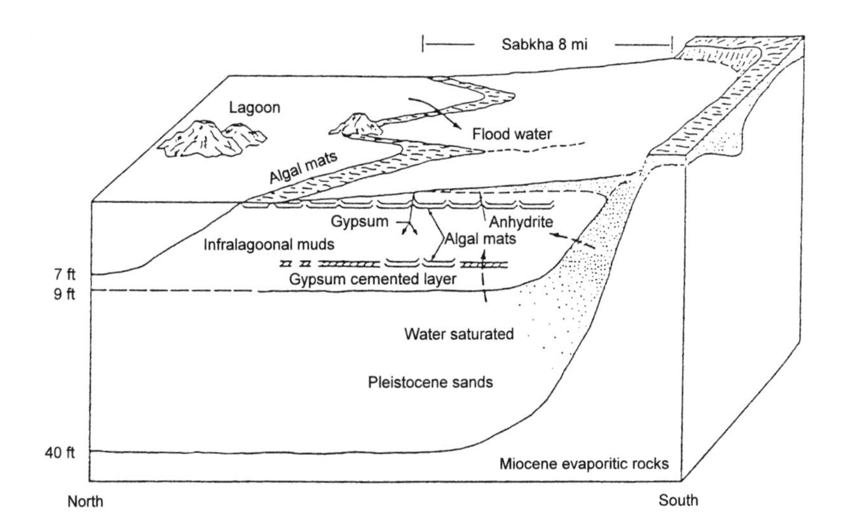

Figure 7.24. Schematic cross section of the Persian Gulf sabkha near Abu Dhabi. Note vertical exaggeration. (Butler 1969)

Many evaporite sequences are nearly devoid of detrital contamination, so that the salt and the potassium ores are relatively pure and high grade. This is because the optimal conditions for maximum evaporation are provided in desert climates by bays or basins that are not fed by a river and that have only a limited connection to the ocean. Seawater enters these bays only in small amounts and does not dilute the brines, yet it provides a steady supply for further evaporation and precipitation of economically important salts. It also helps if the floor of the basin gradually sinks, so that room is created for the deposition of later evaporite minerals.

Table 7.6. Production and Reserves of Potash, 1992 (in millions of metric tons of K_2O equivalent)

Country	*Production*	*Reserves*	*Reserve Base*
USSR (former)	8.2	3,600	3,800
Canada	7.1	4,400	9,700
Germany	3.4	750	900
United States	1.8	83	290
Israel	1.3	53	600*
France	1.1	15	35
Subtotal	22.9	8,901	15,325
Other countries	2.1	499	1,675
Total	25.0	9,400	17,000

Source: *Mineral Commodity Summaries*, 1993.

* The total reserve base in the Dead Sea is equally divided between Israel and Jordan.

World Resources: Estimated domestic potash resources total about 6 billion tons. Most of this lies between 6,000 and 10,000 feet deep, in a 1,200 square-mile area of Montana–North Dakota as an extension of the Williston Basin deposits in Saskatchewan. The Paradox Basin in Utah contains approximately 2 billion tons, mostly at depths of more than 4,000 feet. An unknown quantity of potash resources lies about 7,000 feet under central Michigan. Estimated world resources total about 250 billion tons. The potash deposits in the former USSR account for most of the centrally planned economy countries' resources. Very large resources, about 10 billion tons and mostly carnallite, occur in Thailand.

The world production and reserves of potassium are listed in table 7.6. The ratio of the reserve base to the current production is about 1,000. As in the case of halite, we need not fear that we will deplete the reserves of this commodity. Even if we were to mine out the world's potash reserves, the oceans hold an essentially infinite supply of K^+, retrievable at relatively modest cost. This result is quite important, because potassium is a major component of fertilizers.

Brines rich in magnesium are recovered from some evaporite deposits and are processed to obtain magnesium metal. However, most magnesium metal and compounds are obtained directly from seawater and from magnesium-rich lake brines in desert climates, such as the Great Salt Lake in Utah (Petersen 1994). Magnesium is largely used to produce lightweight alloys with aluminum. The demand for these two metals has been increasing faster than for most other metals.

Today, seawater is processed only to obtain freshwater, magnesium, salt (halite), and bromine. In the future it may also be used by maritime countries to obtain sodium sulfate or carbonate ("soda ash"), potassium, chlorine, sulfur, gypsum, borax ($Na_2B_4O_7.5H_2O$), lithium, strontium, iodine, fluorine, and perhaps even uranium (Petersen 1994). Much of the sodium sulfate is used to make soap, detergents, pulp, and paper. The sodium carbonate is important for the production of glass, chemicals, soap, and detergents. Seawater is unlikely to become a major source of iron, aluminum, base metals (copper, lead, zinc), and precious metals (gold, silver, platinum).

7.7 The Magnesium Problem, Black Smokers, and Manganese Nodules

Shallow and deep-water carbonates were easily identified as the major oceanic sinks of the river input of Ca^{2+}. The major sinks of Mg^{2+} in the oceans have been much more elusive. Ancient marine carbonate sediments contain large amounts of the mineral dolomite. Much of this was probably formed by the dolomitization reaction described in the previous section. Relatively few modern marine carbonates are dolomitized, and the present-day dolomitization of older carbonate sediments does not seem to account for the removal of more than a small fraction of the river input of Mg^{2+}. Silicate minerals are potentially important marine sinks for Mg^{2+}, but an intensive search has turned up little evidence for the formation of large quantities of Mg^{2+} clays in the oceans. Late in the 1960s the search for Mg^{2+} sinks turned to the midocean ridges (MORs).

Very hot brines had been extracted from boreholes on the Reykjanes Peninsula on Iceland, a large volcanic island athwart the Mid-Atlantic Ridge in the North Atlantic Ocean (see fig. 7.25). Although these brines appeared to have started as seawater, their composition on emerging from the boreholes was quite different from that of seawater, presumably because they had reacted with the hot basaltic rocks of the Reykjanes Peninsula. Laboratory experiments confirmed this hunch: Mg^{2+} was removed almost completely from seawater that was heated to a few hundred degrees centigrade in the presence of basaltic rocks. The composition of the solutions at the end of these experiments was almost identical to that of the Reykjanes brines. Here, clearly, was a previously unrecognized way to remove Mg^{2+} from the oceans.

But is removal into hot basalts a major sink for oceanic Mg^{2+}? Models of the flow of seawater through hot basaltic rocks of the MORs indicated that the process could be important for the marine geochemistry of Mg^{2+}. The exit temperature of seawater near the center of the MORs was estimated to be as high as 170°C. In 1976 the first warm seawater on the floor of the oceans was found near

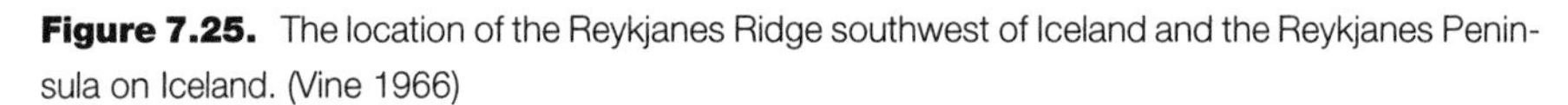

Figure 7.25. The location of the Reykjanes Ridge southwest of Iceland and the Reykjanes Peninsula on Iceland. (Vine 1966)

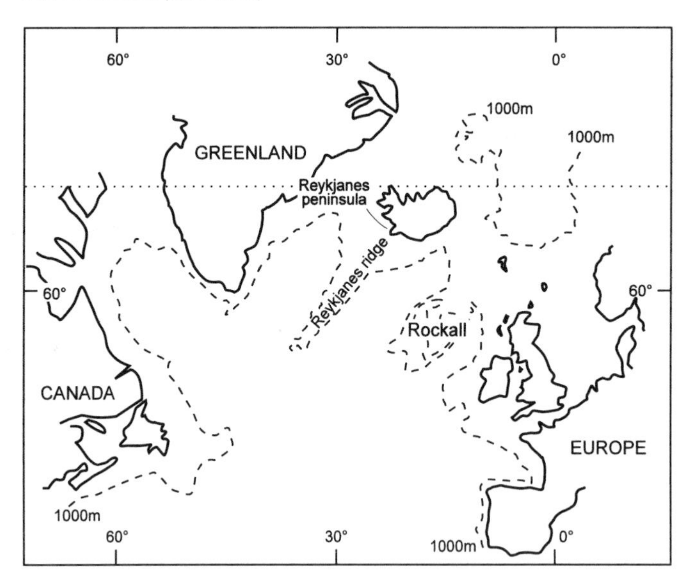

the Galapagos Islands. Water temperatures were only about 25°C, but there were indications that these warm waters were mixtures of normal seawater with very hot solutions, perhaps with initial temperatures as high as 300°C. In 1979, the first undiluted hot-spring waters were discovered on the East Pacific Rise at latitude 21°N. Their discovery by Robert Ballard in the deep submersible *Alvin* (fig. 7.26) opened an exciting new chapter in ocean exploration. The hot-spring vents were dubbed "black smokers" (plate 18) because a black cloud of copper, iron, and zinc sulfide minerals precipitated from the hot solutions as soon as they mixed with cold seawater. The temperature of the hot springs ranged up to 350°C. Chimneys and larger edifices consisting largely of sulfide and sulfate minerals formed around the hot vents and served as conduits for the hydrothermal fluids (fig. 7.27). Ingenious samplers were designed to capture the fluids. Their composition turned out to be similar to the composition of the Reykjanes brines and to the products of the laboratory experiments in which seawater was reacted with basalt at several hundred degrees centigrade (Mottl and Holland 1978). Of particular interest was the observation that the vent fluids were essentially Mg^{2+}-free.

Figure 7.26. The deep submersible *Alvin* on board the *R.V. Lulu* during the second cruise to the submarine hot springs at 21°N on the crest of the East Pacific Rise. H. D. Holland is pondering the sampling equipment.

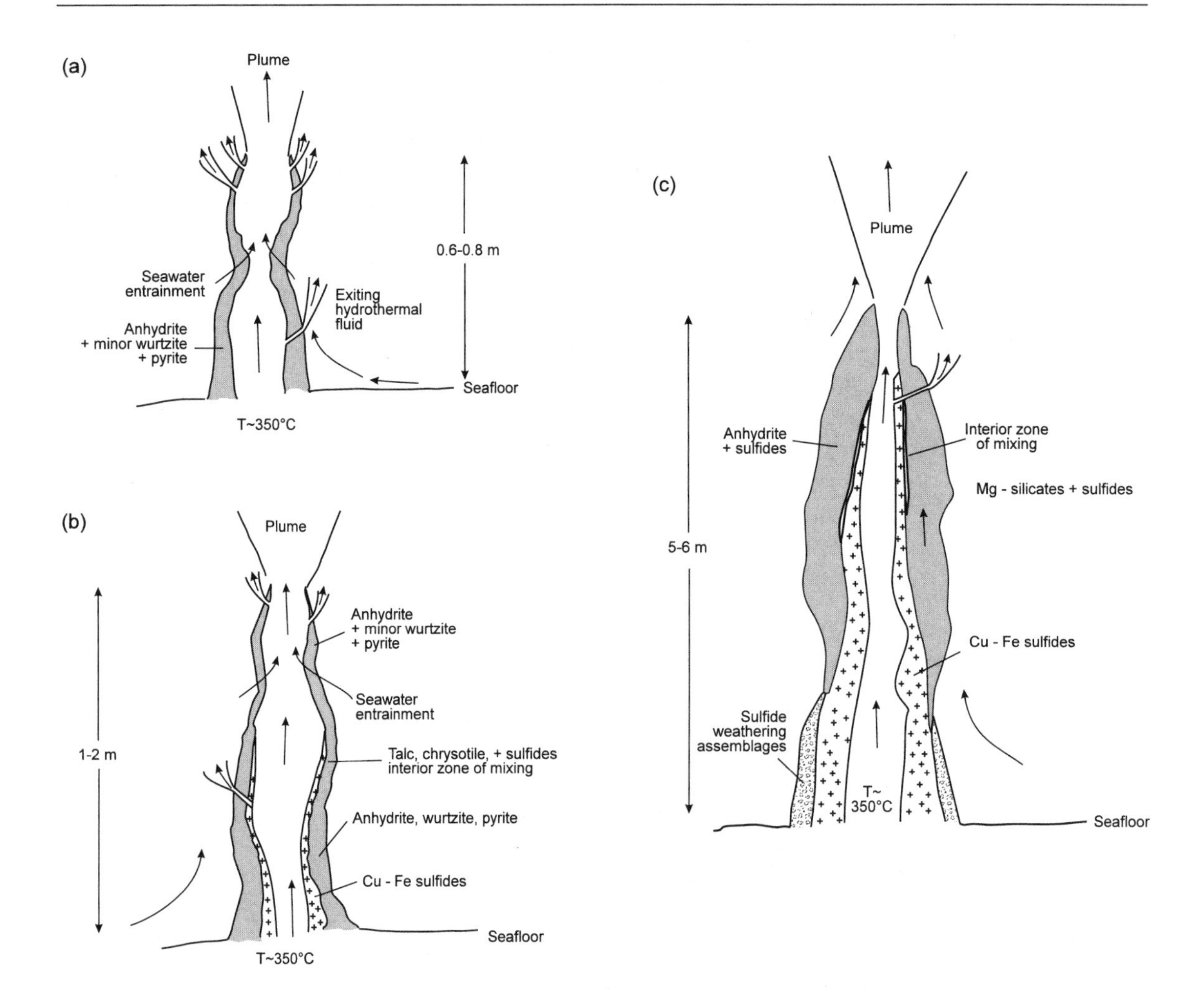

Figure 7.27. The stages of hydrothermal chimney growth are represented in these idealized cross sections of active chimneys. (Goldfarb et al. 1983)

(a) *Stage 1*. During the initial growth stage, the chimney wall is composed dominantly of anhydrite and minor quantities of sulfides, which precipitate during mixing of seawater with hydrothermal fluid. The wall is highly permeable and permits ready passage of both hydrothermal fluid and seawater. Growth rates during this stage may reach 30 cm/day.

(b) *Stage 2*. With continued mixing of seawater and hydrothermal fluid in the chimney wall and consequent mineral precipitation, the permeability of the wall is reduced. The temperature of the hydrothermal fluid inside the chimney rises, leading to precipitation of an inner layer of Cu-Fe sulfides. Mg silicates precipitate in a mixing zone along the boundary between the Cu-Fe sulfide layer and the dominantly anhydrite layer. The upper part of the chimney is identical to the young chimney shown in (a).

(c) *Stage 3*. Cross section of a mature chimney. The upper part of this chimney is equivalent to the chimney illustrated in (b). The lower part matures with increased sealing of the chimney wall. This leads to the separate development of a high-temperature hydrothermal mineral assemblage in the chimney interior and a low-temperature weathering mineral assemblage on the chimney exterior. Prolonged weathering of the chimney walls may lead to the dissolution of anhydrite and the development of a mineral assemblage that resembles those formed by the weathering of sulfides at the Earth's surface (see chapter 6).

Anhydrite = $CaSO_4$; wurtzite = ZnS; pyrite = FeS_2; talc = $Mg_3Si_4O_{10}(OH)_2$; chrysotile = $Mg_3Si_2O_5(OH)_4$.

Plate 1. Sunset, lit by Earth's atmosphere; photograph taken by cosmonauts V. Kovalyonok and A. Ivanchenkov on board the orbital space complex *Salyut 6-Soyuz 29-Soyuz 31*, September 1978 (see chapter 3).

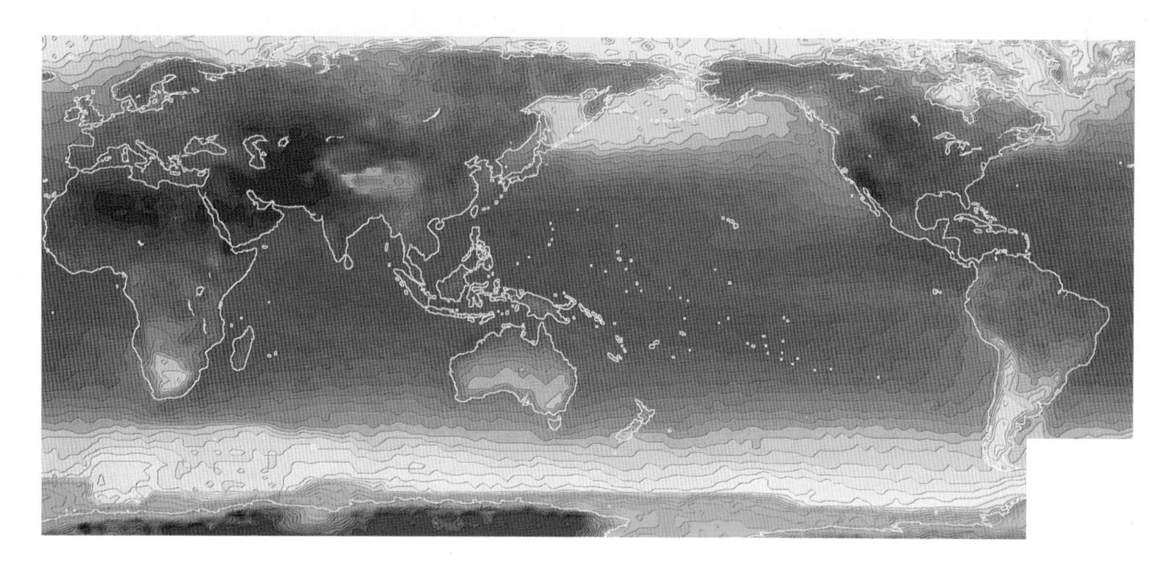

Plate 2. The mean surface temperature for June 1988 as obtained from the infrared and microwave radiometers (HIRS/MSU) aboard the NOAA series of meteorological satellites; the temperature increases from blue to green to yellow to red (see chapter 4).

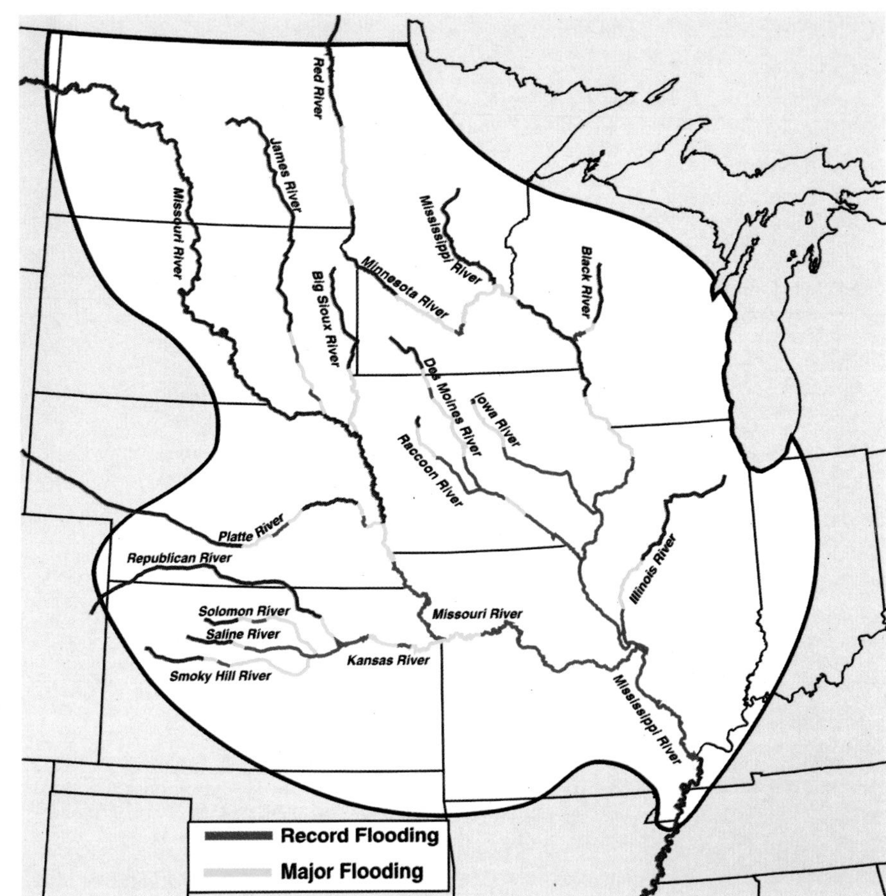

Plate 3. Major river systems in the nine-state Midwest flood region of 1993. Green shading indicates regions of major flooding; red shading indicates regions of record flooding (see chapter 4).

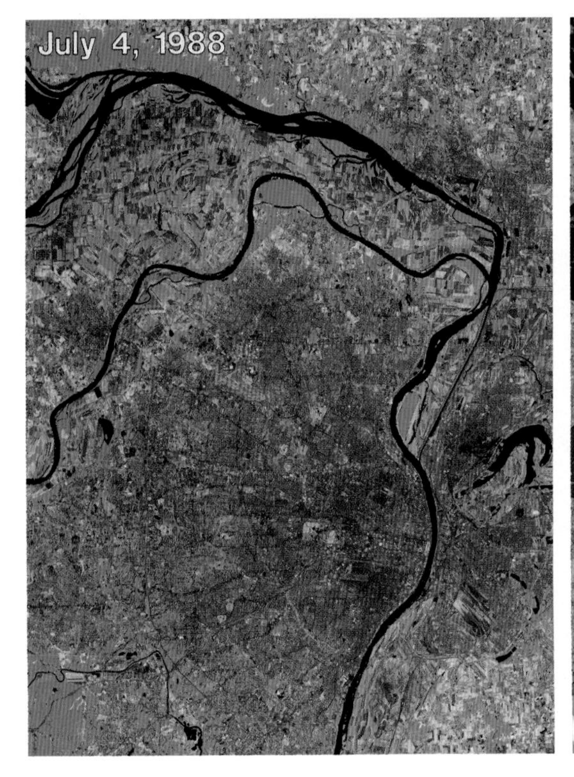

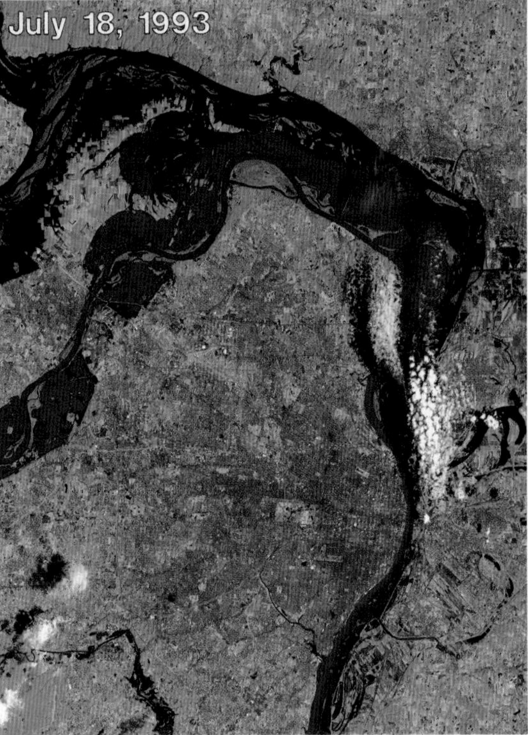

Plate 4. The junction of the Mississippi River with the Illinois and Missouri Rivers at normal river levels and during the 1993 flood (see chapter 4). St. Louis is just south of the junction of the Mississippi and Missouri Rivers.

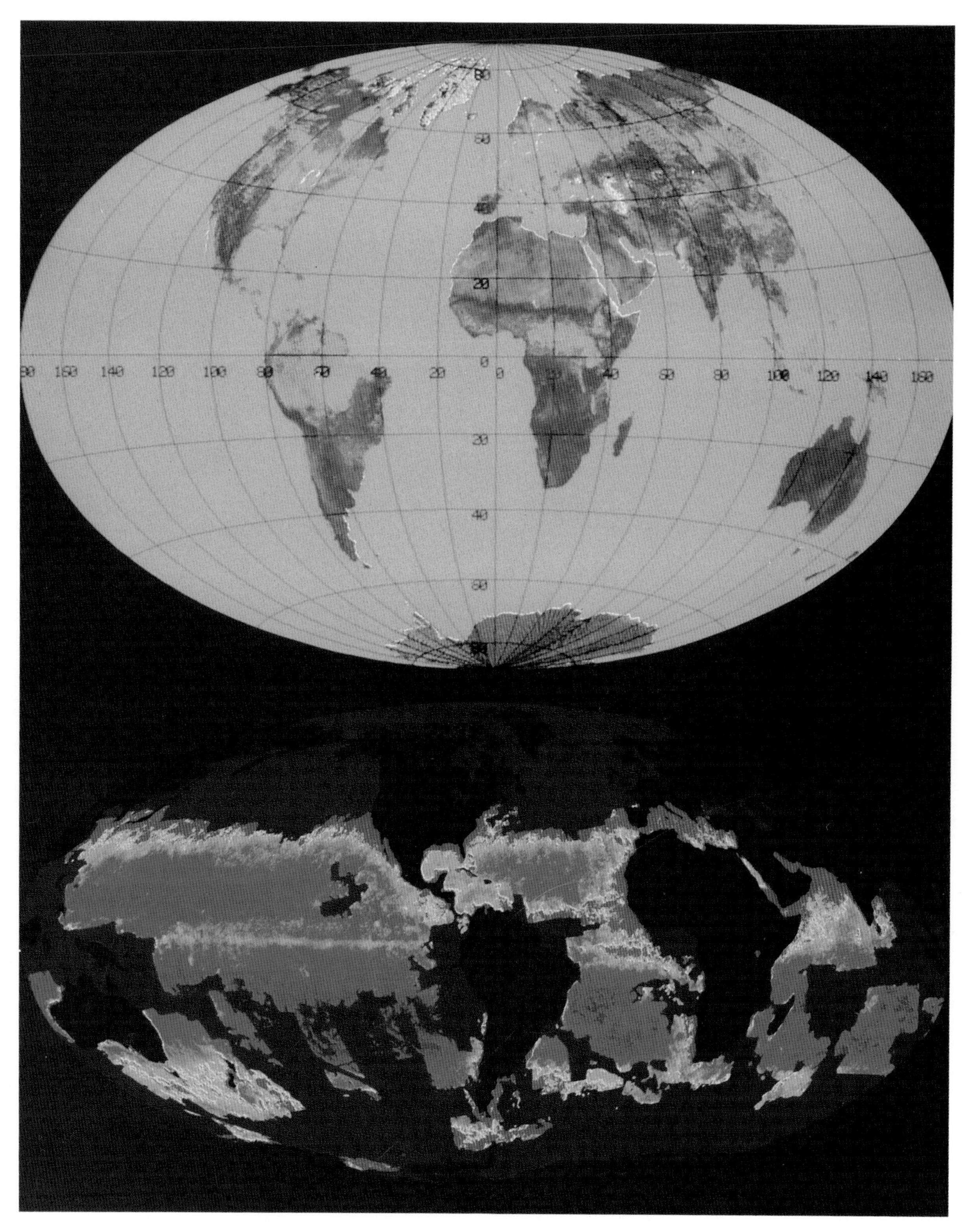

Plate 5. *Upper image:* Worldwide vegetation patterns revealed through a color index derived from environmental satellite observations. *Lower image:* Distributions of chlorophyll-producing marine plankton mapped by space observations of ocean color. Purple to yellow to red: increasing chlorophyll concentration. Gray: no data (see chapter 5).

Plate 6. View of Kilauea Crater and Mauna Loa from Volcano House, Hawaii. Mauna Loa is the largest volcano on Earth. Its shape, which is reminiscent of a shield, makes Mauna Loa a classic shield volcano (see chapter 6).

Plate 7. Lava from Kilauea flowing into the Pacific Ocean (see chapter 6).

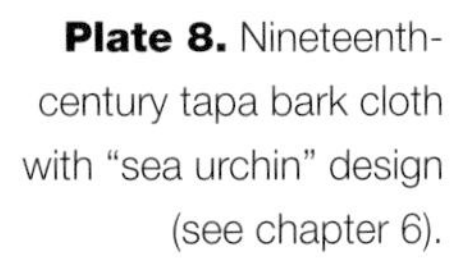

Plate 8. Nineteenth-century tapa bark cloth with "sea urchin" design (see chapter 6).

Plate 9. Crystal of corundum (see chapter 6).

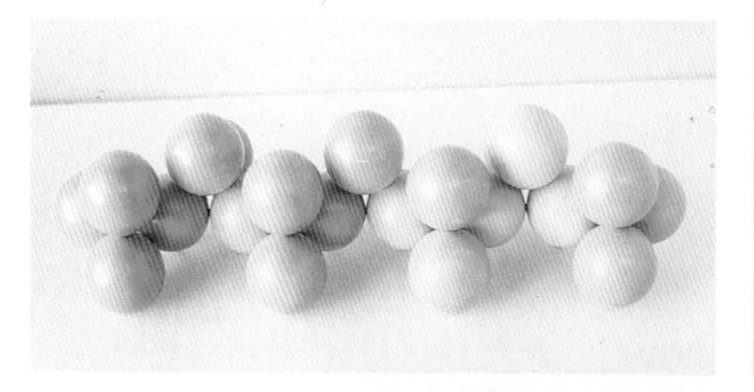

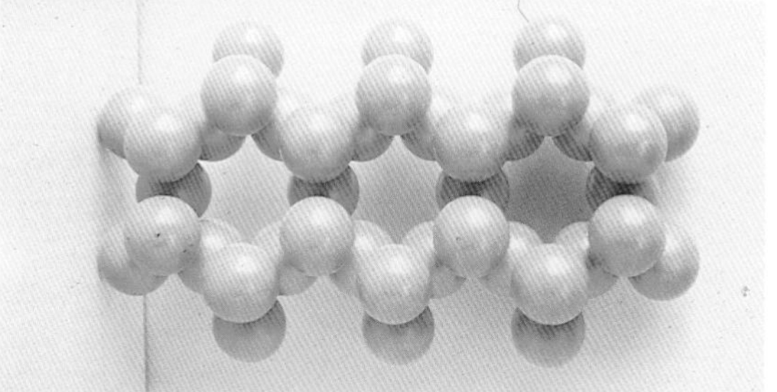

Plate 10. Chains of $[SiO_4]^{4-}$ tetrahedra; pyroxene chain on left, amphibole double chain on right (see chapter 6).

Plate 11. Ohia plant growing in very recent Kilauea lava (see chapter 6).

Plate 12. Soil profile near Wahiawa, Oahu, Hawaii (see chapter 6).

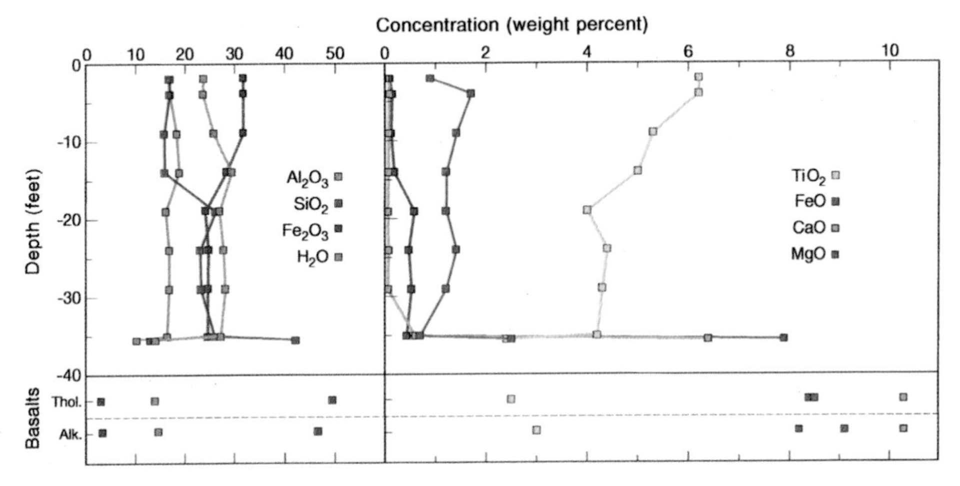

Plate 13. Composition of soil in a vertical soil profile (Ki64) on Kauai, Hawaii (see chapter 6).

Plate 14. Terracing, as practiced by the ancient Incas (see chapter 6).

Plate 15. The contact between Precambrian rocks and the overlying Cambrian Tapeats sandstone in the Grand Canyon, Arizona (see chapter 6).

Plate 16. Holei Sea Arch along the southeastern coast of Hawaii (see chapter 6).

Plate 17. Waipio Valley and sea cliffs, on the north coast of Hawaii (see chapter 6).

Plate 18. Black smoker at latitude 21°N on the East Pacific Rise (see chapter 7).

Plate 19. *Above left:* The *JOIDES Resolution* (see chapter 7).

Plate 20. *Above right:* Tube worms near a hydrothermal vent (see chapter 7).

Plate 21. *Below:* Banded iron formation, Hamersley Basin, Western Australia (see chapter 7).

Plate 22. Open-pit mining at the Mount Tom Price Mine, Western Australia (see chapter 7).

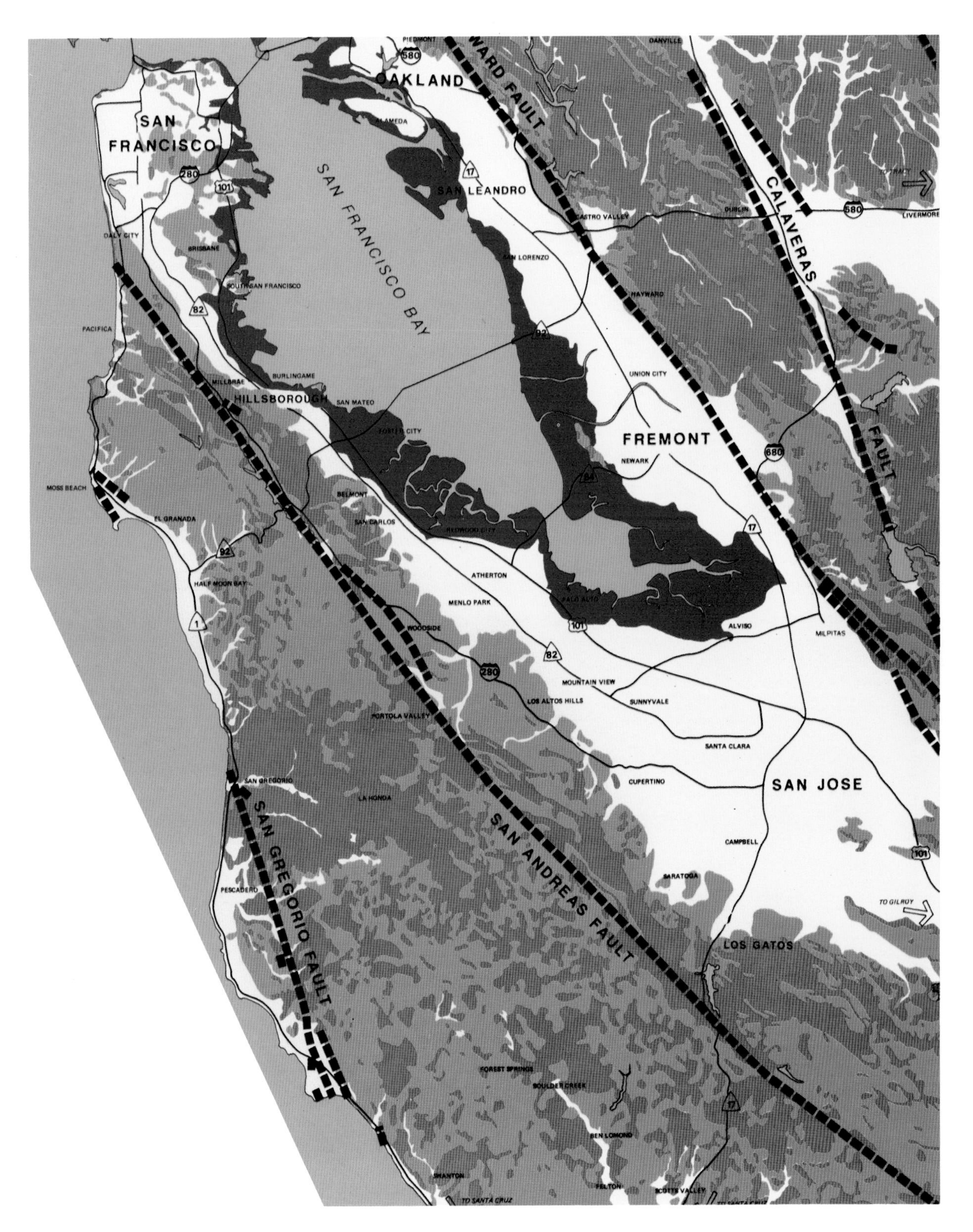

Plate 23. The effect of soil and rock type on earthquake risk in the San Francisco area. Areas in green are composed of stable bedrock (shaking not increased, ground failure unlikely). Areas in orange are composed of unstable bedrock (shaking slightly increased, prone to landsliding if present on steep slopes, decomposed, or water saturated). Yellow areas represent unconsolidated soil (shaking increased, especially if thick and water saturated) and red areas represent mud and fill (shaking strongly increased, prone to ground failure, including liquefaction) (see chapter 8).

Plate 24. One of two collapsed structures at the Golden State Freeway (I-5)-Antelope Valley Freeway (California State Route 14) interchange. Two spans collapsed due to the crushing of a column. Photograph shows superstructure shear failure as a result of the column collapse (see chapter 8).

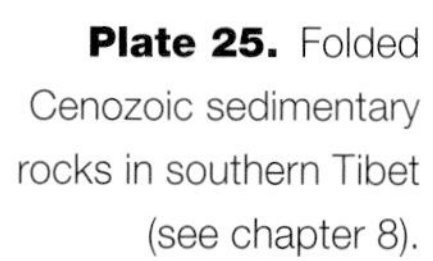

Plate 25. Folded Cenozoic sedimentary rocks in southern Tibet (see chapter 8).

Plate 26. Tightly folded Precambrian sedimentary rocks on the Great Escarpment overlooking the Namib Desert on Africa's southwest coast (see chapter 8).

Plate 27. Late in 1993 Shell Oil Company's $1.2 billion "Auger" tension leg platform was towed from Freeport, Texas, to its permanent location in 2860 feet of water 255 miles southeast of Houston. The deck section of the platform has dimensions of 290 feet by 330 feet. Under tow, the entire structure extends above the water line to the height of a 26-story building. In 1994 oil and natural gas production started from the Auger field, which is forecast to have an estimated gross ultimate recovery of 220 million barrels of oil and gas equivalent (see chapter 8).

Plate 28. Pillow lava on the East Pacific Rise; the photograph was taken from the deep submersible *Alvin* (see chapter 9).

Plate 29. Variation of seismic velocity in the Earth's interior. The section shown in this plate is through the equator. Higher velocities are shown in blue, lower velocities in red. The thin dashed line indicates the 670 km discontinuity in the mantle. The heavy inner line represents the core-mantle boundary. The map, of the Earth's surface from the North Pole, replaces the core. The white line in the map represents the equator, through which the interior seismic velocity section is drawn (see chapter 9).

−1.5%

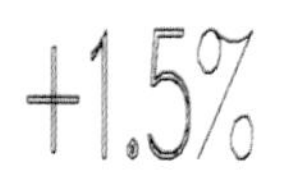

Plate 30. Lava fountain at Kilauea Iki on the island of Hawaii (see chapter 9).

Plate 31. Mount St. Helens before the volcanic eruption on May 18, 1980 (see chapter 9).

Plate 32. Cloud of volcanic debris during the eruption of Mount St. Helens, May 18, 1980 (see chapter 9).

Plate 33. Mount St. Helens after the eruption of May 18, 1980 (see chapter 9).

Plate 34. The Manaslu granite intruded into country rocks of the high Himalayas (see chapter 9).

Plate 35. Polished granite facing at the Shawmut Bank in Winchester, Massachusetts. Note xenolith. Scale is 15 cm long (see chapter 9).

Plate 36a. *Above left:* Watermelon tourmaline from the Dunton pegmatite mine in Maine (see chapter 9).

Plate 36b. *Above right:* Aquamarine crystals (see chapter 9).

Plate 37. *Below:* Chromite layers (black) in the lower part of the Bushveld Complex (see chapter 9).

Plate 38. A geyser in Iceland (see chapter 9).

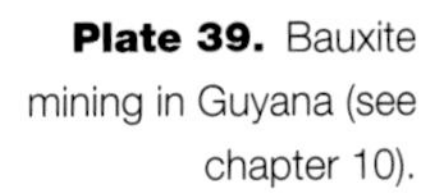

Plate 39. Bauxite mining in Guyana (see chapter 10).

Plate 40. Cattle grazing on a former bauxite mine in Jamaica (see chapter 10).

Plate 41. Global distribution of sulfur emissions with 4° X 5° resolution. Data are yearly means for 1980 except for anthropogenic emissions from the former Soviet Union (1989), North America (1985), and Japan (1983). The maxima reflect the eruption of Mount St. Helens in Washington, smelting activities in Norilsk (Siberia), and coal combustion in eastern Germany and the former Czechoslovakia (see chapter 11). *On opposite page:* Scale of emission flux (kg S km^{-2} yr^{-1})

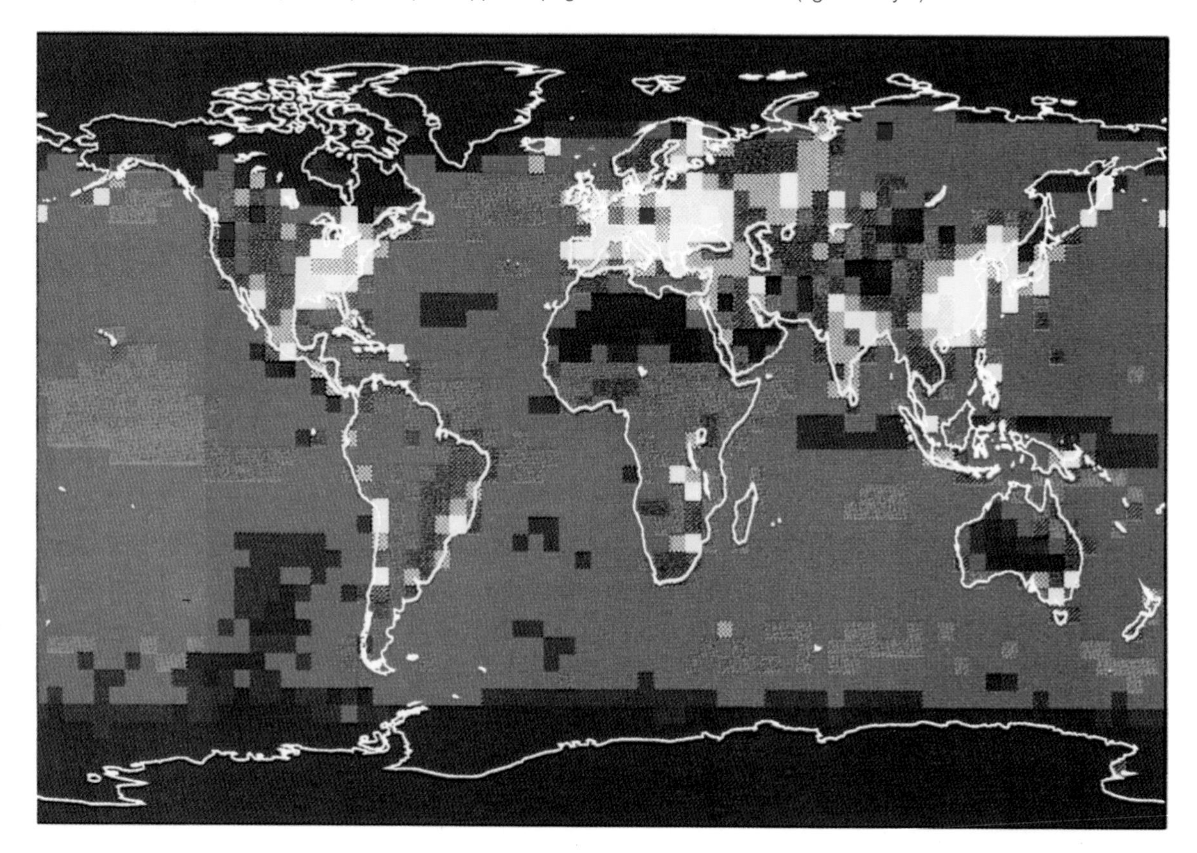

Plate 42. The Yankee Rowe 185 MW pressurized water reactor in Rowe, Massachusetts. Commercial operation started in 1960 and ended in 1992. At the time of closing, the Yankee Rowe reactor was the oldest and smallest commercial reactor in the United States. The cost of dismantling the reactor will be about $370 million (see chapter 11).

Plate 43. The stricken nuclear power station at Chernobyl in the Ukraine (see chapter 11).

Scale for **Plate 41.**

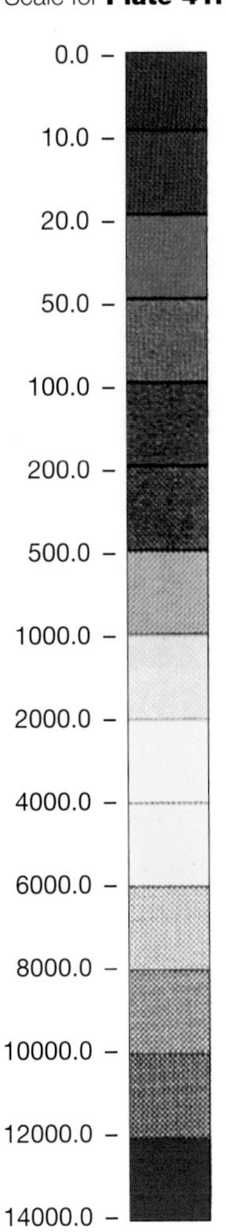

Plate 44. Yucca Mountain, Nevada, candidate site for the first U.S. geologic repository of high-level nuclear waste (see chapter 11).

Plate 45. The Glen Canyon Dam, above the Grand Canyon on the Colorado River (see chapter 11).

Plate 46. Lake Havasu, a vacationer's dream and a nature lover's nightmare (see chapter 11).

Plate 47. Satellite mosaic (1972–90) of the Nile River Delta. The geologic record of the delta provides a basis for understanding the impact of population growth and natural factors on land use. Accelerated loss of agriculture (vegetation shown in red) results from urban expansion (yellow), development of large sand dunes (pink), coastal erosion, and formation of salt pans (white, far right). Reclamation projects are draining the remaining lagoons and marshes (green) (see chapter 11).

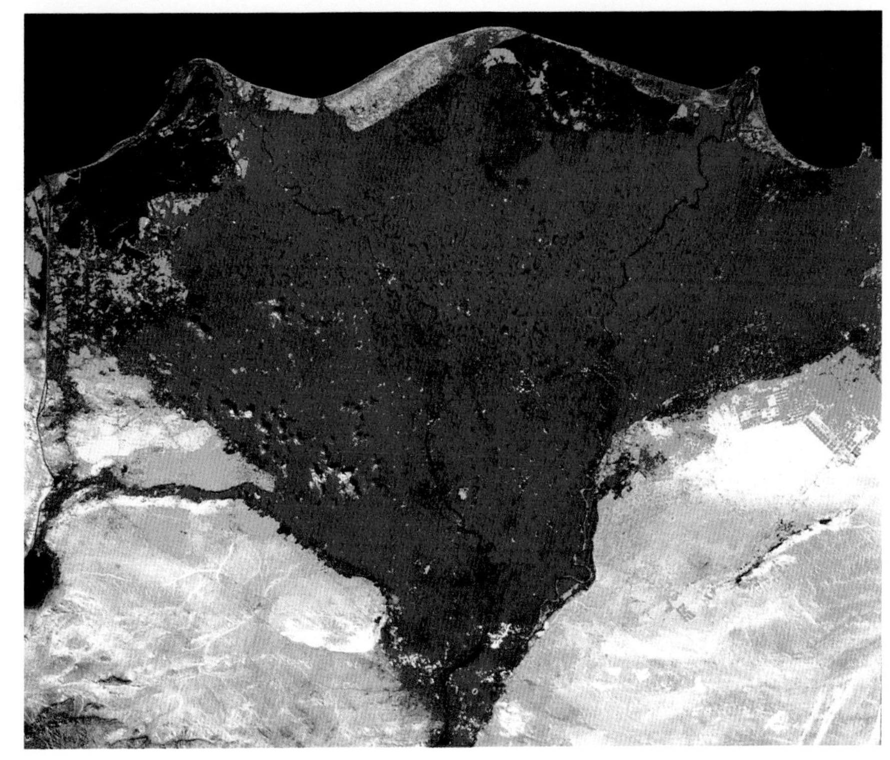

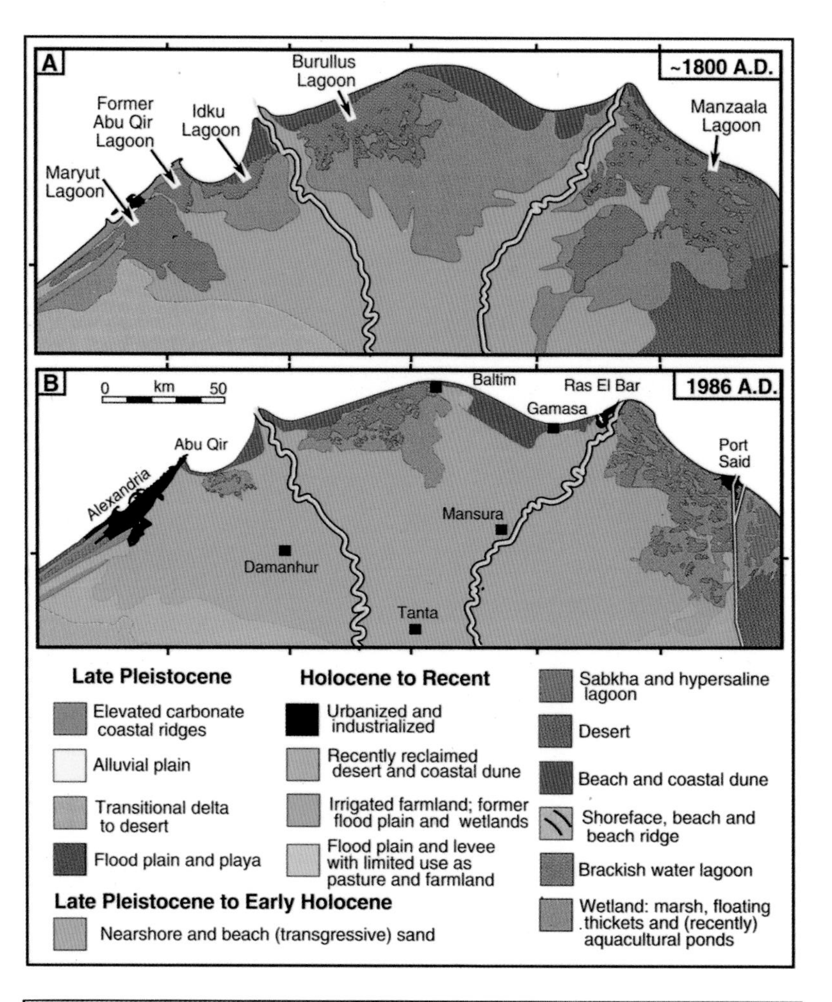

Plate 48. Geography of the northern Nile Delta (A) about A.D. 1800, and (B) at present (see chapter 11).

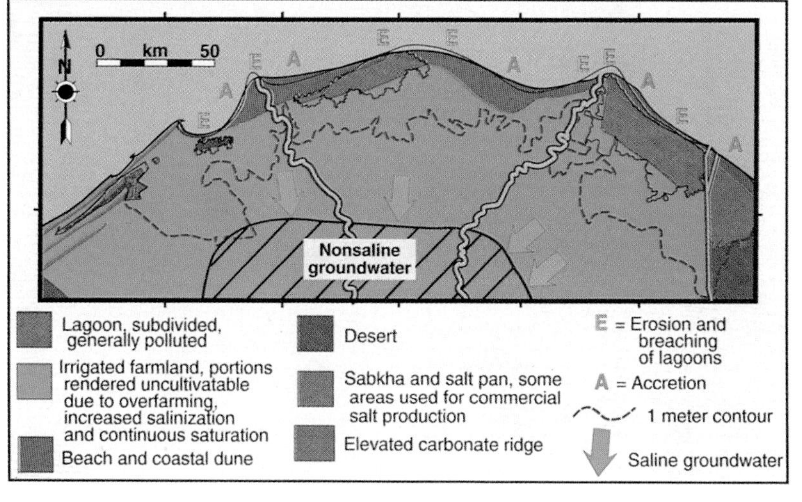

Plate 49. Projected changes in the northern Nile Delta in A.D. 2050 (see chapter 11).

Plate 50. *Right:* The Geysers, California. Drilling rig and steam vent; Unit 12 (110 mw) in the background (see chapter 11).

Plate 51. *Below left:* The 2-megawatt photovoltaic power plant of the Sacramento Municipal Utility District in Sacramento, California. The plant was constructed in two stages, in 1984 and 1986. The cooling towers in the background belong to the Rancho Seco nuclear generating station, which is now closed (see chapter 11).

Plate 52. *Below right:* Wind turbines at Altamont Pass, California (see chapter 11).

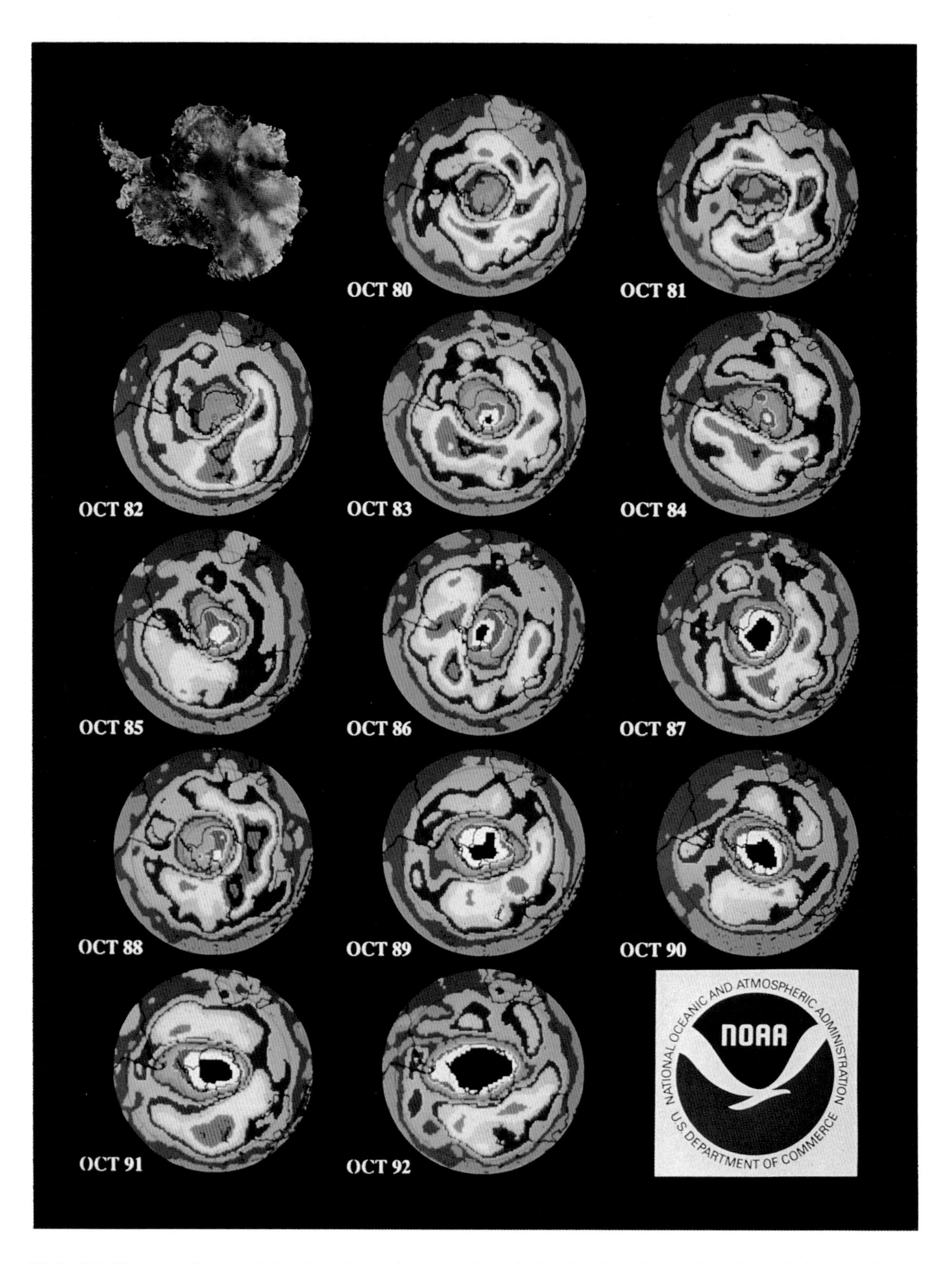

Plate 53. The Antarctic ozone hole attains its maximum depth in mid-October. Data from NOAA's Polar Orbiting TOVS (Tiros Operational Vertical Sounder) are used to display the rapid decline in stratospheric ozone over Antarctica between 1980 and 1992. The growing black spot illustrates the increase in the lowest total ozone values (see chapter 12).

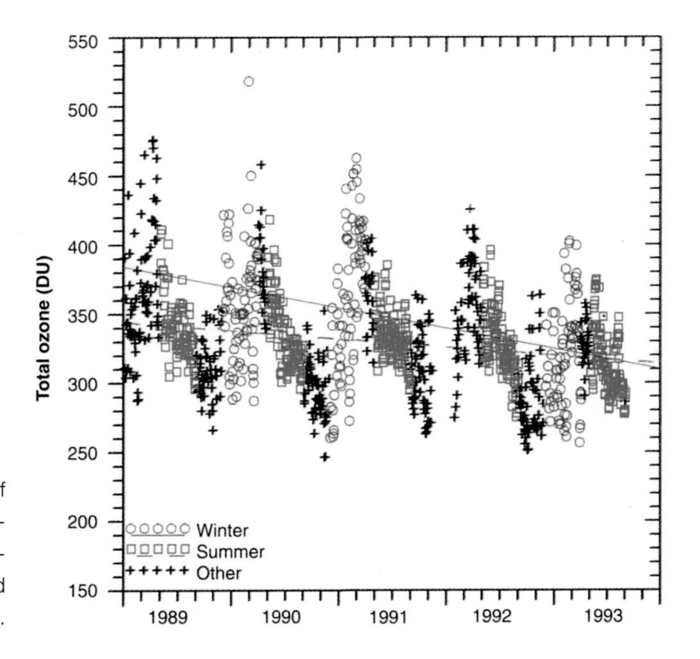

Plate 54. Record of daily ozone measurements at Toronto, Canada, between 1989 and 1993 (see chapter 12).

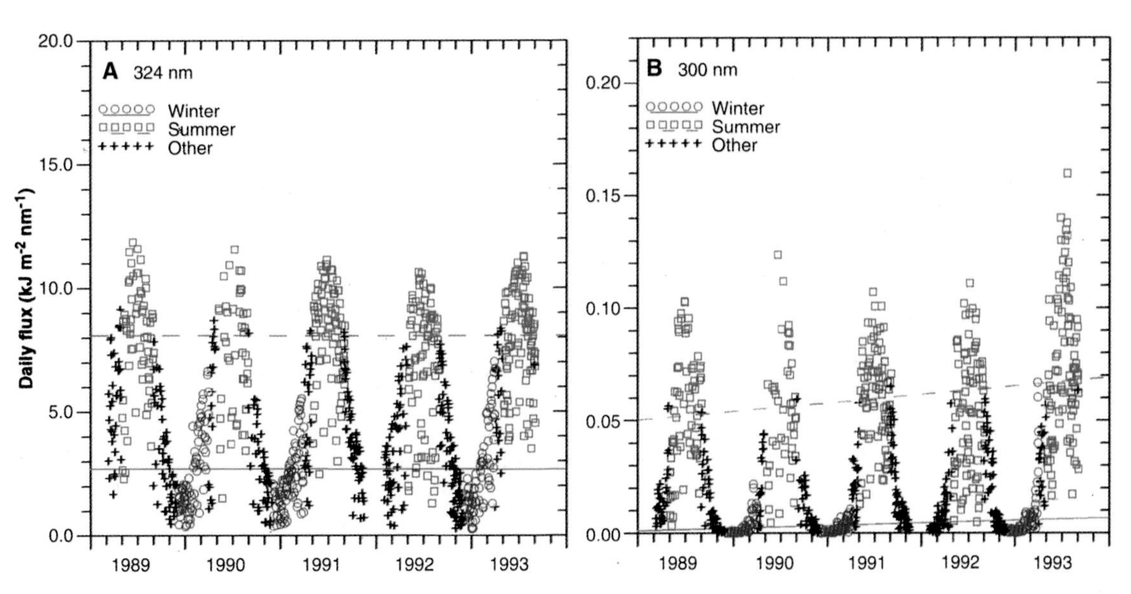

Plate 55. Record of integrated daily total UV radiation at (A) 324 nm and (B) 300 nm at Toronto, Canada, between 1989 and 1993. The light intensity at 300 nm is about 1% of that at 324 nm; the difference is largely due to UV absorption by ozone (see chapter 12).

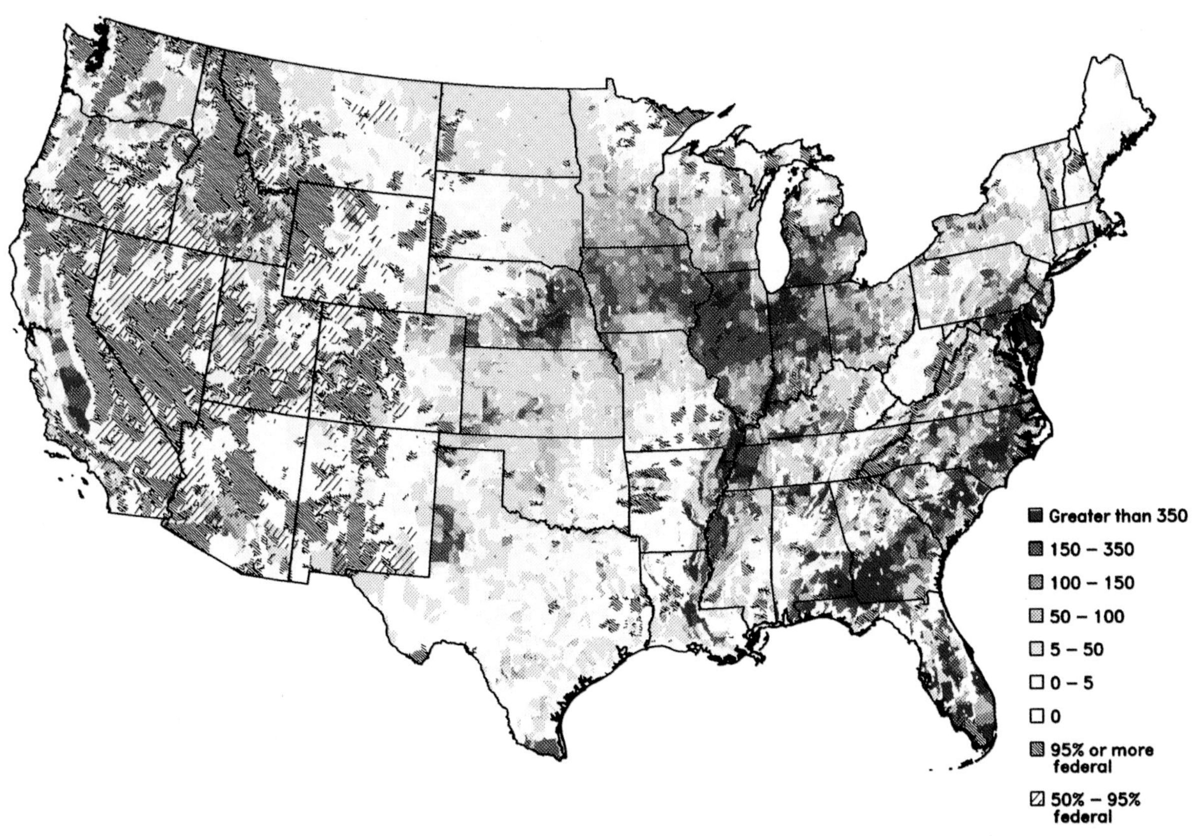

Greater than 350
150 – 350
100 – 150
50 – 100
5 – 50
0 – 5
0
95% or more federal
50% – 95% federal

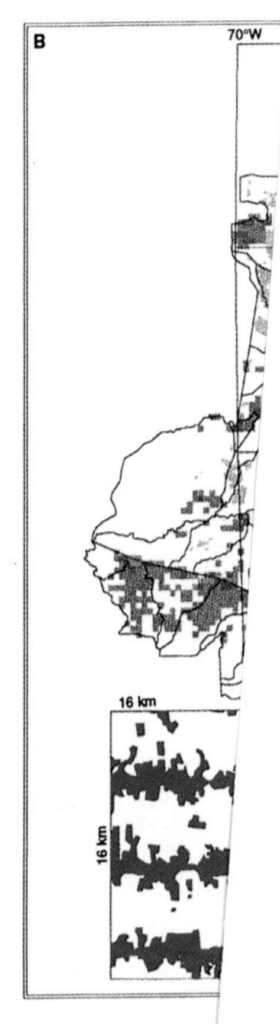

Opposite page:

Plate 56. *Above:* Average groundwater vulnerability index for pesticides in the United States (see chapter 12).

Plate 57. *Below:* Tundra east of Kuujuaq, northern Quebec, Canada (see chapter 12).

This page:

Plate 58. *Above right:* Sawgrass marsh in the Florida Everglades (see chapter 12).

Plate 59. *Below right:* Intermediate egret couples display reassurances and affection by gently preening each other (see chapter 12).

Plate 60. *Rigl*
Landsat Thematic Ma
per color compos
image of south
Rhondonia State, Bra
The area identified
"Isolated Fores
about 3 x 15 km
chapter

Plate 61A.
Represent
deforestatio
Amazon B
Brazil in 19
cha

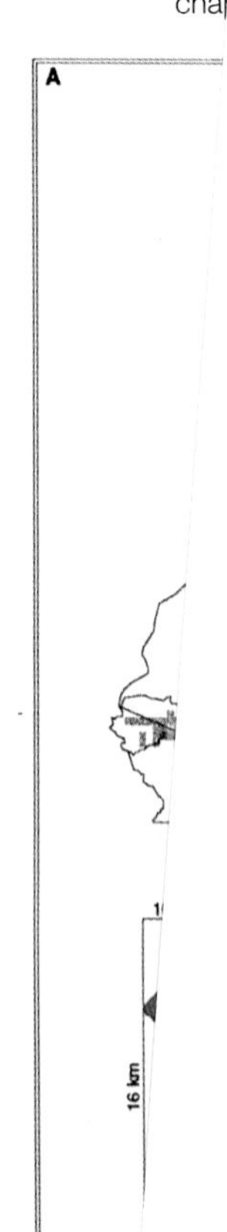

Plate 63. Rain forest in the Sierra de Luquillo, Puerto Rico (see chapter 12).

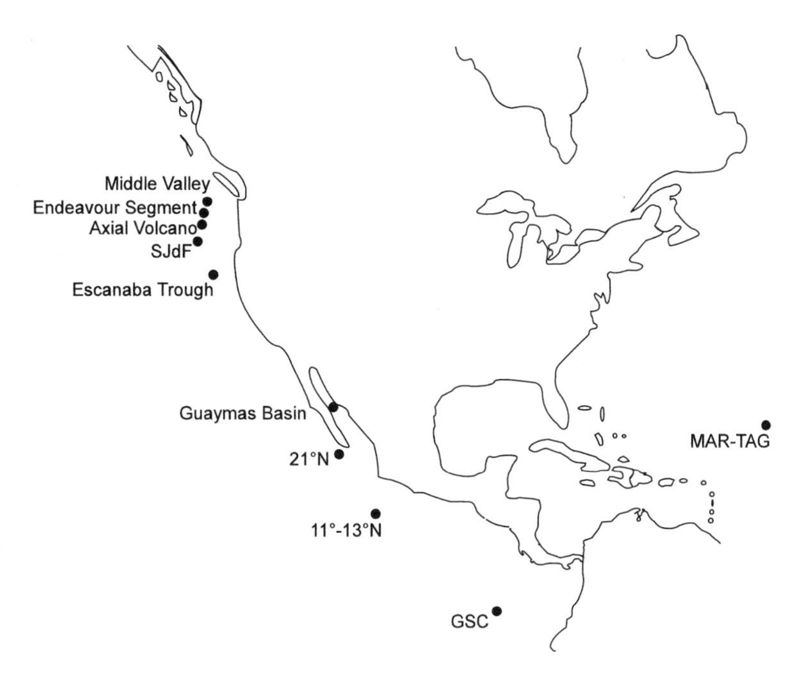

Figure 7.28. Map of some of the sampled seafloor hydrothermal systems. (Von Damm 1990)

Since 1979, hydrothermal vents have been discovered in numerous places along midocean ridges and in several non-ridge settings. The current record for the highest vent temperature, slightly over 400°C, is held by vents on the Juan de Fuca Ridge (SJdF in fig. 7.28) in the northeast Pacific Ocean. It is clear that there are many vent fields still waiting to be discovered, and there is still considerable controversy about the total flux of seawater through the crest of MORs. However, various geophysical and geochemical arguments suggest that this flux and the attendant transfer of Mg^{2+} from seawater to the oceanic crust does not seem to account for the removal of a major fraction of the river flux of Mg^{2+} into the oceans. The "magnesium problem" is therefore still unsolved, but some fascinating discoveries have been made in the search for the solution, and answers have been found to several other perplexing questions. The black smokers and their lower-temperature relatives, the white smokers, have yielded a wealth of new insights. They are excellent examples of hydrothermal ores. A good quantity of metals is extracted from the oceanic crust during the reaction of basalt with seawater at 400°C. Iron and manganese are the most abundant of the large suite of metals in the black smoker fluids (see table 7.7). The H_2S content of the vent fluids is enough to precipitate most of these metals during cooling and mixing with seawater. Very sizable quantities of the sulfides of iron, copper, and zinc accumulate at and in the vicinity of many of the vent sites. The drill ship *JOIDES Resolution* (plate 19) has drilled holes into a number of these. If they were on dry land, they would be worth mining. Their location some 2.5 kilometers below the surface of the oceans makes them commercially unattractive targets at this time, and they may never become ores; but they are superb analogues of ancient zinc and copper ores that formed under similar conditions and that were later lifted above sea level, a process that has made them accessible and attractive for mining.

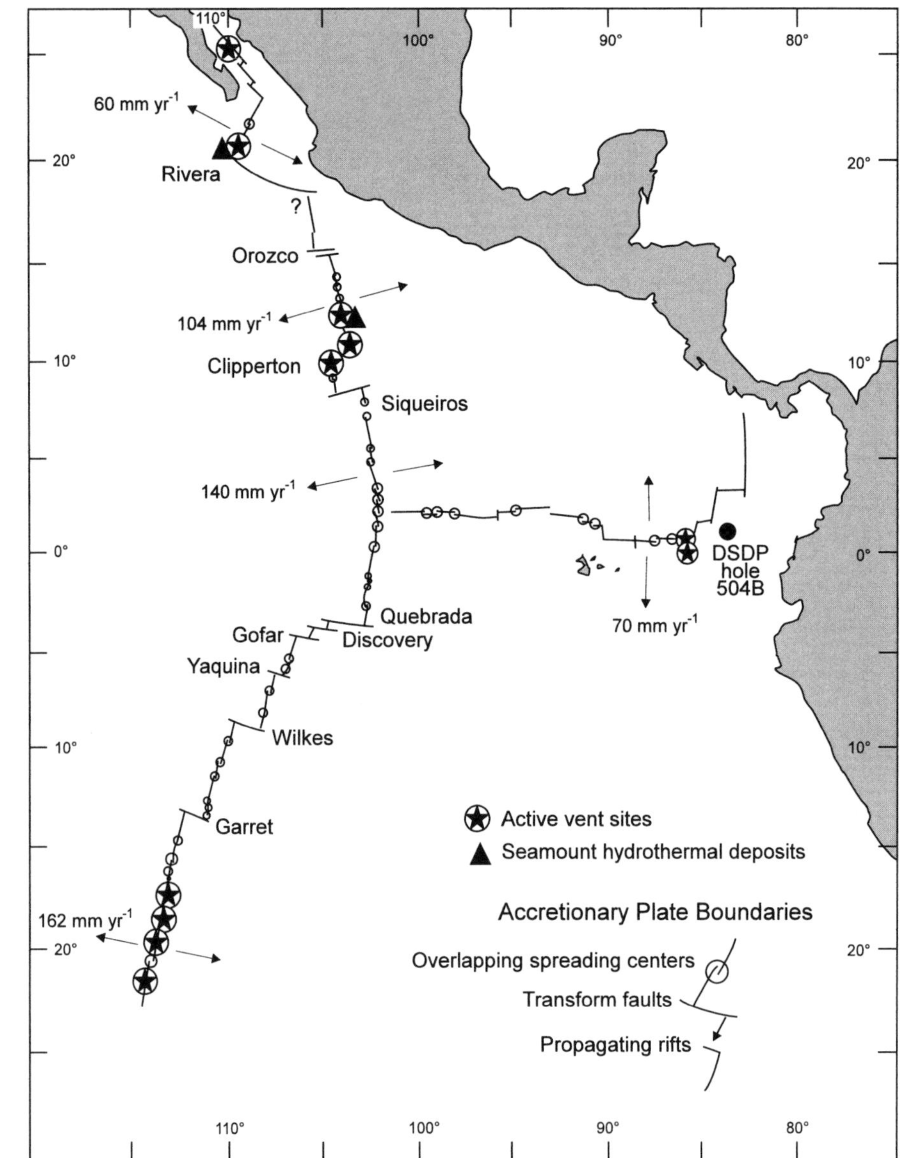

Figure 7.29. Distribution of known sites of active hydrothermal venting along the East Pacific Rise and Galapagos Rift. Solid stars mark sites of active venting; solid triangles mark locations of off-axis seamounts associated with hydrothermal mineral deposits. Solid circle shows location of DSDP Hole 504B. (Haymon 1989)

Some modern vent fields are covered and surrounded by a strange biota. Sunlight does not penetrate even a quarter of the way to the crests of MORs. Energy for biosynthesis, therefore, comes from the bacterially mediated oxidation of reduced compounds in the vent fluids. H_2S, the most important of these compounds, is oxidized bacterially to H_2SO_4. Oxygen dissolved in seawater is the oxidant. The sulfur-oxidizing bacteria are the base of a significant food chain, which includes the tube worms in plate 20.

Much of the manganese and a good deal of the iron in black smoker solutions escapes precipitation close to the vents. The iron is gradually oxidized from Fe^{2+}

Table 7.7. Measured Compositions of End Member Hydrothermal Vent Fluids

Sample		*EPR, 21°N*	*EPR, 13°N*	*Galapagos Rift*	*Seawater*
Alkalis					
Li	u	891–1,033	820	689–1,142	26
Na	m	432–510	450	+,–	464
K	m	23.2–25.8	24	18.7–18.8	9.79
Rb	u	27–33	26	13.4–21.2	1.3
Alkaline earths					
Be	n	10–37	NA	11–37	0.02
Mg	m	0	0	0	52.7
Ca	m	11.7–20.8	16	24.6–40.2	10.2
Sr	u	65–97	90	87	87
Ba	u	7	NA	17.2–42.6	0.14
$^{87}Sr/^{86}Sr$		0.703019–0.703345	0.7041	NA	0.7091
pH		3.3–3.8	3.8	NA	7.8
Alk	meq	(–0.5)–(–0.19)	NA	0	2.3
Cl	m	489–579	550	+,–	541
SiO_2	m	15.6–19.5	22	21.9	0.16
Al	u	4.0–5.2	NA	NA	0.005
SO_4	m	0	0	0	28.6
Trace metals					
Mn	u	699–1,002	610	360–1,140	<0.001
Fe	u	750–2,429	1,800	+	<0.001
Fe/Mn		0.9–2.9	2.9		
Co	u	22–227	NA	NA	0.03
Cu	u	0.02–44	NA	0	0.007
Zn	u	40–106	NA	NA	0.01
Ag	u	1–38	NA	NA	0.02
Cd	u	17–180	NA	0	1
Pb	u	183–359	NA	NA	0.01
Gases					
ΣCO_2	m	5.7–6.4	NA	NA	2.33
CH_4	m	0.04–0.08	NA	NA	0
H_2S	m	6.5–8.7	NA	NA	0
H_2(aq)	m	0.36–1.69	NA	NA	0
He	u	3.75	NA	NA	0.002

Source: Haymon 1989.
Units: m = millimoles/kilogram; u = micromoles/kilogram; n = nannomoles/kilogram; meq = milliequivalents/kilogram; Galapagos data is/liter; + = gain, – = loss; EPR = East Pacific Rise.

to Fe^{3+} in the water column and settles out slowly on the flanks of MORs. Manganese is present as Mn^{2+} in the vent fluids. This cation is oxidized to Mn^{4+} much more slowly than Fe^{2+} is oxidized to Fe^{3+}, but manganese does finally settle out as oxides of Mn^{4+}. Some of the iron and manganese from vent sources are joined by iron and manganese from sediments on the ocean floor to form what are called

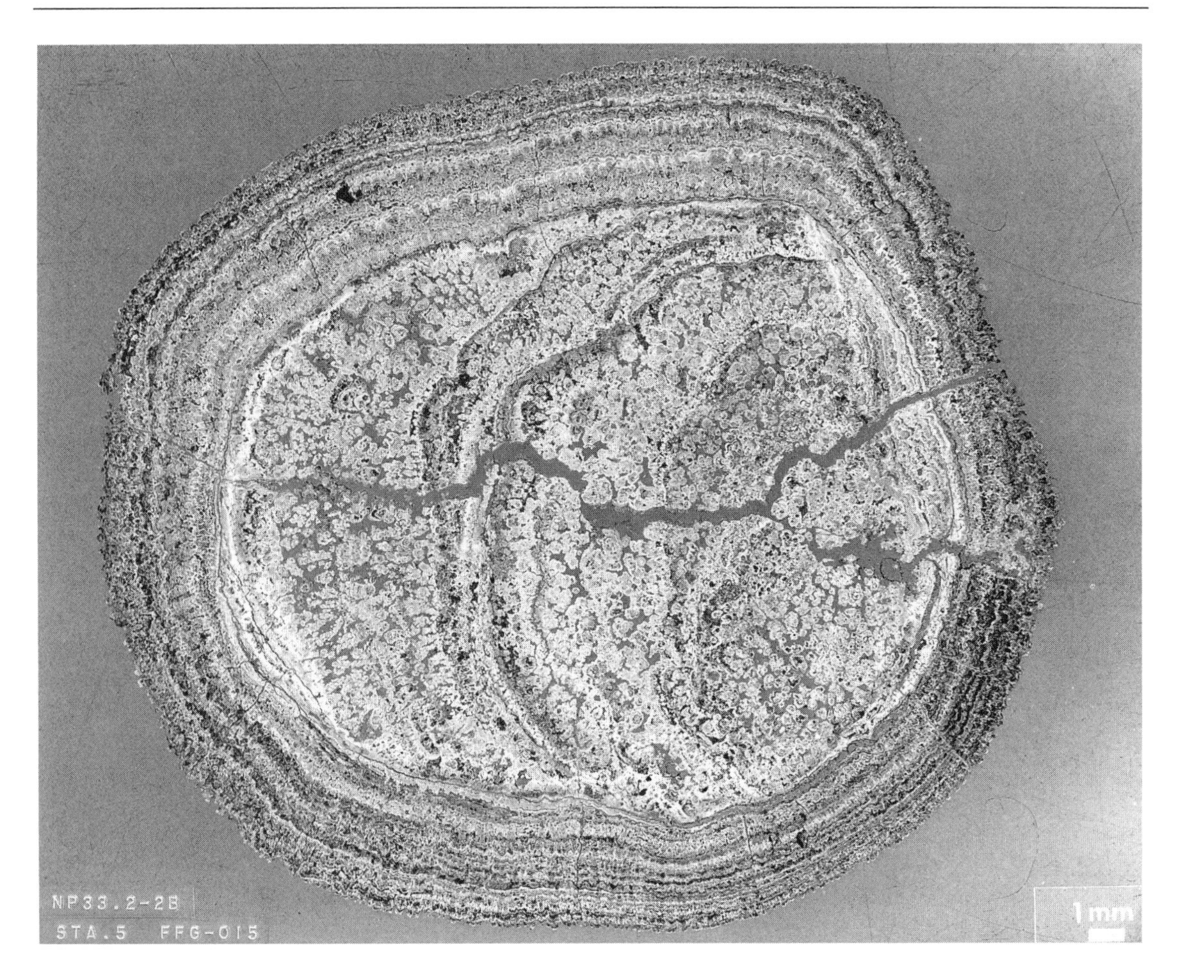

Figure 7.30. Section through manganese nodule NP 33.2–2B from the equatorial Pacific Ocean. This nodule, like many others, consists of a fragment of an older nodule that has been overgrown by later manganese and iron minerals. Bulk composition: Mn, 31.2%; Fe, 9.49%; Cu, 1.40%; Ni, 1.26%; Co, 0.15%. (Courtesy R. K. Sorem)

manganese nodules (see fig. 7.30). These consist largely of iron and manganese oxides, but they also contain enough nickel, cobalt, and copper to make them interesting targets for mineral exploration (see table 7.8). Considerable thought was given to mining manganese nodules on the floor of the Pacific Ocean in the 1970s, when metal prices were high. Some of the richest nodule areas are fairly close to the Hawaiian Islands (fig. 7.31). Understandably, the prospect of nearby ocean floor mining and onshore processing was greeted with mixed feelings in Hawaii. The nodules were to be dredged off the ocean floor. Such dredging would have destroyed large areas of bottom-dwelling communities and would have put large quantities of mud into suspension. These environmental concerns and questions of ownership played a role in discouraging the mining ventures. The most important impediment, however, was economics. The drop in metal prices in the 1980s simply made the proposition unattractive. Land-based sources were able to supply copper, nickel, and cobalt more cheaply. Interest in

Table 7.8. Average Abundance of Mn, Fe, Ni, Co, and Cu in Ferromanganese Oxide Concentrations from the Atlantic, Pacific, and Indian Oceans and the Oceanic Average (in wt %) (Cronan 1976)

	Atlantic			*Pacific*		
	Average	*Maximum*	*Minimum*	*Average*	*Maximum*	*Minimum*
Mn	16.180	37.690	1.320	19.750	34.60	9.870
Fe	21.200	41.790	4.760	14.290	32.73	6.470
Ni	0.297	1.410	0.019	0.722	2.37	0.161
Co	0.309	1.010	0.017	0.381	2.58	0.052
Cu	0.109	0.884	0.022	0.366	1.97	0.034

	Indian			*Oceanic*
	Average	*Maximum*	*Minimum*	*Average*
Mn	18.030	29.16	11.670	17.990
Fe	16.250	26.46	6.710	17.250
Ni	0.510	2.01	0.167	0.509
Co	0.279	1.04	0.068	0.323
Cu	0.223	1.38	0.029	0.233

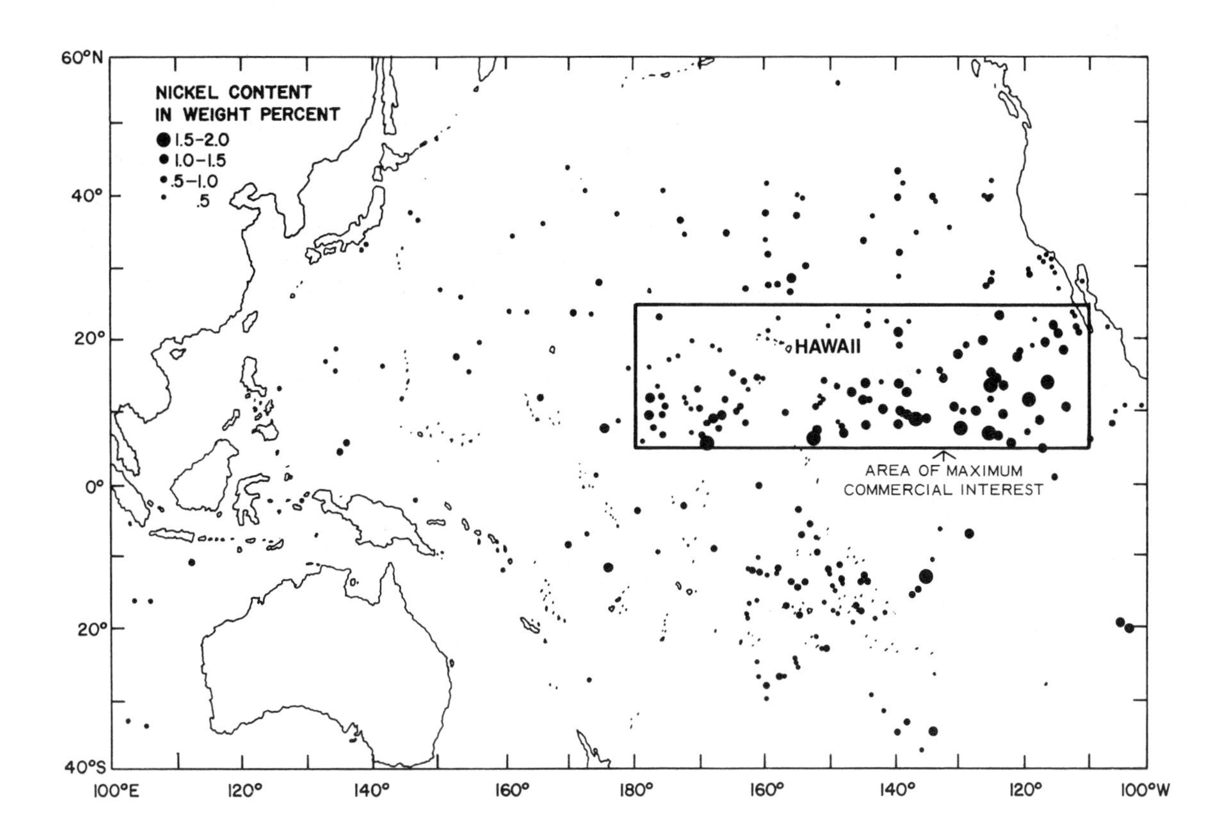

Figure 7.31. Ferromanganese deposits in the Pacific Ocean. (Jenkins et al. 1981)

mining manganese nodules may well revive in the next century, when the deposits of copper, nickel, and cobalt that are now being mined have been exhausted, especially now that the Law of the Sea treaty has been modified and has won the support of the U.S. government.

In the meantime, manganese nodules will continue to be of great scientific interest. They grow extremely slowly; it has taken some nodules several million years to grow to their present diamater of a few centimeters; yet they lie on top of sediments that accumulate at a rate of about one millimeter per 1,000 years. Somehow the nodules have avoided being buried in these sediments. It has been suggested, but with a wink and a twinkle, that there are marine animals who like to roll nodules along the ocean floor. Ocean bottom currents may provide a more prosaic answer.

7.8 Banded Iron Formations

Most of the world's great iron ore deposits are in sedimentary rocks deposited between 3,800 and 1,800 m.y.a. During these 2 billion years of Earth history, great thicknesses of iron-rich sediments were precipitated in several parts of the world (see fig. 7.32). The oldest of these deposits is in the oldest-known sequence of sedimentary rocks, at the edge of the ice sheet near Isua in western Greenland. The youngest, the Sokoman Iron Formation, is in the Labrador Trough in northeastern Canada. The world's major iron-ore-producing countries are listed in table 7.9.

The Hamersley Basin in Western Australia (fig. 7.33) is currently one of the most prolific sources of iron ore. Plate 21 shows a rather typical outcrop of banded iron formation in this basin. Its banding is due to an alternation of layers of quartz (SiO_2) and iron oxides, usually a mixture of magnetite (Fe_3O_4) and hematite (Fe_2O_3). The composition of these oxide facies BIFs (banded iron formations) is surprisingly uniform the world over, as shown in figure 7.34. They

Figure 7.32. The location of some of the major iron deposits in the world. The banded iron formations constitute the world's major iron reserves. (Craig, Vaughan, and Skinner 1988)

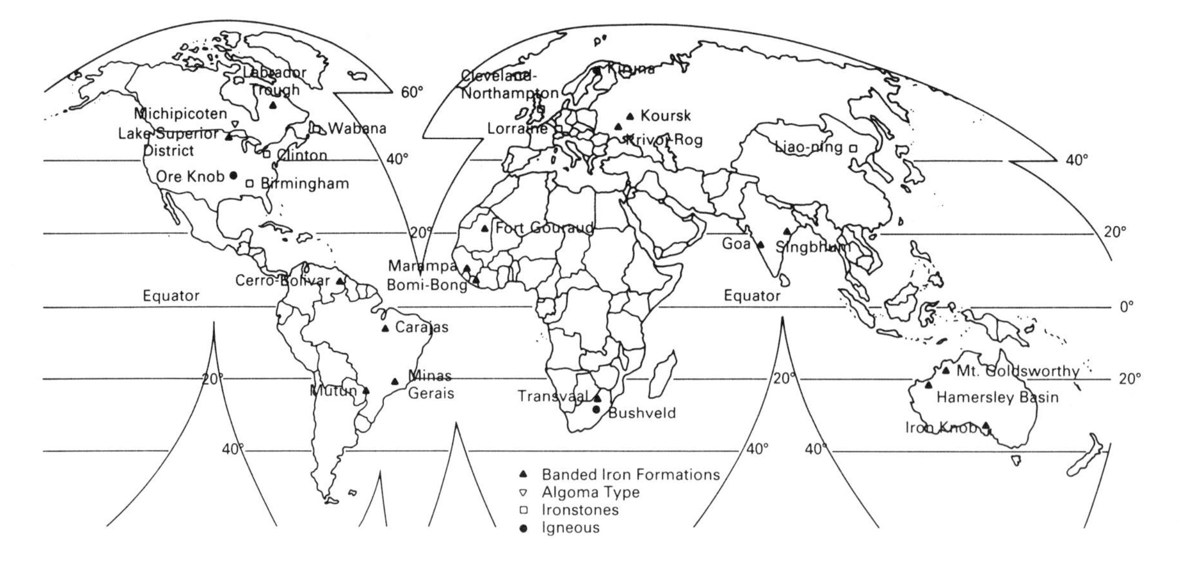

Table 7.9. Production and Reserves of Iron Ore (in millions of metric tons of usable ore)

Country	Production	Crude Ore Reserves	Crude Ore Reserve Base	Iron Content Reserves	Iron Content Reserve Base
USSR (former)	200	63,700	78,000	23,500	29,000
Brazil	156	11,100	17,300	6,500	10,100
China	120	9,000	9,000	3,500	3,500
Australia	118	16,000	28,100	10,200	17,900
India	58	5,400	12,100	3,300	6,300
United States	56	16,100	25,200	3,800	6,000
Subtotal	708	121,300	169,700	50,800	72,800
Other countries	137	28,700	60,300	14,200	27,200
World total	845	150,000	230,000	65,000	100,000

Source: *Mineral Commodity Summaries*, 1993.

World Resources: World resources are estimated to exceed 800 billion metric tons of crude ore containing more than 230 billion tons of iron. U.S. resources are estimated at about 110 billion metric tons of ore containing about 27 billion tons of iron. U.S. resources are mainly low-grade taconite-type ores from the Lake Superior district that require beneficiation and agglomeration for commercial use.

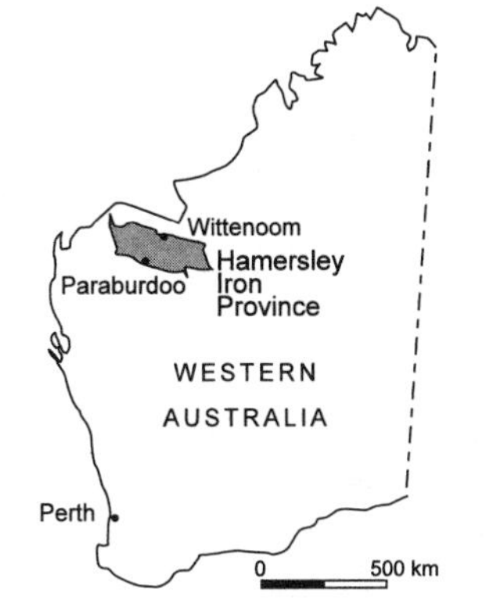

Figure 7.33. Locality map of the Hamersley Basin iron formations. (Ewers 1980)

usually contain about 45% by weight (wt %) iron oxides which together account for about 30% iron.

In some areas these BIFs have been changed into higher-grade iron ores by the removal of nearly all of their quartz. The mechanism by which the quartz has been removed is poorly understood. Since quartz is the only major diluent of the iron oxides in oxide facies BIFs, the quartz-free ores consist almost entirely of iron oxides. Such ores were mined extensively in the Lake Superior district of the United States during the nineteenth and early twentieth centuries. The most famous of the "iron ranges" was the Mesabi. After these rich ores were exhausted,

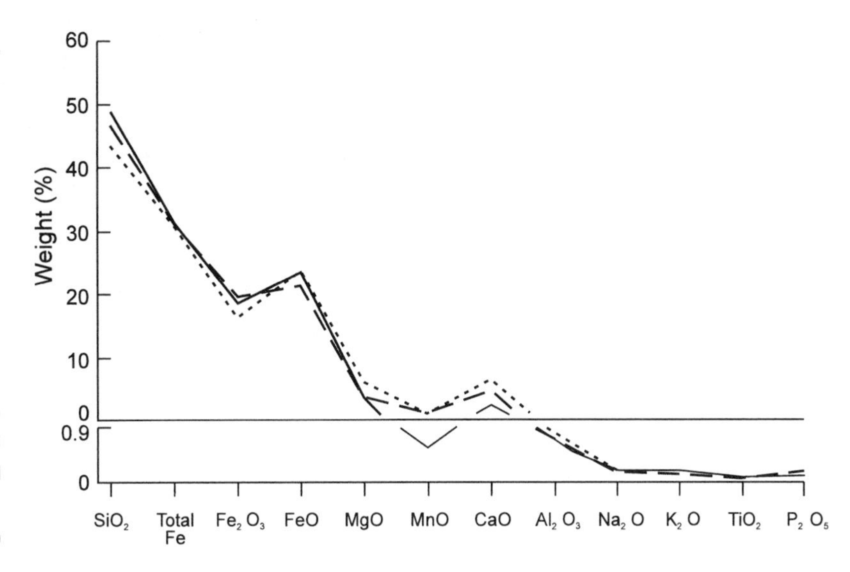

Figure 7.34. The average chemical composition of several banded iron formations. (Gole and Klein 1981)

Table 7.10. Chemical Composition of Sedimentary Rocks

	Average Igneous Rock	*Average Shale*	*Average Sandstone*	*Average Limestone*	*Average Sediment**	*Average Sediment†*	*Average Sediment‡*
SiO_2	59.14	58.10	78.33	5.19	58.49	59.7	46.20
TiO_2	1.05	0.65	0.25	0.06	0.56	—	0.58
Al_2O_3	15.34	15.40	4.77	0.81	13.08	14.6	10.50
Fe_2O_3	3.08	4.02	1.07	0.54	3.41	3.5	3.32
FeO	3.80	2.45	0.30	—	2.01	2.6	1.95
MgO	3.49	2.44	1.16	7.89	2.51	2.6	2.87
CaO	5.08	3.11	5.50	42.57	5.45	4.8	14.00
Na_2O	3.84	1.30	0.45	0.05	1.11	0.9	1.17
K_2O	3.13	3.24	1.31	0.33	2.81	3.2	2.07
H_2O	1.15	5.00	1.63	0.77	4.28	3.4	3.85
P_2O_5	0.30	0.17	0.08	0.04	0.15	—	0.13
CO_2	0.10	2.63	5.03	41.54	4.93	4.7	12.10
SO_3	—	0.64	0.07	0.05	0.52	—	0.50
BaO	0.06	0.05	0.05	—	0.05	—	—
C	—	0.80	—	—	0.64	—	0.49
	99.56	100.00	100.00	99.84	100.00	100.00	100.13

Source: Mason and Moore 1982.
* Shale 80, sandstone 15, limestone 5.
† Garrels and Mackenzie 1971.
‡ Ranov and Yaroshevsky 1969.

mining turned to the unenriched "taconite" ores. Australia still possesses large reserves of very high-grade ore, which is mined by low-cost open pit methods (plate 22), and that is largely shipped to Japan for smelting.

Very large quantities of iron ore are mined each year (see table 7.9), but the reserves of iron ore are truly enormous. The tonnage of taconite in the Hamersley Basin and in the other BIF regions of the world exceeds eight hundred times the present annual production of iron ore. This is important. Most of the world's mining bill pays for ores of three metals: aluminum, iron, and copper (see chapter 10). We have already seen (chapter 6) that the cost of aluminum ore will most likely stay close to its present value for a long time, certainly during the coming century, because there is so much aluminum ore at or close to the current mine grade. The same now turns out to be true for iron ore.

There is another similarity between the two metals: both are abundant elements in the Earth's crust. The average Al_2O_3 content of igneous rocks is about 15 wt %, their average content of FeO + Fe_2O_3 is 7 wt % (see table 7.10). This means that the gap between the iron content of the leanest minable iron ore today (FeO + $Fe_2O_3 \approx 40\%$) and the iron content of average igneous rocks is only about a factor of six. Hence the cost of iron oxide in real terms will probably never exceed the cost of iron oxide in taconite by much more than a factor of six.

The origin of the BIFs is still somewhat of a puzzle. How were such enormous quantities of iron concentrated in such relatively small areas on the world's continents? To what do the BIFs owe their banding? Why did nearly all of the BIFs form between 3,800 and 1,800 m.y.a.? Why did a few, somewhat different BIFs, form more recently, between about 600 and 900 m.y.a.? There are answers to these questions, but few of them are very firmly established. Black smokers were probably the source of the iron in BIFs. The strongest evidence in favor of this view is the fingerprint of basaltic mantle rocks in the isotopic composition of the element neodymium, one of the rare earths, which is present in trace amounts in BIFs. It is also difficult to find another source of iron that could be sufficiently large to account for the iron in the giant BIF deposits.

This answer does, however, raise another serious question: How were the iron oxides concentrated in chemical sediments, some clearly deposited in rather shallow water, when today iron from black smokers is oxidized in the deep ocean and is deposited over large areas of ocean floor flanking midocean ridges? We can take at least a stab at answering this question. Evidence from the behavior of iron in paleosols (chapter 6) and from the resistance of pyrite and uraninite to weathering in conglomerates that are more than 2,200 m.y. old (see sec. 7.3) indicates that the O_2 content of the atmosphere more than 2,200 m.y.a. was very much less than at present. This suggests that very little O_2 was dissolved in surface seawater more than 2,200 m.y.a., and that the deep oceans were probably O_2-free, i.e., anoxic. Black smoker solutions entering such an ocean would have deposited sulfide minerals to the extent that H_2S was available. Metals that were not deposited as sulfides would have accumulated in the deeper parts of the oceans until they welled up into the shallower parts, perhaps as indicated very schematically in figure 7.35. Deposition of iron oxides in shallow waters could have followed the oxidation of Fe^{2+} by trace amounts of O_2 dissolved in surface ocean waters, or the conversion of Fe^{2+} to Fe^{3+} by short wavelength solar ultraviolet radiation that was able to penetrate to the surface of the Earth in the absence of a stratospheric ozone UV screen. In some BIFs, iron is present either as an iron carbonate, an iron silicate, an iron sulfide, or a mixture of these. Such different mineral assemblages can be explained in terms of differences in the location and chemistry of the deposition areas. However, the reason for the alternation of iron minerals and quartz is still somewhat of a mystery.

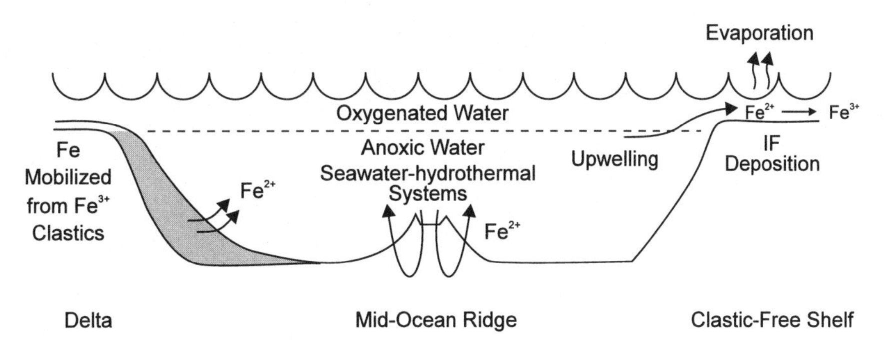

Figure 7.35. Model for the deposition of Lake Superior–type iron formations (after Drever 1974). Deeper water, enriched in Fe^{2+} from either volcanic or diagenetic sources, moves up onto a shallow shelf, where iron minerals and SiO_2 are precipitated as a result of oxidation, mixing, cooling, and possibly evaporation.

The period of major BIF deposition probably ended when the deeper parts of the oceans became oxygenated. If our admittedly slim store of evidence is interpreted correctly, the oxidation of the deep oceans occurred about 1,800 m.y.a. The brief reappearance of BIFs in late Precambrian sequences might be related to a recurrence of anoxia in significant parts of the deeper oceans.

7.9 From Continental Drift to Seafloor Spreading and Plate Tectonics

Anyone enamored of globes and world maps is apt to be struck by the complementarity of the western edges of Europe and Africa with the eastern edges of North and South America. The two continental areas look like adjacent pieces of a global jigsaw puzzle (fig. 7.36). Could they once have been parts of a single supercontinent that somehow broke apart? The idea is surely intriguing, but is it right? After all, one can imagine geologic processes that could have created these matching shorelines in other ways. One test of the idea is to see whether the fossils and rocks on the west side of the Atlantic Ocean match those on the east side when the two land masses are fitted together, as in figure 7.36. Alfred

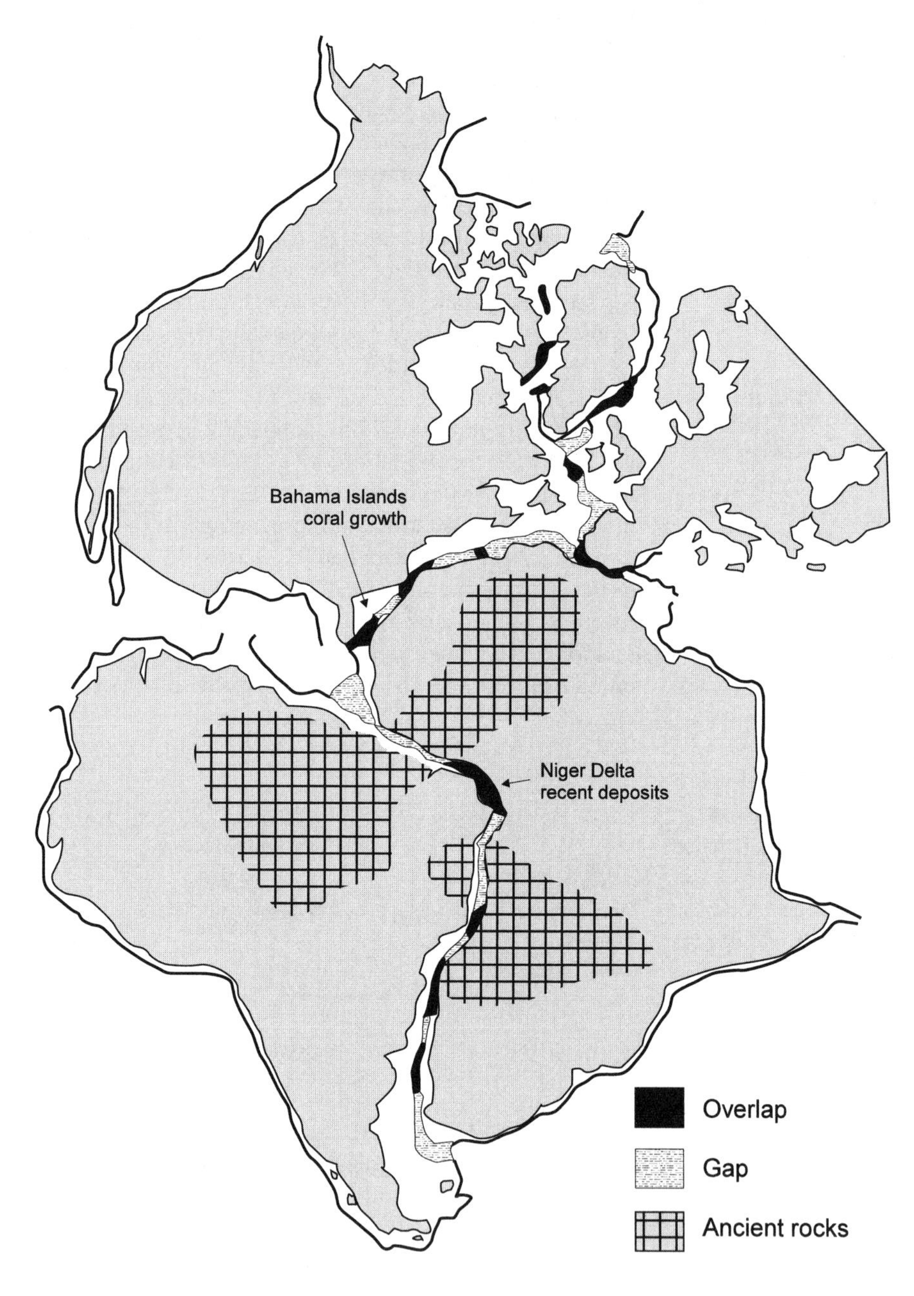

Figure 7.36. Continents on both sides of the Atlantic fit along the 500 m depth contour. Areas of overlap are sites of coral reef growth or sediment deposition since the opening of the Atlantic Basin. Areas of rocks older than 500 million years were split apart when the Atlantic opened. Several areas of Precambrian rocks are not shown. (After Bullard et al. 1965)

Wegener, a German meteorologist, championed the idea of continental drift and looked into the evidence carefully early in the twentieth century (Wegener 1929). He became convinced that the match was excellent, and that these land masses had once been part of a supercontinent. In 1912 he first proposed his theory of continental drift. The notion was rather variously received. In some countries, geologists accepted Wegener's ideas enthusiastically, because they explained so much of their geology. In other countries, continental drift was rejected, in part because the mechanism Wegener advocated to explain the movement of the continents was physically implausible (Jeffreys 1929). In Wegener's theory, the continents moved through the rocks of the Earth's mantle much like icebergs move through polar oceans. The forces that drive icebergs are easy to appreciate, those that could drive continents through mantle rocks are not. Opposition to the theory of continental drift was particularly virulent in the United States (Van Watterschoot van der Gracht 1928).

There matters stood until the end of World War II. In the late 1940s, Maurice Ewing and his associates, at what is now the Lamont-Doherty Geological Observatory of Columbia University, showed that the layer of sediments in the middle of the oceans is very thin. If these sediments represented the accumulated debris of 4,500 m.y. of Earth history, their rate of accumulation was unbelievably slow. Could they have accumulated in much less than 4,500 m.y.? This alternative demanded that the ocean floor is much younger than the age of the Earth. Could this be due to continental drift? If the Americas really split away from Europe and Africa not so long ago, the floor of the Atlantic Ocean would be no older than the separation of these land masses. The validity of this view has now been established by the Ocean Drilling Program, first with the drilling ship *JOIDES Challenger*, which began operating in 1968, and more recently with the *JOIDES Resolution* (plate 19). Sediments are very thin and geologically very young on the flanks of the midocean ridges, including the Mid-Atlantic Ridge. Away from the MORs, in the direction of the continents, the thickness of sediments increases, as does the age of the oldest sediments encountered by the drill bit before it enters the oceanic crust. The oldest sediments in any of the oceans are less than 200 million years old. Two hundred million years is a small fraction of the age of the Earth, small enough to explain the bothersomely small thickness of deep-sea sediments found by Maurice Ewing.

These observations are best explained by way of seafloor spreading, a theory developed about 1960 by Harry H. Hess (Hess 1962). If new ocean floor is created along the crest of the MORs by the injection of basaltic magmas, the oceanic crust and the first sediments deposited on it become progressively older as they move away from the MORs. The creation of new ocean floor at the ridge crests moves older ocean floor away from the ridges. In Hess's words: "A more acceptable mechanism is derived for continental drift whereby continents ride passively on convecting mantle instead of having to plow through oceanic crust." In the process, continental masses that were once joined together become separated. The continents are carried along as parts of large plates, hence the expression "plate tectonics" (see chapter 8).

Seafloor spreading and plate tectonics were accepted before the overwhelming sedimentologic evidence from deep-sea drilling was assembled. Most of the skeptics were converted in the 1960s by evidence from measurements of rock magnetism. The Earth behaves somewhat as if there were a large magnet inside the Earth. Actually, the magnetic field of the Earth is generated by the movement of molten material, probably largely iron, in the outer part of the Earth's core some 2,900 kilometers below the Earth's surface. The exact position of the magnetic poles of the Earth shifts sufficiently rapidly, so that precise compasses need to be adjusted every few years; but the magnetic poles have stayed fairly close to the position of the geographic poles, that is, close to the rotation axis of the Earth. At irregular intervals of some 10^5 to 10^6 years, the poles switch positions: the North magnetic pole becomes the South magnetic pole, and vice versa. Although the "whys and wherefores" of these switches are still a matter of debate, their occurrence and their timing are well recorded in the magnetic properties of igne-

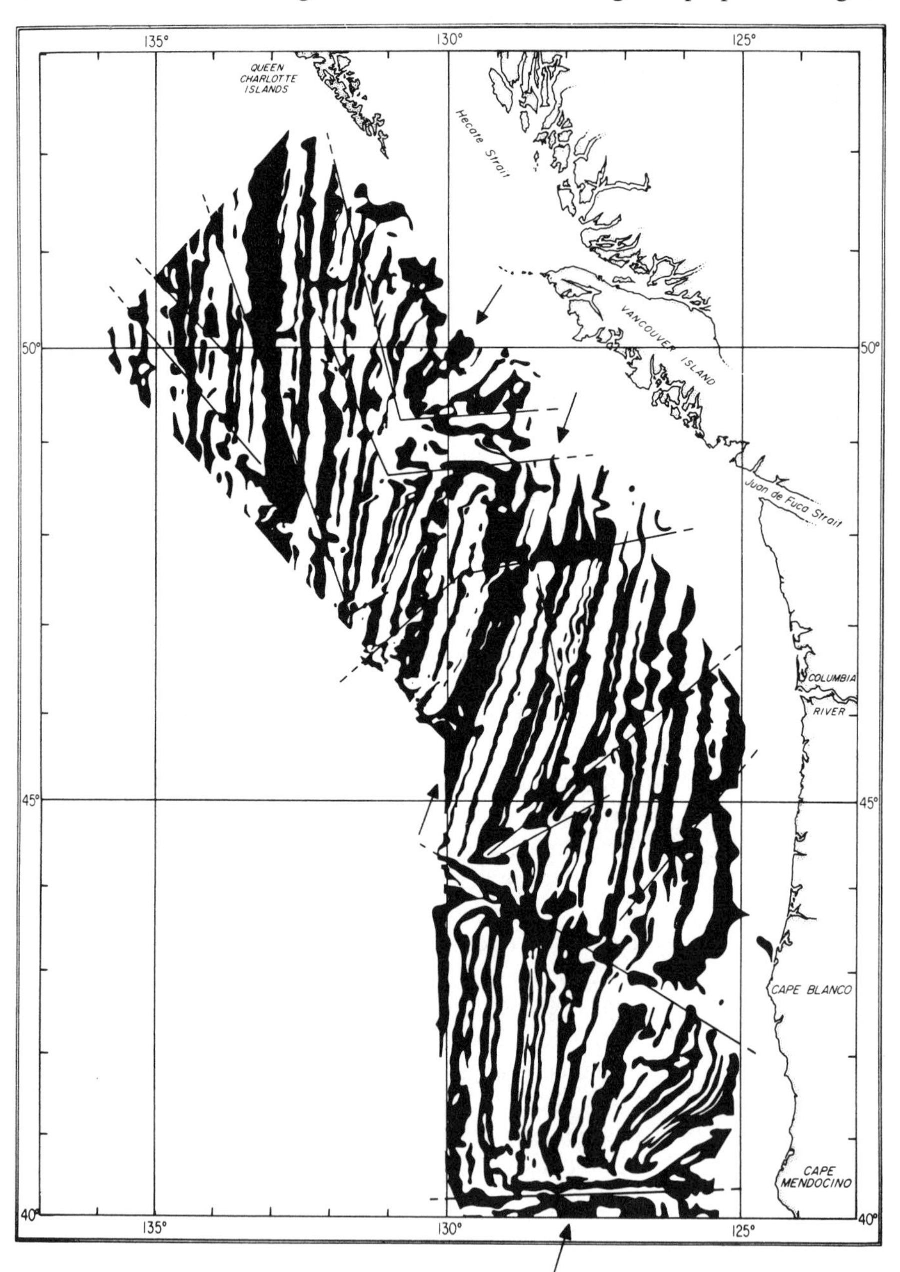

Figure 7.37. Magnetic striping on the floor of the eastern Pacific Ocean. (Mason and Raff 1961)

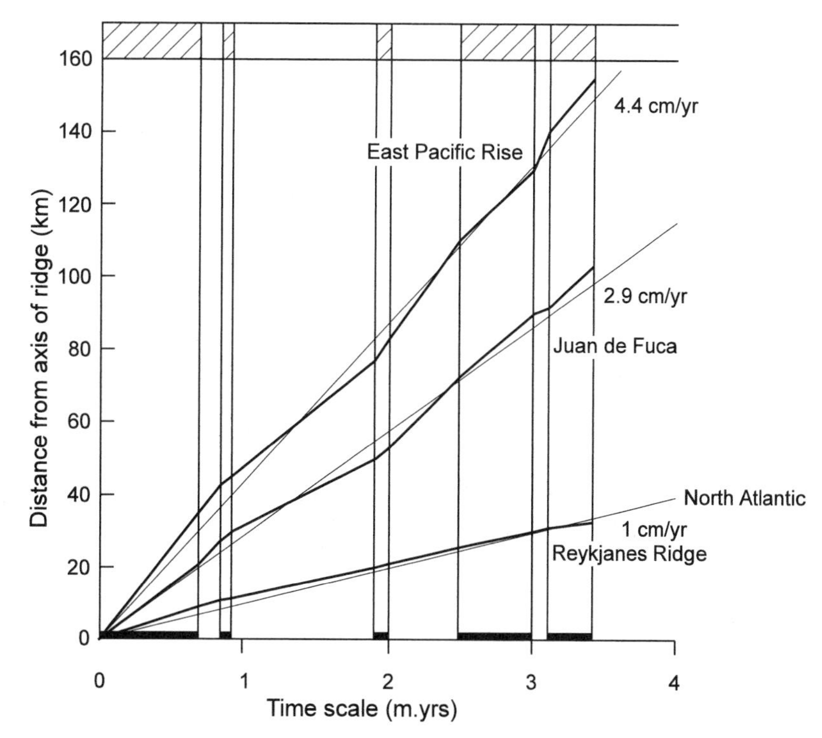

Figure 7.38. Normal-reverse magnetic orientation boundaries within the crust plotted against the timescale of magnetic reversals. (Vine 1966)

ous and sedimentary rocks. When basaltic melts begin to crystallize, the minerals that form are initially too hot to retain a magnetic orientation. However, as they cool, they pass through what is called their *Curie temperature*. Below their Curie temperature, most minerals assume a magnetic orientation parallel to the Earth's magnetic field. The magnetic orientation of a lava flow that cools today is parallel to the present-day field at the point where the flow cools. This is called "normal" magnetic polarity. If the magnetic field reverses direction, later lava flows will have a reversed magnetic polarity, but the existing lava flows will retain their original polarity.

The timing of reversals of the Earth's magnetic field was determined by studying the polarity of the magnetic field of well-dated lavas. In the late 1950s a peculiar "striping" of the magnetic properties of rocks in the oceanic crust was observed with shipboard magnetometers (fig. 7.37). The connection between the two phenomena was pointed out independently by Morley and Larochelle (1964) and by Vine and Matthews (1963), who proposed that the striping was due to the effect of reversals in the Earth's magnetic field. When these authors plotted the distance from the crest of MORs to rocks on the ocean floor that recorded successive magnetic reversals, they obtained a very good correlation with the age of the reversals (see fig. 7.38). These data showed that the ocean floor has acted somewhat like a magnetic tape that recorded the alternations in the direction of the Earth's magnetic field.

The slope of the lower-most line in figure 7.38 shows that the continental masses bordering the North Atlantic Ocean have been moving away from the Mid-Atlantic Ridge at a rate of about 1 cm/yr during the last million years. The

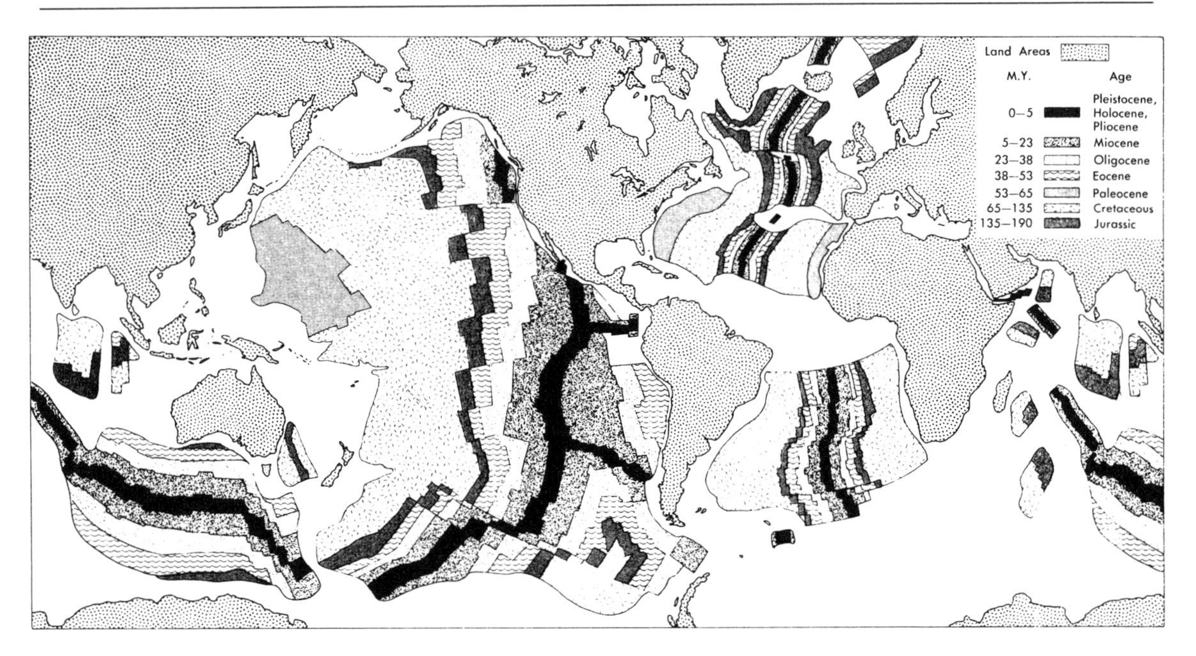

Figure 7.39. Age of the ocean floor based on magnetic anomaly patterns. (From Pitman et al. 1974; Kennett 1982)

distance between the continents has therefore been increasing at about 2 cm/yr. The distance between Halifax, Nova Scotia, and Dakar is 6,400 kilometers (see fig. 7.2). The age of the initial rifting between these continental masses is 200 million years. Therefore, the mean spreading rate since then has been about 3.2 cm/yr, or about 50% higher than the recent rate.

Seafloor spreading is neither constant nor simple. The spreading rate of the East Pacific Rise (see fig. 7.38) is currently somewhat less than 10 cm/yr but was apparently more rapid between 80 and 115 m.y.a. While the pattern of magnetic stripes on the floor of the Atlantic Ocean is reasonably simple, their pattern in the Pacific Ocean is quite complex (see fig. 7.39). This is because the geometry of movements on the Pacific Ocean floor has changed fairly frequently. The Emperor Sea Mounts–Hawaiian Island chain is a good illustration of one such change. The source of the Hawaiian lavas seems to have remained in approximately the same position for the past 70 m.y. As was pointed out in chapter 6, the floor of the Pacific Ocean has moved over this source of lava much like a piece of cloth moves through a sewing machine. Each island is the equivalent of a stitch. The bend in the guyot-island chain some 40 m.y.a. (see fig. 6.6) reflects a change in the motion of the ocean floor over the source of lava.

About 3 km^2 of new ocean floor are created annually along the world's MORs. It has been suggested that the volume of the Earth is increasing, somewhat like a balloon that is being inflated to accommodate the new seafloor; but if that were true, geologically old seafloor should exist somewhere. None has been found, except for some small pieces that have become perched on continents. Nearly all ocean crust that was formed more than about 200 m.y.a. has somehow been destroyed. The processes accompanying the destruction of seafloor account for much of the mountain building, rock deformation, and volcanism at the edges of continents. They are also responsible for the accumulation of many of our natural resources. The next chapter is devoted to these matters.

References

Anikouchine, W. A., and Sternberg, R. W. 1973. *The World Ocean: An Introduction to Oceanography*. Prentice Hall, Englewood Cliffs, N.J.

Bally, A. W. 1979. *Continental Margins, Geological and Geophysical Research Needs and Problems*. A. W. Bally, Chairman. National Academy of Sciences, Washington, D.C.

Berner, E. K., and Berner, R. A. 1987. *The Global Water Cycle: Geochemistry and Environment*. Prentice Hall, Englewood Cliffs, N.J.

Betts, J. N., and Holland, H. D. 1991. The Oxygen Content of Ocean Bottom Waters: The Burial Efficiency of Organic Carbon, and the Regulation of Atmospheric Oxygen. In *Global and Planetary Change*, ed. L. R. Kump, pp. 5–18.

Bezrukov, P. L., Baturin, G. N., and Bliskovskiy, V. Z. 1979. The Material Composition of Oceanic Phosphorites. In *Veshchestvennyy Sostav Fosforitov*, ed. A. I. Vanshin, pp. 65–79. Izd. Nauka. Sib. Otd. Novosibirsk, USSR.

Borchert, H., and Muir, R. O. 1964. *Salt Deposits: The Origin, Metamorphism and Deformation of Evaporites*. Van Nostrand, New York.

Braitsch, O. 1971. *Salt Deposits: Their Origin and Composition*. Springer-Verlag, Berlin.

Broecker, W. S., and Peng, T.-H. 1982. *Tracers in the Sea*. Eldigio Press, Palisades, N.Y.

Bullard, E., et al. 1965. The Fit of the Continents around the Atlantic. *Philos. Trans. Roy. Soc. London* 1088:41–45.

Burnett, J. C. 1974. Phosphorite Deposits from the Sea Floor off Peru and Chile. Hawaii Institute of Physics Report No. 74–3.

Butler, G. P. 1969. Modern Evaporite Deposition and Geochemistry of Coexisting Brines, the Sabkha, Trucial Coast, Arabian Gulf. *J. Sedimen. Petrol.* 39:70–89.

Cook, P. J. 1982. World Availability of Phosphorus—An Australian Perspective. Internat. Geol. Corr. Program Project 156 (Phosphorites). IUGS-UNESCO.

Craig, J. R., Vaughan, D. J., and Skinner, B. J. 1988. *Resources of the Earth*. Prentice Hall, Englewood Cliffs, N.J.

Cronan, D. S. 1976. Manganese Nodules and Other Ferro-Manganese Oxide Deposits. In *Chemical Oceanography*, ed. J. P. Riley and R. Chester, chap. 28, vol. 5. Academic Press, New York.

Cronan, D. S. 1980. *Underwater Minerals*. Academic Press, London.

Damberger, H. H. 1991. Coalification in North American Coal Fields. In *Economic Geology U.S.*, pp. 503–522, vol. P-2 of *The Geology of North America*, ed. H. J. Gluskoter, D. D. Rice, and R. B. Taylor. The Geological Society of America.

Dingle, R. V. 1974. Agulhas Bank Phosphorites: A Review of 100 Years of Investigation. *Trans. Geol. Soc. S. Afr.* 77 (part 3):261–64.

Drever, J. I. 1974. Geochemical Model for the Origin of Precambrian Banded Iron Formations. *Bull. Geol. Soc. Amer.* 85:1099–1106.

Ewers, W. E. 1980. Chemical Conditions for the Precipitation of Banded-Iron Formations. In *Biogeochemistry of Ancient and Modern Environments*, ed. P. A. Trudinger, M. R. Walter, and B. J. Ralph, pp. 83–92. Springer-Verlag, New York.

Ferm, J. C. 1976. Depositional Models in Coal Exploration and Development. In *Sedimentary Environments and Hydrocarbons*, ed. R. S. Saxena, pp. 60–78. AAPG/NOGS Short Course, New Orleans Geol. Soc.

Ferm, J. C., Milici, R. C., and Eason, J. E. 1972. Carboniferous Depositional Environments in the Cumberland Plateau of Southern Tennessee and Northern Alabama. *Tenn. Div. Geol. Rep. Invest. No. 33.*

Garrels, R. M., and Mackenzie, F. T. 1971. *Evolution of Sedimentary Rocks*. Norton, New York.

Gehman, H. M., Jr. 1962. Organic Matter in Limestones. *Geochim. Cosmochim. Acta* 26:885–97.

Goldfarb, M. S., Converse, D. R., Holland, H. D., and Edmond, J. M. 1983. The Genesis of Hot Spring Deposits on the East Pacific Rise. *Econ. Geol. Monogr.* 5:184–97.

Gole, M. J., and Klein, C. 1981. Banded Iron-Formations through Much of Precambrian Time. *Jour. Geol.* 89:169–83.

Haymon, R. M. 1989. Hydrothermal Processes and Products on the Galapagos Rift and East Pacific Rise. In *The Eastern Pacific Ocean and Hawaii*, pp. 125–144, vol. N, of *The Geology of North America*, ed. E. L. Winterer, D. M. Hussong, and R. W. Decker. The Geological Society of America.

Heezen, B. C. 1962. The Deep Sea-floor. In *Continental Drift*, chap. 9, vol. 3, of International Geophysics Series, ed. S. K. Runcorn. Academic Press, New York.

Heezen, B. C., Tharp, M., and Ewing, M. 1959. *The Floors of the Oceans: 1, The North Atlantic*. Special Paper 65. The Geological Society of America.

Hess, H. H. 1962. History of the Ocean Basins. In *Petrologic Studies: A Volume in Honor of A. F. Buddington*, ed. A.E.J. Engel, H. L. James, and B. F. Leonard, pp. 599–620. The Geological Society of America.

Holcombe, T. L. 1977. Ocean Bottom Features. *Geojournal* 6:25–48.

Holland, H. D. 1978. *The Chemistry of the Atmosphere and Oceans*. Wiley, New York.

Holland, H. D. 1984. *The Chemical Evolution of the Atmosphere and Oceans*. Princeton University Press, Princeton, N.J.

Jeffreys, H. 1929. *The Earth: Its Origin, History and Physical Condition*. Cambridge University Press, Cambridge, U.K.

Jenkins, R. W., Jugel, M. K., Keith, K. M., and Meyland, M. A. 1981. *The Feasibility and Potential Impact of Manganese Nodule Processing in the Puna and Kohala Districts of Hawaii*. NOAA, Washington, D. C.

Kennett, J. P. 1982. *Marine Geology*. Prentice Hall, Englewood Cliffs, N.J.

Klitgord, K. D., and Hutchinson, D. R. 1985. Distribution and Geophysical Signatures of Early Mesozoic Rift Basins beneath the U.S. Atlantic Continental Margin. In U.S. Geol. Survey Circ. 946, pp. 45–61.

Klitgord, K. D., Hutchinson, D. R., and Schouten, H. 1988. U.S. Atlantic Continental Margin: Structure and Tectonic Framework. In *The Atlantic Continental Margin, U.S.*, pp. 19–55, vol. I-2, of *The Geology of North America*, ed. R. E. Sheridan and J. A. Grow. The Geological Society of America.

Lefond, S. J. 1975. *Industrial Minerals and Rocks*. 4th ed. American Institute of Mining, Metallurgical, and Petroleum Engineers, New York.

Mason, B., and Moore, C. B. 1982. *Principles of Geochemistry*. Wiley, New York.

Mason, R. G., and Raff, A. D. 1961. Magnetic Survey of the West Coast of North America, 40°–52°N Lat. *Bull. Geol. Soc. Amer.* 72:1267–70.

McArthur, J. M. 1974. The Geochemistry of Phosphorite Concretions from the Continental Shelf off Morocco. Ph.D. thesis, University of London.

McCaffrey, M. A., Lazar, B., and Holland, H. D. 1987. The Evaporation Path of Seawater and the Coprecipitation of Br^- and K^+ with Halite. *Jour. Sedimen. Petrol.* 57:928–37.

Mero, J. L. 1965. *The Mineral Resources of the Sea*. Elsevier, New York.

Morley, L. W., and Larochelle, A. 1964. Paleomagnetism as a Means of Dating Geologic Events. In Royal Soc. Canada Spec. Publ. No. 8, pp. 39–51.

Mottl, M. J., and Holland, H. D. 1978. Chemical Exchange during Hydrothermal Alteration of Basalt by Seawater, I. *Geochim. Cosmochim. Acta* 42:1103–16.

Parker, R. J. 1971. The Petrology and Major Element Geochemistry of Phosphorite Nodule Deposits on the Agulhas Bank, South Africa. M.Sc. thesis, University of Cape Town, South Africa.

Parker, R. J., and Siesser, W. G. 1972. Petrology and Origin of Some Phosphorites from the South African Continental Margin. *Jour. Sedimen. Petrol.* 42:434–40.

Patterson, R. J., and Kinsman, D.J.J. 1977. Marine and Continental Groundwater Sources in a Persian Gulf Coastal Sabkha. *Amer. Assoc. Petr. Geol., Stud. Geol.* No. 4:381–97.

Petersen, U. 1994. Mining the Hydrosphere. *Geochim. Cosmochim. Acta* 58:2387–2403.

Pitman, W. C. III, Larson, R. L., and Herron, E. M. 1974. *The Age of the Ocean Basins.* Map and Chart Series MC-6. The Geological Society of America.

Ronov, A. B., and Yaroshevsky, A. A. 1969. Chemical Composition of the Earth's Crust. In *The Earth's Crust and Upper Mantle*, ed. P. Hart, pp. 37–57. Amer. Geophys. Mem. 13.

Sheridan, R. E. 1989. The Atlantic Passive Margin. In *The Geology of North America: An Overview*, pp. 81–96, vol. A, of *The Geology of North America*, ed. R. E. Sheridan and J. A. Grow. The Geological Society of America.

Sorem, R. K., and Fewkes, R. H. 1979. *Manganese Nodules: Research Data and Methods of Investigation.* Plenum, New York.

Starinsky, A. 1974. Relationship between Ca Chloride Brines and Sedimentary Rocks in Israel. Ph.D. thesis, The Hebrew University of Jerusalem, Israel (in Hebrew).

Strakhov, N. M. 1962. *Osnovy Teorii Litogeneza*, vol. 3, *Zakonomernosti Sostava Irazmeshcheniya Aridnykh Otlozhenii.* Akad. Nauk. S.S.S.R., Geol. Inst., Moscow.

Tissot, B. P., and Welte, D. H. 1984. *Petroleum Formation and Occurrence.* 2d ed. Springer-Verlag, Berlin.

Toggweiler, J. R. 1994. The Ocean's Overturning Circulation. *Physics Today* 47:45–50.

Tucker, M. E. 1981. *Sedimentary Petrology: An Introduction.* John Wiley, New York.

van Krevelen, D. W. 1961. *Coal.* Elsevier, Amsterdam.

Van Watterschoot van der Gracht, W.A.J.M., et al. 1928. *Theory of Continental Drift: A Symposium.* American Association of Petroleum Geologists.

Vine, F. J. 1966. Spreading of the Ocean Floor: New Evidence. *Science* 154:1405–15.

Vine, F. J., and Mathews, D. H. 1963. Magnetic Anomalies over Oceanic Ridges. *Nature* 199:947–49.

Von Damm, K. L. 1990. Seafloor Hydrothermal Activity. In *Ann. Rev. Earth and Planet. Sci.*, pp. 173–204, vol. 18. Annual Reviews, Palo Alto, Calif.

Wegener, A. 1929. *Die Entstehung der Kontinente und Ozeane* (The Origin of Continents and Oceans), trans. John Biram (1966) from the 4th (1929) rev. German ed. Methuen, London.

White, D. 1925. Progressive Regional Carbonization of Coals. *Trans. Amer. Inst. Mining Metall. Engineers* 71:253–81.

Wood, M. 1985. *In Search of the Trojan War.* British Broadcasting Corporation, London.

Zherebtsova, I. K., and Volkova, N. N. 1966. Experimental Study of Behavior of Trace Elements in the Process of Natural Solar Evaporation of Black Sea Water and Sasyk-Sivash Brine. *Geochem. Internat.* 3:656–70.

8 Mountains and Fossil Fuels

8.1 Introduction: Ancient Solar Energy

The last chapter introduced the notion of seafloor spreading and noted that new ocean floor is created at midocean ridges at a rate of about 3 km^2/yr. This does not seem to be a particularly rapid rate. However it is sufficiently fast, so that the entire 350 million square kilometers of ocean floor can be created in about 120 million years. The most ancient ocean floor discovered to date is only about 200 million years old. This shows that today's rate of seafloor spreading is about average for the past 200 million years. Ocean floor older than 200 million years has been largely destroyed. At the present-day rate of seafloor spreading, the floor of the oceans would have been renewed approximately thirty times during the 4,500 million years of Earth history. This chapter deals with the processes that consume the ocean floor, and with the geologic consequences of these processes. Some of the consequences are important for human society: we owe to them much of our oil and natural gas, most of our geothermal energy, and many important metallic ore deposits (chapter 9).

Ocean floor can be consumed within the oceans and at the boundaries between oceans and continents. Of the two locales, the second is currently the more important. In the South Atlantic, ocean floor is created along the Mid-Atlantic Ridge. South America and the floor of the Atlantic Ocean west of the Mid-Atlantic Ridge are moving westward. At the same time ocean floor is being created at the crest of the East Pacific Rise, and the floor of the Pacific Ocean east of the rise is being carried eastward (see fig. 8.1). These westward- and eastward-moving segments of the Earth's crust collide along the west coast of South America. Along this coast Pacific Ocean floor is disappearing under South America. In the process, the western edge of South America is being crumpled and fractured. This process is largely responsible for the Andes, the spectacular mountain chain that runs the length of South America. In the North Pacific, ocean floor moves northward. Along the collision zone with Alaska, ocean floor disappears under the Aleutian Islands. In the western Pacific, ocean floor disappears beneath Japan and under a series of superb island arcs. The Atlantic Ocean is clearly growing today while the Pacific Ocean is shrinking, largely due to the subduction of ocean floor beneath the surrounding continents.

The circumference of the Pacific Ocean has been called the Ring of Fire, and for good reason: the subduction of ocean floor along the edges of this ocean is accompanied by extensive volcanism. However, volcanism is conspicuously absent in the world's highest mountain range, the Himalayas. These mountains have been raised by the collision of India with the Asian continent. Before the collision, as these land masses came closer, ocean floor between them was gradually destroyed. Since their collision, about 1,400 kilometers of lateral continental extent have been lost, and the area of continental crust has decreased, largely by thickening. The consequences of this continent-continent collision are spectacular, and they differ in some ways from those of ocean-continent collisions.

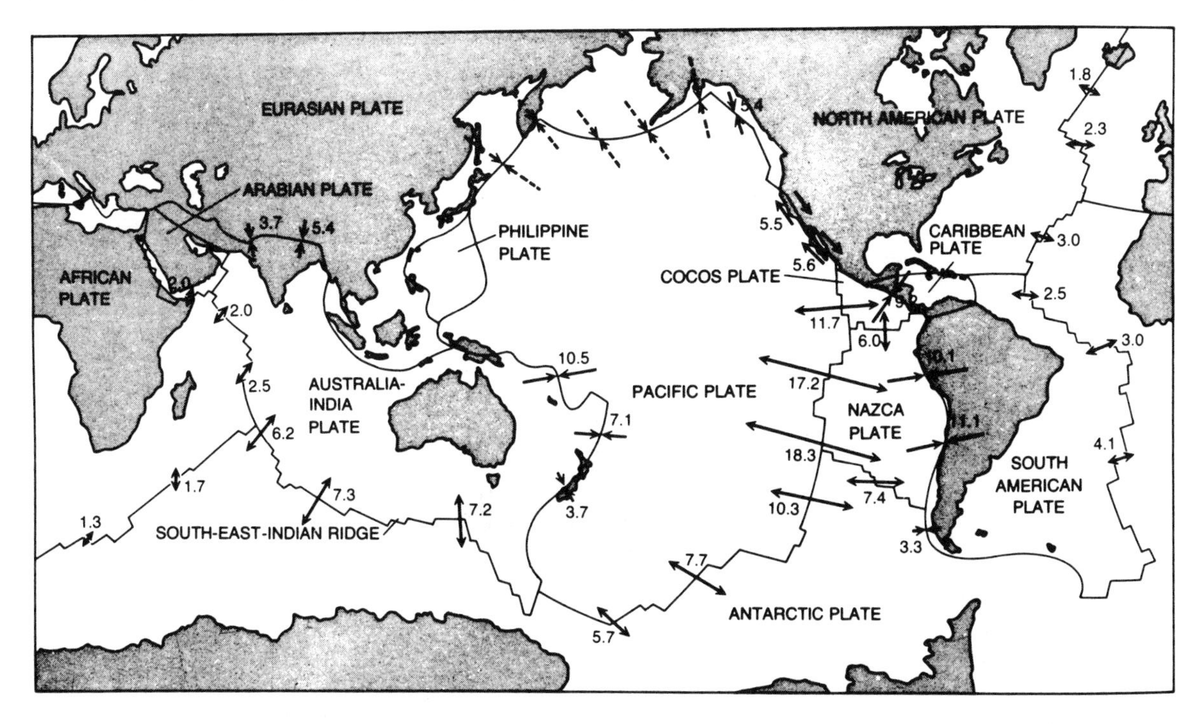

Figure 8.1. Evidence for convection in the mantle is supplied by the motions of the dozen or so large plates into which the Earth's crust is divided. The lines with arrowheads pointing outward show where plates are moving apart at ridges; numbers indicate their relative velocity in centimeters per year. Lines with opposed arrowheads show where plates are moving toward each other, usually at trenches. Plates can slide past each other along transform faults such as the San Andreas fault on the West Coast of North America. (McKenzie and Richter 1979)

The density of oceanic crust is greater than that of continental crust. Oceanic crust therefore tends to sink under continents in ocean-continent collisions. When two continents collide, their crust is apt to be shortened and thickened by thrusting.

Collisions of both types are accompanied by earthquakes. Many of these rank among the largest natural disasters that have beset mankind. In both types of collisions rocks in the collision zone are bent, twisted, broken, and lifted well above sea level. However, only ocean-continent collisions are accompanied by extensive melting of rocks in the deeper parts of subduction zones. These melts rise toward the surface. Some erupt from volcanoes as lavas, others solidify en route and can accumulate below the Earth's surface as very large bodies of igneous rocks.

8.2 Earthquakes

Earthquakes come in all sizes. Most, blessedly, are so small that they are barely noticeable. A teacup may rattle, or a sharp report may raise eyebrows. Some earthquakes, however, are enormously destructive; death tolls in excess of 100,000 are not uncommon in highly populated areas. Hurricanes, floods, earthquakes, and plagues are clearly the three most severe natural disasters that are visited on mankind. A number of scales have been devised as measures of earthquake intensity. The most commonly used, but perhaps not the best, is the Richter scale invented in 1935 by Charles F. Richter, a California seismologist. It is based on the intensity of earthquakes as recorded on seismograph records such as that in figure 8.2. *Seismographs* are instruments that record ground motions. The greater the earthquake, the more intense the ground motion and the larger the amplitude of the seismograph record. Of course, the amplitude of seismometer records also depends on the sensitivity of the seismograph and on the distance of the seismograph from the source of the earthquake. A great deal of effort has gone into relating seismographic information to the energy that is released during earthquakes (see, for instance, Bullen and Bolt 1985, 1993).

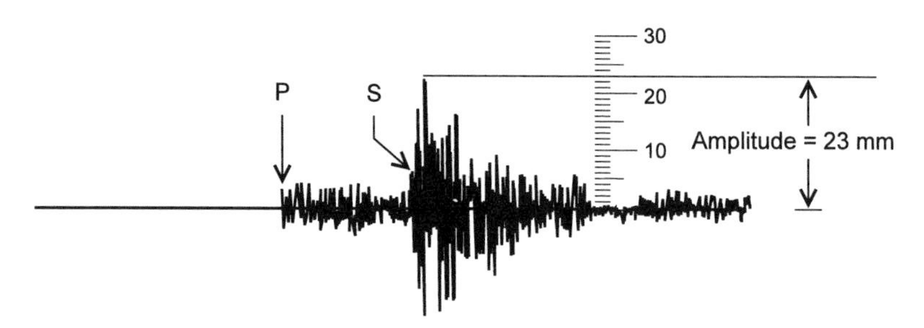

Figure 8.2. Earthquake record from a seismograph. (Bolt 1978)

The amount of energy released during an earthquake is related to its magnitude on the Richter scale in a fairly simple way. Every increase of 1.0 on this scale implies an increase of about a factor of thirty in earthquake energy. A difference of 2.0 between two earthquakes indicates that during the more intense of the two earthquakes some $30 \times 30 = 900$ times as much energy was released than during the less intense of the two earthquakes. The difference between the energy released during an earthquake of magnitude 8.5 and one of magnitude 6.5 is therefore very large. In this century, two or three earthquakes recorded on seismographs had Richter magnitudes of 8.9. The 1906 San Francisco earthquake had a magnitude of 8.25; the great Chilean earthquake of 1960 had a magnitude of 8.5, and the great Alaskan earthquake on Good Friday in 1964 had a magnitude of 8.6.

One of the worst earthquakes in European history occurred in Lisbon, Portugal, on All Saints' Day, November 1, 1755. Many of Lisbon's 250,000 inhabitants were in church when the first shock came. Suddenly the city began to shudder violently. According to one survivor, the city's tall medieval spires "waved like a cornfield in a breeze." Hundreds of people rushed out onto a new marble quay along the shore of the Tagus River, but there was no escape. With the first shock the river receded until the sandbar at its mouth was exposed. At about the time of the second shock, the water swept back in a raging 50-foot crest that surged over the quay and up into the city. By the time the earthquake and the resulting fires had subsided, some 60,000 people had lost their lives (Walker 1982).

Such disasters bring out the worst in some people, the best in others. Robbing, looting, and killing were rampant in Lisbon. Sebastião de Carvalho, secretary of state to the Portuguese king, hastened to nearby Belem, where the king and the royal family were in a state of considerable alarm, surrounded by priests who pointed out that none of this would have happened had it not been for the king's manifold sins. The only recourse now, they said, was for him to pray for belated forgiveness. The king greeted Carvalho with the desperate plea, "What can be done? What can be done to meet this infliction of divine justice?" Carvalho's straightforward reply has gone down in Portuguese history: "Sire, we must bury the dead and feed the living" (Verney 1979). The extent of the Lisbon disaster was contained by Carvalho's coolheadedness and great organizational skills.

Shortly after the Lisbon earthquake, on November 18, 1755, an earthquake shook Boston and its vicinity. The quake was much less intense, but many chimneys came down, brick buildings were damaged, and stone fences were wrecked. The quake was felt from Chesapeake Bay in the south to Nova Scotia in the north and demonstrated that even seismically quiet regions like the East Coast of the United States could be visited by significant earthquakes. Perhaps of most importance, John Winthrop IV, an astronomer and professor of mathematics at Harvard College, observed the effect of an aftershock on his fireplace. He reported that "it was not a motion of the whole hearth together, either from side to side or up and down; but each brick separately by itself." He described the motion as "one small wave of earth rolling along." A century later the wavelike nature of seismic motion became the key to understanding earthquakes (Walker 1982).

Three very destructive earthquakes struck the area near New Madrid, Missouri, between December 16, 1811, and February 7, 1812 (Penick 1981). Like Boston, Missouri is hardly earthquake country, but even in this normally quiet part of the world violent earthquakes can occur. The magnitudes of the quakes were probably about 7.5, 7.3, and 7.8 on the Richter scale (Bolt 1978). The town of New Madrid was destroyed, and the quake produced extensive changes in ground and river configurations, including the course of the Mississippi River. Remarkably few people were killed during the New Madrid earthquakes. Today the area is much more populated, and the death toll of similarly large earthquakes would be very large.

The best-remembered American earthquakes occurred in what is clearly earthquake country: California and Alaska (see fig. 8.3). The San Francisco earthquake of 1906 and the subsequent fire destroyed much of the city (see fig. 8.4a). Seven hundred persons were killed; the damage, much of it due to the fire, was estimated to be about $400 million. The campus of nearby Stanford University in Palo Alto was considerably damaged. Figure 8.4b shows the statue of Louis Agassiz, a former Harvard great and one of the first outstanding geologists in the United States, after it was shaken loose from its pedestal during the earthquake and embedded head-first in pavement. The president of Stanford University was supposed to have remarked: "My dear Agassiz, I have often contemplated you in the abstract, but this is the first time I have seen you in the concrete."

The most recent very large American earthquake shook much of the area around Prince William Sound along the southern coast of Alaska on Good Friday, 1964. It was the same area that was visited by the disastrous *Exxon Valdez* oil spill a quarter of a century later (see chapter 11). In Anchorage, moderately tall structures were damaged extensively. Landslides and slumps demolished many buildings. Docks were destroyed in several ports, and there was extensive sea-wave damage along the western U.S. coast and elsewhere. Altogether, 131 lives were lost. A panel appointed by the U.S. National Academy of Sciences to study and report on the effects of this earthquake (NAS 1973) documented the extraordinary susceptibility of certain water-saturated soils and loose accumulations of river sediments (alluvium) to liquefaction and hence to large-scale

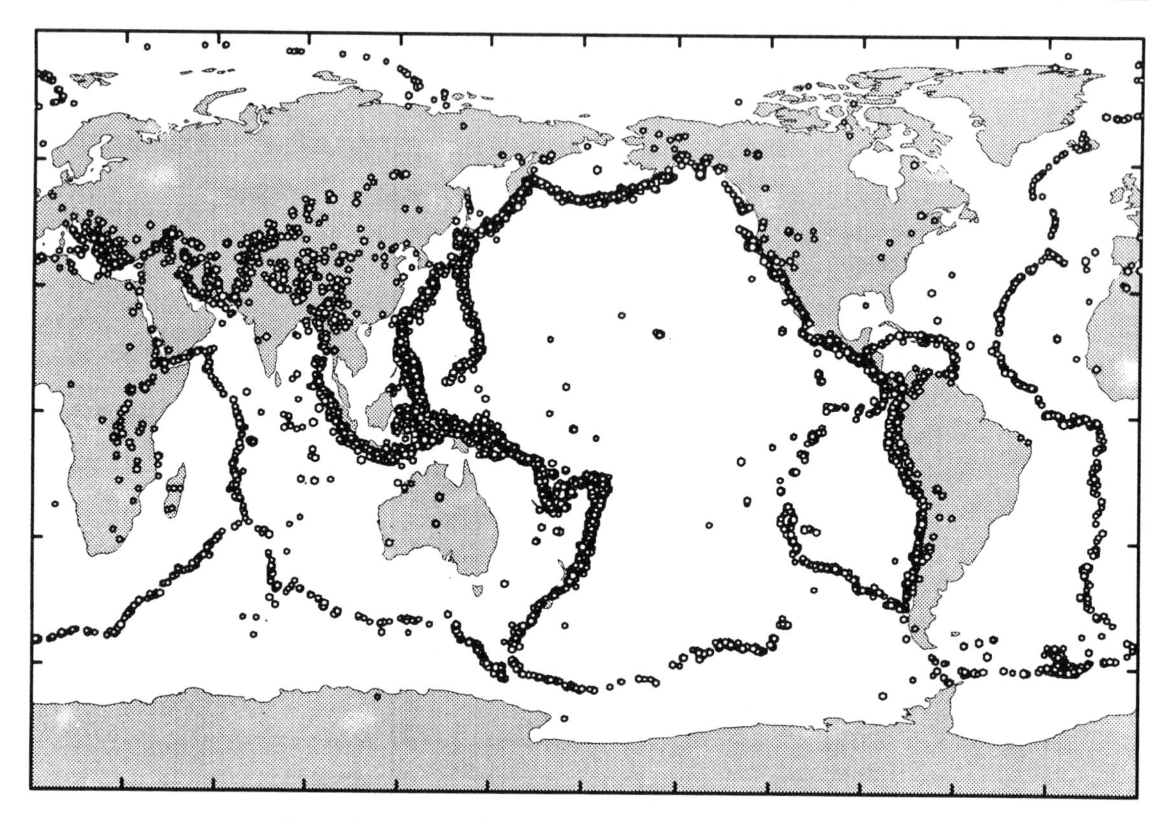

Figure 8.3. A seismic map of the world, showing the location of earthquakes of magnitude greater than 5.0 that occurred at a depth less than 70 km between 1977 and 1994; collectively they clearly show the principal earthquake zones of the world. (Courtesy of G. Ekstrom)

sliding under the influence of seismic shocks. Earthquake damage to buildings reflected not only the magnitude of the earthquake but the physical properties of the ground on which they were built. This was also observed in the 1989 Loma Prieta earthquake, which registered 7.1 on the Richter scale (see plate 23). Only the Marina district in San Francisco, where houses were built on landfill, were damaged significantly by this earthquake. If there is a repetition of the 1906 earthquake in the San Francisco area, buildings built on loose landfill will be in much greater danger than those built on solid rock.

The extent of damage to buildings also depends on their architectural design and on the materials used in their construction. In the 1964 Alaska earthquake, even some structures built in accordance with accepted practice and conforming to the requirements of building codes for seismically active areas performed poorly. Since 1964, significant advances have been made in the design of earthquake-resistant buildings and bridges, and these improvements should reduce damage caused by earthquakes in the future. On January 17, 1994, the Northridge section of the San Fernando Valley in Los Angeles was subjected to an earthquake of magnitude 6.7. Of the two thousand bridges in the epicentral region, six failed and four were so badly damaged that they had to be replaced. Plate 24 shows one of the collapsed connector structures at the Golden State Freeway interchange. The failures were primarily due to fracturing of the sup-

Figure 8.4a. On foot and in carriages and wagons, San Franciscans crowd Market Street after the 1906 earthquake to view the still-smoldering ruins of the city's business district. In the distance, atop Nob Hill, is the famed Fairmont Hotel, its handsome marble façade intact but its interior severely damaged. An unknown photographer captured this image from the Ferry Building's 235 ft tower. (Walker 1982)

Figure 8.4b. During the San Francisco earthquake of 1906, this statue of Professor Louis Agassiz was ignominiously thrown from its niche at Stanford University. (Stanford University Archives)

porting columns that had been designed and constructed before 1971. After the San Fernando earthquake of 1971, the standards for earthquake design for bridges and roads began to be toughened considerably, and bridges constructed since then sustained only minor damage and were easy to repair (Cooper et al. 1994). This is encouraging, but—as shown by the extensive damage caused by the 1995 Kobe earthquake—very large earthquakes will always produce a great deal of damage, particularly in heavily populated areas. A recent estimate suggests that a major earthquake in Tokyo could cause $1.2 trillion dollars in damage. The last very large earthquake in the Los Angeles area occurred in 1857, more than 135 years ago. An equally large earthquake is apt to occur there again, perhaps sooner rather than later. Planning for a potential disaster of such magnitude is a daunting enterprise; the emplacement of effective mitigation policies requires that there be an agreed-upon problem, a generally acceptable plan of action, people interested in working on the realization of the disaster mitigation plan, and the will to put such a plan into action (Alesch and Petak 1986).

Even if all these requirements are met, there is still one very major problem in planning earthquake disaster relief: the difficulty of predicting the time of occurrence. The earthquake-prone regions of the world are well identified (see fig. 8.3), and the stresses responsible for earthquakes are quite well understood. In a rough way, the periodicity of earthquake occurrences is also known. The great Chilean earthquake of 1960 was preceded by large earthquakes in 1575, 1737, and 1837 (see fig. 8.5). Along coastal southwestern Japan large earthquakes have also occurred reasonably regularly. But in neither region has the recurrence of large earthquakes been sufficiently regular to serve as the basis of usable predictions for the timing of future earthquakes.

Measurements of the deformation of earthquake-prone areas are more promising as indicators of impending earthquakes, but so far they have not proved particularly reliable. A large area near Palmdale in southern California (see fig. 8.6) seems to have risen about 25 centimeters between 1961 and 1974 (Raleigh 1984). This became known as the Palmdale Bulge. It was thought that the bulge presaged a large earthquake. It did not. The problem of using this type of data is somewhat like that of predicting the moment when a stick that is gradually being bent will finally break. Earthquake prediction is even more complicated, because it is a three-dimensional problem, where the properties of the stressed medium are poorly known in all three dimensions.

Earthquake predictions based on seismic evidence alone have not been very successful either. Some large earthquakes are indeed preceded by foreshocks, but others are not. In areas where foreshocks were observed, it is difficult to know whether a minor quake is really a foreshock, or whether it is simply an isolated minor quake. The success of Chinese geophysicists in predicting the 1975 earthquake in the Haicheng-Yingchow region saved thousands of lives, but it has not been followed by a string of other successes. Many unpredicted earthquakes have occurred, and many predicted earthquakes have not. The day of reliable earthquake prediction still seems a good way off.

That being so, it is fair to ask whether there might not be some way to prevent large earthquakes. During earthquakes, accumulated strain in the solid earth is suddenly released. Large earthquakes follow the accumulation of large amounts of strain. If the bodies of rock that move past each other in tectonically active regions could be separated slightly, sliding would be continuous, there would be no major hangups, and hence no large earthquakes. Several experiments, not all intentional, to test this notion have been successful. Water has been pumped into areas of rock slippage, and this has been followed by numerous small earthquakes. These quakes stopped when the water was pumped out again; therefore the basic idea is sound. However, the social and legal problems associated with the application of this remedy in areas of great earthquake hazard are formidable. If the opening of a fault plane were to set off a large earthquake, the financial liability could be enormous. Fault plane "lubrication" might be appropriate after

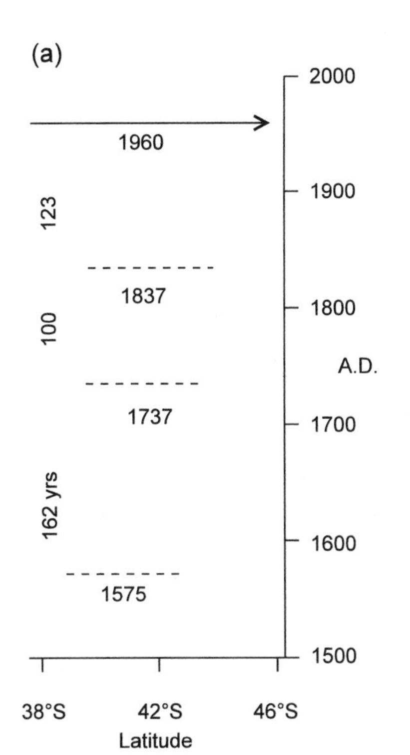

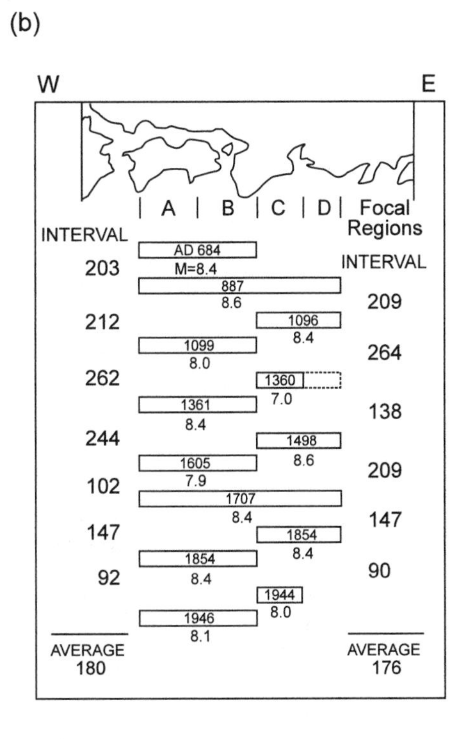

Figure 8.5. (A) Three large Chilean earthquakes in 1837, 1737, and 1575 preceded the great 1960 earthquake. (B) A remarkable series of large earthquakes occurred between A.D. 684 and 1946 in coastal southwestern Japan. Both the Chilean and Japanese localities demonstrate regularity in style and timing of large earthquakes. (Sieh 1981)

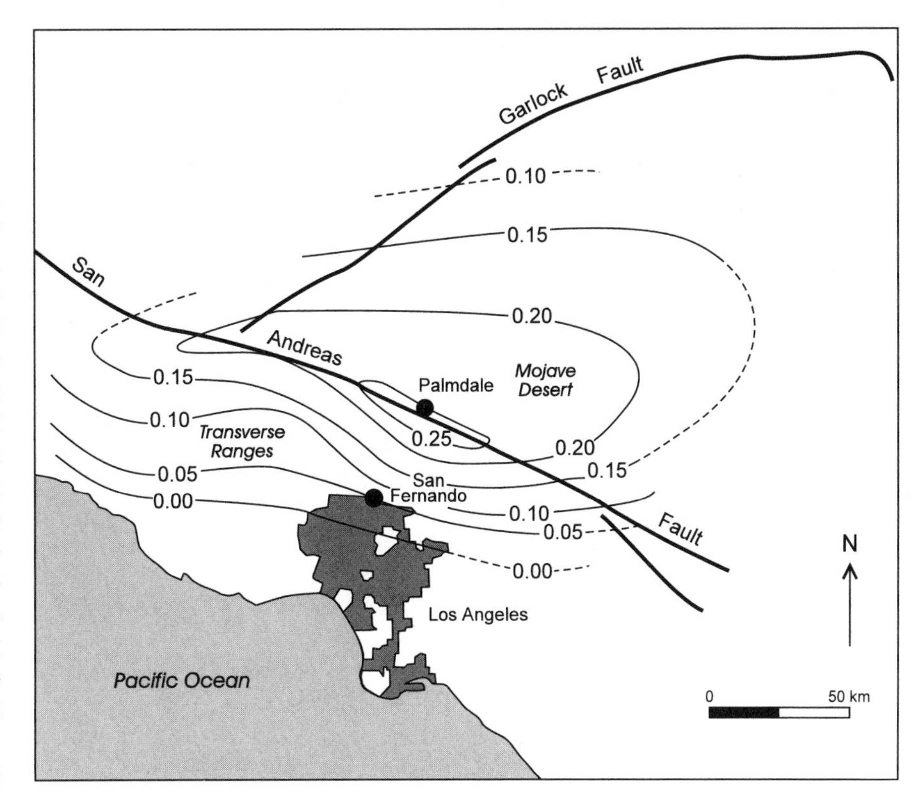

Figure 8.6. Map of a portion of southern California showing contours of the uplift of the ground surface (in meters) as indicated by leveling during the period 1959–74. The maximum uplift of over 25 cm occurred near Palmdale between the Transverse Ranges and the Mojave Desert. The heavy lines show the general positions of the San Andreas and Garlock faults. (Bolt 1978)

much of the accumulated strain in a region has just been relieved by a major earthquake, but even then the possibility of suddenly releasing a large amount of residual strain is apt to dampen the enthusiasm of most engineers for this approach to earthquake mitigation.

8.3 Deformation of the Crust

The major earthquake zones in figure 8.3 clearly coincide with the location of midocean ridges and plate convergence zones. Earthquakes associated with subduction zones are much more frequent and intense than those along midocean ridges. This difference is due in part to differences between the process of seafloor spreading and the mechanisms of seafloor subduction. In part these differences are also due to the different behavior of earth materials at ridges and in subduction zones. Rocks can deform by flowing and by breaking. Lava flows, but solidified lava breaks; yet there are all gradations between flowage and breakage, and earth materials that break under one set of conditions may well flow under others. Silly putty is a good example of a substance with highly variable mechanical properties. A ball of silly putty bounces enthusiastically when it is dropped on the floor or when it is thrown against a wall; but when squeezed slowly, silly putty changes shape readily, much like a lump of butter. The behavior of rocks is similarly variable. Rocks that break when hit with a hammer in the field can deform by flowing when they are deformed slowly at high temperatures or in the presence of fluids at high pressures. Cheeses display a range of mechanical properties similar to those of rocks. At room temperature ripe brie is decidedly runny, while feta cheese is decidedly solid and responds to deformation by breaking and crumbling. If one is so inclined, one can describe the behavior of earth materials in terms of the feta-brie coefficient invented by P. England. Rocks that behave like feta cheese have an f-b coefficient of 100, those that behave like brie an f-b coefficient of zero.

The abundance of earthquakes in subduction zones shows that rupture and the movement of rock masses past each other are important there. This is not surprising. Figure 8.7 is a diagrammatic representation of the major features of subduction zones at ocean-continent boundaries such as the western margin of South America. The ocean floor in figure 8.7 is moving to the left, the continental mass to the right—i.e., the view of South America is to the south. The depth of the ocean increases as one approaches the subduction zone from the ocean side. The maximum ocean depth is reached in the axial trench, which frequently serves as a receptacle for sediments. Landward from the trench, the ocean shallows, and ocean crust descends under an area of intensely deformed sediments, the accretionary prism. The continents on the landward side of this prism usually form a buttress or rock framework. This is often penetrated by rock melts (magmas) that feed volcanoes or that congeal into masses of igneous rocks below the surface of the Earth.

Rocks break in many parts of subduction zones. Figure 8.8 shows the distribution of earthquakes along an east-west line across the northeastern part of the Japan arc in the West Pacific. Many of the earthquakes in this subduction zone occur close to the top of the down-going slab of Pacific ocean floor, because most of the relative movement of the two plates takes place there. There is, however, a second zone of earthquakes within the down-going slab. There is also a great concentration of quakes in the accretionary prism, and shallow earthquakes are

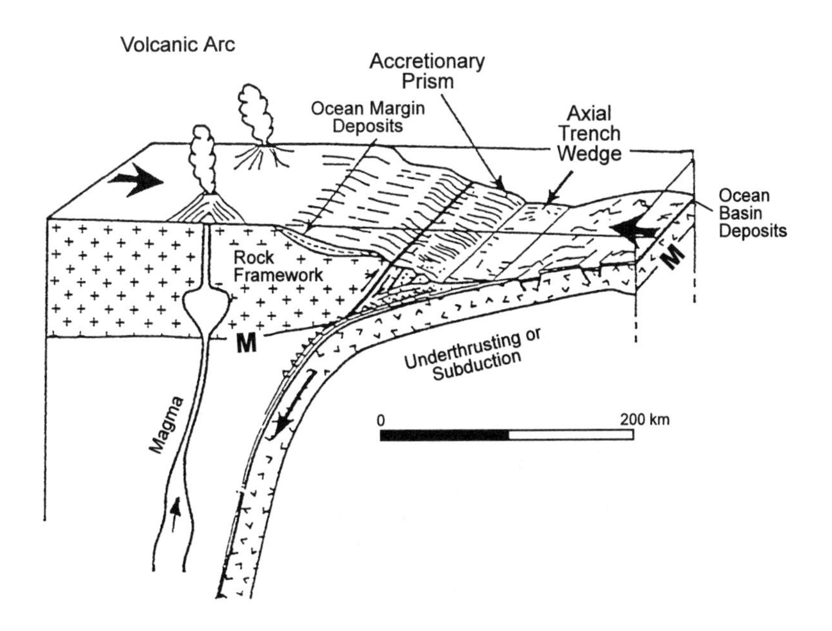

Figure 8.7. Diagrammatic portrayal of the principal sedimentary deposits and morphologic elements of a convergent ocean margin and the sediment-associated subduction processes operating there. Facing arrows indicate that the continental or terrestrial crust of the upper plate (left) converges orthogonally with the underthrusting oceanic crust of the lower plate. Seaward of the margin's framework rock, the process of subduction accretion has formed a small accretionary pile or prism of offscraped trench floor deposits. The companion process of sediment subduction is shown conveying trench and ocean basin deposits beneath the seaward base of the framework and potentially to mantle (M) depths. The diagram is only roughly to scale (vertical exaggeration 2:1). (Von Huene and Scholl 1991)

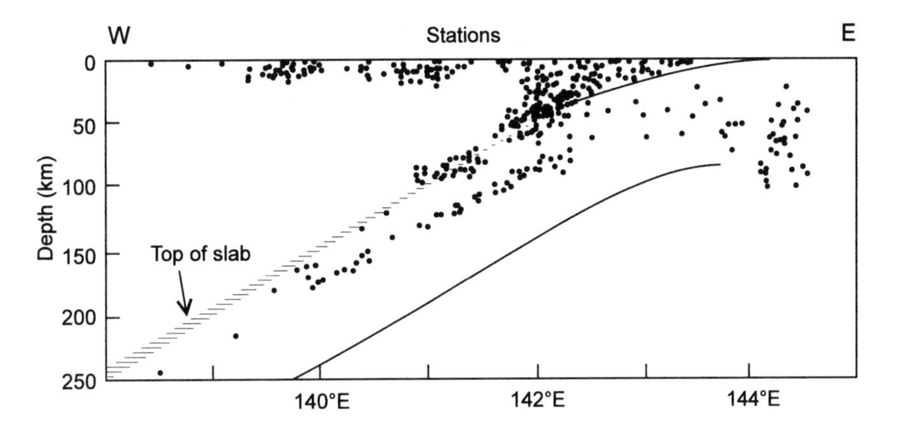

Figure 8.8. Section showing the focal depth distribution of small earthquakes along an east-west line across the northeastern part of the Japan arc and the double-planed deep seismic zone detected by the Tohoku University seismic network. The hatched zone shows the position of the boundary between the descending slab and the overlying mantle. (Bott 1982)

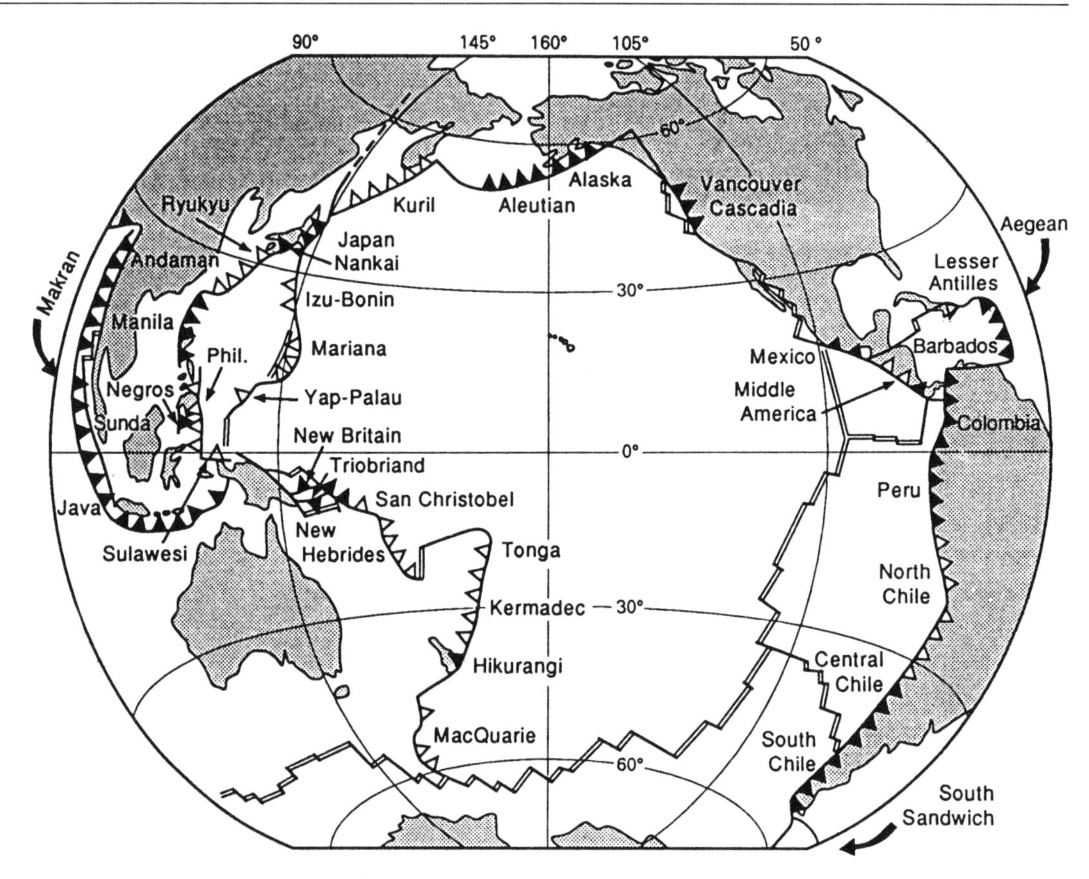

Figure 8.9. Trenches (barbed lines) and convergent margins of the Pacific, Caribbean, and eastern Indian Ocean regions. Not shown, but indicated, are the short trench systems of the Makran region of the Gulf of Oman, northwestern Indian Ocean, the South Sandwich Trench connecting South America and Antarctica, and the Aegean region of the Mediterranean. Filled barbs are accreting trenches; open barbs are trenches at which effectively no accretionary prism forms. (Von Huene and Scholl 1991)

frequent landward of the subduction zone. In some subduction zones, for example in the subduction zone below the Tonga arc (see fig. 8.9), earthquakes occur down to a depth of about 650 kilometers. Neither the mechanisms that produce these deep quakes nor the reasons for their absence below 650 kilometers are well understood.

The size of the accretionary wedge in subduction zones is quite variable. Figure 8.9 shows the global distribution of two types of convergent margins. Large accretionary prisms form in Type 1 margins. Little accretion takes place in Type 2 margins. Sediments can accumulate in subduction zones or can be subducted directly with down-going oceanic plates. In either case, much of the material in ocean-continent collision zones is ultimately subducted. It is still not clear how much of the subducted material is added to the adjacent continental blocks and how much is carried down into the mantle.

The geology along the coast of California is quite different from that along the western coast of South America. The East Pacific Rise disappears into the Gulf

Figure 8.10. Horizontal displacement of the ground during an earthquake in the Imperial Valley of California disrupted the regular pattern of trees in citrus groves. In this aerial photograph of an orchard seven miles east of Calexico, made shortly after the earthquake in 1940, the path of the San Andreas Fault can be clearly traced diagonally across the groves west of the Alamo River. North is to the right. (Boore 1977)

of California (see fig. 8.9). North of this area the Pacific plate moves northward past the North American plate. Figure 8.10 illustrates the horizontal displacement of the ground along a small segment of the San Andreas fault in the Imperial Valley of California, where this fault has cut across the groves west of the Alamo River. The San Andreas fault is one of the major breaks along the western coast of California, but it is hardly alone. Figure 8.11 is a simplified fault map of California. The heavy black lines identify faults along which major earthquakes have occurred since 1836. The pattern of faulting in this region is complicated,

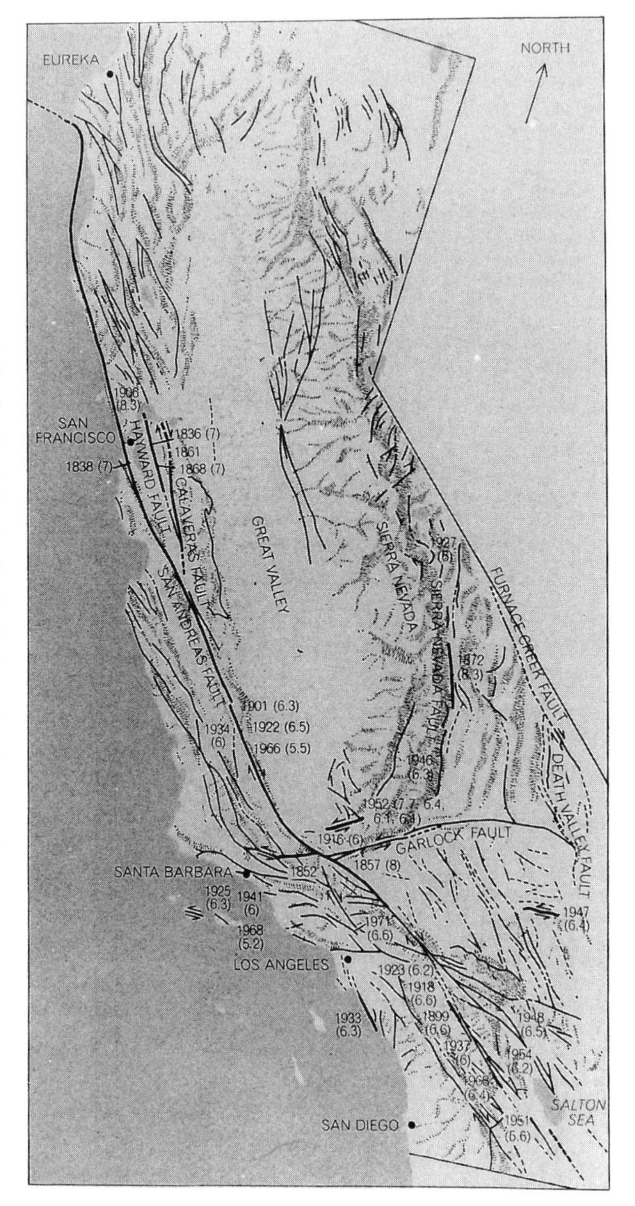

Figure 8.11. Simplified fault map of California; the heavy black lines indicate the faults that have given rise to major earthquakes since 1836. The magnitude of all but two of the earthquakes is given in parentheses next to the year of occurrence. The magnitude of events that predated the introduction of seismological instruments is estimated from historical accounts. For two major events, the earthquakes of 1852 and 1861, information is too sparse to allow a magnitude estimate. Arrows parallel to the faults show relative motion. (Anderson 1971)

as might be expected in an area where complex physical forces are acting on geologically heterogeneous crust (McPhee 1993).

Southern California west of the San Andreas fault is moving majestically but jerkily northward relative to the North American plate at an average rate of about 6 cm/yr. Along some fault segments, the opposing plates slide quite freely past each other, a process called *creep*. Along other segments they tend to lock. Stress can accumulate there for decades or even centuries until the lock finally breaks. Figure 8.12 shows the cumulative displacement of southern California west of the San Andreas fault system during the past 200 years. Periods of quiescence along this fault have been followed by essentially instantaneous displacements during earthquakes. The longer the period of quiescence between earthquakes,

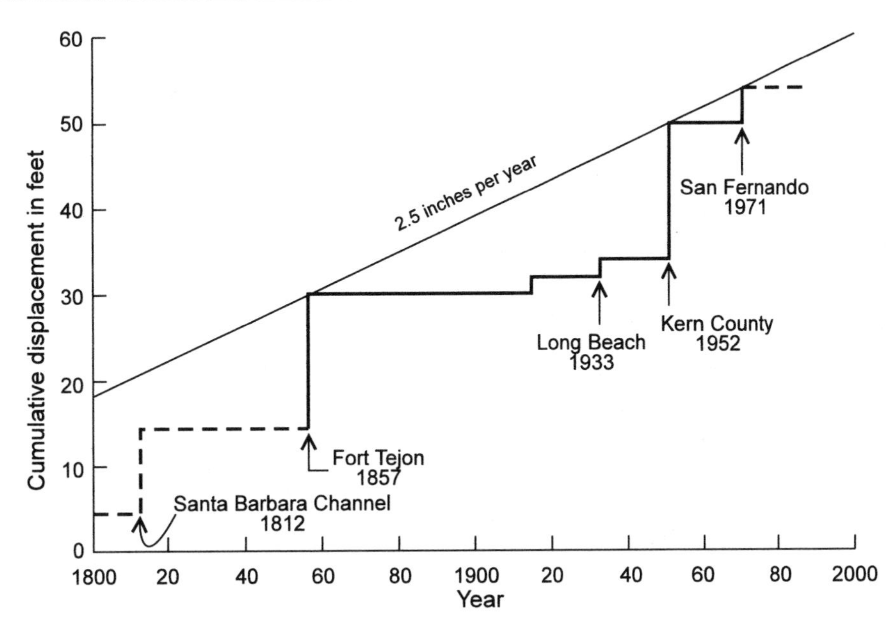

Figure 8.12. Cumulative displacements directly related to earthquakes indicate that southern California west of the San Andreas fault system is sliding northwestward at an average rate of 6 cm per year. Major earthquakes relieve stresses that have built up over decades. (Anderson 1971)

the larger the sudden ground displacement and the magnitude of the associated earthquakes.

When two continents collide, geologic structures tend to differ both from those along the coast of South America and along the California coast. The Himalayas are a spectacular example of a continent-continent collision. India appears to have drifted some 5,000 kilometers northward with respect to Eurasia in the 20 to 30 million years before its collision with Eurasia about 45 million years ago. The probable path of India is shown in figure 8.13. A subduction zone somewhat like that in figure 8.14 probably existed prior to collision. Since the collision, there has been no ocean floor between India and Eurasia. Thrusting has continued as shown in figure 8.14, and the continental crust has thickened, somewhat like the thickening of cars involved in a high-speed, head-on collision.

Not all of the faulting in the Himalayas has thrust Eurasia southward over India. In parts of southern Tibet, ocean floor that once was probably several kilometers below sea level is now exposed 4 kilometers above sea level. This provides an excellent opportunity to study the nature of the seafloor without a deep submersible. Seafloor here flaked off the Indian continent and was thrust upward rather than downward. This is a fine example of obduction, or what has also been called *flake tectonics* (J. Zimmerman, pers. comm.).

The tectonics of the Himalayas are very complex. A series of faults wrap around the northern part of India and extend for several thousand kilometers to the northwest, north, and northeast of the collision zone. The complexities due to the collision seem to be compounded today by collapse features. Anyone who has tried to build a very large sand castle at the beach and has seen it collapse under its own weight can appreciate the problem of holding together the world's highest mountain system. The range actually seems to be collapsing somewhat like a sandpile, but of course much more slowly and imposingly.

We can now summarize the kinds of crustal deformation that we have encountered in collision zones. Rupture, which has dominated the descriptions, can be

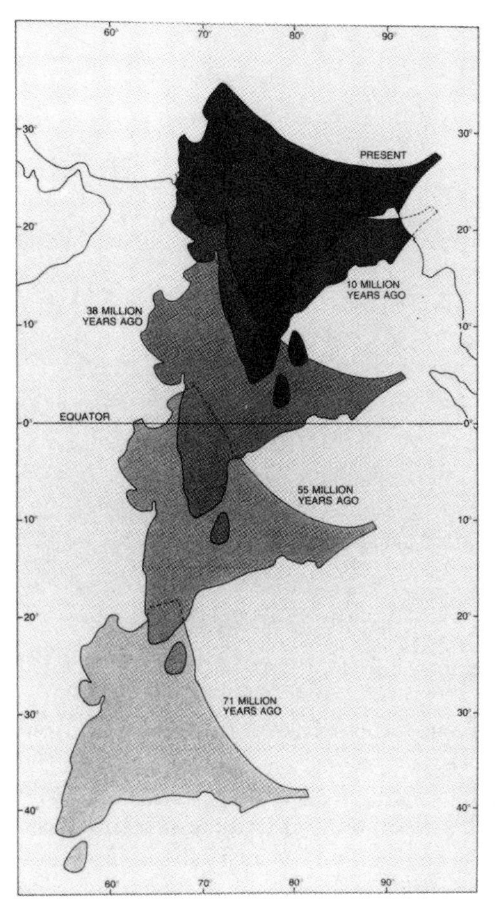

Figure 8.13 (*left*). India's northwest drift during the past 71 million years. (Molnar and Tapponier 1977)

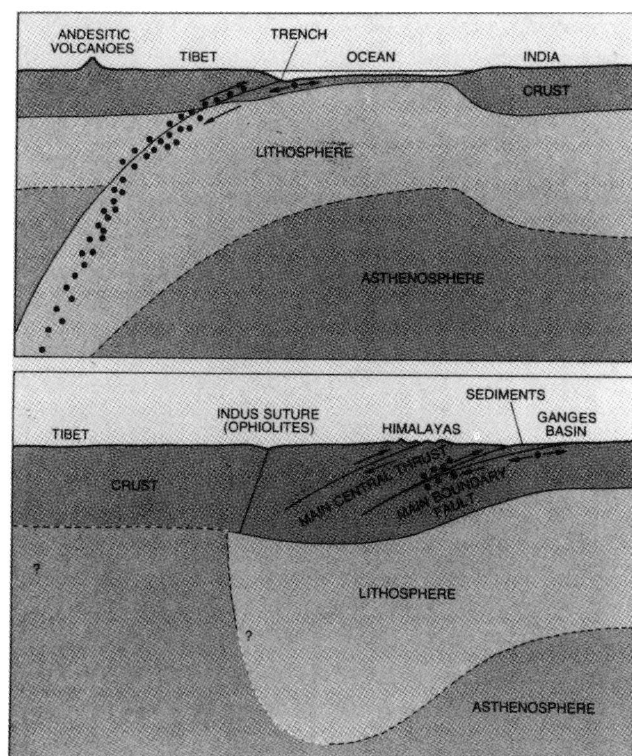

Figure 8.14 (*above, right*). Cross section of the collision zone between the Indian and Eurasian plates shown schematically. The upper diagram (A) shows a cross section through the lithosphere and the asthenosphere about 2 million years before actual contact, when the land masses were still separated by about 200 kilometers of ocean. The black dots show how earthquakes tend to cluster along the boundary between plates and within the descending plate. The suture line, the Indus suture, is marked by the presence of ophiolites: sequences of rocks containing ocean sediments and showing other characteristics of having been formed in a suboceanic environment. Earthquakes are more diffusely distributed and shallower than they were before the collision. The Himalayas are slices of old Indian crust that have overthrust the rest of India to the south, creating new faults that migrate southward. Active faulting seems to occur on the main boundary fault. The crust under Tibet appears to be unusually hot. (Molnar and Tapponier 1977)

described in terms of the block diagrams in figure 8.15. A rock fracture along which motion has occurred is a fault plane. The attitude of fault planes is defined by their strike, i.e., by the compass direction of the fault line; and by their dip, i.e., the inclination of the fault plane with respect to the horizontal. A vertical fault plane, such as the fault plane in figure 8.15d, has a dip of 90°; the fault plane in figure 8.15a has a dip of about 45°.

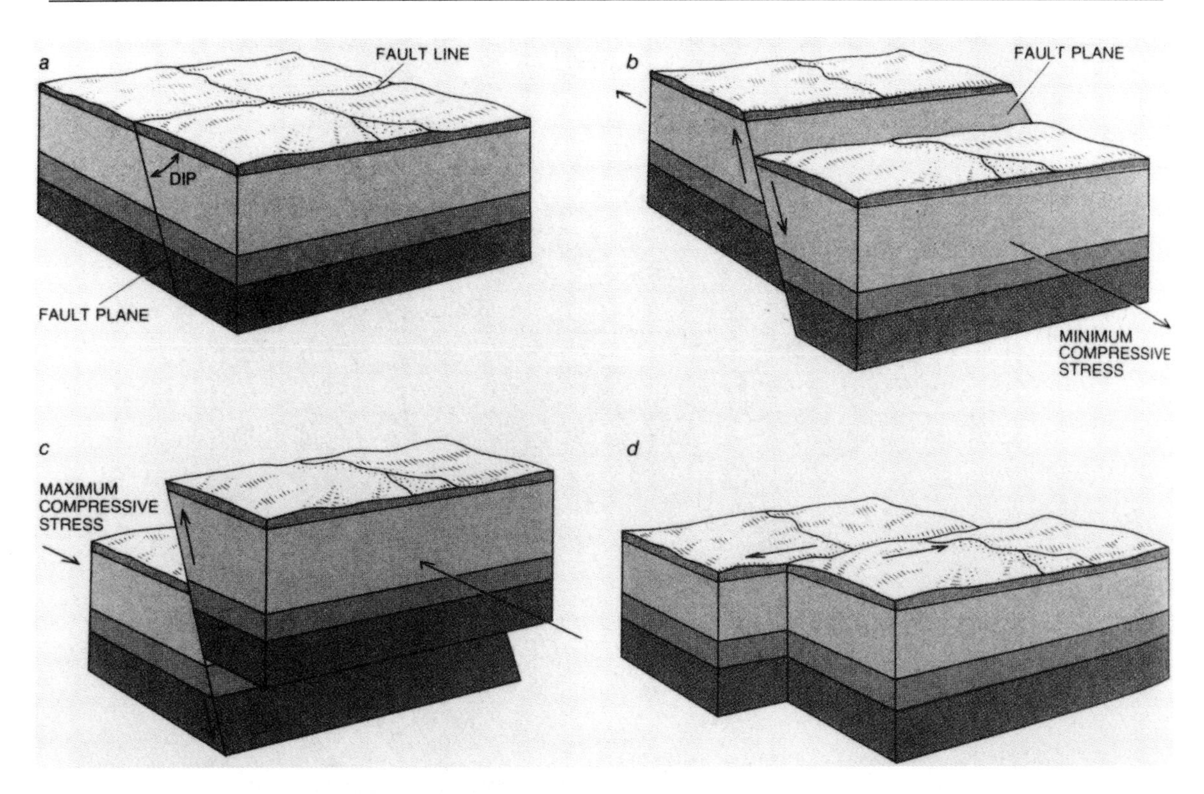

Figure 8.15. During an earthquake, one block of the earth's crust slips with respect to an adjacent block along a fault plane. In a normal fault (b) the blocks act as if they were being pulled apart. The overlying block slides down the dip of the fault plane. In a thrust fault (c) the overlying block is forced up the dip of the fault plane, because the maximum compressive stress is horizontal. In a strike-slip fault (d) the two blocks slide past each other horizontally. Lateral slippage can be combined with normal or thrust faulting. From an analysis of the seismic waves generated by an earthquake, one can tell what kind of fault motion has occurred. (Molnar and Tapponier 1977)

Motion on fault planes can be in any direction. If the motion is at right angles to the fault line and down the dip direction as in figure 8.15b, the fault is called a *normal fault*. This type of faulting occurs in areas of crustal extension and is common at midocean ridges. If motion along a fault plane is at right angles to the fault line but up dip as in figure 8.15c, the fault is called a *thrust fault*. Such faults are caused by compressional forces and are common in collision zones such as those along the west coast of South America and in the Himalayas. If motion on the fault plane is horizontal, the fault is called a *strike-slip fault*. The San Andreas fault and many of the other faults in California are of this type. Other directions of movement are, of course, possible. Motion on a fault can, for instance, be normal with a strike-slip component, or a thrust fault with a strike-slip component. The motion on a fault can also change direction during its lifetime. For example, if an area that has been under compression begins to spread under tension, faults that were once thrust faults can become normal faults.

In areas where rocks and sediments deform without breaking or with only minor breaks, folds tends to develop. Plate 25 shows folded sedimentary rocks in

Figure 8.16. Features of simple folds. Note that the strata dip away from the axes of anticlines but toward the axes of synclines. (Skinner and Porter 1992)

(a) Neighboring anticline and syncline; the axis of the anticline is horizontal

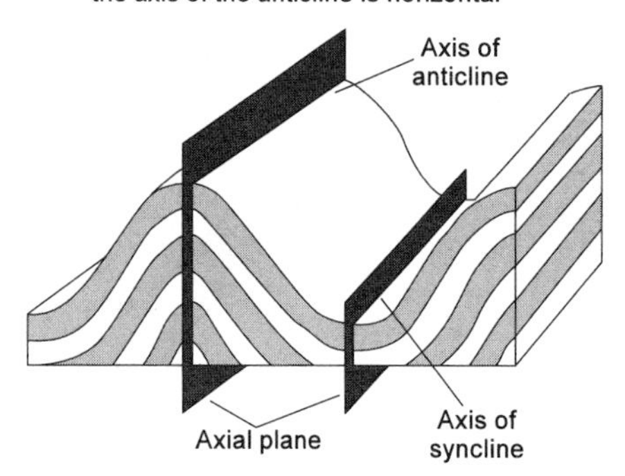

(b) Plunging anticline

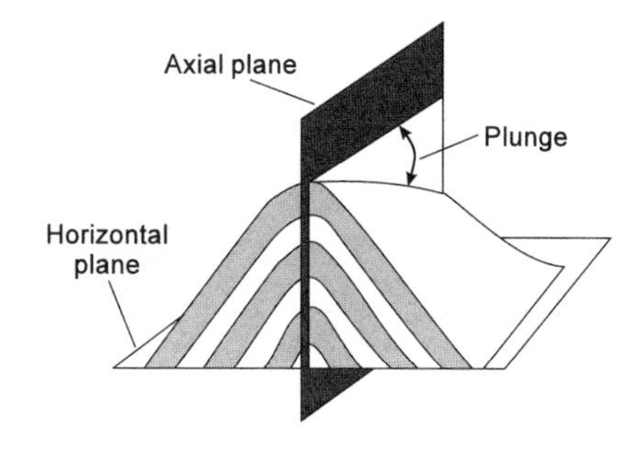

(c) Symmetrical folds

Both limbs dip equally away from the axial plane.

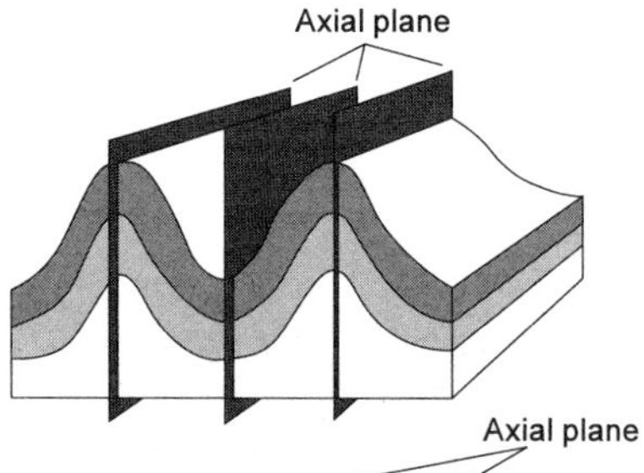

(d) Asymmetrical folds

One limb of the fold dips more steeply than the other.

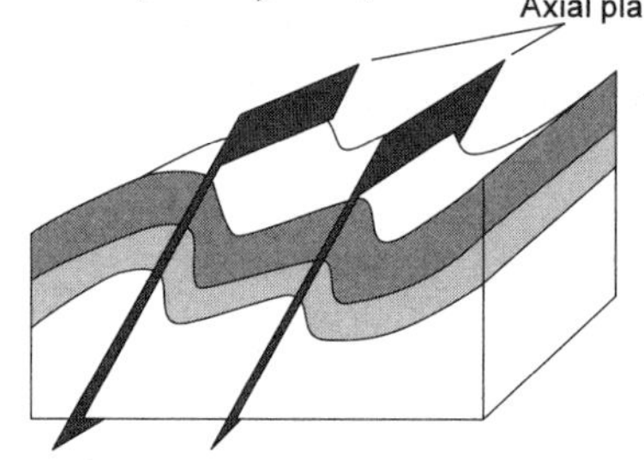

(e) Overturned folds

Strata in one limb have been tilted beyond the vertical. Both limbs dip in the same direction but not at the same angle.

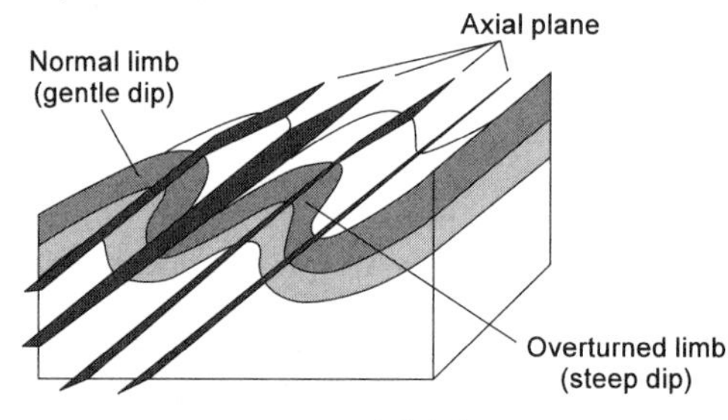

(f) Recumbent folds

Axial planes are horizontal or nearly so. Strata on the lower limb of anticlines and upper limb of synclines are upside down.

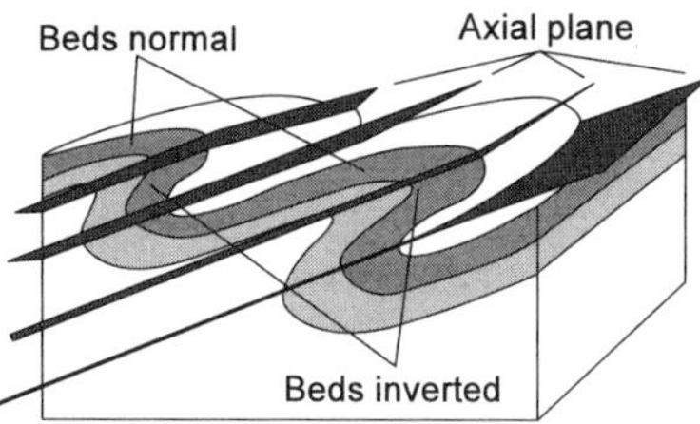

southern Tibet. The folds are not entirely regular, but deformation in this particular area has clearly occurred without major faulting. Folds in rocks are similar to folds in blankets and sheets pushed to the foot of a bed. The geometry of these folds can be contemplated while deciding whether it is worth getting up in the morning. They can be divided into *anticlines* and *synclines*, as in figure 8.16a. Usually the crests of the blanket folds dip downward. Such folds are said to have a *plunge*. You may find that many of the folds are symmetrical, but that others are not, and that the axial plane of the folds is inclined toward the foot of the bed (fig. 8.16b). If you have been particularly energetic, you may have created folds whose axial planes are almost horizontal (fig. 8.16c,d and plate 26).

If these complexities are not enough for you, you may want to try kicking the crumpled blankets from a second direction. This will produce folded folds. If your sheets and blankets are old and threadbare, all these exertions may create tears and thus introduce the additional complexities of faulting. You will then have attained something close to the geometry of many real mountain ranges, and you will be well on the way to appreciating that geological mapping in highly deformed, mountainous terrain tends to be mentally as well as physically demanding.

8.4 Oil and Natural Gas Accumulations

The study and the analysis of geologic structures has been important for understanding the history and the workings of the Earth. They have also been important economically. Most of the world's reserves of oil and natural gas are contained in structurally deformed rocks, and the success of exploring for these resources depends a good deal on understanding geologic structures. However, a knowledge of rock deformation is not sufficient for successful oil and gas exploration. The formation of an oil or gas field can be compared to the baking of a cake. Certain ingredients are essential; these must be mixed in the proper sequence, and the mixture must be subjected to a carefully controlled thermal regimen.

The first essential ingredient in the formation of an oil or gas field is a suitable source rock. Oil and gas are produced when the remains of microorganisms and higher land plants are thermally decomposed. A few large oil fields have been generated from lacustrine (i.e., lake) source rocks, but in general only sediments that contain a good deal of marine organic matter yield large quantities of oil and gas. Such sediments tend to form where the biological productivity of the oceans is high, and where the organic matter that rains down from surface waters is buried rapidly (see chapters 5 and 7). Most of the important source rocks for oil and natural gas are shales that were sediments rich in organic matter deposited in inland seas or on continental shelves or slopes.

The accumulation of sediments rich in organic matter must be followed by suitable thermal processing. Methane, CH_4, the major component of natural gas, can be produced by bacteria very early in the burial history of marine sediments. Much, if not all, of this methane escapes from sediments; it is squeezed out together with seawater during compaction by the weight of overlying sediments. The later generation of oil and nonbiogenic natural gas takes place between about 60° and 220°C (see fig. 8.17). By the time the temperature of sediments has risen into this range, they have usually been buried to a depth of 2,000 to 5,000

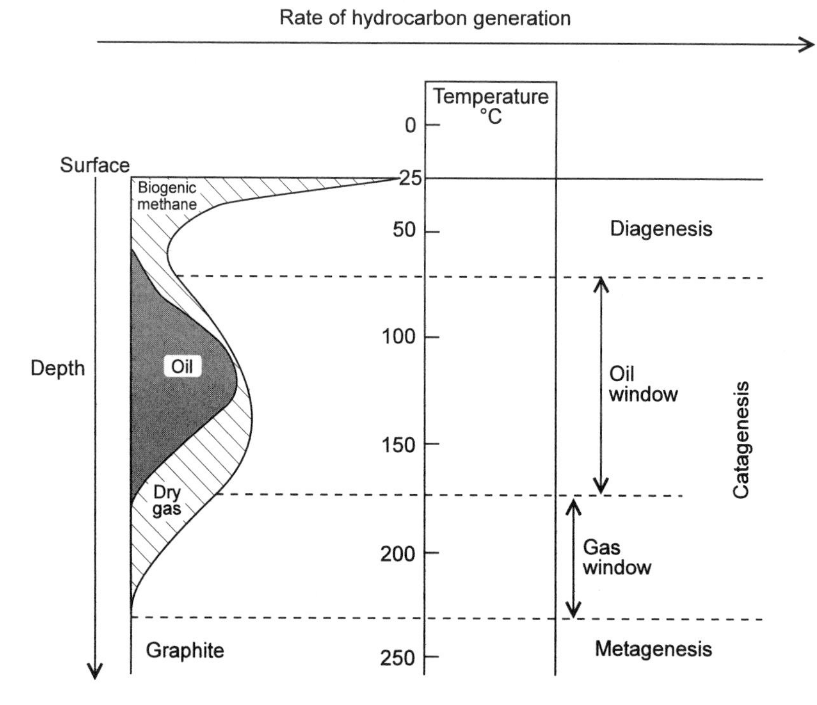

Figure 8.17. Correlation between temperature and the rate of hydrocarbon generation during the burial of organic matter in marine sediments. (Selley 1985)

meters (see fig. 8.18). At these burial depths most of the seawater they contained initially has already been squeezed out. The porosity of shales decreases rapidly during burial to 2,000 meters, and rather little additional water is expelled during passage through the oil and gas "window."

Oil that is produced in the "window" can remain in its source rocks, or it can move into other quarters. If it remains in place, the source rocks gradually turn into shales that contain significant quantities of oil. The lacustrine oil shales in the Green River Formation of Utah and Colorado are important examples of such sediments. Since oil shales are quite impermeable, the oil they contain cannot be released by drilling alone. Several techniques have been developed to release oil from oil shales. They can be quarried, crushed, and then heated to release their contained oil. Oil can also be extracted in situ. Closely spaced, shallow boreholes can, for instance, be drilled into oil shales. Heating elements are then inserted, and the temperature of the rock is raised to about 400°C. Oil released at this temperature is pumped from the wells. In another procedure, oil shales are artificially fractured underground to increase their permeability. Compressed air is then pumped into the fractured rocks and ignited. The high temperature gases move through the shale and retort the oil, which is then recovered from oil-producing wells.

The extraction of oil from oil shales has had a long and somewhat checkered history. Advances in drilling for liquid oil have continued to make the extraction of oil from oil shale economically unattractive. The rapid increase in the price of petroleum in the 1970s renewed interest in recovering oil from oil shale, but this interest ebbed with the drop in the price of oil in the 1980s. Oil shales may become an important source of oil in the next century as the reserves of more

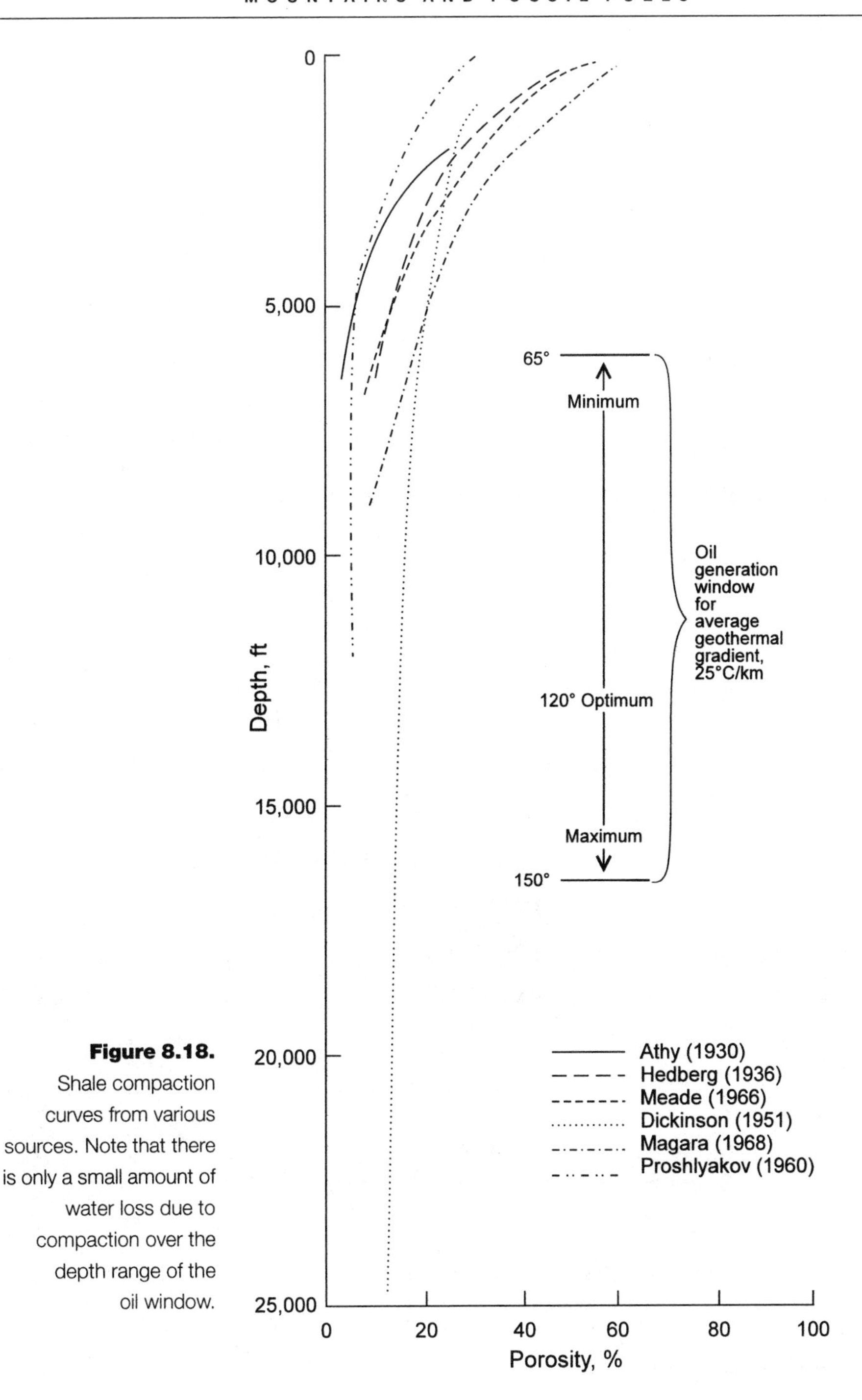

Figure 8.18. Shale compaction curves from various sources. Note that there is only a small amount of water loss due to compaction over the depth range of the oil window.

easily extractable oil become depleted. The quantity of shale oil is very large and may well play an important role in the world's fuel economy during the twenty-first century unless the climatic effects of increasing atmospheric CO_2 levels due to fossil fuel burning turn out to be so severe (see chapter 11) that the world must turn to non-fossil sources of energy.

Some oil and gas generated during the gradual burial of marine sediments manages to escape from their source rocks. Just how they escape is not well understood (see, for instance, Selley 1985). Oil and gas that escape into more permeable sediments move through these carrier rocks, driven by gravitational forces related to the difference between the density of oil, gas, and formation waters, which are the aqueous fluids in subsurface sediments. Formation waters frequently have a salinity and density similar to that of seawater (ca. 1.024 gm/cc). However, in sedimentary sequences that contain marine evaporites, formation waters tend to be highly saline. Some formation waters have dissolved so much salt, that they are saturated with respect to halite and have densities that are quite a bit higher than that of seawater (ca. 1.25 gm/cc).

Once liberated from their source rocks, oil and gas can migrate all the way to the surface and issue there as oil seeps. For much of human history such seeps were mankind's only source of oil. In the nineteenth century the oil seeps near Titusville, Pennsylvania, were well known. In 1859, the area became the site of

Figure 8.19. (a) Edwin Drake (*right*) in front of his oil well on the banks of Oil Creek in Titusville, Pennsylvania, in 1861; this well marked the beginning of the modern extraction of oil. (b, *on opposite page*) The success of Drake's first well resulted in the drilling of large numbers of closely spaced wells as shown here in 1861 on the Benninghoff Farm along Oil Creek. (Drake Well Museum, Titusville, Pennyslvania)

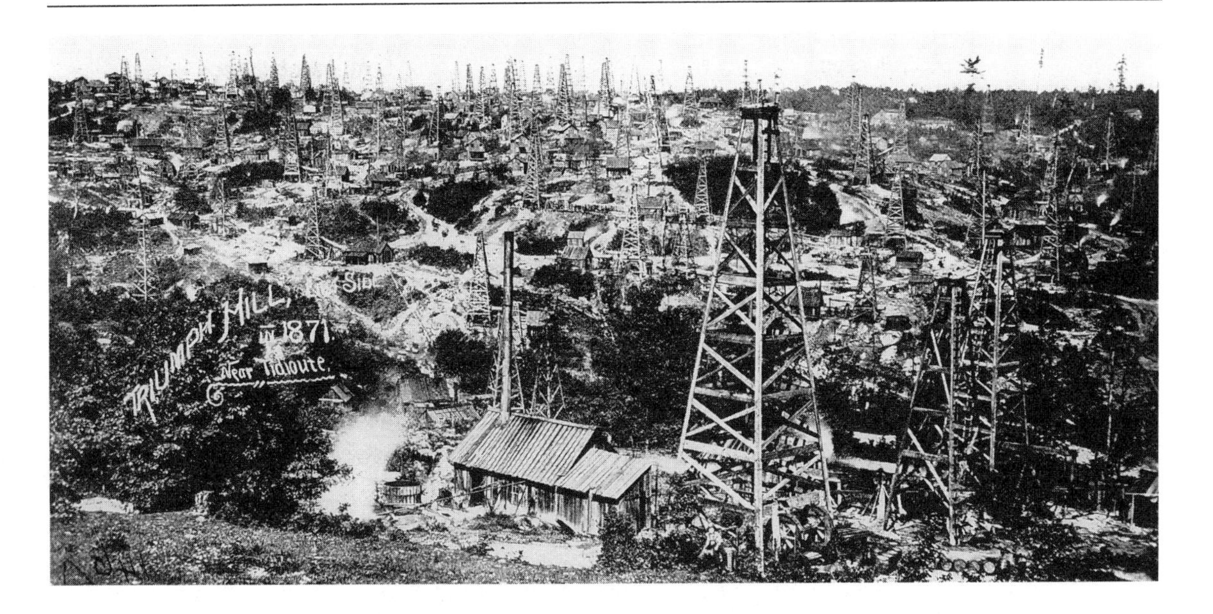

the first successful oil well in the United States (fig. 8.19). The successors of this oil well gradually evolved into the giant oil rigs of the late twentieth century (plate 27).

Oil in seeps can age to tar and to pitch largely as a result of surface biodegradation. The famous tar pits of La Brea in California contain the remains of animals and have become a treasure trove for paleontologists interested in reconstructing the history of the fauna in this area. Pitch from such seeps has been used for thousands of years to make boats waterproof. According to biblical tradition, Noah used pitch in the construction of the ark (Genesis 6:13–14). References to the use of oil, pitch, and asphalt are common in Greek literature composed as long ago as 450 B.C.

Most of the oil that seeped out of the subsurface has been dispersed, metabolized, burned, or otherwise lost. Therefore, most of the oil and gas that have been important sources of energy for mankind during the course of this century were trapped underground, and have remained there until tapped for human use. The bewildering array of underground oil and gas traps that have been discovered to date reflects the ingenuity of oil geologists in ferreting out undiscovered oil and gas fields (see, for instance, Jenyon 1990). The stakes in oil and gas exploration are often very high, and new ideas for locating oil and gas fields are highly prized. It has been well said that if you want to find elephants, you must go to elephant country. But sometimes it is possible to discover new and unsuspected elephant country. The financial rewards for discovering large oil and gas fields are very substantial. This lends a good deal of excitement to the chase (see, for instance, Yergin 1991).

Oil and gas traps can be divided into two classes: structural traps and stratigraphic traps (fig. 8.20). The simplest of the structural traps and by far the most important are anticlines (fig. 8.20a), and domes. Natural gas, the least dense of the subsurface fluids, can migrate into a trap, or it can separate from oil that is moving upward and then accumulate in a trap. Oil accumulates beneath the gas, and the pore space beneath the oil is filled with formation waters. The quantity

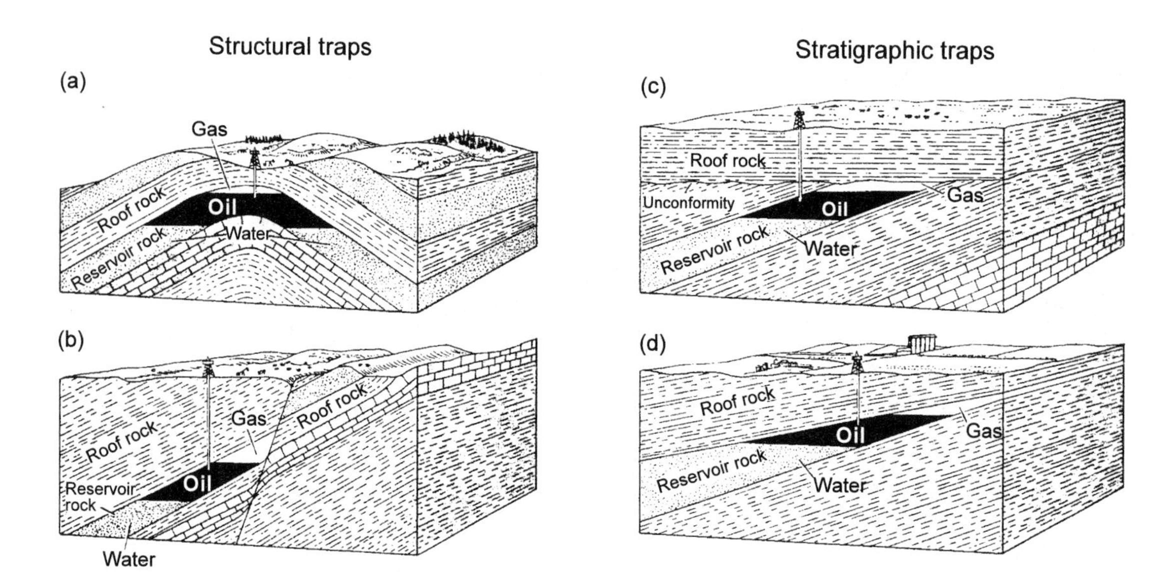

Figure 8.20. Four kinds of oil traps. (a) and (b) are structural traps; (c) and (d) are stratigraphic traps. Folds (a) are traps formed by ductile deformation. They are the most important of all oil traps. Faults (b) are traps formed by fracturing of brittle rocks. In (c), an unconformity marks the top of the reservoir; in (d), a porous stratum (reservoir) thins out and is overlain by an impermeable roof rock. Gas overlies oil, which floats on groundwater, saturates the reservoir rock, and is held down by a roof of shale. Oil fills only the pore spaces in the reservoir rock. (After Skinner and Porter 1989)

of oil and gas that can accumulate in a dome depends on its size. Oil and gas in excess of the quantity that fits within a dome therefore leak into neighboring structures and either accumulate there or fail to be trapped and escape to the surface. Figure 8.21 shows the location of the giant Burgan oil field in southern Kuwait. Figure 8.22 shows the very gentle dip of sediments in the anticline. In spite of these low dips the Burgan field is estimated to contain 72 billion barrels of oil; this makes it the second largest oil field in the world. The world's largest gas field, the Urengoy field in the northern part of western Siberia, has a similar domal structure (fig. 8.23). The structure contains several gas-bearing strata and one commercial accumulation of oil. During 1984, this field was responsible for one-third of the very large total production of natural gas in what was then the USSR.

Some structural oil and gas traps are fault related. The simplest of these is shown in figure 8.20b, where faulting has moved a rock that is impermeable to gas and oil opposite a permeable reservoir rock. If the fault is tightly sealed, so that fluids cannot move along the fault plane, the upward movement of oil and gas stops at the fault. Many important oil and gas fields owe their origin to this type of geometry. Very complex structures can form in areas like the Gulf Coast, which is underlain by thick evaporites. The density of salt is significantly less than that of most sedimentary rocks. A stratigraphic section such as that in figure 7.7 is therefore intrinsically unstable, and small disturbances of the salt layer are apt to produce synclines and anticlines at the salt-sediment boundary. Rocks can flow, albeit slowly. Salt flows more easily than most rocks and tends to move

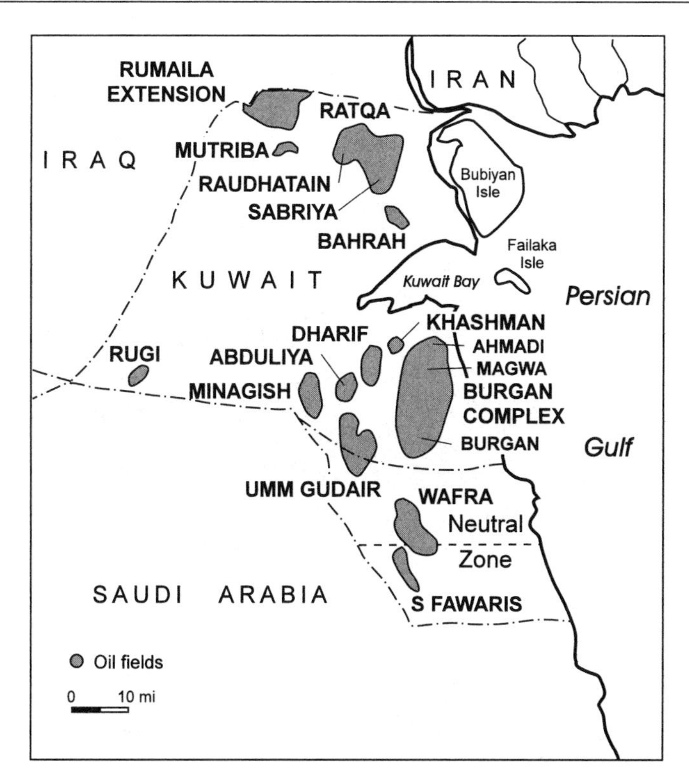

Figure 8.21. The oil fields of Kuwait. (*World Oil*, January 1992)

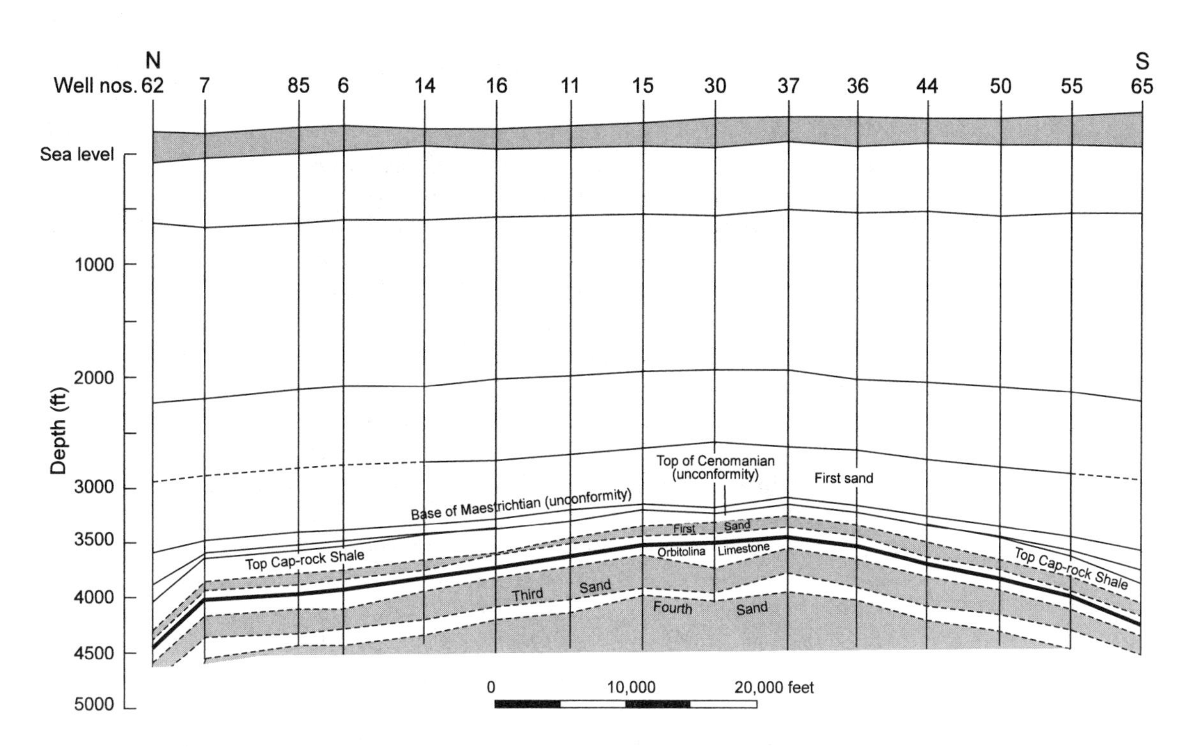

Figure 8.22. Geology of the Burgan structure, with eight times vertical exaggeration. (Jenyon 1990)

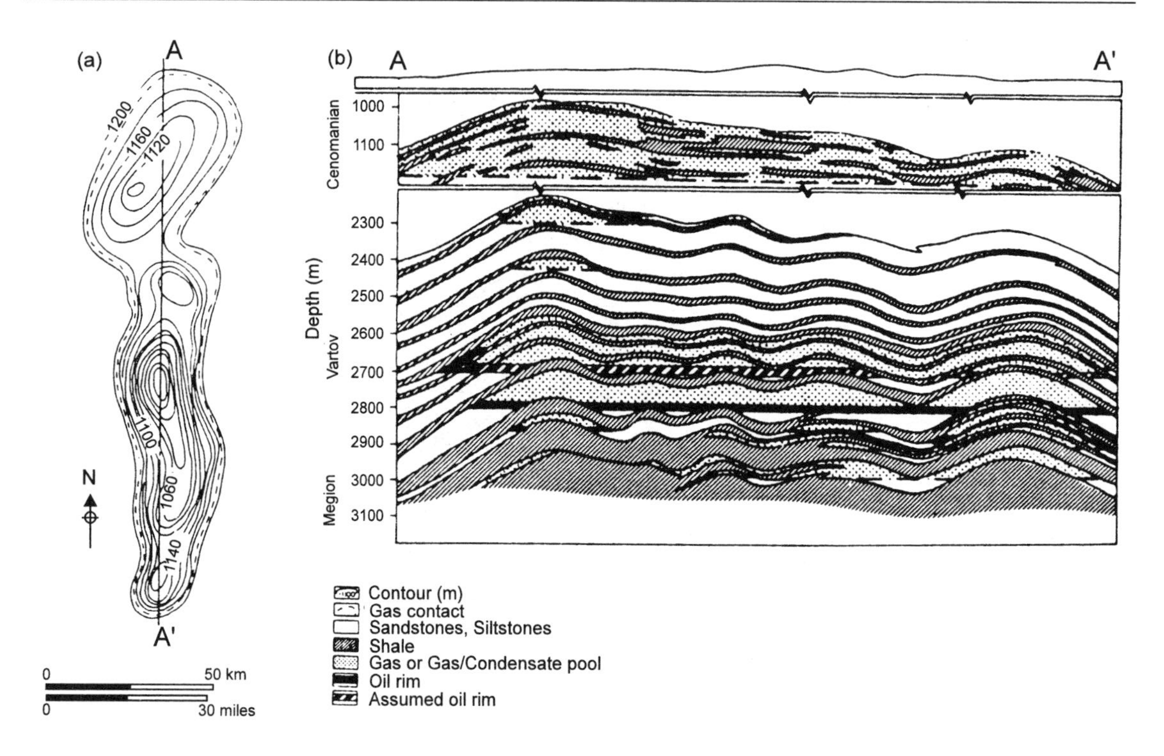

Figure 8.23. (a) Depth-structure contour map on the top of the Pokur Suite (Cenomanian); (b) cross section (location shown by line A-A′ in map (a) approximately along the axis of the elongate Urengoy gas field structure, northern part of West Siberia, former USSR. (Grace and Hart 1986)

into the anticlinal structures. These salt accumulations rise and pierce the overlying sediments as shown in the seismic section of figure 8.24. Continued rise of the salt can produce salt domes such as the one shown in figure 8.25. A variety of structural traps can develop at the sides and above the top of these domes, which frequently develop shapes bordering on the grotesque.

The second class of oil and gas traps are stratigraphic in nature. Two traps in this family are shown in figure 8.20c and d. In figure 8.20d, the reservoir rock gradually pinches out up dip. Oil and gas in such reservoir rocks accumulate in the region of the pinchout and remain there if the surrounding sediments are sufficiently impermeable. The geologic evolution of the block in figure 8.20c is somewhat different. The older sediments represented in this block were tilted, partially eroded, and covered by younger sediments. Oil then migrated into the reservoir rock, moved upward, and accumulated below the unconformity between the older and the younger sediments. If oil and gas had moved into the reservoir rock before the deposition of the sediments above the unconformity, the oil and gas would have been lost by leakage at the surface. Changes in the porosity and permeability of sedimentary rocks can occur in areas other than stratigraphic pinch-outs. Many coral reefs and limestones that have been dolomitized (see chapter 7) are sufficiently porous to serve as reservoir rocks. If they are surrounded by impermeable strata, they can also serve as efficient traps for oil and gas.

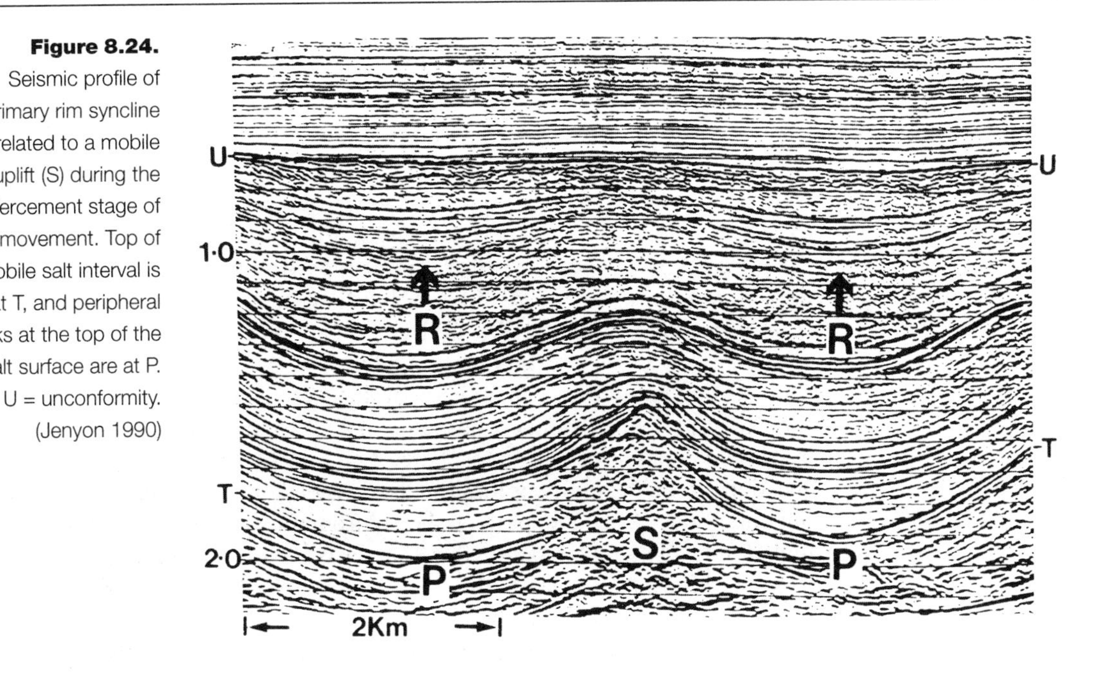

Figure 8.24. Seismic profile of primary rim syncline (at R) related to a mobile salt uplift (S) during the pre-piercement stage of salt movement. Top of the mobile salt interval is at T, and peripheral sinks at the top of the salt surface are at P. U = unconformity. (Jenyon 1990)

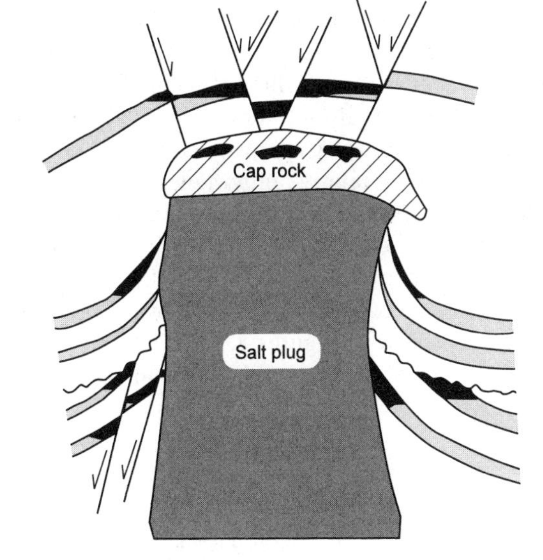

Figure 8.25. Schematic representation of a typical Gulf Coast salt dome, indicating various trap types, including leached secondary porosity traps in a cap-rock. (Jenyon 1990)

The potential complexity of oil and gas traps is almost infinite. Some traps involve both structural and stratigraphic elements, others owe their complexity to settings in highly folded and complexly faulted terrains that have suffered more than one period of deformation. During such a complex geological evolution, oil and gas traps can be emptied as well as filled. It is interesting, and perhaps heartening, that many of the world's largest oil and gas accumulations are in geologically rather simple structures. The accumulation of oil and gas even in these structures depends on the operation of the "petroleum system." A source of oil and gas is required, carrier beds must be available, and a trap must exist.

The presence of all three parts of the system in a single region is quite unlikely. Only a small part of the oil and gas that is generated in source rocks has therefore become trapped and available for human use (Demaison and Huizinga 1991).

8.5 Fossil Fuels and World Energy

Although fossil fuels supply most of the world's energy, it has not always been so. Wood, a renewable energy source, was the world's dominant energy source until the second half of the nineteenth century. In the not-too-distant future, as measured in terms of recorded human history, fossil fuels will cease to be the dominant source of energy. We are living in an unusual era, drawing heavily on solar energy that has been stored during the past several hundred million years in coal, oil, and natural gas. The release of this stored energy has fueled an unprecedented growth of economic well-being; it has also produced significant changes in the composition of our atmosphere. Two major and somewhat contradictory questions cloud the future of fossil fuels at the end of this century: "When will we run out?" and "What are apt to be the environmental consequences of their continued use?" The remainder of this chapter is devoted to answering the first of these questions. An answer to the second question will be ventured in chapter 11.

Figures for energy production and consumption in the United States between 1949 and 1993 are shown in figure 8.26. Breakdowns of the major sources of U.S. energy production and consumption are shown in figures 8.27, 8.28, and 8.29. U.S. energy production and consumption rose rapidly between 1949 and 1973, but only modestly between 1973 and 1993. In 1993, U.S. energy consumption amounted to about 82 quadrillion Btu (British thermal units). About 88% of this was derived from fossil fuel burning. Nuclear energy, hydroelectric power, geothermal and all other energy sources combined accounted for the remaining 12%. The changes in the relative importance of the several energy sources have been significant. Coal has claimed a progressively greater share of total U.S. energy production; nuclear energy made its first appearance in the 1950s and has claimed a small but increasing share of the U.S. energy budget.

The world's energy production in 1990 was approximately 345 quadrillion Btu ("quads"). As shown in figures 8.30, 8.31, and 8.32, fossil fuels accounted for nearly the same percentage of this amount as for U.S. energy production.

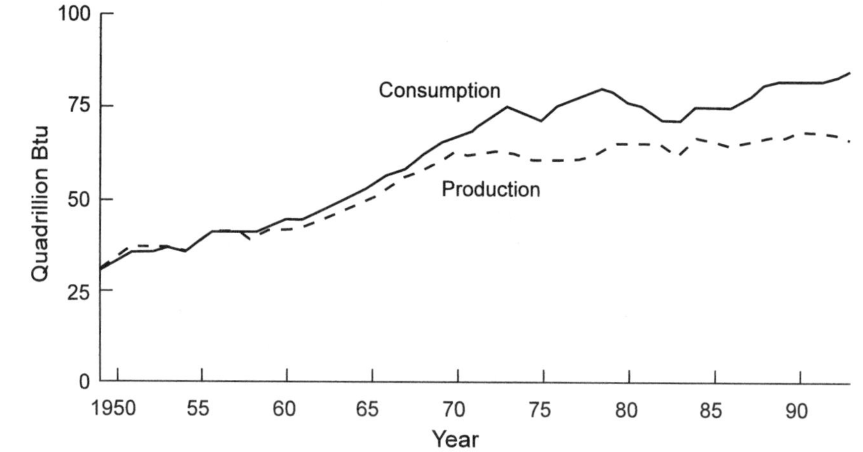

Figure 8.26. The production and consumption of energy in the United States between 1949 and 1993. (*Annual Energy Review 1993*)

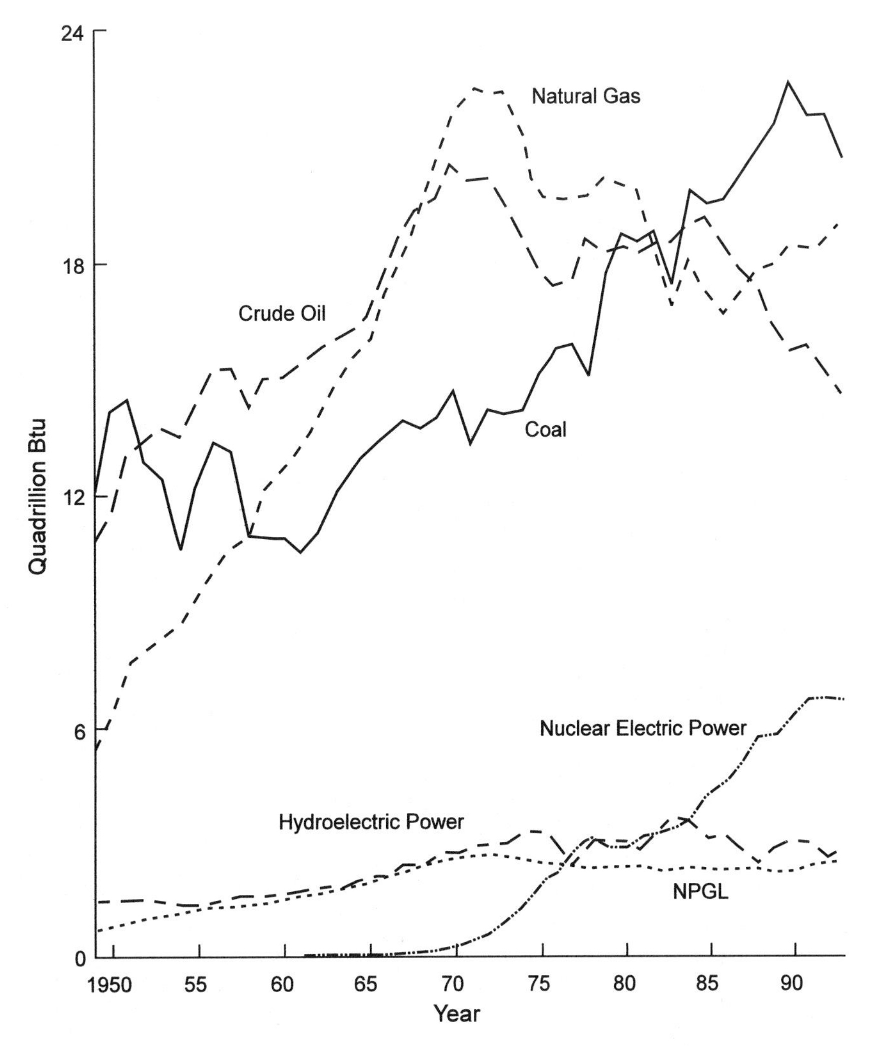

Figure 8.27. The major sources of energy produced in the United States between 1949 and 1993. NPGL = natural pressurized gas liquid. (*Annual Energy Review 1993*)

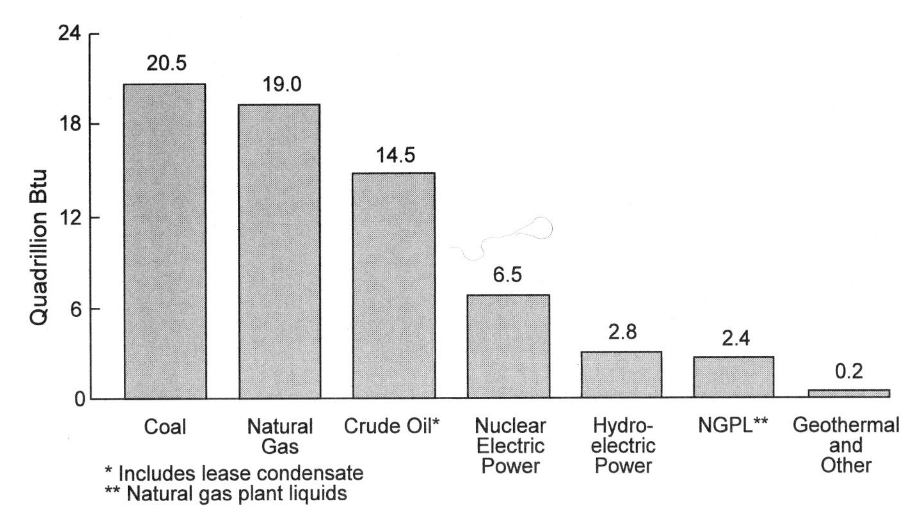

Figure 8.28. The major sources of energy produced in the United States during 1993. NGPL = natural gas pressurized liquid. (*Annual Energy Review 1993*)

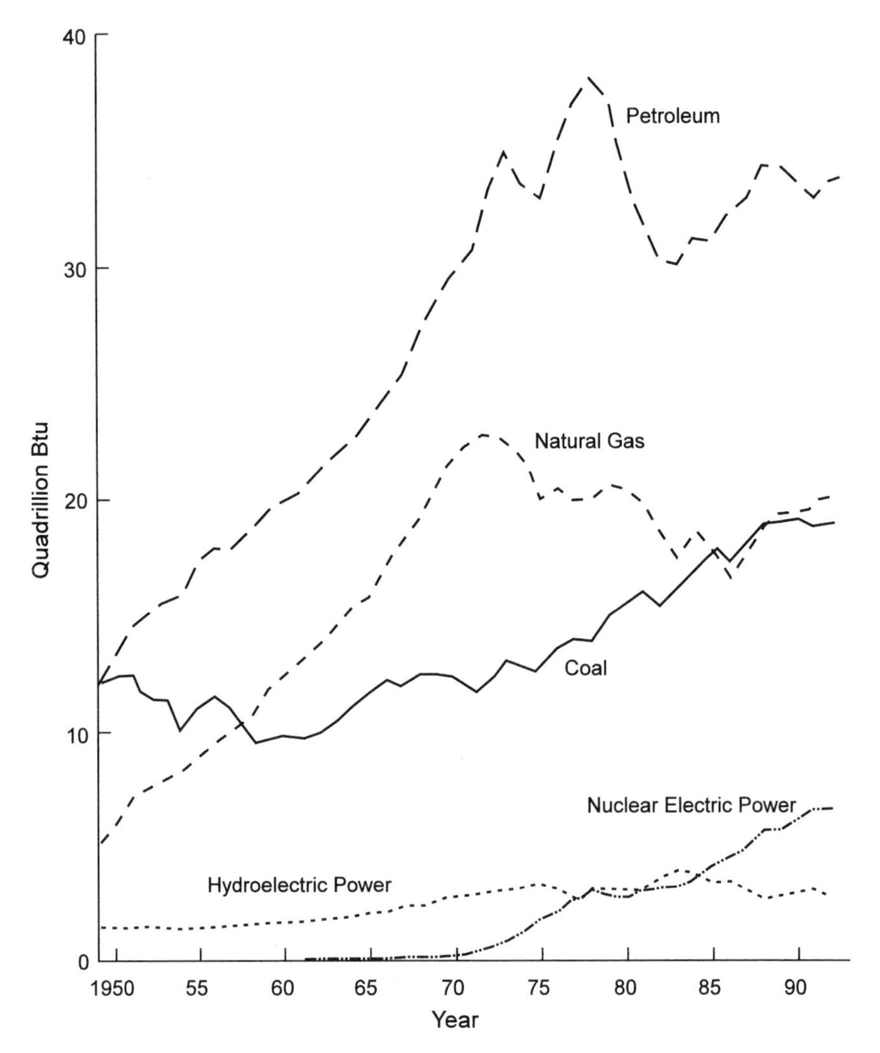

Figure 8.29. The major sources of energy consumed in the United States between 1949 and 1993. (*Annual Energy Review 1993*)

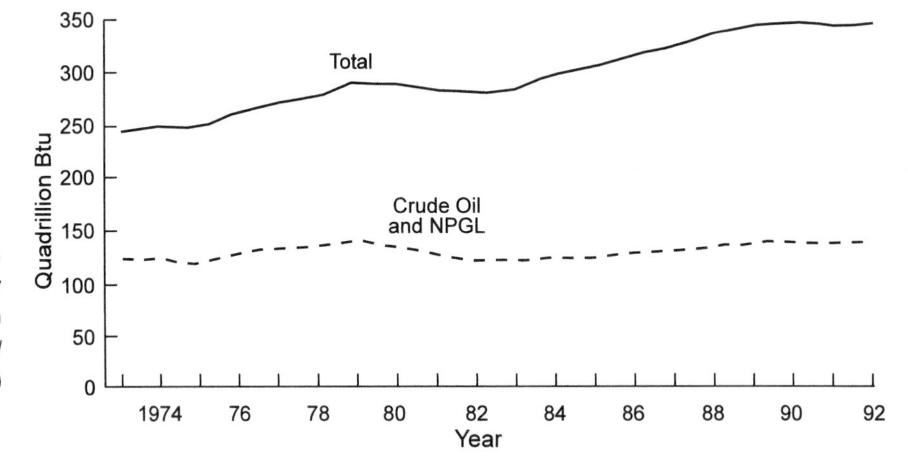

Figure 8.30. World primary energy production between 1973 and 1992. (*Annual Energy Review 1993*)

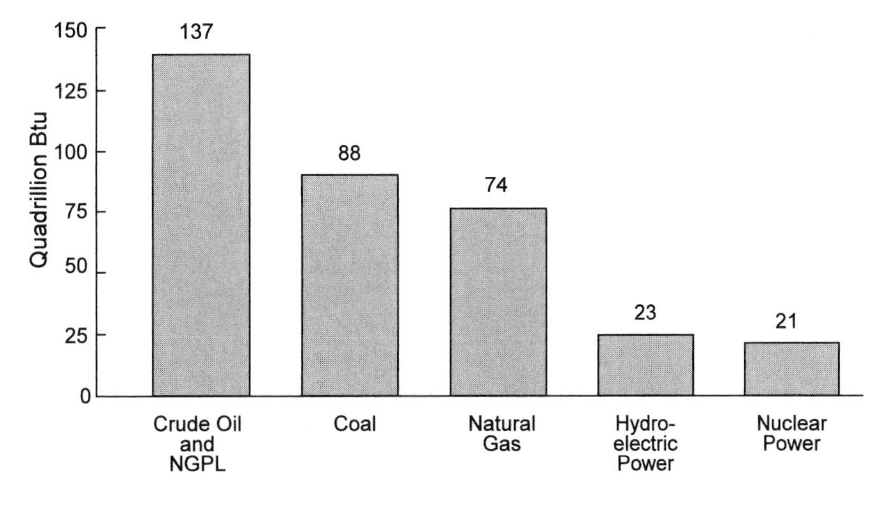

Figure 8.31. The major sources of world energy in 1992. (*Annual Energy Review 1993*)

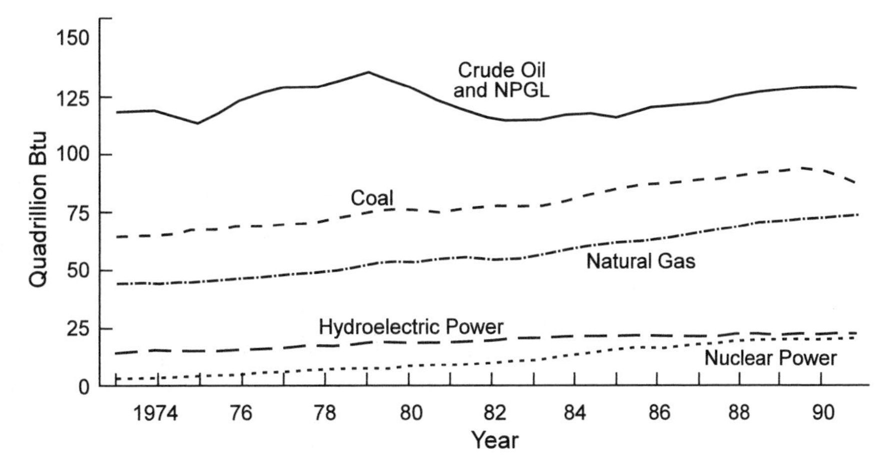

Figure 8.32. World primary energy production by source between 1973 and 1992. (*Annual Energy Review 1993*)

Fossil fuels are therefore by far the most important sources of energy for the world as a whole. Very significant geographic differences in the relative importance of the several sources of energy do, however, exist. France, for instance, relies heavily on nuclear energy. In Norway, nearly all electricity is hydroelectric. Some of these differences are geographical. Norway is particularly well endowed with mountains and streams. The choice of energy sources in France has been largely political.

Energy production in the world is extremely uneven. Figure 8.33a shows that the United States, with approximately 5% of the world's population, produced nearly 20% of the world's energy. Although China's energy production is increasing very rapidly (see fig. 8.33b), China, with approximately 20% of the world's population, currently produces less than 10% of the world's energy. These differences extend to the use of most raw materials and are indicators of the great differences in per capita wealth in the two countries.

Although quads are a convenient unit for dealing with and comparing the importance of various types of energy, it is somewhat difficult to relate to them. To endow them with reality, they need to be converted into the units of mass, volume, and energy that are commonly used in coal mining, oil and gas production,

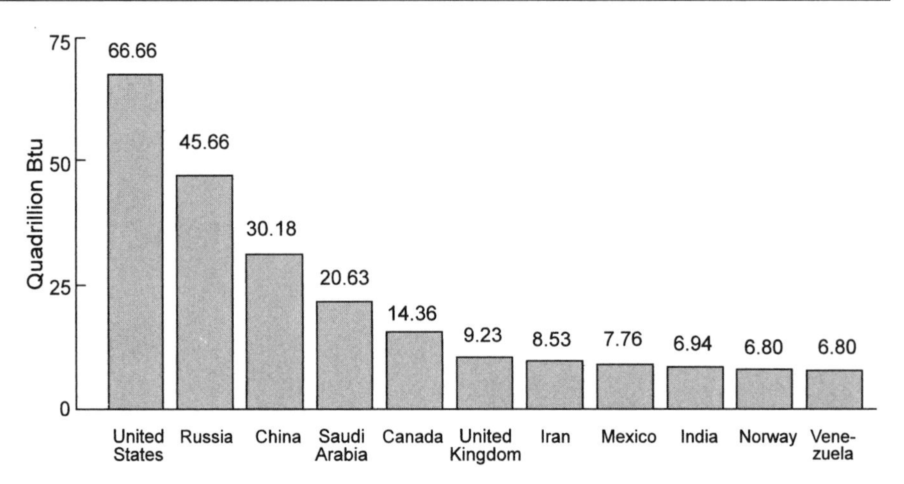

Figure 8.33a. The eleven major primary energy-producing countries in 1992. (*Annual Energy Review 1993*)

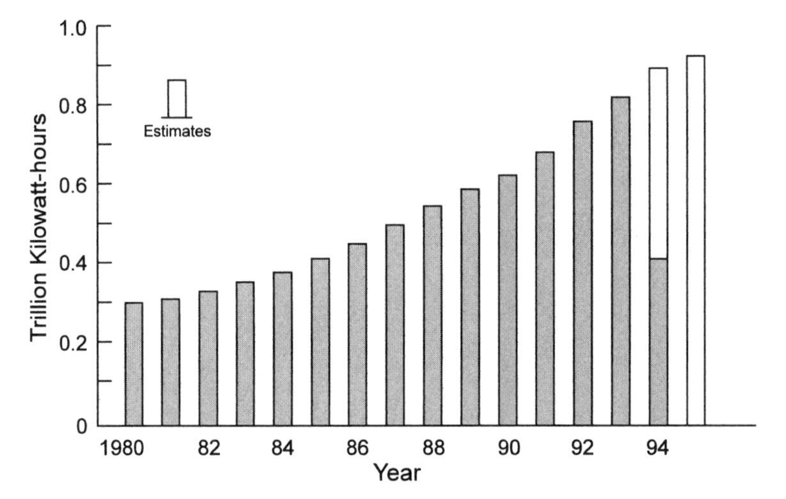

Figure 8.33b. Electricity generation in China since 1980. (*New York Times*, November 7, 1994)

and electric energy generation and consumption. Some conversion factors are listed in appendix B. The Btu is one of the somewhat quaint set of historical English units. It is defined as the energy required to raise the temperature of one pound of water by one degree Fahrenheit. The more commonly used calorie is the energy required to raise the temperature of one gram of water by one degree centigrade. Since one pound contains 453 grams, and 1°C is equal to 1.8°F, 1 Btu is the equivalent of 252 calories. Fossil fuels are used so extensively to generate electricity that it is convenient to relate the Btu to units of electrical energy. The joule is defined as the electrical energy needed to maintain a current of one ampere for one second at a potential of one volt. One Btu is the equivalent of 1,053 joules. The consumption of one joule of electrical energy per second is defined as an energy flow of one watt. One 1,000-watt light bulb burning for one hour consumes one kilowatt-hour (kwh) of energy; this is equivalent to 3,410 Btu. One quad, defined as 1 quadrillion Btu, is equal to 10^{15} Btu and hence is the equivalent of 1.055×10^{18} joules and 2.9×10^{11} kwh.

Since coal is a fuel that is mined, quantities of coal are normally reported in tons. Tons are rather old units of weight. There are several of them, much as there used to be several different sorts of miles, whose appearance on old maps tends both to confuse and to charm collectors. Two thousand pounds make up one short ton, 2,240 pounds one long ton, and 2,204 pounds one metric ton, which is defined as 1,000 kilograms. The heat energy that is released when coal burns depends on its rank: one short ton of bituminous coal yields about 6,500 kwh of electricity or 22 million Btu. In 1990, 5,200 million short tons of coal were mined. These accounted for the production of some 93 quads of energy.

Petroleum is pumped as a liquid whose volume is usually measured in terms of barrels (bbls). One barrel has a volume of 42 U.S. gallons; on burning, one barrel of oil releases about 1,700 kwh. The world production of petroleum is approximately 60 million bbls/day, or 22 billion bbls/year. These account for the release of 136 quads of energy/year.

The quantity of natural gas is usually measured in terms of cubic feet. On burning, one cubic foot of natural gas releases about 1,035 Btu or 0.300 kwh. The annual world production of ca. 75 trillion cubic foot (TCF) of natural gas is responsible for generating 74 quads of energy.

The average price of energy from fossil fuels in 1991 was approximately $1.60/million Btu. The total cost of supplying the world with 345 quads of fossil fuel energy was therefore approximately $500 billion. This figure alone indicates the great importance of energy costs for the world's economy and accounts for the great interest in discovering new coal deposits, and oil and gas fields.

The distribution of the fossil fuels is anything but uniform. Some parts of the world are blessed with abundant fossil fuel supplies, others are almost devoid of coal deposits, and of oil and gas fields. Some Israelis bemoan the fact that, after 40 years of wandering in the wilderness, Moses led them to one of the few areas in the Middle East that is not rich in oil and gas. The production of fossil fuels differs greatly from country to country, partly because of differences in their natural endowments, but also due to differences in the extent of coal, oil, and gas exploration, and differences in the extent to which these commodities have already been exploited.

8.6 Production and Reserves of Coal

Figure 8.34 lists the production of coal in 1992 in the most prolific coal-producing countries. China, the United States, and the former USSR produced by far the largest tonnage of coal. However, the combined coal production in the countries of eastern and western Europe is comparable to that of the United States. Northern Africa and South America are conspicuously absent from the list of major coal-producing areas.

The consumption of coal in 1992 by the ten most consuming countries is detailed in figure 8.35. Most of the coal was used locally. We have to drop to sixth place, Australia, among the major coal-producing countries before we see a significant difference in ranking between coal-producing and consuming countries. The difference between the position of Australia among the producer and consumer nations is due to the export of a large fraction of Australian coal production to Japan, which has virtually no fossil fuel reserves of its own. The United States also exports a good deal of coal to Japan, but the quantity is a small fraction of the total U.S. production.

We mine coal at a prodigious rate. To formulate a sensible energy policy for the next century, or at least for the next several decades, we need to know how

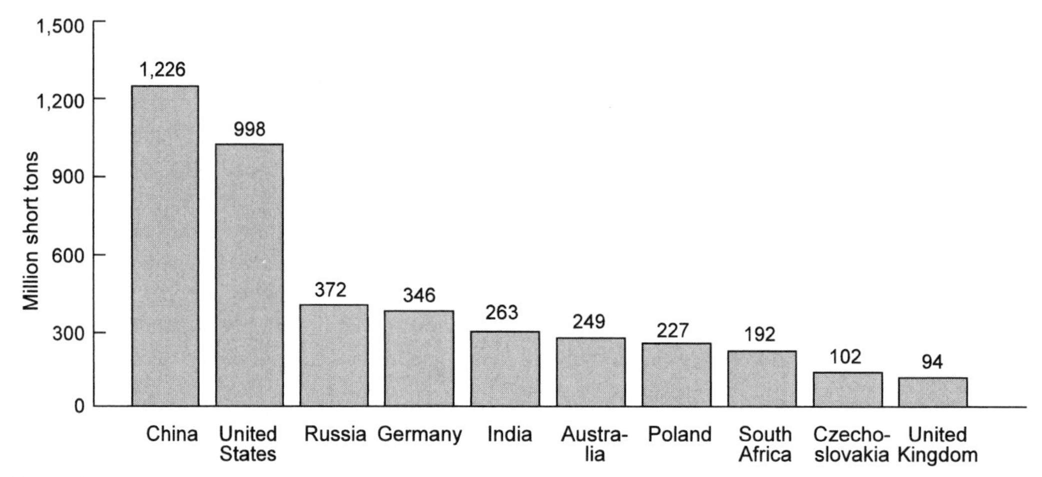

Figure 8.34. 1992 production of coal in the ten leading countries. (*Annual Energy Review 1993*)

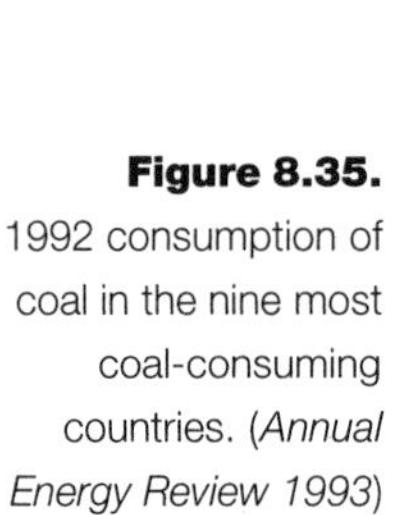

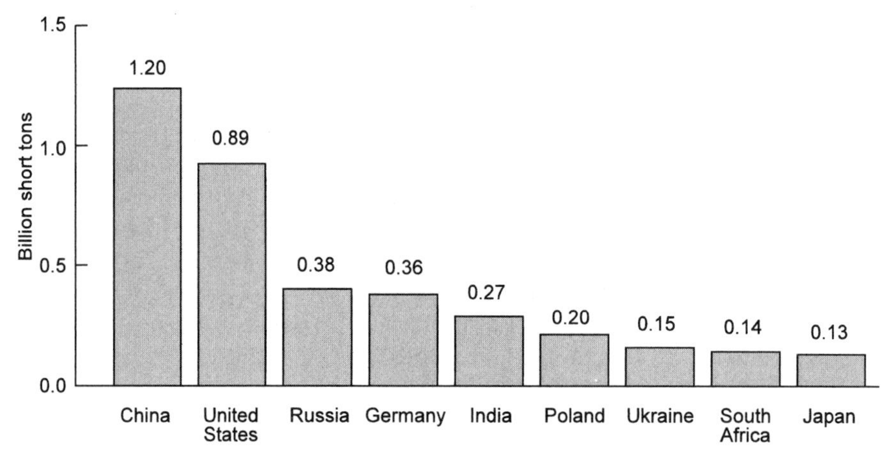

Figure 8.35. 1992 consumption of coal in the nine most coal-consuming countries. (*Annual Energy Review 1993*)

long we can continue to mine this resource at the present or at an accelerating rate. The answer to this question is not simple. We do not have a complete inventory of the world's coal, and we cannot estimate precisely the minimum thickness of coal seams and the maximum thickness of overburden to which coal mining can and will be extended in the future. Figure 8.36 is a diagram that has proved very useful in circumscribing the amount of a given natural resource that is available today and the quantity that is apt to be available in the future. The diagram is usually called a *McKelvey diagram* in honor of its popularizer, Dr. Vincent McKelvey, a former director of the U.S. Geological Survey. The axes of the diagram indicate the degree of assurance to which a given natural resource is known and the degree of economic feasibility with which a given natural resource can be mined or otherwise recovered.

The reserves of a given commodity are usually defined as the quantity that can with reasonable certainty be recovered from known deposits under existing economic and operating conditions. Figure 8.37 shows two independent estimates of the world's coal reserves. They are quite similar for most parts of the world. The

Figure 8.36. The McKelvey Diagram (Adapted from U.S. Bureau of Mines, *Mineral Commodity Summaries*, 1993, p. 200)

Reserve Base

Inferred reserve Base

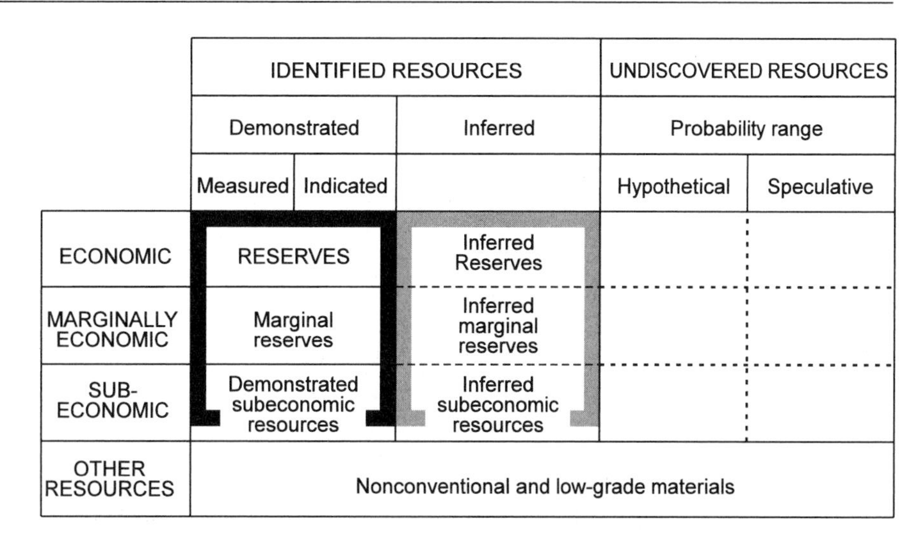

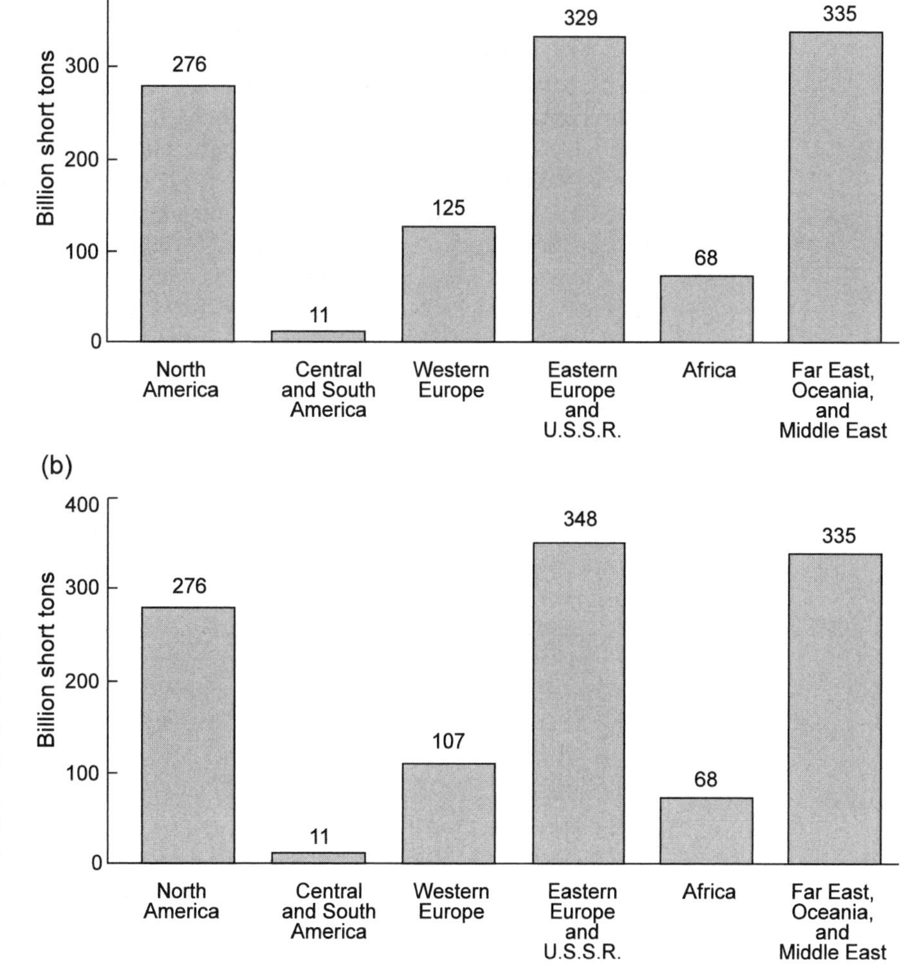

Figure 8.37. World reserves of coal estimated by (a) the World Energy Council, 1991; (b) British Petroleum, 1992. (*Annual Energy Review 1993*)

economic development of China in the next century is apt to depend on the availability of cheap electrical power. The most important source of this power will probably be coal.

The World Energy Council estimates that the world coal reserves currently stand at about 1,800 billion short tons. At the 1990 rate of mining, this implies that our supply of coal would last about

$$\frac{1{,}144 \text{ billion tons}}{5.2 \text{ billion tons/yr}} = 220 \text{ years.}$$

This is a comfortably long period of time; but before taking too much comfort from it, we should remember that both the numerator and the denominator of the fraction are quite uncertain. The rate of coal production is increasing steadily and will probably continue to increase, unless it is demonstrated that very significant environmental problems will be created by continued fossil fuel burning. A major increase in the rate of coal mining during the next century could well drop the "run-out time" to less than 100 years, if the world coal reserves remain at 1,800 billion tons.

This is unlikely. Additional coal deposits will almost certainly be discovered, and improvements in mining technology will almost certainly make it possible to mine coal deposits that are currently subeconomic. In terms of the McKelvey diagram, this implies that resources that are as yet unknown and/or uneconomic will gradually be added to the reserve corner. Estimates of the quantity of coal that will be added to world reserves are, understandably, uncertain. At Mesa Verde National Park in southwestern Colorado, park rangers are frequently asked how many undiscovered Indian remains there are in the park. Estimating the quantity of undiscovered coal deposits is similarly difficult and uncertain, but even rough estimates are useful, and it is worth remembering that most official resource estimates have turned out to be too conservative. Recently, the quantity of ultimately recoverable coal has been estimated to be 6,700 billion short tons (Fulkerson et al. 1990). If this figure is anywhere near correct, coal will certainly be available at least until the year 2100, even if the rate of mining increases very considerably during the twenty-first century.

8.7 Production and Reserves of Oil and Natural Gas

There are many similarities and some differences between the resource picture for coal and that for oil and natural gas. The world distribution of oil and natural gas is very uneven. Unlike coal, a significant fraction of the oil has already been extracted from some heavily explored countries, such as the United States.

Figure 8.38 shows that the world production of petroleum increased rapidly during the 1960s. The share of world crude oil production contributed by OPEC (Organization of Petroleum Exporting Countries), whose members are listed in table 8.1, climbed from 41% in 1960 to 56% in 1973. The sudden rise of oil prices imposed by OPEC in 1973 had a very significant effect on oil exploration and production. World oil production in 1990 was the same as in 1977, but OPEC's share has gradually dropped back to just under 40%.

Figure 8.39 shows the very different history of oil production in four of the leading oil-producing countries since 1960. Oil production in the United States peaked in 1970, and will probably never again exceed the 1970 figure of

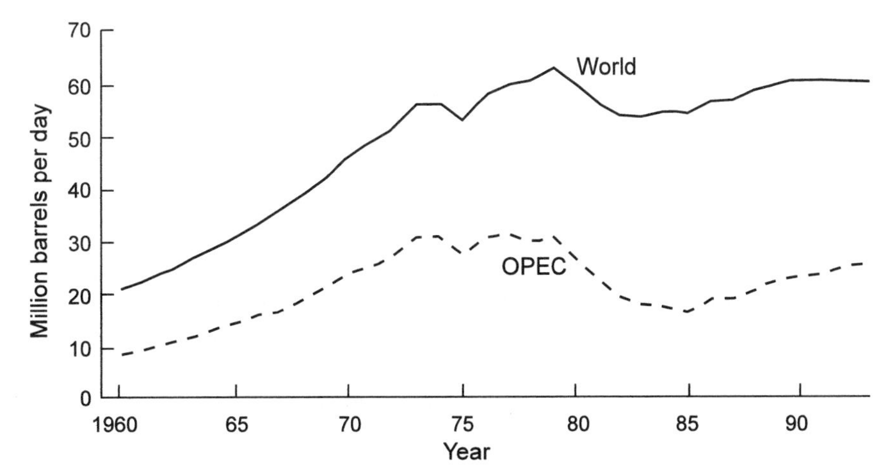

Figure 8.38. Crude oil production, 1960–93; the world and OPEC. (*Annual Energy Review 1993*)

Table 8.1. Membership of the Organization of Petroleum Exporting Countries

Algeria
Gabon
Indonesia
Iran*
Iraq*
Kuwait*
Libya
Nigeria
Qatar
Saudi Arabia*
United Arab Emirates
Abu Dhabi
Fujairah
Sharjah
Dubai
Ras el Khaimah
Ajman
Umm al Qaiwain
Venezuela*

* Organizing members in 1960.

9.6 million bbls/day. Oil exploration in the United States has been very intense, and most of its giant oil fields have probably been discovered, although Alaska and deep-water areas of the Gulf of Mexico offer some hope. Without new discoveries of such fields, additions to U.S. oil reserves will not be enough to offset the depletion of the known oil fields. Oil production will therefore decline, unless it becomes economically attractive to recover much of the oil remaining in reservoirs that have been tapped with past and current oil field technologies.

Oil production in the former USSR increased rapidly until 1980, was nearly constant during much of the '80s, and decreased rapidly during its political disintegration in the '90s. The recent decrease in oil production can almost certainly be reversed, since its causes are political rather than geological.

Saudi Arabia's production has been very irregular. The oil reserves of the country are huge, ca. 26% of world reserves, and the rapid fluctuations in its annual oil production are largely due to political unrest in the Middle East and to the role which Saudi Arabia has played as a leader in OPEC's struggle to maintain the price of oil while meeting the currency demands of its member nations.

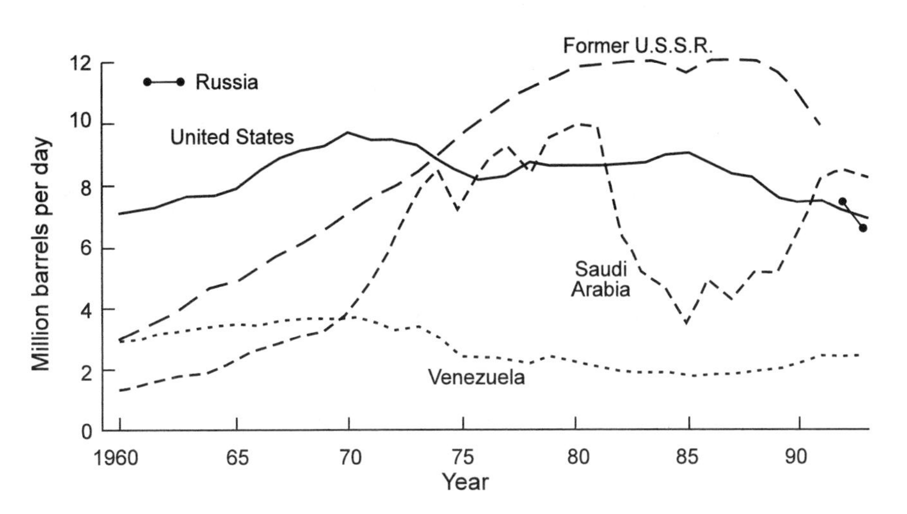

Figure 8.39. Production by the leading crude oil producers, 1960–93. (*Annual Energy Review 1993*)

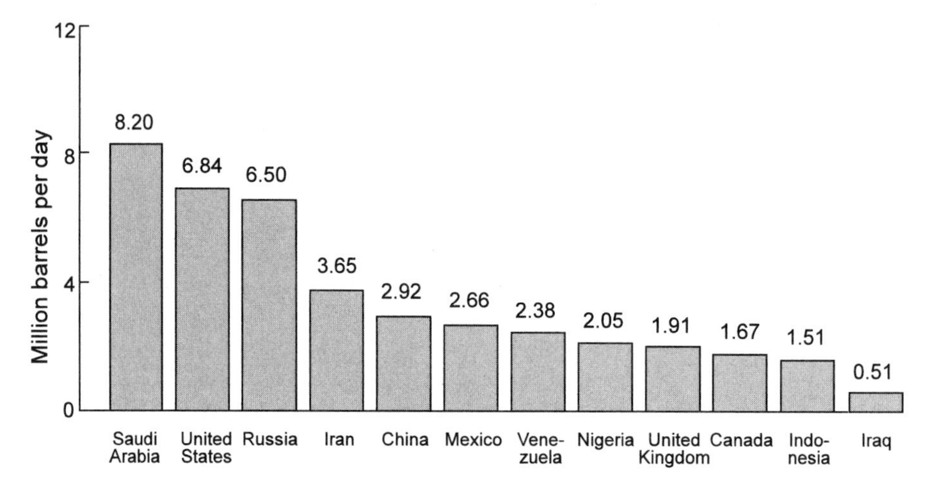

Figure 8.40. 1993 oil production by the twelve most important oil-producing countries. (*Annual Energy Review 1993*)

The twelve major oil-producing countries and their daily production for 1993 are shown in figure 8.40. The relative importance of oil production in the various parts of the world is quite different from the relative importance of their oil reserves. Two rather similar reserve estimates are shown in figure 8.41. Both estimates show how strongly the world's oil reserves are concentrated in the Middle East. About 66% of the known reserves of oil are in the Middle East. Saudi Arabia alone owns 26% of the world's reserves. Iran, Iraq, Kuwait, and the United Arab Emirates each own about 10% of the current world petroleum reserves; by comparison, the United States owns a mere 2.6%. The sheer magnitude of the oil and gas reserves in the Middle East guarantee the strategic importance of this area for at least the first half of the next century.

Even rather modest levels of oil production can have a major effect on the economy of a country. In 1993 the United Kingdom was a distant ninth in oil production (see fig. 8.40). But the discovery of British oil and gas in the North Sea was a major boon for Britain in the 1960s and has been important ever since.

The world's crude oil reserves amount to about 1,000 billion barrels. World oil production has been between 50 and 60 billion bbls/yr during the past decade. If no additional oil were found, we would deplete our petroleum reserves in about 20 years. However, more oil will surely be found, much of it probably in the Middle East, the former USSR, China, and off the coasts of Asia. The quantity of ultimately recoverable oil is much more difficult to gauge than the quantity of ultimately recoverable coal. Much of the new oil will be in rather deep water off shore and in rather deep onshore reservoirs. There are no really reliable techniques for determining how much, if any, oil is present in these potential oil structures. It is likely, however, that the total recoverable reserves of oil are less than 2,000 billion barrels. Unless there are major surprises, we will run out of petroleum by the middle of the next century.

In 1990 the world production of natural gas totaled 74 trillion cubic feet. This represented 21% of the energy contained in fossil fuels produced during this year. The major gas-producing and gas-consuming countries are shown in figures 8.42 and 8.43. The USSR and the United States were by far the largest producers and consumers. Production in the USSR exceeded consumption,

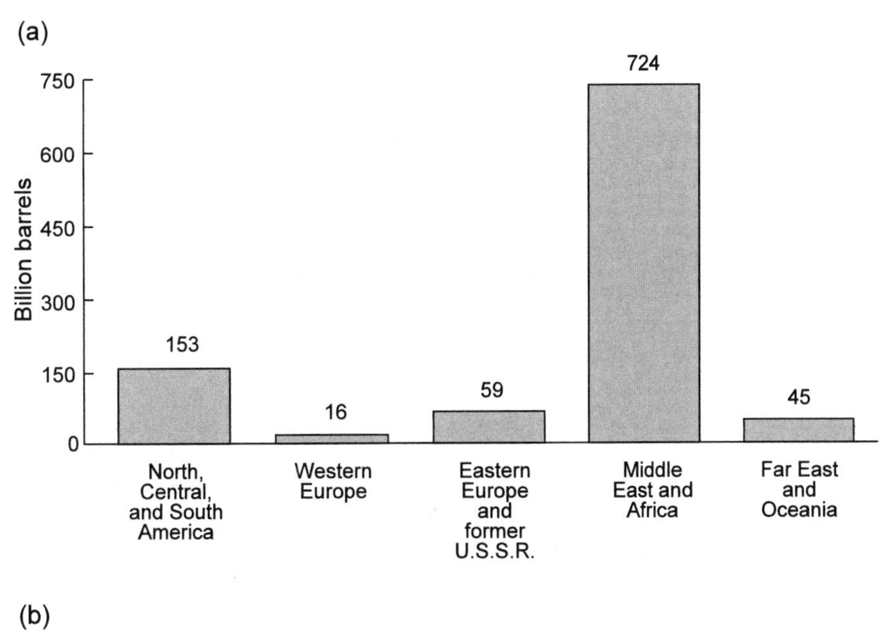

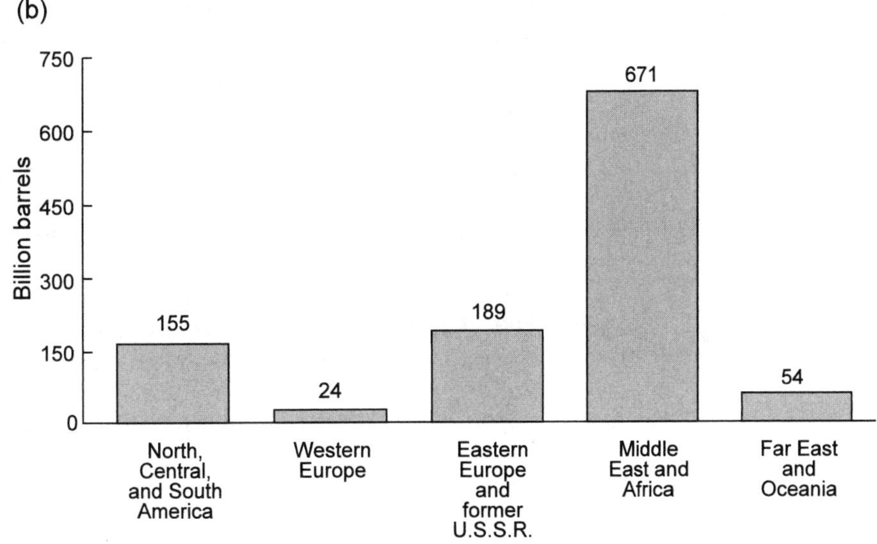

Figure 8.41. The distribution of world oil reserves in 1993. Sources (a) *Oil and Gas Journal*; (b) *World Oil*. (*Annual Energy Review 1993*)

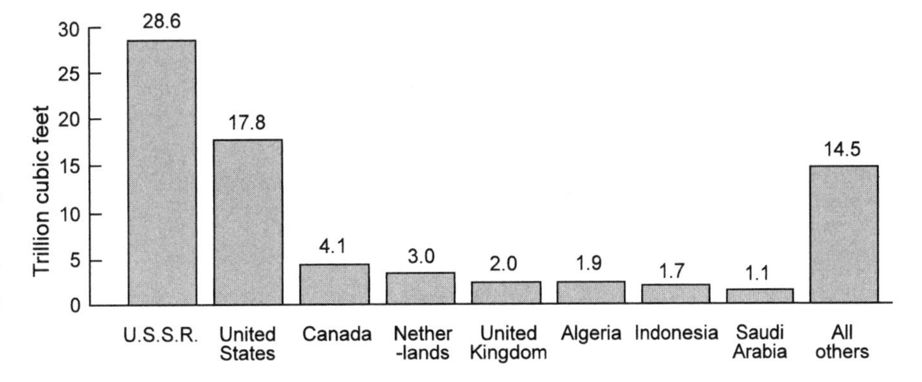

Figure 8.42. The major producers of natural gas in 1991. (*Annual Energy Review 1993*)

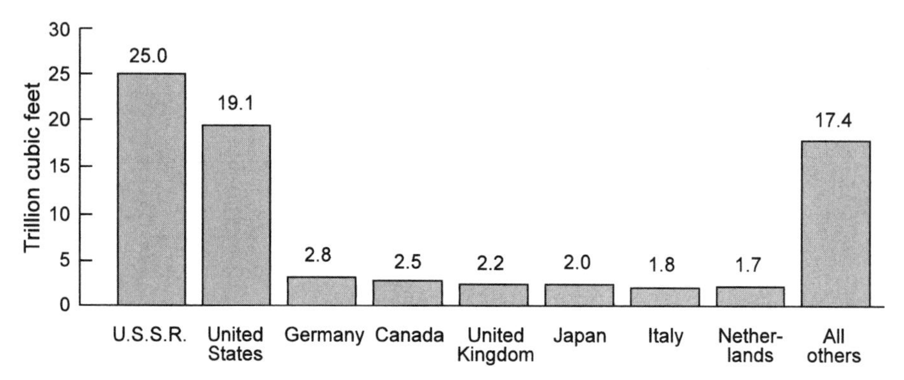

Figure 8.43. The major consumers of natural gas in 1991. (*Annual Energy Review 1993*).

allowing the USSR to export a large quantity of natural gas. The situation was the reverse in the United States, although the fraction of imported natural gas was much smaller than the fraction of imported oil. The "All Others" category among producers includes the countries of the Middle East. These produced 6 trillion cubic feet of natural gas, i.e., less than 10% of total world production. Their importance for the natural gas market is clearly much less than for the petroleum market. Figure 8.44 shows that the world production of natural gas increased quite steadily during the 1980s. Natural gas production increased in the USSR but fell in the United States, for the same reasons that oil production decreased.

Much natural gas has been flared off in the Middle East and elsewhere, because a lot of it is currently uneconomic. This is unfortunate. Per calorie of energy generated, natural gas adds a much smaller amount of CO_2 to the atmosphere than coal. Simply from the point of conservation it would be sensible to pump natural gas that is not currently needed back into the tops of reservoirs and to save it for future use as a fuel or as a feedstock for petrochemicals.

The world reserves of natural gas are estimated to be 4,200 TCF. Of these about 1,600 TCF are in the former USSR. The Middle East is estimated to have reserves of 1,300 TCF, primarily in Iran, the United Arab Emirates, Saudi Arabia, and Qatar. The U.S. reserves are estimated to be only 169 TCF. At the present rate of production, the world is assured of natural gas for approximately

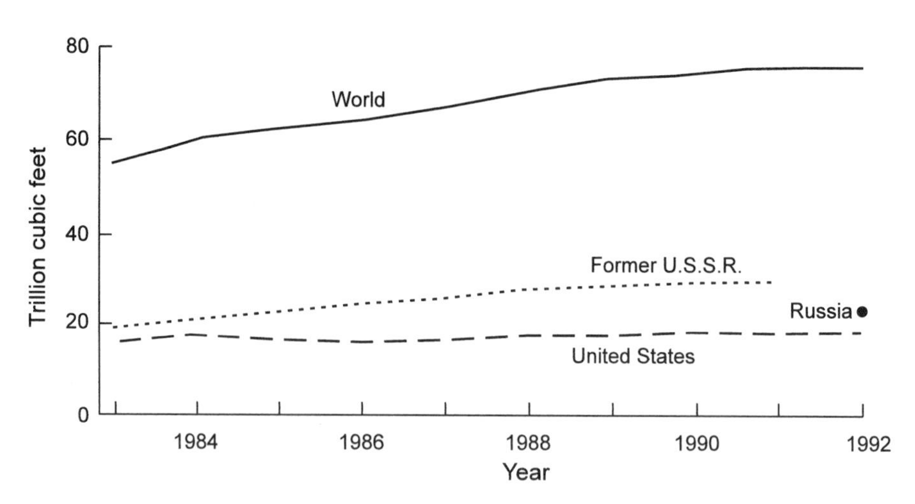

Figure 8.44. Natural gas production in the United States, the former USSR, and the world as a whole between 1983 and 1992. (*Annual Energy Review 1993*)

$$\frac{4{,}200 \text{ TCF}}{75 \text{ TCF/yr}} = 56 \text{ years.}$$

The actual run-out time depends on the amount of natural gas that remains to be discovered and on changes in the rate of natural gas production, and on the quantity of natural gas that is flared off, i.e., that is wasted.

At the present rate of consumption, the United States has known reserves for only about another decade, in part because the public's fear of spills has delayed or prevented the development of new gas fields. In the Middle Eastern countries the reserves of natural gas are apt to last for 200 years at the present rate of production. All these figures suggest that the world's reserves of natural gas will last longer than the world reserves of oil but not nearly as long as the world reserves of coal.

8.8 Subeconomic Sources of Fossil Fuels

Commercial sources of coal, oil, and natural gas account for a minute fraction of the organic carbon that is preserved in the Earth's crust. Unfortunately, nearly all of this carbon is dispersed in shales that contain on the average something like 1% carbon. The cost of extracting carbon from such shales is much greater than the value of the carbon itself. This will almost certainly continue to be so, because the energy required to extract the carbon from such shales is greater than the energy released by burning the extracted carbon. Between such highly uneconomic sources of carbon and commercially viable coal, oil, and natural gas reserves, there are a number of carbon reservoirs which are currently subeconomic but that will be potentially important sources of fossil fuels during the next century. The most important of these are the heavy oils, tar sands, and oil shales.

Heavy oils are oils that are too viscous and too dense to be pumped out of their reservoir rocks by conventional techniques. The high viscosity of these oils is due to a large concentration of high-molecular weight hydrocarbons and organosulfur compounds. In tar sands, the quantity and the composition of heavy hydrocarbons gives the organic matter the consistency of tar, a substance that is clearly unpumpable. A variety of techniques have been developed to extract oil from these sources. Enhanced recovery techniques are being used to soften or liquefy heavy oils and tar by direct heating or by heating with steam injected under high pressure. In other techniques, natural gas is injected into reservoirs. The natural gas mixes with the heavy oil, and the mixture has a viscosity that is sufficiently low so that it can be pumped.

Two major production methods are available for extracting oil from oil shale. One involves open-pit mining of the shales, grinding the rock to a powder, and heating the powdered rock to about 500°C in a retort. The oil is volatilized and leaves the powder as a gas, which condenses at lower temperatures. In the second technique, underground tunnels are driven into the oil shale. The oil shale is then shattered by underground explosions and heated to about 500°C by fire that is controlled with a stream of air, diesel fuel, and steam. Much of the high molecular weight bitumen in the oil shales is vaporized, driven ahead of the combustion zone, and recovered in wells or drains.

The major challenge to the recovery of oil from these sources has been economic, but environmental problems, particularly those created by the recovery of oil from oil shales, have also proved daunting. The 1970s opened a considerable window of opportunity for these alternative sources of petroleum. The enormous increase in the price of pumped oil during that decade came close to making the recovery of oil from tar sands and oil shales economically feasible. The thought of producing enough oil from U.S. sources to make the country energy self-sufficient and independent of the whims of foreign producers was highly appealing. A good deal of effort was therefore expended on improving the technology of oil recovery from oil shale. The subsequent drop in oil prices in the 1980s dampened interest in these technologies, and it seems unlikely that they will be developed and applied on a large scale until the world's reserves of easily pumped oil become significantly depleted. Even then it may be cheaper and environmentally sounder to convert coal to petroleum. As early as 1913 synoil (synthetic oil) was produced by adding hydrogen to coal in a solvent slurry at elevated temperatures and pressures. In the 1970s and 1980s large-scale demonstrations of three processes were undertaken. Although all were technically successful, they were not economic. Even today the cost of producing synoil from coal is significantly higher than the cost of producing oil from heavy oils, tar sands, and oil shale.

Table 8.2. World Resources of Heavy Oil, Tar Sands, and Oil Shale, 1990

Heavy Oil

	Billion Barrels		
	Proved Reserves	*Undiscovered Resources*	*Total Recoverable*
North America	23	30	65
Central and South America	280	16	309
Western Europe	8	0	9
USSR and Eastern Europe (former)	7	21	33
Africa	4	1	5
Middle East	115	22	169
Far East and Oceania	13	4	19
World total	450	94	609*

Tar Sands

	Billion Barrels		
	Measured Resources	*Speculative Resources*	*In-Place Resources*
United States	21	41	~60
Canada			~1,700
Venezuela			~700
World total			~4,000

Oil Shale

		*Billion Barrels of Oil**
United States		630
Western	460	
Eastern	170	
South America (Brazil)		300
USSR (former)		40
Africa (Zaire)		40

Source: Data from compilation by Kulp 1990.
* Recovery = 38% of estimated in-place resource.

The reserves and resources of oil in heavy oils, tar sands, and oil shales are enormous. The data in table 8.2 suggest that the sum of the oil resource in these categories is on the order of 5,000 billion barrels. This is five times the current estimated reserves of oil and about 2.5 times the estimated volume of ultimately recoverable petroleum. Most of the world resources of heavy oil are in Central and South America and in the Middle East. Canada is the most important potential source of oil in tar sands. In 1994 commerical production of oil from tar sands in Canada was a small but not negligible 0.4 million barrels per day. The United States and Brazil own the lion's share of the world resources of oil shale. Although the figures in table 8.2 are uncertain, it is clear that oil in these resources will play a significant role in energy availability and energy politics during the twenty-first century.

8.9 A Forecast for Fossil Fuels

The description of the several types of fossil fuel sources, and the reserves and likely resources of each, has shown that there will be no worldwide shortage of fossil fuels during the next century. Furthermore, it is likely that the cost of producing energy from fossil fuels will not increase dramatically until well into the twenty-first century. The price of fossil fuel energy may, however, fluctuate widely. The experience of the 1970s offers many salutary lessons in this regard. OPEC is rather weak at present; but the politics of the Middle East are volatile, and the relationship of the Middle East to the rest of the world is difficult to foresee. Shortages of oil and large price fluctuations are to be expected. However, strong price excursions are not apt to last for more than a decade or so. The world knows how to economize, how to switch to other fuels, and how to develop alternative sources of petroleum on a decadal time scale.

On a timescale of hundreds of years, the outlook for the continued availability of fossil fuels is quite bleak. Figure 8.45 shows the course of the energy from fossil fuel burning as predicted in 1967 by M. King Hubbert, one of the outstanding American geologists of this century. The rapid rise in fossil fuel use during this century is followed by an equally rapid decline. The total curve has come to be known affectionately as "Hubbert's pimple." It illustrates one of the properties of nonrenewable energy sources like fossil fuels: their lifetime is apt to be short on the timescale of recorded human history. They must be replaced by other sources of energy if the economic welfare of humankind is to be improved or

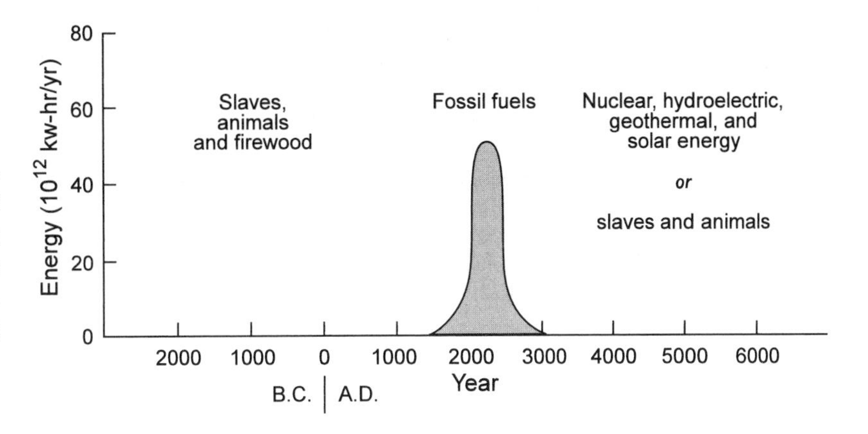

Figure 8.45. The energy sources of humankind, illustrating the brevity of the fossil-fuel epoch. (After Hubbert 1967)

even maintained at its present level. Fortunately, alternatives to fossil fuels are available and others are in the offing (see chapter 11).

We may wish to call on these sources long before our store of fossil fuels is exhausted. At some point the value of the remaining oil and gas may become much more valuable as petrochemical feedstocks than as fuel. Your grandchildren may ask you in amazement: "You *burned* them, those nice molecules, you just burned them?" The burning of fossil fuels also creates local, regional, and global environmental problems. Potentially most vexing is the modification of the Earth's climate due to the increase in the CO_2 content of the atmosphere. The burning of the world's known coal reserves alone would add an amount of CO_2 to the atmosphere that is more than twice its current CO_2 content. An increase of a factor of five in the CO_2 content of the atmosphere by the year 2200 is quite possible. We are clearly altering the composition of the atmosphere to a significant extent. The possible consequences of these alterations are discussed in chapter 11. Suffice it to say that the consequences could be significant and deleterious, but that at this time neither their significance nor their potential harm can be predicted with certainty.

References

Alesch, D. J., and Petak, W. J. 1986. *The Politics and Economics of Earthquake Hazard Mitigation*. Institute of Behavioral Science, University of Colorado, Boulder.

Anderson, D. L. 1971. The San Andreas Fault. *Scientific American* 225 (Nov.):52–68.

Ando, M. 1975. Source Mechanisms and Tectonic Significance of Historical Earthquakes along the Nankai Trough, Japan. *Tectonophysics* 27:119–40.

Athy, L. F. 1930. Density, Porosity and Compaction of Sedimentary Rocks. *Amer. Assoc. Petrol. Geol. Bull.* 14:1–35.

Bolt, B. A. 1978. *Earthquakes: A Primer*. W. H. Freeman, San Francisco.

Boore, D. M. 1977. The Motion of the Ground in Earthquakes. *Scientific American* 237, (Dec.):68–78.

Bott, M.H.P. 1982. *The Interior of the Earth: Its Structure, Constitution and Evolution*. Elsevier, New York.

Bullen, K. E., and Bolt, B. A. 1985. *An Introduction to the Theory of Seismology*. 4th ed. Cambridge University Press, Cambridge, U.K.

Cooper, J. D., Friedland, I. M., Buckle, I. G., Nimis, R. B., and Bobb, N. M. 1994. The Northridge Earthquake: Progress Made, Lessons Learned in Seismic-resistant Bridge Design. *Public Roads* 58 (no. 1):26–36.

Craig, J. R., Vaughan, D. J., and Skinner, B. J. 1988. *Resources of the Earth*. Prentice Hall, Englewood Cliffs, N.J.

Demaison, G., and Huizinga, B. J. 1991. Genetic Classification of Petroleum Systems. *Amer. Assoc. Petrol. Geol. Bull.* 75:1626–43.

Dickinson, G. 1951. Geological Aspects of Abnormal Reservoir Pressures in the Gulf Coast Region of Louisiana, U.S.A., *3rd World Petrol. Congr. Proc.*, sec. 1, and *Amer. Assoc. Petrol. Geol. Bull.* 10:40–430.

Energy Information Administration. 1994. *Annual Energy Review 1993*. U.S. Dept. of Energy, Washington, D.C.

Fulkerson, W., Judkins, R. R., and Sanghvi, M. K. 1990. Energy from Fossil Fuels. *Scientific American* 263 (Sept.):128–35.

Grace, J. D., and Hart, G. F. 1986. Giant Gas Fields of Northern West Siberia. *Amer. Assoc. Petrol. Geol. Bull.* 70:830–52.

Hasegawa, A., Umino, N., and Takagi, A. 1978. Double-Planed Deep Seismic Zone and

Upper-Mantle Structure in the Northeastern Japan Arc. *Geophys. J. (R. Astronom. Soc.)* 54:281–96.

Hedburg, H. 1936. Gravitational Compaction of Clays and Shales. *Amer. J. Sci.* 31:241–87.

Hubbert, M. K. 1967. Application of Hydrodynamics to Oil Exploration. *7th World Petrol. Congr. Proc.* 1B:59–75.

Illing, V. C., ed. 1953. *The Science of Petroleum*. Vol. 6, part 1, of *The World's Oil Fields: The Eastern Hemisphere*. Oxford University Press, London.

Jenyon, M. K. 1990. *Oil and Gas Traps: Aspects of their Seismostratigraphy, Morphology and Development*. John Wiley, Chichester, U.K.

Kelleher, J. A. 1972. Rupture Zones of Large South American Earthquakes and Some Predictions. *Jour. Geophys. Res.* 77:2087–2103.

Kulp, J. L. 1990. *Energy Strategy, 1990–2030*. A Report to the Weyerhaeuser Company.

Levorsen, A. I. 1954. *Geology of Petroleum*. W. H. Freeman, San Francisco.

Magara, K. 1968. Compaction and Migration of Fluids in Miocene Mudstone, Nagaoka Plain, Japan. *Amer. Assoc. Petrol. Geol. Bull.* 52:2466–2501.

McKenzie, D. P., and Richter, F. 1980. Convection Currents in the Earth's Mantle. In *Earthquakes and Volcanoes: Readings from Scientific American*, pp. 136–45. W. H. Freeman, San Francisco.

McPhee, J. A. 1993. *Assembling California*. 1st ed. Farrar, Straus and Giroux, New York.

Meade, R. H. 1966. Factors Influencing the Early Stages of the Compaction of Clays and Sands—Review. *J. Sedimen. Petrol.* 36:1085–1101.

Molnar, P., and Tapponnier, P. 1977. The Collision between India and Eurasia. *Scientific American* 236 (April):30–41.

Morgan, W. J. 1972. Deep Mantle Convection Plumes and Plate Motions. *Amer. Assoc. Petrol. Geol. Bull.* 56:203–13.

National Academy of Sciences. 1973. *The Great Alaska Earthquake of 1964*. National Academy Press, Washington, D.C.

Penick, J. L., Jr. 1981. *The New Madrid Earthquakes*. University of Missouri Press, Columbia.

Proshlyakov, B. K. 1960. Reservoir Properties of Rocks as a Function of Their Depth and Lithology. *Geol. Neft. i. Gaza* 4:34–29.

Raleigh, C. B. 1984. The Southern California Uplift: Review of Recent Results. In *Earthquake Prediction*, pp. 35–45. F. F. Evison, Chair. Terra Scientific Publishing, Tokyo.

Selley, R. C. 1985. *Elements of Petroleum Geology*. W. H. Freeman, New York.

Sieh, K.E. 1981. A Review of Geological Evidence for Recurrence Times of Large Earthquakes. In *Earthquake Prediction: An International Review*, ed. D. W. Simpson and P. G. Richards, pp. 181–207. American Geophysical Union, Washington, D.C.

Skinner, B. J., and Porter, S. C. 1992. *The Dynamic Earth: An Introduction to Physical Geology*. 2d ed. John Wiley, New York.

U. S. Bureau of Mines. 1993. *Mineral Commodity Summaries 1993*. Dept. of the Interior, Washington, D.C.

Verney, P. 1979. *The Earthquake Hand Book*. Paddington Press, New York.

von Huene, R., and Scholl, D. W. 1991. Observations at Convergent Margins Concerning Sediment Subduction, Subduction Erosion, and the Growth of Continental Crust. *Rev. of Geophys.* 29:279–316.

Walker, B. 1982. *Earthquake*. A vol. of *Planet Earth*. Time-Life Books, Alexandria, Va.

Yergin, D. 1991. *The Prize: The Epic Quest for Oil, Money and Power*. Simon and Schuster, New York.

Yonekura, N. 1975. Quartenary Tectonic Movements in the Outer Arc on Southwest Japan with Special Reference to Seismic Crustal Deformation. *Bull. Dept. Geogr. Univ. Tokyo* 7:19–71.

9 Magmas, Water, and Ores

9.1 Introduction: From Ordinary Rocks to Ores

We have encountered magmas and volcanoes on two previous occasions. In chapter 6 volcanoes were introduced as the builders of the Hawaiian Islands. In chapter 8 they reappeared as parts of the machinery of the Earth, as by-products of mantle convection and plate tectonics. In this chapter they will be treated as subjects worthy of study in their own right, and as sources, both direct and indirect, of many of the world's major ore deposits and of heat for geothermal systems. Somewhat in passing, we will look at volcanoes as natural hazards on a variety of distance and timescales.

9.2 The Origin of Magmas

In Hawaii, the legends say, Pele came first to islands without names and then to Kauai, Oahu, and Maui. In each place Pele dug for volcanic fire, but the flames went out. She traveled southeastward until she came to the island of Hawaii. There she went up the mountain, thrust down her magic digging stick, and brought up volcanic fire. Satisfied, she decided to make her home in the volcano Kilauea and became the goddess of volcanoes.

Many are the legends of Pele and her dealings with the people of Hawaii since that time. They still warn geologists not to take rocks away from Kilauea lest Pele's anger be aroused. Undaunted, geologists have removed a good many rocks for study in faraway places. Their samples have included "Pele's tears," solidified droplets of volcanic glass, and "Pele's hair," long strands of blonde-bronze threads of volcanic glass that form during some volcanic eruptions. Studies of the full range of lavas and magmas have revealed a great deal about the history of the Hawaiian volcanoes and about the origin of Hawaiian magmas.

On cooling, nearly all Hawaiian magmas yield basalts. Their composition was described in table 6.1. They are much like the basalts that make up most of Iceland and like the basaltic magmas that flow from the volcanoes along the submerged parts of midocean ridges. They are representatives of one of the most common and widespread types of magmas. The volcanoes on Hawaii are now heavily instrumented. Seismographs listen to their rumblings. Much like stethoscopes, they tell us about movements below the surface and can be used to predict the onset of volcanic eruptions.

Magmas are clearly the products of rock melting. Below Hawaii, magma moves up from magma chambers to the surface and erupts as lava. A moment's reflection is apt to raise several questions about these lavas: what is it that melted? Whatever it was, did it melt completely or are basalts products of partial melting? Where does the melting take place? Is it shallow, or did the magma make its way up from the deep parts of the mantle? We can answer some of these questions with a good deal of confidence. Others still belong to the category of "good questions."

There is excellent evidence that basalts are products of partial rather than complete melting. The speed of earthquake waves depends on the composition of the rocks through which they pass. Very thorough studies of the speed of seismic waves in the Earth's mantle have shown that the upper mantle does not have the composition of basaltic lavas. Seismic velocities in the upper mantle correspond much more closely to seismic velocities through rocks which consist largely of the mineral olivine, $(Mg,Fe)_2SiO_4$. This mineral is present in many Hawaiian basalts, but only in small amounts. The major minerals in cooled Hawaiian lavas are pyroxenes and plagioclase feldspars. Both were described in chapter 6. Lavas of basaltic composition cannot be produced by the melting of pure olivine. But the partial melting of olivine with a small admixture of pyroxene and feldspar will produce melts of basaltic composition. The composition of the uppermost parts of the mantle is probably close to that of peridotite (see table 9.1), a rock named for peridot, an olive-green, gemmy variety of olivine. Laboratory experiments have shown that when peridotite starts to melt, some components are strongly concentrated in the liquid. These include SiO_2, Al_2O_3, CaO, Na_2O, and K_2O. The first four are essential components of some pyroxenes and all plagioclase feldspars. A very good case can be made for the origin of Hawaiian magmas by the partial melting of such mantle material.

The foregoing proposition is strengthened by two important observations. Chunks of peridotitic rock are present in volcanic materials that have been brought to the surface explosively from considerable depths. These chunks have not crystallized from magmas at or close to the surface, as shown by the presence of minerals and of mineral assemblages that are stable only at high pressures. Their bulk chemistry is consistent with evidence from seismology and measurements of gravity for the nature of the upper mantle. In some parts of the world they contain small quantities of diamonds. One of the most famous diamond-bearing volcanic structures is the diamond pipe at Kimberley in South Africa. Diamonds are present there in very small quantities, but they have been mined extensively (see fig. 9.1) from a pipelike body of kimberlite, a volcanic rock that contains many fragments of mantle material. Most of the natural—as contrasted with synthetic—diamonds come to the surface in such pipes. When these pipes weather, diamonds are released from their matrix. Since diamonds are quite re-

Table 9.1. The Composition of Three Peridotites from Israel (major components in wt %, minor components in ppm)

Constituent	*Weight Percent*			*Constituent*	*Parts per Million*		
SiO_2	44.90	44.70	44.90	Cr	3,089	2,425	2,581
TiO_2	n.d.	0.11	0.09	Ni	2,037	2,052	2,077
Al_2O_3	2.10	2.60	3.10	Co	105	n.d.	116
FeO	8.03	8.01	8.60	Mn	1,037	973	941
MgO	42.10	42.00	35.80	Sc	12	10.1	11
CaO	2.00	2.40	3.30				
Na_2O	0.26	0.28	0.19				
K_2O	0.04	0.04	0.04				
Total (wt %)	99.43	100.14	96.02				

Source: Stein and Katz 1989.

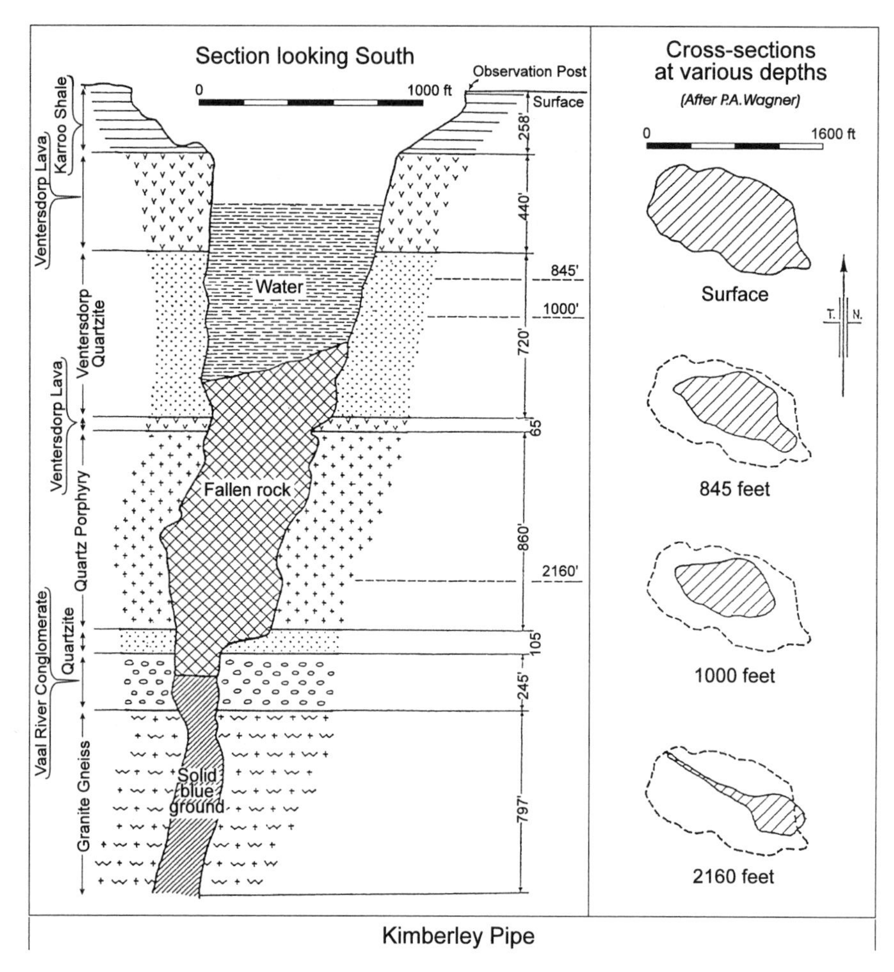

Figure 9.1. Geology of the diamond mine at Kimberley, South Africa. (Pelletier 1964)

sistant to weathering and denser than many silicate minerals, they can accumulate in river gravels. Diamonds are recovered commercially both from diamond pipes and from such sediments.

Diamond is not the stable form of carbon at the Earth's surface. At low pressures, the stable form of carbon is graphite, the writing constituent of "lead" pencils. The conversion of diamond to graphite at Earth surface temperatures is very slow indeed. Diamonds may not be "forever," but they are nearly so, even on a geologic timescale. At high temperatures, however, diamonds are converted quite readily to graphite. The presence of diamonds in kimberlites implies their origin at great depths in the mantle and an upward journey to the surface that was so rapid that they did not have time to be converted to graphite en route. The composition and structure of the minerals associated with diamonds in kimberlites agree well with these inferences. In fact, a good deal has been learned about the temperature structure of the upper 200 kilometers of the mantle from detailed studies of the minerals in kimberlites.

There is a second important line of evidence which supports the conclusion that basalts are the products of the partial melting of peridotitic mantle material.

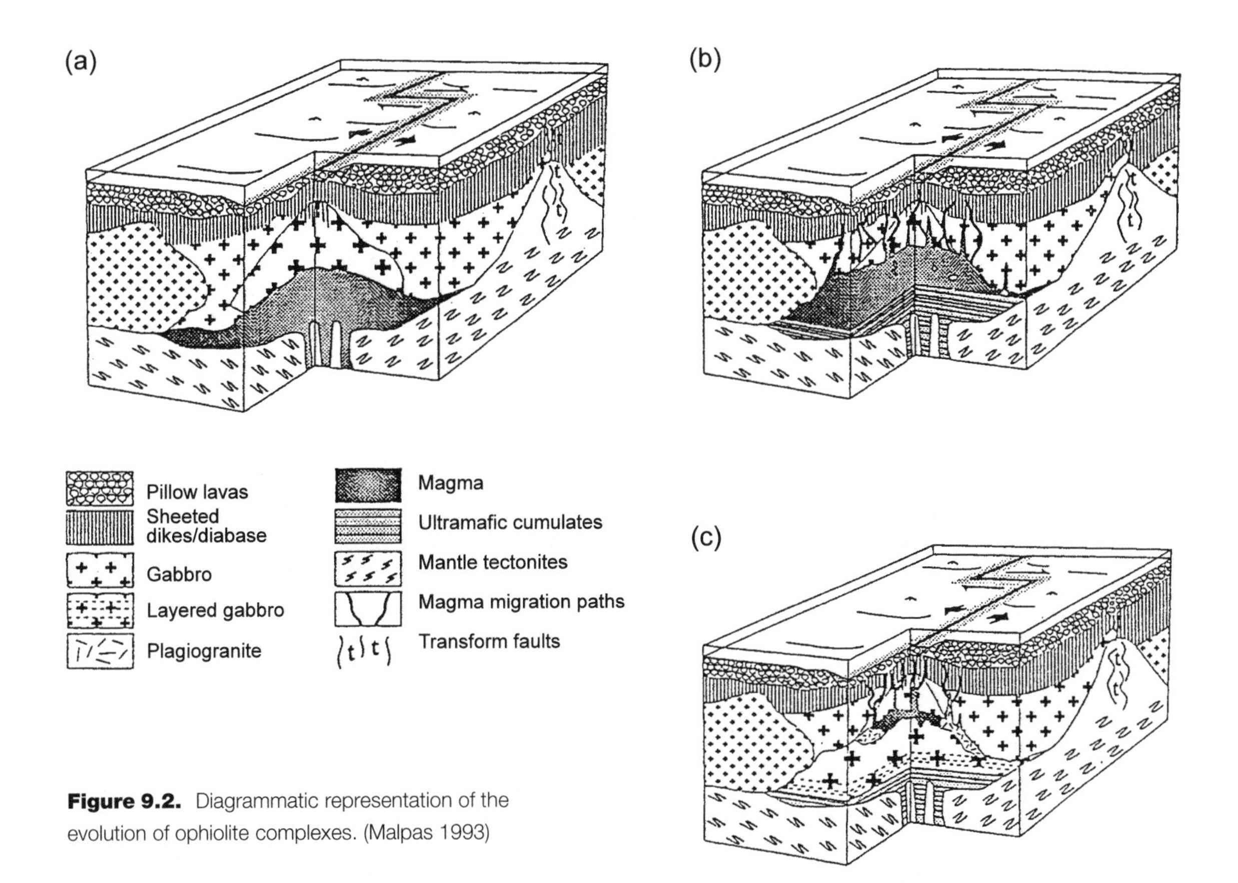

Figure 9.2. Diagrammatic representation of the evolution of ophiolite complexes. (Malpas 1993)

In some parts of the world, pieces of oceanic crust (ophiolites) have been thrust up onto the continents rather than subducted beneath them. Notable examples of ophiolite complexes occur, for instance, in California, on the island of Cyprus in the Mediterranean, in Oman on the Arabian Peninsula, and on the Tibetan Plateau, north of the high Himalayas. Figure 9.2 is a diagrammatic representation of the evolution of an ophiolite complex. Its base consists of mantle material. This has generally been tectonically deformed, hence the name *mantle tectonite*. Above this there is a magma chamber (fig. 9.2a) which gradually cools and solidifies (fig. 9.2b, c). During this process, early formed crystals settle to the bottom of the chamber to form what are called *cumulates*. These usually contain large amounts of olivine and are given the name *ultramafic cumulates*. The main body of magma in the chamber cools to form igneous rocks of basaltic composition. Since the chamber cools slowly, the size of crystals in these rocks is much larger than in rapidly cooled basaltic lava flows. They are usually called *gabbros*, a name applied to rocks of basaltic composition that have cooled slowly beneath the Earth's surface.

During the cooling of basaltic magma bodies, the rocks above the magma chamber frequently crack. Magma moves into and through these cracks, and either emerges as lava on the ocean floor or solidifies in the cracks to form *dikes*, which frequently intrude earlier lava flows. Submarine basaltic flows tend to assume forms like those in plate 28. When such sausage-shaped bodies of basalt are piled up and exposed in vertical cuts, they look rather like mounds of pillows, hence the name *pillow lavas*.

The depths at which basaltic magmas originate is still a matter of considerable debate. Some mantle fragments in kimberlites and in other explosively emplaced volcanic material have probably come from depths as great as 200 kilometers. The origin of some basaltic magmas may lie even deeper. The observation that the source of magmas for the Hawaiian chain remained essentially in one place while the Pacific plate itself moved thousands of kilometers (see chapter 6) suggests that the magma source is below the base of the plate. In fact, the source may be as deep as the base of the mantle, but direct evidence for such a deep origin is still lacking. It is very difficult to detect what is probably a rather small region of melting at great depth.

We do, however, know a good deal about the vertical structure of the mantle (see fig. 9.3). Most of our knowledge has come from studies of the propagation of seismic waves through the Earth's interior. The mantle can be divided into an upper and a lower portion. The upper mantle extends to a depth of approximately 670 kilometers. The lower mantle extends to the core-mantle boundary at a depth of 2900 kilometers. The core extends from the base of the mantle to the center of the Earth at a depth of 6,371 kilometers. It is liquid down to 5,100 kilometers, and solid below that. The core probably consists largely of iron and nickel, much like metallic meteorites, but other elements are surely present. Likely candidates include oxygen and sulfur.

During the 1980s, the three-dimensional structure of the mantle finally started to be defined. Since the velocity of seismic waves in the mantle changes quite rapidly with depth, it was relatively easy to define the vertical structure of the mantle. However, the velocity of seismic waves at a particular depth in the mantle varies laterally by only a few percent. Mapping such small differences throughout the globe requires very large amounts of seismic data and the use of

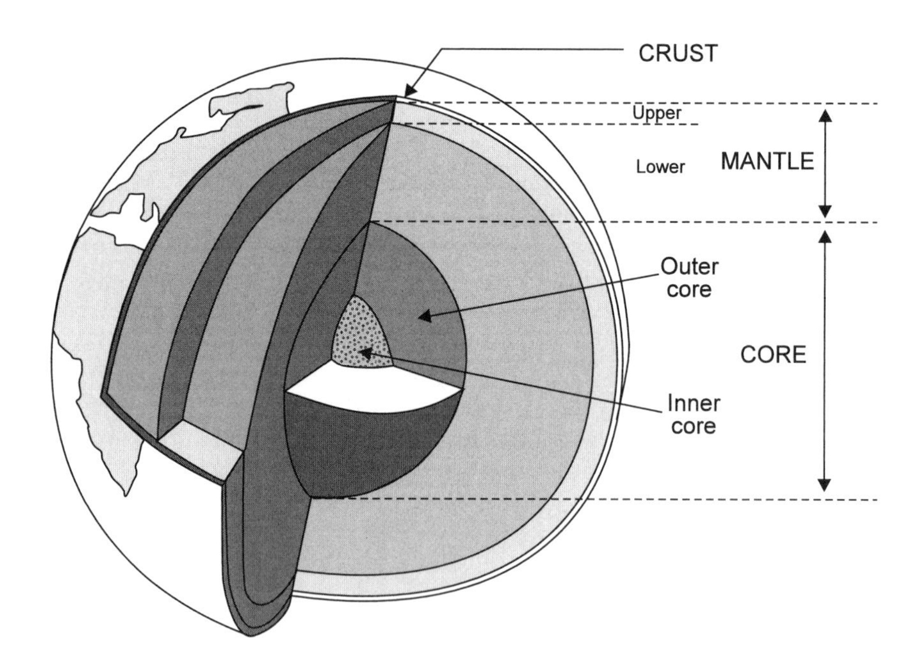

Figure 9.3. The internal structure of the Earth. (After Allègre 1988)

very high speed computers. Even today, the best 3-D maps of the mantle, those prepared by Adam Dziewonski and his group, are still only generalized (see plate 29). Mantle tomography is in its infancy. Much more detail will be required from the geophysical equivalents of medical CAT-scans before they define the patterns of mantle convection. Even then it may be impossible to locate the sources of the Hawaiian lavas and those of the other magmatic hot spots with seismic techniques alone.

By far the largest effusions of basaltic lavas occur in the oceans—at hot spots such as the Hawaiian Islands and along midocean ridges like the Mid-Atlantic Ridge and the East Pacific Rise. However, great masses of basaltic lava are sometimes extruded on land. In the northwestern United States, in much of eastern Washington and Oregon, and in adjoining parts of western Idaho, enormous amounts of lava flowed out of volcanoes between 17 and 13 million years ago on what is now the Columbia Plateau. In central India a similarly large amount of basaltic lava erupted 65 million years ago to form the famous Deccan Traps. This was just at the end of the Cretaceous period, when the last of the dinosaurs died out and when a large number of other animals became extinct. It has been suggested that the eruption of the Deccan Traps caused or was instrumental in causing the extinction events at the "K/T boundary" (the Cretaceous [Kreta]-Tertiary boundary). However, at the same time a very large meteorite hit the Earth, probably on the Yucatan Peninsula of Mexico. This impact is currently favored as the cause of the "K/T extinction," although it is only fair to say that the biological consequences of large impacts have not yet been worked out in detail.

Most basaltic eruptions are rather gentle events. Some lava fountains in Hawaii rise to heights of some 400 meters (plate 30), but there are few truly explosive events. Basaltic lavas are quite fluid and tend to form gently sloping shield volcanoes such as Mauna Loa (see plate 6). Towns built near basaltic volcanoes are apt to be overwhelmed by majestic lava flows rather than by dust and bombs from violent volcanic explosions. Attempts have been made, some successful at least on a modest timescale, to stop lava flows by cooling them with intense jets of water, by erecting retaining walls, and by diverting them with explosive charges set against partly solidified crusts.

Volcanism is not always gentle. Many volcanoes erupt explosively, spreading death and destruction for distances of tens and even hundreds of kilometers. Recent examples include the 1980 eruption of Mount St. Helens in the Cascade Range of the state of Washington, the 1982 El Chichón eruption in Mexico, and the 1992 eruption of Mount Pinatubo in the Philippines. The last century saw the great 1883 explosive eruption of Krakatoa in the Sunda Strait between Java and Sumatra. Perhaps the most famous eruption of antiquity is that of Mount Vesuvius near Naples, Italy. In A.D. 79 the debris from this eruption buried the towns of Pompeii and Herculaneum. The event was recorded by the Roman author and statesman Pliny the Younger. His uncle, Pliny the Elder, the author of a famous book on natural history, insisted on observing the eruption and died during the event. During the present century parts of both Pompeii and Herculaneum have been excavated, and their death has been immortalized by two movie versions of *The Last Days of Pompeii.*

Lava and ash from volcanoes that erupt explosively are usually not basaltic in composition. Lavas from volcanoes in island arcs, such as the Aleutians off the southern coast of Alaska (see fig. 9.4), are dominantly andesites or basaltic an-

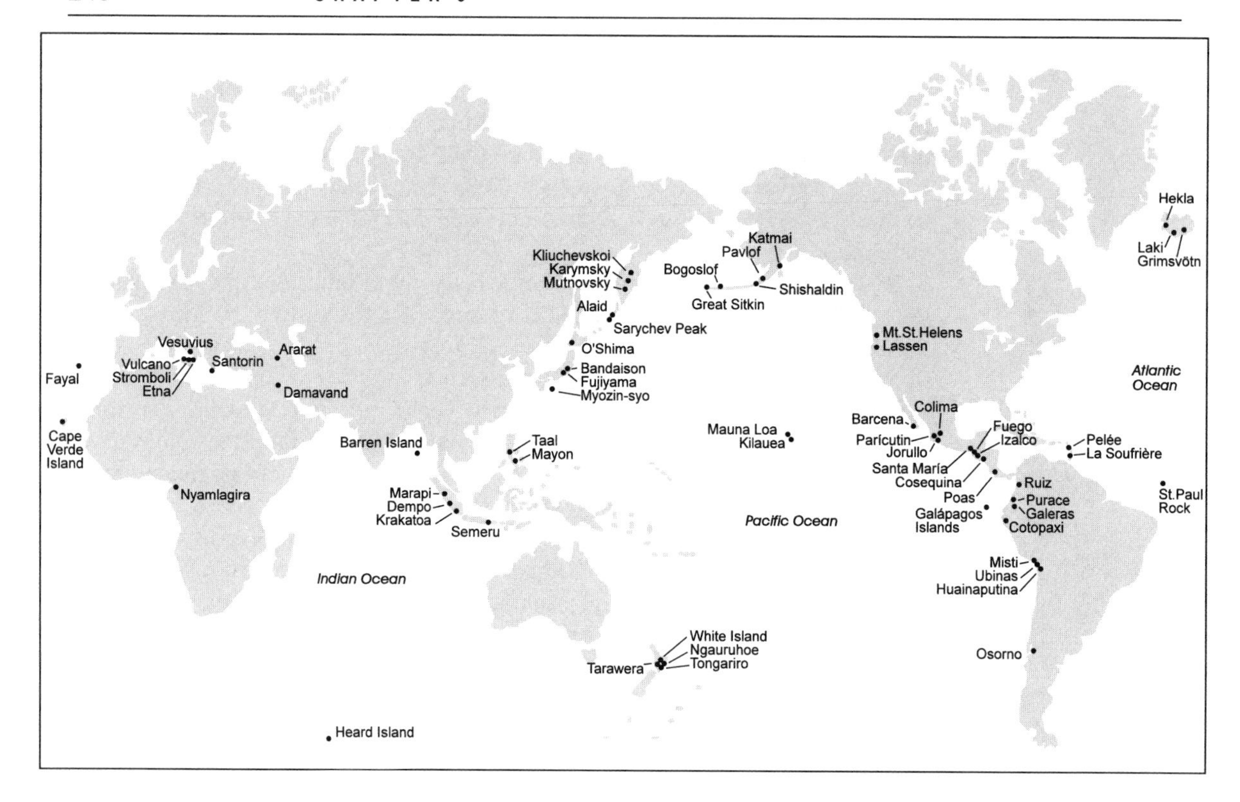

Figure 9.4. Location of some active volcanoes. (Judson and Kauffman 1986)

Table 9.2.
Average Composition of Five Classes of Volcanic Rocks (in wt % oxides)

Constituent	*Rhyolite*	*Dacite*	*Andesite*	*Basalt*
SiO_2	73.66	63.58	54.20	50.83
TiO_2	0.22	0.64	1.31	2.03
Al_2O_3	13.45	16.67	17.17	14.07
Fe_2O_3	1.25	2.24	3.48	2.88
FeO	0.75	3.00	5.49	9.05
MnO	0.03	0.11	0.15	0.18
MgO	0.32	2.12	4.36	6.34
CaO	1.13	5.53	7.92	10.42
Na_2O	2.99	3.98	3.67	2.23
K_2O	5.35	1.40	1.11	0.82
P_2O_5	0.07	0.17	0.28	0.23
H_2O	0.78	0.56	0.86	0.91
Total	100.0	100.0	100.0	100.0

Source: Carmichael, Turner, and Verhoogen 1974.

desites. Representative chemical compositions of andesitic lavas are shown in table 9.2. Basaltic andesites are compositionally intermediate between basalts and andesites. Andesites contain less MgO, CaO, and total iron ($FeO + Fe_2O_3$) than basalts, but more SiO_2, Na_2O and K_2O. Volcanic rocks of andesitic composition are not restricted to island arcs. The word "andesite" itself comes from its type locality in the Andes of South America. Volcanoes throughout central America and up the west coast of North America also emit a great deal of andesite. In fact, lavas from most volcanoes that are located above subduction zones are dominantly andesitic.

The difference in the eruptive style of basaltic and andesitic volcanoes is related to differences in their chemical composition. The viscosity of andesitic lavas is much greater than that of basaltic lavas, because the higher SiO_2 content of andesites allows a greater degree of linking of $[SiO_4]^{4-}$ tetrahedra in these magmas. Andesitic magmas also tend to have a higher volatile content than basalts. The pressure exerted by water, CO_2, and other gases dissolved in andesitic magmas increases as these melts crystallize. The pent-up gas pressure can be released suddenly in major explosions, which frequently blow off large parts of volcanoes. Before its spectacular eruption on May 18, 1980, Mount St. Helens had a nearly perfect cone rising to almost 10,000 feet, i.e., about 3,000 meters (plate 31). Its sides were much steeper than those of Mauna Loa and other basaltic shield volcanoes. Earthquake activity started on March 20, 1980, and continued steadily, along with minor eruptions and the growth of a large bulge on the northern flank of the volcano. On May 18, without a warning increase either in seismic activity or in the rate at which the bulge was expanding, a magnitude 5.1 earthquake shook the weakened cone and triggered one of the largest landslides of historic times (Harris 1988). As the avalanche began to move, a small plume of inky ash shot upward from the bisected summit crater, while lighter ash clouds overtook the avalanche. In a hurricane of hot ash and rock fragments, the clouds rolled nearly 19 miles north from the crater, devastating a fan-shaped area 23 miles across from east to west. The area within 8 miles of the crater looked as if it had been cleared by a giant sandblaster.

Before the lateral eruption reached its maximum extent, the jet plume from the summit crater began to expand into a titanic mushroom cloud that rose to an altitude of 18 kilometers (plate 32). Ash from this cloud settled as far east as Minnesota. The eruption subsided after about 9 hours. Plate 33 is an aerial view of the mountain after the May 18 eruption. The mountain erupted several more times during 1980. Since then, most of the eruptions have been quiet, dome-building events. Mount St. Helens has changed its form and appearance repeatedly during the past 2,000 years; there is every reason to expect similar changes during the next 2,000 years.

The changes in the chemical composition of volcanic rocks as one passes from basalts through basaltic andesites to andesites continue beyond andesites to dacites and rhyolites. In this sequence (see table 9.2), the SiO_2 content of the lavas rises from about 51% to about 74%, the percentage of ($FeO + Fe_2O_3$), CaO, MgO, and TiO_2 continues to drop, and that of K_2O to rise. Rhyolites are much less common than andesites, but some volcanic explosions associated with rhyolitic volcanism have been truly enormous. Rhyolitic volcanics erupted in the Yellowstone area of Wyoming 2.2, 1.2, and 0.6 m.y.a. These eruptions completely dwarf those of Mount St. Helens. The volume of ejecta from the

Table 9.3.
Loss of Life Due to Volcanic Eruptions during Recorded History. (After Nuhfer, Proctor, and Moser 1993)

Year	*Volcano*	*Location (Agent of Mortality)*	*Casualties*
1500 B.C.	Santorini	Greece (explosion, tsunami)	Unknown
A.D. 79	Vesuvius	Pompeii & Herculaneum (pyroclastic flow)	3,360
1631	Vesuvius	Naples & Resina (pyroclastic flows)	3,500
1783	Skaptar (Laki)	Iceland (tephra and starvation)	9,500
1792	Unzen	Kyushu, Japan (avalanche and tsunami from eruption-related earthquake)	15,000
1815	Tambora	Indonesia (tephra, tsunami, and starvation)	92,000*
1822	Galunggung	Indonesia (pyroclastic flows and mudflows)	4,000
1883	Krakatau	Indonesia (tsunami)	36,400
1902	Pelée	Martinique (pyroclastic flows)	29,000
1902	Santa Maria	Guatemala (pyroclastic flows)	6,000
1919	Kelut	Java (mudflows)	5,110
1951	Lamington	New Guinea (debris avalanche, pyroclastic flows)	2,942
1977	Nyiragongo	Tanzania (lava flows)	<100
1980	Mt. St. Helens	Washington (lateral blast, debris avalanche, and mudflows)	57
1982	El Chichón	Mexico (pyroclastic flows)	1,877
1985	Nevado del Ruiz	Colombia (mudflows)	23,000
1986	Nyos	Cameroon (CO_2 gas cloud erupted from beneath lake in old volcano)	1,746
1990	Kelut	Java, Indonesia (tephra)	32
1991	Unzen	Shimabara, Japan (pyroclastic flows)	38
1991	Pinatubo	Luzon, Philippines (tephra and disease)	932
1993	Mayon	Mayon, Philippines	60

* 70,000 deaths from famine due to crop destruction.

2.2 m.y.a. eruption at Yellowstone was some one thousand times greater than that of the Mount St. Helens ejecta. A similarly large eruption in the Yellowstone area could well occur during the next few hundred thousand years.

The loss of human life due to explosive volcanic eruptions has been significant. Table 9.3 shows that death tolls in excess of one thousand have not been unusual. This is still small compared to the loss of life due to major floods and earthquakes, but it is not negligible. Living close to an active volcano that is apt to erupt explosively can have its exciting moments. Regions that have been devastated by explosive eruptions have rarely remained uninhabited for any great length of time, and modern geophysical and geochemical techniques are helping to decrease the element of surprise in the timing of volcanic explosions. Seismic monitoring has proved useful, and careful geodetic surveying can now detect ground movements and premonitory tell-tale swellings of volcanic edifices. Geochemistry has contributed as well, because the composition of volcanic gases emitted by volcanoes tends to change before eruptions. The efficacy of these warning techniques will almost certainly continue to improve, but predicting the time and the size of explosive eruptions precisely will probably continue to be difficult.

Not all of the effects of explosive volcanic eruptions are local. Many of the deaths caused by the eruption of Krakatoa in 1883 (see table 9.3) were caused by the tidal wave generated by the explosion. Even farther afield, it is thought that the explosion of Tambora in 1815 was responsible for "the year without a summer." Crop failures during 1816 created widespread famine in New England and northern Europe. The number of deaths due to famine and the related typhus epidemic that began to spread from Ireland to the rest of the British Isles in late 1816 was probably much greater than the immediate effects of the eruption itself (Stommel and Stommel 1983). Climatic cooling due to explosive volcanic eruptions is due to the injection of dust and sulfur dioxide into the stratosphere. These effects are still not well understood; like the straw that broke the camel's back, they tend to be most destructive when—as in 1815 and 1816—they reinforce global cooling due to other causes.

The highly explosive nature of the prehistoric eruptions in the Yellowstone area was surely due to the high SiO_2 and volatile contents of the lavas. Their high SiO_2 content and that of dacitic and rhyolitic lavas in general is in need of explanation. Specifically, are these lavas derived from the mantle or the crust, and in either case, what accounts for the differences between their composition and that of basalts? These questions are gradually receiving answers, although our understanding of the origin and evolution of volcanic rocks is still somewhat hazy.

Fortunately, rocks in the Earth's mantle tend to have isotopic compositions (see chapter 2) that are different from those of crustal rocks. During partial melting of the mantle, many radioactive elements are separated from their daughter elements. As an example, rubidium is concentrated in magmas much more strongly than strontium. Magmas moving upward from the mantle into the crust therefore have an Rb/Sr ratio greater than that of unmelted mantle; the residual, depleted mantle remaining after the removal of partial melts has an Rb/Sr ratio that is lower than that of unmelted mantle. As a result, the crust develops a higher Rb/Sr ratio than the mantle. ^{87}Rb slowly decays to ^{87}Sr (see chapter 2). Hence crustal rocks contain strontium with a higher proportion of ^{87}Sr than mantle rocks. For similar reasons, the isotopic compositions of lead, neodymium, and several other elements in crustal rocks differ significantly enough from their isotopic composition in mantle rocks to be useful as tracers for the origin of volcanic rocks.

As might be expected, the isotopic signature of these elements in oceanic basalts is essentially the same as that of the mantle. The origin of magmas above subduction zones is more complex. Some volcanic magmas there have isotopic signatures indicative of a mantle origin. Others clearly have come from crustal rocks. Still others bear the imprint of a mixed origin, either because mantle-derived magma was contaminated by crustal rocks during passage to the surface, or because both mantle and crustal rocks were present in the source area.

One of the most spectacular confirmations of the presence of young crustal materials in magmas above subduction zones has been provided by the discovery of minute quantities of ^{10}Be in recent volcanic rocks. ^{10}Be, an isotope of beryllium with a half-life of 2.5 million years, is produced in the Earth's atmosphere by interaction with cosmic rays from space. The ^{10}Be rains out of the atmosphere, much of it into the oceans. There it is scavenged by marine sediments. These are subducted below continents. The presence of ^{10}Be in volcanic rocks shows that some ^{10}Be is caught up in magmas produced by melting below continents and reappears at the surface in a sufficiently short period of time, so that not all of the original ^{10}Be has decayed.

The presence of crustal material in the melting regions of subduction zones affects the melt composition. Two other processes are also important: fractional melting and fractional crystallization. We have already seen how partial melting of the mantle can yield basalts, which differ chemically from the parent mantle. The reason for the difference is that the first melt to form when mantle material is heated contains a disproportionately large share of some elements, for instance Si, Al, Na, and K, and a disproportionately small share of others, for instance Mg. Since these melts usually have a lower density than the solid mantle residue, they tend to rise toward the surface, leaving depleted mantle material behind. The more complete the melting process before the melt becomes separated from the residual solids, the closer the composition of the melt will be to the composition of the mantle. If melting is complete, the composition of the melt will, of course, be the same as that of the original mantle material.

Basaltic magmas form after only a few percent of mantle material have melted. If tens of percent of mantle melt, highly magnesian magmas are formed. After these solidify, they are called *komatiites*. These were first identified in the Archean rocks of the Komati Valley of the Barberton area in South Africa. They are much less common in modern terrains than in Archean terrains, i.e., in rock units that formed more than 2.5 billion years ago. Higher mantle temperatures are required to generate komatiites than to generate basalts. The greater abundance of komatiites in ancient rocks is thought to be related to somewhat higher temperatures in the upper parts of the Earth's mantle during the early part of Earth history.

Fractional crystallization is a process that is complementary to fractional melting. As magmas rise toward the surface, they pass into cooler regions. Some magmas rise to the surface so rapidly that their temperature has decreased only slightly when they pour out on the Earth's surface. Thereafter they cool so rapidly that their composition changes very little during solidification. That is not the fate of magmas that cool slowly. At the end of the 1959–60 eruption of Kilauea Iki, a lava lake formed at the edge of the main crater of Kilauea volcano on the island of Hawaii. This lake has since cooled and is covered by a progressively thickening crust of solidified magma. Numerous drill holes have sampled this crust, and have shown that the composition of the solidified lava is quite variable. The composition of the lava lake was originally homogeneous, so these variations are related to the slow cooling of the lava within the crust of solidified magma. During the crystallization of the lava, the elements that were fractionated into the basalt during mantle melting tend to stay in the melt. Elements that largely remained in the mantle during the formation of the basalt tend to crystallize from the lava first. The composition of the remaining melt therefore becomes progressively enriched in SiO_2, Na_2O, and K_2O, and depleted in MgO and ($FeO + Fe_2O_3$). As a result, silica-rich rocks have formed within the crust of the lava lake, including rocks of rhyolitic composition (see table 9.2) (Helz 1987).

Many melts that form in the Earth's crust and upper mantle move upward so slowly that they begin to crystallize en route. This is particularly common during the rise of highly viscous, SiO_2-rich melts. Once crystallization has begun, the upward progress of these magmas becomes even slower. This is probably the major reason for the scarcity of highly siliceous lavas compared to the abundance of intrusive siliceous igneous rocks, i.e., rock bodies that solidified below the surface. Granites are the intrusive chemical equivalents of rhyolites. Granites in huge bodies called *batholiths* are common in areas of mountain building and in the deeper parts of stable continental areas. Plate 34 shows the contact between the Manaslu granite and the rocks which this body intruded in the high Himalaya. The mineralogy of most granites is relatively simple. Quartz, Na-feldspar (plagioclase), and K-feldspar (orthoclase) each make up about 30% of many granites. Micas, amphiboles, and minor accessory minerals account for the remaining 10%. Plate 35 shows a small part of a polished slab of granite, a favored facing stone for many banks and public buildings. Crystals of the major constituents of granites are frequently in the centimeter range. Many granites contain xenoliths, i.e., strange rocks (see plate 35), which are not formed as part of the crystallization sequence. In most cases these are pieces of the rock formations through which the granitic magma traveled on its way upward, or pieces that were detached from the roof of the magma chamber and sank through the granite magma until they were trapped in partly solidified granite.

Many granitic magmas contain several percent water. Some of this water is incorporated in minerals such as micas and amphiboles, but much of it is concentrated in the last bits of melt to crystallize. In the presence of such a hydrous melt, very large crystals can grow in what are called *granite pegmatites*. Some single crystals weighing more than a ton have been recorded; in a case of extreme gigantism, 13-meter-long crystals of spodumene, $LiAlSi_2O_6$, have been found in pegmatites in the Black Hills of South Dakota. Pegmatites frequently contain abnormally high concentrations of elements that do not fit readily into the crystal structure of the common minerals of granites. Beryllium is one of these elements. The mineral beryl, $Be_3Al_2Si_6O_{18}$, is rare in granites but common in pegmatites, and pegmatites have been important as commercial sources of beryllium.

Water in pegmatites sometimes becomes segregated into pockets. Crystals of many minerals growing into such pockets are clear and transparent. Many pockets are therefore important sources of gems. They often contain transparent, light-blue beryl, the gemstone aquamarine (plate 36b), and pink spodumene, the gemstone kunzite. Large, beautiful, clear crystals of tourmaline are to be found there in a variety of colors. Watermelon tourmalines are among the most delightful varieties of this gemstone. Their pink center is surrounded by a colorless zone, which is followed by a "rind" of dark green tourmaline (see plate 36a). Pegmatites are one of a group of rock units made commercially valuable, and—not coincidentally—very beautiful by the presence of water.

9.3 Magmatic Oxide Ores

Most igneous rocks are not particularly highly valued, except for use as building stones or for aggregate mixes. A few, however, are much sought after, either because they contain unusually high concentrations of certain chemical elements, unusual compounds, or valuable forms of such compounds. Ores of chromium and vanadium belong to the first category. Chromium and vanadium belong to the group of elements that are usually referred to as "minor," but both are sometimes strongly concentrated in minerals of igneous rocks. The concentration of chromium in average crustal rocks is quite small, about 100 ppm (1 part per million = 1 gm per metric ton). It is almost always present in the +3 valence state. The radius of the Cr^{3+} ion is about 0.62 Å, only a little larger than that of Fe^{3+} (0.55 Å), and Al^{3+} (0.54 Å). Cr^{3+} therefore substitutes quite readily for Fe^{3+} and Al^{3+} in a variety of minerals. Substitutions in the spinel group are quite typical. Pure spinel has the formula $MgAl_2O_4$, but the mineral is rarely pure. Fe^{2+}, Mn^{2+}, and Zn^{2+} readily substitute for Mg^{2+}; Fe^{3+}, Cr^{3+}, and Mn^{3+} readily sub-

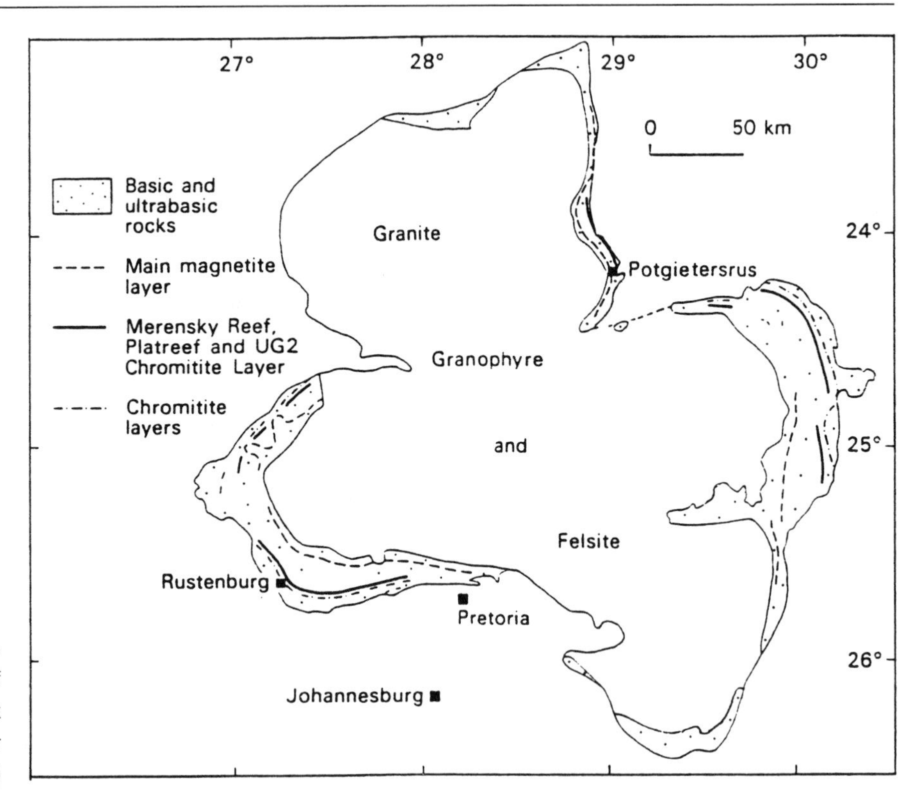

Figure 9.5. Sketch map of the Bushveld Complex in South Africa. (After von Gruenewaldt 1977)

stitute for Al^{3+}. The mineral chromite is a member of the spinel group and is the major ore mineral of chromium. In chromite, Fe^{2+} replaces much of the Mg^{2+}, and Cr^{3+} replaces a good deal of the Al^{3+}. The composition of chromites can therefore be written as $(Fe^{2+}, Mg^{2+})(Cr^{3+}, Al^{3+})_2O_4$. Chromite must contain more than 45% Cr_2O_3 to be of metallurgical and chemical grade. To be used as a refractory material in applications such as furnace linings, chromite must contain more than 30% Cr_2O_3. Such high-Cr spinels are found in some intrusive rocks. In a very few rocks, chromite occurs as the only mineral in layers or in pods. The concentration of Cr in these accumulations is often orders of magnitude greater than in the intrusions as a whole. If the layers are sufficiently thick or the pods sufficiently large, both can be mined at a profit; they are then chromium ores.

Much of the world's chromite ore comes from the Bushveld Complex in South Africa (see fig. 9.5). This is a very large, funnel-shaped, layered intrusion. Its overall composition is gabbroic. Layers of massive chromite (see plate 37) from a few centimeters to 2 meters in thickness occur in the lower parts of the intrusion and can be traced laterally for as much as tens of kilometers. Ore bodies may consist of a single layer or of a number of closely spaced layers. The ore reserves in the Bushveld are enormous. Down to a mining depth of only 300 meters, the quantity of ore amounts to about 2.3 billion tons. If we include lower-grade deposits down to a mining depth of 1,200 meters, the quantity of ore is about ten times larger. The 1993 figures for chromium-producing countries are shown in table 9.4. The reserves and estimated reserve base of South Africa are an impressive fraction of the world total. Second only to the reserves of chromium in the Bushveld are those of the Great Dyke of Zimbabwe (see fig. 9.6). This very

Table 9.4. Production, Reserves, and Reserve Base of Chromium, 1993 (in thousands of metric tons)

Country	*Mine Production*	*Reserves*	*Reserve Base*
Kazakhstan	3,500	130,000	130,000
South Africa	2,300	960,000	5,500,000
India	900	59,000	77,000
Turkey	850	8,000	20,000
Zimbabwe	550	140,000	930,000
Finland	450	29,000	29,000
Subtotal	8,550	1,326,000	6,686,000
Others	1,000	36,000	60,000
Total	9,550	1,362,000	6,746,000

Source: *Mineral Commodity Summaries*, 1994.

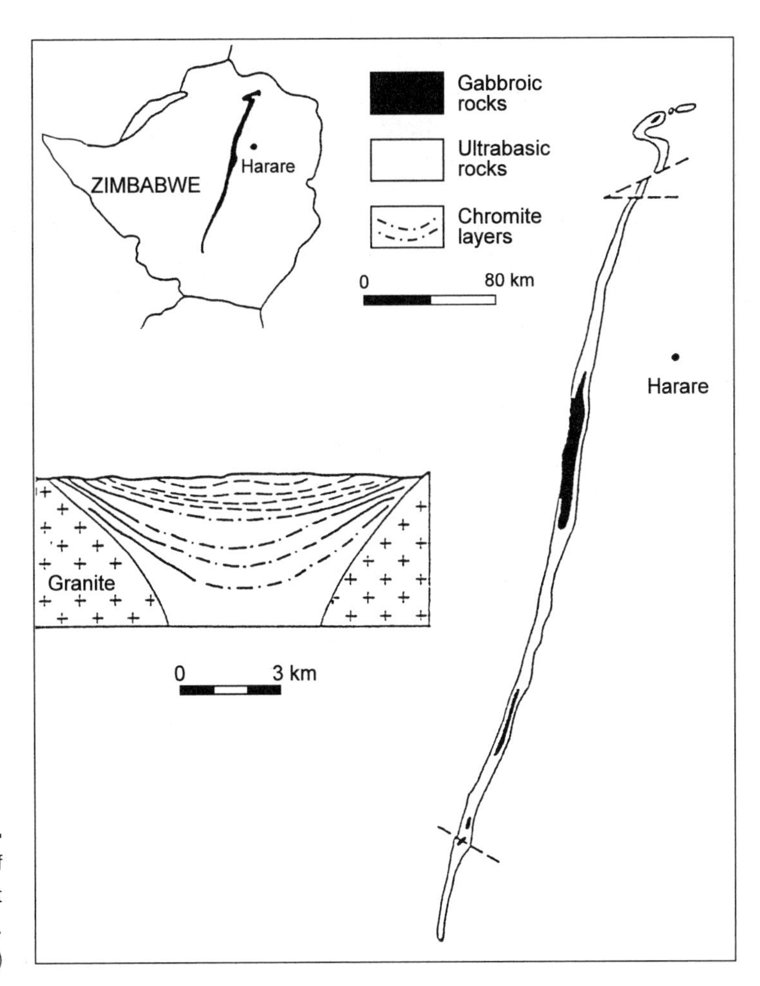

Figure 9.6. The occurrence of chromite in the Great Dyke of Zimbabwe. (Evans 1993)

olivine-rich intrusion is 532 kilometers long and 5 to 9.5 kilometers in width. Chromite layers occur along most of its length.

In 1993 Kazakhstan ranked first among the major producers of chromite. Most of its ores are of the "podiform" variety. Their shape is irregular and unpredictable. Most podiform chromite ore bodies are relatively small. There are, however, exceptions. Some of the chromium ores in the Urals are in very large podiform ore bodies.

Layered chromites were clearly formed during the crystallization of large intrusive bodies. They probably grew at what was then the base of the magma chamber. Just why and how they formed is still a matter of debate. The most likely explanation involves shifts to chromite precipitation when a new pulse of magma was added to a body of magma undergoing fractional crystallization in a large magma chamber.

The geopolitical implications of table 9.4 are important. Chromium is an important alloy in stainless steel and in steel for armaments. The world's reserve base of chromium ore is reassuringly large compared to the current rate of production; their ratio

$$\frac{6{,}746{,}000{,}000 \text{ tons}}{9{,}550{,}000 \text{ tons/yr}} = 706 \text{ years}$$

indicates that we will not run out of high-grade chrome ore for several centuries. The extreme concentration of the world's chrome reserves in South Africa and Zimbabwe gives these countries a strategic importance for chromium comparable to the importance of the Middle East for oil. The United States is not well endowed with chromium. The Stillwater Complex in Montana is a large intrusive that is quite similar to the Bushveld Complex; unfortunately, it does not contain high-grade chromium ore. Some low-grade chromite ore was mined there during World War II. But even the reserve base of this type of ore (10 million tons) is very small compared to the South African and Zimbabwan reserves. Unless unforeseen technological innovations supersede chromium steels, chromium will continue to be an important industrial metal, and the African ores will continue to dominate the market.

Higher up in the Bushveld Complex, where there are no chromite layers, the dominant member of the spinel group is the mineral magnetite. Pure magnetite has the composition $Fe^{2+}Fe_2^{3+}O_4$. Magnetite in the upper layers of the Bushveld Complex contains up to 19% TiO_2 and up to about 1.6% vanadium oxide, V_2O_3. Vanadium, a metal that is also important in steel making, has a crustal abundance of about 135 ppm. It is therefore only slightly more abundant than chromium and is very strongly concentrated in the Bushveld magnetites. The valence of vanadium in igneous rocks is almost always +3. Since the ionic radius of V^{3+} is about 0.64 Å, it can substitute readily for Fe^{2+} (r = 0.61 Å) in magnetite. About half the world's production of vanadium comes from the monomineralic layers of magnetite in the upper parts of the Bushveld Complex. This, together with its chromium and platinum group element ores (see sec. 9.4), makes the Bushveld one of the major mining areas of the world. These ores, together with its gold ores in the Witwatersrand Basin, its coal, and its iron and manganese ores, combine to make the Republic of South Africa a nation that is superbly endowed with mineral wealth.

Most of the non-South African production of vanadium comes from settings that are geologically similar to the Bushveld ores, though the rather minor U.S. production comes from a very different type of deposit, where vanadium is associated with uranium ores (see sec. 9.5). Like the reserves of chromium, vanadium reserves are very large compared to the current rate of vanadium production. If shortages of vanadium do occur, they will almost certainly be due to political or social rather than geologic reasons.

9.4 Liquid Immiscibility and Magmatic Ores

When magmas rise into regions of lower temperatures and pressures, they usually simply crystallize. Some magmas, however, first split into two liquids. In some instances one of the liquids contains abnormally high concentrations of one or more valuable metals. When such liquids crystallize, they can give rise to important ore deposits. From an economic point of view, oxide, sulfide, and carbonate-rich liquids are the most important products of magmatic unmixing. Oxide melts are responsible for many titanium and some iron ore bodies. Sulfide melts are important as the parents of ores of nickel, copper, and the platinum group elements. Carbonate melts are important as hosts of economic concentrations of niobium, tantalum, rare earths, copper, thorium, vermiculite, phosphates, and carbonates and are the source of the materials that bring the red and green phosphors on color television sets. The phosphates can be used in fertilizers, especially in countries lacking cheaper sedimentary phosphate deposits. The carbonates are used in the manufacture of cement in regions lacking adequate sources of limestone.

9.4.1 *Magmatic Ores of Titanium*

Titanium may well be one of the major metals of the future. TiO_2 has replaced lead oxide as a major pigment in white paint. The metal itself is relatively light and has attractive physical properties. In a variety of alloys it imparts a high strength-to-weight ratio, a high melting point, and great resistance to corrosion. It is sufficiently abundant in the Earth's crust to be reported as one of the major components in rock analyses. At present titanium metal is relatively expensive. It is therefore used rather sparingly compared to iron, aluminum, and copper.

In the Earth's crust and mantle, titanium occurs almost exclusively as the tetravalent ion Ti^{4+}. The radius of this ion, 0.61 Å, is just a shade less than that of Fe^{3+} (0.64 Å). Ti^{4+} therefore readily replaces Fe^{3+}. TiO_2 itself occurs in three polymorphs: rutile, anatase, and brookite. In combination with ferrous oxide it forms the common mineral ilmenite, $FeTiO_3$, and the rather rare mineral ulvöspinel, Fe_2TiO_4. The structure of ilmenite is the same as that of hematite, Fe_2O_3, and solid solution between the two minerals is extensive at magmatic temperatures. The structure of ulvöspinel is the same as that of magnetite, Fe_3O_4, and solid solution is extensive between these minerals at magmatic temperatures. In both the $FeTiO_3$-Fe_2O_3 and the Fe_2TiO_4-Fe_3O_4 series, one Fe^{2+} plus one Ti^{4+} ion together replace two Fe^{3+} ions.

In the previous section we saw that magnetite is a common mineral in gabbroic rocks, and that it forms by direct crystallization from gabbroic magmas. That is also true for ilmenite. However, oxide melts rich in titanium sometimes form as droplets in these magmas. Since these droplets are denser than gabbroic magmas, they tend to sink and to coalesce into puddles of oxide melt at the base of magma chambers. They usually crystallize there; however, they may also descend along

Table 9.5. Mine Production, Reserves, and Reserve Base of Ilmenite, 1993 (in thousands of metric tons of contained TiO_2)

Country	*Mine Production*	*Reserves*	*Reserve Base*
Australia	850	24,000	66,000
South Africa	800	36,000	45,000
Canada	600	27,000	73,000
Norway	300	32,000	90,000
United States	W	7,800	33,300
Malaysia	190	—	1,000
India	160	31,000	31,000
USSR (former)	150	5,900	13,000
Subtotal	3,050	163,700	352,300
Others	200	39,900	76,500
Total	3,250	203,600	428,800

Source: *Mineral Commodity Summaries*, 1994.
W = withheld.

cracks in the floor of magma chambers or be squeezed out and crystallize with geometries that betray their origin as oxide melts (Force 1991).

Large ilmenite ore bodies have been found in many parts of the world (see table 9.5). Together with ilmenite in sands and gravels derived from the erosion of normal igneous and metamorphic rocks, they supply about 90% of the world's demand for titanium. The remaining 10% are supplied by TiO_2, largely the mineral rutile. Figures for the 1993 mine production, reserves, and reserve base of TiO_2 in ilmenite are shown in table 9.5, as are the locations of major resources of ilmenite. The ratio of the reserve base to 1993 production is 132 years. World resources of TiO_2 in ilmenite are estimated to be about one billion tons. Titanium will certainly be available in ores close to current grade during all of the twenty-first century.

9.4.2 *Magmatic Ores of Nickel, Copper, and the Platinum Group Elements*

Magmas, particularly gabbroic magmas, can become saturated in both sulfide and oxide melts. If the concentration of sulfur is sufficiently high, melts consisting largely of iron, sulfur, and some oxygen (Fe-S-O melts) can form as droplets within silicate magmas. These droplets can sink, coalesce, be extruded, and solidify into large bodies of iron sulfide, mainly pyrrhotite ($Fe_{1-x}S$), and some magnetite (Fe_3O_4). If that were the entire story, these bodies would only be of minor economic interest. They owe their importance to the fact that a number of metals that are present in the parent magma in minor or trace quantities are very strongly scavenged into Fe-S-O droplets. As sulfide melt droplets sweep through a magma reservoir, they can concentrate large amounts of nickel, copper, cobalt, and platinum group elements (Naldrett 1989a). On cooling, these sulfide liquids crystallize and go through a set of fairly complex mineralogical changes. At room temperature the major minerals are usually pyrrhotite, pentlandite $(Fe,Ni)_9S_8$, chalcopyrite ($CuFeS_2$), and pyrite (FeS_2).

The most famous and important of the magmatic Ni-Cu ore deposits are those of the Sudbury district in southern Ontario, Canada. Ore was discovered there in

1856. At that time the area was very remote, and there was little interest in its mineral potential. The ore was in a sense rediscovered in 1883 along the right-of-way of the then-new Trans-Canadian Railway. Not long thereafter, nickel steel was found to be useful for toughening armor plate. The increasing demand for the metal and the large size of the ore deposits in the Sudbury area allowed the International Nickel Company of Canada and the Mond Nickel Company effectively to control the world's nickel industry for nearly a quarter of a century. Inco Limited, the successor of the International Nickel Company of Canada, continued to produce more nickel than any other company in the world. Most of Inco's Canadian production is still derived from the Sudbury area. The tonnage of nickel produced at Sudbury has risen since World War II; however, its share of the world nickel market has decreased very significantly (Pye, Naldrett, and Giblin 1984).

The generalized geology of the Sudbury Basin is shown in figure 9.7. In cross section, the basin is funnel shaped. The sulfide ore bodies and the nickel-copper mines are concentrated along the wall of the funnel. The average compositions of sulfides in four of the Sudbury mines are summarized in table 9.6. Nickel and copper are both present to the extent of a few percent, and cobalt to a fraction of one percent. Platinum and the other platinum group elements are present in the parts per billion range. Their low concentration is not due to a lack of enthusiasm

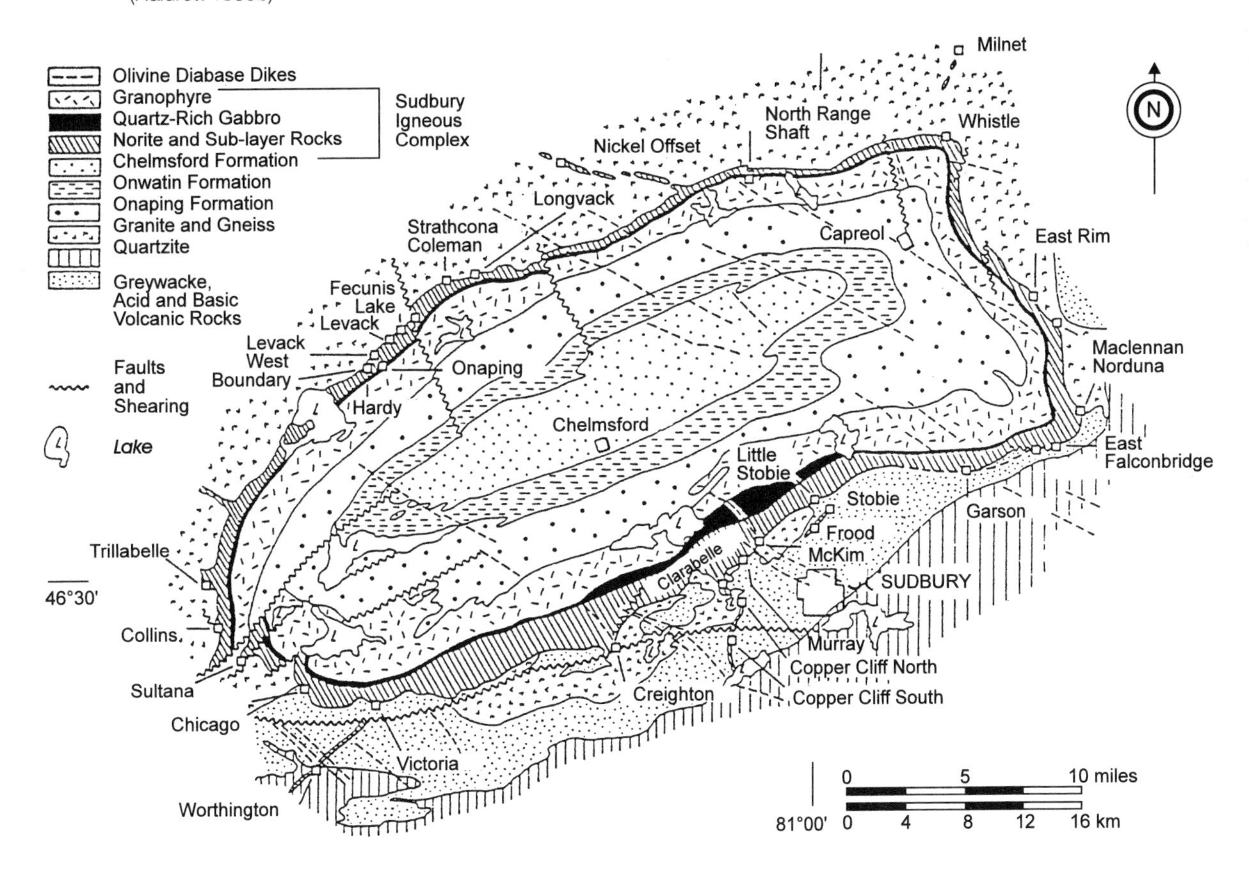

Figure 9.7. Geologic map of the Sudbury District. The small squares mark the location of mines. (Naldrett 1989b)

Table 9.6. The Average Ni, Cu, and Platinum Group Elements (PGE) Content of Sulfides from Four Mines in the Sudbury District

	Wt %			*ppb*							
Mine	*Ni*	*Cu*	*Co*	*Pt*	*Pd*	*Rh*	*Ru*	*Ir*	*Os*	*Au*
Strathcona	3.6	1.2	0.15	420	372	3.0	21	12	8	54
Levack West	5.7	3.7	0.16	1154	1253	186	60	47	22	150
Little Stobie No. 1	3.8	4.4	0.19	1930	2120	119	123	61	29	862
Little Stobie No. 2	4.0	3.6	0.17	2130	3170	303	247	113	46	860

Source: Pye et al. 1984.

for partitioning into sulfide liquids but rather to the extremely low concentration of these elements in gabbroic magmas and in the Earth's crust and mantle as a whole. That they are worth mining at all is due to their great value in a number of industrial applications and to the desirability of platinum jewelry.

The origin of the Sudbury Basin has been hotly debated during the course of the twentieth century. A number of geologic features have been found around its

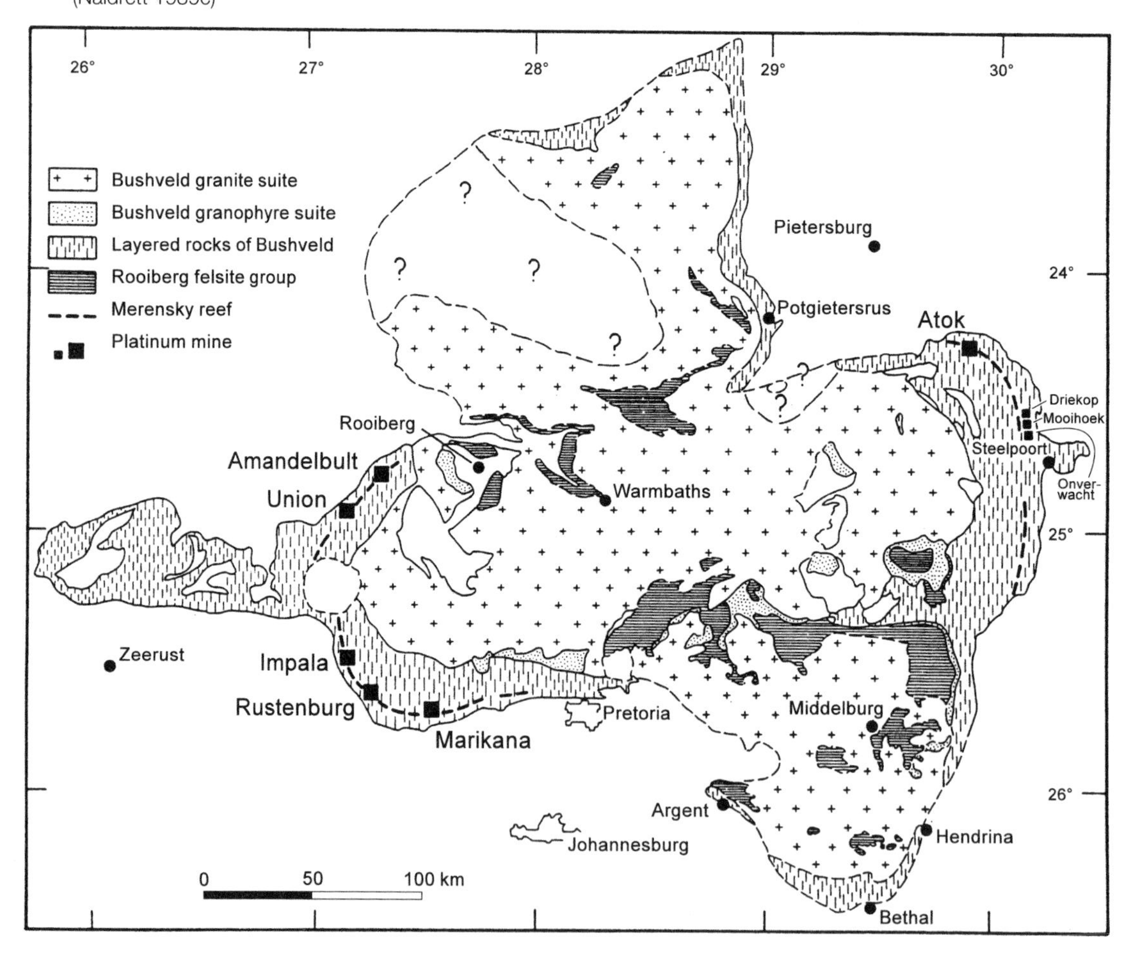

Figure 9.8. Generalized geological map of the Bushveld Complex showing the location of the main platinum group element (PGE) mines. (Naldrett 1989c)

perimeter which suggest that the filling of the basin was preceded by a major impact. The most likely source of these features is the impact of a very large meteorite, but this idea has met with resistance from some geologists who favor a volcanic origin for the extreme pressures that seem to be required to explain the field evidence for intense shock metamorphism.

Unusual concentrations of the platinum group elements (PGE) have also been found in the large layered body of basic intrusive rocks that contain chromium ores: the Bushveld Complex, the Great Dyke, and the Stillwater Complex. Figure 9.8 shows the location of the main PGE mines in the Bushveld complex. The most important PGE containing layer is somewhat above the chromite ores and well below the vanadium ores. The PGE layers are quite inconspicuous, because the PGEs are largely concentrated in thin layers of sulfide minerals. The ratio of the concentration of the PGEs to that of nickel and copper in these layers is much greater than in the ores at Sudbury, and the total quantity of PGEs in the Bushveld is more than one hundred times greater than in the Sudbury ores. It seems likely that the PGEs were scavenged from the Bushveld magma in much the same way as in the Sudbury intrusion, but there is still some question about the role that high-temperature aqueous fluids played in defining the present-day distribution of the PGEs in these intrusions.

9.5 Magmas, Water, and Ores

Magmas always contain some volatile compounds. We have already encountered the role of magmatic volatiles in Hawaiian volcanism, where they are largely responsible for spectacular lava fountains, and we have encountered them in pegmatites, where they are responsible for pockets that frequently contain gemstones. In Hawaii, gases fizz off lavas at the surface somewhat like CO_2 from champagne. In explosive volcanism, the pressure of the gases builds up due to the high viscosity of granitic magmas, eventually leading to their catastrophic release. In deep pegmatites the volatiles are largely contained as dense, highly saline fluids. At intermediate depths, confining pressures are too high to allow volatile compounds to escape as low-density gases, but are low enough so that volatile compounds can escape with reasonable ease from magmas as dense, supercritical fluids. At these depths, magmatic volatiles are often responsible for the formation of very large ore bodies of a number of metals, particularly copper and molybdenum, but also of gold, tin, tungsten, zinc, and lead. The geologic setting of these ore deposits is illustrated in figure 9.9. They are typically associated with granitic-granodioritic batholiths and with their volcanic edifices. Copper ores are usually found in altered porphyries. These are igneous rocks that are a mixture of sizable crystals embedded in a matrix of very small crystals, a consequence of a two-stage cooling history. In porphyry copper and molybdenum deposits (fig. 9.9) such rocks have been altered by aqueous fluids, which brought these metals with them. Most of the ores consist either of copper and iron sulfides or of molybdenum and iron sulfides. Some porphyry ores contain significant quantities of both copper and molybdenum sulfides, and most contain small but significant quantities of gold. The dominant copper sulfide is chalcopyrite, $CuFeS_2$; the dominant molybdenum sulfide is molybdenite, MoS_2.

The ores are of the disseminated type. They occur in thin veinlets that permeate large volumes of rock and are usually low in grade. In most porphyry copper deposits that are currently being mined, the concentration of copper, i.e., the

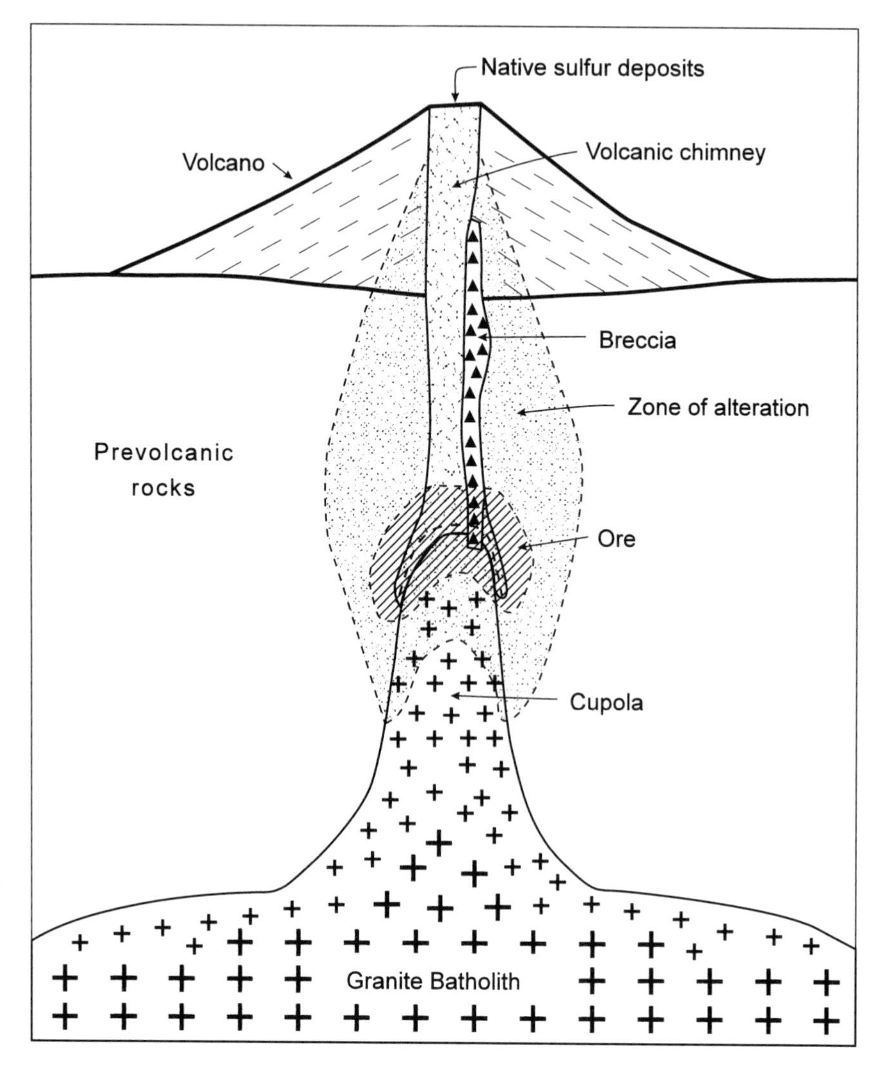

Figure 9.9. Diagrammatic representation of a porphyry copper-molybdenum system at the boundary between volcanic and plutonic environments.

grade, is between 0.5 and 2.0 wt %. What makes such low-grade deposits economic is their very large size, which allows sufficient economies of scale to make them as attractive or more so than smaller, high-grade deposits (see chapter 10).

Controversy has raged about the evolution of these ores. Initially all of the solutions that transported the metals and altered the igneous rocks were thought to have been dissolved in magmas at depth and to have been released during their crystallization. This view was challenged, because the isotopic composition of hydrogen and oxygen in many of the clay minerals in the alteration zone of porphyry copper deposits showed that they had formed by the interaction of igneous rocks with local rainwater heated to several hundred degrees centigrade by contact with volcanic and intrusive rocks. It now looks as if both interpretations are partially correct. The initial concentration and deposition of the ores was largely accomplished by magmatic fluids; the ores were later permeated and redistributed by solutions that evolved from local rainwater.

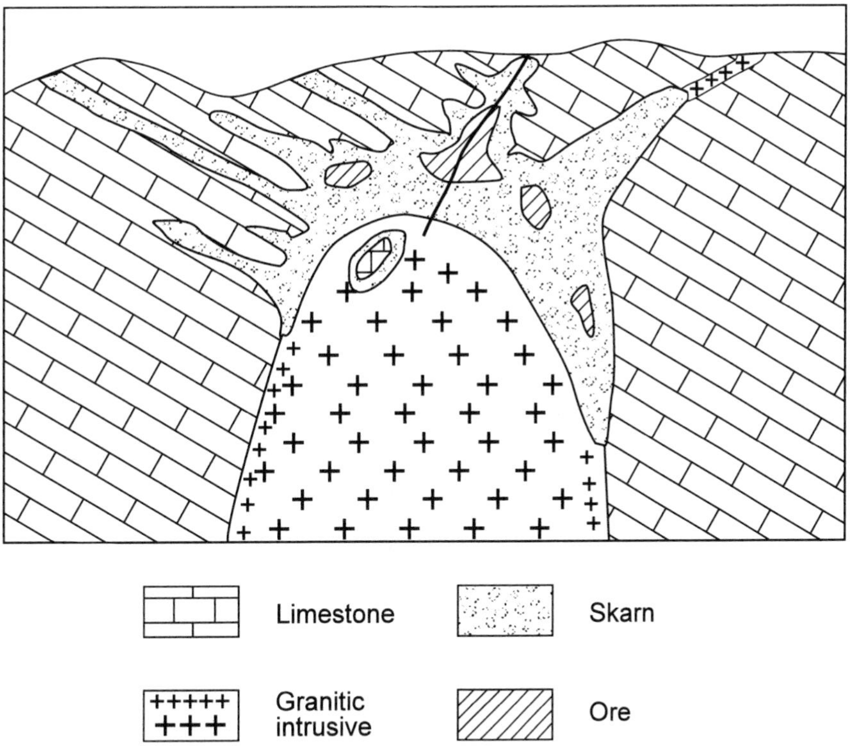

Figure 9.10. Diagrammatic representation of ore in calc-silicate rocks ("skarn") developed around a granitic intrusion. Note extensions of the skarn along favorable sedimentary rocks, fractures, and dikes.

A class of ore deposits that is usually called *skarns* forms in areas where hydrothermal solutions pass from granitic porphyries into limestones or dolomites. These solutions are usually acidic. They dissolve carbonate rocks and replace calcite and dolomite with a variety of silicate minerals, together with sulfides of iron, copper, zinc, lead, and silver; with oxides of iron, tin, and tungsten; and with gold (fig. 9.10). In some large mining districts, such as Bingham Canyon, Utah, disseminated ores are present within porphyries, and high-grade skarn deposits are located in adjoining carbonate rocks.

The regions of shallow granitic and granodioritic intrusives are the natural home of copper and molybdenum porphyries and associated skarn deposits. These regions are particularly common along the western ranges of North, Central, and South America and in the island arcs of the southwestern Pacific Ocean. A good deal of the mineral wealth of the United States is derived from these deposits. The enormous open pit at Bingham Canyon gives eloquent testimony to the size of the porphyry copper deposit that has been mined there. The mountain at Butte, Montana, has been called the richest hill on Earth. Within it a large number of copper, gold, zinc, silver, lead, manganese, arsenic, antimony, and bismuth-bearing veins overlie and surround a porphyry copper deposit. Young miners, from Cornwall, in Great Britain, were said to have been admonished by their mothers: "When you get to the other side of the ocean, don't stop in the United States, go straight to Butte." Many took this advice, and the mining camp at Butte was the richer for the presence of its "Cousin Jacks."

Table 9.7. Production, Reserves, and Reserve Base of Copper, 1993 (in thousands of metric tons of copper content)

Country	*Mine Production*	*Reserves*	*Reserve Base*
Chile	2,020	88,000	140,000
United States	1,770	45,000	90,000
USSR (former)	800	37,000	54,000
Canada	750	11,000	23,000
Zambia	420	12,000	34,000
Poland	390	20,000	36,000
China	380	3,000	8,000
Peru	370	7,000	25,000
Australia	300	7,000	21,000
Indonesia	270	11,000	17,000
Philippines	130	7,000	11,000
Zaire	90	10,000	30,000
Subtotal	7,690	258,000	489,000
Others	1,590	50,000	98,000
Total	9,280	308,000	587,000

Source: *Mineral Commodity Summaries*, 1994.

The figures in table 9.7 for the production, reserves, and reserve base of copper reflect the concentration of porphyry copper ores along the American Cordillera from Canada, through the United States, and into South America. The copper mines and the copper reserves of Chile are particularly noteworthy. The ratio of the reserve base of copper to the annual production of copper is about 63 years. The lowest economically minable grade of porphyry copper ore has gradually been dropping (see chapter 10). There are very large bodies of mineralized porphyry with copper concentrations somewhat below the current cutoff grade. These will almost certainly be mined during the next century unless recycling and the substitution of aluminum and glass fibers for copper reduce the demand for copper very significantly.

9.6 Igneous Rocks, Water, and Ores

Many important ore deposits are closely associated in time and space with volcanism and igneous intrusions but do not owe their metallic riches to magmatic processes per se. In the formation of these ores, magmatic rocks serve primarily as a source of heat and secondarily as a source of metals that can be extracted by hot aqueous fluids. When magmas appear at the surface as lavas or when they form shallow intrusive bodies, their temperature is usually between about 800°C and 1200°C. Surface water that comes in contact with rocks at such high temperatures flashes into steam. In Hawaii, for instance, spectacular steam vents often form where basaltic lava encounters seawater. Groundwater that comes in contact with hot igneous rock at considerable depths below the surface may turn into hot, pressurized water. The boiling point curve of water is shown in figure 9.11. At the Earth's surface near sea level, the pressure of the atmosphere is 1 atm, or 14.7 pounds per square inch (psi); at this pressure, water boils at 100°C. At high

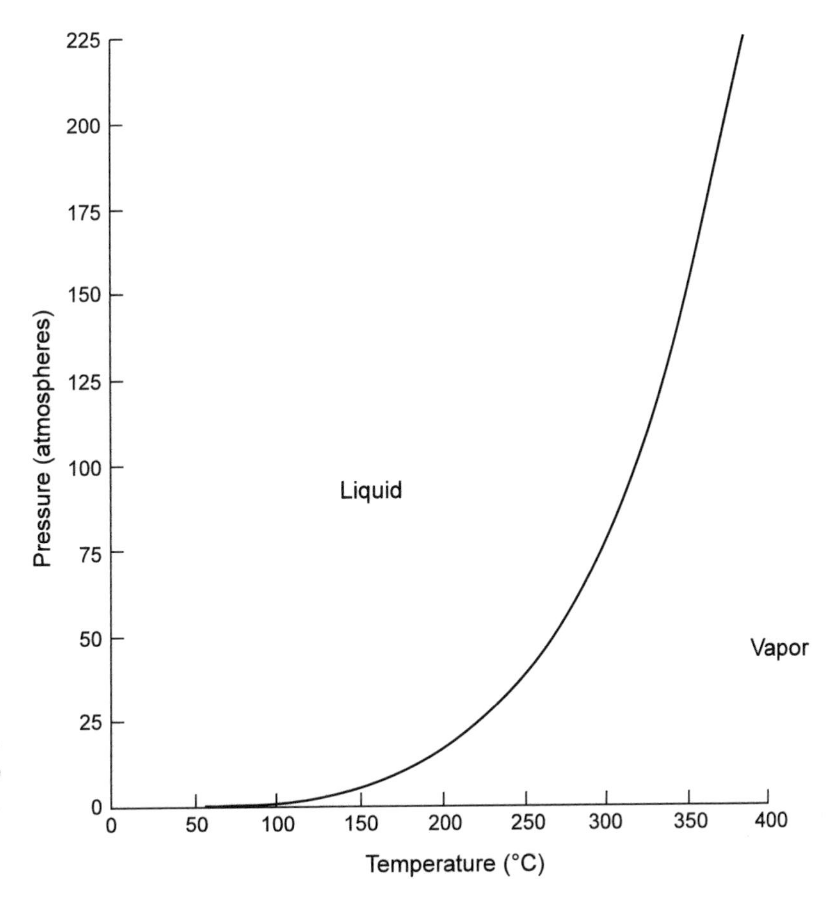

Figure 9.11. The boiling point curve of water.

elevations in mountainous regions, atmospheric pressure is less than 1 atm, and water boils at temperatures below 100°C. As many campers can certify, boiling eggs at an elevation of 3,000 meters (10,000 ft) takes considerably longer than at sea level. Below the surface of the Earth, pressures are always greater than 1 atm. Diving, even to a depth of only 10 meters, creates pressures on one's eardrums that can be quite uncomfortable. At a depth of one kilometer the pressure on a submarine is 100 atm. On land, the pressure exerted by overlying rock at a depth of one kilometer is typically between 250 and 300 atm.

In areas like the Yellowstone Park region of Wyoming (fig. 9.12), much of the rock below the surface is quite hot. Yellowstone has experienced intense igneous activity during the past 2 million years, and subsurface temperatures are still several hundred degrees centigrade. The area is strongly fractured. Some of the rainwater that falls in the region therefore percolates downward and is heated during its passage through the young igneous rocks. Since the density of hot water is less than that of cold water, hot water in these rocks tends to rise and to be replaced by cooler water from the edges of the volcanic area. If rock temperatures are in excess of 100°C, groundwater may be converted into steam. Hot water and steam can emerge at the surface as hot springs and steam vents, or—in particular settings—as geysers like Yellowstone's famous Old Faithful, which fountains about once an hour, or similar geysers in Iceland (plate 38).

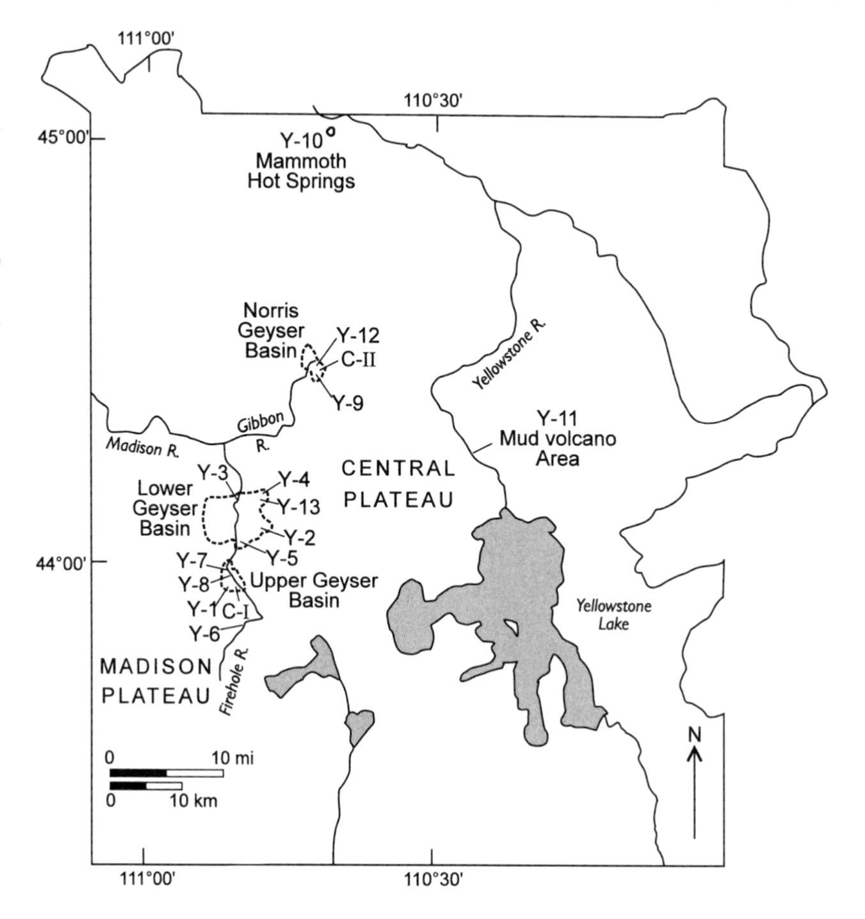

Figure 9.12. Map of Yellowstone National Park showing the location of major geothermal areas and sites of research drill holes. U.S. Geological Survey drill holes are numbers Y-1 to Y-13. (White et al. 1975)

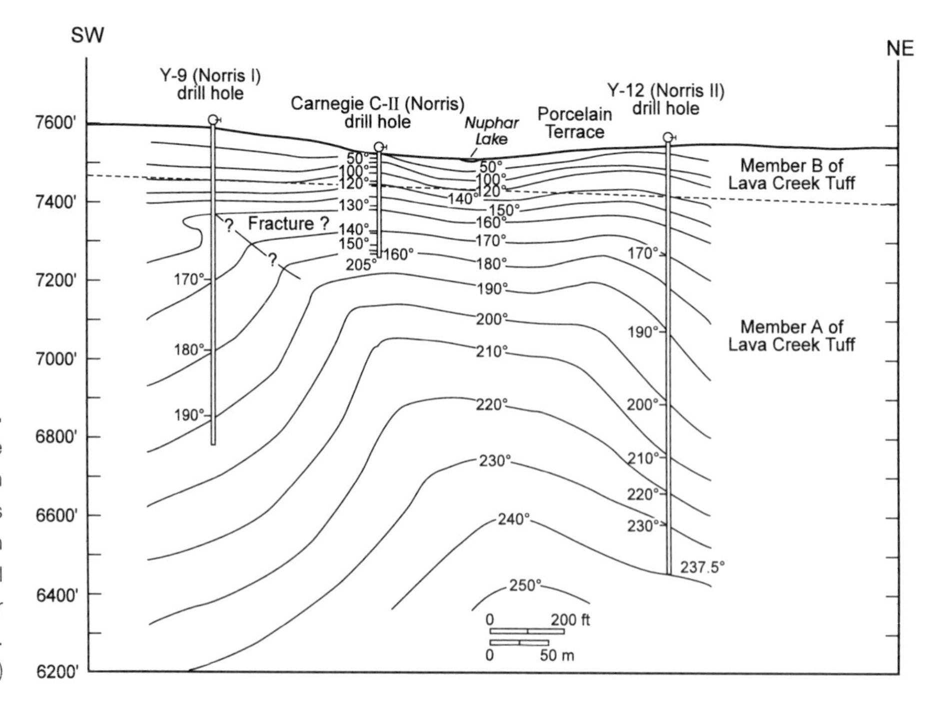

Figure 9.13. Subsurface temperature along a SW–NE section through the Norris Geyser basin in Yellowstone National Park. See figure 9.12 for location of drill holes. (White et al. 1975)

The temperature of many of the hot springs in Yellowstone Park is close to 100°C, and this is not unusual for hot springs in active or recently active volcanic regions. Below the surface at Yellowstone the temperature of groundwater rises rapidly. As illustrated in figure 9.13, temperatures in excess of 200°C are frequently encountered at depths of only a few hundred meters. The flow path of these hot groundwaters can be quite complicated. In many geothermal areas groundwaters circulate repeatedly through porous volcanic rocks. In other geothermal areas, the reservoir rocks are filled with steam or with a mixture of steam and water.

In many instances, the residence time of groundwaters in geothermal systems is relatively short. This is shown most convincingly by the concentration of short-lived radioisotopes like tritium, ^3H, in these hot groundwaters. Tritium is produced naturally in the atmosphere by cosmic ray bombardment and reaches the surface as a normal constituent of rainwater. Atmospheric testing of hydrogen bombs has also added tritium to the atmosphere. Tritium decays with a half-life of 12.5 years. Since it is not replenished beneath the Earth's surface, its concentration in groundwaters decreases by a factor of two every 12.5 years. After 100 years, less than 1% of the original tritium remains. Most of the waters in Yellowstone hot springs contain much more than 1% of the original tritium. Their residence time underground is therefore much less than a century.

A great deal has been learned about the complex hydrology of geothermal systems during exploration for geothermal energy, which is an important source of energy in areas of recent and current volcanism (see chapter 11). In the United States, a large amount of geothermal energy is produced in The Geysers area of California. The island of Hawaii is a large potential source of geothermal energy, but the development of this resource has been strongly opposed by some local residents, in part for environmental reasons.

Figure 9.14. Fluid inclusions in sphalerite from the hydrothermal ore deposit at Creede, Colorado. (Photo courtesy of E. Roedder)

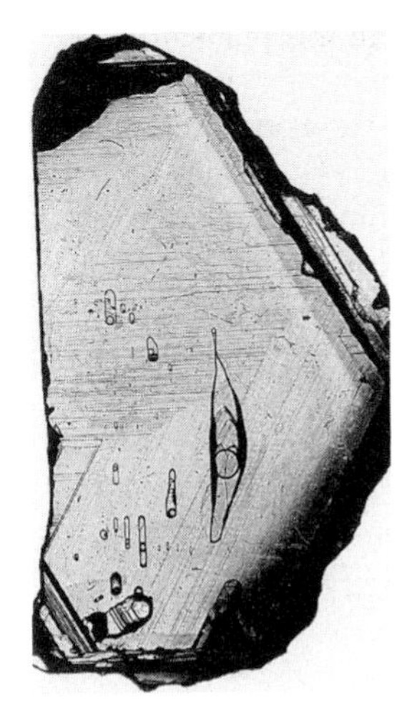

Hot water in geothermal areas usually contains significant amounts of gold. The metal can precipitate in ponds where the solutions cool, but the quantity of gold that is deposited in such settings is rarely of commercial interest. After the death of geothermal systems by cooling, their plumbing systems can be explored by mining. Many large ore bodies of copper, zinc, lead, silver, and gold have been discovered in such settings. The ore minerals and the associated gangue, i.e., worthless diluents such as calcite and quartz, contain tiny inclusions of the fluids from which they precipitated. Examples of such fluid inclusions are shown in figure 9.14. Inclusion fluids can be analyzed chemically and isotopically. Their composition can tell us much about the nature of ore-forming fluids and about the temperature and pressure of ore formation (Roedder 1984). The concentration of the ore metals in these fluids is usually quite small, a few parts per million to a few hundred parts per million. The formation of large ore bodies therefore requires the precipitation of metals from enormous quantities of fluid.

In most geothermal systems, fluids move through large bodies of relatively porous rock. Metals precipitated in such settings tend to be rather widely dispersed and are rarely of ore grade. High-grade ore deposits tend to form where large volumes of fluid flow through restricted channels. Finding such ancient channel ways is a challenge for exploration geologists. In many hydrothermal ore deposits, the metals are arranged in zonal patterns. Copper ore is frequently found at greatest depth. Toward the surface, ores of copper tend to give way to ores of zinc, then to ores of lead and of silver (see fig. 9.15). This sequence

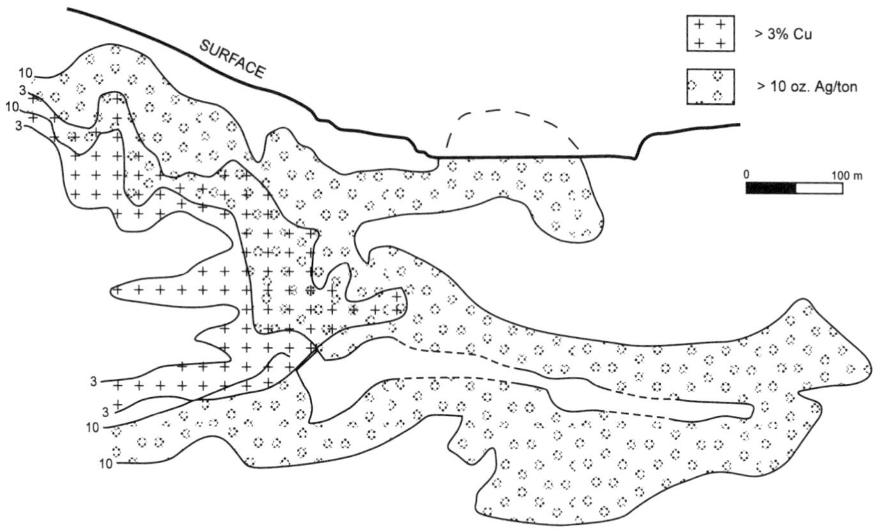

Figure 9.15. Vertical section through the Docenita vein, Julcani District, Peru. The solutions flowed from left to right, first precipitating high-grade copper ore (> 3% Cu) and then high-grade silver ore (> 10 oz Ag/ton).

reflects the progressive precipitation of the sulfide minerals of these metals during the rise of the ore-forming fluids. Once the zoning pattern of metals in a mining district has been established, its geometry can often be used as an effective tool in further mineral exploration (see chapter 10).

The solutions in fluid inclusions in ore and gangue minerals are often quite saline. Salinities comparable to that of seawater (3.5 wt % salt) are common, and solutions saturated with respect to NaCl (ca. 25 wt %) are not at all rare. Clearly, such solutions are not simple rainwater. They may be rainwater that has become saline by dissolving salt in evaporites or by mixing with very saline solutions. They may also be derived either directly or indirectly from seawater. A glance at the world map of the oceans shows that there are more than 50,000 kilometers of midocean ridges. Basaltic volcanism is active along all of these, and all are potential giant underwater geothermal systems. Cold seawater enters their flanks, is heated, picks up metals and hydrogen sulfide as it moves through the oceanic crust, and deposits sulfides of iron, copper, and zinc when it emerges at or close to the crest of the ridges. The sulfide accumulations formed in this manner were described in chapter 7. Their location at water depths of about 2.5 kilometers makes them economically unattractive at the present time. However, some ore deposits formed in such settings have since been elevated above sea level. On the island of Cyprus, for instance, a piece of oceanic crust containing copper ores is exposed to view and has been mined since antiquity. The product of these mines was known as the metal from Cyprium, hence the name copper. There are many large ore deposits of this type. As a group, these are known as volcanogenic massive sulfide deposits.

The Kuroko ores of Japan form a related group of copper-lead-zinc-silver ores. They were also formed in marine settings at a depth of about 2 kilometers, but the igneous rocks with which they are associated are much more silica-rich than basalts. Figure 9.16 shows an idealized cross section through one of these ore deposits. Sulfide ores were deposited in conduits below the ocean floor and at the

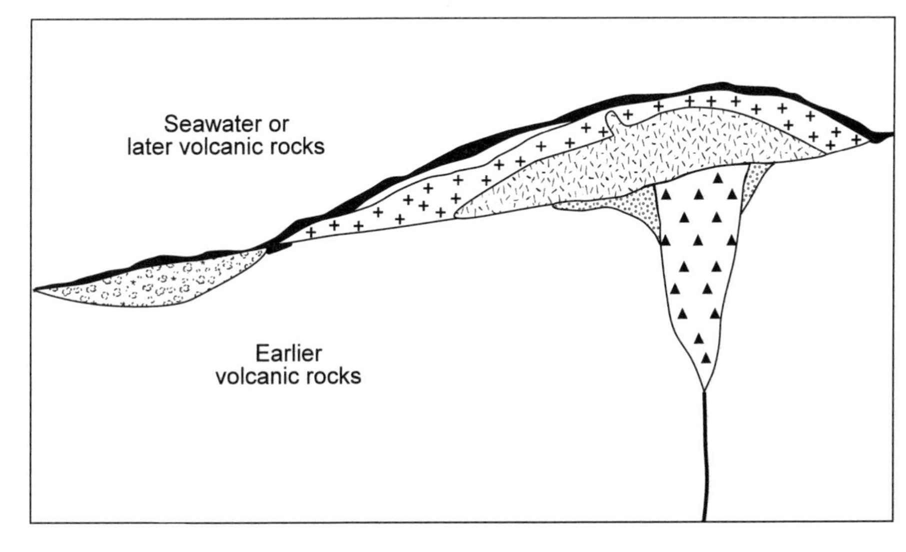

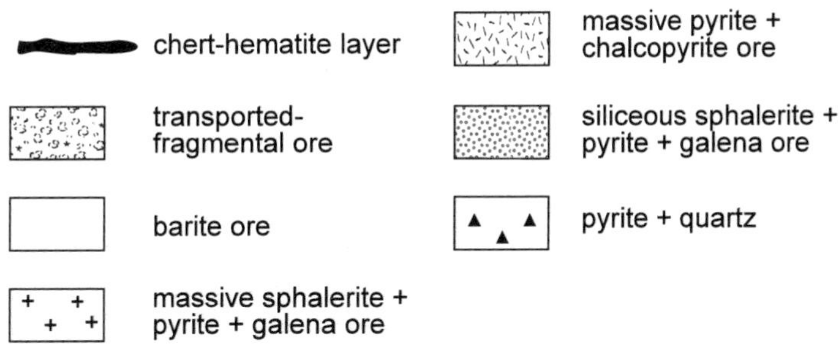

Figure 9.16. An idealized representation of an undisturbed Kuroko deposit.

water-sediment interface. They were preserved under a later cover of sediments and lavas, and were uplifted tectonically to their present position well above sea level.

9.7 Water and Ores

The metallic ore deposits that have been described so far in this chapter are related either directly or indirectly to magmatic processes. However, the connection between magmas and ores is not universal. Several types of ore deposits, some very large and economically important, have no visible connection with magmatic processes. During the last few decades a good deal of progress has been made in understanding how these deposits formed. However, many questions regarding their origin are still unanswered, or more accurately, there are many answers to these questions, and it is not clear which, if any, are correct.

The Mississippi Valley Type (MVT) zinc and lead deposits, named in honor of the large number and size of this class of ore deposits in the Mississippi River basin, are among the largest ore deposits of these metals. This metallogenic province extends from northern Arkansas and Missouri to Kentucky and Wisconsin (see fig. 9.17), and includes the world's largest accumulation of lead around the St. François Mountains of Missouri (see fig. 9.18).

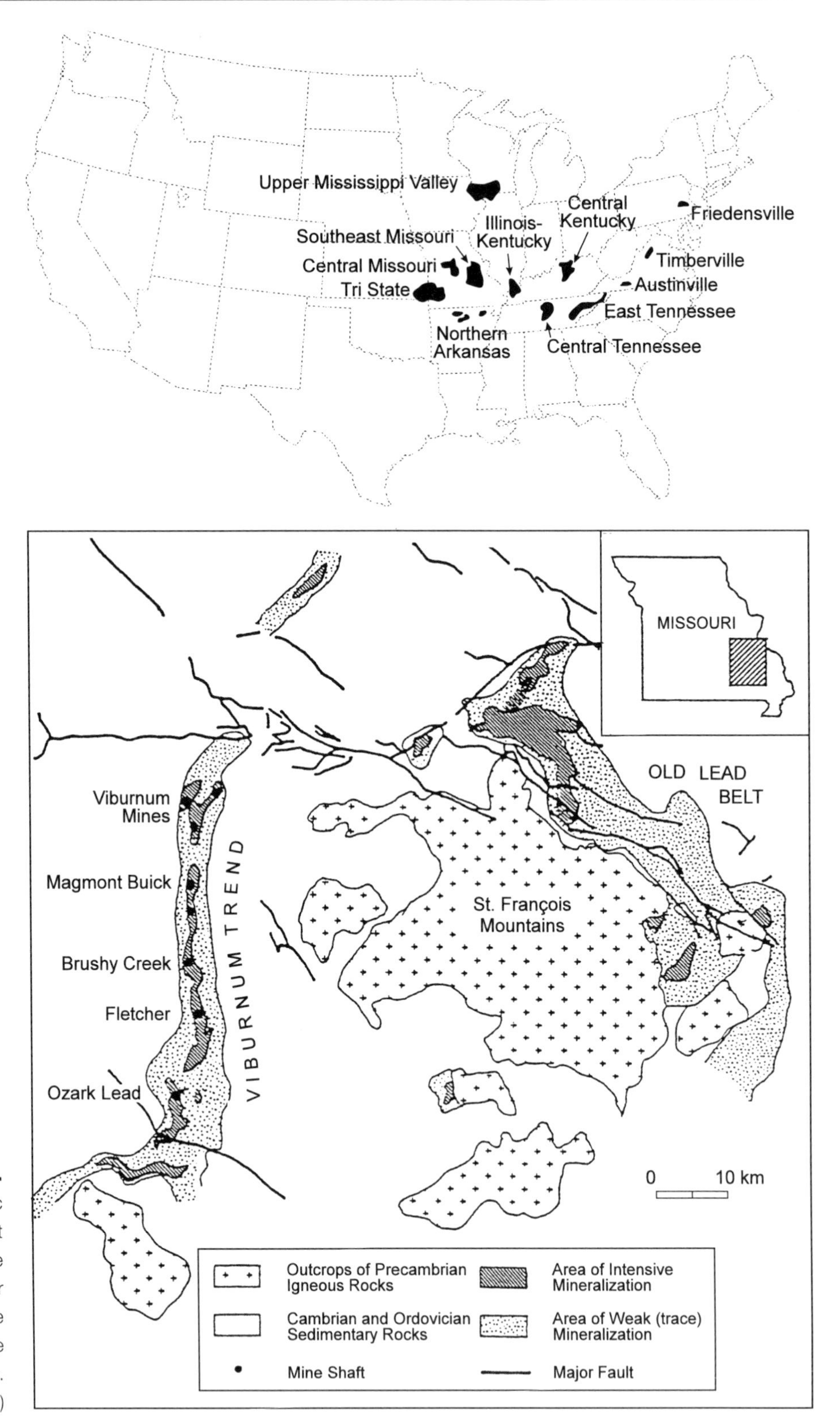

Figure 9.17. Location of the principal Mississippi Valley–type lead-zinc regions in the United States. (Craig, Vaughan, and Skinner 1988)

Figure 9.18. Generalized geologic map of southeast Missouri showing the location of the major Mississippi Valley ore districts near the St. François Mountains. (Sawkins 1984)

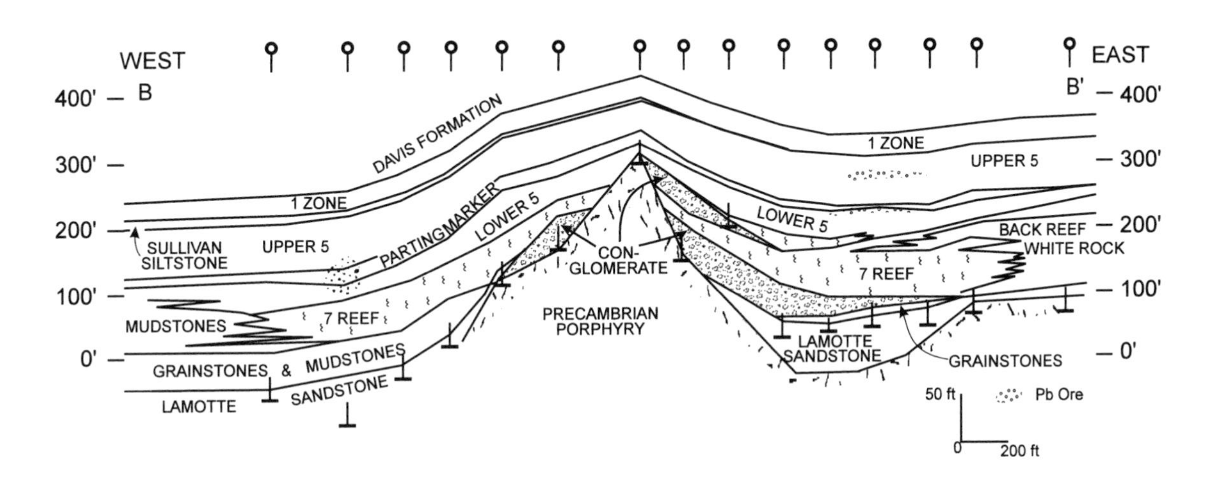

Figure 9.19. Geologic cross section through the Brushy Creek Mine in the Viburnum Trend of southeast Missouri. (Evans 1977)

The mineralogy of these deposits is rather simple. The only major lead mineral is galena, PbS; the only major zinc mineral is sphalerite, ZnS. Copper, when present, is usually in chalcopyrite, $CuFeS_2$; barium in barite, $BaSO_4$; and fluorine in fluorite, CaF_2. The ores are invariably found in dolomites and/or limestones, frequently filling cavities that formed during the prior dissolution of the carbonate host rocks. Superb mineral specimens of the ore and gangue minerals are often found in large, partly filled cavities. The ore-forming fluids were highly saline, often close to saturation with respect to halite, NaCl. Their temperature at the sites of ore deposition was usually between 70°C and 150°C. Chemically, the ore-forming fluids were similar to brines in sedimentary rocks; they are particularly closely related to brines associated with petroleum in oil fields. The isotopic composition of the lead in the ores is quite variable and is generally enriched in ^{206}Pb, ^{207}Pb, and ^{208}Pb, the isotopes derived from the radioactive decay of ^{238}U, ^{235}U, and ^{232}Th, respectively. The sulfur in the sulfide ore minerals also varies isotopically. The very large lead deposits around the St. François Mountains are in carbonate rocks that are stratigraphically above the very extensive and permeable Lamotte sandstone. This formation thins and pinches out against the flanks of the Precambrian core of the St. François Mountains (see fig. 9.19). The ore solutions probably moved through this sandstone and then up into the overlying carbonates, where ore deposition took place. The cause or causes of ore deposition are still a matter of debate. It is likely that the ore-forming fluids contained sulfate, SO_4^{2-}. This ion was reduced to sulfide, S^{2-}, perhaps by petroleum or natural gas. The S^{2-} then served to precipitate sulfide minerals of the ore metals.

The large expanse of midcontinent carbonate rocks was apparently flooded with similar brines. Local sources for the brines are unlikely. One intriguing possibility is that a very large body of brines migrated northward out of the Arkoma Basin in Arkansas during the Carboniferous and Permian deformation and uplift of the sediments in this basin. These brines may well have flooded the rocks in much of the midcontinent area; lead and zinc ores probably formed

Table 9.8. Production, Reserves, and Reserve Base of Zinc, 1993 (in thousands of metric tons)

Country	*Mine Production*	*Reserves*	*Reserve Base*
Canada	1,200	21,000	56,000
Australia	1,000	17,000	65,000
China	700	5,000	9,000
Peru	580	7,000	12,000
United States	530	16,000	50,000
Mexico	270	6,000	8,000
Subtotal	4,280	72,000	200,000
Others	2,700	72,000	130,000
Total	6,980	144,000	330,000

Source: *Mineral Commodity Summaries*, 1994.

Table 9.9. Production, Reserves, and Reserve Base of Lead, 1993 (in thousands of metric tons)

Country	*Mine Production*	*Reserves*	*Reserve Base*
Australia	600	10,000	35,000
United States	400	10,000	22,000
China	400	7,000	11,000
Canada	300	6,000	13,000
Peru	200	2,000	3,000
Mexico	200	1,000	2,000
South Africa	100	2,000	3,000
Sweden	100	500	1,000
Morocco	100	500	1,000
Subtotal	2,400	39,000	91,000
Others	800	24,000	36,000
Total	3,200	63,000	127,000

Source: *Mineral Commodity Summaries*, 1994.

where large masses of galena and/or sphalerite precipitated following the reduction of sulfate in the brines.

Ore deposits that are similar to those in the Mississippi Valley have been found in Canada, northern Africa, Australia, the former Soviet Union, and Europe. Most of the lead ore produced in the United States is mined from MVT deposits. The production, reserve, and reserve base figures for zinc and lead are shown in tables 9.8 and 9.9. The ratio of the reserve base to the current mine production of both metals is in the range of 40 to 47 years.

If the deposition of galena and sphalerite in MVT deposits was due to SO_4^{2-} reduction to S^{2-} in the ore-forming fluids, then they are one of a rather large family of ore deposits that owe their origin to the reduction of one or more of the components of aqueous ore solutions. Several metals are among these components; the most important are uranium, vanadium, and molybdenum. We have encountered one type of ore deposit of each of these three metals already. The very large reserves of uranium in the Witwatersrand Basin of South Africa were briefly described in chapter 7. The vanadium ores of the Bushveld Complex and molybdenum porphyry deposits were mentioned earlier in this chapter. In this section we shall meet some of the other major ores of these metals.

In the Witwatersrand-type ores, uranium is present largely in the mineral uraninite. Pure uraninite has the composition UO_2, but uraninite in Witwatersrand-type ores also contains significant quantities of thorium and rare earths. Its composition is similar to that of uraninite found in granites and granite pegmatites, and it is very likely that the Witwatersrand uraninites were released from such rocks during weathering, transported by streams into the Witwatersrand basin, and accumulated there in sands and gravels. The survival of uraninite during weathering is almost certainly due to the very low oxygen content of the atmosphere more than 2,200 million years ago. No ore deposits of the Witwatersrand type have been found in rocks younger than 2,200 million years. Indeed, uranium ores of a very different kind have formed during the past 2,000 million years. In these ores uranium was typically deposited from aqueous solutions as pitchblende, a very fine grained variety of uraninite, which is frequently somewhat oxidized and contains very little thorium and rare earths. Most of the pitchblende was deposited at temperatures between 100° and 300°C, often in the vicinity of organic matter and other reducing materials. The uranium in the solutions was almost certainly carried in the hexavalent state, i.e., as U^{6+} and its complexes. Pitchblende precipitated when the uranium was reduced to the tetravalent state, i.e., to U^{4+}.

U^{6+} was almost certainly produced during the weathering of rocks under an oxidizing atmosphere. Several lines of evidence indicate that the oxygen content of the atmosphere rose dramatically about 2,200 million years ago (Holland 1994). It is therefore not surprising that the earliest of the "new" kind of uranium ores, the ore deposits in Gabon, West Africa, are 2,000 million years old. These ores are contained in ca. 2,100 million-year-old conglomerates and coarse sandstones, some of which are cemented by asphaltic organic matter. The ore deposits lie close to the contact between these sediments and the underlying granitic basement. The ores are famous not only for their size and high grade, but also because they acted as natural nuclear reactors. During this process ^{235}U was preferentially destroyed by fission (see chapter 11). The ores are therefore depleted in this isotope of uranium and are enriched in the fission products of ^{235}U. A good deal of research has been done on the distribution of the fission products within and in the vicinity of these natural reactors, because they are thought to be natural analogues of repositories for modern high-level nuclear wastes (see chapter 11). Unfortunately, there is reason to doubt that the analogy is sufficiently close to make the comparisons useful.

The largest and currently most productive hydrothermal uranium deposits are those in the Athabasca Basin of northern Saskatchewan, Canada (see figs. 9.20, 9.21) and in the Northern Territory of Australia. These deposits contain about half of the Western world's low-cost uranium reserves. Most of the ore occurs at or just below unconformities, where these unconformities overlie or are cut by faults passing through carbonaceous schists in the basement. The uranium in the ore-forming fluids was probably picked up during the reaction of oxygenated groundwaters with the rocks through which they passed. U^{4+} was oxidized to U^{6+} and probably traveled in the ore-forming fluids as complexes of the uranyl ion, UO_2^{2+}. U^{6+} was reduced to U^{4+} when these solutions encountered organic matter, sulfides, or other reductants. UO_2 was deposited following reactions such as

$$2\ UO_2^{2+} + C + 2\ H_2O \rightarrow 2\ UO_2 + CO_2 + 4\ H^+ \qquad (9.1)$$

at temperatures between about 100° and 300°C.

Similar reactions are probably also responsible for the uranium ores of the Colorado Plateau in the western United States. Groundwater flowing through permeable sediments of what were once river channels often encountered the carbonaceous remains of trees. These served as reductants for U^{6+} and for pentavalent vanadium, V^{5+}, in the solutions. Reduced compounds of uranium and vanadium were deposited in the vicinity of these reductants. Later, many of these minerals were oxidized and transformed into the canary-yellow carnotite ores of

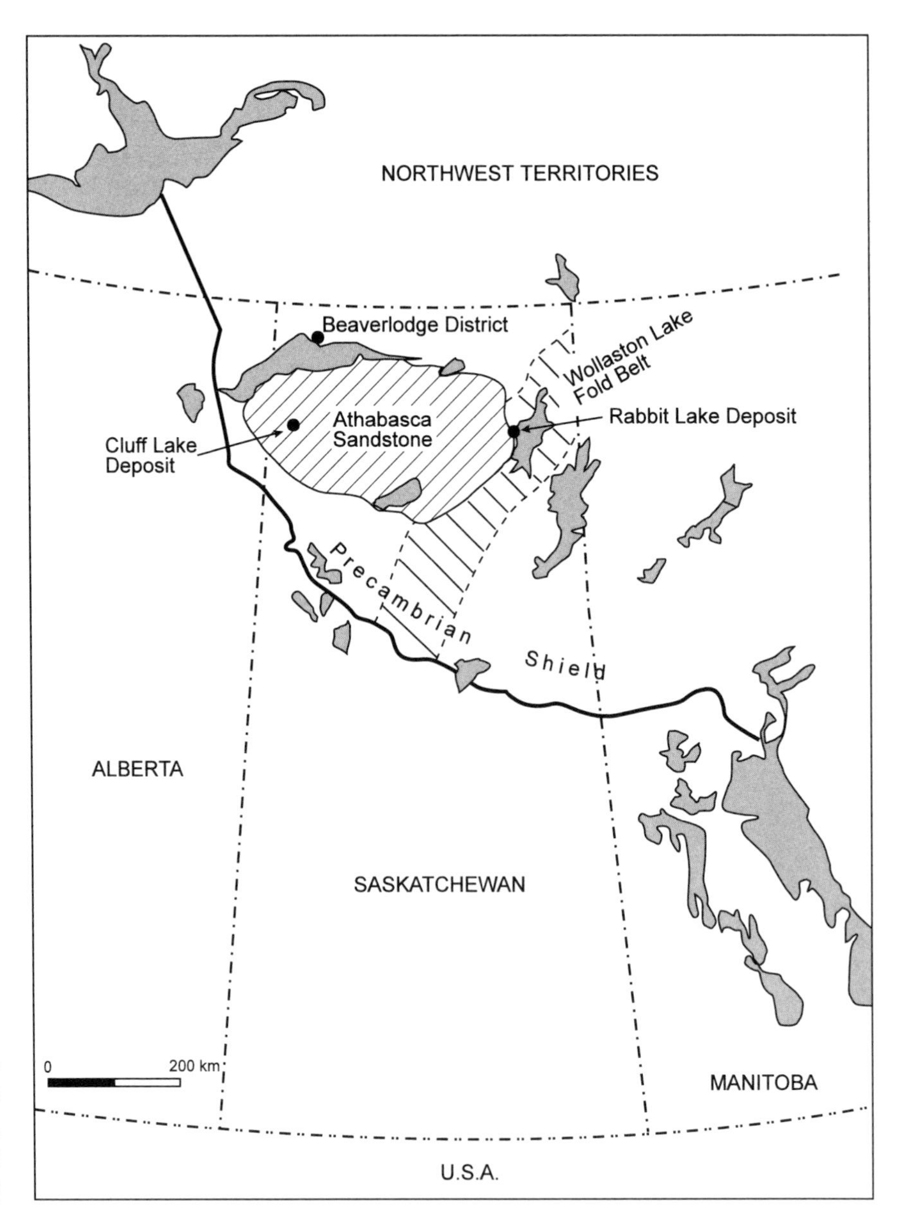

Figure 9.20. The Athabasca Basin in Saskatchewan, Canada. (Rich, Holland, and Petersen 1977)

the Colorado Plateau. For a long time only these yellow ores were known in some parts of the Colorado Plateau, and exploration was aimed at the discovery of additional ores of this type. In the Big Indian Wash area of Utah, black uranium ore containing the rare uranium silicate mineral coffinite was first discovered by mistake. In the early 1950s Charles Steen, an impoverished prospector, found some odd-looking black material in his very last drill hole. His geiger counter was no longer operational. On a chance, he threw the odd piece of drill core into the back of his truck and took it to the assay office in nearby Moab. The counter went wild, and a new type of uranium ore deposit had been discovered.

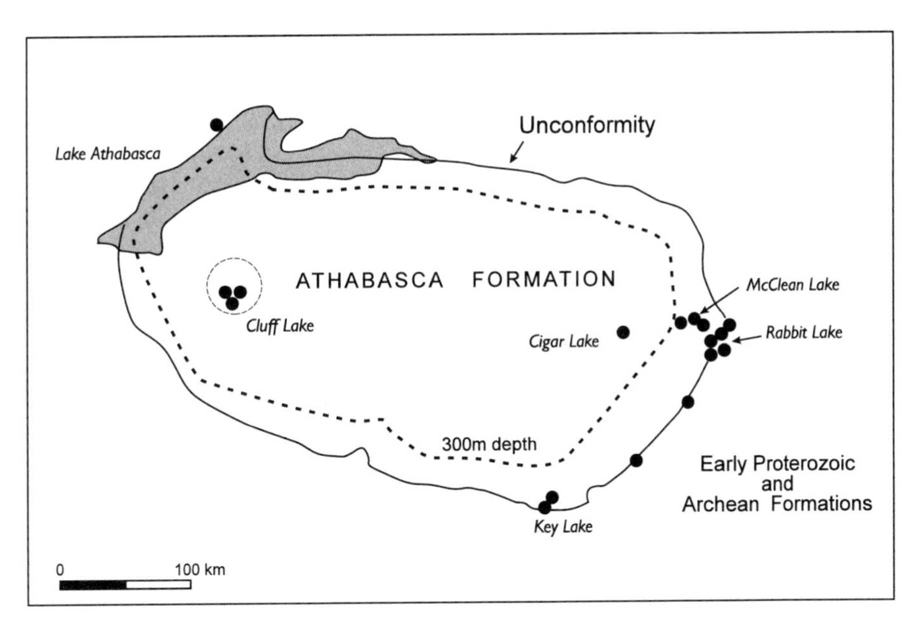

Figure 9.21. Outline of the Athabasca Basin and distribution of the associated uranium ore deposits. The basement rocks outside the basin are Early Proterozoic and Archean in age (Evans 1993). The 300 m contour indicates the location of the contact between the Athabasca sandstone and the basement rocks at a depth of 300 m.

Charles Steen threw a day-long party for the entire town of Moab, and proceeded to become wealthy from the proceeds of his Mivida ore body.

A closely related type of uranium ore deposit has been aptly called the *roll front type*. Figure 9.22 shows its main features. Oxygenated groundwater entering permeable sand or sandstone oxidizes uranium in these sediments and moves the uranium down dip until the dissolved oxygen in the groundwater is exhausted. Thereafter, as the groundwater encounters additional reduced sediments, the U^{6+} again reverts to U^{4+}, and pitchblende precipitates close to the contact between the oxidized and reduced sediments along the flow path of the

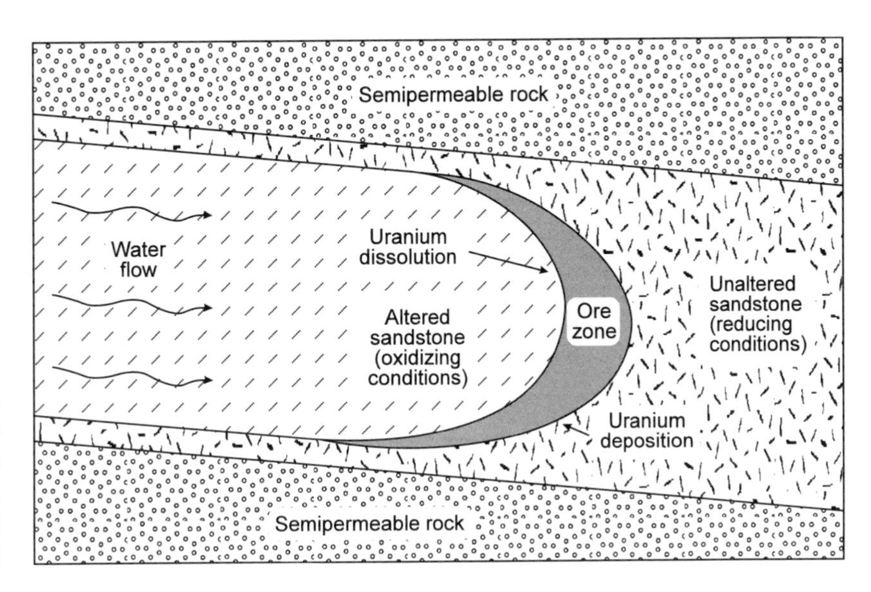

Figure 9.22. Cross section of a roll front uranium ore deposit. (Craig, Vaughan, and Skinner 1988)

groundwater. Continued flow of oxidized groundwater gradually shifts this contact down dip. The location of the pitchblende moves along with the contact.

Abnormally high concentrations of uranium have also been found in ancient carbonaceous marine sediments. Seawater contains only a few parts per billion of uranium, but uranyl ions can diffuse into sediments. If these sediments are highly reducing and if they accumulate very slowly, the concentration of uranium can rise from an original value of a few parts per million to as much as 300 ppm. Vanadium, molybdenum, and a number of other metals can be similarly enriched in highly carbonaceous shales. Although the metal content of many metalliferous shales can be explained by the proposed diffusive transfer of metals from seawater into sediments followed by their precipitation as oxide and sulfide minerals, the origin of some metalliferous shales, especially those containing abnormally high concentrations of the platinum group elements, is still a mystery.

Carbonaceous sediments are normally enriched in phosphorus. When such sediments are oxidized and winnowed, phosphorites are often formed (see chap-

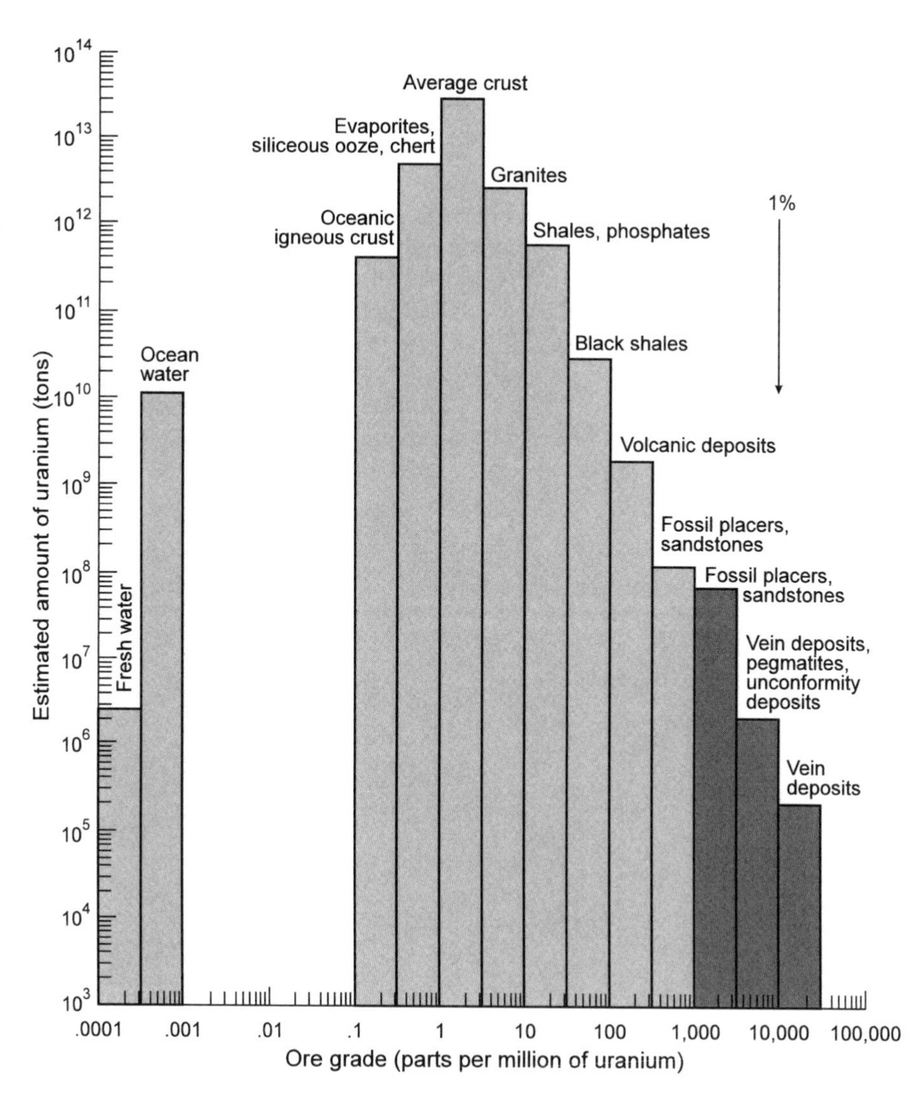

Figure 9.23. The distribution of uranium in the Earth's crust; a plot of the estimated amount of uranium versus ore grade in parts per million of uranium. The bars represent various categories of uranium deposits or repositories of uranium in descending order of uranium content. The three bars on the right represent deposits of the type now being mined specifically for uranium. For approximately every tenfold decrease in grade there is a thirtyfold increase in the amount of recoverable uranium. (Deffeyes and MacGregor 1980)

Table 9.10. World Production of Uranium, 1992 (in million pounds equivalent U_3O_8)

Western Production	
Canada	24.2
Niger	7.7
Australia	6.1
France	5.6
United States	5.1
Namibia	4.3
South Africa	4.3
Gabon	1.4
Others	2.0
Total	60.7
Eastern Production	ca. 34

Source: NUEXCO, 1992.

ter 7). These are the primary sources of fertilizer phosphate. They consist largely of the mineral apatite, $Ca_5(PO_4)_3(OH,F)$, and often contain up to 100 ppm uranium, perhaps a remnant from the original carbonaceous sediments. Although phosphorites are not worth mining for their uranium content alone, uranium can be a valuable by-product of phosphate mining. More often, it is a large environmental headache.

Uranium is clearly an element that is widely distributed and is concentrated in a variety of rocks by a number of different mechanisms. These have changed during Earth history in response to major increases in the oxygen content of the atmosphere. The distribution of uranium in the Earth's crust is of more than passing interest, because nuclear reactors may become the preferred source of energy during the next century, particularly if changes in the Earth's climate due to anthropogenic changes in the composition of the atmosphere turn out to be significant and unpleasant. Figure 9.23 is an attempt to define the future availability of uranium. The concentration of uranium in average crustal rock is a few ppm (gm U/ton of rock). The highest-grade ores contain between 10% and 30% uranium. Although they are so radioactive, that they are difficult to mine, they are the preferred sources of uranium; but they account for less than a millionth of the crustal reservoir of uranium. The amount of uranium in the several types of lower-grade uranium deposits increases progressively as the grade decreases. For every tenfold decrease in grade there is approximately a thirtyfold increase in the amount of recoverable uranium. The data in table 9.10 show that at the present rate of mining the amount of uranium available in ores with more than 0.1% U (1,000 ppm) will last more than 1,000 years. Uranium resources with grades between 100 and 1,000 ppm will probably last for a million years. Clearly, the supply of uranium in the crust is practically inexhaustible. Economic and environmental issues will decide how much of the energy contained in the world's supply of uranium will actually be used (see chapter 11).

9.8 The Future of Mineral Resources

We owe our mineral wealth to many of the complex workings of the Earth system. Weathering is responsible for virtually all of the ores of aluminum and a large share of the world's nickel ores. Placer deposits of gold, tin, diamonds, and uranium have accumulated along river channels. Marine evaporites account for much of the world's gypsum, salt, and potassium ores. Massive, largely sedimentary iron ores and phosphate ores contain most of the world's reserves of iron and phosphorus. Magmas have been responsible both directly or indirectly for the ores of a large suite of metals, and groundwater has been the scavenging agent, transport medium, and source of many others.

Since we will need an increasing supply of mineral resources, it is useful to inventory the mineral stores in our supply cabinet, and to determine whether and to what extent shortages of mineral commodities are apt to limit the material well-being of the human family. Table 9.11 lists the 1992 reserve base, the 1992 production, and the ratio of the two for a large number of mineral commodities. The reserve base of a mineral commodity includes those resources that are known and currently economic, those that are marginally economic, and some that are currently subeconomic. The reserve base is therefore a fair measure of the quantity of a mineral commodity that will be recoverable at a cost that is not much greater than current costs. Figure 9.24 is a plot of the reserve base of the

Table 9.11. The Ratio (T) of Reserve Base to Annual Production of Mineral Commodities, 1992

Commodity	*Reserve Base (tons)*	*Production (tons/yr)*	*T (yr)*
Bromine	VVL	400,000	VVL
Magnesium metal	VVL	303,000	VVL
Salt	VVL	205,000,000	VVL
Silicon	VVL	3,600,000	VVL
Gypsum	VVL	108,000,000	VVL
Peat	510,000,000,000	186,730,000	2,731
Rare Earths	110,000,000	52,000	2,115
Lithium	8,400,000	5,600	1,500
Mg compounds	3,400,000,000	3,090,000	1,100
Vanadium	27,000,000	32,100	841
Yttrium	560,000	700	800
Potash	17,000,000,000	25,035,000	679
Graphite	380,000,000	613,000	620
Iodine	9,700,000	16,400	591
Chromium	6,800,000,000	12,800,000	531
Rutile	170,000,000	410,000	415
Cobalt	8,800,000	24,800	355
Rhenium	10,000	29	345
Helium	5,534	17	323
Columbium	4,200,000	14,000	300
Bauxite	28,000,000,000	105,000,000	267
Boron	630,000,000	2,400,000	262
Manganese	4,800,000,000	18,800,00	255
Phosphate rock	34,000,000,000	141,000,000	241
Platinum group met.	66,000	294	224
Ilmenite	430,000,000	3,200,000	134
Nickel	110,000,000	916,000	120
Iron Ore	100,000,000,000	844,500,000	118
Molybdenum	12,000,000	108,00	111
Barite	500,000,000	5,200,000	96
Fluorspar	310,000,000	3,600,000	86
Tungsten	3,400,000	39,800	85
Tantalum	35,000	410	85
Antimony	4,700,000	60,600	78
Bismuth	250,000	3,200	78
Zirconium	58,000,000	765,000	76
Selenium	130,000	1,800	72
Copper	590,000,000	8,900,000	66
Sulfur	3,500,000,000	52,700,000	66
Mercury	240,000	4,800	50
Tin	10,000,000	200,000	50
Strontium	12,000,000	238,400	50
Cadmium	970,000	20,000	48
Zinc	330,000,000	7,365,000	45
Lead	130,000,000	3,200,000	41
Thallium	640	16	41
Arsenic	1,500,000	45,000	33
Indium	4,600	140	33
Diamond	380	12	33
Silver	420,000	13,700	31
Gold	51,000	2,170	24

Source: *Mineral Commodity Summaries*, 1993.

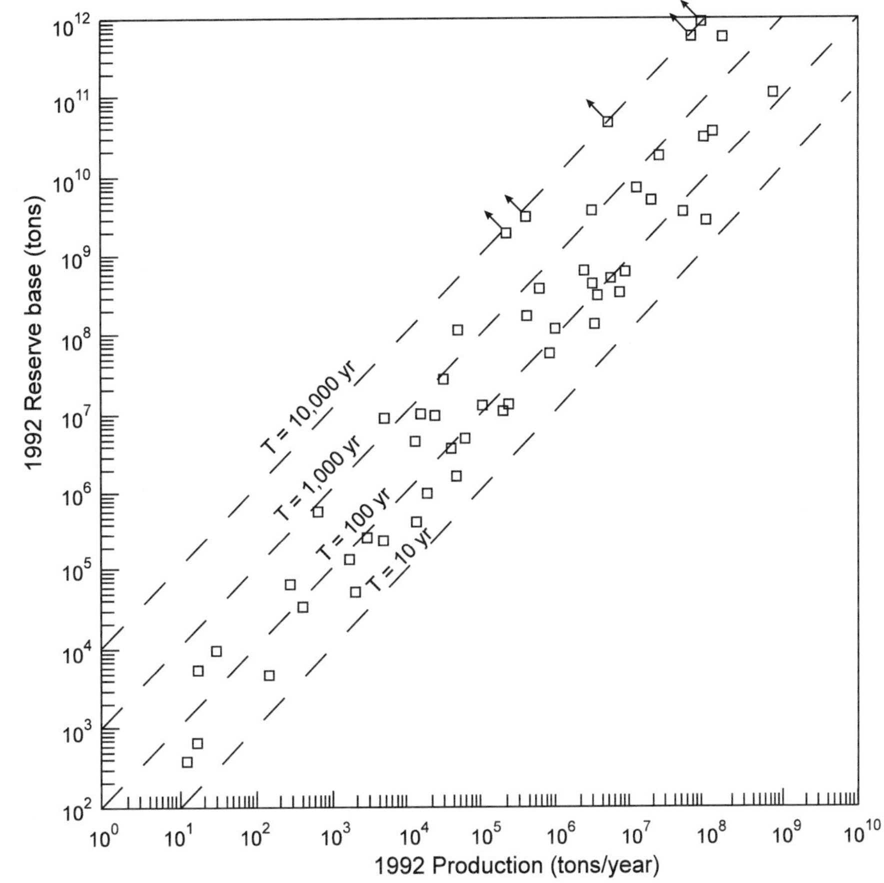

Figure 9.24. Plot of the 1992 reserve base of mineral commodities against their 1992 production. T = ratio of reserve base to production. (Data from *Mineral Commodity Summaries*, 1993)

mineral commodities in table 9.11 against their 1992 production. The two parameters are roughly proportional, i.e., mineral commodities which have a large annual production tend to have a large reserve base. However, the ratio, T, of the reserve base to annual production is not really constant. As discussed earlier, this ratio is a good measure of the continuing availability of a mineral commodity. A commodity for which the ratio of the reserve base to the 1992 production is 100 will probably be available at close to current costs for about a century, if no additional reserves are found and if the production rate remains constant. These "ifs" are very large. Mineral exploration will almost certainly turn up additional ores, and the production of most mineral commodities will almost certainly increase during the next century. Since the effects of new mineral discoveries and technological progress counteract the effects of increasing production on T, the values of T that have been plotted in figure 9.25 can be taken at least as a rough indicator of the time during which ores of approximately present grade will be available.

The range of T's in figure 9.25 is very large. Salt, magnesium, and bromine are currently extracted from seawater. Since an essentially infinite supply of salt, bromine, and magnesium is present at constant grade in the oceans, T is essentially infinite. Silicon is the most abundant metal in the Earth's crust. Nearly pure

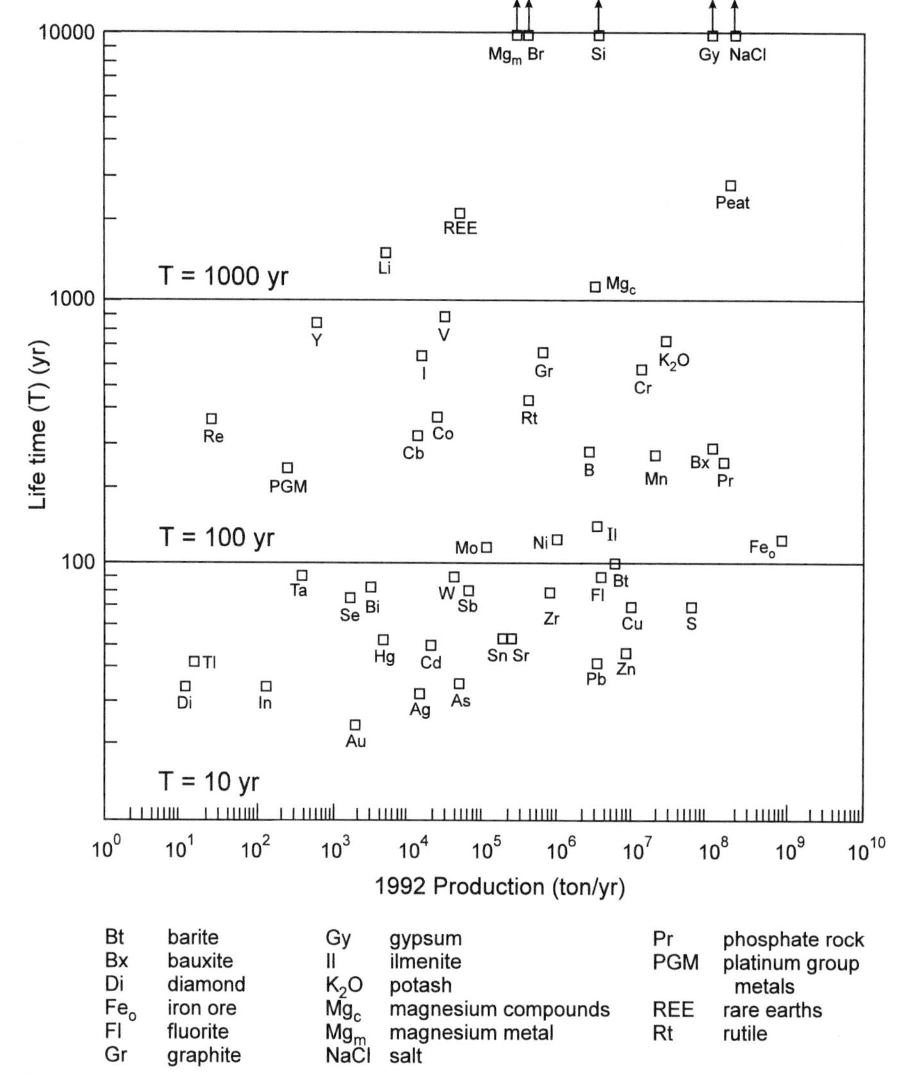

Figure 9.25. The ratio, T, of the 1992 reserve base of mineral commodities to their 1992 production plotted as a function of their 1992 production. (Data from *Mineral Commodity Summaries*, 1993)

quartz sands are so widespread that T for this element is also extremely large. Since gypsum is present in large quantities in evaporite deposits and could be recovered from seawater, its T is similarly very large.

The value of T for many important mineral commodities is in the range of 100 to 1,000 years. Perhaps the most important of these are iron and aluminum (bauxite), and the fertilizers phosphate and potash. The T values of a few mineral commodities are less than 50 years. The most striking of these are gold, silver, and diamonds. Lead and zinc are less glamorous but probably more important metals in this category. Unless large new deposits of high-grade ores of these metals are discovered in the next few decades, the grade of their mined ores will almost certainly decrease considerably during the next century, as has recently occurred for gold.

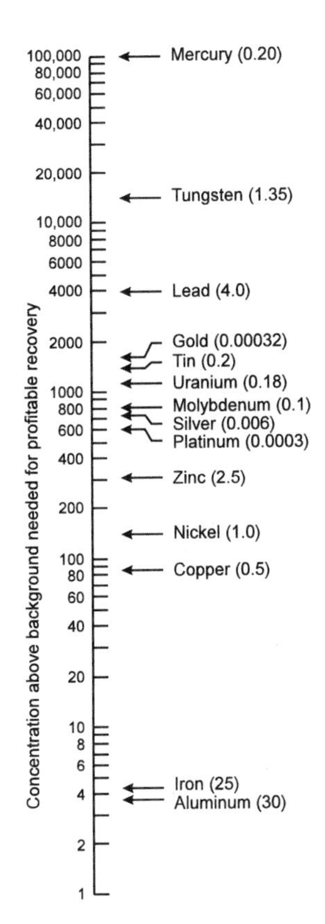

Figure 9.26. The ratio of the concentration of metals in their leanest minable ore to their concentration in average crustal rocks (Skinner 1986). The numbers in parentheses indicate the concentration of metals in wt % in their leanest minable ores.

The long-term trends of the ore grade of many metals are reasonably well defined and can be extrapolated with some degree of confidence. Metals like uranium will be mined from progressively leaner ores, and the grade of uranium ores will gradually approach the uranium content of average crustal rock. The amount of uranium that is available in average crustal rocks is so enormous (see fig. 9.23) that mining will never extend to such low-grade "ore" unless uranium can be produced very cheaply as a by-product. The lowest-grade uranium ore that is currently being mined contains about 0.1% (1,000 ppm) uranium. Average crustal rocks contain about 3 ppm uranium. Roughly, therefore, the cost of extracting uranium from average crustal rocks is about three hundred times the cost of extracting uranium from the lowest-grade ore, all other things being equal, which, of course, they seldom are. Nevertheless, it is safe to say that the cost of extracting uranium will never exceed three hundred times the current cost, and is much smaller in all reasonable forecasts of uranium needs during the next several centuries.

Uranium is an element for which the ratio of the minimum concentration in minable ore to crustal abundance is quite large. As shown in figure 9.26, this ratio is even greater for some elements—mercury, for instance. But for many of the most important metals the ratio is much smaller. It is only five for iron and about four for aluminum. This means that the cost of mining and extracting these metals from average crustal rocks is only some four to five times greater than current costs. It is most unlikely that average crustal rocks will ever be mined for these metals. The cost of iron and aluminum ore during the next several centuries will therefore almost certainly stay within a factor of two to three of current costs, again, assuming that other factors remain more or less constant.

For some commodities the progressive decrease in grade will come to a sudden stop when it becomes economically feasible to mine them from seawater or from other very large reservoirs of constant grade (Petersen 1994). The essentially infinite availability of salt, magnesium, and bromine (see fig. 9.25) is due to the fact that the oceans are already the cheapest source of these commodities. Potassium will almost certainly follow suit, especially if its recovery from seawater is coupled with the production of freshwater in desalination plants.

Such linkages will also become more common in other settings and for other ores. Black, carbonaceous shales are currently unattractive as ores of uranium. Their uranium content is simply too low. However, these shales frequently carry a large complement of other metals. With few exceptions the grade of these metals is also too low to permit mining at present, but multimetal recovery from carbonaceous shales will probably become attractive during the next century.

As metals become significantly more expensive, the tendency to recycle becomes irresistible. Most people throw away aluminum cans with barely a twinge, but those who throw away a gold or silver mug are considered irresponsible, sinister, out of their mind, or extremely wealthy. Aluminum would not have to become much more expensive before recycling of aluminum cans becomes both ecologically and economically highly attractive.

Increases in the cost of producing some mineral commodities would undoubtedly stimulate substitution by others. The increase of the price of tin during the last decades reduced its use in "tin" cans, which are now made largely of aluminum. Aluminum has also made inroads into the preserves of copper, which has suffered from substitution by glass in the transmission of signals via fiber optics.

It seems likely that ceramics and plastics will replace metals in many applications. Aluminum has been the preeminent "new" metal of the twentieth century. Titanium, magnesium, and possibly silicon may be important "new" metals during the next century. All three are abundant and have physical properties that make their alloys very attractive for a number of technical applications.

Altogether, the outlook for the availability of mineral commodities during the next century is reasonably good. The most serious problems may turn out to be organizational and political rather than geological or technological. Like oil and gas, ores are rather unevenly distributed in the world. A major upheaval in South Africa, for instance, could have important consequences for the availability of chrome, gold, and the platinum group elements. Cartels to protect and to advance the price of particular metals have not been very successful during this century. Only the De Beers diamond cartel has managed to continue controlling its market; but the next century may be different. We shall return to this topic in the next chapter, when we consider the likely price of mineral commodities during the twenty-first century.

References

Allègre, C. 1988. *The Behavior of the Earth: Continental and Seafloor Mobility*. Harvard University Press, Cambridge, Mass.

Carmichael, I.S.E., Turner, F. J., and Verhoogen, J. 1974. *Igneous Petrology*. McGraw-Hill, New York.

Chandler, B., ed. 1994. *Treasures of Art and Science at Harvard University*. Harvard University, Cambridge, Mass.

Corcoran, T. 1985. *Mount St. Helens*. KC Publications, Las Vegas.

Costa, J. E., and Baker, V. R. 1981. *Surficial Geology: Building with the Earth*. John Wiley, New York.

Craig, J. R., Vaughan, D. J., and Skinner, B. J. 1988. *Resources of the Earth*. Prentice Hall, Englewood Cliffs, N.J.

Deffeyes, K. S., and MacGregor, I. D. 1980. World Uranium Resources. *Scientific American* 242 (Jan.):66–76.

Eldridge, C. S., Barton, P. B., Jr., and Ohmoto, H. 1983. Mineral Textures and Their Bearing on Formation of the Kuroko Orebodies. In *The Kuroko and Related Volcanic Massive Sulfide Deposits*, pp. 241–81. Economic Geology Monograph 5.

Evans, A. M. 1993. *Ore Geology and Industrial Minerals*. 3d ed. Blackwell Scientific, Boston.

Evans, L. L. 1977. Geology of the Brushy Creek Mine, Viburnum Trend, Southeast Missouri. *Econ. Geol.* 72:381–90.

Force, E. R. 1991. *Geology of Titanium Mineral Deposits*. Special Paper 259. The Geological Society of America.

Harris, S. L. 1988. *Fire Mountains of the West: The Cascade and Mono Lake Volcanoes*. Mountain Press Publishing, Missoula, Montana.

Helz, R.T. 1987. Differentiation Behavior of Kilauea Iki Lava Lake, Kilauea Volcano, Hawaii. An Overview of Past and Current Work. In *Magmatic Processes, Physiochemical Principles: A Volume in Honor of Hatten S. Yoder, Jr.*, ed. B.O. Mysen, pp. 241–58. Special Publication No. 1. The Geochemical Society.

Holland, H. D. 1994. Early Proterozoic Atmospheric Change. In *Early Life on Earth*, ed. S. Bengston, pp. 237–44. Columbia University Press, New York.

Judson, S., and Kauffman, M. E. 1990. *Physical Geology*. 8th ed. Prentice Hall, Englewood Cliffs, N.J.

Kisvarsanyi, G. 1977. The Role of Precambrian Igneous Basement in the Formation of the Stratabound Lead-Zinc-Copper Deposits in Southeast Missouri. *Ec. Geol.* 72:435–42.

Lowell, J. D., and Guilbert, J. M. 1970. Lateral and Vertical Alteration-Mineralization Zoning in Porphyry Ore Deposits. *Econ. Geol.* 65:373–408.

Malpas, J. 1993. Deep Drilling of the Oceanic Crust and Upper Mantle. *GSA Today* 3 (no. 3):53–57.

Naldrett, A. J. 1989a. Sulfide Melts: Crystallization Temperatures, Solubilities in Silicate Melts, and Fe, Ni, and Cu Partitioning between Basaltic Magmas and Olivine. In *Ore Deposition Associated with Magmas*, ed. J. A. Whitney and A. J. Naldrett, chap. 2, vol. 4, of *Reviews in Economic Geology*. Society of Economic Geologists.

Naldrett, A. J. 1989b. Contamination and the Origin of the Sudbury Structure and Its Ores. In *Ore Deposition Associated with Magmas*, ed. J. A. Whitney and A. J. Naldrett, chap 7, vol. 4, of *Reviews in Economic Geology*. Society of Economic Geologists.

Naldrett, A. J. 1989c. Stratiform PGE Deposits in Layered Intrusions. In *Ore Deposition Associated with Magmas*, ed. J. S. Whitney and A. J. Naldrett, chap. 8, vol. 4, of *Reviews in Economic Geology*. Society of Economic Geologists.

Pelletier, R. A. 1964. *Mineral Resources of South-Central Africa*. Oxford University Press, Cape Town.

Petersen, U. 1994. Mining the Hydrosphere. *Geochimica et Cosmochimica Acta* 58:2387–2403.

Pye, E. G., Naldrett, A. J., and Giblin, P. E., eds. 1984. *The Geology and Ore Deposits of the Sudbury Structure*. Ontario Geological Survey, Special Volume 1.

Rich, R. A., Holland, H. D., and Petersen, U. 1977. *Hydrothermal Uranium Deposits*. Elsevier, Amsterdam.

Roedder, E. 1984. *Fluid Inclusions*. Vol. 12 of *Reviews in Mineralogy*. Mineralogical Society of America, Chelsea, Michigan.

Sawkins, F. J. 1984. *Metal Deposits in Relation to Plate Tectonics*. Springer-Verlag, New York.

Sillitoe, R. H. 1973. The Tops and Bottoms of Porphyry Copper Deposits. *Econ. Geol.* 68:799–815.

Skinner, B. J. 1986. *Earth Resources*. 3d ed. Prentice Hall, Englewood Cliffs, N.J.

Stein, M., and Katz, A. 1989. The Composition of the Subcontinental Lithosphere beneath Israel: Inferences from Peridotite Xenoliths. *Israel Jour. Earth Sci.* 38:75–87.

Stommel, H., and Stommel, E. 1983. *Volcano Weather: The Story of 1816, the Year without a Summer*. Seven Seas Press, Camden, Maine.

Su, W. J. 1994. *Geotimes* 39 (no. 6), cover.

U.S. Bureau of Mines. 1993. *Mineral Commodity Summaries*. U.S. Dept. of the Interior, Washington, D.C.

Von Gruenewaldt, G. 1977. The Mineral Resources of the Bushveld Complex. *Miner. Sci. Eng.* 9:83–95.

Von Gruenewaldt, G., Sharpe, M. R., and Hatton, C. J. 1985. The Bushveld Complex: Introduction and Review. *Econ. Geol.* 80:803–12.

Wenkam, R. 1987. *The Edge of Fire: Volcano and Earthquake Country in Western North America and Hawaii*. Sierra Club Books, San Francisco.

White, D. E., Fournier, R. O., Muffler, L.J.P., and Truesdell, A. H. 1975. Physical Results of Research Drilling in Thermal Areas of Yellowstone National Park, Wyoming. U.S. Geol. Survey Prof. Paper 892.

10 The Cost and the Price of Mineral Commodities

10.1 Introduction: How It All Adds Up

The last chapter showed that most mineral commodities will be available at close to their present grade during the next several decades if not for the entire twenty-first century, and that an essentially inexhaustible supply of some mineral commodities is available at their current grades. In this chapter we will explore the likely costs of producing mineral commodities during the next century as well as their likely prices. Their cost will include monetary as well as nonmonetary items; their price will reflect these costs but will also be influenced by a wide range of economic, environmental, and political factors.

To bring a mineral commodity to market, it has to be found, extracted, and—usually—purified. For most metals, prospecting and general exploration are followed by the development of mines, by mining itself, by milling, smelting, and finally by refining. These processes vary from metal to metal, but the underlying principles are similar.

10.2 Prospecting and General Exploration

An unkempt, grizzled old man slowly climbs a narrow trail leading to a mountain pass. His companion is a mule, laden down with picks, shovels, pots, pans, and a bedroll. The old-timer is clearly a prospector, a member of the fabled clan that found so much of America's mineral wealth during the nineteenth century. He is one of a long line of men who scoured the surface of the Earth in search of riches. Most were uneducated, but relatively few outcrops of ore deposits escaped their notice. Their importance in finding ore has been eclipsed during the course of the twentieth century, but only because many of the newly found ore deposits did not crop out at the surface or were detectable only with instruments.

Despite the diligence and success of prospectors, some of the most famous mineral discoveries of the nineteenth century were made by chance or almost by chance. John A. Sutter, a Swiss immigrant to California, settled beside the Sacramento River in the 1830s to pasture cattle and to grow grain. He prospered in the 1840s, and his holdings grew into a valley empire. In July 1847 he dispatched James Wilson Marshall and a companion to the mountains on the American Fork to select a site for a sawmill. In the new mill race, Marshall noticed some yellow particles mixed with the reddish earth. Marshall knew about pyrite and its similarity to gold, but all of the physical tests that he applied showed that he had found real rather than "fool's" gold. Once detected, gold turned out to be widespread in the area. Word of this soon spread. During 1848 great hosts of men made for these hills. The tranquil days of Sutter's farming had ended. Stampeding men and trampling feet had caught up with him (see, for instance, Jackson 1970). During the following year the Forty-Niners arrived, a hundred thousand strong. They sang "Oh, Susannah" and changed the last line of the popular tune to "I'm off to California with my washbowl on my knee." They were immortalized by what they found, and of course by the song "Clementine":

In a cavern, in a canyon
Excavating for a mine,
Dwelt a miner, forty-niner.
And his daughter Clementine.

Ten years later the California gold rush had subsided. Across the Sierras, in Nevada, placer mining for gold proceeded at a modest rate. In 1857 some fair gold placer ground was discovered just below the future Virginia City. The next spring two miners began to work for gold a little higher on the side of the mountain. What they found there was a quantity of strange black sand. To their surprise and delight this turned out to be fabulously rich in both silver and gold. By 1859 they had shown that the great, barren-looking, quartz-rich vein that extended along the face of the high, bare hills contained large, extremely rich ore bodies of both metals. They had discovered the Comstock Lode. Much of the ore in the Comstock Lode was quite shallow. The enormous fortunes that were made in Virginia City during the next twenty years were largely based on ore mined from a depth of less than 500 meters (see, for instance, Smith 1966). Between 1859 and 1882 ore worth nearly $300 million was mined from the Comstock Lode. At current metal prices that amounts to about $5 billion.

The name most closely associated with the Comstock Lode is that of John W. Mackay, by all accounts an extraordinary man, a benefactor of many individuals and institutions (including the University of Nevada's Mackay School of Mines) and husband of Marie Hungerford Mackay, a charming and energetic lady who dazzled society in Paris and London (see, for instance, Lewis 1947). The early days of Virginia City were recorded most memorably by a young reporter on the Territorial Enterprise, one Samuel Langhorne Clemens, better known to history as Mark Twain. His two years (1862–64) in the new boom town occupy a good deal of *Roughing It*. The lode apparently never took him to its heart; that he was not entirely enamored of Virginia City can be gathered from his description of the town:

> Vice flourished luxuriantly during the hey-day of our "flush times." The saloons were overburdened with custom; so were the police courts, the gambling dens, the brothels and the jails—unfailing signs of high prosperity in a mining region—in any region for that matter. Is it not so? A crowded police court docket is the surest of all signs that trade is brisk and money plenty. Still, there is one other sign; it comes last, but when it does it establishes beyond cavil that the "flush times" are at the flood. This is the birth of the "literary" paper. [*Roughing It*, chapter LI]

Geology has changed a great deal since the discovery of the Comstock Lode, and so has the search for ore deposits. Prospecting by individuals has been replaced almost completely by exploration programs involving teams of specially trained geologists with access to a wide range of geophysical and geochemical tools. Most modern exploration is guided by the adage: "If you want to hunt elephants, go to elephant country." "Elephant country" for porphyry copper deposits are regions like the western Cordillera of North, Central, and South America, where volcanic edifices rise above large granitic and granodioritic intrusives (see chapter 9). Porphyry copper deposits have not been found in regions where such edifices have been removed by erosion. "Elephant country" for bauxite ores are regions where the climate is or has recently been hot and humid. The polar regions are not likely target areas for finding such aluminum ores. The application of insights like these has led to the discovery of many major ore deposits. Some, like the discovery of the great Henderson molybdenum porphyry deposit in Colorado, followed painstaking mapping and the application of insights gained from the study of known porphyry molybdenum deposits. The search was almost abandoned when early drilling did not locate ore, but the persistence of the exploration team ultimately paid off handsomely.

The elephant country adage does, however, have its limits. If the elephants in a country have been hunted to extinction, further searches will be fruitless. By analogy, regions that have already been heavily prospected may not be the best place to look for new ore deposits. The elephant country adage also implies that we know how to identify elephant country. Sometimes we don't. If a new habitat is discovered for elephants or for ores, a new era of successful exploration may begin. The recognition of unconformity-type uranium deposits (see chapter 9) rapidly led to major uranium ore discoveries. The recognition of the large, low-grade Carlin-type gold deposits has had an important effect on gold mining, particularly in the United States. It has been said that "gold is where you find it." This dictum has turned out to be rather too pessimistic, but in designing exploration programs for this and for other metals it is well to remember Hamlet's admonition:

> There are more things in heaven and earth, Horatio,
> than are dreamt of in your philosophy. (*Hamlet*, act 1, scene 5)

Sometimes accumulated wisdom can be not only stultifying but positively misleading. At the end of World War II the silver mine at Julcani, high up in the Peruvian Andes, was worked by Cerro de Pasco Corporation (CdeP) under a lease arrangement, and was apparently nearing exhaustion. The chief geologist of CdeP advised his company against purchasing the mine, because he believed that, as in many silver mines, the ore would pinch out at depth. He also thought that lateral extensions of the ore were unlikely. However, Alberto Benavides, CdeP's chief of exploration, believed that the lateral and depth extensions of the ore had not been adequately tested. He formed a new company, Cia. de Minas Buenaventura, and bought the mine. CdeP retained a partial ownership in compensation for the existing installations. Benavides's optimism paid off. Since 1950 Julcani has been one of the major silver producers of Peru and the cornerstone of a large and prosperous Peruvian mining company.

The prospectors of the nineteenth century were limited largely by what their eyes could see and by what their simple chemical tests could analyze. The vision of modern prospectors has been expanded by the availability of many new techniques. Airplanes and satellites permit rapid reconnaissance of large, often inaccessible areas with spectrometers that can detect slight differences in the color of rocks, soils, and vegetation that may indicate mineralization. Sensitive chemical techniques can now detect anomalies in the concentration of ore metals and of elements associated with ores in soils, rocks, and vegetation above ore deposits that are not directly exposed at the surface, and in rivers that drain areas underlain by ore deposits. Physical techniques can also sometimes supply valuable information. In the search for uranium and thorium ores, radiation counters are invaluable adjuncts to the hammer and the hand lens. Counters are normally hand carried, but airborne radiometric surveys are frequently used to locate promising areas. In the search for ores that are either magnetic, or associated with magnetic minerals, aeromagnetic surveys are commonly used. Electrical techniques have often proved useful, because sulfide minerals have properties that differ from those of the common silicates. In some instances, the density contrast between ore bodies and the surrounding country rocks is sufficiently great, so that gravity surveys are useful as ore indicators. Seismic techniques, which have been so important in defining oil and gas structures, are usually less effective in the search for mineral deposits, because ore bodies offer very much smaller and more irregular targets. The usefulness of data from many of the new techniques has been enhanced by the application of multivariate statistical analysis and the use of high-speed computers.

Many parts of the Earth's surface have been explored quite thoroughly with these techniques. Areas that are difficult of access will surely be found to contain valuable ores, but the main hope for large new mineral deposits seems to lie at greater subsurface depths, such as in the Precambrian basement of the central United States. In much of Canada, the Precambrian basement is exposed at the surface and contains an abundance of ores. Similar ores probably exist in the Precambrian shield in the United States, which is largely covered by younger sedimentary rocks. When the cover of these rocks is more than 1,000 meters, mineral exploration is difficult and mining would be expensive, but such targets will probably become attractive during the next century.

Ocean floor mining for copper, zinc, nickel, cobalt, and manganese may also become feasible. Sizable accumulations of copper and zinc sulfides are already known in midocean ridge settings. Copper, nickel, and cobalt in manganese nodules (see chapter 7) were nearly economic in the 1970s and may well become attractive in the twenty-first century, although the environmental costs of mining the deep ocean floor may turn out to be severe.

10.3 Mine Exploration and Development

Roughly speaking, only one promising outcrop in a thousand develops into a mine, because there are so many requirements for a project's economic health and environmental adequacy. The first steps along the path from initial discovery to the beginning of mining usually involve detailed surface geological mapping, trenching, and pitting. These must be preceded by laying legal claim to or acquiring the area, a procedure that can be long and difficult. If mapping and shallow exploration indicate that significant quantities of ore may be present at depth, the subsurface geology of the area is explored by drilling. A wide range of drilling techniques is available to mining geologists. They will, of course, choose those that reveal the most about the mineralized area for the least expense. The pattern of drilling will depend on the likely geometry of the potential ore body. If a large quartz vein crops out at the surface, the ore may be expected to lie within a more or less tabular inclined body such as that in figure 10.1. Drill holes designed to intersect ore at depth may then be poked through the mineralized area as indicated. The number of drill holes and their location will depend on the detail to which the reserves of potential ore must be defined and on the grade and thickness of the ore that is encountered early on.

Defining the size, shape, and grade of a potential porphyry copper ore body poses somewhat different challenges, because porphyry copper ore bodies are large in all three dimensions (see fig. 9.9) and vary considerably in overall geom-

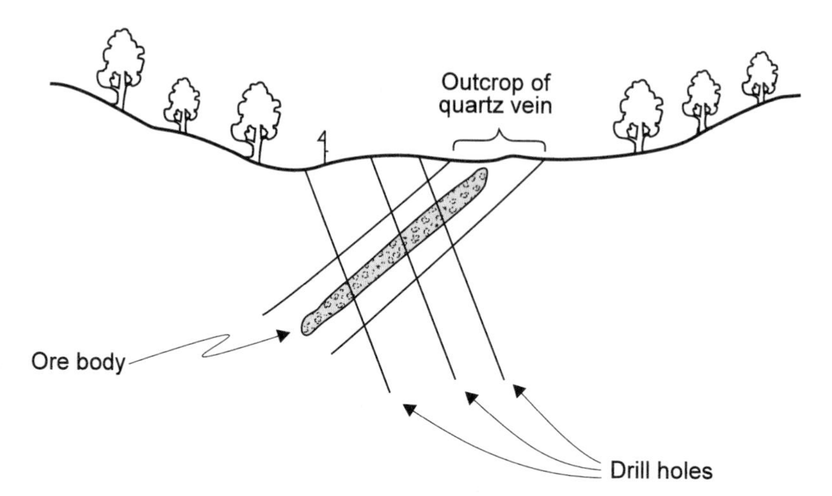

Figure 10.1. Drill pattern for assessing the extent of a tabular, inclined ore body.

etry. Drilling typically yields a rather incomplete picture of the distribution of ore in a mineralized body of rock. The process is somewhat like poking toothpicks into a Schwarzwälder Kirschtorte to assess the number, size, and distribution of cherries. The geographic range of information obtainable from drill holes can, however, be extended by lowering instruments that determine the physical and chemical properties of rocks surrounding the holes. Statistical techniques have been developed to estimate the distribution of ore grades between adjacent and somewhat more distant drill holes.

Sometimes closely spaced drilling is more expensive than underground development. In that case, access to the mineralized area by shafts and/or adits (horizontal tunnels) is used to define the geometry of ore bodies (see fig. 10.2). This type of development is expensive, and the cost is nonrecoverable if, in the end, the project is abandoned as uneconomic. In fact, the entire process of mine exploration is among the riskiest parts of mining ventures. It is therefore normally carried forward in stages and with considerable caution.

At the end of the exploration process a fair but incomplete description of the distribution of potential ore will be available, and the infrastructure required to develop and mine the property can then be designed. This is accomplished with a feasibility study. For large mines, like those required to mine porphyry copper ore bodies, an extensive and expensive infrastructue is needed. If the ore body is in a sparsely inhabited area, roads will have to be built, possibly a railroad, and perhaps even a port to handle the ore. A village or a town may have to be built, complete with schools and a hospital, and procedures to minimize environmental damage will have to be developed. The total cost can amount to several billion dollars. The decision to undertake a venture of this magnitude is not taken lightly. Even mines requiring only one-tenth as much capital are hardly minor investments.

If the feasibility study demonstrates that a profitable mine can be developed, the next step is financing the operation. Most of the money for developing mines is borrowed, usually from banks, but also from equipment suppliers and potential customers. The viability of many mining projects therefore depends as much on

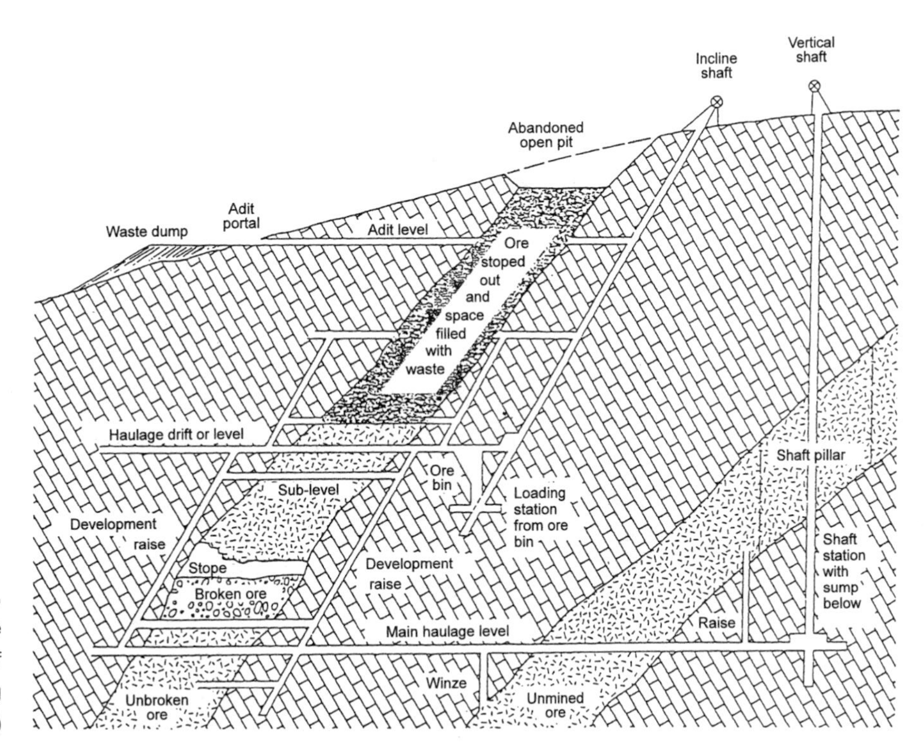

Figure 10.2. A diagram to illustrate the meanings of commonly used mining terms. (Evans 1993)

the terms of the required loans as on the grade and tonnage of the ore. It also depends on the taxes and on the royalties that the mining company is required to pay. In many countries the rules that determine these matters are spelled out in a mining law or tax code. In some countries individual agreements are reached, especially if the ore body is large and the sums involved are substantial. The negotiations can be very delicate, because they tend to involve political as well as economic considerations. The engineering and economic aspects are easy to spell out. Mining companies want to maximize revenues from the operation. The host countries want not only the monetary benefits of the operation, but also its benefits as a source of employment for their citizens. Developing countries view mines as a means of acquiring advanced technology, as a means of developing new or depressed regions, and as a stimulus to related industries, to educational institutions, and to future investment. They also want to make as certain as possible that the political opposition does not construe the permission to mine an ore body as a giveaway of public property.

Each side has to decide whether the development of a mine meets its minimum requirements. If it does not, the project fails, at least for the present. If the development of an ore body does meet the minimum requirements of both parties, then a deal can be hammered out. The key will be the division of the profits that go beyond the minimum expectations of each party. The tendency on the part of the mining company will be to maximize its share, so as to reduce the payback time, to increase its total returns, and to shorten the time during which it is exposed to the risks of operating in a country where it has little or no control over changes in taxation policies and where nationalization and expropriation may be a very real threat. Experience has shown that an equitable division of "excess" profits, both anticipated and unanticipated, is to be preferred. A mining company that recoups its development costs very rapidly because an unexpectedly large volume of rich ore was discovered early in the course of mining is vulnerable to the charge of sharp dealing, and its negotiating partners in the government are vulnerable to charges of stupidity and/or bribery.

Once the decision to proceed has been reached, the development work required to start mining is undertaken. Thereafter, further exploration and development go hand in hand with mining to extend the mine life. Even the best-laid plans for a mining venture can go awry for economic and/or political reasons. Mines where costs are high and the ore grade is low are particularly vulnerable to decreases in metal prices and to increases in labor costs. The fate of large mines that have lifetimes of many decades may be sealed by political changes that were unforeseen when the mines were opened. Many mines have played important roles in the development of the economy of numerous countries. The role that mining companies from technologically advanced countries have played in the economy of developing countries has generally been more helpful than the role of state-run enterprises, although there are exceptions to this rule.

10.4 Mining

The subject of mining can be divided conveniently into surface mining and underground mining. *Surface mining* involves the removal of ore and waste from the surface on down. *Underground mining* involves the construction of shafts and/or adits that give access to ore at depth before mining can begin. The two types of mining are not mutually exclusive. Ore bodies that are initially mined from the surface may be more profitably mined by subsurface techniques as a mine deepens. Conversely, ore bodies that were once mined by subsurface methods may be "open pitted" later, if, for instance, the mining of large quantities of low-grade ore near the surface becomes economically attractive.

Historically, there has been a gradual shift from underground to surface mining. This is particularly true of the mining of coal, copper, and gold, three important mineral commodities. Some underground miners tend to look down on surface mining as simply "dirt moving." It usually involves a good deal more than that, but there is enough truth in the slur to make it sting. The simplest form of surface mining is probably placer mining (see chapter 7) for gold, tin ore, gems, and several other mineral commodities. The equipment for moderate-size operations on land consists of bulldozers and front-end loaders. Drag lines, power shovels, and bucket wheel excavators are used in larger mines. In hydraulic caving, giant hydraulic monitor nozzles are directed against banks that may be as much as 50 meters high.

Placer mining in streams, ponds, and offshore is done by dredging (see fig. 10.3). This is a very long step beyond the panning practiced by old-time miners. Large bucket-line dredges can handle about 1,500–2,000 tons of gravel per hour and can operate in water that is 45 meters deep. Although large dredges are expensive, operating costs are low, some fifty cents to a dollar and fifty per ton of gravel dredged. This is about one-tenth the cost of mining hard rock, and is the reason why placer mining can show a profit from mining even very low-grade ores.

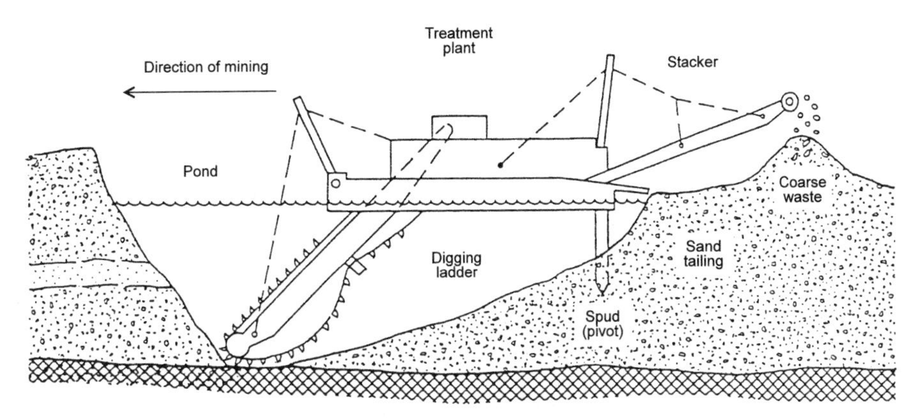

Figure 10.3. Placer mining by bucket line dredge. (Peters 1987)

Nearly all of the sand and gravel dredged in placer operations is moved only short distances, just far enough to extract the usually small amounts of valuable minerals they contain. Surface mining of bauxite is quite different. Essentially all of the bauxite is removed for processing, and large mining operations such as those in Guyana (see plate 39) usually leave unsightly scars on the landscape. This blight is not, however, inescapable. On the island of Jamaica, agriculture does not do well on bauxite, because the ore is so poor in plant nutrients. The difference between the plant cover in areas underlain by bauxite ore and in non-bauxite areas on this island is so large that bauxite deposits can be spotted from the air. The Reynolds Metals Company acquired areas underlain by bauxite, removed the aluminum ore, replaced the topsoil, allowed vegetation to return, and stocked the former mine areas with cattle. After bauxite removal, the cover of grass did very well, and the mining company—somewhat to its embarrassment—discovered that it had become the owner of a lucrative cattle business (see plate 40). This, surely, is a recipe for good environmental stewardship.

The same solution cannot be applied in all areas of major surface mining. The great iron mine at Mount Tom Price in Western Australia (plate 22) and the enormous open pit mine at Bingham Canyon, Utah are located in areas that are too dry for a Jamaican solution. The iron ore mined at Mount Tom Price and in similar mines is very rich. Since there is little waste, the large hole that is produced by mining is more or less permanent. At Bingham Canyon most of the rock that is mined is waste. It could be returned to the open pit, but not before the end of mining, because mining would be seriously hampered by the presence of waste. It is probably unreasonable to consider filling the pit after all of the mining has been completed, except possibly with municipal waste. Proposing this option would probably give the Utah Department of Health an advanced case of the vapors.

In areas of high rainfall, mining must usually be accompanied by extensive pumping. Open pits tend to fill with surface water, and underground mining has often been limited by the inability of pumps to keep up with the influx of groundwater. After surface mining has finished, mines in wet areas fill with water. In this way abandoned quarries have become "swimming holes" since time immemorial, and large mines can become lakes with a variety of recreational uses.

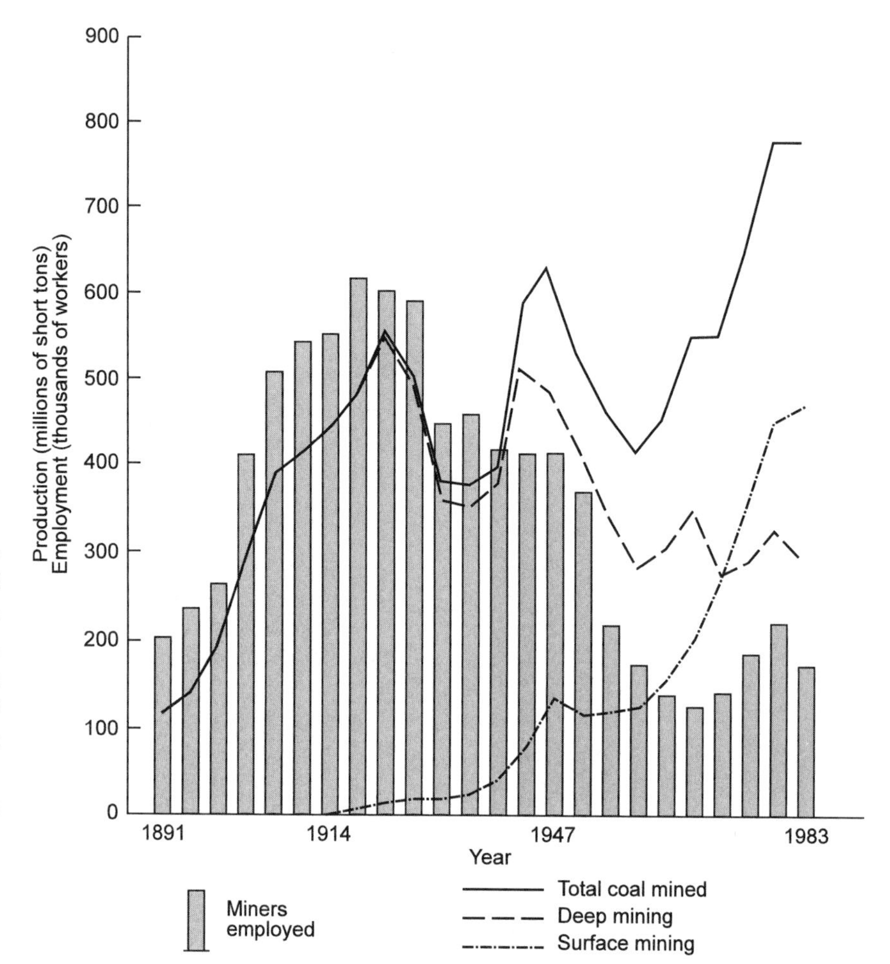

Figure 10.4. Coal production and employment in the United States, 1891–1983. Production is measured in millions of short tons, employment in thousands of miners. (Miernyk 1986)

Unfortunately the disturbance of aquifers and of the water table can have serious environmental consequences. At the beginning of the twentieth century coal was largely mined underground. Now, surface mines account for the major portion of the coal mined in the United States (see fig. 10.4). Extremely large equipment has been developed for mining coal that occurs in flat-lying beds that are exposed at or just below the surface. The capacities of dragline bucket and shovel dippers may exceed 100 cubic meters. One dragline, "Big Muskie" in the Ohio coal fields, weighs 13,500 tons and delivers a 325-ton bucket load in one-minute cycles. A great deal of land is affected by mining on this scale. The land area involved in surface mining of coal by the year 2000 is estimated to be between about 350 and 2,700 square kilometers (COMRATE 1975). The larger number is an area corresponding to a square some 52 kilometers on edge. From the point of view of the United States as a whole, this is a very small square. But on a local level the environmental damage can be very severe. Fortunately, much of this damage can be contained. Overburden, i.e., soil and rock between the surface and the coal seam, can be replaced, and the area as a whole can be restored and revegetated. In this way the recurrence of the worst excesses of past environmental degradation due to surface mining of coal can be prevented.

Some environmental effects of coal mining are difficult to eliminate, or even to contain. Coal and many other ores often contain pyrite. When the atmosphere and/or oxygenated groundwater gain free access to pyrite, the mineral is oxidized. The products of the oxidation are $Fe(OH)_3$ and sulfuric acid

$$4FeS_2 + 14H_2O + 15O_2 \rightarrow 4Fe(OH)_3 + 8H_2SO_4. \quad (10.1)$$

$Fe(OH)_3$ usually forms as a brownish, gelatinous precipitate nicknamed "Yellow Boy." Sulfuric acid increases the solubility of the compounds of many trace metals. As a result, it can make groundwaters quite acid, unfit for human consumption, and a threat to the biota where it mixes with surface drainage.

In any particular area, the economic feasibility of surface mining depends in large part on the *stripping ratio*, i.e., the ratio of the amount of overburden that must be removed to the amount of the ore that can be mined. This ratio is essentially constant for a given deposit or district in flat country that is underlain by horizontal coal beds. In hilly country the stripping ratio may be prohibitively large under hills (fig. 10.5a). When a coal bed has a significant dip, the stripping ratio increases progressively as mining proceeds downdip (fig. 10.5b); surface mining stops when the stripping ratio becomes so large that the cost of removing the waste together with the costs of mining and treating the coal exceeds the value of the coal.

The feasibility of switching to underground mining depends on the economics of this very different way of extracting ore. Underground mining can be divided into three parts: gaining access to the ore, mining it, and bringing it to the surface. Gaining access to ore can be done horizontally, vertically, or at some intermediate angle. All three methods are illustrated schematically in figure 10.2. Horizontal tunnels, or adits, are usually driven into the side of hills containing an ore body. The technique is an ancient one. Figure 10.6 illustrates the technique as it was practiced in the middle of the sixteenth century. The illustration comes from Georgius Agricola's classic 1556 treatise, *De Re Metallica*. The author—Georg Bauer prior to latinization of his name—was a doctor in the Erzgebirge, one of

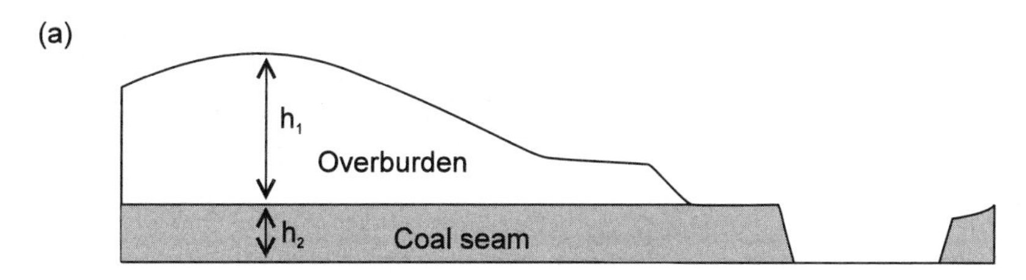

Figure 10.5a. In this mining area the stripping ratio h_1/h_2 is much greater under the hill than in the valley.

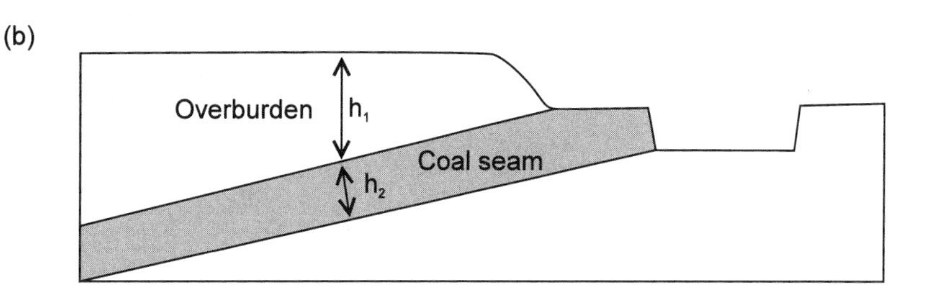

Figure 10.5b. The stripping ratio h_1/h_2 increases down dip in a coal seam that is not flat-lying.

Figure 10.6. Shafts and an adit in mid-sixteenth-century mines. (From Book 5 of Agricola 1556)

the most prolific and long-lived mining regions in eastern Germany and in the western part of the former Czechoslovakia. His treatise served as the definitive exposition of mining and metallurgy for nearly 200 years. The translation by Herbert Hoover (a mining engineer and later president of the United States) together with his wife, Lou Henry Hoover, made the treatise accessible to the English-speaking world and extended the historical perspective of all who are interested in the history of mining and metallurgy in the Western world. Figure 10.6 also illustrates the nature of shafts in the sixteenth century. These provide vertical access to ore bodies. Inclines or inclined shafts descend at intermediate angles; they often run parallel to the dip of orebodies (see fig. 10.2).

A variety of techniques is in use for mining ore. The choice of a mining technique for a particular mine depends on the nature, the inclination, and the distribution of the ore as well as on the nature of the host rocks. In room-and-pillar mining (fig. 10.7a), ore is removed in a series of rooms, leaving pillars to support the roof rocks. This technique is used, for instance, for mining flat coal seams and the large lead ore bodies of southeastern Missouri (see chapter 9). A major ad-

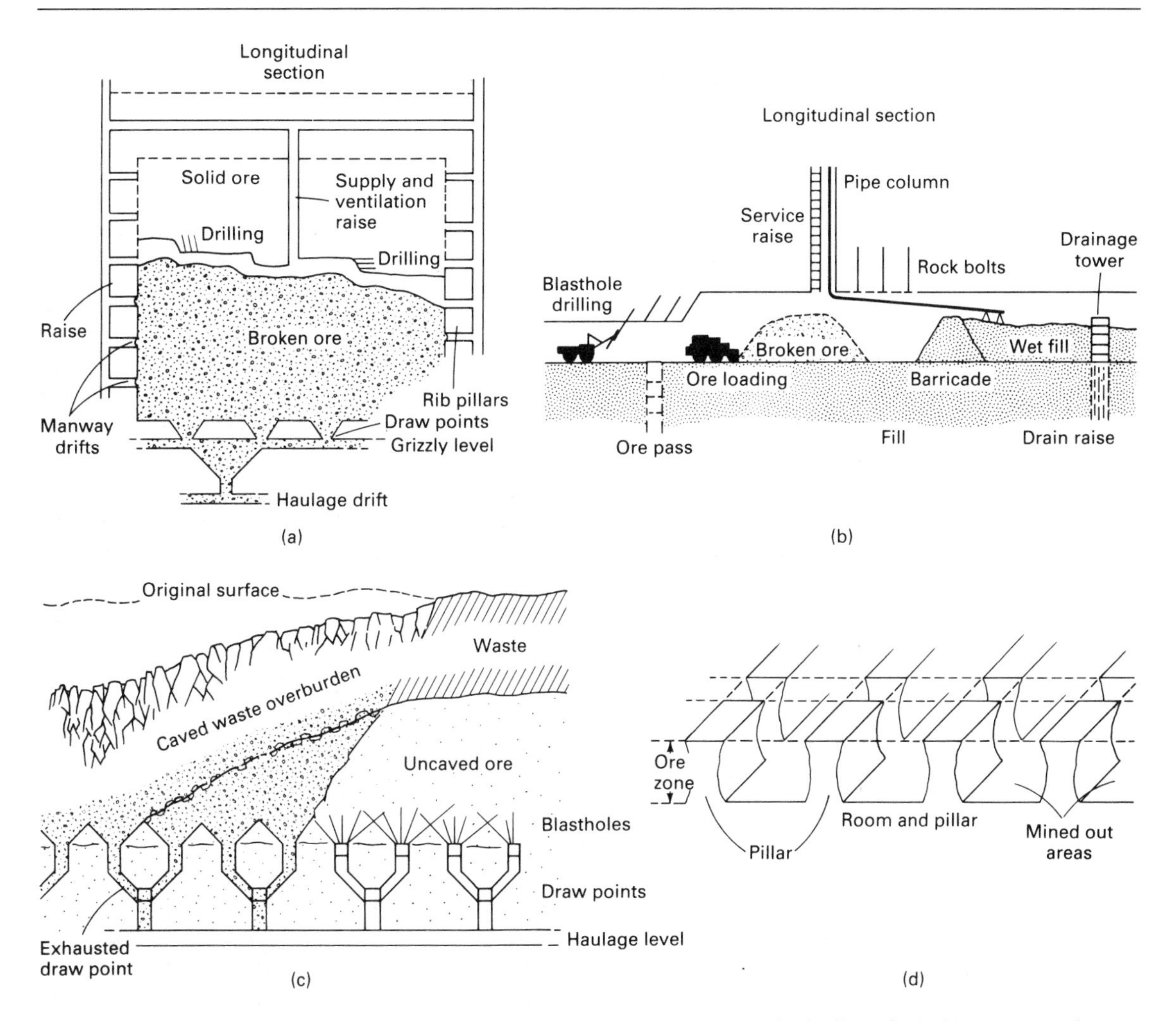

Figure 10.7. Four common underground mining methods. (After Craig, Vaughan, and Skinner 1988)

vantage of room-and-pillar mining is that the roof is stable and does not collapse. A major disadvantage of the technique is that as much as half the ore can remain unmined in the pillars. In the other common mining methods, few or no pillars are left. In cut-and-fill stoping (fig. 10.7b), the open space left when ore is removed is refilled with waste material. The method is particularly suitable for steep veins, but it is expensive. Nevertheless, it has the advantage of allowing the disposition of mining wastes below ground rather than in mine dumps on the surface. In the other common mining methods, such as shrinkage stoping and block caving (fig. 10.7c,d), the ore is broken and induced to move downward to haulage drifts. It is then transported along these drifts to a shaft, and from there to the surface. This type of mining tends to produce surface collapse, an undesirable consequence, especially when a town has been built directly above an ore body.

Gravity can be a very real asset in mining, especially in shrinkage stoping and block caving, where the ore is blasted and moves downward under the force of gravity. But gravity is also a source of danger. Rock falls are the cause of most mine accidents, and are in large part responsible for making mining such a dangerous occupation. Coal mining has claimed an undue number of lives. The total annual number of U.S. coal mine deaths from accidents during this century is shown in figure 10.8. The rapid decrease in the number of deaths during the course of the century is gratifying. This decrease is partly due to increased mine safety; in part it reflects the decrease in the total number of miners from a high of nearly 900,000 in the early 1920s to its present level of about 150,000, and the shift from underground to surface mining. Figures 10.9 and 10.10 are more reliable indicators of the increase in safety during the twentieth century. Figure 10.9 shows that the number of fatalities per million man-hours worked has decreased significantly during the course of the century. Figure 10.10 shows how the number of accidental fatalities per million tons of coal mined in the United States has decreased. This parameter is a good measure of the cost in human lives of producing this important source of energy (see chapter 11).

However we choose to measure risk to human life, underground mining of coal is more dangerous than surface mining. This is due only in part to the incidence of rock falls. Underground mining for coal is intrinsically more dangerous than "hard rock" mining, because coal beds frequently contain methane, i.e., natural gas. This gas is easily ignited. Underground fires and explosions are therefore a constant threat in many coal mines. Mine accidents are not the only occupational hazard for coal miners. Coal dust is a ubiquitous product of coal mining. Unless precautions are taken, coal dust is inhaled by miners and can produce pneumoconiosis, commonly known as black lung. This can be fatal. Black lung used to be a major cause of death among coal miners. Government regulations now set stringent limits on the amount of coal dust permitted in underground mines, and the incidence of death from black lung should decrease greatly (see, for instance, the President's Commission on Coal 1980).

Mine safety regulations and their enforcement have come slowly during the nineteenth and twentieth centuries. The history of mining is full of mankind's inhumanity to man (see, for instance, *How Green Was My Valley* by Richard Llewellyn, 1939). Brutality was not confined to the mining industry, but it was particularly visible there; conditions underground were invitations to misery. The history of the mine workers' unions is burdened with some of the fiercest labor struggles of the nineteenth and early twentieth centuries (see, for instance, Long 1989), and produced some of the most colorful union leaders. "Mother" Mary Jones (1837–1930), an organizer for the United Mine Workers of America (UMWA), used the battle cry: "Pray for the dead, and fight like hell for the living." John L. Lewis, the powerful, long-term (1920–60) president of the UMWA ruled his union arbitrarily and capriciously. His methods and motives have been questioned, but no one can doubt the favorable benefits he negotiated for working miners.

Coal is not the only commodity whose mining is beset with serious environmental hazards. Respiratory problems due to the inhalation of silica can produce silicosis, and the health effects of inhaling asbestos have been debated furiously during the past two decades. It is very clear that inhaling "blue asbestos," the mineral crocidolite, can be fatal. Inhaling "chrysotile asbestos" is clearly much less dangerous, and it has been argued persuasively that the large-scale removal of this type of asbestos from buildings does little, if anything, to improve the

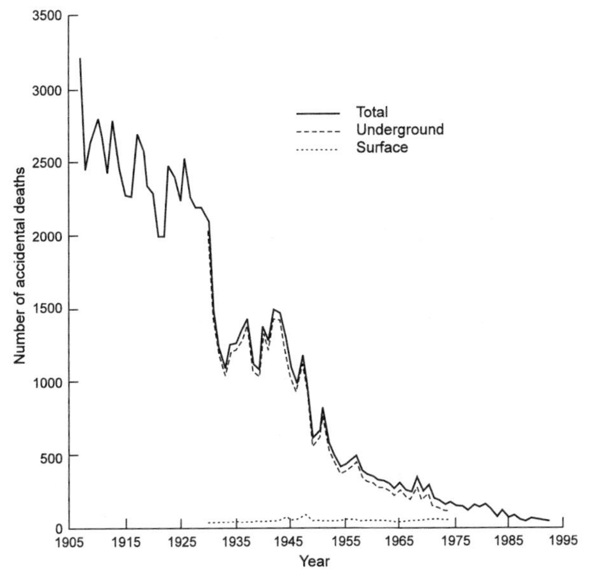

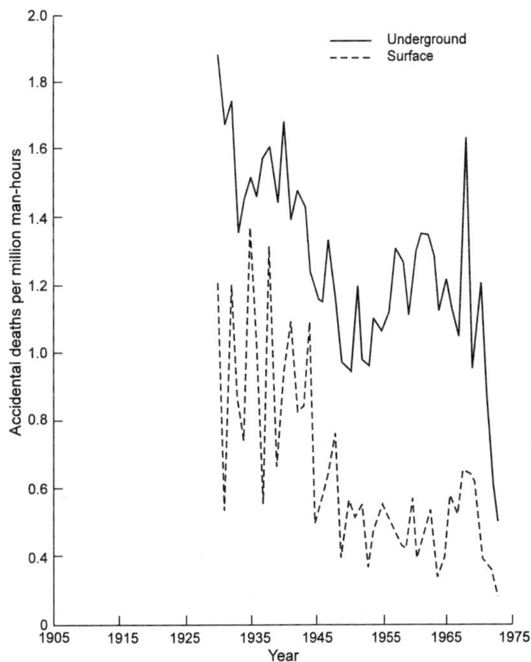

Figure 10.8 (*above, left*). Total U.S. coal mine deaths from accidents, 1907–73. (COMRATE, 1975)

Figure 10.9 (*above, right*). Fatal accidents in U.S. coal mines between 1907 and 1973 per million man-hours worked. (COMRATE, 1975)

Figure 10.10 (*below, right*). Accidental fatalities per million tons of coal mined in the United States, 1907–73. (COMRATE, 1975)

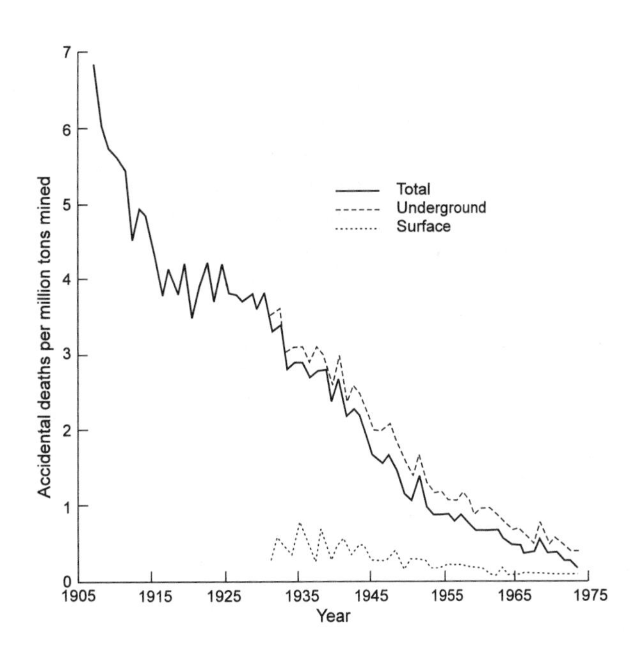

health of the occupants (see, for instance, Solid-Earth Sciences and Society 1993, pp. 218–19; Ross 1981, 1982).

Uranium miners face somewhat different threats. In high-grade uranium mines, miners are exposed to dangerously high levels of radiation from the ore itself. They also inhale radon, a gas in the radioactive decay chain of uranium, which leaks out of uranium ore into mine air. Since the gas is radioactive, it irradiates the miners' lungs. Its decay products are not gases. They tend to be deposited in the respiratory system of miners and to continue the process of irradiation. Ventilation of uranium mines reduces the danger from radon inhalation, but even with this precaution some of the richest uranium mines cannot be mined safely except by robots.

The inhalation of mercury vapor, which produces hydrargyrism, was particularly severe in the Almadén mine in Spain, where native mercury is abundant. The king of Spain decreed that the miners had to refrain from mining one week a month and to grow crops on plots of land especially provided by the authorities. The idea was to sweat out the inhaled mercury. Human nature being what it is, the miners sublet their plots, basked in the sun, and enjoyed the additional revenue from their plot leases.

10.5 Beneficiation

Not everything that is taken out of mines is ore. Many shafts, adits, and mine workings are driven in rock that has no commercial value; in strip mining, the volume of the waste rock often exceeds that of the ore. Waste rock is usually piled on mine dumps. In some cases waste is reintroduced into the mine for support and to reduce the environmental effects of mining operations.

Mined ore is transported to a mill, unless it is so rich that it is already of "shipping grade" or "direct smelting ore." Mills are designed to separate the ore minerals from associated, useless gangue minerals. The first steps in this process involve crushing and grinding. For this purpose a number of techniques have been developed, most of them quite energy intensive. When the ore minerals are very fine grained, less than approximately 0.01 millimeters (10 microns) in diameter, the mills, like those of the gods, must grind "exceeding fine." Sometimes they cannot grind fine enough. Therefore, although the metal content of some rocks is very high, they cannot qualify as ore, because the grain size of the valuable minerals they contain is too small for their separation from the associated gangue minerals with available milling techniques. The beneficiation of these ores is an obvious challenge for the future.

After crushing and grinding, the valuable minerals are separated from the gangue minerals by a variety of methods based on differences in their density, their magnetic and electrical properties, and their surface properties. Differences in the surface properties of minerals are used in *flotation*, a mineral separation technique that revolutionized the beneficiation of ores when developed early in the twentieth century (see, for instance, Kelly and Spottiswood 1982). Most ore minerals are hydrophilic: water attaches itself readily to their surfaces. Flotation of these minerals is achieved by coating their surfaces selectively with collectors, organic compounds that are heteropolar. One end of these molecules sticks to the surfaces of a particular mineral. The other end of the molecules is usually an organic chain or group that is hydrophobic, i.e., that does not mix with water. The coated mineral grains, somewhat like M & M's, are hydrophobic, but they do attach themselves to air bubbles that are coated with *frothers*, water-soluble organic reagents that absorb at the water-air interface. During flotation, air is passed through a slurry of ore and gangue minerals to which collectors and frothers have been added. The coated ore minerals preferentially attach themselves to the air bubbles of the froth, and are carried to the top of flotation cells, where they are skimmed off, dried, and accumulated as "concentrate." Many ores contain minerals of more than one metal. For instance, sphalerite, ZnS, is often associated with galena, PbS, and with silver minerals. These minerals have to be separated from each other as well as from the associated gangue. The flow sheets of mills are therefore often quite complex.

The tails, i.e., waste from mills, usually accumulate in tailings piles or in tailings ponds. In some operations, the tailings are mixed with cement and are pumped back underground. This is by far the best disposal technique; but the conversion of old tailings piles into golf courses is also acceptable provided the tailings do not contain excessively high concentrations of toxic elements in soluble form.

Uncovered surface tailings piles are usually unsightly. They are subject to erosion and the dispersal of toxic, high-metal dust. They also react with rainwater. Sulfide minerals in the tailings piles tend to oxidize (see eq. 10.1), and the sulfuric acid produced during oxidation leaches metals from tailings. These metals can contaminate both the local surface water and groundwater. Sometimes such reactions can be put to good use. The separation of ore from gangue minerals is never complete, and dump leaching almost always dissolves some of the remaining ore. The metals in the leach solutions are recoverable. In some areas dump leaching has therefore proved to be profitable; more often than not it has also been fraught with environmental problems.

Uranium mill tailings can create particularly severe environmental difficulties, because they may contain much of the radioactivity that was present in the original ores. Groundwater contamination with radionuclides and with a number of other species has been particularly bothersome. Some of the difficulties can be contained by stabilizing the tailings piles with vegetated earth cover, but in some areas this was not done. At Grand Junction, Colorado, more than 300,000 tons of radioactive tailings were used as fill material for land on which many buildings were later constructed. Some of the waste was even used to make concrete blocks for foundations. The program to clean up the highly radioactive materials in this area cost many millions of dollars (Craig, Vaughan, and Skinner 1988).

This number pales in comparison with the probable cost of several billion dollars that will be required to clean up the uranium mining area in the Erzgebirge, part of the former German Democratic Republic (GDR). After the Second World War the uranium mines in this region became major suppliers of nuclear materials for the former Soviet Union. Very little care was taken for the health of the miners or for their environment. Approximately 500 million tons of mining waste have seriously contaminated an area of 168 square kilometers. A further 1,500 square kilometers have been classified as possible dump sites, and more than 1.5 million tons of uranium-tainted sulfuric acid have seeped into the subsurface from mining waste dumps.

10.6 Smelting and Refining

In most ores the valuable metals are present either as oxide or as sulfide minerals. Iron and aluminum are generally present as oxides, whereas copper, lead, and zinc occur largely as sulfides. During the smelting of oxide ores, the metals are reduced; during the smelting of sulfide ores, the metals are separated from sulfur. The techniques that are used to decompose oxides and sulfides vary somewhat from metal to metal, but they are sufficiently similar, so that a brief description of the smelting and refining of iron, aluminum, and copper can serve as an adequate introduction to the subject.

10.6.1 *The Smelting and Refining of Iron*

Figure 10.11 is a diagrammatic representation of the interior of a blast furnace. Iron ore, coke, and limestone are added at the top of the furnace. Hot air is introduced through tuyeres at the base of the furnace. The chemistry of the reactions in blast furnaces is complex. The most important reactions are those of coke at ca. 1600°C to produce carbon monoxide

$$\underset{\text{coke}}{C} + \frac{1}{2}O_2 \rightarrow CO \tag{10.2}$$

and the reduction of the iron oxides to liquid iron. If the ore mineral is hematite, the reaction is

$$\underset{\text{hematite}}{Fe_2O_3} + 3CO \rightarrow \underset{\text{liquid iron}}{2Fe} + 3CO_2. \tag{10.3}$$

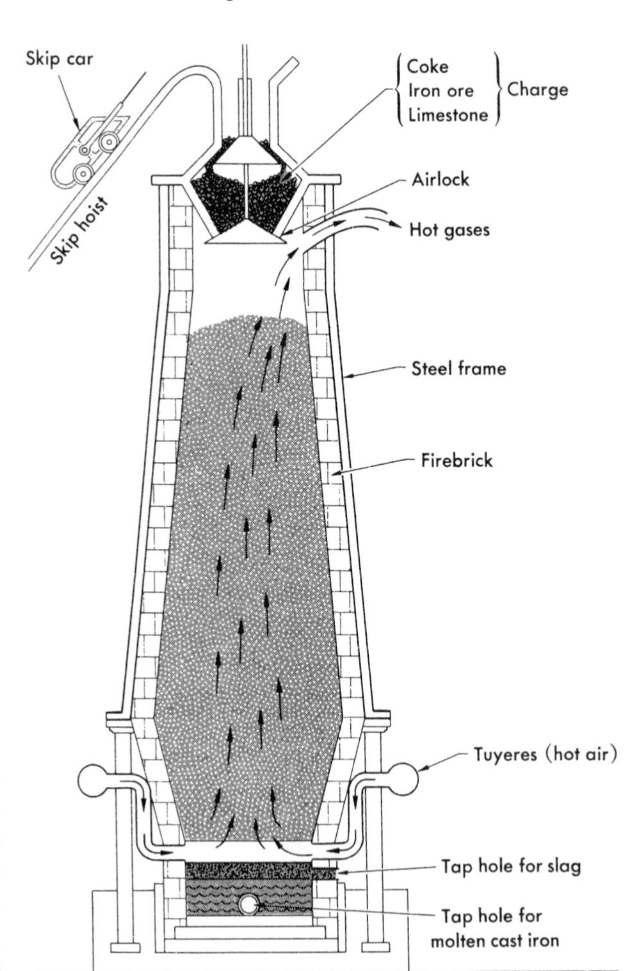

Figure 10.11. Diagrammatic representation of the interior of a blast furnace for the production of pig iron. (Craig, Vaughan, and Skinner 1988)

Effectively, carbon reduces iron oxides to metallic iron and is itself oxidized to CO_2 in the process.

Iron ore usually contains quartz, particularly if its source is an iron formation. Limestone in the blast furnace charge combines with quartz and other impurities to form a silicate melt. These melts are much less dense than liquid iron. Liquid iron therefore sinks to the bottom of the blast furnace. From there it is tapped and solidifies into "pig iron." The slag is tapped separately, and is frequently used as concrete aggregate or as railroad ballast; in Old San Juan, Puerto Rico, cobblestones, which are rather slippery when wet, are made of furnace slag.

Pig iron is the major starting material in the production of steel. It is heated together with scrap iron and fluxes; oxygen is then blown through the melt (see, for instance, Moniz 1992). Some of the undesirable impurities in the pig iron are oxidized and move into the slag. During another stage of smelting, carbon and a variety of metals are added to liquid iron; these give the several types of steel their desirable physical and chemical properties.

10.6.2 The Smelting and Refining of Aluminum

The process by which aluminum in bauxite is converted into aluminum metal is basically similar to the process by which iron in oxide ores is converted into metallic iron. In the smelting of both metals, metal oxides are reduced, but the steps by which this is accomplished are somewhat different. Bauxite usually contains large amounts of impurities—iron and titanium oxides as well as silicate minerals. To remove these, bauxite is ground and mixed with lime and caustic soda, a strong alkali (NaOH). The mixture is pumped into large pressure tanks. Under direct steam, heat, and pressure, the aluminum oxides and hydroxides dissolve by reactions such as

$$Al(OH)_3 + NaOH \rightarrow Al(OH)_4^- + Na^+. \tag{10.4}$$

Most of the impurities in the bauxite remain as solids, which are removed in a series of pressure-reducing tanks and filter presses. The solution containing the dissolved alumina passes through the filters into cooling towers and from there into precipitators, where nearly pure $Al(OH)_3$ is precipitated. The $Al(OH)_3$ is filtered and washed, and is then converted to Al_2O_3 by heating in kilns at about 1,000°C.

Unlike Fe_2O_3, Al_2O_3 cannot be reduced simply by reacting with carbon at high temperatures. In 1886 an efficient method to reduce Al_2O_3 to Al was developed independently by Paul Héroult in France and by Charles Martin Hall in the United States. Hall was a recent graduate of Oberlin College in Ohio and did his pioneering work in a woodshed laboratory behind the family home near the college. Prior to that time aluminum was a rare and prized metal. Napoleon III, nephew of Napoleon Bonaparte, and emperor of France from 1852 to 1871, had an aluminum rattle made for his infant son and used the metal for his most prized eating utensils.

What was probably the first architectural use of aluminum came in 1884 with the casting of a 100-ounce piece of the metal to serve as the cap of the Washington Monument in Washington, D.C. Appropriately, the cap was first exhibited at Tiffany's in New York. It is alleged that a New Yorker bet that he could jump over the top of the Washington Monument. To prove his point, he led his mark to the elegant jewelry house, leaped nimbly over the display case, and collected his winnings—or so the story goes (ALCOA 1969).

In modern aluminum smelters Al_2O_3 is dissolved in a bath of molten cryolite, Na_3AlF_6, in large electrolytic furnaces called *pots*. These are deep, rectangular steel shells lined with carbon. Long rows of pots are connected electrically in series. Carbon anodes are suspended in each pot. High-amperage, low-voltage direct current is passed through the cryolite baths to the carbon walls of the pots. Molten aluminum accumulates at the bottom and is siphoned off periodically to be further purified or to be alloyed with other metals. The oxygen (O^{2-}) in the Al_2O_3 charge combines with the carbon anode and is released as CO_2.

Aluminum smelting is very energy intensive. The relatively high cost of the metal is not due to the cost of mining its ores but to the cost of the electricity required to produce the metal. This is such an important cost factor that bauxite is often transported long distances for smelting to places where electricity is cheap. In the United States, hydroelectric power, particularly in the Pacific Northwest and in the Tennessee Valley, used to be a major source of electricity for aluminum smelting, but today relatively little cheap "excess" hydroelectric power is available. In the future other sources of power are apt to become more important, and aluminum smelting operations will probably be moved to parts of the world where electric power is more abundant and less expensive.

10.6.3 The Smelting and Refining of Copper

Chalcopyrite, $CuFeS_2$, is the most important mineral in many copper ores. Much of the concentrate that arrives at copper smelters consists largely of chalcopyrite, and hence contains about 25% copper and about 30% sulfur. Copper concentrates are roasted in air and are mixed with what are called *flux materials* to assist in the smelting process. Roasting drives off 25%–50% of the combined sulfur as sulfur dioxide, SO_2. The roasted material is then smelted in reverberatory furnaces. During smelting, gangue is eliminated in the form of slag, more sulfur is lost as SO_2, and copper matte containing 35%–45% copper is formed. The matte is converted to blister copper (98% copper) by further oxidation and slag removal, and the metal is finally purified in an anode furnace.

The large quantities of SO_2 that are released during the smelting of copper and other sulfide concentrates set their products apart from those of the smelting of oxide concentrates. Both types of smelters suffer from the generation of large quantities of mineral dust. The development and installation of Cottrell precipitators did much to alleviate the dust problem, but the SO_2 problem has only been solved gradually. The construction in 1717 of the first copper smelter near Swansea in Wales caused considerable controversy and consternation. In order to disperse the smoke more widely, taller chimneys were built, and this has been the favored remedy for a long time. The smelter at Sudbury, Ontario, in Canada is a particularly fine example of smelters whose stack height has been increased steadily to disperse what was originally an insufferable source of local pollution. Copper smelters in the American West have been a source of public controversy since the early 1890s. Some of the earliest smelters were forced to close when irate farmers took the issue of crop damage to court. Smelter companies sometimes settled with farmers out of court simply by buying their crops or even their farms. A particularly severe problem of this type occurred near the Cerro de Pasco Corporation's smelter at La Oroya in Peru. The company purchased a good deal of the affected land, and then proceeded to decrease the SO_2 emissions from the smelter. Thereafter, the Scottish sheep farmers who had settled on the land did very well on their adopted South American range.

Environmental legislation and regulations, notably under the Clean Air Act, have now set tight limits on the permitted level of SO_2 emissions in the United States. The quantity of SO_2 released by smelters to the atmosphere has decreased very significantly, in part due to the conversion of SO_2 to sulfuric acid, in part due to the closing of smelters. Rising environmental costs have contributed to these closings, but declining domestic mine production, obsolete smelting technology, and severe competition for foreign concentrate supplies have also played a major role in reducing U.S. smelter capacity and the self-sufficiency of the U.S. mineral industry (the International Competitiveness of the U.S. Non-Ferrous Smelting Industry and the Clean Air Act, 1982).

The smelting of copper ores containing enargite (Cu_3AsS_4), of gold ores containing arsenopyrite (FeAsS), and of silver ores containing silver sulfosalts, such as tetrahedrite, $(Cu,Ag)_{10}(Fe,Zn)_2Sb_4S_{13}$, has released significant amounts of arsenic and antimony to the atmosphere. Small quantities of cadmium and mercury have been released by the smelting of zinc ores. Such environmental problems have led to the development of processes in which the ore minerals are dissolved in closed autoclaves at high pressure and temperature.

10.7 The Cost of Producing Metals

The cost of producing an ounce of gold or a ton of steel is the sum of the costs of all the steps from mineral exploration to metal refining, coupled with the taxes and royalties levied on the operations. These costs can be divided, at least roughly, into two categories. The first category includes costs proportional to the tonnage of ore required to yield a given quantity of the metal. The second category includes costs related to processing and transporting a given quantity of the metal.

The most obvious costs belonging to the first category are those of mining and milling. Many aluminum ores contain about 30% Al. Therefore 3.3 tons of ore are required to obtain one ton of metal. Many gold ores contain about 5 gm of gold per ton, and 200,000 tons of such ores are required to obtain one ton of gold. If the cost of mining a ton of aluminum ore were exactly equal to the cost of mining a ton of gold ore, the cost of mining a ton of gold would be $\frac{200,000}{3.3} = 60,000$ times greater than the cost of mining a ton of aluminum. This explains in part why gold costs so much more than aluminum. Other factors also contribute to the large difference in the cost of the two metals. The cost of developing a mine depends largely on its size. Two mines of equal size, one an aluminum mine, the other a gold mine, will produce quite different quantities of these metals. Therefore, the cost of developing a gold mine prorated per ton of gold produced will be much greater than the cost of developing an aluminum mine prorated per ton of aluminum produced.

The cost of milling scales in the same way. Since the quantity of gold ore that has to be milled to obtain one ton of gold is so very much larger than the quantity of aluminum ore that has to be milled to obtain one ton of aluminum, the milling costs per unit weight of the two metals are very different.

The second cost category includes smelting and refining. Since at that point in the production most of the gangue has been separated from the ore minerals, the percentage of many metals is similar in their mill concentrates. If taxes and royalties are levied on the value of the finished products, these costs are also independent of the ore grade. Then the cost, C, in U.S. dollars of producing a pound of a particular metal is roughly given by the expression

$$C \approx \frac{A}{G} + B, \tag{10.5}$$

where

A = the sum of the costs that depend on the ore grade (in U.S. dollars per lb of ore)
G = the grade of the ore (in lb of metal per lb of ore)
A/G = the sum of the costs that depend on the ore grade (in U.S. dollars per lb of metal)
B = the sum of the costs that are independent of the ore grade in U.S. dollars per lb of metal).

As an illustration of equation 10.5, let us consider aluminum production. Most bauxite is easy to mine and mill. In 1985 the combined cost of mining and milling a metric ton (mt) of bauxite was approximately \$15.00. In bauxite that contains 0.30 tons of Al per ton of ore, the cost of mining and milling was

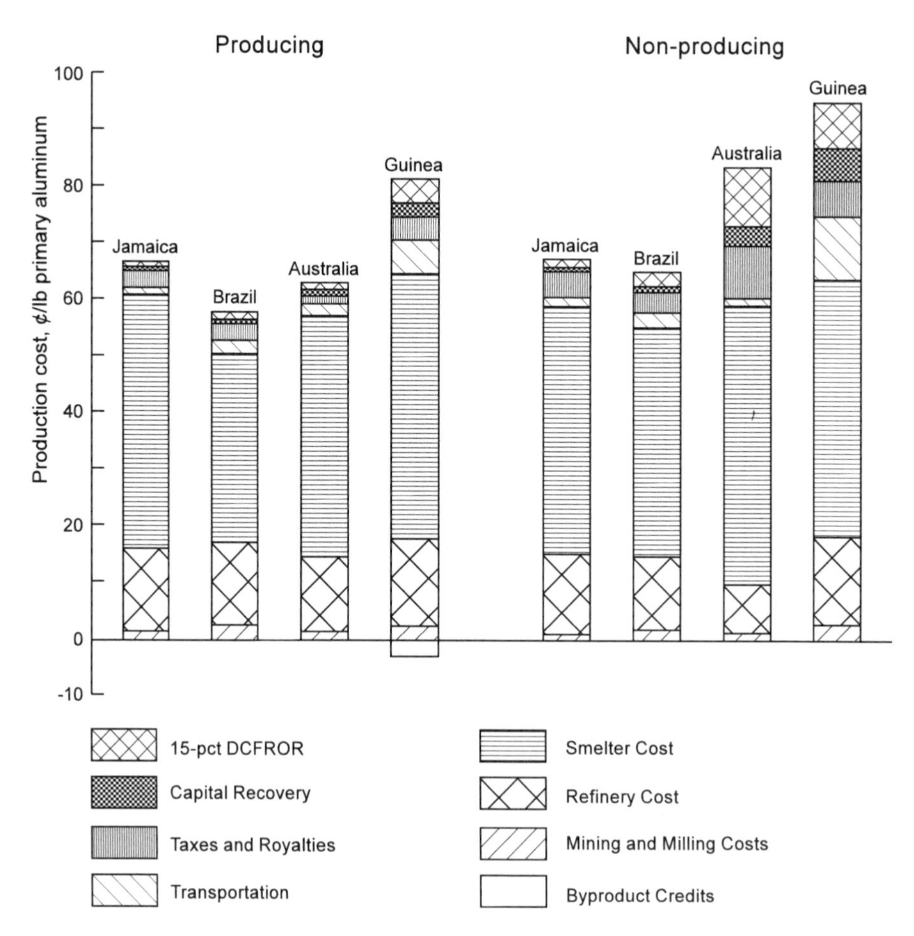

Figure 10.12. Aluminum production costs in January 1985 dollars for producing and nonproducing operations in selected market-economy countries. DCFROR = discounted cash flow rate of return. (*U.S. Bureau of Mines Bulletin* 692)

$$\frac{\$15.00/\text{ton of ore}}{0.30 \text{ tons Al/ton of ore}} = \$50/\text{mt Al}.$$

This is equivalent to 2.3 cents per pound of aluminum. As shown in figure 10.12, this cost is very small compared to the total cost of producing a pound of aluminum. Most of the difference is due to the cost of smelting and refining the metal. This cost depends on the quantity of Al_2O_3 treated, not on the quantity of ore mined to yield the Al_2O_3. The cost of smelting is particularly high for aluminum, because aluminum smelting is very energy intensive (see the previous section), but even in the production of iron and steel the cost of mining and milling iron ore is much less than that of smelting and refining the contained metal.

Gold stands at the other end of the grade spectrum. The cost of mining a ton of gold ore is not very different from the cost of mining a ton of aluminum ore, but the quantity of gold recovered from a ton of ore is so small that the cost of gold is very much greater. Many gold ores run about 4 grams of the metal per ton of ore. If it costs $15.00 to mine a ton of gold ore, mining contributes about $3.75 per gram to the cost of gold, or about $1,330 per pound. The price of gold is usually reported in dollars per troy ounce, which has a weight of 31 grams. Mining a troy ounce of gold therefore requires roughly $120, as illustrated in figure 10.13. Since 1985 the gold industry has changed substantially. Increasing amounts of ore are being mined by open pit methods, and the gold is recovered by low-cost heap leaching. In the United States and elsewhere, this has reduced

Figure 10.13. Gold production costs in January 1985 dollars for producing mines in selected market-economy countries. DCFROR = discounted cash flow rate of return. (*U.S. Bureau of Mines Bulletin* 692)

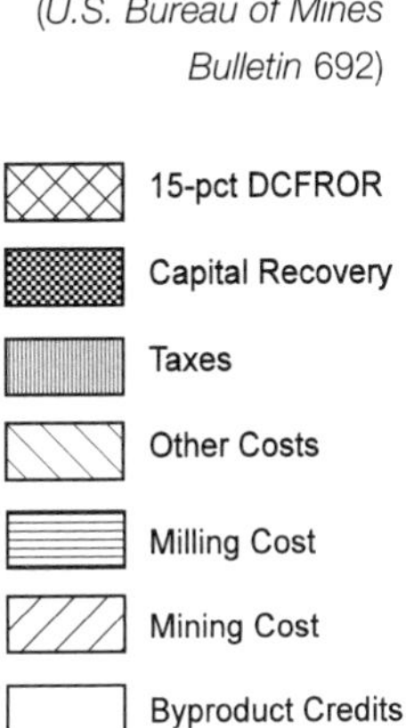

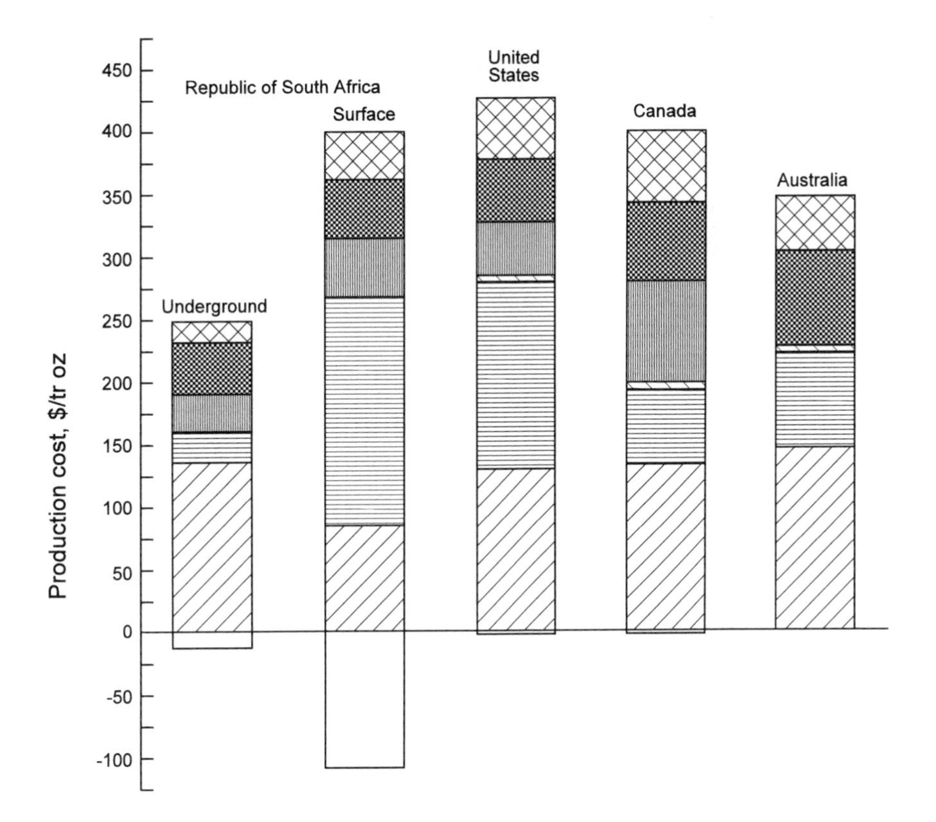

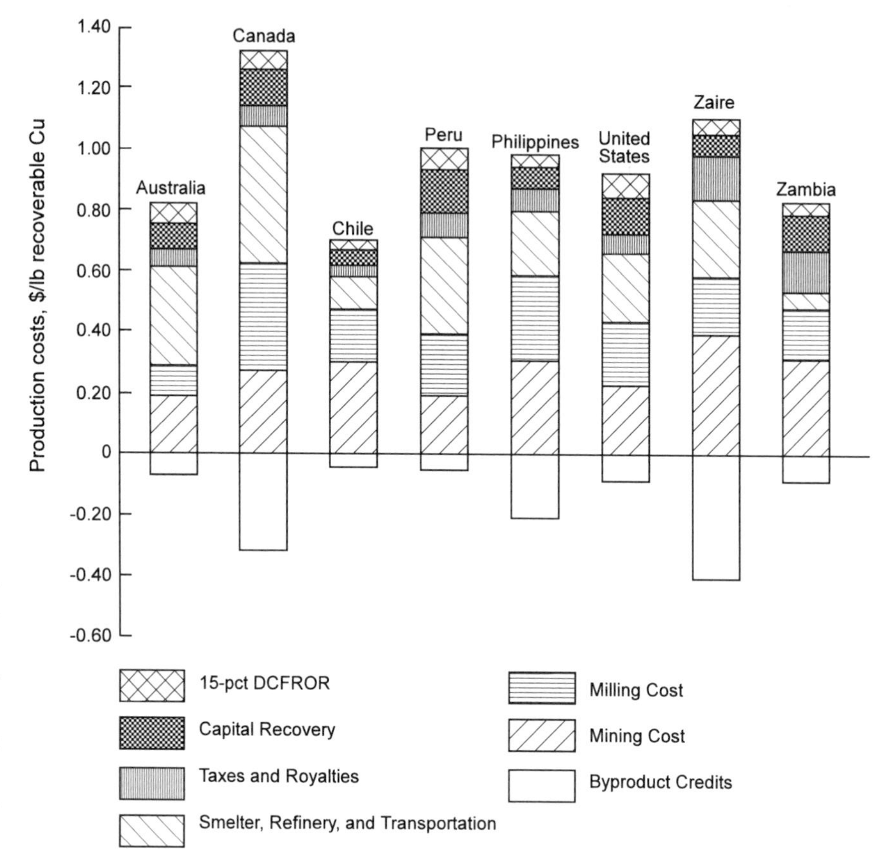

Figure 10.14. Copper production costs in January 1985 dollars for producing mines in selected market-economy countries. DCFROR = discount cash flow rate of return. (*U.S. Bureau of Mines Bulletin* 692)

the gold production costs, so that ores containing 1 gm/ton or even less can be mined for $200–$300 per ounce. Even so, the mining and milling costs of gold together dominate the cost of producing this metal, in complete contrast to the insignificance of these costs in the production of aluminum.

The grade of copper ores is intermediate between those of aluminum and gold ores. In many producing copper mines the ore grade in 1985 was about 0.8%. The cost of mining a ton of ore was approximately only $4.50, because most copper ore was obtained from open pit mines. One ton of such ore contained 16 pounds of copper. The cost of mining was therefore about twenty-eight cents per pound of copper. As shown in figure 10.14, the actual range of this cost is quite large, but it is always much greater than the cost of mining one pound of aluminum, and it is always very much smaller than the cost of mining one pound of gold.

We have shown that equation (10.5) is qualitatively correct. It is interesting that it also does fairly well in a more quantitative test. When the ore grade of a metal is very low, A/G is much greater than B; in that case

$$C \approx \frac{A}{G} \tag{10.6}$$

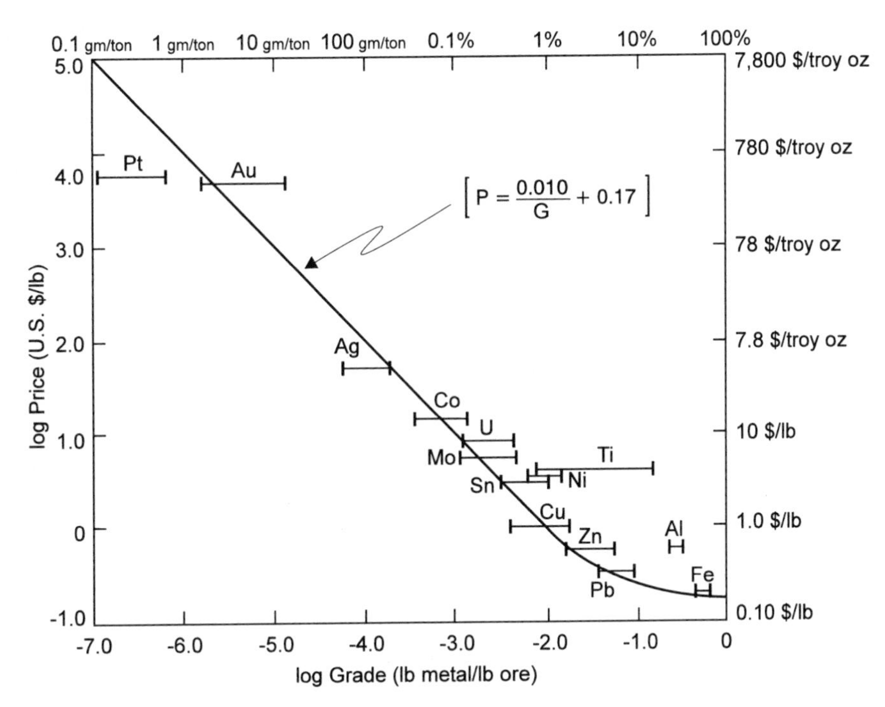

Figure 10.15. The logarithm of the price of metals in U.S. dollars per pound plotted against the logarithm of the common grade range of their ores in pounds of metal per pound of ore. (Metal price data from U.S. Bureau of Mines, *Metal Prices in the United States through 1991*, 1993; ore grade data from Cox and Singer 1986)

and

$$\log C \approx \log A - \log G. \tag{10.7}$$

Figure 10.15 is a plot of the logarithm of the price of a rather large group of metals against the logarithm of their grade. The price of many of the metals that occur in low-grade ores plots reasonably close to a single straight line. The slope of the line is about -1, as predicted by equation (10.7) for the cost-grade relationship. That so many metals fall so close to a single line indicates that in a very rough way the price-grade relation is similar to the expected cost-grade relation, and that the value of A does not vary greatly from metal to metal. It must be remembered, however, that figure 10.15 is a log-log plot, which hides a lot of variability. Some of the scatter in figure 10.15 is due to the fact that more than one metal is extracted from some ores. Many porphyry copper ores contain significant quantities of gold; many vein deposits are mined for copper, zinc, lead, and silver; and Mississippi Valley–type ores are mined for zinc and lead. The cost of mining a ton of these ores and a part of the milling costs can be prorated among the several metals. This decreases the magnitude of the parameter A for each metal and shifts their position downward in figure 10.15.

The cost of the abundant metals is determined largely by parameter B in equation (10.5). For these metals the mining costs are very much less than the cost of smelting, refining, and transportation, and hence

$$C \approx B. \tag{10.8}$$

The cost of these metals is almost independent of the grade of their ore, as suggested by the position of the data point for iron in figure 10.15. The data points for aluminum and titanium fall well above the trend, because the cost of energy to smelt the ores of these metals is greater than that of most other metals.

The data in figure 10.15 have obvious implications for the price of metals in the future. The grade of the ores of most, if not all, metals will decrease during the next century. The data in figure 10.15 suggest that modest decreases in ore grade will have little effect on the price of the abundant metals but will tend to increase the price of the rare metals. Whether these inferences turn out to be correct will depend on future changes in the values of parameters *A* and *B*.

10.8 The Price of Mineral Commodities, Past and Present

The demand for mineral commodities is the major factor that links their cost to their price. Figure 10.16 illustrates this principle for the cost and the price of copper in 1985. The cost of producing copper in a few mining areas was less than \$0.40/lb. If the demand for copper in 1985 had not exceeded the supply from these sources, the price of copper would have been quite low. Since the demand exceeded the supply, copper from higher-cost operations was also used. The data in figure 10.16 show that copper from mines with an operating cost as high as \$0.67/lb had to be used to satisfy a 1985 demand of 4.7×10^6 metric tons (mt) of copper. If the demand had been as high as 6.0×10^6 mt/yr, copper would have had to be used from mines where operating costs amounted to \$1.40/lb. Put another way, at a demand of 4.7×10^6 mt in 1985 all of the mines where the operating costs were less than \$0.67/lb of copper could, at least in principle, have made a profit. At a demand of 6.0×10^6 mt all of the mines where operating costs were less than \$1.40/lb of copper could, at least in principle, have made a profit. The profitability or unprofitability of many copper mines was therefore strongly influenced by the demand for copper.

The actual cost of producing copper is always greater than the operating costs alone. The recovery costs of capital expended in developing a mine, as well as taxes and royalties, have to be included. Then there is the return on investment that mining companies need, or feel they need. In the upper curve of figure 10.16 the recovery costs of capital, taxes, and royalties have been added to the operating cost. When this is done, 3.5×10^6 mt of copper were recoverable at a market price of \$0.67/lb. To recover 4.7×10^6 mt of copper/yr required a market price of about \$0.85/lb. If the cost of a 15% return on investment is added in, the cost of producing much of the world's copper in 1985 is increased by another \$0.10/lb.

These observations say a good deal about the vulnerability of mining enterprises. Mines where the ore is so rich that the production costs are much less than the market price of their metals are reasonably insulated from swings in commodity prices. Mines where the production costs are close to the market price are very vulnerable. Unfortunately, many of the copper mines in the American West are in the latter position.

The price of most metals has varied considerably and rapidly during the twentieth century. Figure 10.17 shows how the price of copper has changed during the course of the century. The lower curve is the price of copper in current dollars, i.e., the actual price in each successive year. The upper curve is the price of the metal in 1987 dollars. This curve allows a better estimate of changes in the "real"

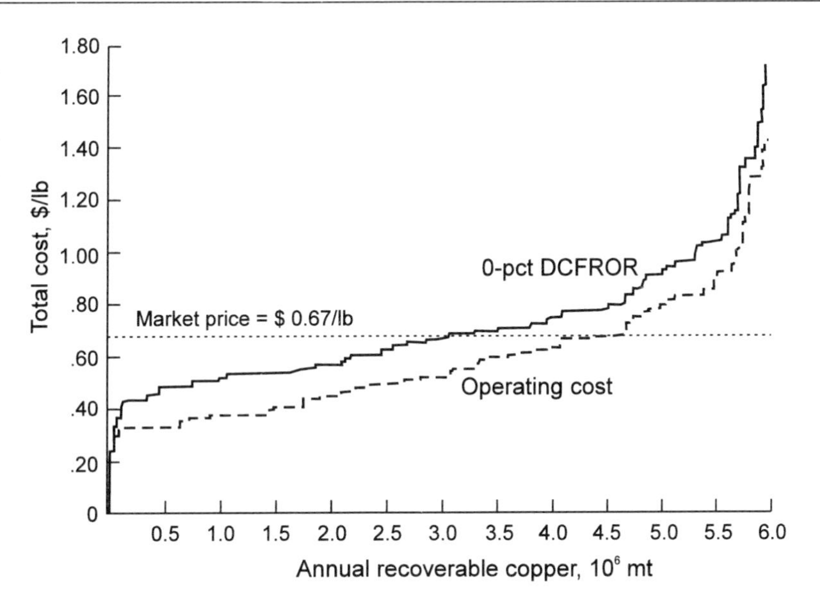

Figure 10.16. 1985 copper capacity from producing mines in market-economy countries (MECs), prices in January 1985 dollars. (*U.S. Bureau of Mines Bulletin* 692)

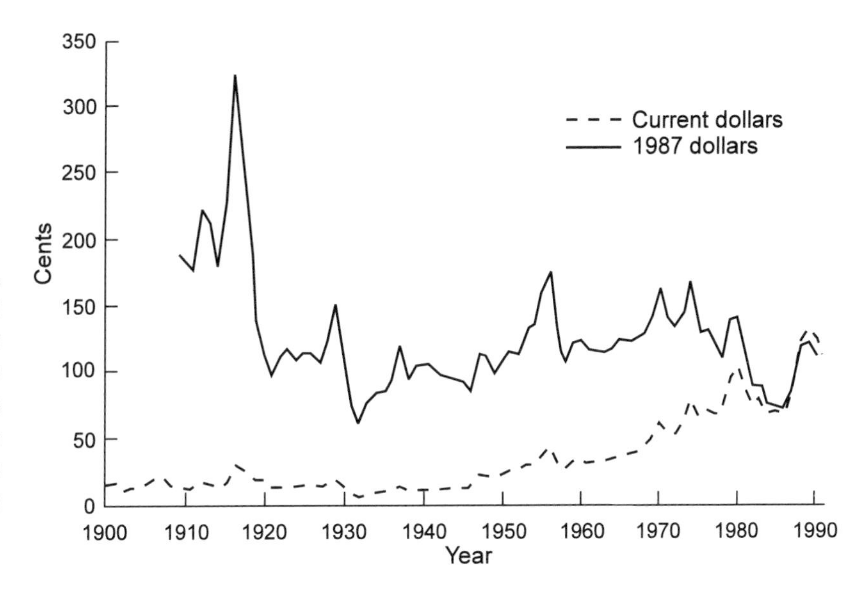

Figure 10.17. The average annual price of copper in cents/lb between 1900 and 1991. (U.S. Bureau of Mines, *Metal Prices in the United States through 1991*, 1993)

price of the metal, because it has removed much of the effect of changes in the value of the dollar during the twentieth century.

Since 1973 the price of copper in 1987 dollars has varied by more than a factor of two. Swings of this magnitude are common for many metals. On a decadal timescale such changes are often due to the swings of the business cycle. During upswings demand is high, exploration is intense, operating mines are expanded, new mines are put into operation, and a general spirit of euphoria prevails. During downturns demand dims, prices fall, and mines become idle. The "boom and bust" cycles of the mining and of the oil and gas industries are legendary.

Table 10.1. Some Important Producer Associations

Years	*Association*
1960–	OPEC: Organization of Petroleum Exporting Countries
1967–	CIPEC: Conseil Intergouvernmental des Pays Exportateurs de Cuivre
1974–	IBA: International Bauxite Association
1974–	APEF: Association des Pays Exportateurs de Minerai de Fer
1975–	PTA: Primary Tungsten Association
1975–	UI: Uranium Association
1950–1985:	ITC: International Tin Council

Source: Gocht, Zantop, and Eggers 1988.

Attempts to control the price of mineral commodities have led to the formation of cartels and a variety of industrial associations. Some of these are listed in table 10.1. OPEC has been the most visible cartel since the early 1970s. It effected a dramatic increase in the price of oil. However, this increase stimulated energy conservation, substitution, and, most important, the search for more oil outside the OPEC domain. In the 1980s OPEC's share of the world market decreased significantly, as did its ability to control the price of oil. Most metal cartels have fared even worse. The collapse of the tin cartel in 1985 illustrates the difficulty of maintaining the price of a mineral commodity in the face of a dramatic decrease in demand. The only cartel that has really stood the test of time in the twentieth century is the De Beers diamond cartel.

The price history of metals has been complicated not only by the existence of cartels but also by stockpiling. The mineral industry maintains stockpiles at all levels of production and trade. For metals, commercial stockpiles are usual in mines, in smelting plants, in the processing industry, and in the licensed warehouses of the metal exchanges. Governments tend to build stockpiles to assure material supplies in times of national emergency. In the United States, stockpiles of critical materials are maintained to decrease the country's dependence on foreign sources of supply. Fluctuations in the size of these stockpiles have had a significant impact on metal prices. Prices have been stimulated during times of buildup and have been depressed in times of sell-off.

Once in a while the actions of a single person or of a small group can affect the price of a mineral commodity quite drastically. In 1979–80, the Hunt brothers of Dallas, Texas, tried to "corner" the silver market. Figure 10.18 shows that they managed to drive up the price of the metal dramatically, but only for a short period of time. Their bid to control the silver market was unsuccessful in the long run, and the price of silver in constant dollars has since returned to values close to the 1970 price.

Long-lasting effects on the price of silver and gold have been exercised by their use as stores of value. During the late eighteenth, the nineteenth, and part of the twentieth centuries the prices of gold and silver were pegged to the value of the dollar. This system of stabilizing currencies was finally abandoned, but both metals, especially gold, still carry a special cachet as possessions whose value is more assured than that of paper currencies. No one has ever said of gold, "It's not worth the paper it's printed on." On a timescale of millennia the value of the precious metals has decreased very significantly as measured by the number of hours an average wage earner has had to work to earn an ounce of gold or silver.

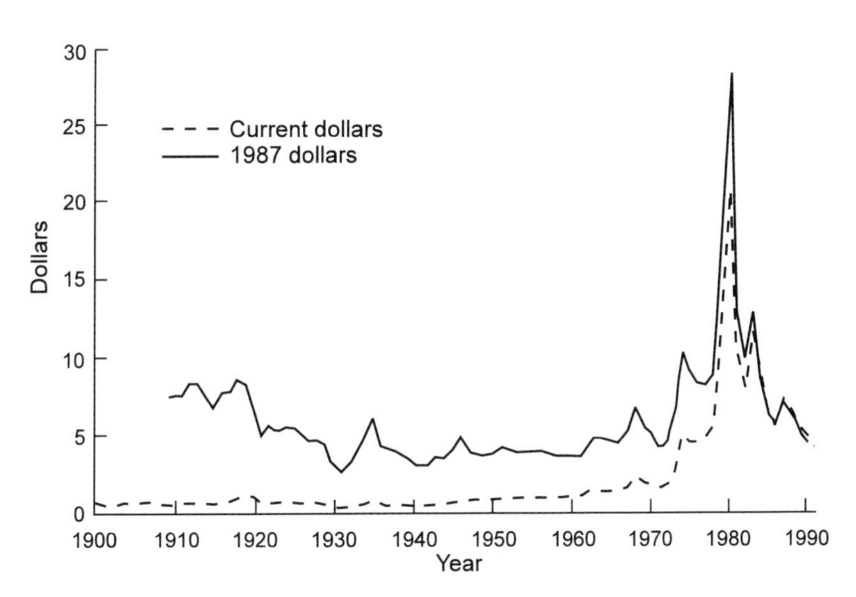

Figure 10.18. The average annual price of silver in dollars per troy ounce between 1900 and 1991. (*U.S. Bureau of Mines, Metal Prices in the United States through 1991*, 1993)

During the twentieth century the price of silver in 1987 dollars has remained roughly constant, although there has been the one major price excursion described above.

Table 10.2. Average Grade of Copper Ore Mined in the United States, 1880–1985

1880	3.0%
1889	3.3
1902	2.7
1906–10	2.1
1911–20	1.7
1921–30	1.6
1931–40	1.6
1941–50	1.0
1951–56	0.8
1985	0.55

Source: Data for 1880–1956 compiled by O.C. Herfindahl (Brooks 1974); 1985 data from *U.S. Bureau of Mines Bulletin* 692.

The price of copper in constant dollars decreased rapidly during the first 20 years of the twentieth century (see fig. 10.17). Since then its price has fluctuated considerably, but at present the price of copper in constant dollars is essentially the same as it was in 1920. This record is rather amazing, because the average grade of copper ore mined in the United States decreased by a factor of three between 1920 and 1985 (see table 10.2). The history of copper prices between 1880 and 1920 is even more spectacular. During those 40 years the average grade of U.S. copper ore dropped by a factor of nearly two, while the price of the metal in constant dollars dropped by a factor of three. The decrease in the price of copper in the face of decreasing ore grades was due to major improvements in the technology of mining and milling. Copper mining shifted from underground mining of rich vein ores to surface mining of very large, relatively low-grade porphyry copper ores. Economies of scale and improved methods of mining and transportation reduced the mining costs of copper. The invention of flotation techniques for preparing mill concentrates of copper minerals did much of the rest.

The economy is engaged in a never-ending struggle between advances that enhance labor productivity and the exhaustion of resources. Over the last two centuries, technology has been the clear winner in the production of nearly all metals (Gordon et al. 1987). Aluminum is a particularly important example. The development of new milling and smelting techniques in the 1880s lowered the price of this element from among those of the precious metals to one allowing it to become a mainstay of industrial society. Figure 10.19 shows how dramatically the price of aluminum in constant dollars has dropped during the twentieth century.

Not all of the drops in the price of metals are due to improvements in technology. Substitutions of one metal for another, and of a nonmetal for a metal, have

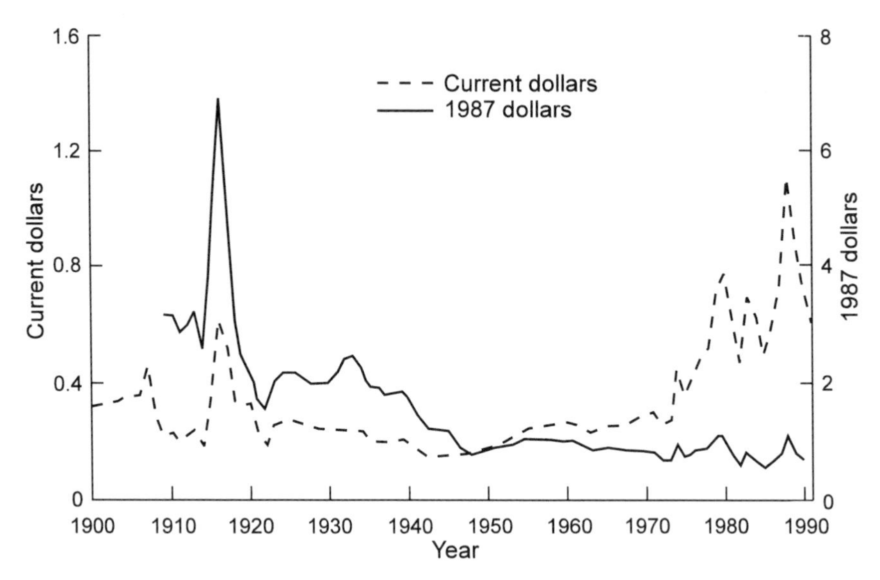

Figure 10.19. The average annual price of aluminum in dollars per pound between 1900 and 1991. (U.S. Bureau of Mines, *Metal Prices in the United States through 1991*, 1993)

been responsible for some price drops. The substitution of aluminum for tin, for instance, has led to a major decrease in the price of tin. Environmental concerns have also had an impact, particularly on the use of lead, mercury, and uranium.

10.9 The Price of Mineral Commodities in the Future

The price of mineral commodities passes through three stages that reflect progressive changes in their cost (Petersen and Maxwell 1979). During the first stage, technological improvements in mining and/or metallurgy are decisive, and the price of mineral commodities in real terms falls progressively. During the second stage, the effect of technological improvements on the cost of mineral commodities is balanced by the effects of decreasing ore grades and increasingly difficult mining. During the third stage, cost decreases due to improvements in technology cannot keep up with cost increases due to increasing scarcity. All, or nearly all, metals are currently in stage 1 or in stage 2 of their life history. Aluminum is a particularly important example of a stage 1 metal; iron and copper are fine examples of stage 2 metals.

The price of mineral commodities in the future will depend on the timing of their transition to stage 3. This transition will depend, at least in part, on their abundance. Iron, aluminum, and magnesium are so abundant in the Earth's crust and/or in seawater that changes in the grade of their ores will probably be slight during the next century. Shifts of these metals into stage 3 will largely be the result of increases in the cost of milling, smelting, and refining. The factor most liable to increase these costs is energy. The last several chapters have touched on this matter briefly, and it is an important theme in chapter 11, where we will argue that the cost of energy is not apt to change in a major way during the twenty-first century. Since so much of the money that is spent on mineral commodities is spent on iron and aluminum, the outlook for the price of the most important metals is reasonably rosy.

Most of the other metals belong to the group of minor and scarce elements. The grade of their ores has decreased during the present century and will surely continue to do so in the future. Copper is a particularly interesting and important member of this group. As shown in table 10.2, the average grade of copper ores mined in the United States dropped from about 3% to 0.55% between 1880 and 1985. Very large porphyries with copper grades between 0.5% and 0.1% Cu are known, and these will probably be mined sometime during the next two centuries (Gordon et al. 1987). It is not clear how long cost reductions due to improvements in mining and mineral technology can keep up with the increase in costs due to the progressive decrease of ore grades. Other types of copper ores, which are not economically attractive today, may replace porphyries as major sources of copper. Among these, metal-rich, black, carbonaceous shales look quite promising. In addition to copper, these often contain significant concentrations of uranium, molybdenum, vanadium, zinc, and lead. Some rather odd black shales are strongly enriched in the platinum group elements, and the carbon content of a few black shales may itself turn out to be of interest as a by-product.

If the price of the scarce metals begins to rise significantly during the twenty-first century, some will surely share the fate of tin in the twentieth century: they will be replaced. Replacement does not need to be by another metal, as shown by the replacement of copper in telephone wires by glass fiber optics. Ceramics and plastics may well replace metals in a number of applications, and it will be interesting to see whether high-temperature superconductors will revolutionize electricity transmission during the next century. Even if none of these factors serve to moderate price rises, increased recycling surely will. It seems likely that we will gradually come to depend more on the superabundant and less on the scarce mineral commodities, and that the key industrial processes will adapt to these new conditions. As long as society does not severely hamper substitution and recycling, the gradual decrease in the importance of the scarce elements should not act as a significant drag on the world's economy.

References

Agricola, G. 1556. *De Re Metallica*. Trans. by H. C. Hoover and L. H. Hoover and republished in 1950. Dover Publications, New York.

ALCOA. 1969. *Aluminum*. Aluminum Company of America.

Brooks, D. B., ed. 1974. *Resource Economics: Selected Works of O. C. Herfindahl*. Resources for the Future, Washington, D.C.

Committee on Mineral Resources and the Environment (COMRATE). 1975. *Mineral Resources and the Environment*. National Research Council, National Academy of Sciences, Washington, D.C.

Cox, D. P., and Singer, D. A., eds. 1986. Mineral Deposit Models. U.S. Geol. Survey Bull. 1693.

Craig, J. R., Vaughan, D. J., and Skinner, B. J. 1988. *Resources of the Earth*. Prentice Hall, Englewood Cliffs, N.J.

Evans, A. M. 1993. *Ore Geology and Industrial Minerals*. 3d ed. Blackwell Scientific, Boston.

Everest Consulting Assoc. Inc. and CRU Consultants, Inc. 1982. *The International Competitiveness of the U.S. Non-Ferrous Smelting Industry and the Clean Air Act.*

Gocht, W., Zantop, H., and Eggert, R. G. 1988. *International Mineral Economics: Mineral Exploration, Mine Valuation, Mineral Markets, International Mineral Policies*. Springer-Verlag, New York.

Gordon, R. B., Koopmans, T. C., Nordhaus, T. C., and Skinner, B. J. 1987. *Toward a New*

Iron Age? Quantitative Modeling of Resource Exhaustion. Harvard University Press, Cambridge, Mass.

Jackson, J. H. 1970. *Anybody's Gold: The Story of California's Mining Towns.* Chronicle Books, San Francisco.

Kelly, E. G., and Spottiswood, D. J. 1982. *Introduction to Mineral Processing.* John Wiley, New York.

Lewis, O. 1947. *Silver Kings; The Lives and Times of Mackay, Fair, Flood, and O'Brien, Lords of the Nevada Comstock Lode.* Knopf, New York.

Llewellyn, R. 1939. *How Green Was My Valley.* M. Joseph, London.

Long, P. 1989. *Where the Sun Never Shines: A History of America's Bloody Coal Industry.* Paragon House, New York.

Miernyk, W. H. 1986. *The Coal Mining Industry in the United States.* Research Series No. 83, International Institute for Labour Studies, Geneva.

Moniz, B. J. 1992. *Metallurgy.* American Technical Publishers, Homewood, Ill.

Peters, W. C. 1987. *Exploration and Mining Geology.* 2d ed. John Wiley, New York.

Petersen, U., and Maxwell, R. S. 1979. Historical Mineral Production and Price Trends. *Mining Engineering* (January): 25–34.

President's Commission on Coal, The. 1980. *The American Coal Miner: A Report on Community and Living Conditions in the Coalfields.* J. D. Rockefeller IV, Chairman. Washington, D.C.

Ross, M. 1981. The Geologic Occurrences and Health Hazards of Amphibole and Serpentine Asbestos. In *Amphiboles and Other Hydrous Pyriboles—Mineralogy*, pp. 279–323, vol. 9A, of *Reviews in Mineralogy*, ed. D. R. Veblen. Mineralogical Society of America.

Ross, M. 1982. A Survey of Asbestos-Related Disease in Trades and Mining Occupations and in Factory and Mining Communities as a Means of Predicting Health Risks of Nonoccupational Exposure to Fibrous Materials. U.S. Geol. Survey Open File Report 82–0745.

Smith, G. H. 1966. *The History of the Comstock Lode, 1850–1920.* lst ed. (rev.). Geology and Mining Series No. 37, The University of Nevada Bulletin.

Solid-Earth Sciences and Society. 1993. *Committee on Status and Research Objectives in the Solid-Earth Sciences: A Critical Assessment.* Commission on Geosciences, Environment and Resources, National Research Council, National Academy of Sciences, Washington, D.C.

U.S. Bureau of Mines. 1993. *Mineral Commodity Summaries 1993.*

U.S. Bureau of Mines, Branch of Metals and Branch of Industrial Minerals. 1993. *Metal Prices in the United States through 1991.*

U.S. Bureau of Mines, Staff. 1987. *An Appraisal of Mineral Availability for 34 Commodities.* U.S. Bureau of Mines Bull. 692.

11 Energy Options

11.1 Introduction: Choosing from the Energy Menu

Some pessimists have proposed that we will shortly run out of energy, and that before long we will be freezing in the dark. This prognosis is clearly wrong. The data in chapter 8 showed that the reserves of fossil fuels alone can see the world through the next several hundred years. The data in chapter 9 showed that world reserves of uranium are very large and could, depending on the choice of nuclear power plants, supply the world energy demand for many centuries. Hydropower is currently underused. Only a minute fraction of the available solar energy is being tapped, and a number of other sources of renewable energy are in the same state. The world is not lacking in energy. The problem, happily, is one of choice: which of the many energy options should we adopt? More specifically, what mix of energy sources is optimal for the several parts of the Earth and for the planet as a whole?

Historically, the agenda for energy providers has been to supply energy in the forms required to support economic growth and social progress. They have invested in projects and programs to deliver energy and to receive consumer payments for underwriting their capital and running expenses. Consumers have set specifications by their needs and value systems, and have sought to pay the lowest possible price for energy. Government has served as a referee, auditor, and regulator of this process, and has intervened to set ground rules, to provide incentives, and to mitigate the consequences of disasters (Kieschnick and Helm 1990). The success of this enterprise has created an interesting dilemma, because the present modes of energy production can create serious environmental problems. Energy planning for the future must consider not only traditional economic and political aims and constraints, but must also identify those energy sources whose use will create the least adverse impact on the environment. This turns out to be difficult, because predicting the environmental impacts of energy sources is an uncertain enterprise. The long-term effects of fossil fuel burning on climate and on the entire world ecosystem are not known. They could either be trivial or severe. We cannot be sure whether the prophets of doom are, to mix metaphors, Chicken Littles crying wolf, or whether we need to pay attention to Machiavelli's counsel:

> It happens then as it does to physicians in the treatment of consumption, which in the commencement is easy to cure and difficult to understand; but when it has neither been discovered in due time nor treated upon a proper principle, it becomes easy to understand and difficult to cure. The same thing happens in state affairs; by foreseeing them at a distance . . . the evils which might arise from them are soon cured; but when, from want of foresight, they are suffered to increase to such a height that they are perceptible to everyone, there is no longer any remedy. [*The Prince*, chapter 3]

11.2 Fossil Fuels

Solar energy stored by plants has been and is by far the largest source of energy for the world's population today. Before the year 1875, biomass, largely in the form of wood, was the dominant energy source (fig. 11.1). Coal dominated the energy picture from 1875 until about 1965, when it was replaced by oil and natural gas plant liquids (NGPL). In 1990 oil plus NGPL accounted for about 36% of the world's primary energy production, coal for 25%, natural gas for 19%, and all other energy sources combined for about 20%. Although wood and other biomass fuels (biofuels) account for less than 10% of the world's energy supply, they still are the main source of energy for about half the world's population (Golob and Brus 1993).

The dominant position of the fossil fuels among the array of energy sources is largely due to economics. However, environmental concerns now cast a shadow over their future. These concerns include those for the environmental effects of fossil fuel acquisition and distribution, and for the consequences of fossil fuel use.

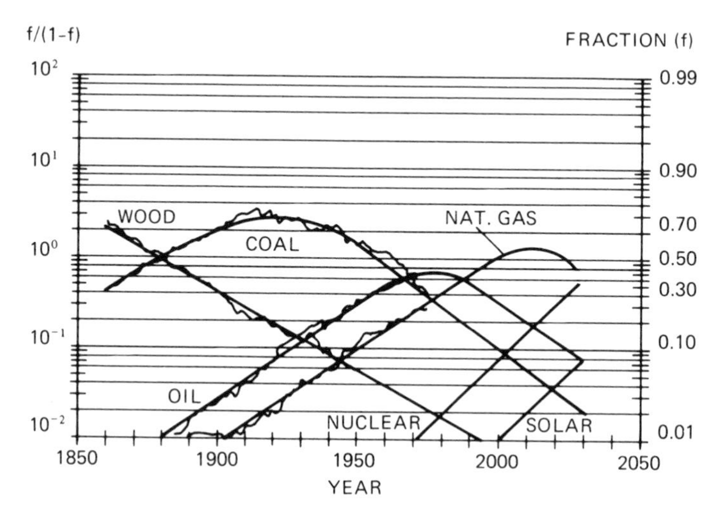

Figure 11.1. The history of the global use of major sources of energy. In 1875 coal replaced wood as the major energy source; in 1965 oil replaced coal as the major energy source. (Data largely from Häfele et al. 1981)

11.2.1 Environmental Effects of Obtaining and Distributing Fossil Fuels

Mining coal, particularly underground mining, is still a hazardous occupation. The health effects of coal mining and the frequency of deaths due to mining accidents were discussed in chapter 9. Drilling for oil and gas is not as dangerous, but the consequences of large blow-outs on oil rigs at sea are very severe for personnel on the rigs and for coastal areas where oil spills come ashore. The demand for restrictions on drilling for oil in U.S. coastal waters has largely been driven by fears of widespread coastal fouling by oil and by the discharge of brines that are produced together with oil.

Some oil field disasters are man-made. The most spectacular and pernicious was the firing of 732 oil wells in Kuwait by Iraqi invasion forces in August 1990. Air pollution was intense, large areas were covered with oil, and roughly 8 million barrels of oil spilled into the coastal and marine environments of the western Gulf. The slow, counterclockwise currents of the Gulf moved the oil along the eastern coast of the Arabian Peninsula, endangering marine life, polluting hundreds of miles of coastline, and threatening water desalination plants (El-Baz 1992).

Things could have been worse. Early predictions of a global meteorological disaster turned out to be wrong. The dire forecasts of a worldwide "nuclear winter" that would block out the sun's rays for months did not materialize. Thanks to unprecedented efforts by twenty-seven international teams of firefighters using a wide range of traditional and highly innovative techniques, the last of the oil-well fires was capped in November 1991. The cost of the well control effort was about $1.5 billion. It is a measure of the enormous oil wealth of Kuwait that less than 2% of the country's recoverable reserves of about 96 billion barrels of oil were burned off, and that at a production capacity of 1.5 million barrels per day it would take only about 50 days of production to pay for the entire clean-up costs of Iraqi ecoterrorism (World Oil 1992).

A great deal of oil is produced from a very small part of the Earth, and oil tankers transport very large quantities of the "liquid gold" from oil-producing to oil-importing countries. Although only about 1/500 of 1% of the oil shipped through U.S. waters is spilled en route, some tanker accidents have been truly spectacular and have raised concerns about the viability of the entire enterprise. Figure 11.2 shows the tonnage of crude oil and petroleum products that passed through U.S. ports in 1988. About 9,000 tons of crude oil and petroleum products are spilled annually in U.S. waters. Spills greater than 30 tons, or about 10,000 gallons, comprise less than 3% of these events, but they account for nearly 95% of the accidental spillage, and they can be devastating. In March 1989 the tanker *Exxon Valdez* grounded in Prince William Sound, Alaska, shortly after leaving the port of Valdez. Nearly 35,600 tons (11 million gal) of oil were spilled. Although this was only 20% of the ship's cargo, the environmental effects were very severe, and some $2 billion were spent on the cleanup. In 1994 Exxon was asked to pay $5 billion in punitive damages for the accident. The *Valdez* spill was a watershed event for the petroleum industry. The climate for domestic offshore drilling changed completely, and the trend toward international exploration by U.S. companies was hastened substantially.

The accident also renewed the long-standing debate over the need for more spill-resistant tank vessel designs, particularly the use of double-hulled vessels in petroleum transport, which will probably prevent half of the accident-related oil

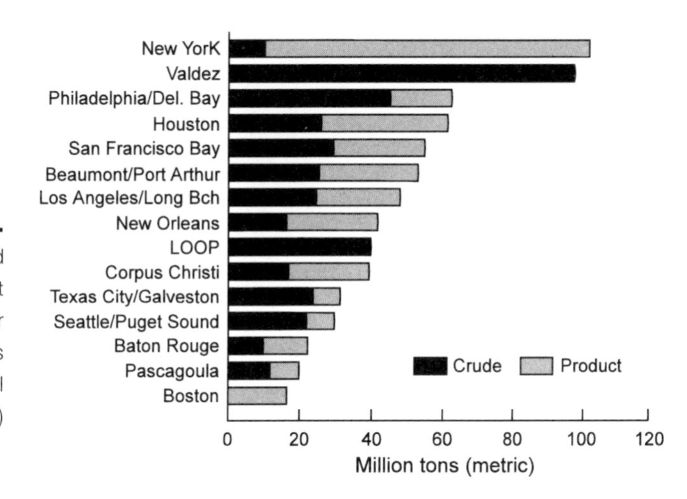

Figure 11.2. Volume of crude oil and petroleum products that passed through major U.S. port complexes in 1988. (National Research Council 1991)

spills at sea. The Oil Pollution Act of 1990 now requires that all new tank vessels operating in U.S. waters or the Exclusive Economic Zone (EEZ) must be fitted with double hulls. In the spring of 1994 the *Eagle*, the first double-hulled, very large crude-oil carrier to be owned by an oil company finally joined the tanker fleet of the Mobil Corporation. Existing single-hull tank vessels are permitted to operate until January 1, 2010. The cost of converting the world's fleet of oil tankers to double-hulled ships is very large, and transport costs are estimated to rise by some $700 million annually. However, this will translate into an incremental cost of less than two cents per gallon (National Research Council 1991).

11.2.2 *Environmental Effects of Using Fossil Fuels*

A very large number of compounds are produced during the burning of fossil fuels. Quantitatively the most important is, of course, CO_2, the gas produced by the complete burning of carbon and hydrocarbons. In simplified form this burning can be written as

$$\underset{\text{coal}}{C} + O_2 \rightarrow CO_2 \tag{11.1}$$

$$\underset{\text{natural gas}}{CH_4} + 2O_2 \rightarrow CO_2 + 2H_2O \tag{11.2}$$

$$\underset{\text{oil}}{C_nH_{2n+2}} + \frac{(3n+1)}{2}O_2 \rightarrow nCO_2 + (n+1)H_2O. \tag{11.3}$$

Currently about 6×10^{15} gm (6 Gt) of carbon are burned annually, and these add about 22×10^{15} gm CO_2/yr to the atmosphere. Among the minor gases produced during fossil fuel burning, carbon monoxide (CO), sulfur dioxide (SO_2), several nitrogen oxides (NO_x), and volatile organic carbon compounds (VOCs) are the most important. Suspended particulate matter is also a frequent and very undesirable by-product of fossil fuel burning. The quantity of these by-products is often reported in units of teragrams (Tg), where 1 Tg = 10^{12} gm. The lifetime of many of these compounds in the atmosphere is relatively short. Some, like SO_2, are removed in times on the order of days; their effect on the environment therefore tends to be local (100 km) or regional (1,000 km) rather than global. CO_2, on the other hand, has a residence time in the atmosphere that is much longer than the mixing time of the atmosphere. The CO_2 content of the atmosphere is therefore changing globally, and will continue to increase as long as fossil fuels are burned at anywhere near the present rate. Since the behavior of CO_2 and the potential environmental problems associated with its buildup in the atmosphere are quite different from those of the minor and trace products of fossil fuel burning, we will discuss these two classes of pollutants separately.

The Effects of Minor and Trace Products of Fossil Fuel Burning

Figure 11.3 shows the history of many important emissions that went into the atmosphere above the United States. Figure 11.3a summarizes the history of emission rates since 1900 for SO_x (mainly SO_2), volatile organic carbon compounds, and NO_x (the mix of nitrogen oxides N_2O, NO, NO_2, and N_2O_5). Figure 11.3b shows the emission record since 1970 for CO, total solid particulates

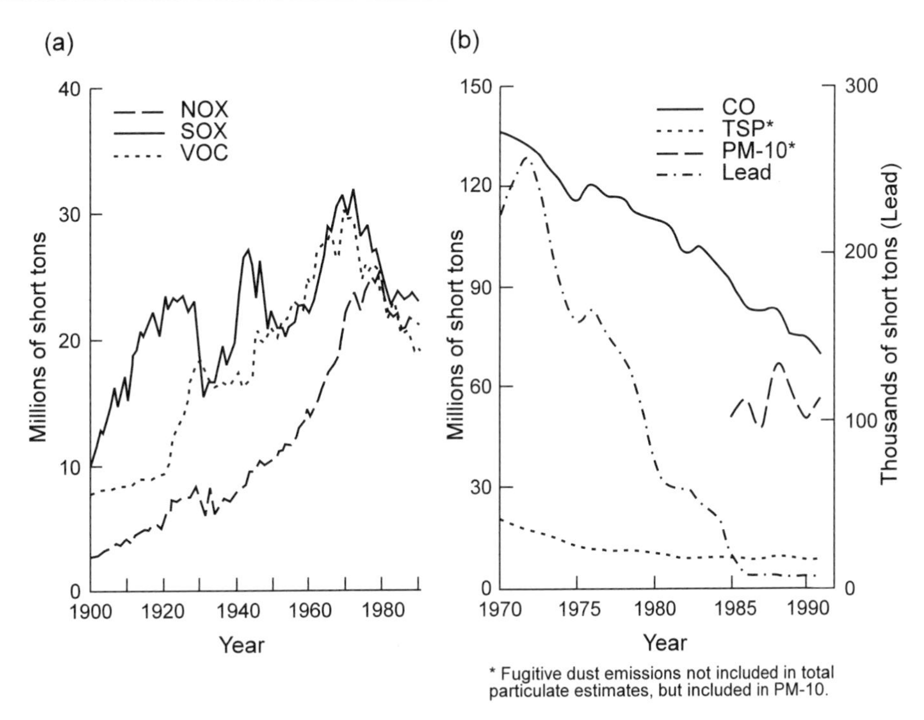

Figure 11.3. Trends of important gaseous and particulate emissions in the United States (*National Air Pollutant Emission Estimates 1990–1991*, EPA 454/R-92–013). (a) Emission rates of nitrogen oxides (NO_x), sulfur oxides (SO_x), and volatile organic compounds (VOC) between 1900 and 1991; (b) emission rates of carbon monoxide (CO), total solid particulates (TSP), particulate matter less than 10 μm in diameter (PM-10), and lead between 1970 and 1990.

(TSP), a newly defined category of particulate matter (PM-10), and lead, which was connected to the others mainly by the burning of leaded gasoline. Broadly speaking, these emission trends can be divided into a pre-1970 and a post-1970 period. Between 1900 and 1970, emissions generally rose. The increases were somewhat fitful, particularly that of the sulfur oxides. The effects of the Great Depression in the 1930s and of World War II in the first half of the 1940s are easy to spot. Since 1970, emission rates have decreased, largely as a result of the implementation of federal regulations. The role of the federal government in the regulation of air pollution began when Congress adopted the Air Pollution Control Act of 1955. This was followed by the Clean Air Act of 1963 and the Air Quality Act of 1967. Congress established a dominant federal presence in the regulation of air pollution with its 1970 legislation. Complex provisions were added in 1977, and attempts were made throughout the 1980s to add further amendments to the Clean Air Act. A major reauthorization measure for the Clean Air Act was signed into law in 1990. The new provisions changed much of the existing regulatory framework and gave a clear signal that Congress was responding to a very strong public mandate to improve the nation's air quality. Figure 11.4 compares the U.S. emission rate of major pollutants in 1970 to those in 1991. Progress has been rather uneven. Lead emissions have virtually disappeared, CO and particulate emissions have been halved, and SO_x emissions have decreased by about 25%. However, NO_x emissions have changed very little.

Ozone, O_3, an important pollutant in the generation of smog, is produced photochemically from other pollutants in the lower atmosphere. Its presence there is both a nuisance and a considerable health hazard. Somewhat paradoxically, we are destroying ozone in the stratosphere where it serves as a major sun screen for solar ultraviolet radiation (see chapter 12), while we are generating

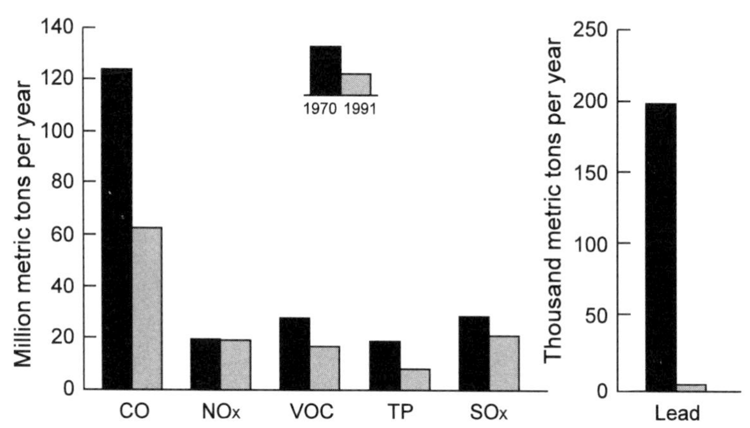

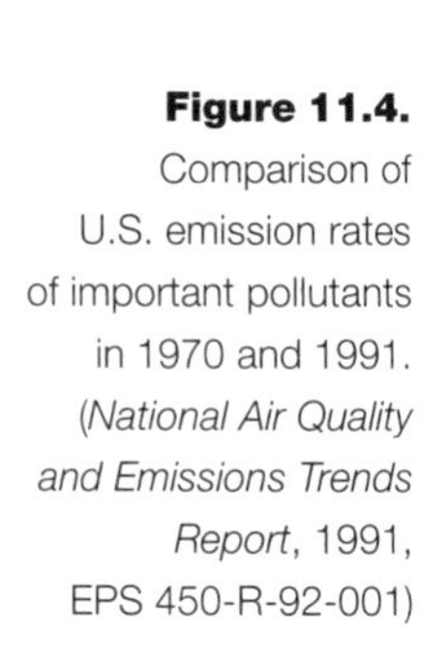

Figure 11.4. Comparison of U.S. emission rates of important pollutants in 1970 and 1991. (*National Air Quality and Emissions Trends Report*, 1991, EPS 450-R-92-001)

ozone in the troposphere where it is a major pollutant. Some of the major O_3^- forming reactions in the troposphere are summarized in figure 11.5. Volatile organic compounds (VOCs) are represented by the symbol RH, where R is an organic group. In the simplest case, where the VOC is methane (CH_4), R is CH_3. VOCs react with hydroxyl radicals (OH) to form organic radicals

Figure 11.5. Some photochemical reactions related to air pollution (from *Rethinking the Ozone Problem in Urban and Regional Air Pollution*, 1991). M is a molecule that acts as a catalyst for reactions in the atmosphere.

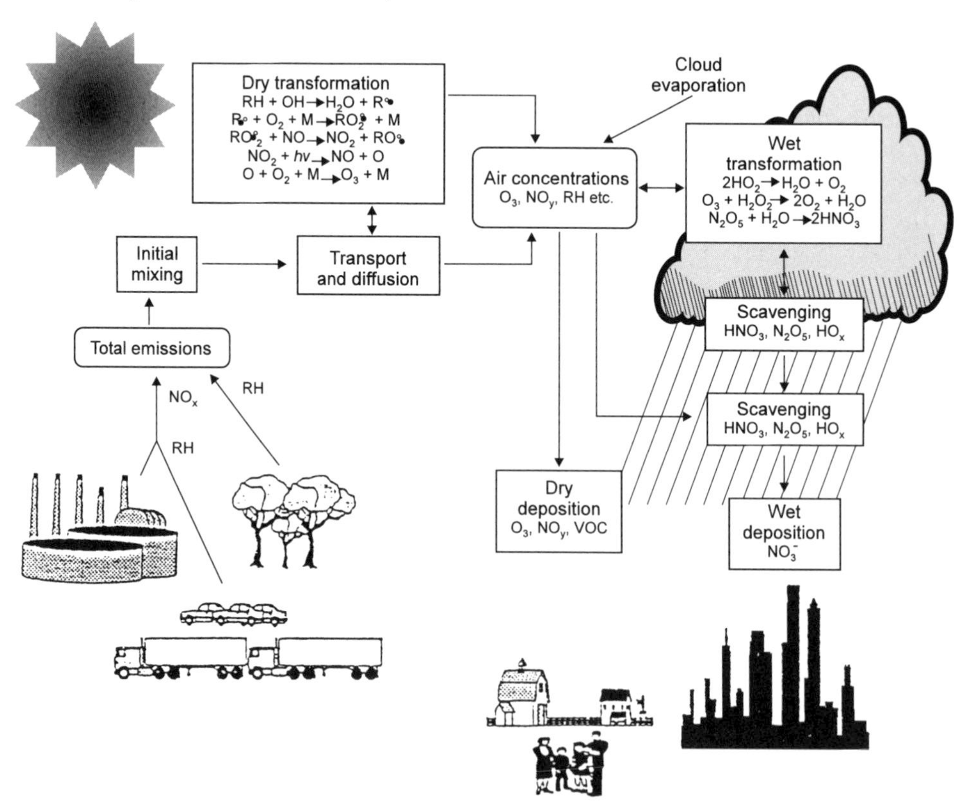

$$RH + \cdot OH \longrightarrow R\cdot + H_2O. \quad (11.4)$$

These can react with O_2 in the presence of an inert third body or catalyst (M) to form peroxy radicals,

$$R\cdot + O_2 \xrightarrow{M} RO\cdot_2. \quad (11.5)$$

These, in turn, react with nitric oxide (NO) to form nitrogen dioxide (NO_2),

$$RO_2\cdot + NO \longrightarrow NO_2 + RO\cdot \quad (11.6)$$

NO_2 is photodissociated by solar radiation to reform NO and atomic O in the ground state,

$$NO_2 \xrightarrow{light} NO + O(^3P), \quad (11.7)$$

which combines with molecular O_2 in the presence of a third body to form ozone

$$O(^3P) + O_2 \xrightarrow{M} O_3. \quad (11.8)$$

Ozone is then photodissociated to form O_2 and atomic oxygen in an excited state,

$$O_3 \xrightarrow{\text{near-uv light}} O_2 + O('D), \quad (11.9)$$

which can react with water vapor to form two OH radicals,

$$O('D) + H_2O \longrightarrow 2OH\cdot. \quad (11.10)$$

These can reinitiate the process. The process is a classic chain reaction, which can lead to the buildup of unhealthy concentrations of ozone in the troposphere. Figure 11.6 shows that despite reasonably strenuous efforts, ozone levels in the U.S.A. did not change significantly between 1982 and 1991. The dashed line marked NAAQS is the upper limit of tropospheric ozone, 0.12 ppm, permitted under the National Ambient Atmospheric Quality Standard.

The United States is a major source of atmospheric pollutants. This is illustrated in plate 41 which shows the worldwide distribution of sulfur emissions in 1980. Sulfur emissions were particularly intense in the northeastern part of the United States, in Europe, and in China. The SO_2 "hot spot" in the state of Washington is the signature of the eruption of Mount St. Helens in May 1980 (see chapter 9). The intense sulfur emissions in Europe and in China reflect the burning of high-sulfur coal, particularly in what were then East Germany and Czechoslovakia.

Several sources contribute to the emission of these atmospheric pollutants. Some, like power plants and factories, are stationary. Others, like automobiles, aircraft, and ships, are mobile. Transportation is the major source of CO emissions, and was the major source of lead emissions until unleaded gasoline became the dominant transportation fuel. NO_x is contributed about equally by sta-

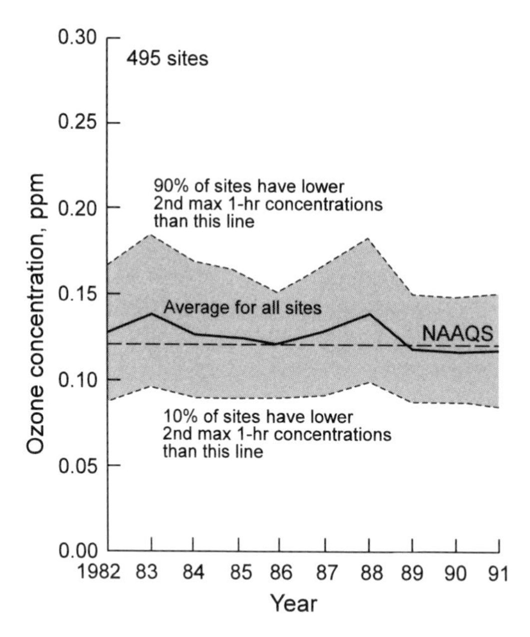

Figure 11.6. The trend in ozone concentrations at 495 sites in the United States from 1982 to 1991. NAAQS = National Ambient Atmospheric Quality Standard.

tionary and nonstationary sources. Stationary fuel combustion accounts for much of the SO_x emissions. Particulate matter is emitted in large part by industrial processes and by stationary fuel combustion.

Should we care? The answer is clearly "yes." Many of the terrible local pollution problems during the early industrial revolution are behind us. Today, few areas are as bad as Coketown, described by Charles Dickens in *Hard Times*:

> Coketown lay shrouded in a haze of its own, which appeared impervious to the sun's rays. You only knew the town was there because there could be no such sulky blotch upon the prospect without a town. A blur of soot and smoke, now confusedly tending this way, now that way . . . a dense formless jumble, with sheets of cross light in it, that showed nothing but masses of darkness.

Pittsburgh, Pennsylvania, a city that was shrouded in soot at the turn of the century, is now respectably clean. Sudbury, Ontario, an early disaster area due to poorly controlled sulfide smelting (see chapter 9), has mended its polluting ways, has reduced SO_2 emissions, and is spreading the remainder in diluted form over a larger area. The infamous London "black fog" episode, which killed several thousand people in 1952, led to stricter emission regulations in Great Britain and has not been repeated.

At the same time, air quality has clearly deteriorated in some large urban areas. Mexico City's location in a topographic depression encourages the accumulation of air pollutants. Figure 11.7 shows that its levels of ozone, SO_2, and total suspended particulates (TSP) are now considerably greater than those in Los Angeles, the U.S. city with the highest levels of atmospheric pollution. Mexico City's unenviable position in the charts is due to its geographic location, to the rapid growth of its population, and to a concurrent increase in automobile and industrial emissions. Large urban areas, especially those in tropical areas, where photochemical ozone production is most intense, are particularly prone to severe atmospheric pollution. With care, repetitions of the London "black fog" episode

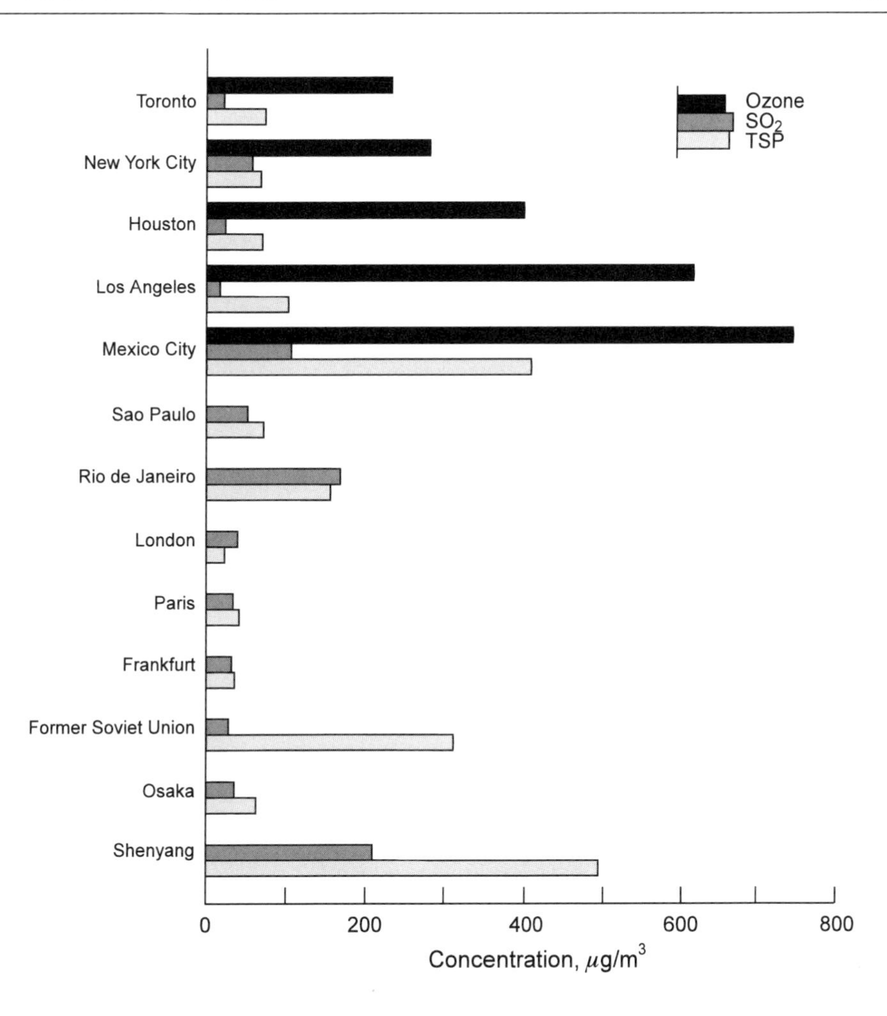

Figure 11.7. Comparison of ambient levels of ozone, total suspended particulate matter, and sulfur dioxide in selected cities. (*National Air Quality and Emissions Trends Report*, 1991; EPA 450-R-92-001)

can be avoided, but less-lethal respiratory problems will continue to be a part of life in towns like Mexico City unless emissions are reduced very significantly.

Two somewhat related aspects of atmospheric pollution are frequently slighted: the effects of biofuels on air quality, and the problem of indoor air pollution. The two topics are related, because a large fraction of biofuel use is for home heating and for cooking in the poorer countries. On the whole, biofuels are more polluting than coal, oil, and natural gas. To many people this comes as a surprise; after all, biofuels are "natural" and are considered, almost by definition, to be benign. It is easy to understand that coal is dirty when it is burned; after all, it is dirty stuff containing contaminants of varying toxicity even before combustion. Wood, by comparison, seems clean, wholesome, and nontoxic; but it is not. Wood smoke has a toxicity similar to that of tobacco smoke. Even their levels of mutagenicity are similar (see, for instance, Smith 1987).

Among the biofuels, dung is the most polluting. Crop residues are next, then wood. All of these fuels are more polluting than the major commercial fuels. Which fuels are used in households depends on income levels and on local circumstances. In some situations, households retreat to dirtier fuels. Deforestation and continued poverty may force people to move down the ladder from wood to crop residues or dung (Smith 1988). Since biofuels are used extensively in cooking, women and children in the poorer countries tend to be more exposed than men to air pollution from biofuels.

Indoor air quality where biofuels are used extensively is often poorer than the quality of air outdoors. Somewhat paradoxically, this has also been one of the consequences of improved home insulation in the wealthier countries. Following the rapid increase in fuel oil prices in the 1970s, thermostat settings were lowered and insulation was improved in many American homes. Fuel bills dropped, much to the relief of bill payers, but indoor air pollution from gas stoves, furnaces, and fireplaces increased significantly. The incidence of potential problems due to the accumulation of radon, a radioactive gas in the ^{238}U decay series, also increased (see chapter 12).

Many of the minor products of fossil fuel burning have residence times in the atmosphere that are long enough to allow for their dispersal over distances of hundreds or even thousands of kilometers. SO_2 and NO_x are the chief among these products. They tend to be deposited with dry precipitation and with rain at considerable distances from their sources. In 1907 this led to a famous law suit between the state of Georgia and the Tennessee Copper Company and Ducktown Sulphur, Copper and Iron Company Ltd. The mines at Ducktown, Tennessee, like those at Sudbury, Ontario, are blessed with rich sulfide ores. At the turn of the century smelting in both areas was done irresponsibly, to the great detriment of the local environment. Winds carried enough SO_2 into Georgia to create problems there. In its judgment the U.S. Supreme Court stated what has remained a basic concern of governments:

> It is a fair and reasonable demand on the part of a sovereign state that the air over its territory should not be polluted on a great scale by sulphurous acidic gas; that the forests on its mountains, be they better or worse, and whatever domestic destruction they may have suffered, should not be further destroyed or threatened by the acts of persons

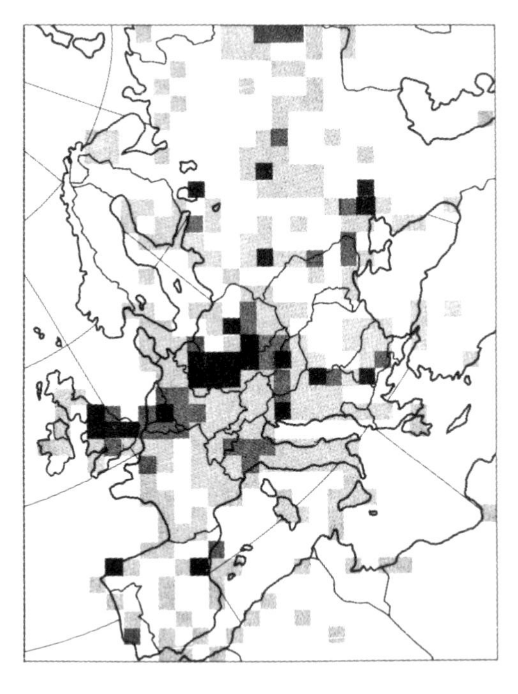

Figure 11.8. Emissions of sulfur dioxide in Europe in 1985. White areas indicate the lowest levels of emissions (fewer than 10,000 metric tons of sulfur per year); gray areas are intermediate (from 10,000 to 299,000 metric tons per year); and black areas represent the highest emission levels (over 300,000 metric tons per year). (Sand 1990)

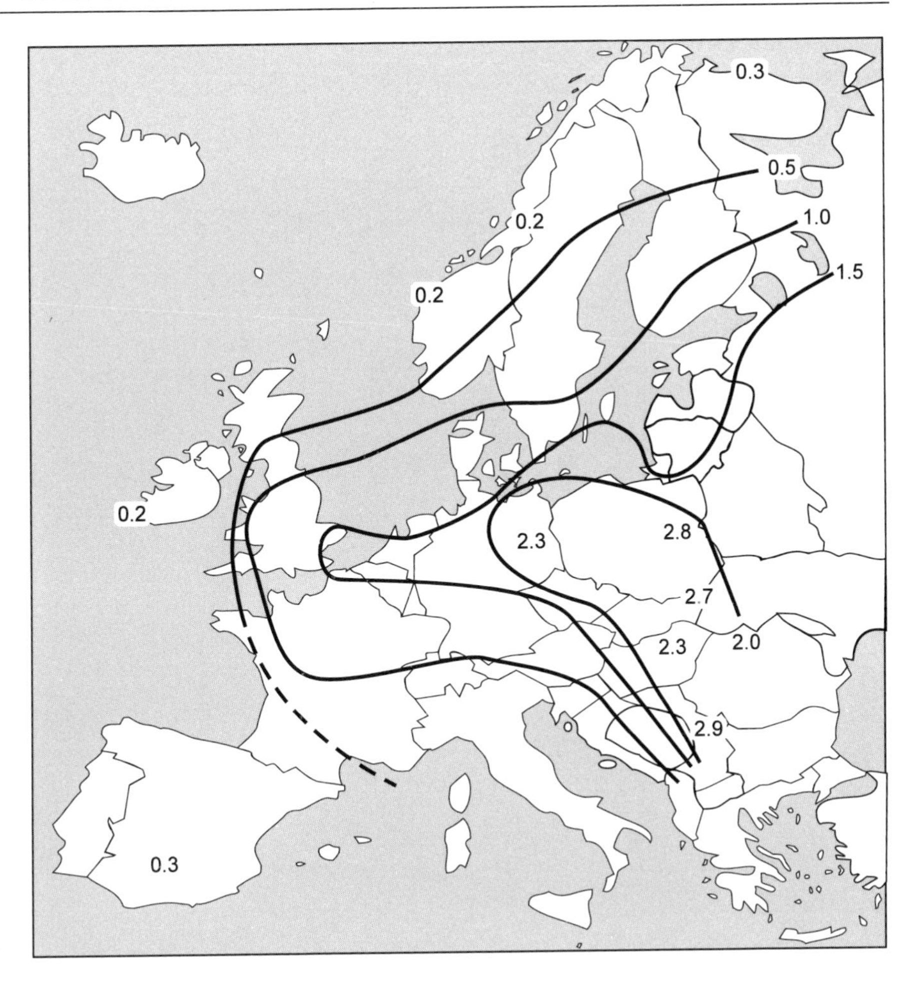

Figure 11.9. Mean annual concentration of sulfate in precipitation in Europe in 1985 (milligrams S per liter). (Sand 1990)

beyond its control; and that the crops and orchards on its hills should not be endangered from the same source.

In Europe, where countries are relatively small, long-range transboundary air pollution is a particularly important concern. Scandinavian reports of acid rain damage due to air pollution from western and central Europe led to the formation of the Cooperative Program for Monitoring and Evaluation of Long-Range Transmission of Air Pollutants in Europe (EMEP). Voluminous and increasingly reliable evidence has now shown that sulfur and nitrogen compounds emitted by a wide range of stationary and mobile pollution sources are indeed dispersed widely through the atmosphere in Europe (Sand 1990). In about half of the European countries, the major part of the total pollutant deposition within their borders is the product of foreign emissions.

Figure 11.8 shows the intensity of European SO_2 emissions in 1985. The most intense SO_2 sources were in the English Midlands, in Belgium, in Poland, and in what was then East Germany. SO_2 emissions from Norway and Sweden were modest. SO_2 is oxidized to SO_3 in the atmosphere and is rained out as sulfuric acid, H_2SO_4. Figure 11.9 shows the mean annual concentration of sulfate in precipitation in Europe during 1985 corrected for SO_4^{2-} from sea salt. The highest

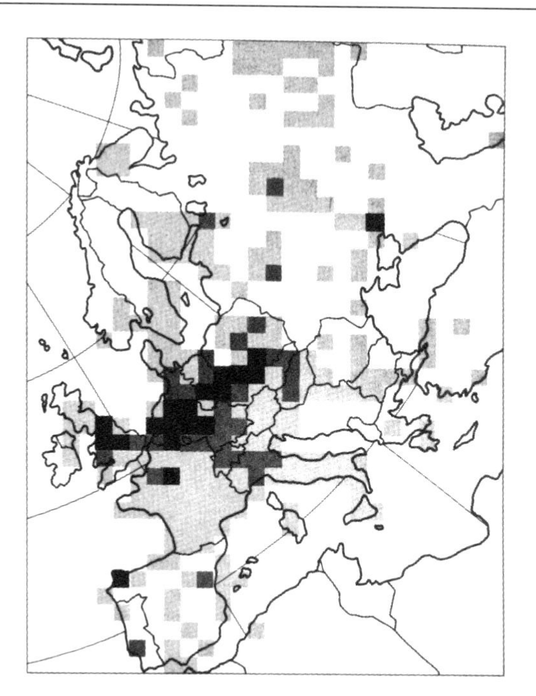

Figure 11.10. Emissions of nitrogen oxides in Europe in 1985. White areas indicate the lowest levels of emissions (fewer than 10,000 metric tons of NO_2 per year); gray areas are intermediate (from 10,000 to 299,000 metric tons per year); and black areas represent the highest emission levels (over 300,000 metric tons per year). (Sand 1990)

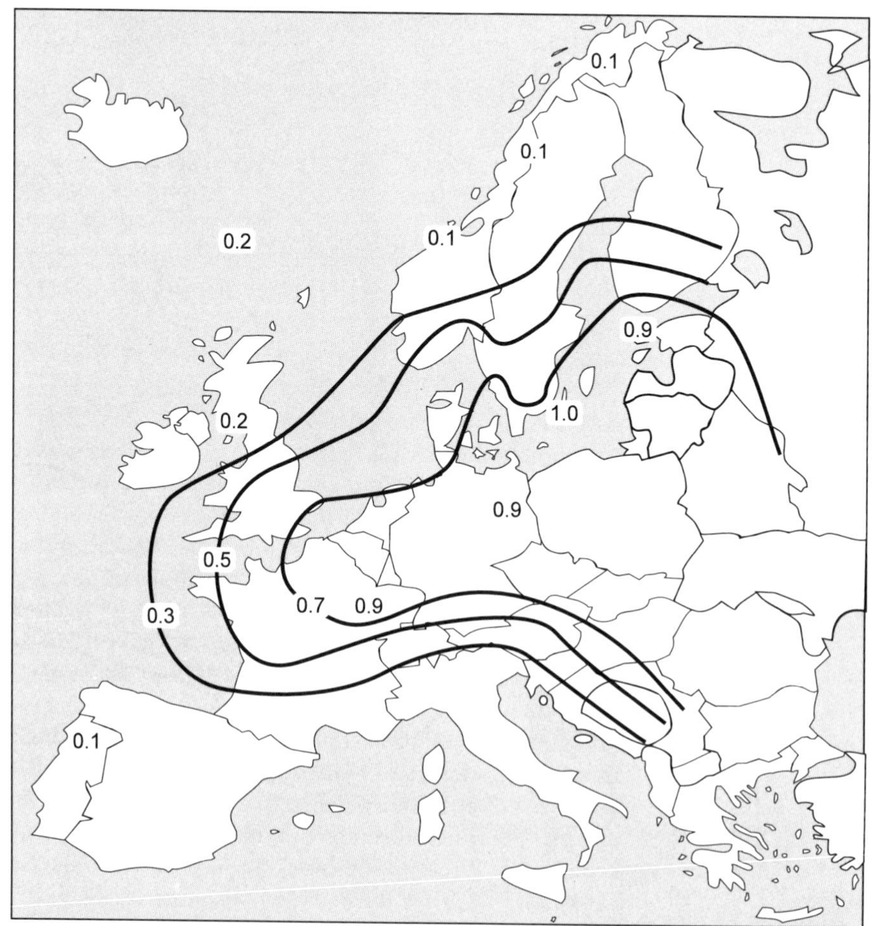

Figure 11.11. Mean annual concentration of nitrate in precipitation in Europe in 1985 (milligrams N per liter). (Sand 1990)

sulfate concentrations were in rain that fell in Poland and in the former East Germany (the GDR). However, Scandinavia clearly received a good deal of sulfate that was not Scandinavian in origin, and less than 15% of the sulfate deposited in Switzerland during 1980 was derived from Swiss sources.

The picture is much the same for NO_x. Figure 11.10 shows the 1985 distribution of European NO_x emissions. The areas of most intense NO_x emissions were essentially the same as the areas of most intense SO_2 emissions. Nitrogen oxides are oxidized in the atmosphere and are removed in part as HNO_3 dissolved in rain. As might be expected (see fig. 11.11), the pattern of nitrate concentrations in European rain was similar to the pattern of sulfate concentrations. The Scandinavian countries received a considerable excess of NO_3^- as well as of SO_4^{2-}.

The concentration of sulfate and nitrate in European rain is actually quite low, even in the most polluted areas. Since both constituents serve as nutrients for vegetation, it is fair to inquire why so much fuss is made about the issue. The answer is in the acidity of the precipitation. Sulfate and nitrate in rainwater are almost entirely present as constituents of H_2SO_4 and HNO_3, respectively. H_2SO_4 and HNO_3 are both strong acids. The H^+ concentration in rain over much of Europe is therefore determined not by H^+ from the ionization of H_2CO_3 but by H^+ from the ionization of H_2SO_4 and HNO_3.

Figure 11.12 shows the distribution of the pH in precipitation in Europe in

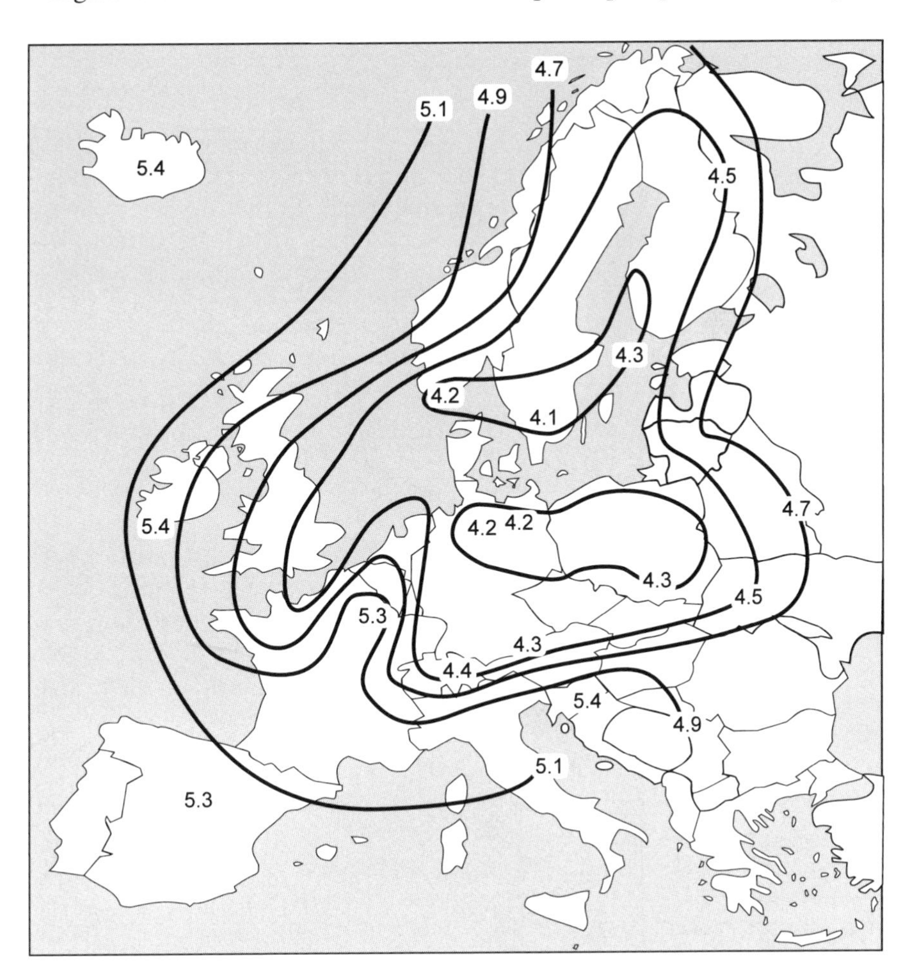

Figure 11.12. Mean annual pH of precipitation in Europe in 1985. (Sand 1990)

Table 11.1. The Acidity of Some of America's Favorite Nonalcoholic Beverages

Beverage	*pH (25°C)*
Coca Cola	
Classic	2.61
Diet	3.32
Pepsi Cola	2.61
Diet Pepsi	3.18
Ginger Ale (Schweppes)	3.12
Sprite	3.44
Root Beer (Hires)	4.38
Coffee (Harvard University)	
Regular	5.05
Decaffeinated	4.98
Poland Spring Water	6.82

Note: Measurements made by R. O. Rye and P. H. Kuo, July 8, 1993, on contents of containers immediately after opening.

Table 11.2. Approximate pH Range in Which Various Fish Species Suffer Reproductive Failure or Mortality

pH	*Species*
6.0 to 5.5	Smallmouth bass (*Micropterus dolomieui*)
	Walleye (*Stizostedion vitreum*)
	Rainbow trout (*Oncorhynchus mykiss*)
	Common shiner (*Notropis comutus*)
	Burbot (*Lota lota*)
5.5 to 5.2	Lake trout (*Salvelinus namaycush*)
	Trout perch (*Percopsis omiscomaycus*)
	Fathead minnow (*Pimephales promelas*)
5.2 to 4.7	Brook trout (*Salvelinus fontinalis*)
	Brown bullhead (*Ictalurus nebulosus*)
	White sucker (*Catostomus commersoni*)
	Largemouth bass (*Micropterus salmoides*)
	Rock bass (*Ambloplites rupestris*)
4.7 to 4.5	Cisco (*Coregonus artedii*)
	Yellow perch (*Perca flavescens*)
	Lake chub (*Couesius plumbeus*)

Source: Restoration of Aquatic Ecosystems, 1992.

1985. Far from sources of atmospheric pollution, as in Iceland, the pH of rainwater is 5.4, close to the pH of unpolluted rainwater that has equilibrated with the CO_2 content of the atmosphere. In 1985 the pH of rainwater in much of Germany and Poland was between 4.1 and 4.3. The pH of dilute solutions such as rainwater is related to the concentration of H^+ ions, m_{H^+}, by the approximate expression

$$pH \approx -\log m_{H^+}. \tag{11.11}$$

A solution with a pH of 4.3 therefore contains ten times as much H^+ as a solution with a pH of 5.3. The much greater H^+ concentration in central European rainwater can be explained quantitatively by pollution with H_2SO_4 and HNO_3. This is also true of the similarly low pH values of precipitation in the northeastern part of the United States.

As acids go, rainwater, even in the highly industrialized areas of Europe and the United States, is very dilute. Table 11.1 lists the pH of some of America's favorite soft drinks. Rainwater clearly cannot compete with Coca-Cola or Pepsi-Cola in terms of acidity. Drinking acid rain with a pH of 4.0 is not detrimental to human health. It is, however, detrimental to the health of a number of aquatic animals. Most animals in rivers and lakes cannot live in water with a pH less than about 4.5 to 5.0. Estimates of the pH tolerance of various aquatic organisms are listed in table 11.2 and illustrated in figure 11.13. Rivers and lakes whose pH drops below 5.0 lose their fish, and both their fauna and flora change dramatically.

The pH of most rivers and lakes is between about 6.5 and 8.5, i.e., well in excess of rainwater pH values. Weathering reactions (see chapter 6) account for

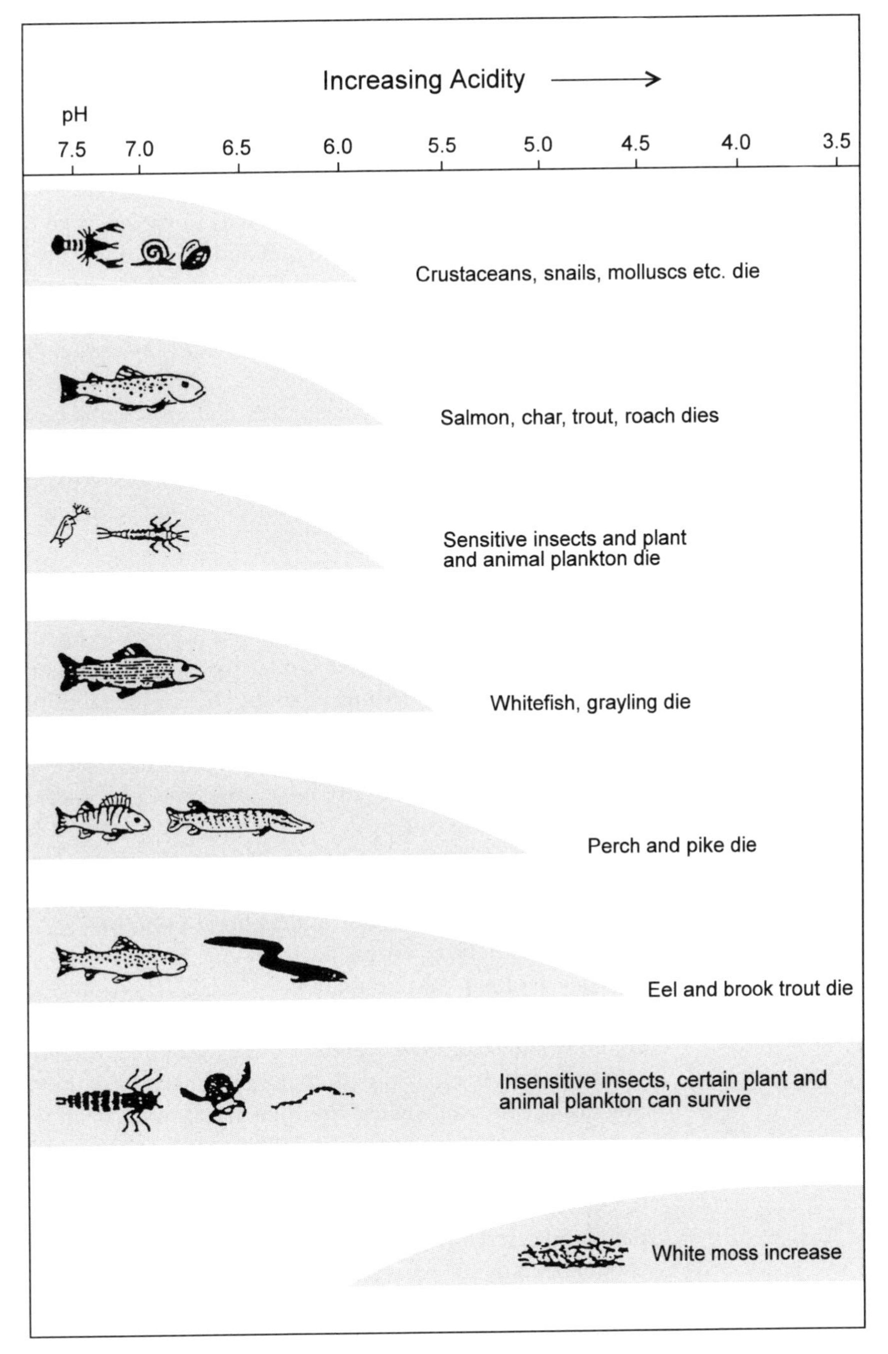

Figure 11.13. The pH tolerance of various aquatic organisms. (Ina Lehmann)

most of this difference. Carbonic acid, H_2CO_3, in rainwater reacts with rocks, converts them into soils, and is neutralized in the process (see chapter 6). The same reactions serve to neutralize other acids in precipitation. Where these reactions are allowed to proceed to completion, the presence of H_2SO_4 and HNO_3 in rainwater does not disturb the pH of rivers and lakes. Where an area is underlain by limestones or dolomites, carbonate minerals react swiftly and completely with H_2SO_4 and HNO_3. Acidification of rivers and lakes is therefore hardly, if ever, a problem in carbonate terrains. In common parlance, carbonate rocks are said to buffer the pH of rainwater.

In areas underlain by silicate rocks the situation can be quite different. Acid rain reacts much more slowly with feldspars and ferromagnesian minerals than with carbonate minerals. It tends to be neutralized only slightly during its passage through soils into rivers and lakes in regions like the Canadian shield and the Adirondack Mountains. Rivers and lakes in such regions are therefore susceptible to acidification. The pH history of lakes can be reconstructed quite well by studying changes in the assemblages of diatoms preserved in sediment cores. Diatoms are microorganisms which use SiO_2 to build their skeletons. The diatom assemblages in lakes are strongly affected by the pH of the lake water. When the pH changes, so do the species in the diatom populations. This type of evidence from sediment cores has shown that the pH of many lakes in granitic rocks of the northeastern United States has decreased markedly during the course of the twentieth century in response to the increasing acidity of the precipitation falling in their watersheds.

The extent of fish losses in these lakes has been very difficult to quantify. Some early studies clearly exaggerated the losses (National Research Council 1992), but the National Acid Precipitation Assessment Program (1990) concluded "with reasonable confidence" that acidification has resulted in the loss of one or more fish populations in about 16% of the lakes in the Adirondack region of New York State. Fish population declines have also been associated with increasing acidification of lakes and streams in Norway, Sweden, Scotland, and in the Canadian provinces of Ontario and Nova Scotia (Meybeck, Chapman, and Helmer 1989).

The effect of intense local deposition of SO_2 on the land biota can be devastating, as has been demonstrated repeatedly in the vicinity of large smelters. On a regional scale the effects of atmospheric pollution on trees is still a matter of considerable debate. Forest declines can be caused by complex interactions of multiple biologic and nonbiologic factors. There have been numerous studies of the effects of specific air pollutants on specific plants, but understanding the complex nature of many forest declines usually requires intensive, multilayered investigations of the community at risk. There are few examples where air pollution has clearly contributed significantly to forest decline in the United States. In only two U.S. cases has air pollution been implicated unequivocally in forest damage: ozone has been implicated in the decline of the mixed conifer forests of the San Bernardino Mountains of southern California and of the Eastern white pine in eastern North America. It has been suggested that a third and more recent decline, that of the red spruce at high elevations in the eastern United States, is due to atmospheric deposition in association with natural causes. Finally, recent growth-rate reductions in southern pines has led to speculation that regional air pollution is a primary cause (Bartuska 1990).

Although the effects of atmospheric pollutants are minor in the United States, they seem to be serious in central and eastern Europe. Various estimates have suggested that between 30% and 60% of forests are damaged in Austria, the former Czechoslovakia, Germany, Poland, and Switzerland (see fig. 11.14). Forests closest to major sources of pollutants, particularly SO_2, lead, and zinc, are particularly damaged. However, great differences in the degree of forest damage

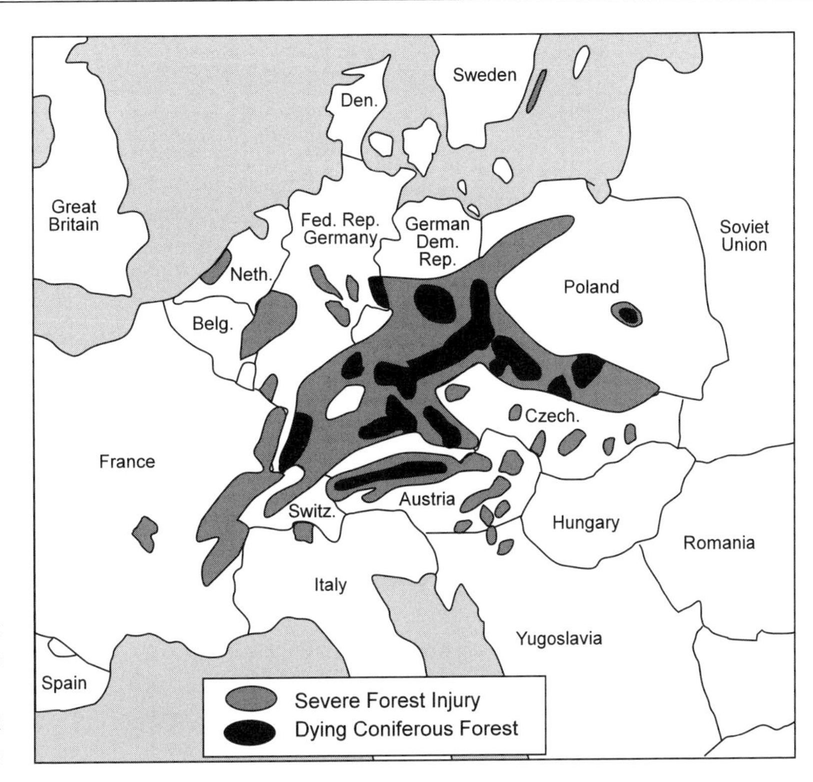

Figure 11.14. Distribution of forest injuries in some European countries. (Wentzel, 1987; Godzik and Sienkiewicz 1990)

exist among areas that have received similar doses of atmospheric pollution. This suggests that forest damage is due to more than one cause, and that multiple factors, including the combined effect of several atmospheric pollutants, and the underlying geology are responsible for the injury and decline of forests in central and eastern Europe (Godzik and Sienkiewicz 1990).

Atmospheric pollutants affect agricultural crops as well as forests. In the United States crop damage by ozone is significant. Damage by other atmospheric pollutants is comparatively minor, and is probably due to interaction with O_3. Most crops show growth, biomass, and yield reduction when grown under elevated ambient air concentrations of O_3. Estimates of annual U.S. economic losses from O_3 damage to all crops range from one to seven billion dollars. These losses vary from year to year depending on O_3 levels and on meteorological conditions (Heck 1990).

Similarly careful and detailed studies of crop damage have not been made in eastern Europe, but significant reductions in crop yields have been observed near sources of intense air pollution in Upper Silesia. Figure 11.15 shows the correlation between the atmospheric SO_2 concentration and the yield of beans, barley, and potatoes near such sources. Major decreases in yield occur when the SO_2 concentration exceeds 100 $\mu g/m^3$. Figure 11.7 shows that very few cities in the world suffer from such a high degree of SO_2 pollution. Fortunately, crop losses at more usual SO_2 concentrations near 10 $\mu g/m^3$ are minor, probably between 3% and 4% (Godzik 1990). There are no reliable estimates for crop losses due to atmospheric pollution in central and eastern Europe as a whole, but the losses may well be in this general range.

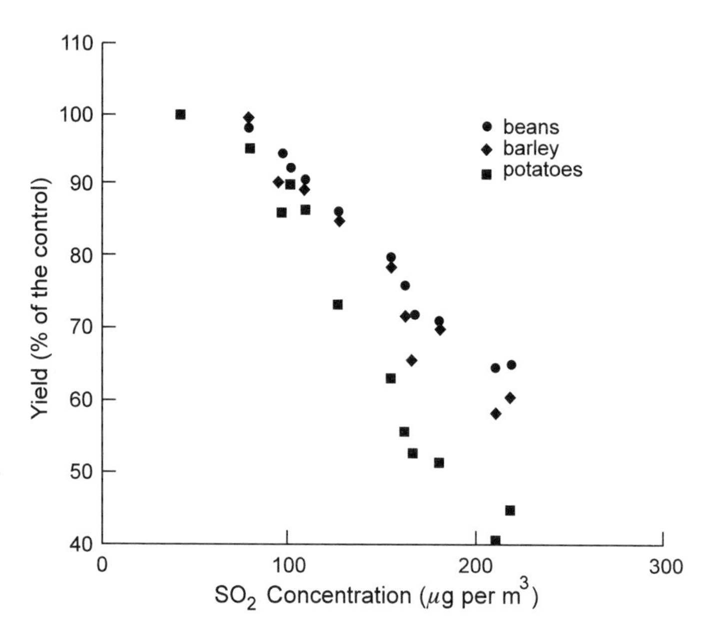

Figure 11.15. Effects of sulfur dioxide on the yield of plants in the field; means of three years of experiments. Sulfur dioxide concentration (in μg/m^3) is calculated from the sulfation rate. (Godzik 1990)

CO_2 Inputs and Impacts

If asked the meaning of the phrase "global change," many Americans would answer: "That's greenhouse warming and the ozone hole." The warnings of global warming have been particularly dire and are at the heart of the controversy regarding the wisdom of the continued use of fossil fuels as the world's major energy source. The CO_2 content of the atmosphere has increased significantly since the start of the industrial revolution and could increase by roughly an additional factor of six before the world runs out of fossil fuels. The effects of such a major increase in atmospheric CO_2 on the world's climate and biota are difficult to predict. We do not know whether the continued use of fossil fuels is an invitation to disaster, or whether it is simply mankind's most magnificent geochemical experiment.

Measurements of the CO_2 content of the atmosphere have been made for well over a hundred years, but systematic, highly precise measurements were only begun in the 1950s. Figure 11.16 is a summary of the detailed, long-term measurements of the CO_2 content of air at Mauna Loa on the island of Hawaii. The Mauna Loa site is particularly well suited for measuring the composition of uncontaminated air. Disturbances from volcanic vents do occur, but they can be readily identified and excluded from the record. Between 1958 and 1990 the CO_2 content of air on Mauna Loa increased by 12% from 316 parts per million by volume of dry air (ppmv) to 354 ppmv. On average, the CO_2 content of the atmosphere has risen 1.2 ppmv per year during this 32-year period. During the past decade the rate of increase has been 1.5 ppmv/yr.

The annual zigzag pattern of the CO_2 concentration in figure 11.16 is due to imbalances in the rate of removal of CO_2 from the atmosphere by photosynthesis and the rate of resupply of CO_2 by the decay of organic matter. The size of these annual swings depends on the intensity of photosynthesis in the region where the measurements are made. At the South Pole, for instance, the annual excursions

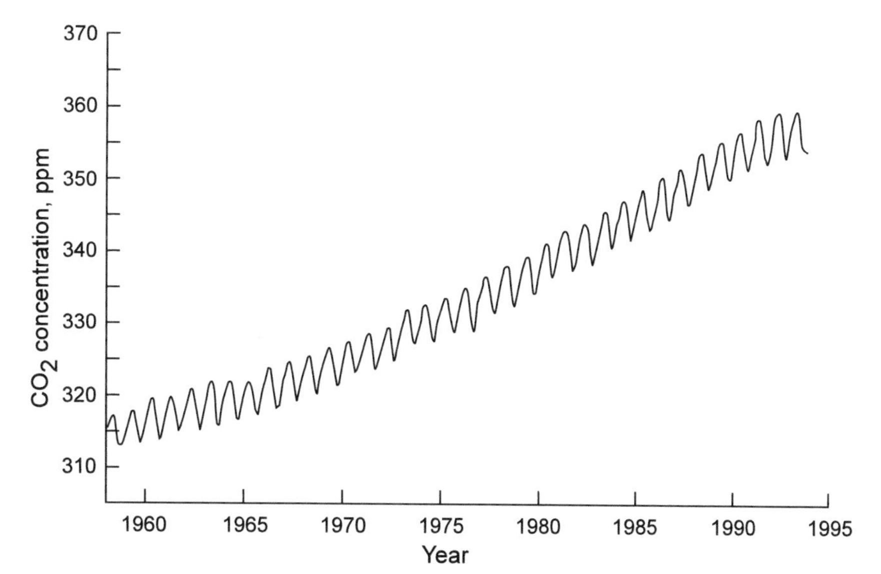

Figure 11.16. The CO_2 content of air on Mauna Loa, Hawaii, from 1958 to 1993. The measurements were made by Charles D. Keeling and his associates. (Halpert et al. 1994)

about the mean CO_2 content of the atmosphere are much smaller than in Hawaii; but in Antarctica, as well as in the several other stations where the CO_2 content of the atmosphere is now being monitored, the long-term trends are the same.

The most reliable pre-1957 indications of atmospheric CO_2 come from analyses of air that was trapped in snow and ice in Greenland and Antarctica. In these cold regions, snow becomes gradually compacted and turns into ice. Snow, being quite fluffy, contains a large amount of air. During compaction and conversion to ice under the weight of the snowfalls of later years, much of the original air is expelled. However, some air is trapped in tiny bubbles in the ice. The air in these bubbles can be extracted and analyzed chemically. Figure 11.17 shows the Mauna Loa data together with results of analyses of CO_2 in air bubbles in an ice core taken by a joint U.S.-Swiss team at Siple Station in Antarctica. The analyses show that the concentration of CO_2 has been rising steadily since 1740. The

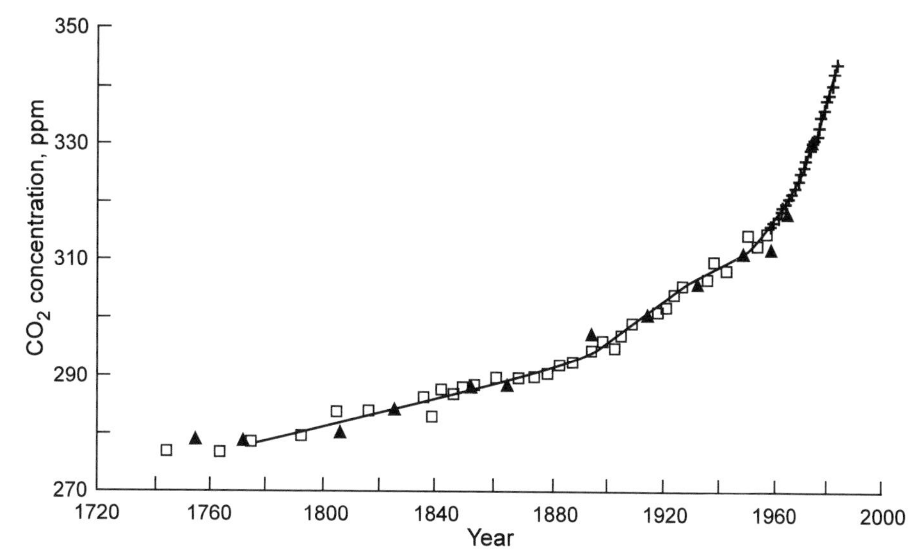

Figure 11.17. The CO_2 concentration in the atmosphere from 1740 to 1990. Squares and triangles represent measurements on the ice core from Siple Station, Antarctica; crosses represent data from Mauna Loa, Hawaii. (Oeschger and Siegenthaler 1988)

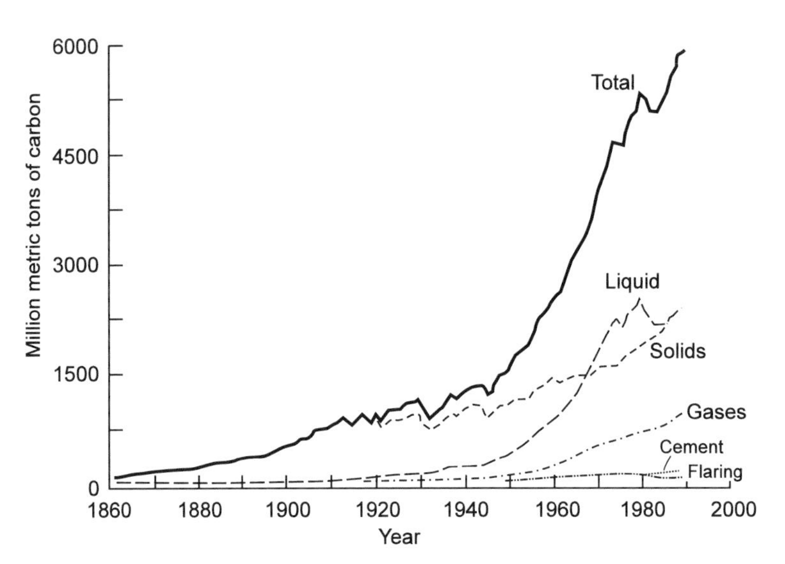

Figure 11.18. Global CO_2 emissions from fossil fuel burning and cement manufacture, 1860–1989. (From *Trends '91: A Compendium of Data on Global Change*)

increase was modest until about 1900, and has accelerated dramatically during the twentieth century. Data from very long ice cores that sample thousands of years of Earth history have shown that the CO_2 content of the atmosphere was remarkably constant during the long interval between the end of the last ice age about 10,000 years ago and the beginning of the industrial revolution.

Three processes are responsible for most of the increase in atmospheric CO_2 since 1700: fossil fuel burning, cement making, and deforestation. The history of the first two processes since 1860 is documented in figure 11.18. Deforestation, particularly by the American pioneers of the nineteenth century, accounted for most of the CO_2 increase before 1860. Since then, fossil fuel burning has dominated the CO_2 increase. Today, cutting of the South American rain forests is thought to account for the burning of about 1,000 million metric tons of carbon per year. This represents about 15% of the combined anthropogenic CO_2 inputs to the atmosphere.

It is easy to show that the CO_2 content of the atmosphere would be rising by 3.3 ppmv/yr if all of the anthropogenic CO_2 were accumulating in the atmosphere. The current annual increment is only half as large. The difference is undoubtedly real. It implies that about half of the CO_2 that is added annually to the atmosphere is somehow removed on a short timescale. Upward escape into interplanetary space is not possible. CO_2 can escape downward by dissolving in the oceans and by participating in the growth of the biosphere. There are indeed areas where the biosphere is growing. New England is one of these. In the eighteenth and nineteenth centuries much of the region was deforested and converted into farmland. A large fraction of this farmland reverted to forest after farming had moved into the Midwest and into California. However, the uptake of CO_2 in New England and in other areas like New England is almost certainly too small to account for more than a small part of the difference between the anthropogenic inputs of CO_2 into the atmosphere and the observed growth of its CO_2 content. The oceans are almost certainly the major sink of the "missing" CO_2 (see, for instance, Sundquist 1993).

The CO_2 content of the atmosphere will surely rise during the next century.

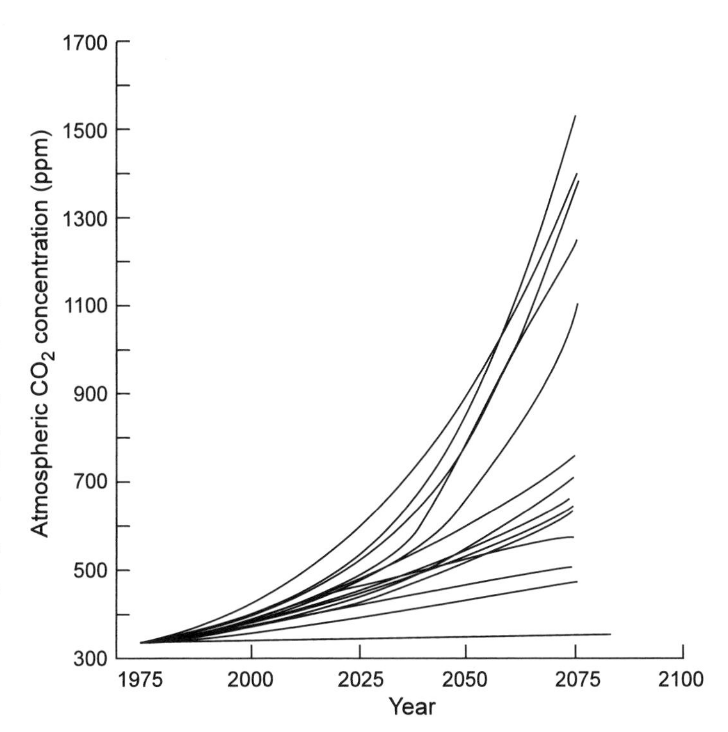

Figure 11.19. Projections of the atmospheric CO_2 concentration between 1975 and 2075. The major variables are the growth of world population, per capita energy consumption, and the fraction of energy that will be supplied by fossil fuel burning. (Häfele 1981)

How much will depend on the rate of anthropogenic CO_2 injection and on the rate at which the injected CO_2 is removed from the atmosphere by natural and by engineered processes. A wide range of scenarios can and have been imagined. A number of these are shown in figure 11.19. In scenarios where fossil fuel energy is replaced by non-CO_2-producing energy sources, the CO_2 content of the atmosphere increases very little. In scenarios where fossil fuel burning continues to supply most of the energy to a world with a rapidly increasing population and significantly increasing per capita energy use, the CO_2 content of the atmosphere rises steeply and could perhaps reach values as high as 1,600 ppmv by the year 2075. A quadrupling of atmospheric CO_2 by the year 2100 is unlikely, but not impossible. If the world population triples, and if the per capita fuel consumption also increases by a factor of three, the rate of CO_2 production could be nine times greater than the present rate. The annual increment in atmospheric CO_2 would then be about 13 ppmv. At that extreme rate the CO_2 content of the atmosphere would increase by 1,300 ppmv in one century. The Monte Carlo Error Analysis figure of 700–800 ppm is a more likely estimate for the CO_2 content of the atmosphere in the year 2075.

Ultimately, the CO_2 content of the atmosphere will be limited by the amount of available fossil fuel. Using the data in chapter 8, one can show that burning the inferred reserves of fossil fuels could increase the CO_2 content of the atmosphere by about a factor of ten. If half of this CO_2 dissolves in the oceans or enters the biosphere, the maximum CO_2 content of the atmosphere would be about six times the present value. The number six is rather uncertain, but it is quite clear that burning the world's inventory of fossil fuels during the next 200 years would increase the concentration of CO_2 in the atmosphere several fold.

The consequences of a several-fold increase in the CO_2 content of the atmosphere are poorly known. Intense efforts have been made during the past decade to develop general circulation models (GCMs) of the atmosphere to predict the climatic effects of doubling atmospheric CO_2. There is now general agreement regarding the effect of such a doubling in "clear sky" models, but there is great uncertainty about the effect of doubling atmospheric CO_2 on the extent of the world's cloud cover. Various models predict temperature increases that range from less than 1°C to as much as 5°C. A mean temperature increase of 1°C would only be twice the temperature increase that the Earth has apparently experienced since the middle of the nineteenth century. A temperature increase of 5°C would almost certainly influence world ecology in a major way. An increase of this magnitude would be similar to the temperature increase during the transition from the last ice age to the present regime.

The problem of predicting changes in cloud cover is being attacked vigorously, but it may be some time before the problem is solved. There is a story, probably apocryphal, about a cloud physicist who always pulled down the window shade next to his seat when he was flying cross-country, because he became so daunted contemplating the complexity and variability of cloud formations. The problem is indeed formidable. As the temperature of the atmosphere rises, more water evaporates from the oceans and global rainfall increases, but it is not clear whether or how much the cloud cover will increase. The mean cloud cover today is close to 50%. An increase in cloud cover would almost certainly cool the Earth. Sunbathers are well aware of the chilling effect of clouds that obscure the sun as they pass overhead. Even a rather modest increase in cloud cover could act as a strong restraint on global warming due to an increase in the concentration of CO_2 and other greenhouse gases.

In the absence of convincing predictions of greenhouse warming based on global modeling, we must turn to the historical record for guidance. The CO_2 content of the atmosphere has increased by about 30% since the beginning of the industrial revolution; perhaps we can already discern a pattern of change that can be extrapolated to a doubling of atmospheric CO_2. Figure 11.20 shows a compilation of temperatures in the Northern and Southern Hemispheres and for the entire globe between 1880 and 1987 relative to a 1951–80 reference period. Figure 11.21 is a similar compilation for the period between 1861 and 1988 relative to a 1950–79 reference period. The diagrams in the left-hand column of figure 11.21 summarize the raw temperature data; in the diagrams on the right, these data have been corrected by subtracting the temperature effect of the El Niño/ Southern Oscillation (ENSO) events in the Pacific Ocean. The data in figures 11.20 and 11.21 suggest that the global Earth surface temperature has increased by about 0.5°C during the past 130 years. Some doubt has been cast on the reliability of these data due to the lack of a temperature trend in the satellite observations of global temperatures between 1979 and 1993 (fig. 11.22).

Nevertheless, the data in figures 11.20–11.22 look like a useful beginning for a long-range forecast; unfortunately, they are not. Doubts assail us from several directions. First, the course of the increase in the average surface temperature does not match that of the increase in atmospheric CO_2. The average global surface temperature was apparently steady or even declined somewhat between 1940 and 1970, while the CO_2 content of the atmosphere was increasing at an accelerating rate. Second, the increase in temperature could be interpreted as a

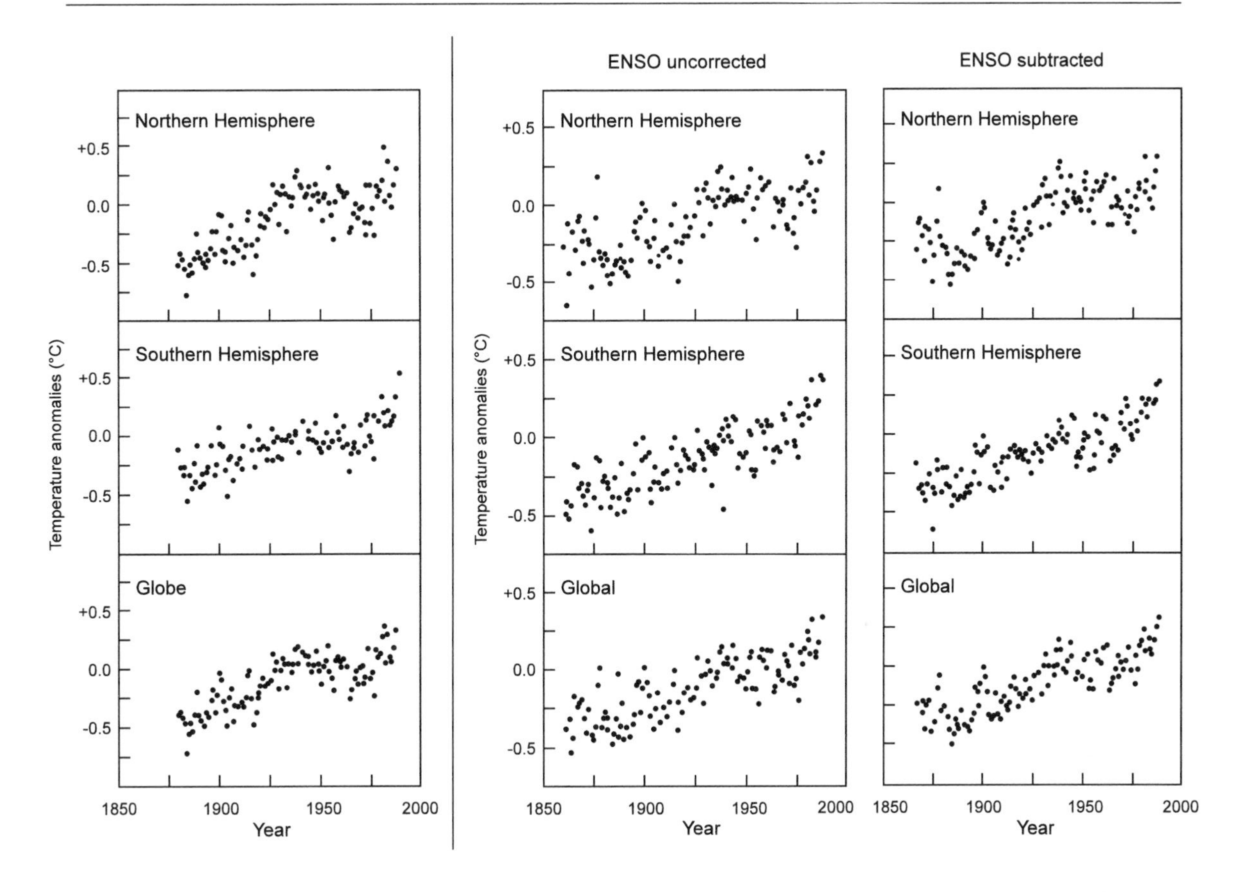

Figure 11.20 (*left*). Annual surface air temperature anomalies in the Northern and Southern hemispheres and for the Earth as a whole, 1880–1987. (Based on data by J. Hansen and S. Lebedeff; *Trends '90: A Compendium of Data on Global Change*)

Figure 11.21 (*right*). Annual global and hemispheric temperature anomalies, 1861–1988. (Based on data by P. D. Jones, T.M.L. Wigley and P. B. Wright; *Trends '90: A Compendium of Data on Global Change*)

Figure 11.22. Mean global tropospheric temperature anomalies from satellite data (85°S–85°N) from 1979 to 1993. (Data from R. W. Spencer and J. R. Christy; Halpert et al. 1994; see also Kerr 1995)

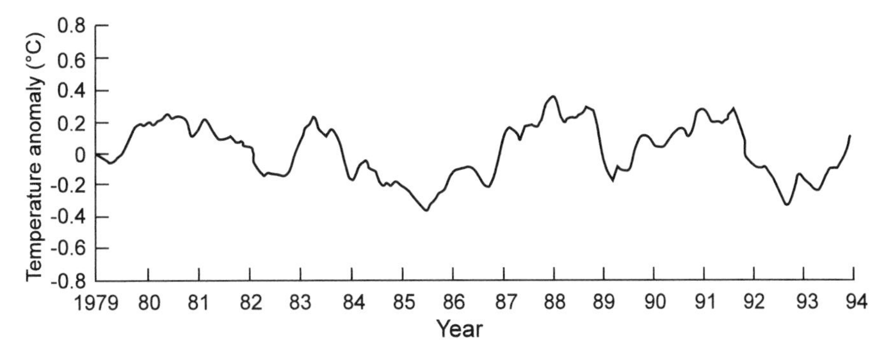

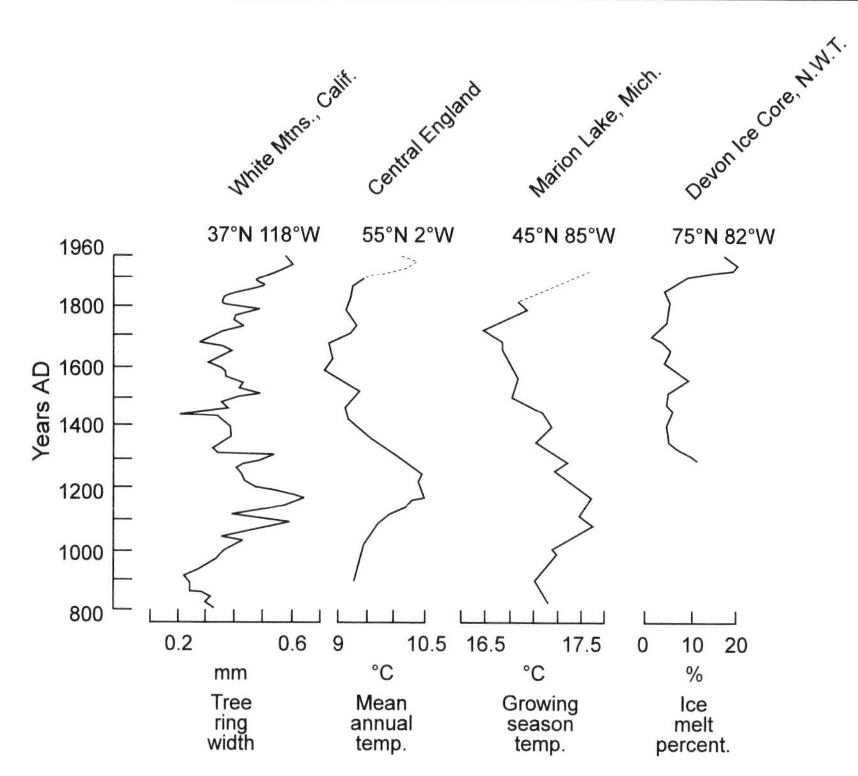

Figure 11.23. Comparison of independently derived indicators of environmental and climatic change since A.D. 800. (Crowley and North 1991)

recovery from the "little ice age" between ca. 1500 and 1800. Some indicators of temperature fluctuations since A.D. 800 are shown in figure 11.23. Temperatures have fluctuated by 1°C or so, and it can be argued that the warming trend of the past 130 years is simply the warming part of a cycle that is unrelated to the CO_2 content of the atmosphere. To demonstrate this, we would need to know the cause or causes of these climatic cycles. We do not . It is therefore impossible to extract meaningful information regarding the influence of the rising CO_2 content of the atmosphere on global climates from temperature indicators like those in figures 11.20–11.22. If all the global warming since 1850 has been part of a recovery from the little ice age, then the increase in the CO_2 content of the atmosphere has exerted no detectable climatic effect. On the other hand, if the little ice age would have deepened progressively in the absence of anthropogenic additions of CO_2, then the rise in atmospheric CO_2 during the past 130 years has saved the Earth from a very cold century, and further additions of CO_2 are apt to have a very large heating effect on the atmosphere. Clearly, policy decisions regarding the use of fossil fuels that are driven by fears of global warming are still based on a very incomplete understanding of the influence of CO_2 additions on Earth surface temperatures.

Temperature increases may not in fact be the most important cause for concern. Many Americans move to warmer climates when they retire. Svante Arrhenius, the famous nineteenth-century Swedish chemist, looked forward to global warming as a welcome source of relief from the cold Stockholm winters and as a hedge against the possible onset of the next ice age. In many parts of the United States the pain of higher air-conditioning bills in the summer would be offset by the joy of lower fuel bills in the winter. Other consequences of a major CO_2

increase may be more serious. Rainfall will probably increase, but rainfall patterns may well shift. There is still considerable doubt about the existence of changes in stream flow in the United States since 1940 (Lins and Michaels 1994), but if the breadbaskets of the world turn into dust bowls, the result would be disastrous for the affected regions. If the decrease in food production in these regions is not compensated by increases in other regions, the effects could well be disastrous for the world as a whole. Some compensation can be expected due to the fertilizing effect of additional atmospheric CO_2, but a high CO_2 environment is no substitute for rain.

A good deal of thought has gone into predicting the effect of global warming on sea level. Intuitively, sea level should rise in response to a rise in the average surface temperature of the Earth. Ice sheets in Greenland and on the Antarctic continent should melt, the volume of the oceans should increase, and sea level should rise. After all, it is now well established that at the height of the last ice age sea level was about 100 meters lower than it is today, and that the sea rose to its present level in response to the melting of the glaciers at the end of the last ice age. The matter is more complicated, however. As temperatures increase, the evaporation and the precipitation of water also increase. An increase in precipitation on Antarctica will produce an increase in the ice volume (Bentley and Giovinetto 1992) unless melting at the edges of the ice sheet more than compensates for this increase. It is difficult to predict which process will dominate during the next century. Sea level has risen slightly during the past century. Figure 11.24 summarizes the average sea level changes in six of the world's regions. In most

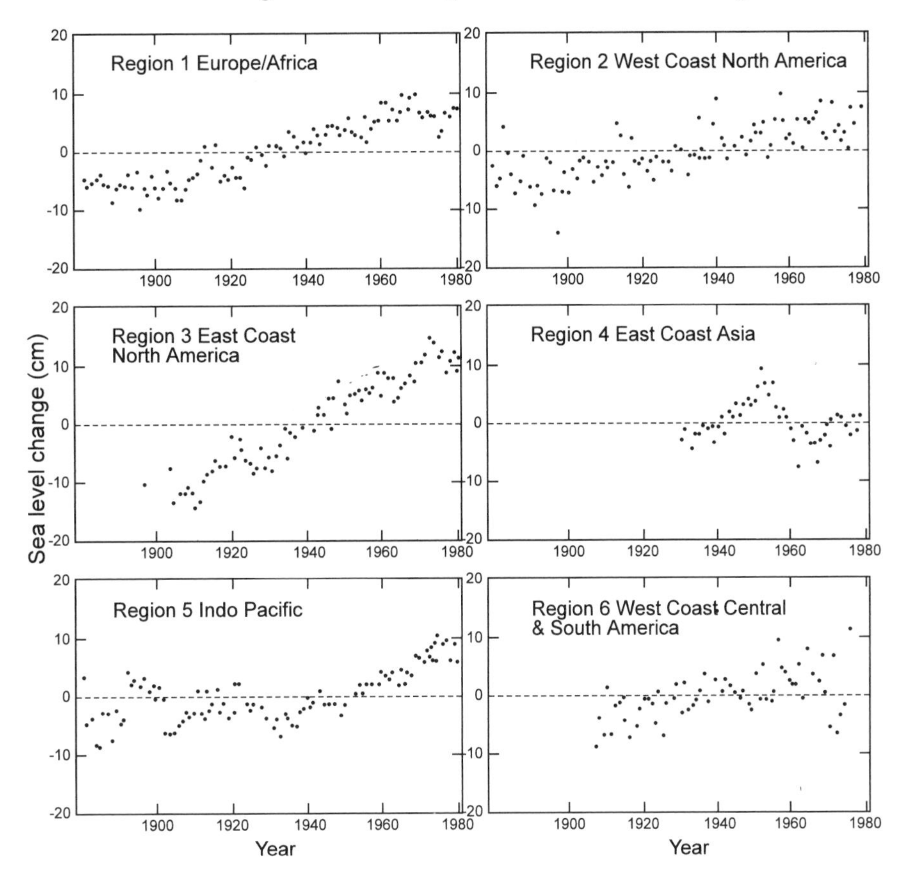

Figure 11.24. Regional averages of sea level change, 1880–1980. (Barnett 1990)

of these regions, sea level has risen since 1880. For the Earth as a whole, sea level seems to have risen at a rate of 0.1 to 0.2 mm/yr (Barnett 1990). At this rate, sea level would be only a very modest and unimportant 10–20 centimeters higher in 2100 than it is today. Some of this increase is due to the mining of water from aquifers and its transfer to the oceans. If the rate of sea level rise were to increase by a factor of ten during the next century, the results could be very serious. Bangladesh is a country with a mean elevation above sea of only a few meters. A sea level rise of one meter, either gradually or by the sudden loss of the west Antarctic ice sheet, could make a very large, densely populated part of this country uninhabitable. At the end of the last ice age, huge armadas of icebergs launched from Canada spread across the northern Atlantic Ocean, triggering a climate response of global proportions (Broecker 1994). Similar armadas from Antarctica or Greenland during the next century could have similar major effects.

A significant increase in atmospheric CO_2 could also have serious consequences for the marine biota. The pH of surface seawater would decrease, and this would drive down the supersaturation of surface seawater with respect to calcium carbonate. Surface seawater could ultimately become undersaturated with respect to calcite and aragonite. The effect of this on marine biologic communities is not known. The effects of CO_2 addition due to fossil fuel burning could become severe during the next two centuries (see also Manabe, Stouffer, and Spelman 1994). Since the effects have been very modest to date, it may be sufficient simply to monitor the most vulnerable parts of the Earth system and to switch to other energy sources if undesirable effects become apparent. Whether or not this is a sensible approach depends on the timescale of environmental change compared to the timescale of energy substitution. Large-scale energy substitution will probably require several decades. The timescale of potential environmental changes due to increasing the CO_2 content of the atmosphere is not known. It seems likely that their timescale is longer than decadal. However, very sudden changes in climate seem to have occurred during the course of the last ice age (Alley et al. 1993), and we know too little about instabilities in the climate system today to rule out the possibility of significant climate changes on timescales of just a few years.

11.3 Nuclear Fuels

During December 1938, on the eve of World War II, Otto Hahn and Lise Meitner in Berlin discovered that uranium nuclei can be split by neutrons. The energy released per atom was about 100 million times greater than the energy released in the most energetic chemical reactions. The potential use of nuclear fission was recognized immediately and led to secret efforts in the United States and in Germany to develop nuclear weapons. In December 1942, at the height of World War II, a group of scientists working under the leadership of Enrico Fermi at the University of Chicago produced the first sustained nuclear reaction. The event was a critical step in the production of plutonium and of the nuclear weapons that were detonated in Japan over Hiroshima and Nagasaki in August 1945. The Chicago reactor was also a forerunner of today's civilian power reactors.

Commercial nuclear power in the United States was born in December 1953 when President Eisenhower made his famous "Atoms for Peace" proposal. Nine years later the first U.S. commercial nuclear power plant, Indian Point I, went into operation 26 miles north of New York City. The cost of electricity from this plant approached parity with competing power sources. At the time the need for electric power was expanding at the remarkable rate of 7% per year. Between 1962 and 1972 the generation of electricity in the United States nearly doubled. The future for the new, cost-effective energy source looked very bright.

The generation of electricity by nuclear power plants has indeed increased impressively since 1962, as shown in figures 11.25 and 11.26. In 1991 nuclear energy accounted for 19% of all the electrical energy generated in the United States and about 16% of the world's electrical energy. These figures are, however, somewhat misleading. In the U.S. no new orders for nuclear power plants have been placed since 1978. In Sweden a referendum has called for the termination of nuclear power generation by 2010. In Switzerland and Germany there is a de facto moratorium on the building of new nuclear power plants. Only France, Japan, North Korea, and South Korea are actively pursuing the nuclear energy option. Nuclear power plants generate nearly 75% of electricity in France, and a number of nuclear power plants are under construction. Only about 30% of Japanese electricity is nuclear, but the percentage is increasing rapidly. Nuclear power plants generate approximately 50% of South Korea's electricity (Häfele 1990). The reasons for the great differences in attitude toward nuclear energy are rooted in the history of the world's countries, their endowment with fossil fuels, and a wide range of political factors.

In the United States most environmentalists are aligned against nuclear energy, even though nuclear power generation contributes neither to acid rain nor to the buildup of greenhouse gases in the atmosphere. Not long ago bumper stickers admonished the public to "Split wood, not atoms." Might it have been better to urge Americans to "Split atoms, save the forests"? The reasons for the current distrust of nuclear energy are many. They need to be appraised before we can assess the desirability of using nuclear energy in the next century.

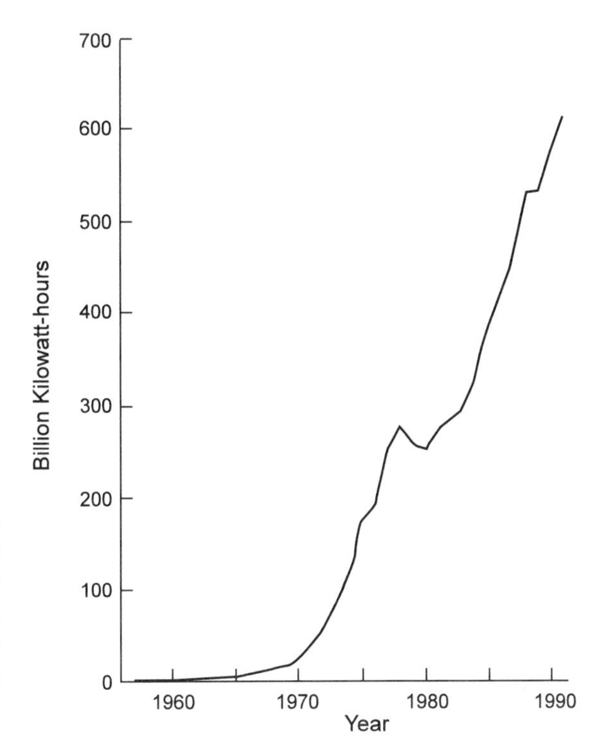

Figure 11.25. Net nuclear generation of electricity in the United States, 1957–91. (*Annual Energy Review 1991*)

Figure 11.26. Nuclear electricity gross generation (*Annual Energy Review 1993*)

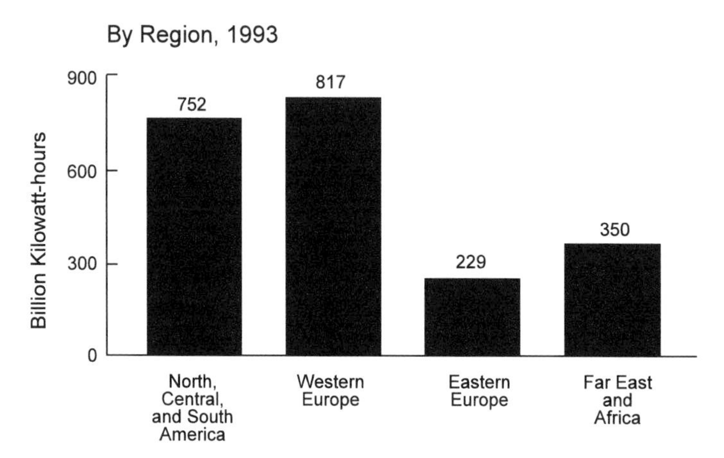

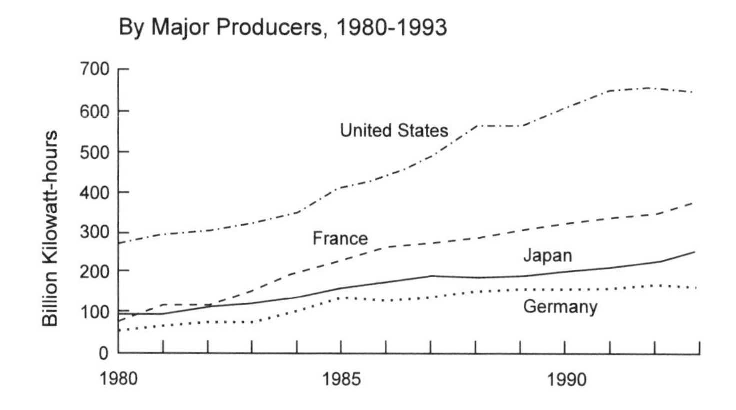

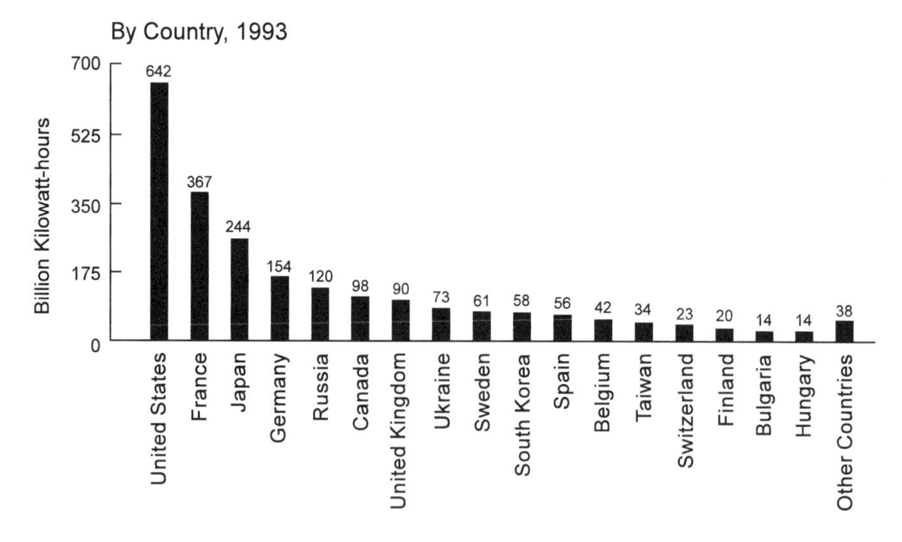

11.3.1 *The Physics of Nuclear Fusion and Fission*

The energy of stars is generated by fusing light nuclei into heavier nuclei. The major source of energy in the Sun is the conversion of hydrogen into helium (see chapter 2). Higher temperatures and pressures are required for the conversion of He into even heavier atoms. However, the energy that is released per combined proton or neutron decreases as the atomic weight of the product nucleus increases. As shown in figure 11.27, the binding energy per proton or neutron passes through a maximum at an atomic weight between 50 and 100. Energy can therefore be released by splitting very heavy nuclei (fission) as well as by combining very light nuclei (fusion). On Earth, nuclear fission has been put to use in atomic bombs and in nuclear power plants. Nuclear fusion has been used only in bombs. Taming nuclear fusion reactions for generating power has proved daunting, because it is extraordinarily difficult to contain a gas at temperatures of several million degrees in vessels that melt at a few thousand degrees.

In 1994 the Tokamak reactor at Princeton University set a world record by generating 10.7 megawatts of fusion energy. A new Tokamak is scheduled to be in operation in 2001, and the $10 billion International Thermonuclear Experimental Reactor is scheduled for operation by 2005. The new machines will surely advance the difficult technology of producing economically viable energy with nuclear fusion.

Beyond element number 83, bismuth, there are no elements with stable isotopes. All of these are radioactive (i.e., unstable), and all gradually change into lighter nuclei. Since the Earth is approximately 4,500 million years old, most of the remaining nuclides have a very long half-life. Among the exceptions are the short-lived intermediates in the radioactive decay of some very long-lived nuclides. Two isotopes of uranium, ^{238}U, ^{235}U, and one isotope of thorium, ^{232}Th, undergo a rather complex sequence of nuclear reactions before they finally attain stability as three isotopes of lead: ^{206}Pb, ^{207}Pb, and ^{208}Pb, respectively. Short-lived nuclides are also produced in the atmosphere and surface rocks by bombardment of various elements with cosmic rays and by the very small flux of neutrons in the Earth's interior.

The gradual conversion of uranium and thorium isotopes into isotopes of lead involves the loss of a number of alpha and beta particles. The losses are not

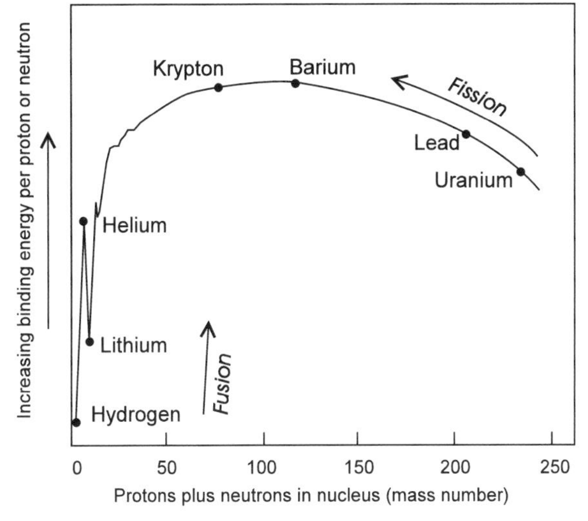

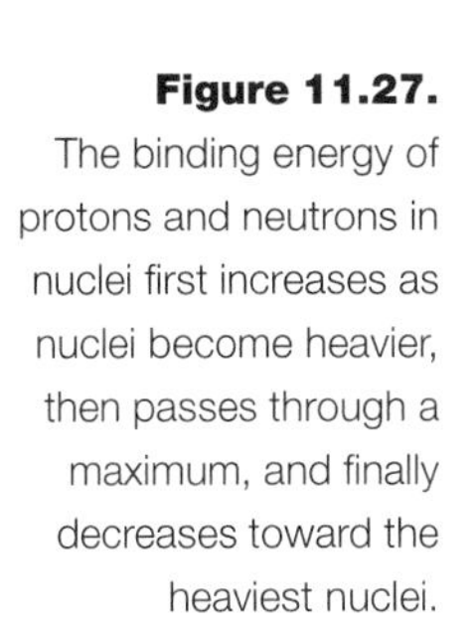

Figure 11.27. The binding energy of protons and neutrons in nuclei first increases as nuclei become heavier, then passes through a maximum, and finally decreases toward the heaviest nuclei.

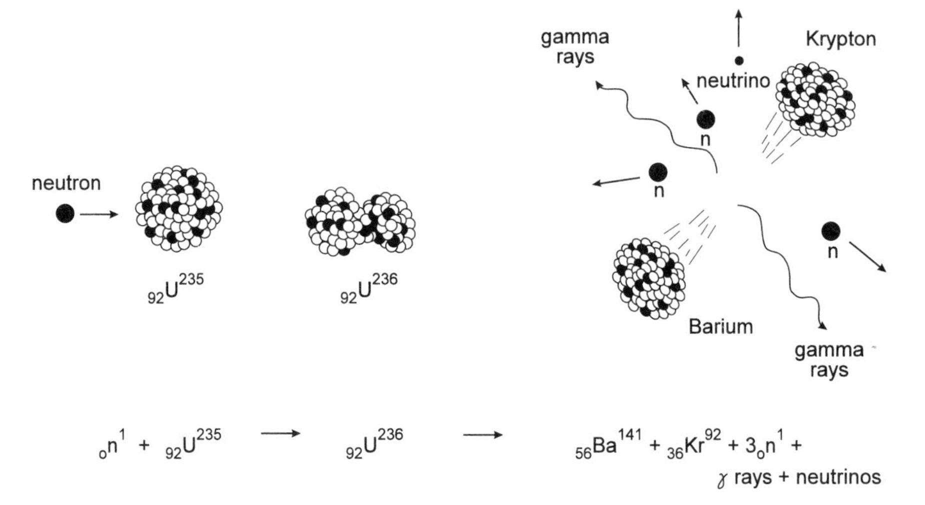

Figure 11.28. Diagrammatic representation of the reaction of a neutron with a ^{235}U nucleus. The reaction produces ^{236}U, which, in this instance, fissions into barium, krypton, neutrons, neutrinos, and gamma rays.

particularly violent events. However, on rare occasions some of these nuclei undergo the much more violent process of spontaneous fission. For some nuclei, fission can be accelerated greatly by bombarding them with neutrons. ^{235}U, ^{233}Th, and ^{239}Pu are particularly susceptible to neutron fission, and are known as *fissile nuclides*. Figure 11.28 illustrates the fission of ^{235}U. A neutron combines with a ^{235}U nucleus to produce unstable ^{236}U. This fissions almost immediately into two nuclei of somewhat different size, and liberates about three neutrons in the process. Not all breakups of ^{235}U nuclei produce barium and krypton. The spectrum of fission products is somewhat like the size distribution in the remains of plates that have been dropped on the floor: there are some large pieces, some smaller ones, and a lot of tiny fragments. If all of the neutrons released during a ^{235}U fission event encounter additional ^{235}U atoms, they will start a chain reaction whose intensity increases explosively until the supply of ^{235}U nuclei is exhausted. The process is illustrated in figure 11.29. It has been compared to the effect of dropping a mousetrap into a box filled with mousetraps that have all been carefully set. Each mousetrap, as it is sprung, springs several additional traps. Chaos ensues until all of the traps have been sprung. In the fissioning of ^{235}U, this process is so rapid that it serves as the basis for the highly effective atom bomb.

To make the fission process suitable for power generation, all but one of the neutrons liberated during the fissioning of a ^{235}U nucleus have to be absorbed by nonfissioning material. The one remaining neutron splits another ^{235}U nucleus. In such an arrangement the flux of neutrons is constant, as is the rate of power generation. The trick is to keep the neutron flux exactly constant. If more than one neutron per fission goes on to split additional ^{235}U atoms, the neutron flux increases very rapidly, and this produces explosive results. If less than one neutron per fission goes on to split additional ^{235}U atoms, the rate of power generation decreases rapidly. Nuclear reactors are therefore designed to monitor and to regulate their neutron flux very precisely. Accidents have happened when the reactor controls have not functioned properly, either due to equipment failure or due to human error.

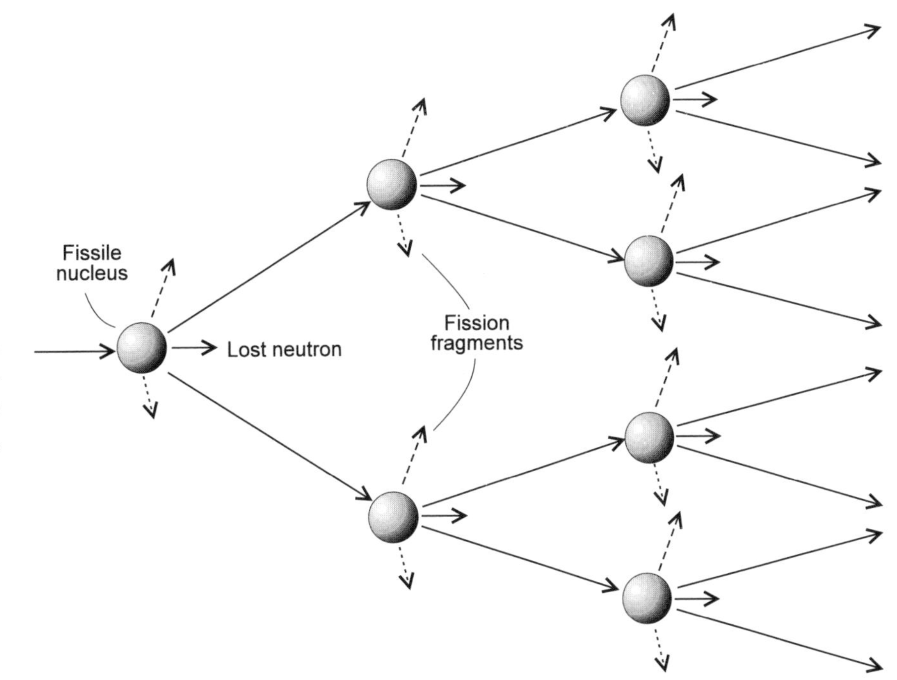

Figure 11.29. Schematic diagram illustrating a chain reaction in a fissile material.

11.3.2 *Nuclear Reactors*

There is now a rather bewildering variety of nuclear power reactors and a correspondingly forbidding jungle of associated acronyms. However, most reactors have a number of basic components in common: fuel, a moderator, coolant, control elements, a reflector, and a containment vessel. The fuel frequently consists of natural uranium, which contains 0.7% fissile ^{235}U and 99.3% nonfissile ^{238}U, or of enriched uranium, i.e., uranium in which the concentration of ^{235}U is higher than 0.7%. In some reactors the uranium is enriched in ^{233}U produced earlier by the conversion of ^{232}Th. It may also be enriched in ^{239}Pu produced earlier by the conversion of ^{238}U.

Neutrons from the fissioning of nuclei are highly energetic. Moderating materials such as water or graphite are used to slow down neutrons, because slow neutrons are more likely to cause fission than fast neutrons. The large amount of energy that is generated in reactors is removed by a coolant. This is usually water, which also serves as a moderator. In liquid metal reactors (LMRs), the coolant is molten sodium metal.

The neutron flux in reactors is usually controlled by inserting or withdrawing control elements consisting of neutron-absorbing materials such as boron or hafnium. To redirect some of the neutrons that would otherwise escape, the core of reactors is often surrounded by a neutron reflector. The nuclear portion of the reactor is contained in a reactor vessel. In reactors where the coolant is at high pressure, the reactor vessel is called a *pressure vessel.*

More than three-fourths of the power reactors that are operational and under construction today are light water reactors (LWRs), i.e., reactors in which hydrogen in the water is the light isotope, ^{1}H. Reactors of this type are expected to dominate the industry during the next several decades. The simplest of the LWRs

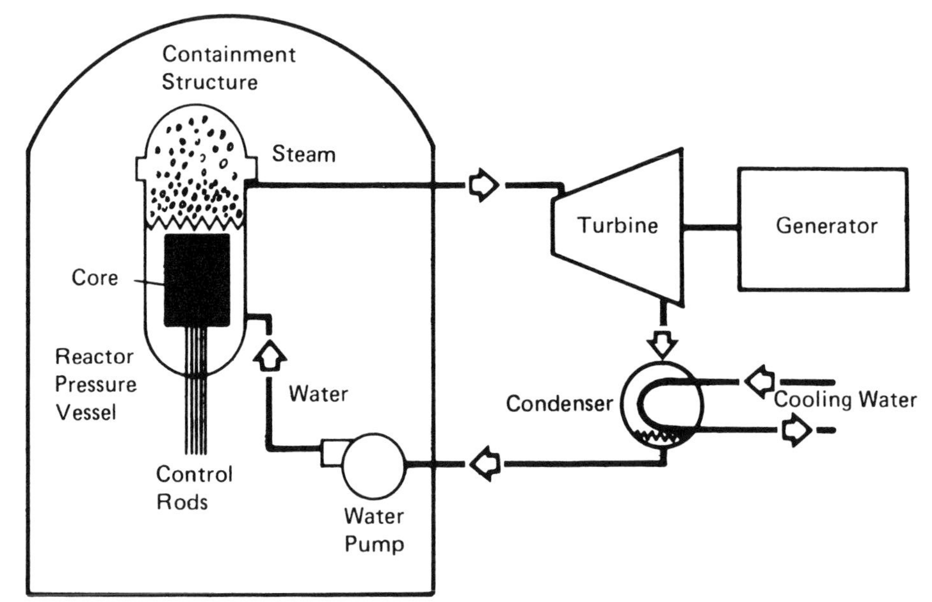

Figure 11.30. Schematic diagram of a boiling water reactor (BWR) power system. (Nuclear Energy Policy Study Group 1977)

are the boiling water reactors (BWRs). A schematic diagram of a BWR power system is shown in figure 11.30. The reactor fuel usually consists of stacks of pellets of uranium dioxide containing about 3% ^{235}U encased in tubular cladding. The heat generated by fission in the fuel rods is transferred through the cladding to the surrounding water, which boils. A mixture of steam and water flows out of the top of the core. After passing through steam separators, the steam leaves the reactor vessel and is piped to a turbine. The turbine drives a generator, which produces electricity. The exhaust from the turbine passes through a condenser and is pumped back to the reactor inlet. BWRs typically operate at pressures of about 70 atm. At this pressure water boils and forms steam at about 280°C.

Since the BWR system operates on a direct cycle, any radioactive contamination of the coolant is spread throughout the system. To prevent this, one can use pressurized water reactors (PWRs) such as the system shown in figure 11.31. The main difference between PWR and BWR systems is that in PWR systems the cooling water that flows through the reactor does not flow through the turbine. The reactor vessel is maintained at a sufficiently high pressure, so that the coolant water does not boil. Boiling takes place in steam generators that are heated by the primary coolant. After passing through the steam generators, the primary coolant is pumped back to the reactor inlet. The turbines and electrical generators are run by the steam produced in the secondary loop. Plate 42 is a view of the Yankee Rowe PWR reactor in the northwestern corner of Massachusetts.

LWRs must be shut down during refueling. This is not necessary in heavy water reactors (HWRs). These use deuterated water, i.e., water in which the hydrogen has a mass of 2 rather than 1. Heavy water is a much more effective moderator than light water. Uranium fuel that has not been enriched in ^{235}U can therefore be used in HWRs, and fuel elements can be inserted and removed automatically while the reactor is at full power, thus avoiding down-time losses.

A schematic diagram of the commercially available CANDU (Canadian-deuterium-uranium) HWR is shown in figure 11.32. The coolant flows through

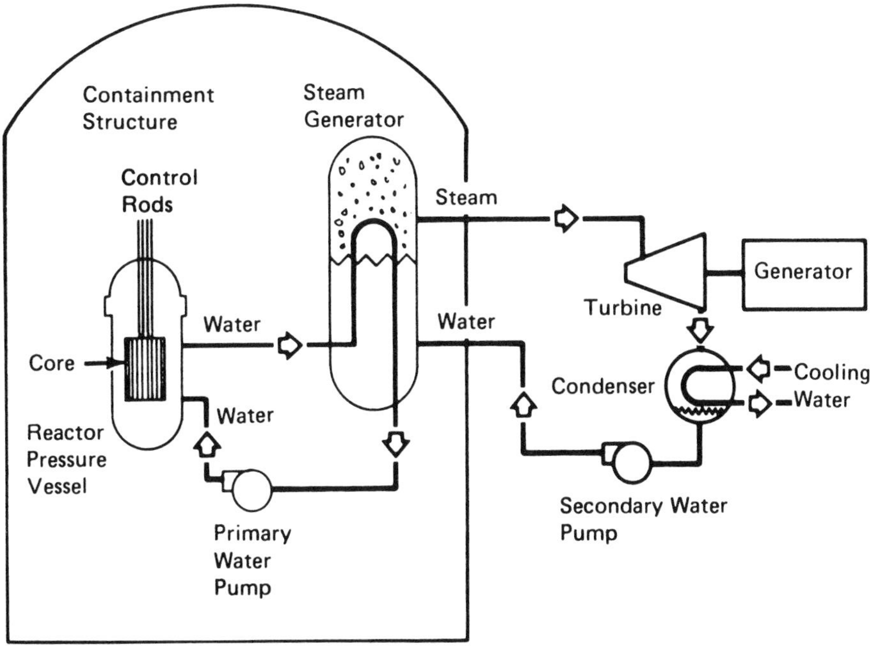

Containment
Structure
Steam
Generator
Control
Rods
Steam
Generator
Turbine
Water
Water
Core
Condenser
Cooling
Water
Water
Reactor
Pressure
Vessel
Secondary Water
Pump
Primary
Water
Pump

Figure 11.31. Schematic diagram of a pressurized water reactor (PWR) power system. (Nuclear Energy Policy Study Group 1977)

Containment
Structure
Steam
Generator
Control
Rods
Fuel
Heavy Water
Moderator
Vessel
Steam
Turbine
Generator
Heavy
Water
Heavy
Water
Water
Heavy Water
Condenser
Cooling
Water
Moderator
Primary
Heavy
Water
Circulating
Pump
Secondary Water
Pump
Heavy Water

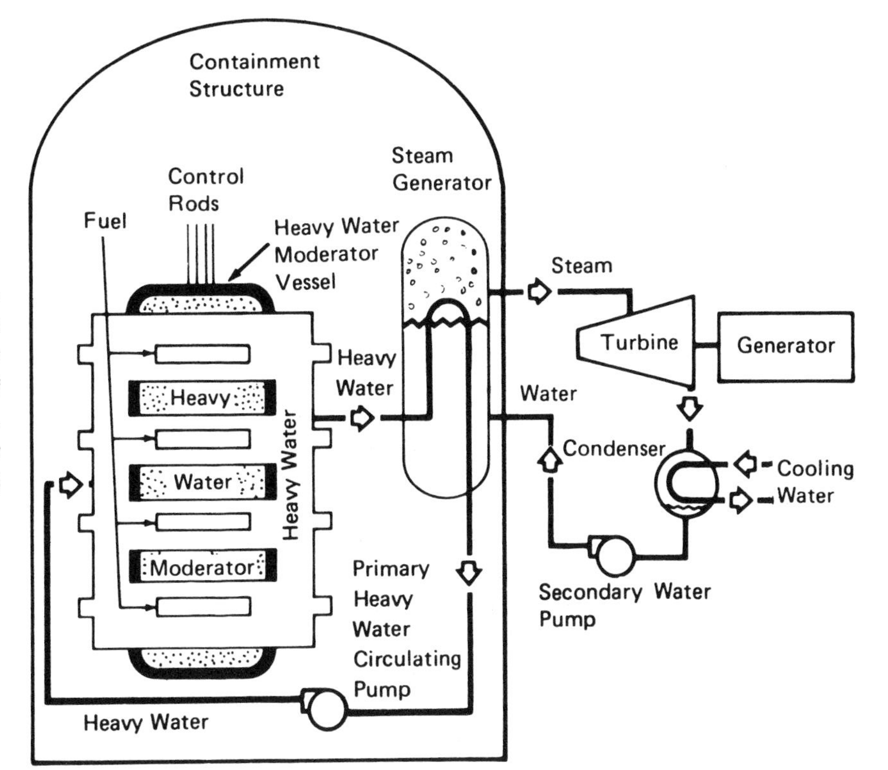

Figure 11.32. Schematic diagram of a heavy water reactor power system (HWR) of the CANDU type. (Nuclear Energy Policy Study Group 1977)

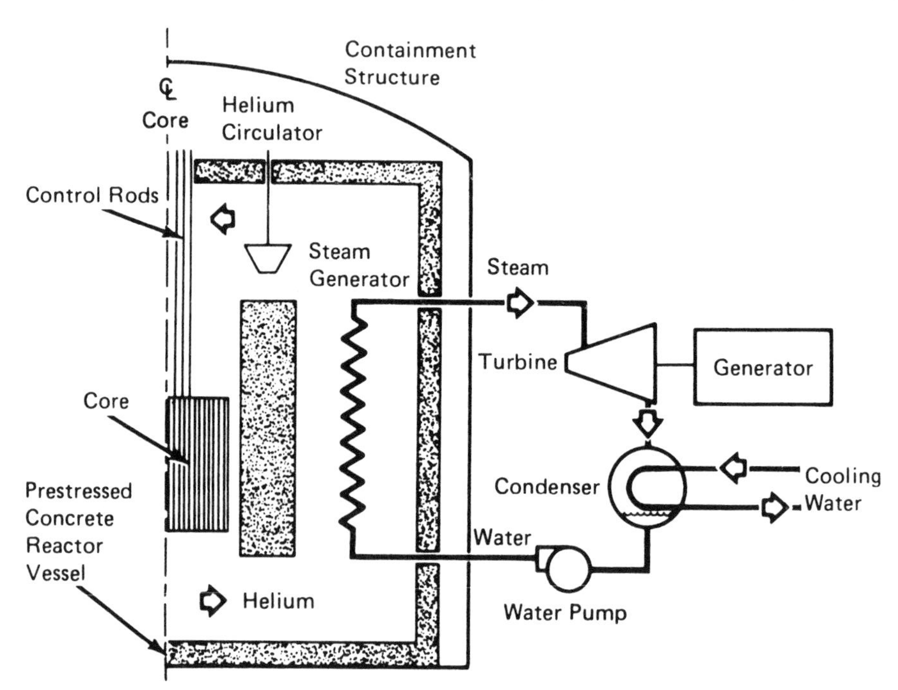

Figure 11.33. Schematic diagram of a high-temperature, gas-cooled reactor power system (HTGR). (Nuclear Energy Policy Study Group 1977)

several hundred individual pressure tubes that contain the fuel. The pressure tubes are immersed in unpressurized heavy water at room temperature. This serves as a moderator. The core is surrounded by a large cylindrical vessel. The CANDU system employs an indirect cycle similar to that of pressurized-water reactor power systems (PWRs). Ordinary light water is used as the secondary coolant, which is piped into the turbine.

Water is not the only possible coolant for nuclear reactors. In high temperature gas-cooled reactors (HTGRs), helium under high pressure is used as the coolant. The fuel materials consist of highly enriched (over 90%) ^{235}U, ^{233}U, and ^{232}Th. The mix of fuels changes during the lifetime of the reactor. Figure 11.33 is a schematic diagram of an HTGR power system. The gaseous helium leaves the reactor at 800°C and 50 atm, circulates through the tubes in the steam generator, and then returns to the reactor. Water is converted into steam in the steam generator, passes through the turbine and condenser, and returns as a liquid to the steam generator. Only one commercial HTGR reactor has operated in the U.S.; it was a notable economic failure.

The reactors described above all consume nuclear fuel. Reactors can, however, be used to generate new fuel by using unneeded neutrons to convert nonfissile nuclides into fissile nuclides. Fast breeder reactors can obtain about sixty times more energy from a given amount of uranium than the previously described nuclear reactors. The liquid metal fast breeder reactor (LMFBR) converts ^{238}U into ^{239}Pu at a rate faster than it consumes fissile fuel. LMFBR fuel rods contain a mixture of plutonium dioxide and uranium dioxide depleted in ^{235}U. A blanket of rods containing depleted UO_2 surrounds the core. An LMFBR power system is shown schematically in figure 11.34. Liquid sodium is the coolant in the primary loop. Since the sodium becomes radioactive in passing through the reactor core,

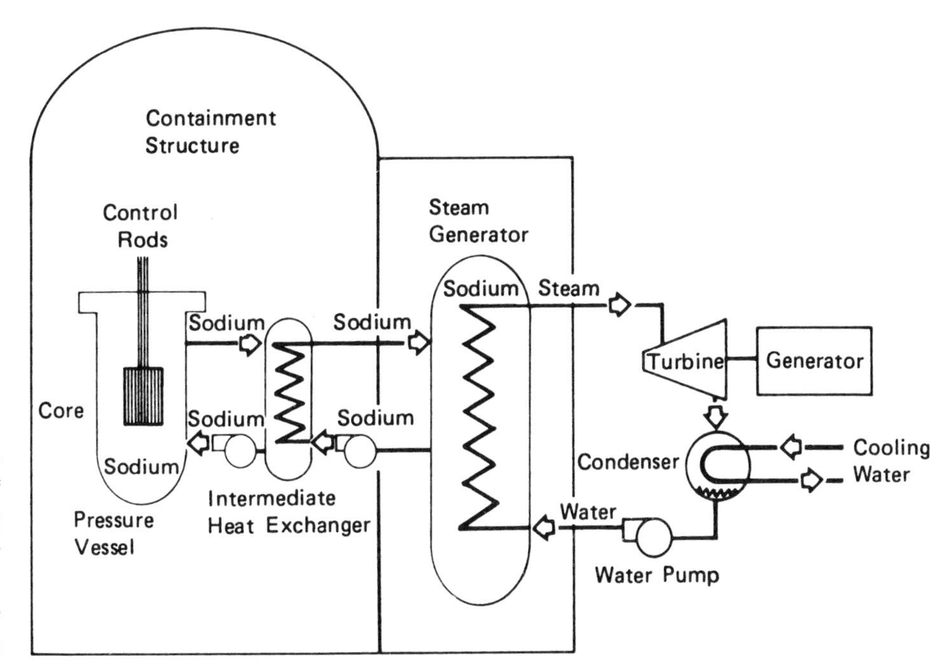

Figure 11.34. Schematic diagram of a liquid-metal, fast breeder reactor power station (LMFBR). (Nuclear Energy Policy Study Group 1977)

a heat exchanger is used to transfer heat from sodium in the primary loop to nonradioactive sodium in a secondary loop. Sodium in the secondary loop passes through a steam generator, in which heat is transferred to water. Thereafter the system is similar to other reactor systems. Breeders can also be based on much of the existing LWR technology. Light water breeder reactors (LWBRs) can produce enough ^{233}U from thorium to refuel themselves repeatedly. The fact that the LWBR is a water-cooled thermal reactor gives it a safety margin over sodium-cooled fast reactors, which are prone to sodium leakage.

The relative advantages of several advanced reactor technologies have been assessed recently by a committee of the U.S. National Research Council (NRC). Table 11.3 summarizes their relative qualitative rankings. Each of the concepts embodied in these technologies can be used to design and operate reactors that will meet or closely approach the safety objectives currently proposed for future, advanced LWRs. There seems to be no single, optimal approach to improved safety. The committee's overall assessment of the reactor technologies in table 11.3 was mostly driven by market suitability. However, the economic projections are highly uncertain, because past experience suggests higher costs, longer construction times, and lower availabilities than projected, and because different assumptions have been made for the several reactor technologies.

The large new light-water reactors are likely to be the least costly to build and to operate on a low cost per kilowatt-hour basis. The high-temperature gas-cooled reactors and liquid metal reactors are likely to be the most expensive. Safeguards against thefts of nuclear materials and security considerations did not offer much in the way of discrimination among the several reactor technologies. However, the CANDU and the LMR will require special attention to safeguards. In the NRC committee's judgment, the large evolutionary LWRs and the mid-

Table 11.3. Assessment of Advanced Reactor Technologies by the Committee on Future Nuclear Power Development of the U.S. National Research Council, 1992

		Evaluation Criteria							
Reactor Designation	*Available Design Information*	*Safety*	*Economy*	*Market Suitability*	*Fuel Cycle*	*Safeguards & Physical Security*	*Maturity of Devel-opment*	*Licensing*	*Overall Assess-ment*[a]
ABWR	○	○	○	○	◑	◑	○	○	○
APWR	○	○	○	○	◑	◑	○	○	○
SYS 80+	○	○	○	○	◑	◑	○	○	○
AP 600	◑	○	◑	○	◑	◑	◑	○	○
SBWR	◑	○	◑	○	◑	◑	◑	○	○
CANDU	○	○	●	◑	◑	◑	○	◑	◑
SIR	◑	○	◑[b]	●	◑	◑	●	●	●
MHTGR	◑	○	●	●	◑	◑	●	●	●
PIUS	●	○	◑[b]	●	◑	◑	●	●	●
PRISM-LMR	●	○	●[c]	●[c]	○	◑	●	●	◑

Legend rating: ○ high ◑ moderate ● low

Reactor designations: ABWR, advanced boiling water reactor; APWR, advanced pressurized water reactor; Sys 80$^+$, system 80$^+$ large evolutionary LWR; AP 600, advanced pressurized 600 water reactor; SBWR, simplified boiling water reactor; a mid-sized LWR with passive safety feature; CANDU, Canadian deuterium uranium reactor; SIR, safe integral reactor; MHTGR, modular high temperature gas-cooled reactor; PIUS, process inherent ultimate safety reactor; PRISM, power reactor, innovative small module, liquid metal reactor.

[a] Overall assessment was mostly driven by market suitability.

[b] Lack of design maturity results in great uncertainty relative to vendor cost projections.

[c] Long-term economy and market potential could be high, depending on uranium resource availability.

sized LWRs with passive safety features rank highest. The large evolutionary LWR reactors could be ready for deployment by 2000; the mid-sized LWRs could be ready for initial plant construction soon after 2000. Both types of reactors are likely to be significantly safer than current reactors.

11.3.3 The Safety of Nuclear Reactors

But how safe are nuclear reactors? Safety management has been built around the concept of "defense in depth," which erects several barriers to block public exposure to reactor accidents and their consequences:

1. Reactors should have remote siting to reduce population exposure in case of radionuclide release.
2. Engineering safety systems should be adequate to prevent the initiation of accident sequences.
3. Operators should be well trained to recognize and contain the development of accident sequences.
4. The reactors should be thoroughly contained to prevent the release of radioactivity after an accident involving the reactor core.
5. Emergency planning should be well developed to reduce exposure.
6. Insurance to compensate human and property losses should be available.

The 109 reactors that are currently operating in the U.S.A. (see fig. 11.35) have a very good safety record. It is not, however, perfect. At 4 A.M. local time on

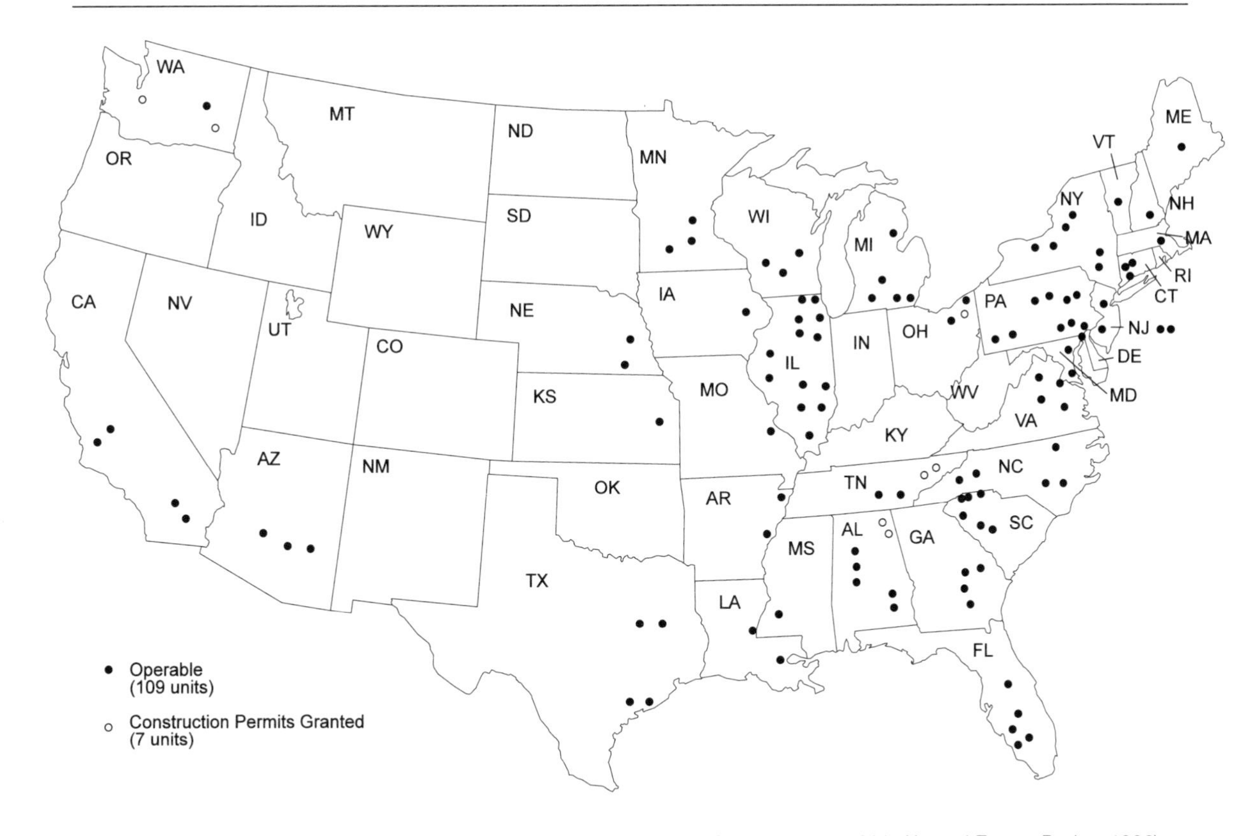

Figure 11.35. Nuclear Generating Units, December 31, 1993. (*Annual Energy Review 1993*)

March 28, 1979, Unit 2, a 850 MWe pressurized water reactor of the Three Mile Island (TMI) nuclear plant near Harrisburg, Pennsylvania, suffered automatic shutdown of its steam generator, because its feed water pumps had stopped. Moments later, as pressure in the steam generator built up, the pressure relief valve on top of the steam generator opened automatically as planned. It should have closed automatically soon thereafter but remained stuck in an open position. During the following 7 hours the reactor core was partially uncovered and melted, and some radioactive noble gases (^{133}Xe and ^{85}Kr) were released into the environment. Misjudgments by plant operators and regulatory officials enlarged the scope of the accident (Kemeny et al. 1979; Hohenemser, Goble, and Slovic 1992).

No one was killed at Three Mile Island, but the event indicated that serious reactor accidents can happen. Decontamination of the reactor has cost about $2 billion, and a large number of lawsuits have been filed by individuals who believe that they were affected by the accident. Since the time of the accident, not a single new nuclear reactor plant has been ordered in the United States. The accident is generally seen as a significant contributor to the continuing decline in the public's confidence in nuclear power.

Seven years after the accident at Three Mile Island, a much more serious accident occurred in the former USSR at Unit 4 of the Chernobyl nuclear plant, about 80 kilometers north of Kiev in the Ukraine (plate 43). The accident was clearly

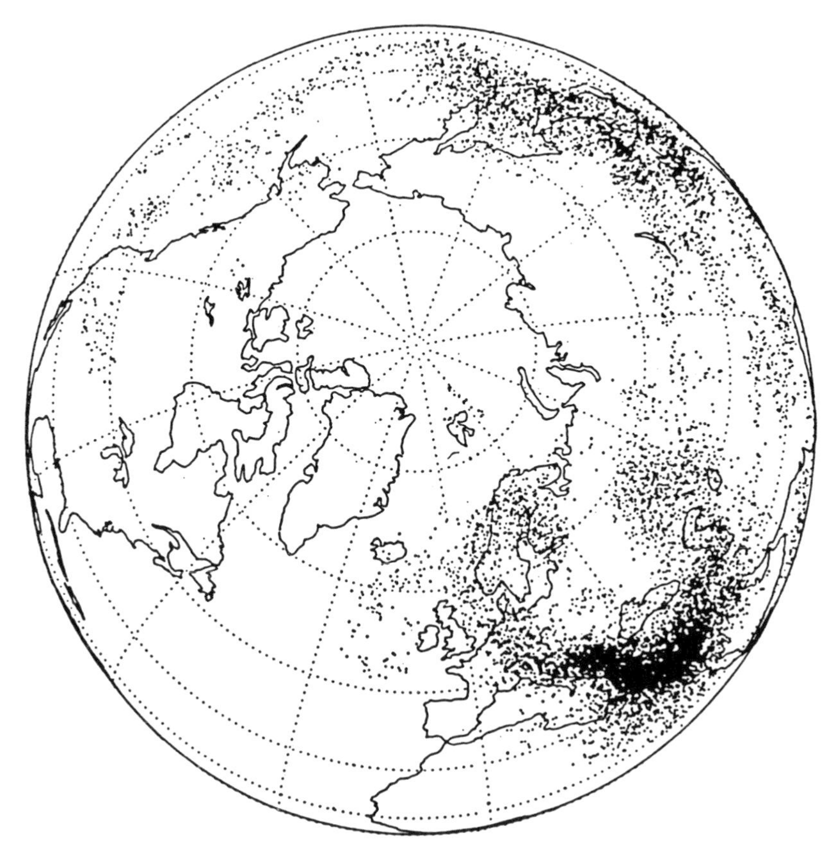

Figure 11.36. Calculated spatial distribution of radioactivity over the Northern Hemisphere 10 days after the Chernobyl accident, as illustrated by the Lawrence Livermore National Laboratory. (Lange, Dickerson, and Gudiksen 1988).

due to a chain of inexcusable operating errors by the staff of the power station. The combination of their fundamental violations of safety rules made the reactor unstable. The technicians lost control of the reactor, which suffered a prompt neutron power burst. At 1:23 A.M. on April 26, 1986, Unit 4 reached one hundred times normal power. Some of the fuel disintegrated and evaporated the cooling water. This caused a steam explosion, which sheared off the top of all 1,661 pressure tubes in the graphite-moderated tube reactor, lifted the 1,000-ton cover off the core, ruptured the inadequate containment, dislodged the refueling crane, discharged hot molten and/or pulverized fuel to an altitude of at least 7.5 kilometers, and started some thirty fires. The release of radionuclides, which was the largest ever recorded in a technological accident, continued for ten days. Dispersed radioactivity reached all of the countries in northern Europe and triggered emergency protection measures in many European countries (May 1990; Hohenemser, Goble, and Slovic 1992).

Dry weather, favorable siting, evacuation of 116,000 people, and—above all—dispersal of radionuclides to high altitudes (see fig. 11.36) contributed to holding the immediate deaths to thirty-one. However, some estimates suggest that the worldwide dispersal of reactor core material may lead to as many as 50,000 to 100,000 cancer fatalities worldwide over the next 50 years (see, for instance, Hohenemser 1988 and Chernousenko 1991). Since these fatalities are spread over such a long time period, the projected cancer fatalities are expected to exceed normal cancer fatalities by more than 1% only for the evacuees from the vicinity of Chernobyl. For non-Soviet Europe the projected increase in the cancer fatalities is only about 0.1%.

Since the accident, the stricken reactor has been encased in a concrete shell, which is cracking. A new containment shell is to be constructed. A second reactor was shut down after a fire in 1991. The other two 1,000-megawatt reactors at the site are still in operation at the center of a restricted zone roughly 30 kilometers in radius, which is managed by the Pripet Research Industrial Association (PRIA). The zone will probably require special management for 100–150 years. PRIA employees live outside the zone, but about one thousand people have returned to live inside the zone, mostly pensioners who expect the health risks of Chernobyl to matter less in old age.

About one million cubic meters of soil have been scraped up and placed in some six hundred trenches. Hundreds of artesian wells have been drilled to replace the contaminated water supply of a large part of the area (Ausubel 1991). Pulses of heavy metals and radionuclides have washed down through the Dnieper River basin, where tens of millions of people obtain their drinking water. The major ecological problem during the next few years will probably be due to the presence of radionuclides—particularly strontium, cesium, and plutonium—in the soil.

The events leading up to the Chernobyl accident involved a dramatic sequence of faulty judgments and arrogance on the part of the operators. It was also a consequence of deliberately overriding safety devices and procedures. As a result, the scale of the accident was equal to "worst case" projections for pressurized water reactors. Since the design of the Chernobyl reactor differed substantially from that of Western LWRs, the accident says little about the likelihood of a Chernobyl-scale accident in the United States. Yet the Chernobyl accident has had a substantial impact here. It demonstrated that "worst case" accidents can happen, and that transnational emergency planning is essential. The accident strongly reinforced public perceptions derived from the Three Mile Island accident, and entered the U.S. political process in the debate concerning the licensing of the Seabrook, New Hampshire, and Shoreham, New York, nuclear power plants. How can one seriously prepare to remove, contain, and bury the topsoil from hundreds of square kilometers? Preparing and publishing a map showing where six hundred trenches would be sited in New Hampshire, or on Long Island, or in Massachusetts is an unimaginable exercise. If it could be done, it would almost certainly foreclose the possibility of siting or operating nuclear power plants in populated areas.

11.3.4 *The Disposal of Nuclear Wastes*

Nuclear waste disposal is almost as contentious an issue as the safety of nuclear reactors. The total quantity of waste is rather modest, but its radioactivity is initially very high, as is the toxicity of some of its constituents. No one wants radioactive waste stored in their "backyard," and there is a general uneasiness about terrorist access to nuclear waste. The problem of finding storage sites for nuclear wastes has been complicated by the somewhat casual attitude of the defense establishment toward the disposal of military radioactive wastes, and by the political use of the radioactive waste storage problem by opponents of nuclear energy.

Can anything be stored safely for long periods of time? From a historical point of view, the answer is a qualified "yes." The Egyptian pyramids have been re-

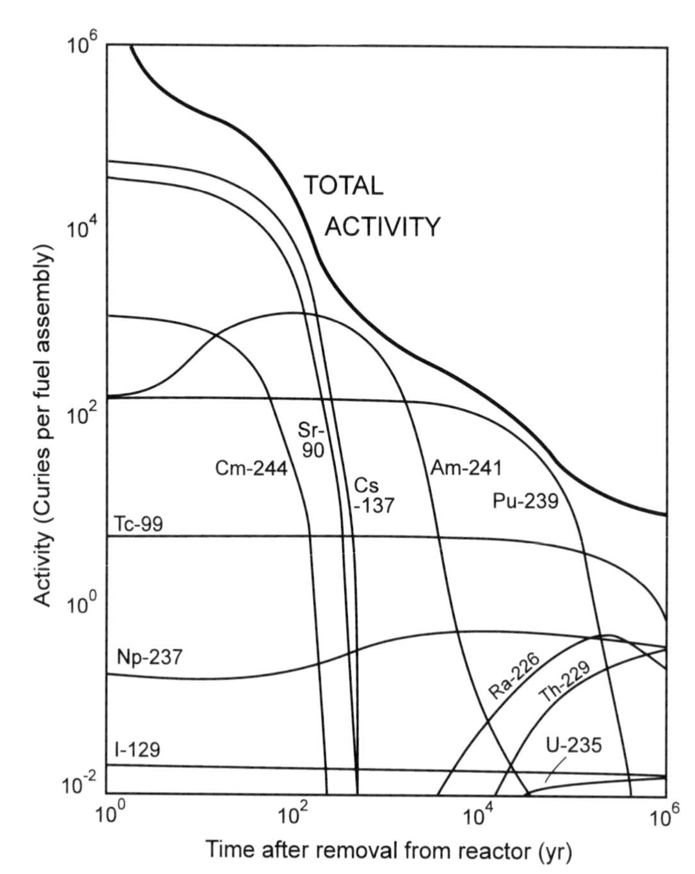

Figure 11.37. Radioactive decay of nuclides in a fuel assembly from a light water reactor. (Brookins 1984)

positories of valuable objects for several thousand years. The spectacular objects in the tomb of Tutankhamen were buried some 3,300 years ago and have been beautifully preserved in the dry climate of Egypt. However, many of the royal tombs have been rifled despite strenuous efforts on the part of their builders to prevent burglaries. The storage of valuables for periods of thousands of years is clearly possible but not always secure.

Uranium, thorium, their decay products, as well as radionuclides produced in the geologically old natural reactor at Oklo in Gabon have been retained completely or nearly so for about two billion years. This is clearly an encouraging observation for the underground storage of nuclear wastes. However, many radionuclides dissolve and are transported quite readily in oxygenated groundwaters. The behavior of radionuclides in nature therefore serves both as a reassurance and as a caution for designers of nuclear waste storage facilities.

Wastes from nuclear power reactors are highly radioactive and contain some very long-lived radionuclides. The activity of a fuel assembly from a light water reactor is shown in figure 11.37. The unit of radioactivity is the curie. One curie is equivalent to the radioactivity of one gram of radium, which undergoes 3.7×10^{10} disintegrations per second. One year after removal from a reactor, the activity of a full assembly is in excess of 10^6 curies. Much of this activity is due to relatively short-lived radionuclides. After 100 years the radioactivity of fuel elements will have decreased by more than a factor of one hundred, i.e., less than 1% of the initial radioactivity will remain. After 10,000 years the radioactivity will have decreased by another factor of one hundred, but thereafter the radioactivity of the assemblies decreases more slowly. A decrease in their activity by another factor of ten requires a wait of a million years. This does not mean that fuel assemblies remain highly dangerous for a million years. Most of the danger has abated after a few thousand years, roughly the time since the death of Tutankhamen.

In France, spent fuel assemblies from power reactors are reprocessed to recover reusable fuels, and the wastes are incorporated in glass. No reprocessing has taken place in the United States since the late 1970s. Several containment forms for waste from processed fuel have been developed, among them synroc, an acronym for synthetic rock. Synroc consists of solid phases that readily accommodate most of the radionuclides in processed wastes. To make synroc, calcined high-level waste is mixed with synroc additives and is hot pressed at 1200–1300°C.

Some military wastes are extremely complex, highly radioactive mixtures. At the Hanford site in the state of Washington (fig. 11.38), a variety of very high-level liquid wastes have been mixed in storage tanks. Many of these tanks are now at the end of their engineered lives and are leaking. The characterization, removal, and processing of the wastes poses formidable scientific and engineering problems; their proximity to the Columbia River adds an element of urgency to a solution. As an alternative to their safe storage it has been suggested that they could be destroyed in accelerators by a process known as *accelerator transmutation of wastes* (ATW). It is not clear whether, or to what extent, this solution is feasible. Until its feasibility is demonstrated, high-level wastes must be sequestered in safe repositories.

To date much spent nuclear fuel has been stored in "swimming pools" on or near nuclear reactor sites. One such facility is shown in figure 11.39. The radioactivity of water in these pools is monitored continuously, so that leaks in the

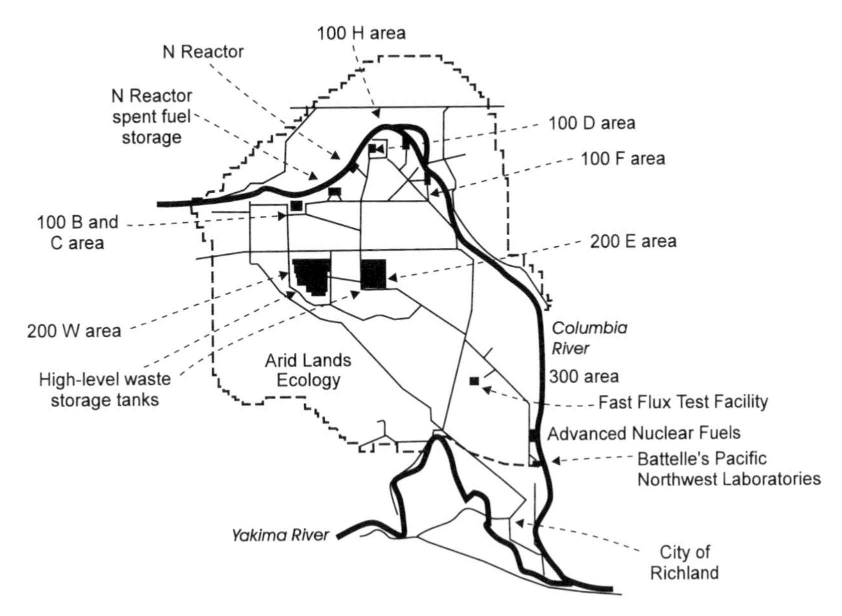

Figure 11.38. The Hanford site borders on the Columbia River in southeastern Washington. The nine plutonium production reactors were built in the so-called 100 area, along the river to the north. High-level waste tanks are buried in the 200 area, where the spent fuel was processed. The 300 area was used mostly for fabrication of reactor fuel. (Levi 1992)

Figure 11.39. Refueling the nuclear reactor at the Pilgrim Station in Plymouth, Massachusetts. The crane in the left rear of the photograph lifts spent fuel assemblies from the reactor and moves them underwater to the spent fuel pool at the right. The reactor is then refueled and is closed with the round reactor top, which is partly visible in the left foreground. (Photo by Frank Keenan, courtesy of Boston Edison Company)

fuel assemblies can be detected quickly. Although this type of storage has proved adequate in the short run, it is not a reasonable solution for the long-term storage of nuclear wastes. The long-term management of nuclear wastes requires either permanent disposal or storage in repositories where they can be monitored and from which they can be retrieved. The United States and most other countries early opted for the first alternative. In the 1950s the possibility of storing high level wastes (HLW) in salt deposits was seriously explored. Since then, the suitability of a variety of environments has been investigated. These include space, the ocean floor, and a wide variety of rock units. Disposal in space was discounted rather early, because launch vehicles are not sufficiently reliable. Disposal on the seabed was considered seriously but was finally abandoned, because the pathways of the wastes to water and from there to the biosphere were considered to be too short.

HLW disposal on land in areas that are very dry, very thinly populated, and tectonically stable is scientifically most attractive, but its implementation has proved politically difficult. No one is anxious to be close to an HLW repository, and proponents of particular sites have found it essentially impossible to demonstrate that HLW can be stored there safely for 10,000 years. Risk assessments for even the most favorable sites must allow for a variety of unlikely events. The risk of failure is never zero, and this jeopardizes the licensing of all permanent repositories. After decades of pushing for a permanent repository for radioactive wastes from power plants, the nuclear power industry is reluctantly concluding that one or more temporary storage sites will be needed before a permanent one can be ready. Financial experts estimate that an Indian tribe or a county willing to rent a patch of dry land about half the size of Central Park in New York City for temporary storage could collect $50 million a year for at least 20 years (*New York Times*, August 27, 1993).

The most likely candidate site for a permanent repository or possibly a monitored retrievable repository of high-level wastes in the United States is at Yucca Mountain in southern Nevada, close to the Nevada-California state line (fig. 11.40). The area is dry and desolate, as shown in figure 11.41. Although it is probably as well suited as any site in the States, objections have been raised to building an HLW repository there. The area is not particularly stable tectonically, and volcanism has occurred there during the last 500,000 years. The water table is currently several hundred meters below the level of the planned repository, but might rise significantly during the next 10,000 years if the climate were to become much wetter. Although this is unlikely, the possibility is difficult to rule out. Natural climatic changes on this timescale approach those of glacial-interglacial periods, and the regional effects of changes in the composition of the atmosphere due to fossil fuel burning are currently impossible to predict.

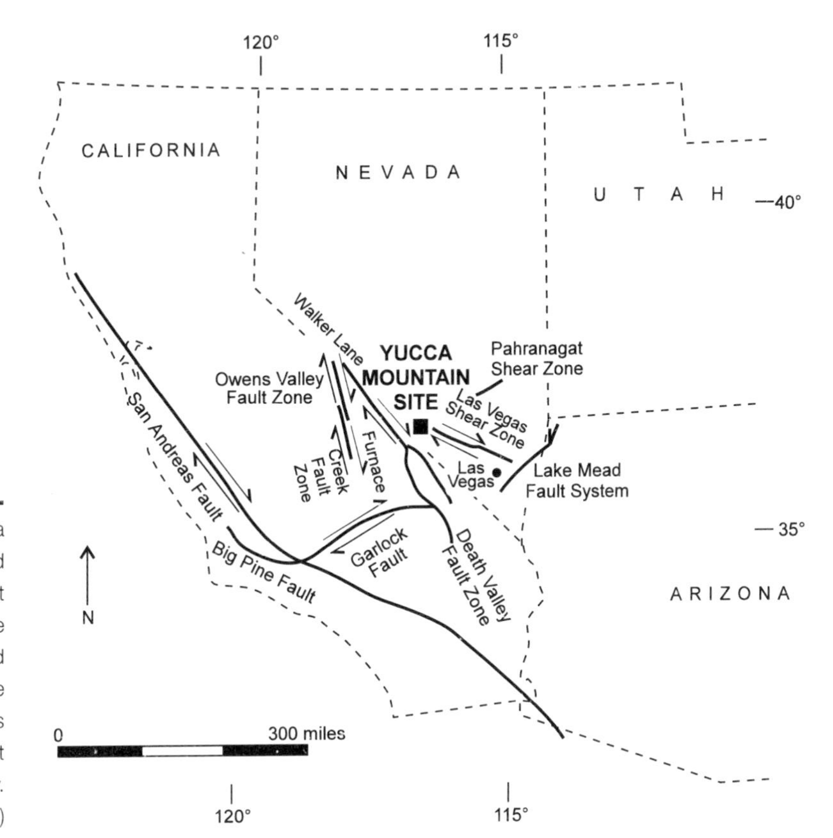

Figure 11.40. Location of Yucca Mountain, the proposed site for a permanent HLW repository in the southwestern United States. Also shown are major strike-slip faults of the southern Great Basin and vicinity. (Tien et al. 1985)

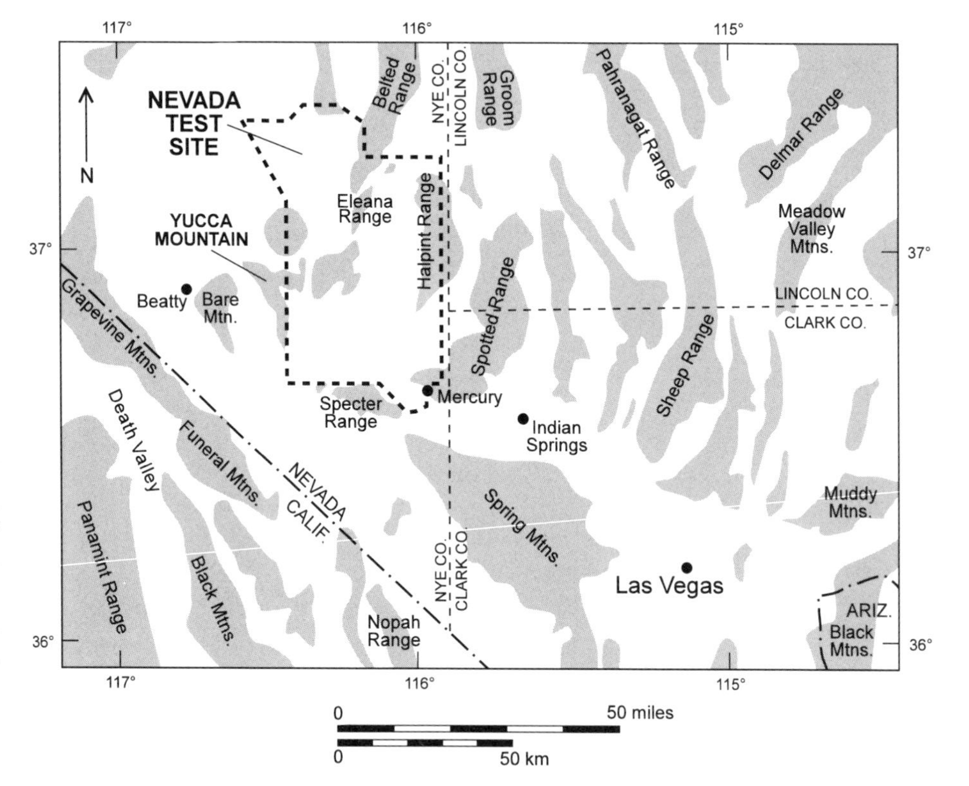

Figure 11.41. Topography of southwestern Nevada and the location of Yucca Mountain and the Nevada Test Site (NTS). Note location of Las Vegas. (Tien et al. 1985)

The location proposed for the repository is 25 miles south of the Nevada Test Site (NTS) at Yucca Flats, an area that has been heavily contaminated by the underground testing of nuclear weapons (fig. 11.41 and plate 44). The products of underground nuclear explosions are not carefully contained. Some radionuclides have already migrated considerable distances from their source at Yucca Flats, and it can be argued that minor additional radionuclide contamination from a repository at Yucca Mountain will be of little consequence for the region. Understandably, this line of argument has carried little weight with opponents of siting an HLW repository anywhere in the state of Nevada.

If and when a repository is opened, the HLW will be isolated from the environment by several barriers. There is, first, the waste form itself. This will be placed in a canister, probably of titanium alloy, which is designed to be leak-proof for 300 to 1,000 years. The canisters will be surrounded by a backfill, which will probably consist of clay minerals that tend to retain radionuclides and limit groundwater access to the canisters. The combination of the waste form, canister, and engineered backfill will probably be sufficient to isolate the radionuclides in HLW from the surrounding environment for periods in excess of 500,000 years. This is well in excess of the required quarantine period. An absolute guarantee of isolation cannot, however, be given. If a volcanic vent were to cut directly through the repository at Yucca Mountain, the consequences could be disastrous. Although the probability of such an event is very small, it is not zero. The notion of designing repositories from which HLW can be retrieved is therefore appealing. Such monitored retrievable storage sites, MRS, would be considerably easier to license because the guaranteed safe isolation time of HLW would be relatively short.

11.4 Water Power

The potential of running water as a source of power was recognized more than 2,000 years ago. Greek historians described how water wheels were used to grind corn during the first century B.C. The Roman Empire relied heavily on water-powered mills. By A.D. 1200, more than five thousand water mills were operating in Britain. Figure 11.42 shows how water wheels were used to dewater mines in the sixteenth century. The use of water power continued to expand during the following centuries. In the nineteenth century, the development of hydroelectric power plants ushered in a new era for water power. The number and installed capacity of these plants increased rapidly in the late nineteenth and early twentieth centuries. The combined efforts of the U.S. Corps of Engineers and the Bureau of Reclamation together with other agencies and private companies built more than two million dams on the nation's rivers. Eighty-seven dams impound

Figure 11.42. Machine for dewatering mines. A stream is diverted into a reservoir from which water is supplied to the water wheel. (Agricola 1556)

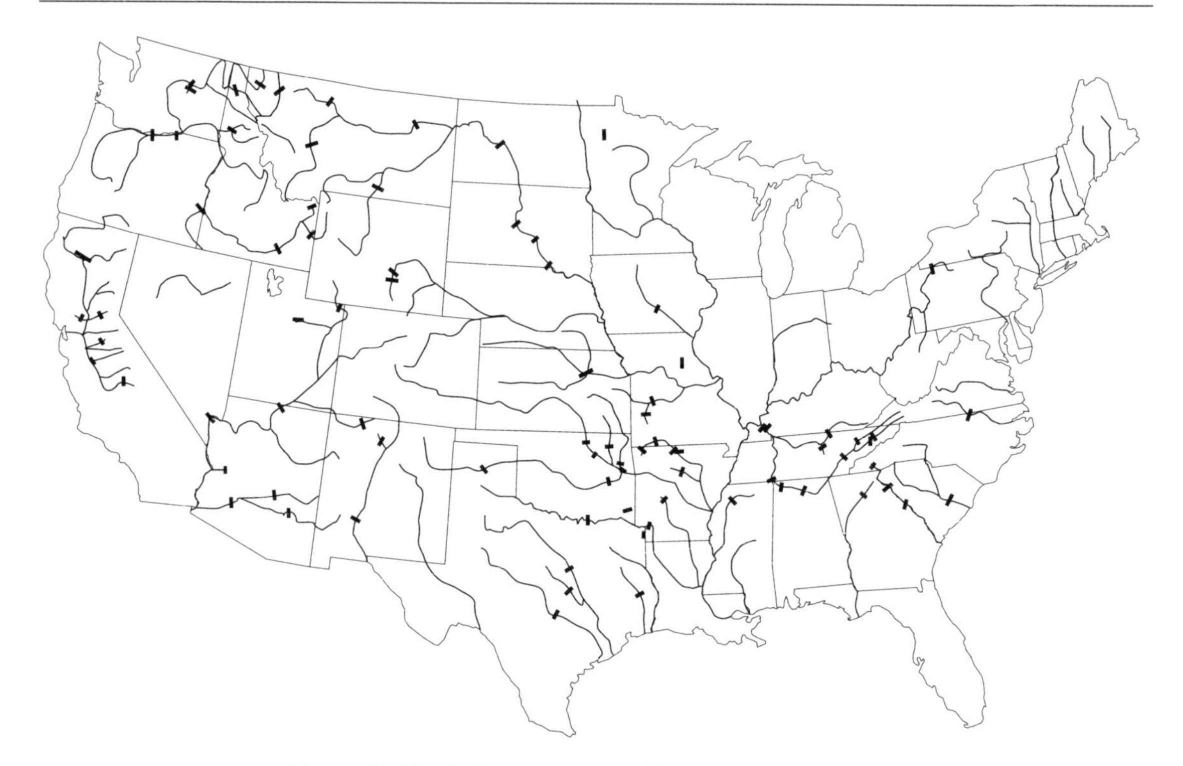

Figure 11.43. Distribution of dams with a reservoir capacity of one million acre-feet or more in the continental United States. (Sources: U.S. Department of the Interior 1986, U.S. Department of the Army 1986, and van der Leeden et al. 1990; from Graf 1993)

reservoirs with a storage capacity in excess of a million acre-feet (see fig. 11.43). Currently only 2% of the nation's rivers remain undeveloped (Graf 1993). Worldwide, hydropower production increased sixfold between 1950 and 1989. In 1991 hydropower provided about 4% of total U.S. energy and some 9% of U.S. electrical energy (fig. 11.44). On the world stage, hydropower provided about 6% of all energy in 1989 and about 19% of all electrical energy (fig. 11.45). The total amount of hydroelectric energy generation has increased dramatically during the entire twentieth century, but its share of both the U.S. and world energy production has been declining since midcentury, because the growth of generating capacity from fossil fuel and nuclear sources has far outstripped that of hydropower.

Although the contribution of hydropower to world energy production is modest, it is the dominant electricity source in several countries. At the end of the 1980s, hydropower accounted for over 99% of the total electricity production in Norway, 95% in Nepal, 93% in Brazil, 78% in New Zealand, and 58% in Canada. This enormous range in the relative importance of hydropower is due to differences in the availability of suitable sites for large dams and penstocks, and differences in the availability of other relatively inexpensive sources of energy. Hydropower has a number of important advantages over energy production by fossil fuel burning and nuclear reactions. Hydropower generates neither green-

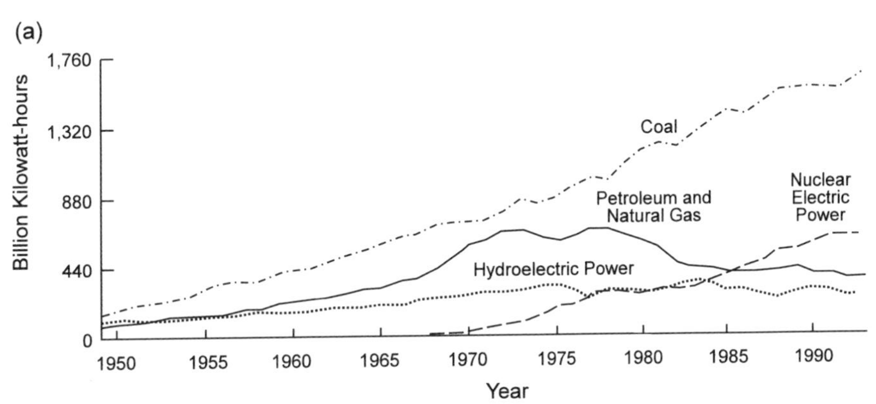

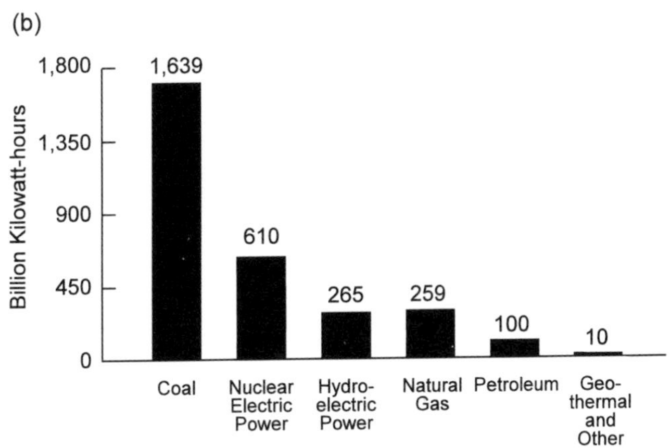

Figure 11.44. (a) The major sources of U.S. net electricity generation from 1949 to 1993; (b) the contribution of the major sources of U.S. net electricity generation in 1993. (*Annual Energy Review* 1993)

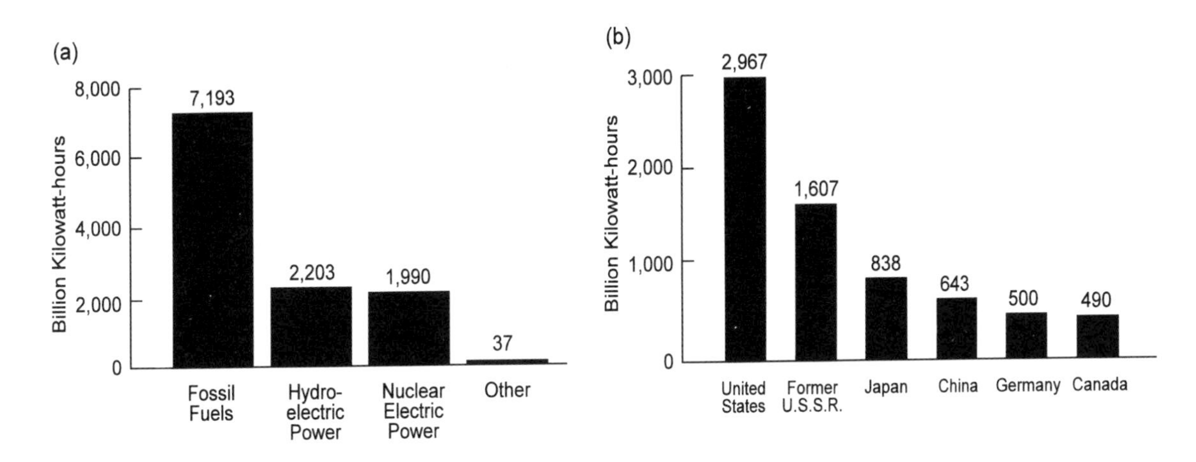

Figure 11.45. (a) World net generation of electricity by type, 1991; (b) world net generation of electricity by leading countries, 1991. (*Annual Energy Review* 1993)

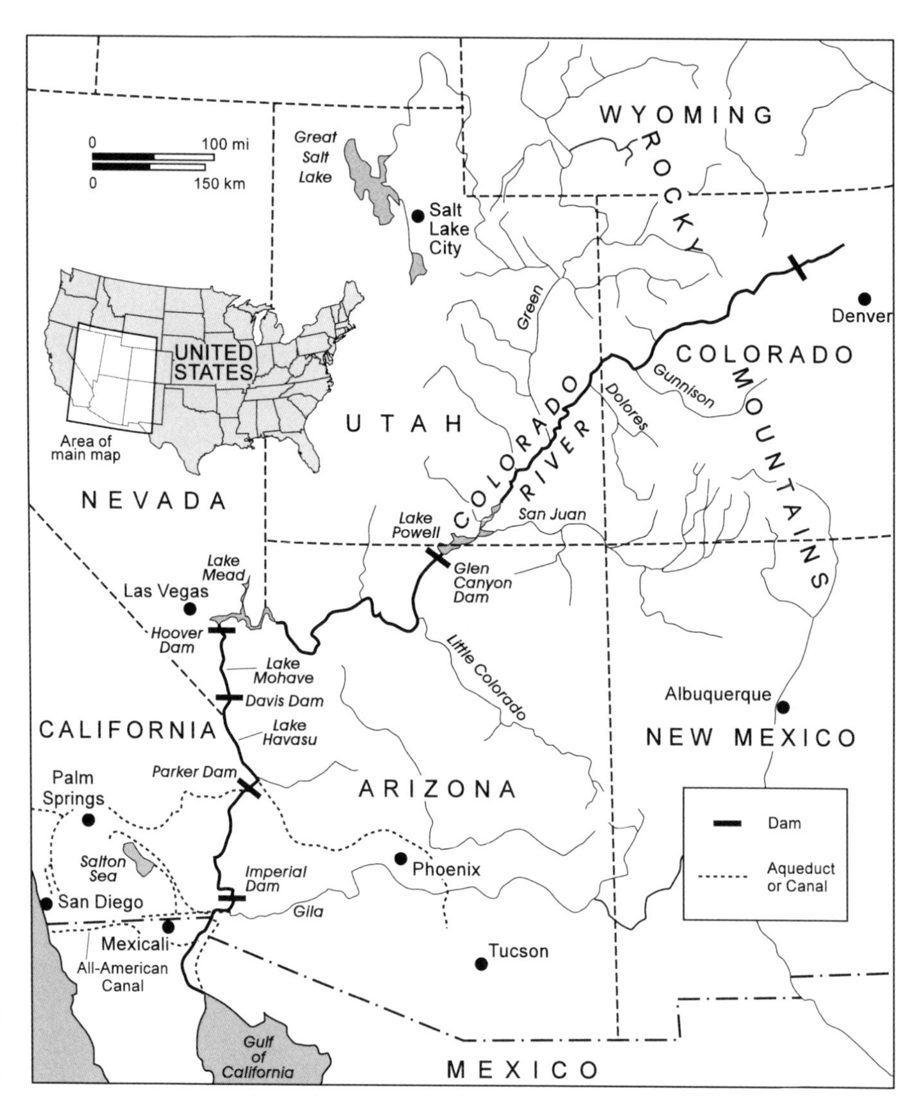

Figure 11.46. The course of the Colorado River. (*Time*, July 22, 1991)

house gases nor other objectionable wastes, and dams serve several useful functions besides power generation. In the American West and in other dry areas, reservoirs behind dams supply water for drinking and irrigation. In the early days of the American West, the availability of water was often a matter of life and death. The importance of water has been celebrated in many Westerns.

No American rivers illustrate the importance and difficulty of water management better than the Columbia (Dietrich 1995) and the Colorado (see fig. 11.46). The Colorado begins high above the tree lines among the glaciers and snowpack on the western slopes of the Rocky Mountains. In Utah it is joined by the Green River and by the San Juan. The river rushes through the Grand Canyon of Arizona (plate 45), and finally trickles into the ocean at the head of the Gulf of California. In the past half-century the Colorado River has been tamed by a spectacular system of dams and reservoirs. It supplies water to 20 million people and 2 million acres of farmland. The six major dams along its course generate 12 million kilowatts of electricity. Some stretches along the river are much as they once were, but people also come to the Colorado to play. Six national parks and recreation areas along the Colorado support a multimillion dollar recreation industry of boating, hiking, fishing, and white-water rafting. These are a dream for vacationers and a nightmare for nature lovers and conservationists (see plate 46).

The 1922 Colorado River Compact divided up the water among the four Upper Basin states (Utah, Wyoming, Colorado, and New Mexico), and the three Lower Basin states (California, Nevada, and Arizona). The Upper Basin states have never used all of the water allotted to them; the surplus was often picked up by the Lower Basin states—mostly California—but the demand for water has risen dramatically since 1922. Powerful interests have staked out claims to the unused portion of this critical resource; its use is complicated by uncertainties about Indian water rights, and the issue promises to be very contentious for a long time to come.

Most large hydroelectric projects involve the inundation of large areas. This is particularly true in relatively flat country, where river gradients are low. The Canadian province of Quebec is one of these regions. Huge areas in Quebec will be flooded if the planned damming of the Great Whale River by Hydro-Quebec is carried to completion. The project would affect the regions east of James Bay and southeast of Hudson Bay, and would be one of the largest hydroelectric complexes in the world. The first phase of the project, south of the Great Whale Basin, is known as the La Grande River complex and has been completed. In phase 2, Hydro-Quebec plans to push north into the Great Whale area. In phase 3, the Nottaway and the Rupert Rivers are to be diverted into the Broadback River. If and when all three phases are completed, the project would produce 26,400 megawatts of electrical power, of which 15% would be available for export.

Quebec's expansive plans are drawing bitter resistance from native Cree Indians, who are concerned about preserving their land and traditional Indian ways of life, as well as from environmentalists, who warn of impending environmental disasters. On the other hand, many Québecois tend to see hydropower as the province's pathway to economic and possibly political independence from the rest of Canada. The future of the project is somewhat clouded, in part because the demand for additional electricity in New York City and along the eastern seaboard of the United States is not increasing as rapidly as had been projected when Hydro-Quebec laid its plans in the 1970s. For New York the disadvantages of generating electricity locally in fossil fuel burning plants early in the next century are much greater than those of importing Canadian hydroelectric power.

The inundation of large land areas is not always the most serious source of environmental problems created by large hydroelectric projects. The Aswan High Dam on the Nile River now supplies approximately 20% of Egypt's electricity. The water stockpiled in Lake Nasser, the reservoir behind the Aswan Dam, spared Egypt from the devastating effects of the severe East African droughts of the 1980s and from potentially damaging floods when the rains finally returned (Golob and Brus 1993). However, the dam also acts as a barrier for the silt and nutrient-rich sediments that are carried by the Nile. Virtually all of these sediments are now trapped behind the dam; virtually none are brought down to the Nile Delta. The intervention has resulted in a series of responses that threaten the northern Nile Delta (Stanley and Warne 1993; see plates 47–49). Erosion, salinization, and pollution are inducing a marked decline in agricultural productivity and a loss of land and coastal lagoons at a time when the population of Egypt is expanding exponentially. Due to delta subsidence during the next 60 years, the effective sea level is likely to rise at least 1 millimeter per year. The subsidence rate along the coast will range from 1 to 5 millimeters per year. Erosion will dominate along the shoreline. Lagoons will be infilled on their seaward margin, and pollution levels will increase substantially. Nile waters will flow into ever smaller, fragmented lagoons and will deposit their sediment load of municipal, agricultural, and industrial wastes. Northern Egypt has ceased to be a balanced delta system due to human intervention. Measures to reverse the decline can be envisioned, but will probably prove inadequate at the current levels of population growth in Egypt (Stanley and Warne 1993).

Controversy is also swirling around the proposed Three Gorges Dam along the Yangtze River in China. If and when it is completed, this dam will have by far the largest rated electrical capacity of any dam. It will have nearly three times the capacity of the Grand Coulee Dam, the largest in the United States. The dam has been under discussion since 1919. The major reason for this enormous undertaking is flood control rather than the generation of electricity; but the costs are staggering, probably some $17 billion. Some 1.2 million people would be forced to move, and critics claim that silting behind the dam is apt to produce a $30 billion bog that would inundate China's finest natural scenery and that would stand as a monumental and vexing legacy of Chinese communism. There are good reasons for this forecast. When Soviet engineers designed the 350-foot high Sanmenxia Dam on China's Huanghe (Yellow River), they failed to consider adequately the immense amount of silt that this river carries to the sea. Silt deposited behind the dam nearly filled the reservoir in just four years. The dam has since been extensively reengineered to allow silt to continue downstream, but the capacity of its reservoir has been reduced by a factor of three. Nevertheless, there are compelling reasons for building the Three Gorges Dam. The recurrence of the devastating flood of 1870, when some 100,000 lost their lives, is one of China's worst nightmares. If the dikes along the Yangtze were to burst at night, hundreds of thousands of people might well die. If the presence of the Three Gorges Dam could prevent such a disaster, its construction would be worthwhile.

Small hydroelectric power plants, those which supply electricity to individual cities or industrial concerns, tend to have few or negligible environmental impacts. They usually divert only part of a river into a canal or tunnel. Upstream inundation is therefore negligible. The walls used to deflect river water into canals or tunnels do not affect river ecology, and small dams can use spillways that function like rapids. A combination of canals and tunnels takes the water to a penstock where it drops to a hydroelectric power generating plant. The water is then returned to the river. The hydroelectric plants that can be built this way may be quite large. In Peru a hydroelectric plant that uses a 20-kilometer tunnel and a drop of 850 meters generates 798 megawatts of electricity. Such plants make sense when it is unnecessary to supply water to distant regions or to manage the stream flow. Unfortunately, this type of responsible ecological and cultural stewardship precludes the use of the full energy generating capacity of a river.

11.5 Other Sources of Energy

Fossil fuels, nuclear fuels, and water power together account for nearly all of the energy produced in the United States. In 1991 only about 0.3% of the energy generated by electric utilities was derived from geothermal plants and about 0.03% was produced from wood, waste, wind, photovoltaic, and solar thermal sources connected to electric utility distribution systems (Annual Energy Review 1991, table 2). So why should one bother to describe these sources of energy? It turns out there are three cogent reasons for doing so. A good deal of the energy from some of these sources is not connected to electric distribution systems; in some parts of the world, energy production by these globally minor sources is by no means negligible; and—perhaps most important—we need to consider whether these energy sources might become a great deal more important during the twenty-first century.

11.5.1 Geothermal Energy

Geothermally heated springs have been used for bathing and cooking for thousands of years. The Romans and the Japanese bathed in geothermally heated pools 2,000 years ago, and people in Iceland cooked with geothermal heat as early as the ninth century A.D. During the last several centuries many European hot spring areas became fashionable spas, none perhaps more fashionable than Bath, some 110 miles west of London (fig. 11.47). The baths there were important during the Roman occupation of Britain, but the golden age of Bath began at the beginning of the eighteenth century, when Richard "Beau" Nash became master of ceremonies and restored order and good manners to Bath society. On the side of decorum, he did not like the custom of the sexes bathing together; but one day at the Cross Bath, when a certain husband said that his wife, in the water, looked so like an angel that he wished to join her, Nash, to confirm his own reputation as a man of gallantry and spirit, threw the husband in (Connely 1955, p. 29).

The three mineral springs that have brought such fame and wealth to Bath yield more than 500,000 gallons of water per day at temperatures ranging from

Figure 11.47. The King's Bath, filled from a hot spring at its center. At the left, the Pump Room, Bath, U.K. (Connely 1955)

114°F (46°C) to 120°F (49°C). The waters are taken either as a drink or in the form of baths or douches. During the reign of Beau Nash, nobility and royalty lent such glamour to the baths that the assembly rooms came to figure prominently in the pages of Henry Fielding, Tobias Smollett, Fanny Burney, Jane Austen, and Charles Dickens.

On a somewhat more prosaic note, in Iceland the Reykjavik Municipal District Heating Service, which oversees the largest program of its kind in the world, has tapped geothermally heated water since 1930. In 1990 about 85% of all residential buildings in Iceland were heated geothermally. Space heating with geothermally heated water in Reykjavik costs less than one-half as much as space heating with oil. Reykjavik is no longer unique. Geothermal district heating projects are providing substantial amounts of energy in Ferrara, Italy; in the Paris Basin and Bordeaux, France; and in Mons, Belgium.

In volcanically active parts of the world, shallow groundwater can reach temperatures well above 100°C. The distribution of subsurface water temperatures in Yellowstone Park, Wyoming, was described in chapter 9. Drill holes in such areas can encounter very hot water, steam under pressure, or a mixture of the two. When brought to the surface, they can be used to drive turbines that generate electricity. Yellowstone, by virtue of its position as a national park and with its enormous number of visitors, is not in danger of being blessed with a power station.

The Geysers area about 90 miles north of San Francisco is currently the most developed geothermal resource in the world (see plate 50). Twenty-six generating units have been installed there since 1960. These have a total electrical generating capacity of about 2,000 megawatts. They account for about three-quarters of the total U.S. geothermal energy capacity and for some 35% of the world total (Golob and Brus 1993). Among the other countries with large installed geothermal capacity, the Philippines, Mexico, Italy, New Zealand, and Japan rank highest. All of the installations in these countries are located close to areas of current or recent volcanism. None of them contribute more than a small fraction of the electrical energy of their country.

Geothermal energy is frequently classed among the renewable energy sources. This is not quite correct. The steam and water produced from geothermal reservoirs are usually discarded. The reservoirs are therefore gradually depleted; water and steam are in a real sense being mined. The plants at The Geysers have not been generating power at full capacity because of steam depletion in the reservoir. Only about one-quarter of the steam that has been withdrawn from The Geysers has been reinjected into the reservoir; the rest has been vented into the atmosphere. Although natural resupply and reinjection of water and steam underground solve the depletion problem, they also cool the reservoir. Without resupply and reinjection, the life of geothermal systems is limited by the quantity of available steam and water. With reinjection, the life of these systems is limited by the quantity of available heat.

Aside from their occasional noisiness, geothermal power plants can also pollute their environment with hydrogen sulfide, mercury, and arsenic. Reinjection of water and steam into underground reservoirs prevents a good deal of this pollution. In some instances the concentration of gold in geothermal fluids is sufficiently high, so that its recovery is becoming commercially attractive.

11.5.2 Solar Energy

Like geothermal energy, solar energy has been used for heating purposes at least since Roman times. The sudden shock of steeply increasing oil prices in the early 1970s gave a great impetus to energy conservation. Thermostats were lowered, home insulation was improved, and new houses fitted with solar heating were designed and—to a limited extent—built. Passive solar heating involving no additional machinery optimized the use of available solar energy. Active solar heating employed heat pumps and other devices for heating purposes.

Solar energy use in electricity generation using power towers, parabolic trough mirror systems, and parabolic dish collectors of sunlight was explored enthusiastically in the 1970s. Although many of the new technologies have achieved functional success, they have not yet been an economic success. As the price of fossil fuels retreated in the 1980s, government subsidies for alternative energy development dwindled. Supporters of the new power sources point out that they are relatively pollution free and that their use should therefore be encouraged by government policies designed to enhance their competitiveness.

Perhaps the most successful of the new technologies for harnessing solar energy involves the use of photovoltaic energy conversion. Crystals of silicon can be doped with small amounts of impurities. Some, like phosphorus, contain a larger number of labile electrons than do silicon atoms; others, like boron, contain fewer. The two types of doped silicon can be joined to form cells that convert the energy in solar photons into electricity. The cost per watt of electricity generated by photovoltaic (PV) cells has decreased dramatically during the last two decades, but it is still four to five times higher than the cost of conventionally generated electricity. Plate 51 shows the 2-megawatt photovoltaic power plant in Sacramento, California. This and similar photovoltaic power plants have served as testing grounds for new PV technology, but at present the total electrical capacity of all PV models is far less than the capacity of a single coal-fired or nuclear plant. If PV power plants do become economically attractive, a rather large land area will be needed to generate a significant fraction of the current U.S. electrical power. To generate an amount of electrical energy that is equal to the present U.S. needs, an area of about 8,000 square miles would have to be covered with PV cells; that is equal to the area of Massachusetts.

PV cells have already made their mark in applications where the cost of the delivered energy is of minor importance. Calculators, watches, battery chargers, solar outdoor lights, navigational aids, warning signs, and satellites have benefited greatly from PV technology. Solar energy can also be competitive for homes in remote locations that are not served by transmission lines.

11.5.3 Wind Power

Wind is also an ancient source of energy. Sailing ships plied the Nile River 5,000 years ago and traveled the Mediterranean Sea shortly thereafter. Windmills were used more than 2,000 years ago in China, India, and Persia, and they have been used in Europe since the twelfth century. Figure 11.48 is a Rembrandt etching of a seventeenth-century Dutch windmill. Between 1850 and 1970, 6 million windmills were installed on the Great Plains to supply mechanical energy for pumping irrigation and drinking water (Golob and Brus 1993). Rural electrification led to a decrease in the use of wind energy. Interest in wind energy revived briefly in the 1980s, but the expiration of tax credits in the mid-1980s signaled the end of the wind boom and led to a shakeout in the wind energy industry.

Figure 11.48. *The Mill* by Rembrandt van Rijn, 1641; the etching illustrates the use of wind power in the seventeenth century. (Courtesy of the Fogg Art Museum, Harvard University Art Museums; gift of William Gray from the collection of Francis Calley Gray)

Wind turbines are built with either a horizontal or a vertical shaft. They are usually deployed in large wind farms. Plate 52 shows a few of the 7,500 wind turbines at Altamont Pass in California, a particularly windy place, where some 1.1 billion kwh of electricity were generated in 1991. The design of modern wind turbines benefited a great deal from progress in the design of helicopter rotors. The demands on the design of wind turbine blades are in some aspects greater than on helicopter blades, because wind turbines have to operate efficiently under conditions of highly variable wind speed and wind direction. Design improvements during the last two decades have dropped the cost of wind-generated power quite dramatically. At some of the best sites in the United States, the cost is now below seven cents per kilowatt hour, which is comparable to the cost of electricity from new coal-fired and nuclear plants. Nevertheless, in 1991 the total amount of wind-generated electricity in the states, most of it produced in California, was only 2.7 billion kwh, about 0.1% of the total net U.S. electricity generation.

Wind energy could produce much larger quantities of electricity. If all of the available wind power were harnessed, power generation in California, the Great Plains, and the Rocky Mountains could exceed the entire current U.S. production of electricity. Wind energy may, in fact, become the most economical new base-load source for California by the year 2002. It seems unlikely, however, that wind-generated electrical energy will account for more than a few percent of total U.S. energy consumption during the next 40 years. The status and likely future of wind energy in Europe are similar to those in the United States. Currently, Denmark, long a pioneer in generating energy with wind turbines, produces only about 2.5% of its electricity in this fashion. Even so, Denmark's contribution accounts for about 80% of all wind-generated electricity in Europe. The European Community has set 10% as its target for the contribution of wind energy to its total electric consumption by the year 2030. The generation of electricity by wind turbines will probably continue to play a minor role in world energy production during the next several decades, unless there is a great urgency to reduce the use of fossil fuels.

11.5.4 *Ocean Energy*

A wide range of techniques has been proposed to harness the huge amount of potentially tappable ocean energy. Only one technology is currently in use: the harnessing of tidal energy. Two technologies have received a good deal of attention but are not yet commercially viable: the harnessing of wave energy and the use of oceanic thermal energy.

To harness tidal power, a dam is built across the mouth of a bay or estuary. Sluice gates regulate the flow of water through the dam. When the difference in water level inside and outside the basin is sufficiently large, water is released through turbines to generate electricity. Tidal power stations can operate either during falling tides, by allowing water trapped behind the dam to return to the ocean, or during both incoming and falling tides, by reversing the turbine flow during an entire tidal cycle.

The world's first and largest tidal electric facility began operating in 1968 on the La Rance River estuary in France. Its twenty-four turbines generate up to 540 million kilowatt hours of electricity per year, approximately 0.02% of the electricity generated in the United States during 1991. In North America the Bay of Fundy in Nova Scotia boasts a tidal range of up to 16.6 meters (54.5 ft), the highest tidal range in the world. Making use of this energy resource has been a topic of discussion for many decades. In 1984 a modest beginning was made with the opening of a tidal energy plant on the Annapolis River, about 100 miles west of Halifax, Nova Scotia (fig. 11.49). When the tide rises, seawater flows in through sluice gates. The gates are closed when the tide begins to fall. As soon

Figure 11.49. The tidal energy plant on the Annapolis River in Nova Scotia, Canada. (Photo courtesy of Nova Scotia Power)

as the water level in the reservoir is 1.5 meters (4.9 ft) higher than in the river, the gates are opened, and the generators begin to produce electricity. The annual energy output is 30 to 35 million kwh of electrical power, considerably less than the output at La Rance. All of the tidal plants in the former USSR and in the People's Republic of China are considerably smaller still.

Tidal power plants are practical only in areas with a tidal range in excess of about 5 meters (16 ft). This imposes a serious limit on the maximum amount of power that can be generated from tidal power on a worldwide basis. One estimate puts this maximum at only about 1% of the electricity that could be produced by hydropower. Tidal power plants are also expensive. They suffer by comparison with hydroelectric installations, because they cannot generate power continuously, and because potential storm damage requires expensive "hardening" of the facilities. Nevertheless, the United Kingdom is seriously considering building tidal power plants in the estuaries of the Severn and Morsey Rivers.

Harnessing the energy in ocean waves is an attractive idea. The most interesting techniques for doing so involve surface followers, oscillating water columns, and focusing devices. None of the proposed techniques has been used to date in power plants that can compete with conventional sources. Ocean thermal energy conversion (OTEC) has looked attractive for more than a century as a way of extracting solar energy stored as heat in the ocean's surface waters. Two types of OTEC systems have been explored. In open-cycle systems, warm surface water is converted into steam in a partial vacuum. The steam is used to drive a turbine and is condensed by cold water piped from the ocean depths. In closed-cycle OTEC systems, warm surface water is pumped into a heat exchanger and is used to evaporate a liquid with a low boiling point, such as ammonia or chlorofluorocarbons. The vapor then drives a turbine and is condensed by cold, deep-ocean water. Hybrid OTEC systems combine elements of the open and closed systems to produce both electricity and fresh water. OTEC systems can be made to generate electricity in parts of the oceans where the temperature of surface water exceeds that of deep water by 20°C or more, but electricity cannot be generated at competitive prices. Presently there are no operating OTEC plants. If floating cities ever become a reality, OTEC installations might well be called on to supply their energy needs.

11.6 Energy Mixes for the Future

Fossil fuels, hydropower, and nuclear power currently dominate world energy production. Fossil fuels are by far the most important. There are, however, great differences among countries and from region to region in the relative importance of their energy sources. France depends heavily on nuclear energy; most of Norway's electricity is generated in hydroelectric plants; Iceland relies a good deal on geothermal energy; biomass burning supplies a fair fraction of the energy for developing countries.

Virtually all of the world's energy sources are currently in abundant supply. The shortages of the 1970s were largely political, and the price of gasoline in real terms is no greater now than it was 40 years ago. It is most unlikely that fossil fuels will become scarce during the twenty-first century, although there may well be shortages for political reasons. Nuclear fuels are also abundant. The discovery of large, high-grade uranium ore deposits and the availability of uranium stockpiles that were originally intended for nuclear weapons have glutted the world uranium market. Nuclear power could therefore replace fossil fuels as the major source of the world's electric generating capacity during the next century. The use of hydropower increased dramatically during the twentieth century and could expand considerably in the next. Although hydropower cannot become the world's major energy source, it could well be of major importance in some regions. Several of the energy sources that are very minor today could also become major players in the twenty-first century. Solar energy is the most likely to do so. The energy consumed by humanity is a very small percentage of the flux of solar energy to our planet. Capture of even a small part of this energy flux could therefore supply a good deal of the world's energy needs.

All of this demonstrates that there will be no shortage of energy in the twenty-first century. If we "freeze in the dark," it will be because we have been stupid, not because we have run out of energy. The real cost of energy will probably not increase dramatically during the course of the century, unless very expensive energy options are adopted for environmental reasons. The world is in the happy position of being able to choose its preferred energy mix, or—more precisely—the preferred energy mixes in the several parts of the world.

The choice of energy mixes today is largely based on economics. Most of the alternative energy sources that looked attractive in the 1970s were abandoned or relegated to very minor roles when the price of fossil fuels dropped back to pre-1970s levels, even though the cost of energy production with alternative technologies has decreased dramatically. Only the fate of nuclear energy has been strongly influenced by noneconomic forces. The energy policies of the twenty-first century will probably be driven more by environmental considerations than past and current energy policies. If the climatic consequences of fossil fuel burning turn out to be highly undesirable, the world will have to turn to other energy sources. None of these is risk-free, and there will surely be a continuing evaluation of the relative advantages and disadvantages of the available energy options. Then, as now, different parts of the world are apt to favor different energy systems, but some energy choices will have to be made in concert. Fossil fuel burning in one region affects the CO_2 content of the entire atmosphere, and the nuclear accident at Chernobyl showed that radioactive debris do not respect national boundaries. The 120-nation Berlin mandate of 1995 to hold talks on reducing the emission rate of greenhouse gases such as CO_2 is surely welcome.

Fossil fuels may well be exhausted during the course of the next several hundred years. Breeder reactors may then become important, together with solar energy, and quite possibly a range of new energy systems. On the timescale of millennia, still short compared to that of human history, energy systems will have to approach long-term sustainability. Increases in energy efficiency and a greater emphasis on energy conservation will help to delay that day.

References

Agricola, G. 1556. *De Re Metallica.* Trans. H. C. Hoover and L. H. Hoover and republished in 1950. Dover Publications, New York.

Alley, R., Bond, G., Chapella, J., Clapperton, C., Del Genio, A., Keigwin, L., and Peteet, D. 1993. Global Younger Dryas? *EOS Transactions* 74:587–89.

Ausubel, J. H. 1991. Chernobyl amidst Perestroika: Reflections of a Recent Visit. Edited transcript of a seminar at the Massachusetts Institute of Technology, February 21, 1991, Cambridge, Mass.

Barnett, T. P. 1990. Recent Changes in Sea Level: A Summary. Chap. 1 in *Sea-Level Change*. Panel on Sea Level Change, Geophysics Study Committee, National Research Council. National Academy Press, Washington, D.C.

Bartuska, A. M. 1990. Air Pollution Impacts on Forests in North America. In *Ecological Risks: Perspectives from Poland and the United States*, ed. W. Grodzinski, E. B. Cowling, and A. I. Breymeyer, pp. 141–54. Polish Academy of Sciences and National Academy of Science. National Academy Press, Washington, D.C.

Bentley, C. R., and Giovinetto, M. B. 1992. Mass Balance of Antarctica and Sea Level Change (Abs. AGU 1992 Fall Meeting). *EOS Transactions* 73:203.

Broecker, W. S. 1994. Massive Iceberg Discharges as Triggers for Global Climate Change. *Nature* 372:421–24.

Brookins, D. G. 1984. *Geochemical Aspects of Radioactive Waste Disposal*. Springer-Verlag, New York.

Chapman, N. A., McKinley, I. G., and Hill, M. D. 1987. *The Geological Disposal of Nuclear Waste*. John Wiley, Chichester, U.K.

Chernousenko, V. M. 1991. *Chernobyl: Insight from the Inside*. Springer-Verlag, New York.

Committee on Future Nuclear Power Development. 1992. *Nuclear Power: Technical and Institutional Options for the Future*. National Research Council. National Academy Press, Washington, D.C.

Committee on Tropospheric Ozone Formation and Measurement. 1991. *Rethinking the Ozone Problem in Urban and Regional Air Pollution*. National Research Council. National Academy Press, Washington, D.C.

Connely, W. 1955. *Beau Nash, Monarch of Bath and Tunbridge Wells*. Werner Laurie, London.

Crowley, T. J., and North, G. R. 1991. *Paleoclimatology*. Oxford University Press, New York.

Dietrich, W. 1995. *Northwest Passage, the Great Columbia River*. Simon and Schuster, New York.

Edwards, N.E.S. 1977. *Tutankhamen: His Tomb and His Treasures*. Metropolitan Museum of Art and Alfred A. Knopf, New York.

El-Baz, F. 1992. The War for Oil: Effects on Land, Air, and Sea. *Geotimes* 37 (May):13–15.

Godzik, S. 1990. Impacts of Air Pollution on Agriculture and Horticulture in Poland. In *Ecological Risks: Perspectives from Poland and the United States*, ed. W. Grodzinski, E. B. Cowling, and A. I. Breymeyer, pp. 196–214, Polish Academy of Sciences and National Academy of Sciences. National Academy Press, Washington, D.C.

Godzik, S., and Sienkiewicz, J. 1990. Air Pollution and Forest Health in Central Europe: Poland, Czechoslovakia, and the German Democratic Republic. In *Ecological Risks: Perspectives from Poland and the United States*, ed. W. Grodzinski, E. B. Cowling, and A. I. Breymeyer, pp. 155–70. Polish Academy of Sciences and National Academy of Sciences. National Academy Press, Washington, D.C.

Golob, R., and Brus, E. 1993. *The Almanac of Renewable Energy*. Henry Holt, New York.

Graf, W.L. 1993. Landscapes, Commodities, and Ecosystems: The Relationship between Policy and Science of American Rivers. In *Sustaining Our Water Resources*, chap. 2. National Research Council. National Academy Press, Washington, D.C.

Häfele, W. 1990. Energy from Nuclear Power. *Scientific American* 263 (Sept.):136–44.

Häfele, W., Anderer, J., McDonald, A., and Nakicenovic, N. 1981. *Energy in a Finite World: Paths to a Sustainable World*. Ballinger Publishing, Cambridge, Mass.

Halpert, M. S., Bell, G. D., Kousky, V. E., and Ropelewski, C. F., eds. 1994. *Fifth Annual Climate Assessment 1993*. Climate Analysis Center, National Oceanic and Atmospheric Administration.

Hansen, J., and Lebedeff, S. 1990. Global and Hemispheric Temperature Anomalies. pp.

196–197 In *Trends '90: A Compendium of Data on Global Change*, pp. 196–97. Carbon Dioxide Information Analysis Center, Oak Ridge National Laboratory, Tennessee.

Heck, W. W. 1990. Impacts of Air Pollution on Agriculture in North America. In *Ecological Risks: Perspectives from Poland and the United States*, ed. W. Grodzinski, E. B. Cowling, and A. I. Breymeyer, pp. 171–195. Polish Academy of Sciences and National Academy of Sciences. National Academy Press, Washington, D.C.

Hohenemser, C. 1988. The Accident at Chernobyl: Health and Environmental Consequences and Implications for Risk Management. *Annual Review of Energy* 13:383.

Hohenemser, C., Goble, R. L., and Slovic, P. 1992. Nuclear Power. In *The Energy-Environment Connection*, ed. J. M. Hollander, chap 6. Island Press, Washington, D.C.

Jones, P. D., Wigley, T.M.L., and Wright, P. B. 1990. Global Hemispheric Temperature Anomalies. In *Trends '90: A Compendium of Data on Global Change*, pp. 194–95. Carbon Dioxide Information Analysis Center, Oak Ridge National Laboratory, Tennessee.

Kemeny, J. G., Babbit, B., Haggerty, P. E., Lewis, C., Marks, P.E., et al. 1979. *Report of the United States President's Commission on the Accident at Three Mile Island*. U.S. Government Printing Office, Washington, D.C.

Kerr, R. A. 1995. Is the World Warming or Not? *Science* 267:612.

Kieschnick, W. F., and Helm, J. L. 1990. Energy Planning in a Dynamic World: Overview and Perspective. In *Energy: Production, Consumption, and Consequences*, ed. J. L. Helm, pp. 1–17. National Academy Press, Washington, D.C.

Kuwait: the Blowouts are History. 1992. *World Oil* (Jan.):35–44.

Lange, R., Dickerson, M. H., and Gudiksen, P. H. In press. Dose Estimates from the Chernobyl Accident. *Nuclear Technology*.

Levi, B. G. 1992. Hanford Seeks Short- and Long-Term Solutions to Its Legacy of Waste. *Physics Today* 45 (March):17–21.

Lins, H. F., and Michaels, P. J. 1994. Increasing U.S. Streamflow Linked to Greenhouse Forcing. *EOS Transactions* 75:281, 284–85.

Manabe, S., Stouffer, R. J., and Spelman, M. J. 1994. Response of a Coupled Ocean-Atmosphere Model to Increasing Atmospheric Carbon Dioxide. *Ambio* 23:44–49.

May, J. 1990. *The Greenpeace Book of the Nuclear Age: The Hidden Story of the Human Cost*. Victor Gollancz, London.

McPhee, J. 1971. *Encounters with the Archdruid*. Farrar, Straus and Giroux, New York.

Meybeck, M., Chapman, D. V., and Helmer, R. 1989. *Global Freshwater Quality: A First Assessment*. World Health Organization and the United Nations Environment Programme. Blackwell Reference, Cambridge, Mass.

National Acid Precipitation Assessment Program (NAPAP). 1990. *Current Status of Surface Water Acid-Base Chemistry*. State of Science and Technology Report 9, NAPAP Interagency Program, Washington, D.C.

National Research Council. 1991. *Tanker Spills: Prevention by Design*. Committee on Tank Vessel Design. National Academy Press, Washington, D.C.

National Research Council. 1992. *Restoration of Aquatic Ecosystems: Science, Technology and Public Policy*. Committee on Restoration of Aquatic Ecosystems. National Academy Press, Washington, D.C.

Nuclear Energy Policy Study Group. 1977. *Nuclear Power Issues and Choices: Report of the Nuclear Energy Policy Study Group*. Ballinger Publishing, Cambridge, Mass.

Oak Ridge National Laboratory. 1990. *Trends '90: A Compendium of Data on Global Change Highlights*. Carbon Dioxide Information Analysis Center, Oak Ridge National Laboratory, Tennessee.

Oak Ridge National Laboratory. 1991. *Trends '91: A Compendium of Data on Global Change Highlights*. Carbon Dioxide Information Analysis Center, Oak Ridge National Laboratory, Tennessee.

Oeschger, H., and Siegenthaler, U. 1988. How Has the Atmospheric Concentration of CO_2

Changed? In *The Changing Atmosphere*, ed. F. S. Rowland and I.S.A. Isaksen, pp. 5–23. John Wiley, New York.

Office of Air Quality Planning and Standards. 1992a. *National Air Pollution Emission Estimates, 1990–1991*. Publication No. 454-R-92-013. U.S. Environmental Protection Agency, Washington, D.C.

Office of Air Quality Planning and Standards. 1992b. *National Air Quality and Emissions Trends Report, 1991*. Publication No. 450-R-92-001. U.S. Environmental Protection Agency, Washington, D.C.

Ray, D. L., and Guzzo, L. 1993. *Environmental Overkill: Whatever Happened to Common Sense*? Regnery Gateway, Washington, D.C.

Revelle, R. D. 1990. Overview and Recommendations. In *Sea-Level Change*, pp. 3–34. Panel on Sea Level Change, Geophysics Study Committee, National Research Council. National Academy Press, Washington, D.C.

Sand, P. H. 1990. Regional Approaches to Transboundary Air Pollution. In *Energy, Production, Consumption and Consequences*, ed. J. L. Helm, pp. 246–64. National Academy Press, Washington, D.C.

Smith, K. R. 1987. *Biofuels, Air Pollution, and Health: A Global Review*. Plenum Press, New York.

Smith, K. R. 1988. Air Pollution: Assessing Total Exposure in Developing Countries. *Environment* 30 (no. 10):16–35.

Spiro, P. A., Jacob, D. J. and Logan, J. A. 1992. Global Inventory of Sulfur Emissions with $1° \times 1°$ Resolution. *Jour. Geophys. Res.* 97 (no. D5):6023–36.

Stanley, D. J., and Warne, A. G. 1993. Nile Delta: Recent Geological Evolution and Human Impact. *Science* 260:628–34.

Stone, J. L. 1993. Photovoltaics: Unlimited Electrical Energy from the Sun. *Physics Today* 46 (Sept.):23–29.

Sundquist, E. T. 1993. The Global Carbon Dioxide Budget. *Science* 259:934–41.

Tien, P.-L., Updegraff, C. D., Wahi, K. K., Siegal, M. D., and Guzowski, R. V. 1985. *Repository Site Data Report for Unsaturated Tuff, Yucca Mountain, Nevada*. Sandia National Laboratories Report NUREG/CR-4110 SAND84–2668. Albuquerque, N.M.

U.S. Department of Energy. 1992. *Annual Energy Review 1991*. Energy Information Administration, U.S. Department of Energy. U.S. Government Printing Office, Washington, D.C.

U.S. Department of Energy. 1994. *Annual Energy Review 1993*. Energy Information Administration, U.S. Department of Energy. U.S. Government Printing Office, Washington, D.C.

van der Leeden, F.; Troise, F. L.; and Todd, D. K. 1990. *The Water Encyclopedia*. 2d ed. Lewis Publishers, Chelsea, Mich.

Weinberg, C. J., and Williams, R. H. 1991. Energy from the Sun. In *Energy for Planet Earth*, pp. 107–118. Readings from Scientific American Magazine. W. H. Freeman, San Francisco.

Wentzel, K.-F. 1987. Waldschäden—Was ist wirklich neu? *Mensch und Umwelt* 1987 (Sept.):19–28.

12 Global Change

12.1 Introduction: What Have We Wrought?

During his presidency, Franklin Delano Roosevelt was fond of discussing what he called "global problems." Margaret Chase Smith, who was a member of the House of Representatives in the 1940s, frequently disagreed with the president's views and described them pithily as "globaloney." The term "global" has been slow to recover, but it is once again in common usage, particularly as part of "global change," a phrase that encompasses all worldwide anthropogenic changes. These changes are generally considered to be for the worse. In Gro Harlem Brundtland's words (1990):

> When the 20th Century began, neither human beings nor human technology had the power to radically alter the global ecosystem. Today, as the century draws to a close, human beings in ever-increasing numbers have that power, and as a result of their activity on the planet, major unintended changes are taking place in the atmosphere, the biosphere and the hydrosphere. These changes outstrip our present ability to cope; the world's financial and political institutions are out of step with the workings of nature.

On an even gloomier note, the great humanitarian doctor-musician-theologian Albert Schweitzer wrote that "Man has lost the capacity to foresee and to forestall. He will end by destroying the Earth." This chapter seeks to summarize and to evaluate what humankind has done and is doing to the face of the Earth.

Global change did not start with *Homo sapiens*. Geologists and other Earth scientists have been occupied for several centuries in reconstructing the changes that have shaped our planet during the past 4,500 million years. Some of these changes were local, some regional, others global. Some, like earthquakes and meteorite impacts, occurred almost instantaneously; others, like ice ages, occupied a time span of about 100,000 years; some, like the drift of the continents since the breakup of the supercontinent Pangaea, are measured in hundreds of millions of years; some, like the cooling of the Earth's interior, have continued through all of Earth history.

People have been present for only a small part of Earth history. Our effect on the environment before the last ice age was probably minimal, but our influence during the past 5,000 years is easy to identify. Its magnitude has grown enormously since the onset of the industrial revolution because our numbers have increased rapidly, and because the ability of each individual to influence the environment has been multiplied a hundredfold by harnessing the Sun's energy and the energy stored within the Earth. We have altered the atmosphere, surface and ground waters, the oceans, the biosphere, and the uppermost parts of the solid Earth. Of course, not all change is for the worse. In *Candide* (chapter 1), Voltaire's Dr. Pangloss proclaimed, "Dans ce meilleur des mondes possibles . . . tout est au mieux" (In this best of all possible worlds . . . everything is for the best), to which James Branch Cabell replied, "The optimist proclaims that we live in the best of all possible worlds, and the pessimist fears that this is true"

(*The Silver Stallion*, chapter 26, 1926). We need change to improve the world, but we also need to inquire which changes in our environment are for the better and which are for the worse. It is a truism that many of the changes that *Homo sapiens* has wrought have had both good and bad consequences; all of these need to be weighed in assessing the merits of environmental change.

12.2 The Changing Atmosphere

The total weight of humanity is only about one ten-millionth of the weight of the atmosphere. It is not surprising, therefore, that human activities have had virtually no effect on the major atmospheric constituents: nitrogen, oxygen, and argon. We have only altered the quantity of minor and trace gases in the atmosphere. Although these gases account for less than 0.1% of the atmosphere, the changes in their concentrations have had significant local, regional, and global consequences.

12.2.1 Anthropogenic Inputs and Their Residence Times

The actual and potential influence of human activities is best assessed by comparing the rate of anthropogenic emissions of atmospheric gases with their natural, "background" rate of emission. Table 12.1 presents such a comparison. The figures in this table are rather uncertain, because it is difficult to determine the magnitude of global fluxes. It is clear, however, that the anthropogenic fluxes of many of these gases are equal to or greater than their nonanthropogenic fluxes. We are disturbing the atmospheric chemistry of these gases to a very considerable extent.

To assess the consequences of these disturbances, we need to look at the location of the sources of these gases, at the rate of their dispersal throughout the atmosphere, and at the rate at which they are removed from the atmosphere. Table 12.2 shows that in 1991 transportation and fuel combustion were the two

Table 12.1. Estimated Anthropogenic and Total Emissions of Some Important Atmospheric Gases (in millions of metric tons/yr)

Emission	*Anthropogenic Emissions*	*Total Emissions*	*Residence Time in Atmosphere*
Carbon monoxide (CO)	700[a]	2,000[a]	2 months[d]
	less than half of	3,300 ± 1,700[b]	
	total emissions[a]	1,500–4,000[c]	
		1,400–3,100[i]	
NO_x Gases	20–30[a]	30–50[a]	days[a]
Nitrous oxide (N_2O)	6[a]	25[a]	170 years[a]
	4[h]	19–22[i]	
Sulfur dioxide (SO_2)	200–260[g]	300–400[a]	days to weeks[a]
		140–340[i]	
Carbon dioxide (CO_2)	22,000[f]	400,000	ca. 100 years
Methane (CH_4)	≥ 125 ± 50[e]	500 ± 200[e]	10 years[a,e]
	350[h]		
Chlorofluorocarbons (CFCs)	1.6[h]	1.6[h]	ca. 70 years

[a] Graedel and Crutzen (1990).
[b] Seiler and Conrad (1987).
[c] Logan et al. (1981).
[d] Cicerone (1988)
[e] Ehhalt (1988)
[f] Oeschger and Siegenthaler (1988).
[g] Charlson (1988).
[h] Watson et al. (1990) and Stern et al. (1992).
[i] Crutzen and Andreae (1990).

Table 12.2. Sources of Anthropogenic Emissions in the United States, 1991 (in millions of metric tons)

Source	*CO*	NO_x	*VOC*	*TP*	SO_x	*Pb*
Transportation	43.49	7.26	5.08	1.57	0.99	1.62
Fuel combustion	4.68	10.59	0.67	1.94	16.55	0.45
Industrial processes	4.69	0.60	7.86	2.55	3.16	2.21
Solid waste disposal	2.06	0.10	0.69	0.34	0.02	0.69
Miscellaneous	7.18	0.21	2.59	1.01	0.01	0.00
Total	62.10	18.76	16.88	7.41	20.73	4.97

Source: National Air Quality and Emission Trends Report, 1991 (EPA)
Key: VOC, volatile organic compounds; TP, total particulates; Pb, lead.

most important sources of the major atmospheric pollutants in the United States (see chapter 11). The locations of the pollutant sources, concentrated as they are in urban areas, along major highways, and at fossil fuel power plants located within and outside major metropolitan areas, are responsible for the heterogeneous distribution of the pollutants and their decomposition products.

Figure 12.1 shows the distribution of arithmetic mean annual concentrations of sulfur dioxide in the United States in 1991. SO_2 concentrations were high in the heavily populated Midwest and Northeast and near point sources in the West. All of the large urban areas had ambient air quality concentrations lower than the current annual standard of 80 $\mu g/m^3$ (see table 12.3). The SO_2 concentration in the air above cities varies considerably during any given year, in part due to fluctuations in the emission rate, and in part due to meteorologic changes such as temperature inversions during which urban air can be trapped and stagnant for unusually long periods of time. Nevertheless, during 1991 the maximum 24-hour

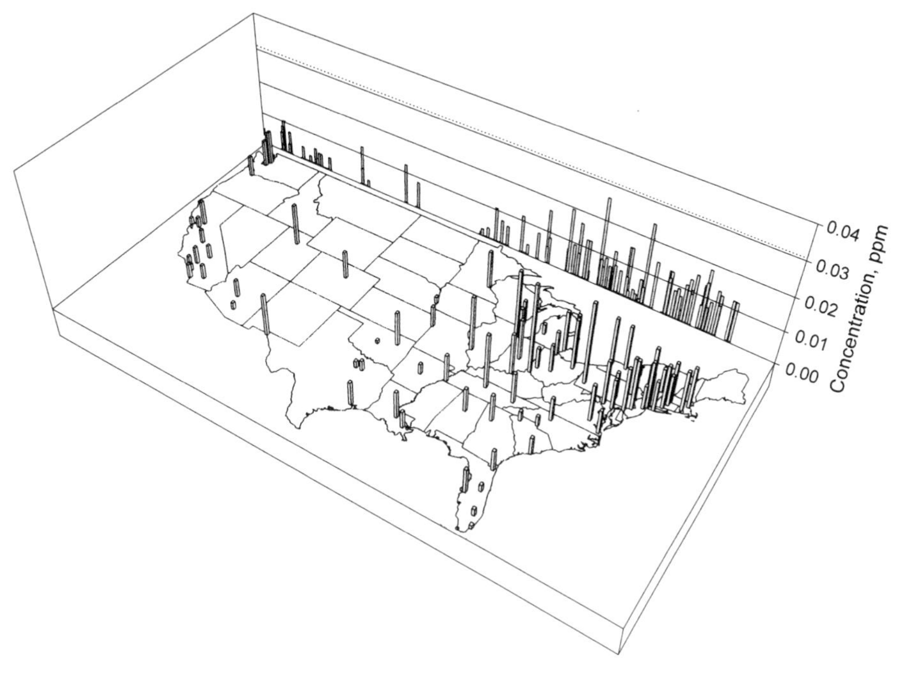

Figure 12.1. 1991 arithmetic mean sulfur dioxide concentrations in U.S. areas with populations greater than one-half million. The map does not reflect air quality in the vicinity of smelters or large power plants in rural areas. (EPA Report 450-R-92-001)

Table 12.3
National Air Quality Standards (NAQS) in Effect in 1992

	Primary (Health Related)		Secondary (Welfare Related)	
Pollutant	*Type of Average*	*Standard Level Concentration*[a]	*Type of Average*	*Standard Level Concentration*
CO	8-hour[b]	9 ppm (10 mg/m^3)	No secondary standard	
	1-hour[b]	35 ppm (40 mg/m^3)	No secondary standard	
Pb	Maximum quarterly average	1.5 $\mu g/m^3$	Same as primary standard	
NO_2	Annual arithmetic mean	0.053 ppm (100 $\mu g/m^3$)	Same as primary standard	
O_3	Maximum daily 1-hour average[c]	0.12 ppm (235 $\mu g/m^3$)	Same as primary standard	
PM-10	Annual arithmetic mean[d]	50 $\mu g/m^3$	Same as primary standard	
	24-hour[d]	150 $\mu g/m^3$	Same as primary standard	
SO_2	Annual arithmetic mean	80 $\mu g/m^3$ (0.03 ppm)	3-hour[b]	1300 $\mu g/m^3$ (0.50 ppm)
	24-hour[b]	365 $\mu g/m^3$ (0.14 ppm)		

Source: EPA 450-R-92-001.

[a] Parenthetical value is an approximately equivalent concentration.

[b] Not to be exceeded more than once per year.

[c] The standard is attained when the expected number of days per calendar year with maximum hourly average concentrations above 0.12 ppm is equal to or less than 1, as determined according to Appendix H of the Ozone NAAQS.

[d] Particulate standards use PM-10 (particles less than 10μ in diameter) as the indicator pollutant. The annual standard is attained when the expected annual arithmetic mean concentration is less than or equal to 50 $\mu g/m^3$; the 24-hour standard is attained when the expected number of days per calendar year above 150 $\mu g/m^3$ is equal to or less than 1; as determined according to Appendix K of the PM NAAQS.

average SO_2 concentration of 365 $\mu g/m^3$ allowed by the National Ambient Air Quality Standards (NAAQS) in 1992 was only exceeded in one metropolitan area: Chicago. During the past 20 years SO_2 emissions have decreased markedly in many developed countries (see fig. 12.2), but there are indications that SO_2 emissions are increasing in developing areas, particularly in China, Mexico, and India.

The concentration of NO_2 in the largest U.S. metropolitan areas during 1991 is shown in figure 12.3. There is an obvious similarity between this figure and figure 12.1. However, NO_2 concentrations were relatively much higher in southern California than SO_2 concentrations. The difference is due to the small fraction of high-sulfur fuels that is used in California. Los Angeles, with an annual NO_2 mean concentration of 0.055 ppm, was the only area in the United States

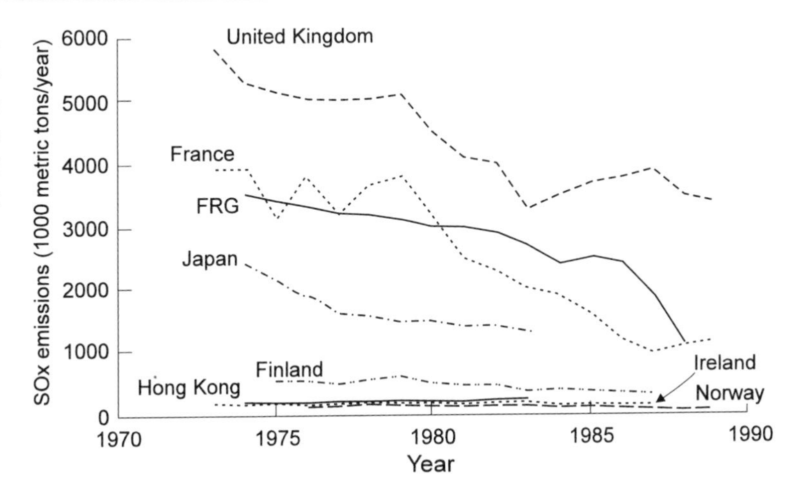

Figure 12.2. Trends of sulfur oxide emissions between 1970 and 1990 in selected developed countries. (EPA Report 450-R-92-001)

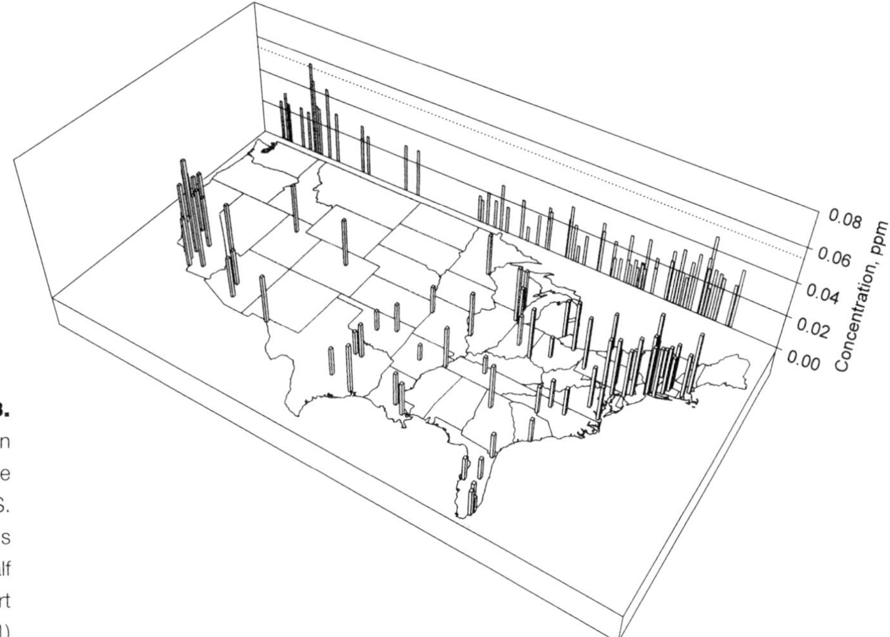

Figure 12.3. 1991 arithmetic mean nitrogen dioxide concentration in U.S. areas with populations greater than one-half million. (EPA Report 450-R-92-001)

that exceeded the NO_2 NAAQS air quality standard of 0.053 ppm (see table 12.3).

Ozone, O_3, is not part of the family of anthropogenic pollutant gas emissions. It is formed both in the stratosphere, where its presence is important in filtering out solar ultraviolet, and in the troposphere, where it constitutes a health hazard for people and for vegetation (see chapter 11). During the 1980s, ozone levels increased approximately 10% throughout the troposphere over Europe. Since tropospheric O_3 is a small part of total atmospheric O_3, the 10% addition to tropospheric O_3 increased the total column abundance of the gas by only about 1%. Tropospheric O_3 is generally attributed to a combination of in situ photochemical production and destruction coupled with the downward movement of

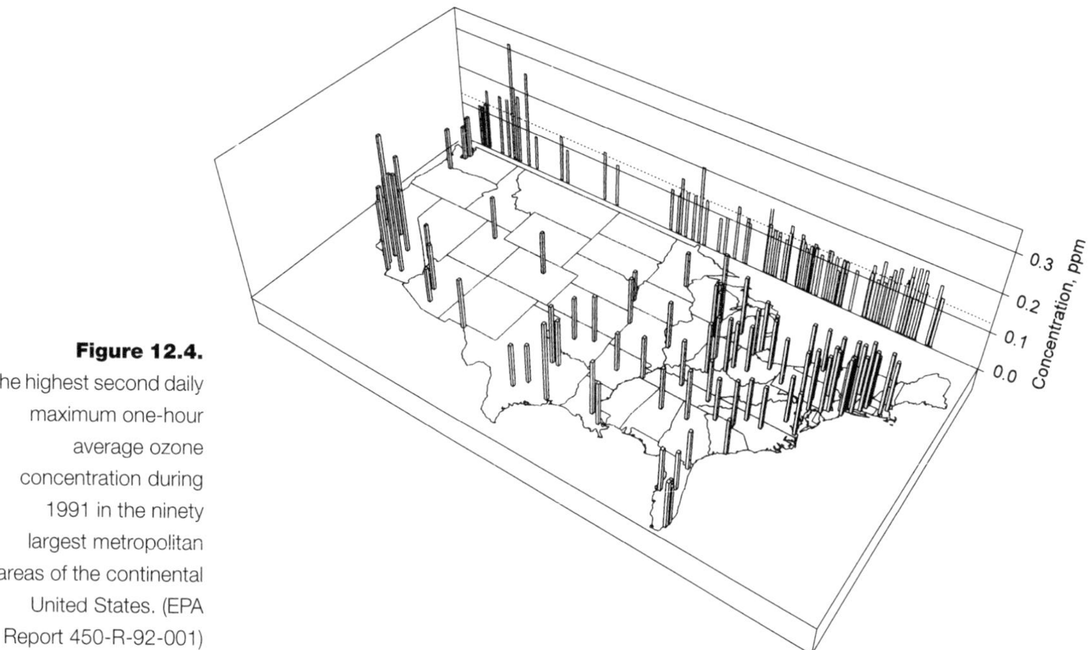

Figure 12.4. The highest second daily maximum one-hour average ozone concentration during 1991 in the ninety largest metropolitan areas of the continental United States. (EPA Report 450-R-92-001)

O_3-rich stratospheric air. The most critical aspect of the tropospheric ozone problem is its formation downwind of large urban areas, where, under certain meteorological conditions, emissions of NO_x and volatile organic compounds (VOC) can result in O_3 concentrations as high as 0.2 to 0.4 ppm (Committee on Tropospheric Ozone Formation and Measurement 1991; chapter 1). The similarity between the data for ozone in figure 12.4 and for NO_2 in figure 12.3 is striking and not unexpected. Thirty-eight of the ninety largest metropolitan areas in the United States did not meet the NAAQS of 0.12 ppm for ozone in 1991. The highest O_3 concentrations were observed in southern California, but high levels also persisted in the Texas Gulf Coast, in the Northeast Corridor, and in other heavily populated regions.

Not all anthropogenic gas emissions are clustered in and around urban centers. The concentration of methane, CH_4, in the atmosphere was about 0.7 ppm before the start of the industrial revolution. Its concentration has been rising at a rate of about 1% per year and is now about 1.7 ppm (see fig. 12.5). Much of this very significant increase is probably due to increasing methane production in rice paddies, flaring of natural gas by cattle, and biomass burning.

The distribution of anthropogenic gases in the atmosphere depends in large part on their residence time, i.e., their mean atmospheric lifetime. Table 12.1 shows that the residence time of some pollutants is measured in days (NO_x gases), some in days or weeks (SO_2), some in months (CO), others in decades (CH_4), and some in centuries (N_2O, CO_2, and CFCs). Pollutants with residence times of days are destroyed before they can be transported very far from their points of origin. Their effects are therefore largely local. Pollutants like SO_2, which have somewhat longer residence times, can be transported considerable distances before they are removed from the atmosphere. Effects such as acid rain

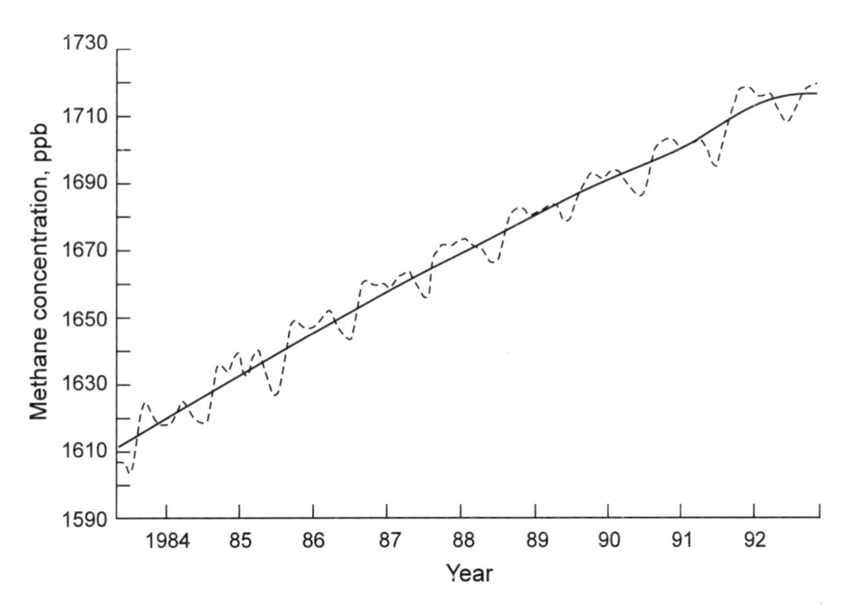

Figure 12.5. Globally averaged, biweekly methane concentration (ppb) by volume determined by the NOAA/CMDL Carbon Cycle Group cooperative air sampling network. Solid line shows the growth with the seasonal cycle removed. (Data provided by NOAA/CMDL; Halpert et al. 1994)

are therefore regional (see chapter 11). Pollutants with residence times in excess of a year or two become distributed over the entire surface of the Earth and can create global problems. CO_2 and the halocarbons, including the CFCs, are particularly important gases in this category.

CO_2 is the only gas with a preindustrial atmospheric concentration in excess of 100 ppm whose concentration has been seriously affected by humanity. Chapter 11 has already dealt with the increase in the concentration of this gas during the past 200 years and with the possible climatic impacts of the potentially large increases in CO_2 due to fossil fuel burning and deforestation during the next two centuries.

A rather different impact on atmospheric chemistry has been produced by the addition of what are, by global standards, minute quantities of chlorofluorocarbon compounds (CFCs) and other halocarbon compounds. Their influence on ozone levels in the Antarctic stratosphere has become dramatic. Their effect was unanticipated, their discovery has taken some ironic turns, and the resulting cooperation among nations in dealing with a global threat to the environment has been encouraging.

12.2.2 *The Ozone Hole*

Many people, when asked to name the most pressing global change problems, voice their concern about global warming and the ozone hole. In some respects the two phenomena are similar; in others they are quite different. The extent of future global warming is still hotly debated, while most of the physics and chemistry of the ozone hole are now well understood. To do something significant about mitigating possible global warming would involve major changes in the world's energy sources that would be politically and economically difficult to achieve. Preventing a serious worsening of atmospheric ozone depletion is not nearly as expensive. Partly for this reason the steps needed to limit ozone depletion proved to be politically feasible. Their implementation represents a major advance in the way nations come together to solve international environmental problems.

Most of the world's ozone is produced in the stratosphere by the action of solar ultraviolet radiation on atmospheric oxygen (see chapter 3). Solar UV dissociates molecular oxygen, O_2, into two oxygen atoms. The subsequent reaction of O with O_2 can produce ozone, O_3. This molecule is decomposed by ultraviolet photons, and the concentration of O_3 in the stratosphere is determined by the balance between the rates of O_3 production and decomposition. Figure 12.6 shows the average global annual mean vertical distribution of atmospheric ozone. The partial pressure of ozone reaches a maximum of about 130 nanobars (0.13 microbars) at an elevation of about 23 kilometers. The maximum ozone mixing ratio, the proportion of ozone in air, lies at a somewhat higher elevation, because total atmospheric pressure decreases rapidly with increasing altitude.

The decomposition of ozone in the stratosphere is speeded up considerably by the presence of NO, OH, and ClO. NO can react with O_3 to form NO_2 and O_2:

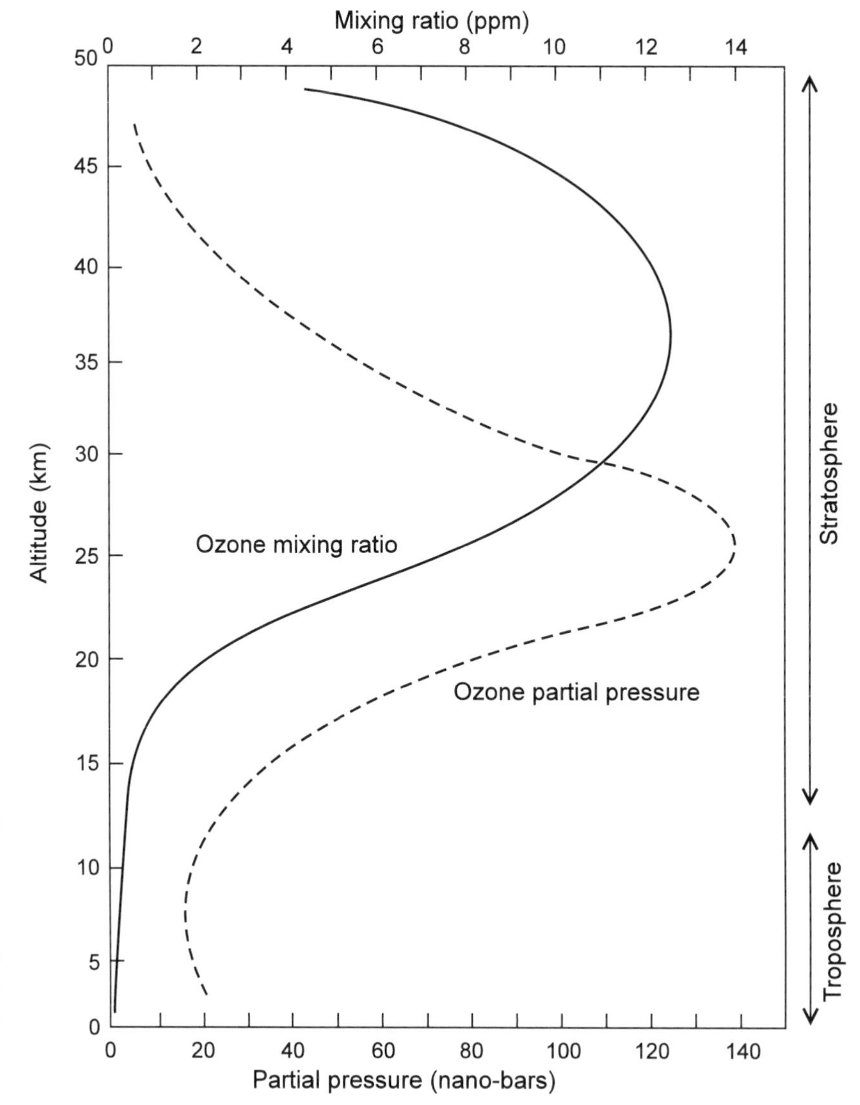

Figure 12.6. The typical global annual mean distribution of ozone in the atmosphere. (Committee on Tropospheric Ozone Formation and Measurement 1991)

$$NO + O_3 \rightarrow NO_2 + O_2. \quad (12.1)$$

NO_2 can then react with O to regenerate NO:

$$NO_2 + O \rightarrow NO + O_2. \quad (12.2)$$

In the process, one O atom and one O_3 molecule are combined to yield two O_2 molecules:

$$\begin{array}{l} NO + O_3 \rightarrow NO_2 + O_2 \\ NO_2 + O \rightarrow NO + O_2 \\ \hline O + O_3 \rightarrow O_2 + O_2 . \end{array} \quad (12.3)$$

In this pair of reactions, NO therefore acts as a catalyst for the decomposing ozone.

The free radicals OH and HO_2 act in a similar fashion. One of the mechanisms for O_3 destruction operates as follows:

$$HO + O_3 \rightarrow HO_2 + O_2 \quad (12.4)$$

$$HO_2 + O \rightarrow HO + O_2 \quad (12.5)$$

$$O + O_3 \rightarrow O_2 + O_2 . \quad (12.6)$$

Several reactions involving Cl atoms and ClO_x, free radicals containing Cl and O, are also important. As an example:

$$Cl + O_3 \rightarrow ClO + O_2 \quad (12.7)$$

$$ClO + O \rightarrow Cl + O_2 \quad (12.8)$$

$$O + O_3 \rightarrow O_2 + O_2 .$$

The higher the stratospheric concentrations of NO, OH, and ClO, the more rapid the decomposition of ozone and the lower the steady-state concentration of ozone in the stratosphere.

A serious decrease in the total O_3 content of the atmosphere would have important consequences for most terrestrial organisms. Ozone is the major absorber of solar ultraviolet radiation between 240 and 320 nanometers. Without a significant ozone shield, the solar ultraviolet radiation reaching the Earth's surface would be sufficient to damage the DNA and RNA of many organisms, and would cause a great deal of skin cancer, some of it fatal. In 1974, Mario Molina and F. Sherwood Rowland pointed out that continued addition of chlorofluorocarbons (CFCs) could add enough chlorine to the stratosphere to lead to a significant reduction in the protective ozone layer by the year 2050. This prediction was taken seriously, and in 1978 the United States banned the use of CFCs in aerosol spray cans.

In 1985, J. C. Farman reported that during the Antarctic spring seasons of 1980 to 1984, the total amount of ozone over the British Antarctic Survey Station at Halley Bay was about 30% lower than it had been during the same season

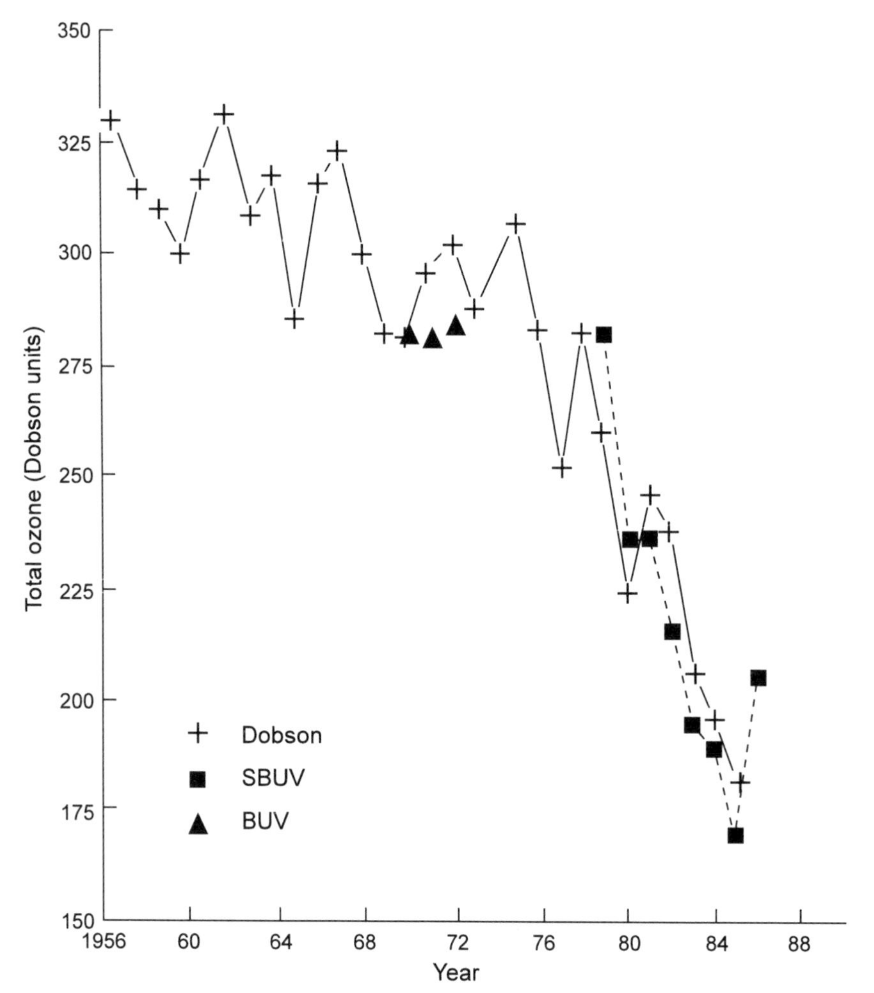

Figure 12.7. October monthly mean total ozone measurements over Halley Bay. Plusses are Dobson measurements by Farman et al., triangles are *Nimbus-4* BUV measurements from 1970 to 1972; squares are *Nimbus-7* measurements from 1979 to 1986. (Stolarski 1988)

between 1957 and 1973. Such a large decrease had not been foreseen in Molina and Rowland's analysis, and was much larger than had been observed anywhere else in the world. Farman's observations agreed with Japanese measurements at Syowa Station. They were, however, received with skepticism, because the reported satellite measurements of ozone over Antarctica did not seem to bear them out. Fortunately, the satellite data were reevaluated. It turned out that the computer which recorded the satellite data had been instructed to eliminate all measurements outside a predetermined range, because it was thought that such measurements would surely be erroneous and probably due to faulty operation of the on-board instruments. When the entire data set was restudied, it was found that while the satellite had been faithfully observing the same decrease in the springtime ozone levels above Antarctica that Farman had reported, the computer had just as faithfully deleted the record of this decrease. As shown in figure 12.7, the corrected satellite data agreed very well indeed with Farman's ground-based measurements.

Ozone levels over Antarctica during the Antarctic spring have been decreasing ever since. The distribution of Antarctic ozone during October of the years 1980 to 1992 is shown in plate 53. The depth of the "hole" has increased progressively. Until the 1970s the lowest ozone levels were about 300 DU (Dobson Units). In the Antarctic spring of 1993 just 90 DU of ozone remained in the Antarctic stratosphere. That is an extremely small amount of ozone. If it were removed and compressed to a pressure of one atmosphere, it would occupy a layer only 90×10^{-6} m thick.

In spite of the 1978 U.S. ban on the use of CFCs in aerosol cans, their use in other applications grew rapidly both in the United States and abroad. By the mid-1980s it was clear that unilateral action by the United States was not enough if there was to be a continued reduction in CFC use. The discovery of the Antarctic ozone hole sparked the Vienna Conference, which led to the Montreal Protocol on Substances that Deplete the Ozone Layer. The protocol was initiated in 1987. It is an environmental treaty between member states of the United Nations that constitutes a commitment to control the use of CFCs and halons (bromine-containing carbon compounds). The signatories to the protocol agreed to limit the annual CFC consumption to their 1986 levels, and within three years to limit their use of halons to 1986 levels as well. From July 1, 1993, annual CFC production and consumption was not to exceed 80% of 1986 levels in any given country. From July 1, 1998, annual CFC production and consumption was not to exceed 50% of 1986 levels. All of the signatories also agreed to exchange information on recycling, recovery, and destruction techniques for the controlled substances as well as on possible alternatives to those substances. Further provisions were made for signatory countries, guaranteeing access to alternative technologies and substitute products for countries that had not already developed their own CFC technology.

Figure 12.8 shows the course of CFC production up to the signing of the Montreal Protocol and during the succeeding 4 years. The continuing increase in the depth of the ozone hole and the discovery of a 3%–5% decrease in ozone levels at lower latitudes (fig. 12.9) led to the 1990 London revisions of the Mon-

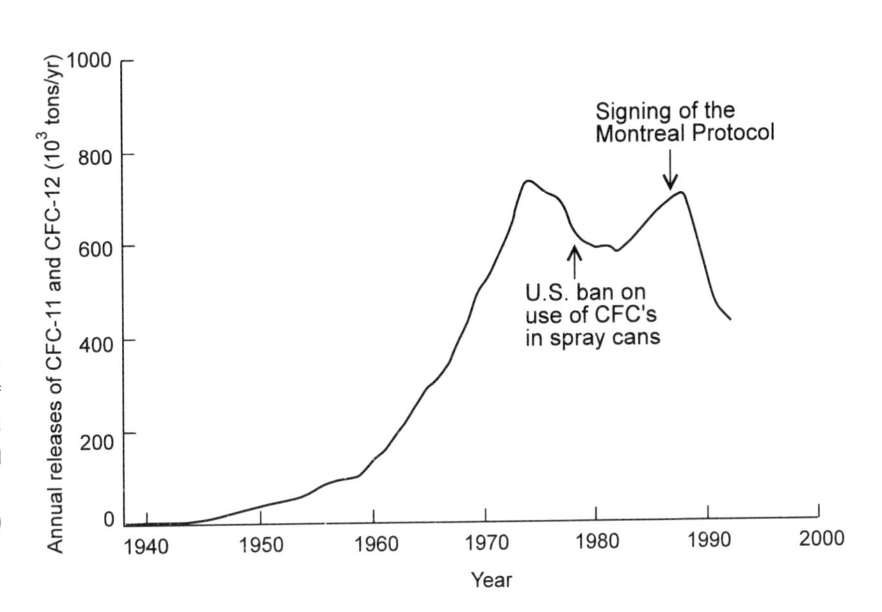

Figure 12.8. Annual release of chlorofluorocarbons between 1938 and 1992. (Boden et al. 1994)

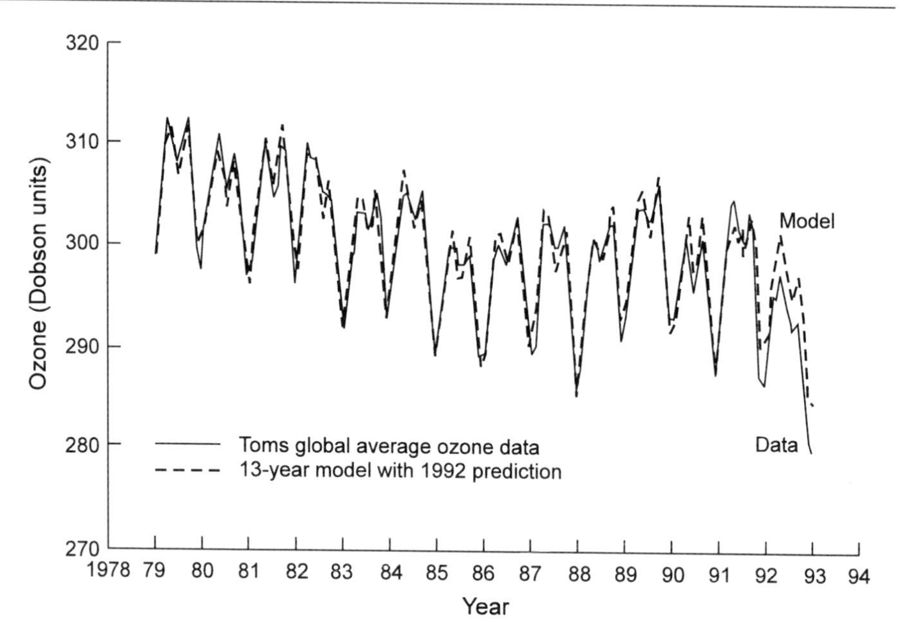

Figure 12.9. The annual average amount of atmospheric ozone for the latitude range 65°S to 65°N between January 1979 and December 1993. (Gleason et al. 1993)

treal Protocol. These revisions extended the scope of the regulations to cover a larger number of CFCs and halons that were not previously designated as controlled substances. The revisions also offered potential alternatives to the controlled substances. Finally, and most significantly, the use of all of the controlled CFCs and halons was banned by the year 2000.

Not everyone is convinced that all of these measures are necessary. In the mid-1980s it was not even clear whether CFCs were responsible for the Antarctic ozone hole. A variety of alternative explanations were advanced; but the deepening of the hole since then and the extensive studies of the photochemistry of the stratosphere during the Antarctic spring confirmed that the CFCs are indeed the major cause of the hole, although aerosols lofted into the stratosphere during major volcanic eruptions do contribute to ozone depletion. A new era in ozone hole exploration is about to begin with the introduction of *Perseus A* and *Perseus B*, small, lightweight unmanned aircraft that will be able to explore the ozone hole much more thoroughly and over greater periods of time than has been possible to date (see fig. 12.10).

Figure 12.10. *Perseus A* roll-out. *Perseus A* is a lightweight, unmanned aircraft designed to fly to 90,000 ft. Its companion aircraft, *Perseus B*, will fly long-duration missions at 65,000 ft. (Photograph courtesy of Lenny Solomon)

During the Antarctic winter, strong circumpolar winds virtually isolate the Antarctic stratosphere from the rest of the stratosphere. In the absence of warming sunlight, the temperature of the Antarctic stratosphere drops below 195°K (–78°C). Under these frigid conditions, nitric acid, HNO_3, and water co-condense as ice particles. During this process HCl and $ClONO_2$, a rather unreactive form of stratospheric chlorine, are decomposed, liberating ClO, which is a very effective catalyst for ozone destruction (see eqs. 12.7 and 12.8). Since ozone destruction requires solar UV, ozone loss begins when the Sun reappears in October, at the end of the Antarctic winter. As spring progresses, the Antarctic vortex that isolates Antarctica in the winter breaks up, and the ozone-depleted air of the Antarctic stratosphere mixes northward with the relatively ozone-rich stratospheric air at lower latitudes.

The absence of an equivalent ozone hole in the Arctic stratosphere is due in large part to differences in the meteorology of the two polar regions: the Arctic polar vortex is not nearly as tight and closed as the Antarctic polar vortex. Nevertheless, further increases in the abundance of chlorine in the stratosphere could lead to significant ozone loss during the Arctic winter and to a thinning of the UV shield in the Northern Hemisphere. The thinning to date appears to be small. Plate 54 documents the changes in ozone levels at Toronto between 1989 and 1993. O_3 levels have clearly been decreasing. The summer ozone maximum declined from ca. 480 DU to ca. 400 DU. However, the flux of UV changed rather little during this 4-year period (see plate 55). In 1993, the daily UV flux at 300 nm was higher than in previous years. At a wavelength of 324 nm the changes were minor. It could be argued that these changes are too small to warrant the drastic actions mandated by the 1990 London revision of the Montreal Protocol. However, the trends in stratospheric ozone levels are clear, and the increasing depth of the Antarctic ozone hole is a considerable warning of future worldwide reductions. Furthermore, adequate substitutes for CFCs and halons are available. The phase-out of the production of CFC and halons is surely a prudent measure.

12.3 The Changing Hydrosphere

Humanity has had a significant impact on the atmospheric concentration of many trace components, one minor component, and none of the major components. These anthropogenic changes in atmospheric chemistry have affected human health, but not in a major way. By contrast, our effect on the terrestrial hydrosphere has been dramatic, and polluted waters annually account for millions of deaths, largely among children in the Third World. If the severity of environmental problems is measured in terms of the number of deaths for which they are responsible, then water pollution is surely our most severe environmental problem.

12.3.1 *Rivers, Lakes, and Groundwater*

Water Use and Salinity Changes

Few rivers in the developed countries have escaped major changes during the past 200 years. Flood control measures have harnessed the flow of most rivers during all but their periods of greatest excess (see chapter 4). The many dams that have been built play an important role in taming rivers and supply a significant percentage of the world's electrical energy (see chapter 11). Extensive irrigation

Table 12.4. The World Water Demand during the Twentieth Century (km³/yr)

Water Users	*1900*	*1940*	*1950*	*1960*	*1970*	*1980*	*1990*	*2000*
Irrigated area (Mha)	47.3	75.8	101	142	173	217	272	347
Agriculture								
A	525	893	1,130	1,550	1,850	2,290	2,680 (68.9)	3,250 (62.6)
B	409	679	859	1,180	1,400	1,730	2,050 (88.7)	2,500 (86.2)
Industry								
A	37.2	124	178	330	540	710	973 (21.4)	1,260 (24.7)
B	3.5	9.7	14.5	24.9	38.0	61.9	88.5 (3.1)	117 (4.0)
Municipal supply								
A	16.1	36.3	52.0	82.0	130	200	300 (6.1)	441 (8.5)
B	4.0	9.0	14	20.3	29.2	41.1	52.4 (2.1)	64.5 (2.2)
Reservoirs								
A	0.3	3.7	6.5	23.0	66.0	120	170 (3.6)	220 (4.2)
B	0.3	3.7	6.5	23.0	66.0	120	170 (6.1)	220 (7.6)
Total								
A	579	1,060	1,360	1,990	2,590	3,320	4,130 (100)	5,190 (100)
B	417	701	894	1,250	1,540	1,950	2,360 (100)	2,900 (100)

Source: Shiklomanov 1991.
Note: Figures in parentheses are percentages.
Key: A, Total water consumption; B, irretrievable water loss.

schemes have diverted so much water from some rivers in desert country that their flow has been reduced to a mere trickle (see chapters 4 and 11).

The expected doubling or tripling of the human population during the next century will increase the demand for freshwater and will apply ever greater pressure for even more complete and efficient use of surface and groundwaters. Table 12.4 documents the upward trend in water use during the twentieth century. Irrigation has been largely responsible for water that is lost irretrievably, and this loss has increased more rapidly than world population. If past trends continue through the twenty-first century, the rate of water use will probably more than triple. The irretrievable water loss will then be a large fraction of the runoff in areas where crop irrigation is most needed.

Rivers have served not only as sources of water and energy, but also as sites for waste disposal. On a sufficiently small scale the effects of this use are benign. Dilution can be a solution to pollution, but only when the pollutant flux is sufficiently small, so that pollutant concentrations do not exceed the limits for safe water use. The world population increase, much of it occurring along or close to rivers, together with the increase in per capita waste production, has turned a number of the world's rivers into what have been aptly described as open sewers. Fortunately, rivers tend to respond quickly to decreases in pollutant inputs. Many highly polluted rivers have been restored within a few years by decreasing the level of waste discharged along their courses. Pollutants such as mercury and selenium that have accumulated in river and lake sediments may, however, remain in place for long periods, and may continue to threaten human and animal health long after their addition to a river system has ceased.

Anthropogenic additives to river waters can be divided conveniently into three groups: inorganic, organic, and biologic. Each category includes a wide variety of substances. Some inorganic compounds are added in such large quantities that they modify the major element chemistry of rivers. This type of pollution is particularly frequent downstream from salt and potash mines. Groundwaters can become heavily polluted by infiltration of saline irrigation waters, with saline brines, and with seawater. Figure 12.11 illustrates the effect of increasing irrigation returns and of a drought on the concentration of total dissolved solids in the Murray River, Australia. The diversion of upstream flows for irrigation and other water uses can have a marked impact on downstream water quality by greatly reducing the dilution of irrigation runoff. For example, upstream water diversion from the Yakima River in the state of Washington is so large that almost the entire summer flow of this river consists of irrigation returns (Meybeck, Chapman, and Helmer 1989).

In northern California the increasingly large amounts of water that have been siphoned from the Sacramento Delta to slake the thirst of farms and fast-growing California cities has brought several fish species to the brink of extinction and has threatened the quality of drinking water in San Francisco's suburbs. The environmental damage is caused by seawater from the Pacific Ocean that backs up through the 1,600-square-mile estuary extending east from San Francisco Bay nearly to Sacramento. An even more severe environmental change has accompanied the withdrawal of water from the Colorado River. Today, 99% of the

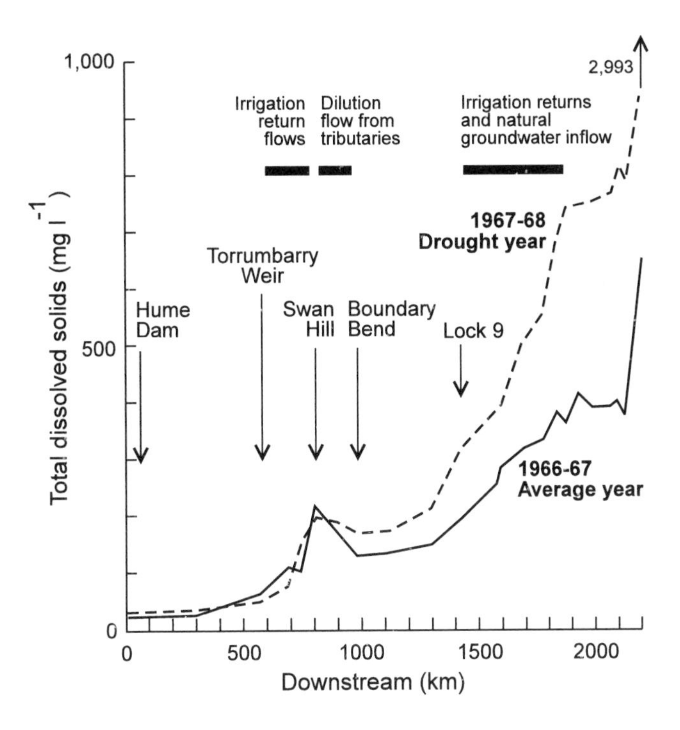

Figure 12.11. Increases in salinity along the course of the Murray River, Australia, during an average and during a drought year. (Meybeck, Chapman, and Helmer 1989; modified from Holmes 1971)

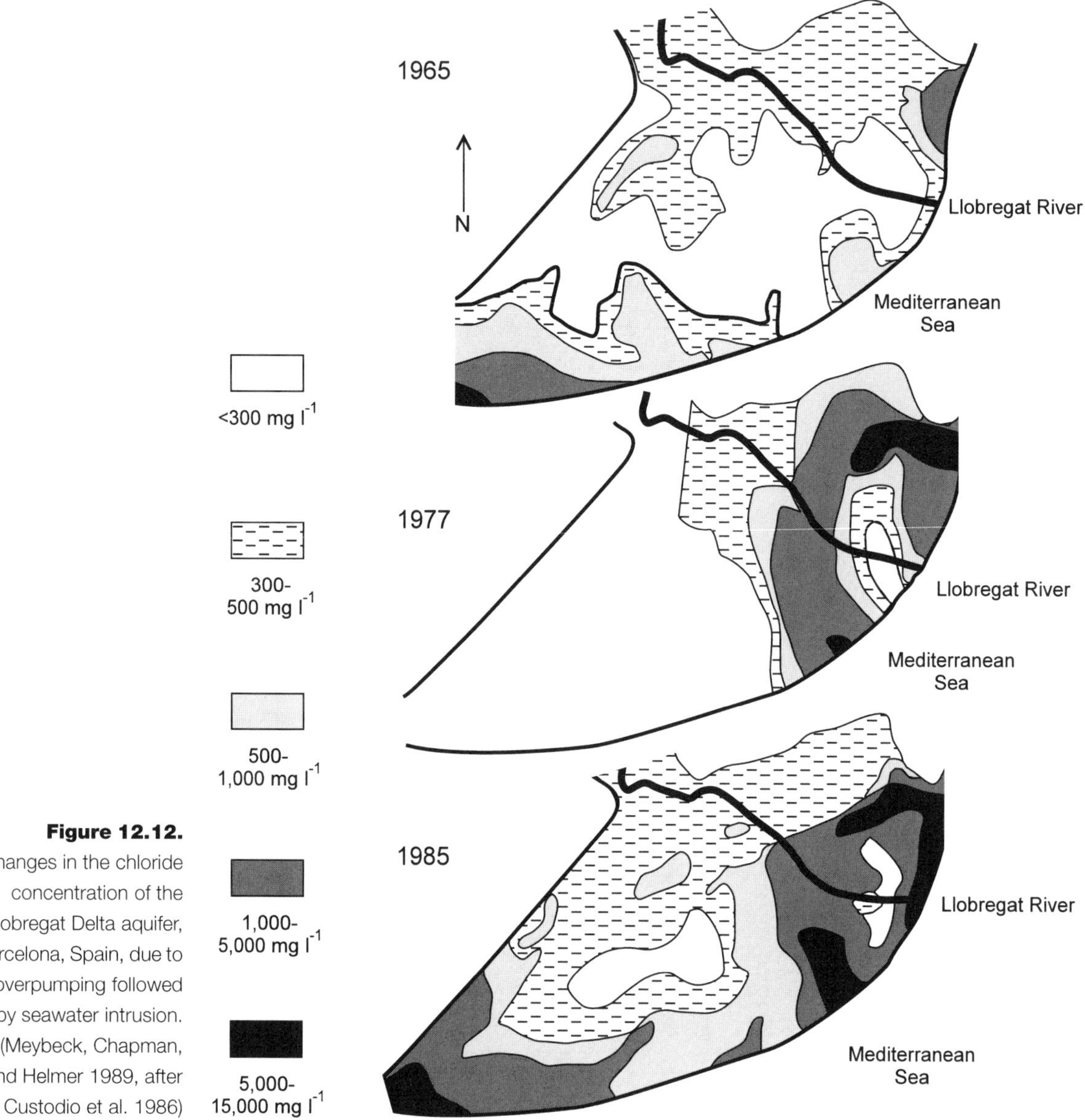

Figure 12.12. Changes in the chloride concentration of the Llobregat Delta aquifer, Barcelona, Spain, due to overpumping followed by seawater intrusion. (Meybeck, Chapman, and Helmer 1989, after Custodio et al. 1986)

river flow is withdrawn, first by Arizona, Utah, and California, and finally by Mexico.

Figure 12.12 shows the salinity increase in the Llobregat Delta aquifer near Barcelona, Spain, between 1965 and 1985. The delta contains many of Barcelona's urban and industrial suburbs and areas of diminishing but still important irrigated agriculture (Custodio et al. 1986). The demand for water in this area was satisfied largely by exploiting the Llobregat Delta aquifer, and this has resulted in declining water levels and in seawater intrusion.

Inorganic Trace Contaminants

Trace contaminants in surface and groundwaters can be divided into three groups (see fig. 12.13). Group A contaminants produce no adverse health effects until their concentration exceeds a threshold value. Above the threshold value their toxicity increases progressively with increasing concentration. Nitrate, NO_3^-, belongs to this contaminant group. The ion is innocuous at concentrations below about 50 mg/l, but is toxic when present at higher levels in drinking water and may cause methaemoglobinaemia, the condition called "blue babies," in bottle-fed infants. For Group B contaminants there is no threshold for the onset of toxicity. Genotoxic substances that cause carcinogenicity, mutagenicity, and birth defects probably belong to this group, which includes some natural and synthetic organic compounds, many chlorinated microorganic compounds, some pesticides, and arsenic. Group C elements are part of the essential dietary intake. Fluorine, iodine, and selenium are crucial dietary constituents. Their absence from drinking water can cause severe health problems, but so can their presence at high concentrations (Thornton 1993).

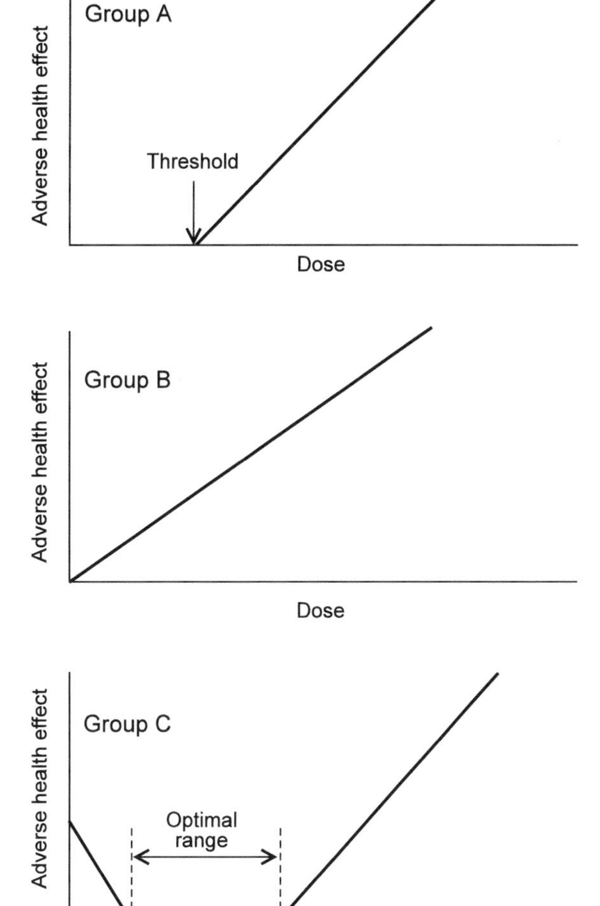

Figure 12.13. Schematic representation of the adverse health effects of three classes of water contaminants (modified from Maybeck, Chapman, and Helmer 1989). (a) Contaminants that produce health effects only above a threshold. (b) Contaminants for which there is no threshold for the onset of adverse health effects. c) Contaminants where adverse health effects occur at very low or high doses.

Table 12.5. Natural and Anthropogenic Fluxes

Element	*1990 World Production (gm/yr)*	*Unpolluted River Flux (gm/yr)*	*Ratio of 1990 World Production to Unpolluted Dissolved Plus Particulate River Flux*
Carbon	$5,000 \times 10^{12}$	400×10^{12} organic 250×10^{12} inorganic	*6*
Iron	500×10^{12}	$1,000 \times 10^{12}$	0.50
Sulfur	58×10^{12}	100×10^{12}	0.58
Nitrogen (ammonia)	110×10^{12}	10×10^{12}	*11*
Aluminum	18×10^{12}	$1,600 \times 10^{12}$	0.01
Phosphorus	22×10^{12}	2×10^{12}	*11*
Copper	9.0×10^{12}	1.1×10^{12}	8.2
Zinc	7.3×10^{12}	1.4×10^{12}	5.2
Lead	3.3×10^{12}	0.26×10^{12}	*13*
Chromium	11.7×10^{12}	2.0×10^{12}	5.8
Nickel	1.0×10^{12}	6.5×10^{12}	0.2
Tin	220×10^{9}	60×10^{9}	*3.7*
Magnesium	380×10^{9}	$400,000 \times 10^{9}$	0.95×10^{-3}
Molybdenum	114×10^{9}	30×10^{9}	*3.8*
Uranium	20×10^{9}	36×10^{9}	0.55
Cadmium	21×10^{9}	4×10^{9}	*5.2*
Mercury	6.0×10^{9}	1.6×10^{9}	*3.7*
Gold	2.0×10^{9}	0.08×10^{9}	*25*
Platinum group metals	285×10^{6}	200×10^{6}	*1.4*

Source: World production figures from *Mineral Commodities Summaries*, 1991.

Most inorganic trace contaminants belong to Group A or B, and many are mined and produced at rates that are comparable to or that exceed their flux in unpolluted rivers (see table 12.5). Humanity is disturbing the geochemical cycle of many elements in a major way. Fortunately, this does not mean that all rivers are heavily polluted. Many of the elements that are mined in great abundance ultimately enter rivers as components of sediments, not as part of the dissolved load, and do not influence water quality significantly. However, in some river environments, minerals containing toxic elements can dissolve and create severe environmental problems. Mercury is such an element. Organometallic mercury compounds can form in rivers both biotically and abiotically. The accumulation of methyl mercury (CH_3Hg^+) and dimethyl mercury ($[CH_3]_2Hg$) in fish is probably the most significant pathway of mercury to humans, and was the cause of several thousand cases of severe mercury poisoning in villages near Minamata Bay, Japan (Takeuchi et al. 1959; Takeuchi 1972).

There are several other reasons why the data in table 12.5 are geochemically intriguing but unreliable measures of the health hazard of inorganic substances. Consider gold, for instance. The ratio of world gold production to the unpolluted river flux of this metal is very large, about 25:1. However, the quantities of gold that are mined are so small that the concentration of gold in streams would be too small to cause alarm even if all mined gold were flushed to the oceans dissolved in streams, which it is not.

A more serious objection to using the data in table 12.5 as an indicator of potential pollution problems is the geographically heterogeneous input of wastes into rivers and lakes. Few wastes enter rivers and lakes in a geographically homogeneous manner. Some enter the hydrosphere from what are essentially point sources, where they can create severe local environmental problems. Mine wastes are a case in point. Cadmium-containing effluents from mine wastes that flooded low-lying rice fields in the Jintsu River area of Japan were responsible for several hundred cases of the painful and debilitating "Itai-Itai" disease among the local inhabitants (Kobayashi 1971; Friberg, Piscator, and Nordberg 1971), and mine-related instances of serious water pollution have been recorded in many other parts of the world, particularly where little attention has been paid to the environmental consequences of mining and milling (Förstner and Wittmann 1979, and chapter 11 in Meybeck, Chapman, and Helmer 1989). But mines are not the only point sources of metals. Effluents from manufacturing sites can be equally or even more polluting. Electroplating plants and pigment and battery manufacturers have been the cause of serious environmental problems. In municipalities where industrial effluents mix with municipal sewage, the concentration of toxic metals in sewage can be so high that fertilizers produced from dehydrated and treated sewage sludge cannot be applied to fields that grow crops for human consumption.

Highly industrialized regions usually harbor many point sources of pollutants. Their combined effluents tend to produce regional sources of pollution. The Rhine River basin illustrates this process. Figure 12.14 shows the location of major urban centers along the Rhine. The river drains an area of 185,000 square kilometers in Switzerland, Germany, France, and the Netherlands, and accumulates a heavy load of pollutants along its 1,320-kilometer passage from the Alps, through spectacularly beautiful wine country that is romantically decorated with castles, and finally to the North Sea. The highly polluted nature of Rhine River water was widely recognized at the beginning of the 1970s, and since then a great effort has been made to improve its quality. The dissolved and particulate load of cadmium and mercury transported from Germany into Holland has decreased dramatically (see fig. 12.15), as have the concentrations of a number of other metals (see table 12.6). Much of this improvement is due to more effective wastewater treatment along the course of the river and to the replacement of toxic metals in several critical applications.

Table 12.6. Trends in Metal Concentrations in the Rhine at Lobith, Close to the Dutch-German Border (μg/l)

	1975	*1980*	*1983*	*1984*	*1985*
Arsenic	4.5	3.0	3.6	0.3	1.8
Cadmium	2.3	1.6	0.4	0.2	0.1
Chromium	35.0	20.0	11.0	5.0	8.0
Copper	20.0	14.0	10.0	5.0	6.0
Mercury	0.4	0.2	0.1	0.07	0.07
Lead	22.0	15.0	7.0	6.0	4.0
Nickel	10.0	9.0	5.0	5.0	5.0
Zinc	135.0	102.0	57.0	37.0	50.0

Source: UNEP 1990.

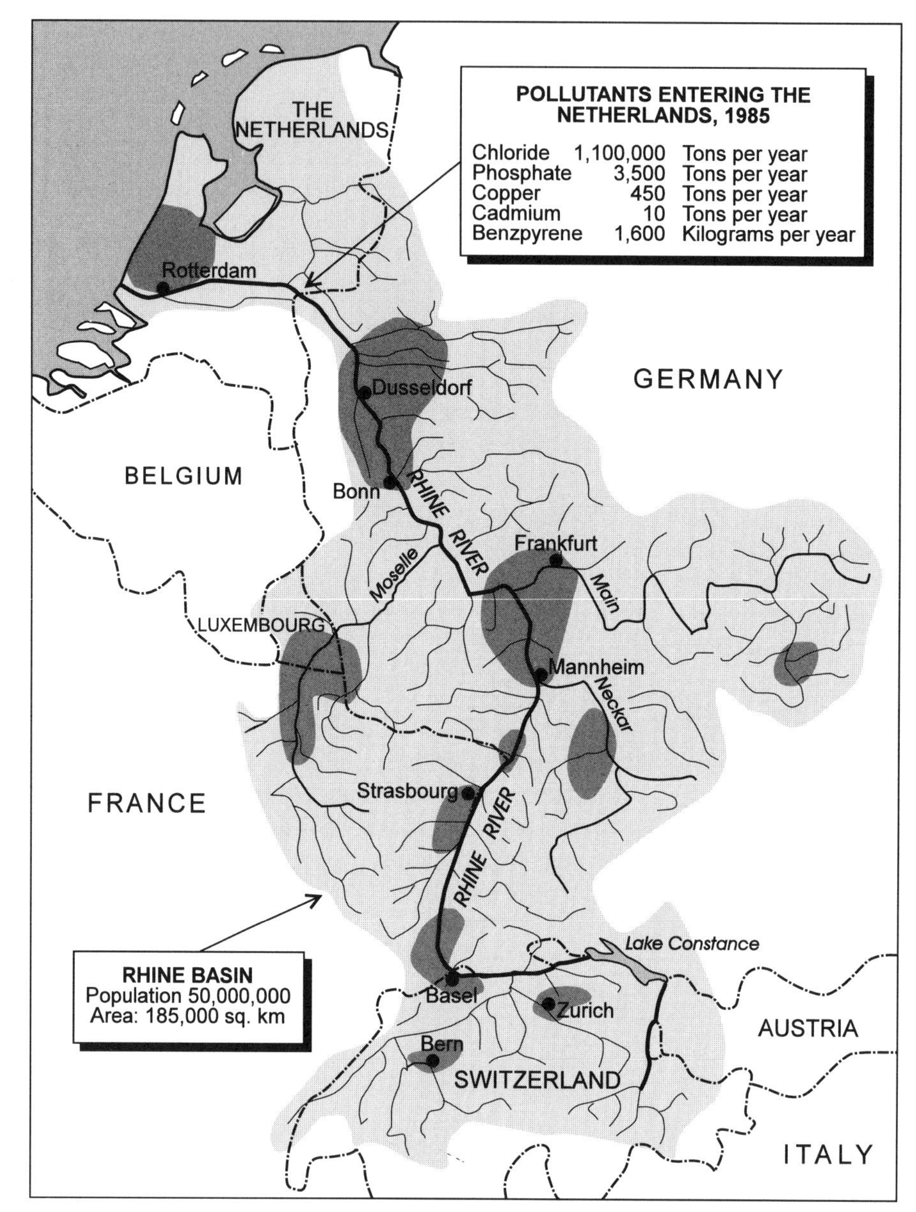

Figure 12.14. The Rhine river basin in western Europe. The major urban areas are indicated in dark grey. (Modified after la Rivière 1989)

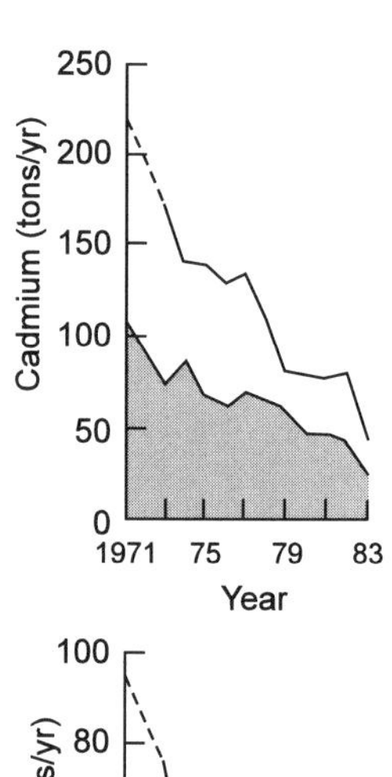

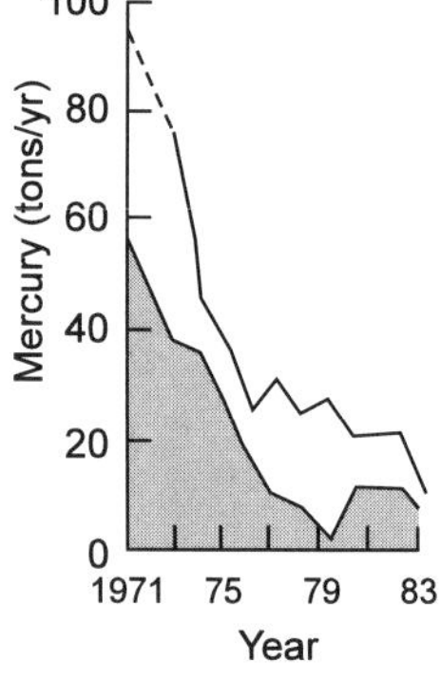

Among the metals listed in table 12.6, lead has had a particularly interesting and turbulent history. The metal is industrially important, and its toxicity has been known for a very long time. In 1859, James Pugh Kirkwood published a *Collection of Reports and Opinions of Chemists in Regard to the Use of Lead Pipe for Service Pipe, in the Distribution of Water for the Supply of Cities*. An 1862 addendum to the preface ventures that "I am satisfied that the only safe course in families, is to avoid entirely the use of lead pipe for service pipe." It has been suggested that lead poisoning of the upper classes contributed to the decline

Figure 12.15. Changes in the mercury and cadmium loads (dissolved and particulate) in the Rhine River at the Dutch-German border between 1971 and 1983. (After Malle 1985)

of the Roman Empire. It has been demonstrated that the ingestion of lead paint from peeling walls and ceilings has contributed materially to the ill-health of poor, inner-city children (Committee on Measuring Lead in Critical Populations 1993).

During the course of the twentieth century, such point sources receded in importance as sources of environmental lead contamination. Tetraethyl lead, $(C_2H_5)_4Pb$, was found to be an excellent anti-knock additive for gasoline. The spectacular increase in the consumption of lead in gasoline between 1930 and 1966 is well documented in figure 12.16. All of this lead was sprayed into the atmosphere as a constituent of automobile exhaust. Concerted lobbying finally led to the replacement of leaded with unleaded gasoline, and to the rapid decline in the use of lead in gasoline during the 1970s. Very hearteningly, in the United States levels of lead in human blood have followed suit. They decreased markedly during the 1970s (see fig. 12.17) and continued to fall in the 1980s. In blood

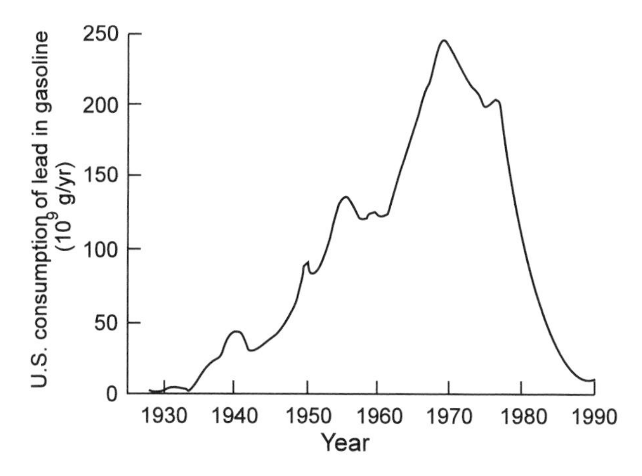

Figure 12.16. The annual consumption of lead in gasoline in the United States between 1930 and 1982. (After Boyle et al. 1986)

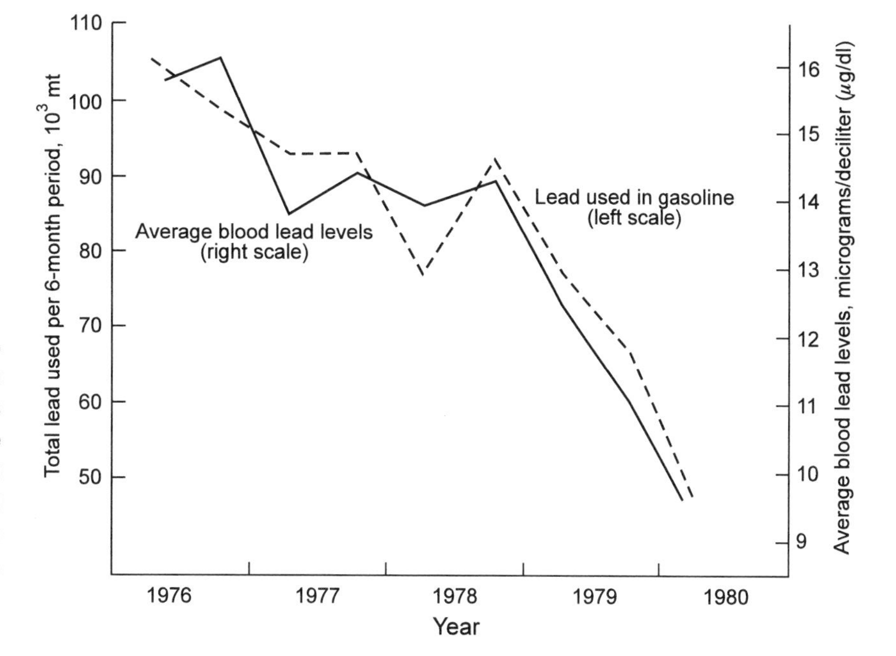

Figure 12.17. Parallel decreases in blood lead levels observed in the NHANES II study and amounts of lead used in gasoline during 1976–80. (EPA 1986)

Figure 12.18. "Baker" underwater nuclear explosion at Bikini, July 25, 1946, a part of Operation Crossroads (Delgado 1993). (Photo-Brown Brothers, Sterling, Pennsylvania)

samples taken between 1988 and 1991 from Americans aged one to seventy-four, lead levels had dropped to 2.8 micrograms per deciliter (Pirkle et al. 1994).

Not all stories of pollution with inorganic compounds have such a happy ending. The residual effects of the Cold War are still with us, and their environmental impacts may become more serious before they begin to recede. The mushroom cloud in figure 12.18 was produced during one of the 1946 atomic bomb tests in the vicinity of Bikini Atoll in the western Pacific. Nearly two dozen nuclear and thermonuclear bombs were detonated there between 1946 and 1958. The natives of Bikini were removed prior to the 1946 tests. Together with two generations of their descendants, they still remain exiled on other islands of the Marshall chain. The sole habitation in Bikini that is active today is a U.S. Department of Energy field station maintained by a transient community of workers and visiting scientists (Delgado 1993). Today, Bikini is remembered largely for the bathing suit by that name, which was inspired by the Operation Crossroad tests, and for the 1954 hydrogen bomb tests that led to the treaty to ban atmospheric testing of nuclear weapons. But the contaminated islands of Bikini Atoll remain, as do the sunken ships in the lagoon. All are parts of the nuclear testing legacy.

Less spectacular American legacies of the Cold War were discussed in chapter 11. U.S. national laboratories that were occupied with the production of nuclear weapons are now devoting a good deal of effort and a great deal of money to cleaning up after themselves and to prevent the spread of radioactive wastes beyond the confines of the federal reservations. It remains to be seen how successful these efforts will be, especially in places like Hanford, Washington, which is uncomfortably close to the Columbia River. The effects of radionuclide contamination from these installations may be dwarfed by contamination from their Russian counterparts at Chelyabinsk, Krasnoyarsk, and Tomsk, from the nuclear weapons testing areas at Novaya Zemlya and Semipalatinsk, from the effects of the 1957 radiation disaster at Kyshtym, and from the nuclear reactor accident at Chernobyl. The map in figure 12.19 shows the location of some of the most important actual and potential sources of widespread radionuclide contamination in the former USSR.

A much less dramatic but nonetheless serious problem has been created by the rapid increase in phosphate and nitrate discharges into the world's rivers and lakes. Anthropogenic phosphate inputs are largely the result of sewage outfalls

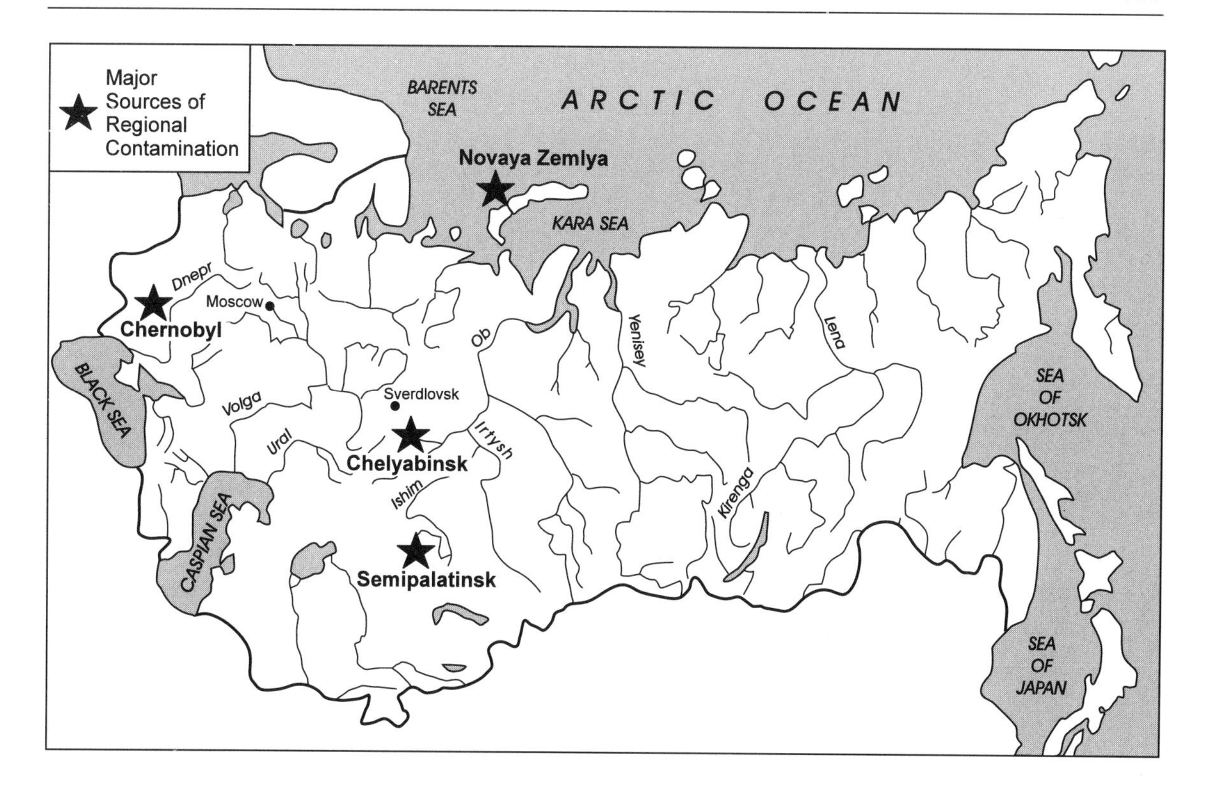

Figure 12.19. Map of the major Russian rivers, showing the locations of major sources of regional contamination with radionuclides in the former USSR. (I. Khodakovskiy 1993, pers. comm.)

and phosphate-containing detergents. Nitrate inputs are largely due to sewage outfalls and fertilizers. Phosphate (PO_4^{3-}) is a limiting nutrient in most lakes. The addition of phosphate therefore stimulates the growth of plants, particularly algae. This creates taste and odor problems in drinking water. Dead algae tend to accumulate at the bottom of lakes, to decay there, and to deplete the oxygen content of bottom waters. Oxygen depletion can have a major effect on fish populations in lakes and on bottom faunas. In the early 1960s, the combination of these processes generated intense public concern regarding entrophication in the Great Lakes region of the United States (see fig. 12.20). Along the shores of Lake

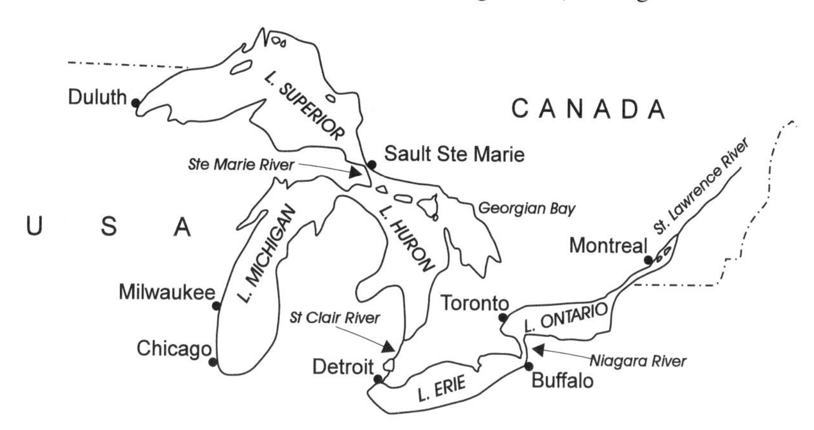

Figure 12.20. The Great Lakes region of the United States and Canada.

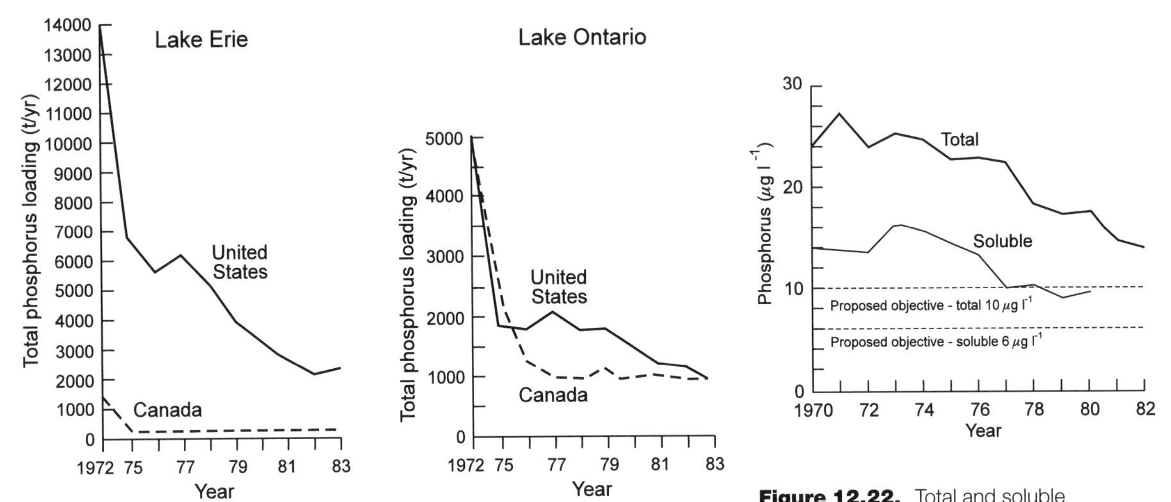

Figure 12.21. Municipal phosphorus loadings to the lower Great Lakes, 1972–83. (Data from the 1983 Report on Great Lakes Water Quality, I.J.C.)

Figure 12.22. Total and soluble phosphorus concentrations in the spring in Lake Ontario, 1970–82. (After Neilson 1983)

Erie, decomposing algae had to be removed with bulldozers. During the summers, the lower part of the water column of the central lake basin became completely anoxic. In 1972, a phosphorus management plan was adopted for the region. This plan limited the allowable phosphorus content of household laundry detergents and—probably of much more importance—reduced the allowable total phosphorus concentration in the effluents from the larger municipal wastewater treatment plants (Meybeck et al. 1989, chapter 15). Figure 12.21 shows the decrease in municipal phosphorus loadings to the lower Great Lakes between 1972 and 1983. The concentration of phosphorus in Lake Ontario followed suit

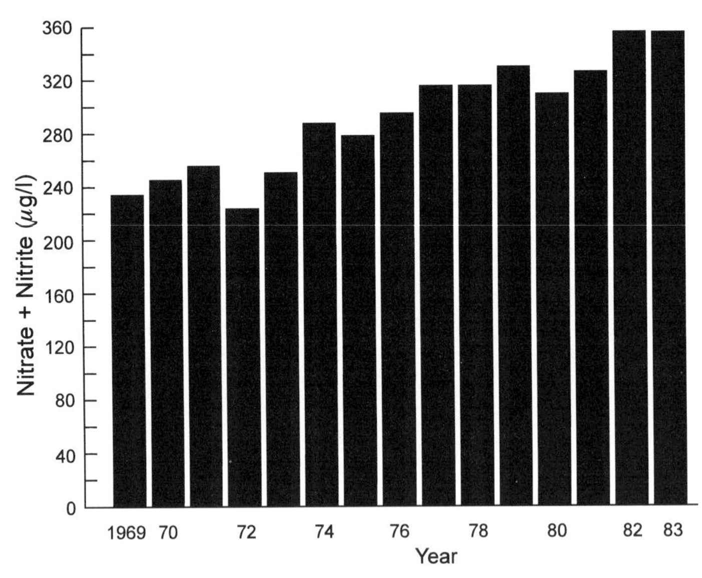

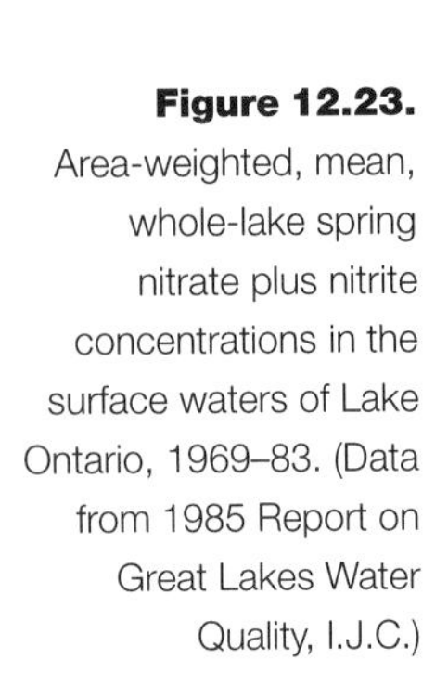

Figure 12.23. Area-weighted, mean, whole-lake spring nitrate plus nitrite concentrations in the surface waters of Lake Ontario, 1969–83. (Data from 1985 Report on Great Lakes Water Quality, I.J.C.)

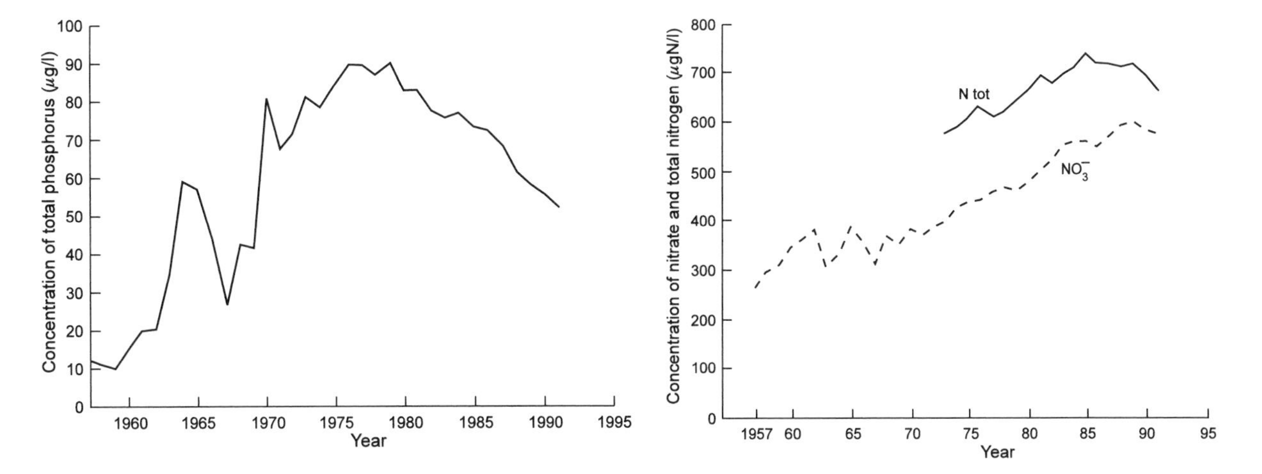

Figure 12.24 (*left*). The concentration of total dissolved phosphorus in Lake Geneva, Switzerland, between 1957 and 1991. (Lake Geneva–Rapport CIPEL 1992)

Figure 12.25 (*right*). The concentration of NO_3^- and of total dissolved nitrogen in Lake Geneva, Switzerland, between 1957 and 1991. (Lake Geneva–Rapport CIPEL 1992)

(see fig. 12.22), and the eutrophication of the lakes has been brought under control, but at a cost of some $7.6 billion for upgrading and constructing municipal waste-water plants throughout the basin.

During this period the nitrate concentration in Lake Ontario continued to increase (see fig. 12.23). These trends in phosphate and nitrate are typical of rivers and lakes in heavily fertilized areas (see figs. 12.24 and 12.25 for phosphate and nitrate trends in Lake Geneva, Switzerland). Figure 12.26 shows the long-term trend in fertilizer use in the United States on a logarithmic scale; figure 12.27 shows similar data on a linear scale for a number of countries in western Europe.

Figure 12.26 (*left*). The use of nitrogen and phosphorus fertilizer in the United States between 1850 and 1985. Note the logarithmic scale. (After Nixon et al. 1986)

Figure 12.27 (*right*). Fertilizer nitrogen use in western Europe between 1940 and 1980. (After Roberts and Marsh 1987)

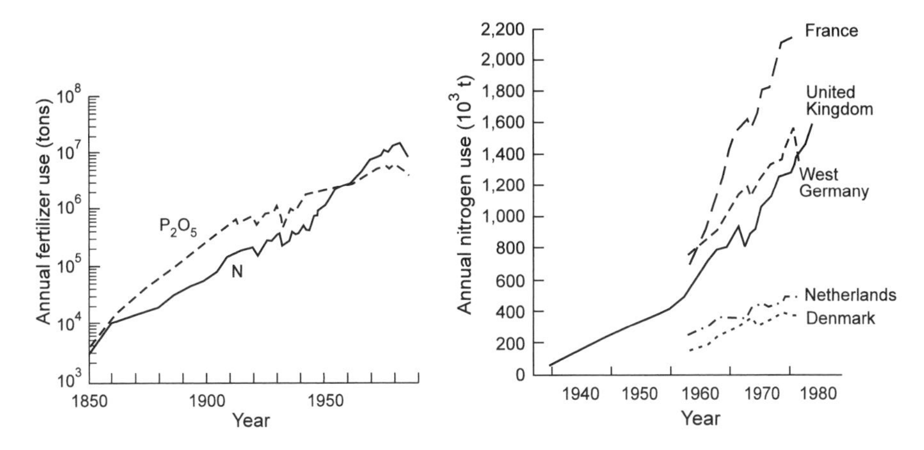

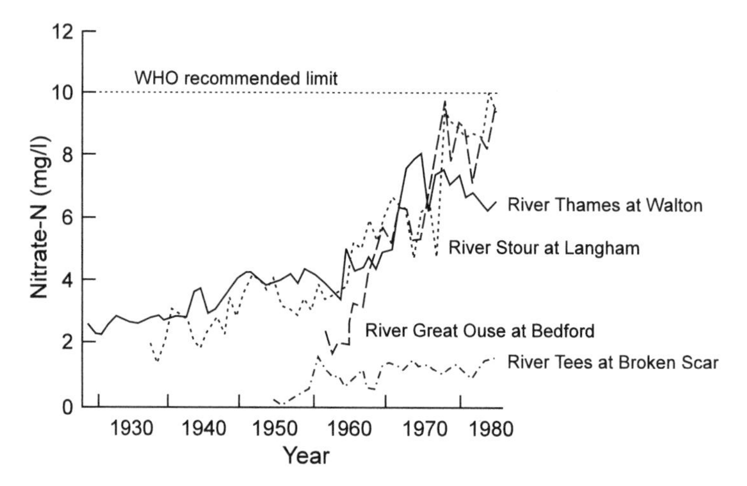

Figure 12.28. Long-term trends in the concentration of nitrate in four U.K. rivers. (After Roberts and Marsh 1987)

Fertilizer phosphate that is not used in plant growth tends to be immobilized in soils as a constituent of phosphate minerals of aluminum, iron, and calcium. Usually, less than 5% of the applied phosphorus finds its way into rivers and lakes in dissolved form. The behavior of nitrate is very different, because nitrate minerals are very soluble. The ion leaves soils largely in solution and is added to the nitrate content of river and lake waters as well as to that of groundwaters. The dramatic increase in the nitrate concentration of some rivers in Great Britain since World War II is documented in figure 12.28. Nitrate concentrations in excess of 50 mg/l are considered a threat to human health. It seems likely that during the next century fertilizers will have to be applied in progressively larger quantities to satisfy the increasing demand for agricultural products. More rivers, lakes, and groundwaters will therefore become heavily polluted with nitrate unless effective countermeasures are applied.

Organic Contaminants

Of all the chemical elements, carbon is by far the most versatile. It can bond to itself and to a wide range of other elements in a somewhat bewildering variety of ways, and it is the building block of several million compounds, including many that are critical for the existence of life. The biosphere consists largely of organic compounds, although a few inorganic compounds are important for the hard parts of animals. We eat mainly organic compounds, and we are healed by them. Some of the most famous medicines are natural products. Penicillin, that wonderful, powerful antibacterial substance, is produced by certain molds of the genus *Penicillium*. Cyclosporin, a compound that has saved many organ transplant patients, is produced by the fungus species *Tolypocladium inflatum Gams*.

Eighteenth- and early-nineteenth-century chemists believed that only living organisms could synthesize organic compounds. The synthesis of urea by Friedrich Wöhler in 1828 ushered in a new era in organic chemistry, and laid the foundation for the synthesis of tens of thousands of useful organic compounds. Many of these do not threaten the environment, but quite a few do. We have already encountered some of the latter. Coal, oil, and gas are sources of much atmospheric pollution (see chapter 11). Volatile organic chlorine-containing compounds, such as the CFCs and carbon tetrachloride (CCl_4), are responsible for deepening the ozone hole (see the earlier part of this chapter). The presence of many other organic compounds in surface and groundwaters creates serious concerns for human health. The EPA has published a list of 129 chemical pollutants as priority risks for the environment (Callahan et al. 1979). Of these, 114 are organic substances. The European Economic Community (EEC) has drawn up a similar list of 129 potentially dangerous substances (EEC, 1982). Theirs includes 118 organic substances that are slated to be eliminated from wastes, particularly from industrial wastes that are discharged directly into the aquatic environment. The seventy-three organic pollutants that are common to both lists can be classified by use and by the manner of their dispersal in the environment. Table 12.7 is one such classification. Group I in this table includes materials that are actively dispersed in the environment to protect agricultural products. They are designed to poison a range of organisms; not surprisingly, they also have nasty effects on many untargeted organisms. The most famous among these materials is the organochlorine insecticide DDT (dichlorodiphenyl trichloroethane). The structure of this compound is shown in figure 12.29. In 1962, Rachel Carson published *Silent Spring*, a book that riveted public attention on the environmental problems created by DDT and by other synthetic pesticides. Her book contributed a great deal to the ecological and conservationist awareness of the 1970s and 1980s, and led to the banning of many organochlorine compounds, including DDT, in many industrialized countries. The subsequent decrease in DDT levels in wildlife is shown in figure 12.30. In some tropical countries these insecticides are not banned, because they are such important weapons in the war against malaria. Their continued use in these countries illuminates the difficult choices that

Table 12.7.
A Classification of Organic Products Considered to be Dangerous to the Aquatic Environment

I. *Pesticides*

1. Insecticides
   a. Organochlorine compounds, such as DDT (see fig. 12.29)
   b. Organophosphate compounds
2. Herbicides
   a. Plant hormones
   b. Triazines
   c. Substituted ureas
   d. Others

II. *Materials for widespread household and industrial use*

1. Halogenated derivatives of methane, ethane, and ethylene
2. Polycyclic aromatic hydrocarbons

III. *Materials used essentially in industry*

1. Carbon tetrachloride
2. Vinyl chloride
3. Chlorinated derivatives of benzene, naphthalene, phenol, and aniline
4. Heat exchange fluids and dielectric substances

Source: After Meybeck, Chapman, and Helmer 1989.

Figure 12.29. The molecular structure of DDT, PCBs, and dioxins.

DDT is one of the chlorophenyls, a group of compounds that also includes DDD, DDE, and methoxychlor.

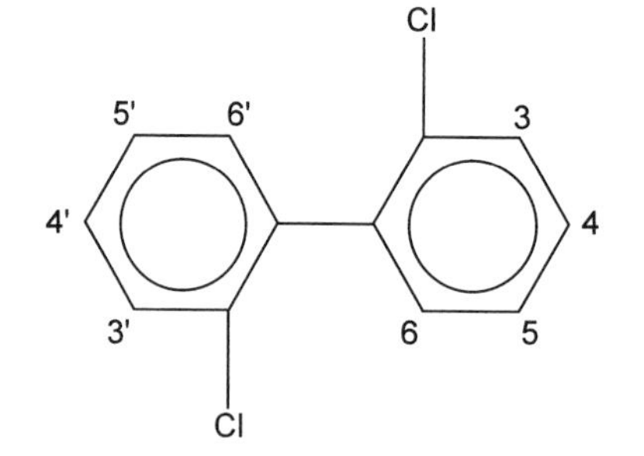

The structure of 2, 2'-dichlorobiphenyl. Other PCBs have similar structures but differ in the number and/or arrangement of their Cl atoms.

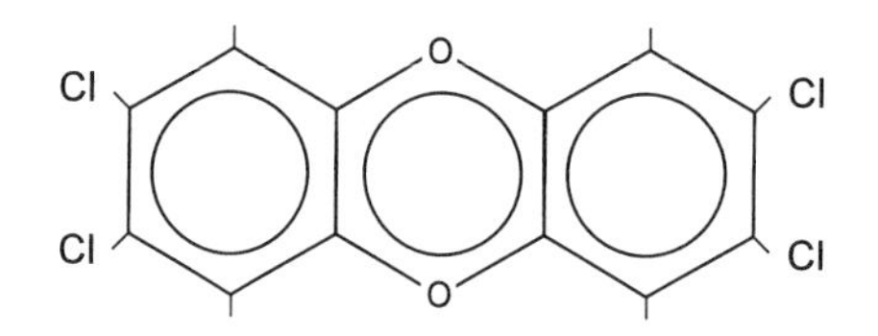

The structure of the chlorinated dioxin 2,3,7,8-TCDD. Other chlorinated dioxins have similar structures but differ in the number and/or arrangement of their Cl atoms.

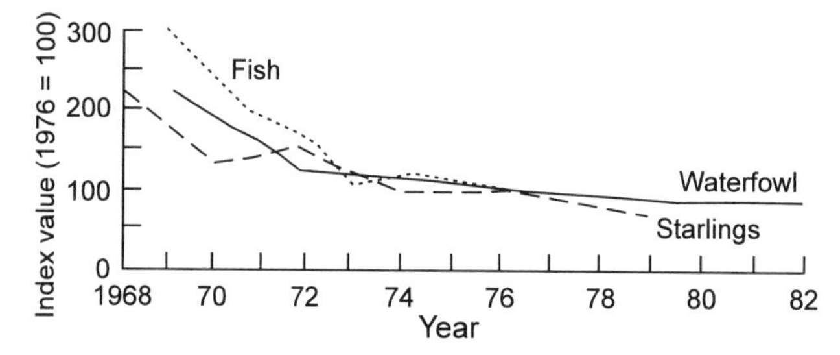

Figure 12.30. Trends in levels of DDT in wildlife, 1968–79. (Data from U.S. Fish and Wildlife Service, *State of the Environment*, 1981, a report from the Conservation Foundation)

sometimes have to be made between human health and the health of the human environment.

The sale of pesticides in the United States has increased to about one billion pounds a year from about 100 million pounds a year at the end of World War II. Worldwide, annual sales of the major categories of pesticides now amount to about $50 billion (see fig. 12.31). Concern about the impact of pesticides has grown with their use. As the sensitivity of the techniques for analyzing pesticide residues has improved, these compounds have been detected in a progressively larger number of water supplies. In many instances, pesticide concentrations are extremely small, and major unresolved battles have been fought over the impact of trace quantities of these substances on human health. Two types of criteria are commonly used to assess pesticide toxicity: tests on animals, and epidemiological studies on human populations.

Rodents are usually used in animal-based evaluations of chemical toxicity. Adverse health effects observed in these animals are then extrapolated to human

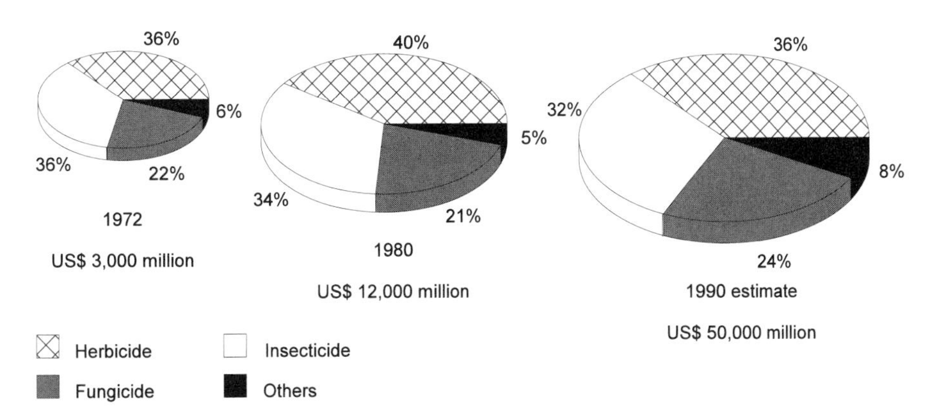

Figure 12.31. Trends in worldwide sales of the major categories of pesticides. (Meybeck, Chapman, and Helmer 1989, chap. 12)

populations. There is considerable uncertainty about the validity of using the effects of what are usually very large doses on animals to predict the effects of much smaller doses on humans. Some toxic effects that occur in animals (e.g., liver tumors in mice) are rarely seen in human populations. Of the roughly four hundred different types of pesticides used on food crops, more than seventy—including most of the top-selling ones—cause cancer in laboratory animals. However, it is not certain that the responses of human physiology to smaller concentrations of these pesticides are sufficiently similar to those of laboratory animals to larger doses to permit meaningful extrapolations from mice to men and women, or—for that matter—from mice to infants and children (Committee on Pesticides in the Diets of Infants and Children 1993).

Studies of human epidemiology are therefore very important. Unfortunately, many of these studies are inconclusive, because they are not well controlled. The causes of illnesses in people are often difficult to pin down, particularly if there are many contaminants in their environment and if the concentrations of these contaminants are close to the threshold for producing adverse human health effects. Nevertheless, as Abelson (1993) has pointed out, it is surely significant that hundreds of thousands of agricultural workers received huge exposures to pesticides during the 1960s and 1970s, and that large numbers of cancers should now be evident in the exposed group if these pesticides were highly carcinogenic. The Council on Scientific Affairs of the American Medical Association has reviewed the evidence on the matter. It reports that a large number of pesticidal compounds have shown evidence of genotoxicity or carcinogenicity in animal and in in-vitro screening tests, but no pesticides—except arsenic and vinyl chloride (once used as an aerosol propellant)—have been proved to be definitely carcinogenic in humans. The wording here is quite careful, and one would clearly like to know more, but it is reassuring that the number of outbreaks, cases, and particularly of deaths in the United States that are ascribable to organic chemicals between 1961 and 1983 (see table 12.8) was very small. Nevertheless, new, inexpensive methods for protecting crops that do not require large-scale applications of highly toxic chemicals and that do not create other environmental hazards would surely be most welcome.

The volatile organic compounds belonging to Group II in table 12.7 are widely used in households and in industry, and tend to be widely dispersed in the environment. Many Group III organic compounds are used as direct or as intermediate agents in industrial chemical syntheses; others are finished products used in

Table 12.8. Waterborne Outbreaks of Illness in the United States, 1961–83

Etiologic agents	*Outbreaks*	*Cases*	*Deaths*
Inorganic chemicals (metals, nitrates, etc.)	29	891	0
Organic chemicals (pesticides, herbicides, etc.)	21	2,725	7
Bacterial			
Shigella	52	7,462	6
Salmonella	37	19,286	3
Campylobacter	5	4,773	0
Toxigenic *E. coli*	5	1,188	4
Vibrio	1	17	0
Yersinia	1	16	0
Viral			
Hepatitis A	51	1,626	1
Norwalk	16	3,973	0
Rotavirus	1	1,761	0
Protozoan			
Giardia	84	22,897	0
Entamoeba	3	39	2
Unidentified agents	266	86,740	0
Total	572	153,394	23

Source: After Meybeck, Chapman, and Helmer, chapter 12.

closed systems, such as heat exchange fluids, and as dielectric substances. PCBs (polychlorinated biphenyls; see fig. 12.29) are prominent among the latter. Less than 5% of most Group III organic compounds are dispersed into the environment, but even relatively minor spills of these substances can do substantial harm on a local scale.

Between the late 1940s and 1977 the General Electric Company discharged PCBs into the Hudson River from its plants in Hudson Falls and Fort Edward, New York (see fig. 12.32). PCBs were an important component of transformers and capacitors during this period, and their toxicity was not widely recognized. The Hudson is a major spawning ground for striped bass, a delicacy that accounted for a large part of the yearly income of hundreds of commercial fishermen in the New York area. The U.S. Congress banned the manufacture of PCBs in 1979, after it was shown that the chemicals are highly toxic to fish and other animals. A ban was imposed on commercial fishing in the New York–Long Island area after the concentration of PCBs in striped bass was found to exceed the federal limit of five parts per million. The concentration of PCBs in the Great Lakes region became a serious concern in the early 1970s. Since these chemicals were banned, the concentrations of PCBs in Lake Michigan trout have fallen dramatically, as has their concentration in herring gull eggs (fig. 12.33).

Industrial accidents tend to be short-lived but may be followed by violent consequences. On November 1, 1986, a major fire broke out in the Sandoz-Basel chemical factory in Basel, Switzerland. Twelve hundred fifty tons of highly toxic compounds, including insecticides, fungicides, herbicides, and mercury-contain-

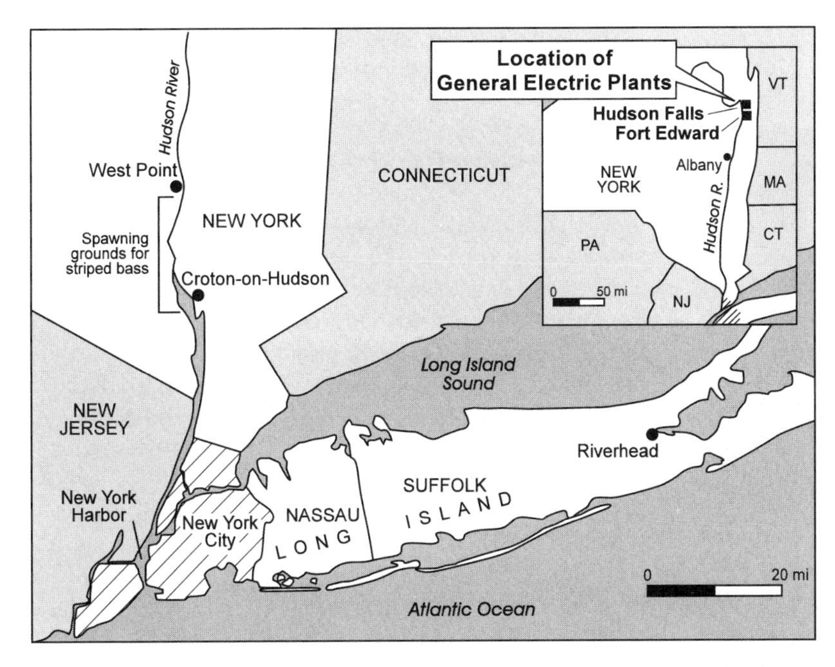

Figure 12.32. The location of the General Electric Company's plants on the Hudson River, which were the source of PCB contamination in the area between the late 1940s and 1977. (*New York Times*, August 13, 1993)

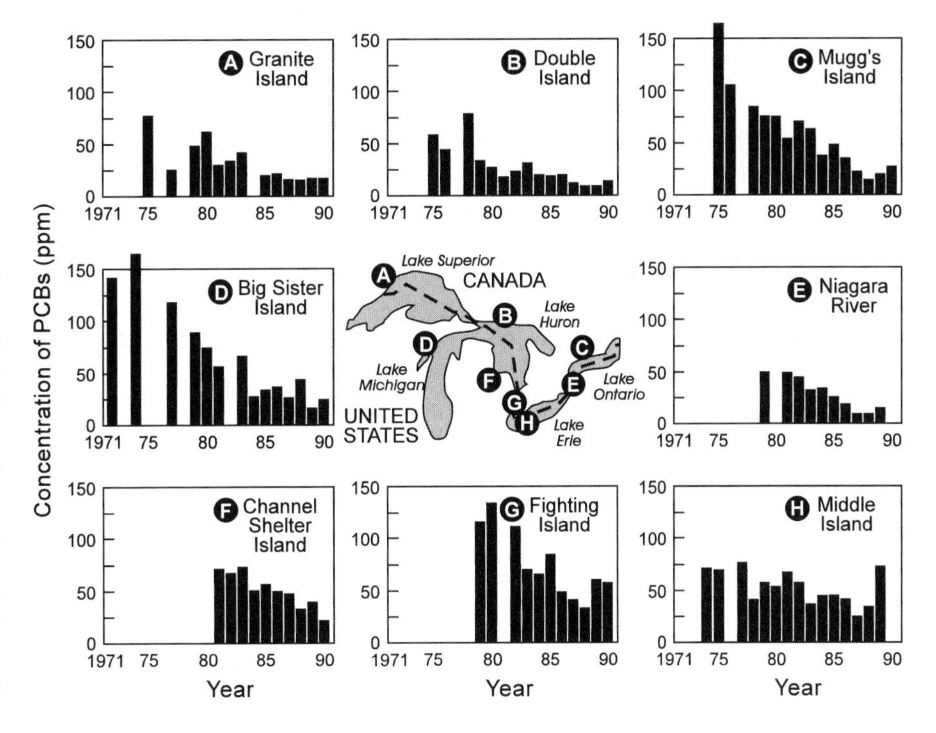

Figure 12.33. Changes in the concentration of PCBs (in ppm) in herring gull eggs at eight sites in the Great Lakes region, United States. (Bishop and Weseloh, *Environment Canada*, 1992, and U.S. Environmental Protection Agency; *New York Times*, May 7, 1994)

ing compounds, were stored there. Part of this stock of chemicals was buried. The rest was washed into the Rhine with the water that was used to extinguish the fire. The spill seriously affected the quality of water in the Rhine from Basel to the sea (Durousseau 1987).

Although accidents like this are serious, contamination of groundwaters with organic chemicals may turn out to be a more serious threat. Once contaminated, aquifers tend to stay that way for long periods of time. There are several reasons for this. The flow of underground waters is orders of magnitude slower than that of surface waters (see chapter 4). Groundwater systems are therefore hard to flush. Many of the common contaminants are organic compounds that have very low solubilities in water and therefore form immiscible lenses of nonaqueous-phase liquids (NAPLs). Some of these contaminants, such as gasoline, are less dense than water. These belong to the light-NAPLs (LNAPLs). Others, such as the common solvent trichloroethylene, are denser than water and are classified as DNAPLs. As NAPLs move through the subsurface, a portion of the contaminating liquids tends to become trapped as small, immobile globules that cannot be removed by pumping but can dissolve in and contaminate the passing groundwater. Removing DNAPLs is complicated by their tendency to migrate deep underground, where they are difficult to detect and where they may remain in pools that slowly dissolve and contaminate the groundwater. A further difficulty is created by the adsorption of NAPLs on solid materials in the subsurface. Adsorbed contaminants can remain underground for long periods of time; they tend to be released when their concentration in groundwater decreases.

The remediation of contaminated aquifers is technically difficult (see, for instance, Water Science and Technology Board 1990, 1993, 1994), and the number of contaminated sites is very large. Estimates of the total number of waste sites in the United States where groundwater and soil may be contaminated range from approximately 300,000 to 400,000. Recent estimates of the total cost of cleaning up these sites over the next 30 years have ranged as high a $1 trillion.

Two major pieces of legislation demand that contaminated groundwaters be restored to drinking water quality. CERCLA, the Comprehensive Environmental Response, Compensation, and Liability Act of 1980, is commonly known by its nickname, "Superfund." The overall goal of the law is to ensure that environmentally dangerous disposal sites are cleaned up, and that the responsible parties bear the cleanup burden to the extent possible. Notwithstanding the large funding for Superfund, neither sufficient money nor human resources are available to attack all of the known problems simultaneously. A hazard ranking system is therefore employed to place the sites on a risk-based priority list. The Superfund Law as passed originally is clearly flawed. Attempts to amend its provisions in 1994 failed in Congress.

The Resource Conservation and Recovery Act (RCRA) was passed four years before CERCLA. While Superfund deals with areas that are already contaminated, RCRA provides the basic framework for the federal regulation of hazardous wastes. The law controls the generation, transportation, treatment, storage, and disposal of hazardous waste through a comprehensive "cradle to grave" system of hazardous waste management techniques and requirements.

There is considerable doubt that existing techniques and innovative new technologies are up to the task demanded by these laws. Most of the decontamination programs involve "pump-and-treat" systems. These require installing wells at strategic locations and pumping contaminated groundwater to the surface for treatment. Although cleanup with this technique is possible at some sites, there are a great many sites where the properties of the subsurface and the contaminants make restoring contaminated groundwater to drinking water standards technically infeasible even in a matter of decades. New, innovative technologies for groundwater cleanup are subject to many of the same limitations as conventional technologies. They can improve the efficiency of groundwater cleanup efforts, but groundwater restoration to health-based goals is impracticable with existing technologies at a large number of contaminated sites (Water Science and Technology Board 1994). In some of the most difficult sites the question arises whether it makes sense to attempt remediation, or whether a contaminated area should simply be declared permanently off-limits to uses that are adversely affected by the presence of contaminated groundwater. This option is not available in areas where aquifers drain into rivers or lakes that serve large populations. Strenuous efforts are then required to prevent the dispersal of the contaminants. The estimated vulnerability index of U.S. groundwaters to pesticides is summarized in plate 56 (Kellogg et al. 1992). If even a small fraction of the highly vulnerable aquifers were to become heavily contaminated, the availability of unpolluted groundwaters would be severely curtailed for a large fraction of the U.S. population.

It helps that nearly all organic compounds are chemically unstable, and that they can be destroyed by microorganisms, by heating, and by incineration. Highly efficient means for destroying organic compounds are already available (Committee on Alternative Chemical Demilitarization Technologies 1993), and the technology will surely improve during the next century (see, for instance, Bradshaw et al. 1992). Destruction in a contained pressure vessel at 400°–600°C in the presence of water and oxygen is very efficient, and the technique is safe, because the decomposition products can be contained until the safety of their release into the environment has been demonstrated. But the method is expensive and will probably be used only for the destruction of extremely toxic substances. Incineration is now widely used. The technique is relatively inexpensive and generally very safe, but in some instances highly toxic compounds, such as chlorinated dioxins (see fig. 12.29) form during the combustion of some organic materials. These and other toxic compounds can contaminate stack gases and fly ash, and can become regional sources of pollution. Needless to say, this has further dimmed people's enthusiasm for siting incinerators in their neighborhoods.

Several innovative techniques have been developed for destroying toxic organic compounds. Among these, bioremediation looks particularly promising. Microorganisms are used to decompose toxic organic chemicals into innocuous compounds. This technique for destroying toxic wastes will probably become very important during the next century.

Landfills are and will continue to be important for sequestering toxic organic compounds, but extremely unhappy experiences with improperly designed landfills such as the Love Canal landfill near Niagara Falls, New York, point up the potential dangers of landfills. The Hooker Chemicals and Plastics Corporation disposed of chemical wastes in the abandoned hydroelectric canal, covered it, and deeded the 16-acre site to the Niagara Falls Board of Education in 1953. The Board built an elementary school on the site, and houses were later built around the school. Nearby residents first noticed chemicals leaking into their basements in 1976. Since then, Love Canal has become a symbol of the nation's environmental problems. The state and federal governments declared an environmental emergency there in 1978, moved 239 families from the neighborhood, and agreed to buy more than five hundred other contaminated homes in 1980. In 1989 families began to return to the area.

Contamination with Pathogens

Pathogens and infectious diseases associated with bodies of water have created much greater health hazards in the United States than inorganic and organic chemicals (see table 12.8). Pathogens are ever-present potential contaminants in municipal water supplies, and the contamination of mountain streams with giardia has led to widespread problems, even in remote areas. In April 1993 the

Table 12.9. Diseases Associated with Water

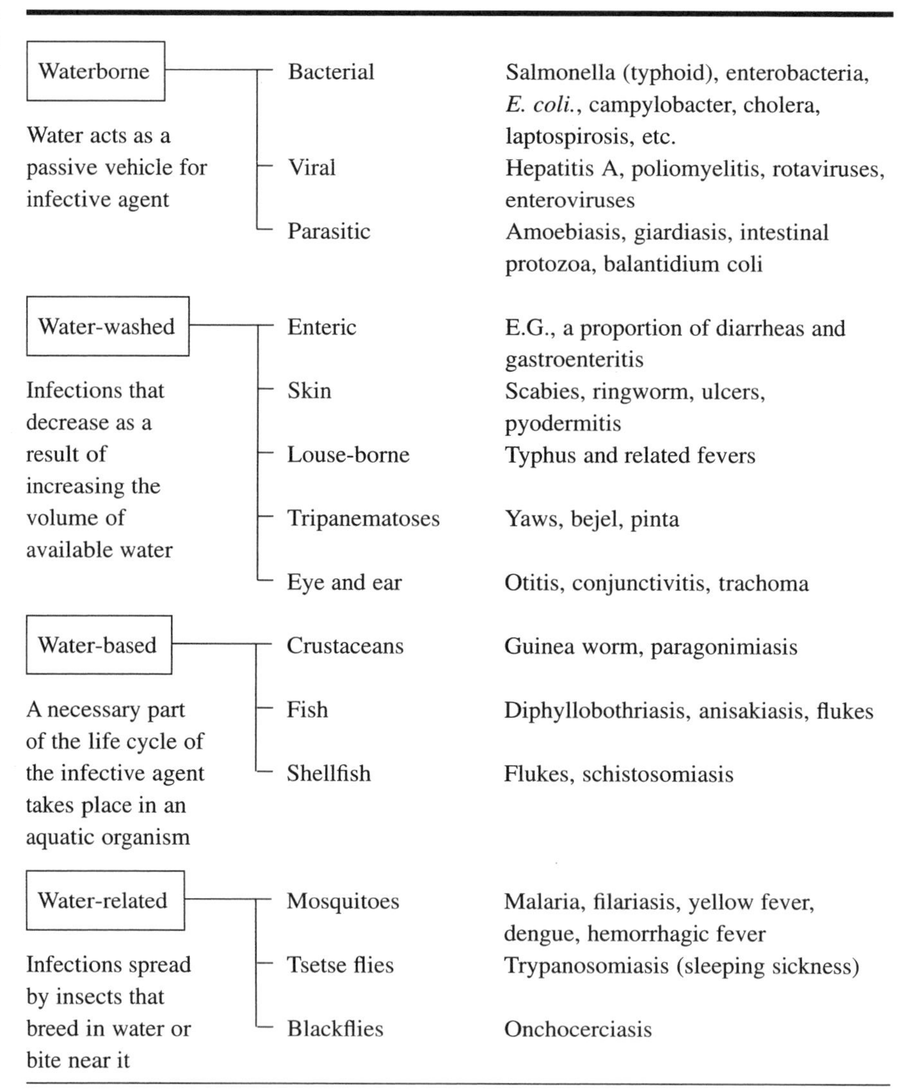

Category	Description	Type	Diseases
Waterborne	Water acts as a passive vehicle for infective agent	Bacterial	Salmonella (typhoid), enterobacteria, *E. coli.*, campylobacter, cholera, laptospirosis, etc.
		Viral	Hepatitis A, poliomyelitis, rotaviruses, enteroviruses
		Parasitic	Amoebiasis, giardiasis, intestinal protozoa, balantidium coli
Water-washed	Infections that decrease as a result of increasing the volume of available water	Enteric	E.G., a proportion of diarrheas and gastroenteritis
		Skin	Scabies, ringworm, ulcers, pyodermitis
		Louse-borne	Typhus and related fevers
		Tripanematoses	Yaws, bejel, pinta
		Eye and ear	Otitis, conjunctivitis, trachoma
Water-based	A necessary part of the life cycle of the infective agent takes place in an aquatic organism	Crustaceans	Guinea worm, paragonimiasis
		Fish	Diphyllobothriasis, anisakiasis, flukes
		Shellfish	Flukes, schistosomiasis
Water-related	Infections spread by insects that breed in water or bite near it	Mosquitoes	Malaria, filariasis, yellow fever, dengue, hemorrhagic fever
		Tsetse flies	Trypanosomiasis (sleeping sickness)
		Blackflies	Onchocerciasis

Source: White et al. 1972.

municipal water supply in Milwaukee, Wisconsin, became contaminated with cryptosporidium, a parasite that causes flulike symptoms in healthy people and is deadly to people with weakened immune systems such as the elderly, AIDS patients, and cancer patients receiving chemotherapy. The source of the contaminated water was quickly identified: a water plant built in the 1950s with a water intake pipe less than 2 miles from the outflow of a sewage treatment plant. Before the source of pollution from the plant had been stopped, an estimated 400,000 people had fallen ill, and forty-two clients of the AIDS project who had contracted cryptosporidiosis had died in what was arguably the largest outbreak of a waterborne disease in recent U.S. history (Terry 1993; Smith 1994).

Before the introduction of chlorine in water purification systems, typhoid, cholera, and dysentery devastated many American cities. These and other waterborne and water-related diseases are still responsible for a large portion of the high infant and under-five mortality rates in developing countries, particularly those in the tropics. Diarrheal diseases alone kill between 4 and 5 million children under five years of age in these countries (Water and Health 1991), 80% of all illness is attributable to unsafe and inadequate water supplies, and half the hospital beds are occupied by people suffering from water-related illnesses (Tolba et al. 1992). On this basis alone, waterborne and water-associated diseases are by far the most serious environmental threat to human health. Waterborne diseases are caused by bacteria, viruses, and parasites (see table 12.9). Water-washed diseases are due in part to a lack of sufficient water for cleaning and cleansing. Water-based diseases are caused by organisms that live in water. Water-related diseases are carried by insects that breed in water or that bite near water. An enormous number of people are affected by these diseases (see table 12.10). More than half the world's population is at risk for malaria, and more than one million people are killed annually by this disease. Of those killed by

Table 12.10. Morbidity and Mortality Caused by Water-Related Diseases in Africa, Asia, and Latin America, 1977–78

Type of Disease	*Infection*	*Infections* (10^3a^{-1})	*Deaths* (10^3a^{-1})
Waterborne	Amoebiasis	400,000	30
	Diarrheas	3–5,000,000	5,000–10,000
	Polio	80,000	10–20
	Typhoid	1,000	25
Water hygiene	Ascariasis (roundworm)	800,000–1,000,00	20
	Leprosy	12,000	very low
	Trichuriasis (whipworm)	500,000	low
	Hookworm	7–9,000,000	50–60
Water-habitat with water-related vectors	Schistosomiasis (bilharzia)	200,000	500–1,000
	African trypanosomiasis (sleeping sickness)	1,000	5
	Malaria	800,000	1,200
	Onchocerciasis (river blindness)	30,000	20–50

Source: Modified after Meybeck et al. 1989.

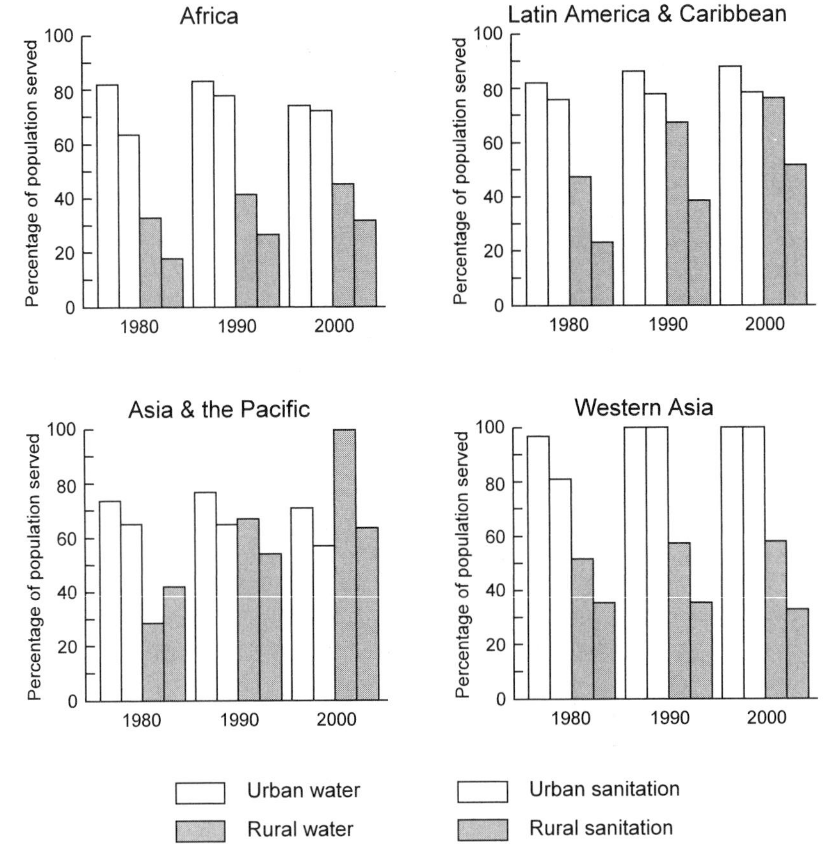

Figure 12.34. Percentages of populations served by clean water supplies and sanitation. (Tolba et al. 1992)

malaria, about 800,000 are children under age five in sub-Saharan Africa. About 600 million people are at risk for schistosomiasis, and about 200 million new infections occur annually (Tolba et al. 1992).

Waterborne diseases are best controlled by providing safe drinking water. To focus global attention and resources on this basic need, the United Nations and the World Health Organization designated the 1980s as the "International Drinking Water Supply and Sanitation Decade." Figure 12.34 summarizes the changes in the percentage of urban and rural populations that are served by clean water supplies and have access to sanitation. In most areas there was a significant increase. However, at the end of 1990 approximately 15% of the urban and 41% of the rural population in the developing countries were still without adequate and safe water supplies. About 26% of the urban populations were estimated to be without access to an appropriate means of excreta disposal, and 60% of the rural populations of developing countries were estimated to be without access to adequate sanitation facilities (The International Drinking Water Supply and Sanitation Decade, WHO/CWS-92/12, 1992). Clearly, a great deal remains to be done. The problem of supplying clean water and adequate sanitation to all of the world's people is daunting because populations are growing so rapidly in many of the countries that are the poorest and have the least adequate water and sanitation systems.

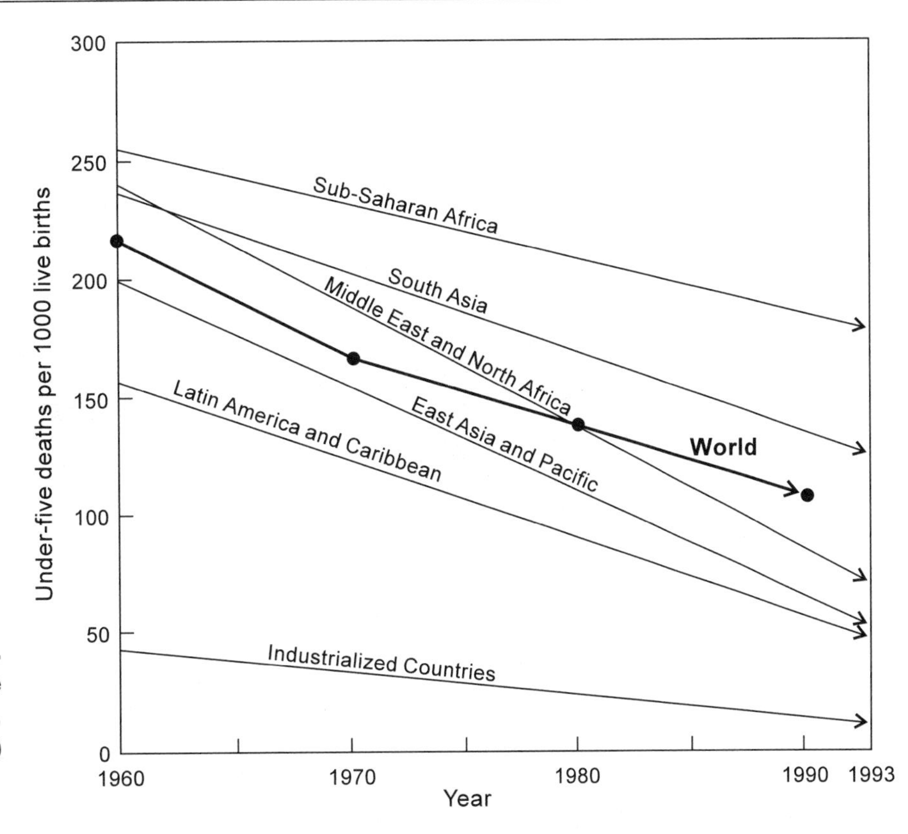

Figure 12.35. Trends in under-five mortality, by region, 1960–90. (Grant 1995)

But there has been progress. Figure 12.35 shows that child mortality has decreased dramatically since 1960 in all parts of the world, even in sub-Saharan Africa. The decreases in child and adult mortality since 1960 are not due to improved water supplies and sanitation facilities alone. To control water-related diseases, the disease-carrying insects must be eliminated or at least greatly reduced in numbers, and medication must be available to treat the infected population. The control of onchocerciasis, or river blindness, illustrates the importance of both factors. The disease is caused by a worm (*Onchocerca volvulus*) that is transmitted by bites of the black fly (*Simulium damnosum*), which breeds in fast-flowing tropical rivers. In its first stage, the infection usually develops unnoticed. The larvae develop slowly, but the worms may live for 12 years, producing millions of embryos that invade the skin. Fibrous, nodular tumors grow around adult worms settled in connective tissues and in deep organs. When they invade the eyes, they can cause partial or total blindness, hence the name of the disease. In some areas, the number of infected people is so high that it is common to see a young child leading a string of adult men through a village (fig. 12.36). An estimated 90 million people are at risk from river blindness; some 17 million are already infected, mainly in Central Africa and Central America (Water and Health 1991).

In 1974, the Onchocerciasis Control Programme was started by the World Health Organization (WHO). Support came from the World Bank and from other sources. The program initially set out to control the black fly population in an

Figure 12.36. A macabre procession of people suffering from onchocerciasis, or "river blindness." (*Water and Health*, 1991)

area occupied by about 20 million people in West Africa by weekly application of a rapidly biodegradable larvicide delivered by aircraft to all of the black fly breeding sites. Although this approach was successful, the sprayed areas were reinvaded by infected black flies from areas up to 400 kilometers away. The area treated with larvicide therefore had to be extended very significantly. The success of the eradication program was then threatened by the development of resistance to the insecticide used to control the black fly populations. Alternating the use of several insecticides helped but did not solve the problem completely (WHO Expert Committee on Onchocerciasis 1987).

At the start of the onchocerciasis program in 1974, no effective and convenient drug existed that did not also have serious side effects. Since then ivermectin, a very effective drug for combating onchocerciasis has been developed. Merck, Sharp, and Dohme, the manufacturers of ivermectin, are providing the drug free of charge to all countries where the disease is endemic. In the WHO target area, onchocerciasis has now been largely contained, and the program has demonstrated how the disease can be eliminated in other parts of the world (WHO Press Release 56, 1992; Holden 1995).

The need to eradicate water-associated diseases has sometimes clashed with the need for additional water resources. Irrigation systems create marvelous breeding grounds for mosquitoes and flies, and enhance conditions that favor the proliferation of snails and other disease vectors. The Diama Dam on the Senegal River was completed in 1986. Since then the area has witnessed one of the most serious outbreaks of schistosomiasis in the history of modern West Africa. Changes in river water salinity favoring the growth of the snail that transmits schistosomiasis and stabilization of water levels leading to more favorable living conditions for the snails are believed to be responsible for the change in transmission patterns in the river basin. The construction of the Aswan Dam in Egypt has had a similar effect. Intestinal schistosomiasis, which has dominated the Nile Delta, is now spreading toward Upper Egypt.

The complete eradication of water-related diseases is not yet a realistic target; but the prevalence of these diseases can be reduced to the point that they are no longer major public health problems by reducing people's contact with infected water, by providing potable water, and by supplying appropriate chemotherapy for the infected populace. These measures are all part of the canon of primary health care: health education, safe water supply and sanitation, maternal and child health care, and drug treatment.

12.3.2 *The Oceans*

Some years ago, one of the authors (HDH) and a colleague went to lunch at their Faculty Club. HDH extolled the virtues of the oyster stew, but his colleague chose another item on the menu. "I know too much," he said, "about the feeding habits of oysters." This comment was not meant to disparage oysters, but rather the degree of pollution in their New England habitats. At the time, Boston could justly claim to possess one of the most polluted harbors in the United States. Sewage treatment was inadequate, and the outfall from the sewage treatment plant on Deer Island entered the ocean just at the edge of the harbor. A multi-billion-dollar program has now upgraded the sewage treatment facilities, a plant has been put into operation for converting sewage sludge into fertilizer, and the outfall has been moved 9 miles out to sea.

Other areas along the U.S. coasts are also being cleaned up. New York City has been the largest source of marine sediments along the eastern seaboard. The 17 million people who live in the New York Metropolitan Zone, together with the population from upstate New York in the Hudson River drainage and in the New Jersey drainage of the Passaic and Raritan Rivers contribute a host of pollutants to the New York Bight. Some of these pollutants are from sewage, some from the many industries in the area, and some from agriculture. As the population of the area has grown, so has the emission of all these pollutants (see table 12.11). Only since the late 1960s has a concerted effort been made to reduce the input of all pollutants into the New York Bight. These efforts are now bearing fruit. Figure 12.37 shows the course of emissions of oil and grease, and of some toxic metals into the bight between 1880 and 1980. Figure 12.38 summarizes the use of chlorinated pesticides, and figure 12.39 the input of total organic materials. A full range of river contaminants has been poured into the bight. Fortunately, their emission rate has been decreasing in response to legislation, and the bight, like other near-coastal areas in the United States, is gradually recovering. Fish that have been absent for decades are returning to many rivers and estuaries, although the recovery is by no means universal.

In 1924, New York and the surrounding cities began dumping their sewage and industrial wastes in 40 meters of water at a site 12 miles out of New York Harbor. In 1987, this dump site was closed by the EPA. Over 8 million wet tons of sludge were then dumped annually at a more distant site, 106 miles out of New York Harbor at a water depth of 2,000 meters (Marshall 1988). Ocean dumping has now been prohibited entirely, a change that is surely good for marine life and possibly for humans as well (see, for instance, Sawyer et al. 1987). However, the prohibition against ocean dumping requires that many municipal wastes must be either destroyed or buried on land. Both alternatives are expensive and politically

Table 12.11. Emissions into the Hudson-Raritan Basin

(a) Total Emissions of Organic Matter and Metals (units = tons)

	Total Organic	*Arsenic*		*Cadmium*		*Chromium*		*Copper*		*Mercury*		*Lead*		*Zinc*	
Year	*Material*	*low*	*high*	*low*	*high*	*low*	*high*	*low*	*high*	*low*	*high*	*low*	*high*	*low*	*high*
1980	262,966	681.4	682.0	52.3	52.3	878.7	878.7	1,120.0	1,120.8	64.2	64.2	5,451	5,458	7,872	7,874
1970	551,803	1,060.6	1,100.9	118.3	18.6	1,526.4	1,526.4	1,241.7	1,250.9	106.4	106.4	9,994	10,143	10,240	10,262
1960	668,050	884.1	1,034.2	104.1	104.8	1,652.4	1,652.5	997.1	1,017.4	56.8	56.8	7,625	8,017	8,754	8,808
1950	607,705	954.5	1,285.3	89.4	90.4	1,497.6	1,497.8	1,186.0	1,228.4	66.7	66.8	5,529	7,138	9,221	9,348
1940	544,722	1,515.2	1,826.9	81.5	82.6	1,593.2	1,593.9	1,384.9	1,418.9	66.7	66.8	3,480	5,685	8,567	8,729
1930	501,677	1.139.5	1,421.5	45.4	56.1	538.3	538.6	656.0	686.7	58.9	58.9	1,144	2,256	7,353	7,481
1920	502,495	722.3	928.2	47.8	76.9	662.9	663.5	877.8	905.7	52.6	52.7	1,342	2,991	5,704	5,905
1900	334,980	243.9	263.0	24.8	26.9	699.3	699.3	723.6	735.4	31.9	31.9	1,562	1,800	1,715	1,855
1880	174,463	56.8	57.8	7.5	7.6	160.8	160.9	236.4	237.8	29.2	29.2	45	53	514	569

(b) Use of Chlorinated Pesticides in the Hudson-Raritan Basin (1945–80) (units = tons)

Year	*DDT*	*BHC/ Lindane*	*Aldrin*	*Chlordane*	*Dieldrin*	*Endrin*	*Heptachlor*	*Toxaphene*
1980	0	0	0	0	0	0	12.6	2.0
1976	0	8.6	0.4	78.8	1.2	0.4	11.6	4.1
1971	48.0	8.6	0.7	295.9	1.7	0.5	13.0	5.0
1966	124.9	18.6	1.2	206.5	4.7	0.7	12.6	5.2
1964	135.3	25.6	0.8	162.5	4.8	0.8	11.6	5.4
1960	187.7	66.8	0.5	159.3	13.8	0.9	10.8	5.3
1955	165.4	333.6	0.6	159.3	13.4	0.9	9.5	5.3
1950	154.2	147.6	0.1	159.3	2.6	0.4	7.0	3.7
1945	84.6	86.9		15.8				

Source: Tarr and Ayers 1990.

(c) Historical Oil and Grease Loading in the Hudson-Raritan Estuary

	Oil and Grease (1,000 t)		
Year	*Point Source*	*Runoff*	*Total*
1980	45	59	104
1970	507	57	564
1960	325	61	386
1950	234	53	287
1940	156	45	201
1930	128	45	173
1920	55	28.5	83
1900	18.5	4.5	23
1880	6.0	2.0	8

unpalatable. For the United States in particular and for the world as a whole, sewage and waste disposal will almost certainly become progressively more difficult during the coming century (see, for instance, Gourlay 1988).

Near-shore pollution of the oceans is not restricted to inputs from land. Spills from oil tankers have also created major environmental problems (see chapter 11). Most of these spills have occurred near shore, where tankers ran aground or collided with other ships. But a good deal of crude oil has also spilled into the oceans from blow-outs in offshore wells (see chapter 8) and from tankers washing out their tanks at sea. The accumulation of tar on many beaches is one of the unhappy consequences of these offshore spills (Butler, Morris, and Sass 1973).

But oil tankers are hardly the only culprits that contaminate the seas. Ships, including oceanographic research vessels, routinely dump their garbage overboard. Some of this garbage washes up on beaches, even in very remote areas. Ducie Atoll, arguably one of the most remote islands in the world, lies more than 3,000 miles northeast of New Zealand. The nearest inhabited island is Pitcairn, the island of "Mutiny on the Bounty" fame, some 293 miles distant from Ducie.

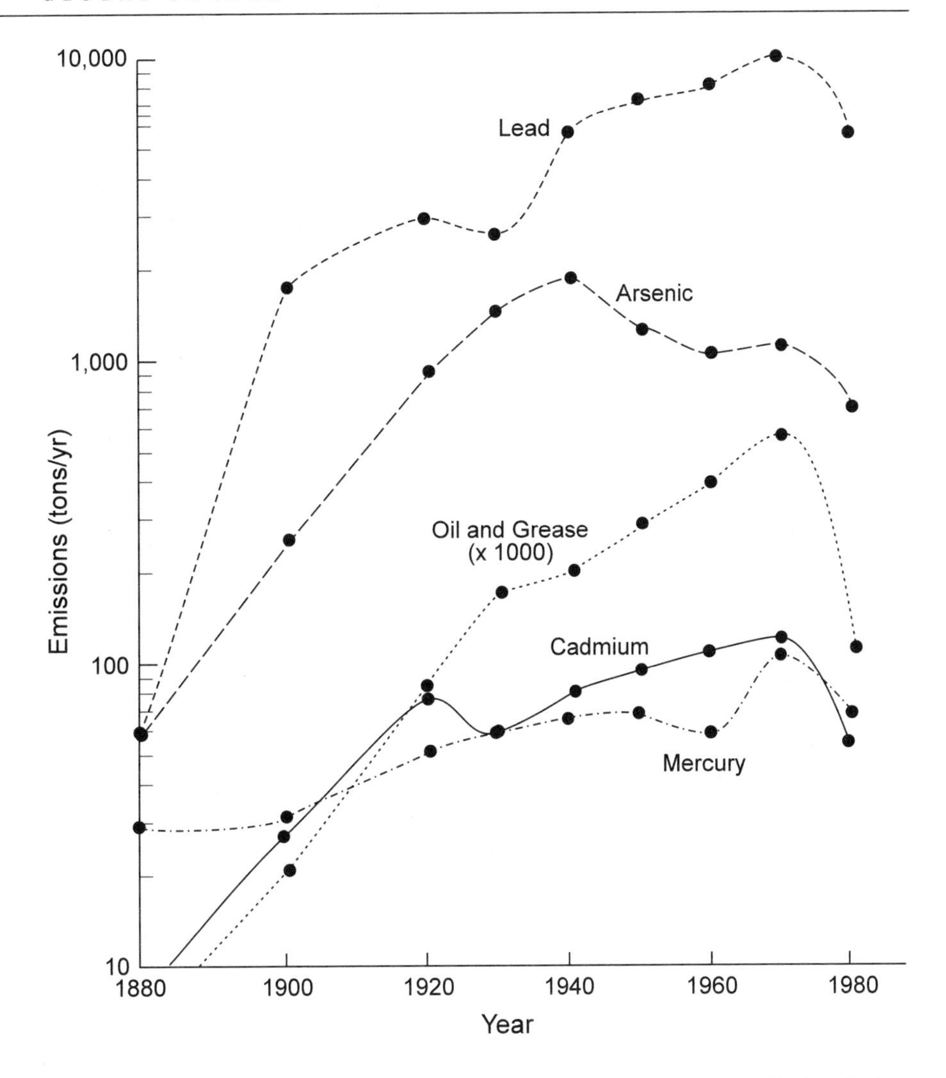

Figure 12.37. Emissions of oil and grease, and of several toxic metals, into the Hudson-Raritan Basin (tons/yr). Note the logarithmic scale. (Data from Tarr and Ayres 1990)

Yet the beaches of Ducie Atoll are covered with junk that would be perfectly at home in a city landfill (*New York Times*, July 16, 1991). One of the bottles recently found on Ducie contained a message. It was dated 1979 and read, "If you find this message, please send a message to"; the rest was obliterated, perhaps appropriately so. The oceans are very large, but humanity's ability to litter is equal to the task of fouling a good deal of its surface.

Humanity also affects the far reaches of the oceans via the atmosphere. The oceans appear to be a major sink for CO_2 from fossil fuel burning on land (see chapter 11). Much of this CO_2 is probably taken up directly from the atmosphere. The SO_2 and NO_x from fossil fuel burning that are wafted out to sea are largely washed into the oceans with rain. The host of other anthropogenic gases generated on land and at sea enter the oceans by one or the other of these mechanisms, or by both, or as constituents of solid particles (see, for instance, Prospero 1978).

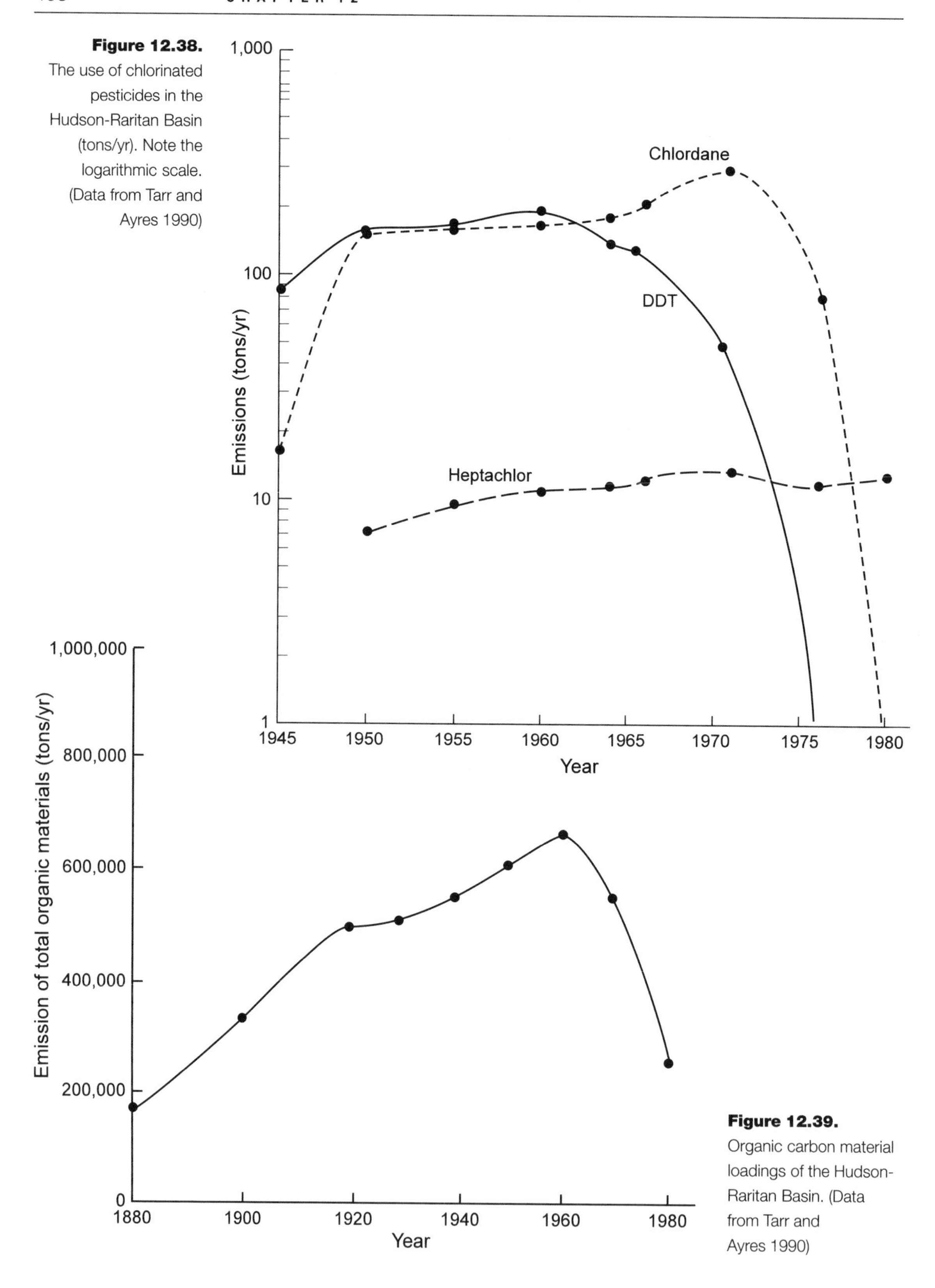

Figure 12.38. The use of chlorinated pesticides in the Hudson-Raritan Basin (tons/yr). Note the logarithmic scale. (Data from Tarr and Ayres 1990)

Figure 12.39. Organic carbon material loadings of the Hudson-Raritan Basin. (Data from Tarr and Ayres 1990)

12.4 The Changing Biosphere

12.4.1 Introduction

If the space traveler of chapter 5 had arrived on Earth 15,000 years ago, he would have found much more ice and many fewer people. If he had arrived 1,000 years ago, he would have found much more wildlife and many fewer lights. If he had arrived 100 years ago, the skies would have been clear of aircraft and the airwaves would have carried no electronic signals, and he might even have been greeted with a rendition of "Home on the Range," a song that has since become a cherished but meaningless chestnut:

Oh give me a home
Where the buffalo roam
Where the deer and the
antelope play,
Where seldom is heard
A discouraging word
And the skies are not
cloudy all day.

Figure 12.40. The evolution of the range beef. (C. M. Russell 1929)

About 60 million buffaloes—or, more correctly, American bison—roamed the western plains for the last several thousand years. Within a mere 15 years, between 1870 and 1885, hunters wiped out nearly all of these vast herds. By 1900 fewer than one thousand and possibly as few as three hundred buffaloes remained in the country. The history of their replacements is illustrated in figure 12.40. Today the buffalo population in the United States has rebounded to about

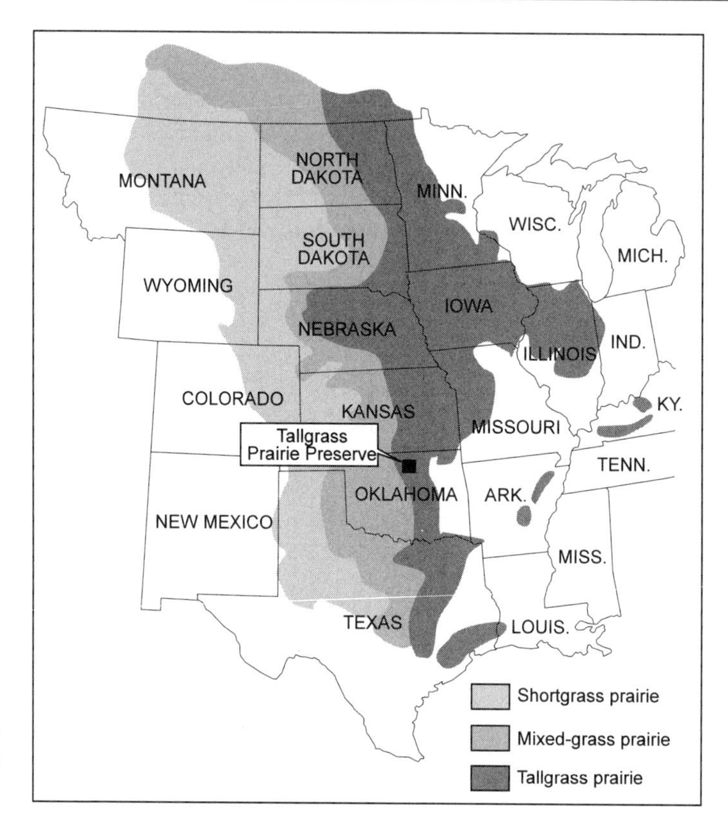

Figure 12.41. Tallgrass prairie, the vanished realm of the bison. (Stevens 1993)

130,000, most of them on private ranches. Ironically, the buffalo is no longer in danger of extinction, in part because its meat, which tastes much like beef but is low in fat, has turned the buffalo into a health food delicacy.

In 1993, three hundred buffaloes galloped headlong onto 5,000 acres of unobstructed tallgrass prairie in the most ambitious attempt so far to revive one of America's formerly vast and robust but now moribund ecosystems (Stevens 1993). Before European settlers and their descendants arrived, more than 220,000 square miles of tallgrass prairie stretched from Canada into Texas and from Nebraska to the Great Lakes (see fig. 12.41). Today more than 90% of this tallgrass prairie has vanished under the plow, and most of the rest is fragmented and heavily degraded. The prairie and its attendant oak savannas have become the rarest of North America's major biomes.

The charming nineteenth-century American poet Emily Dickinson imagined the prairie thus:

> To make a prairie it takes a clover
>   and one bee—
> One clover, and a bee
> And revery.
> The revery alone will do
> If bees are few.

The reality of the twentieth and twenty-first centuries unfortunately requires more than reverie. The new preserve in northern Oklahoma currently encompasses 36,000 acres, which were once part of Oklahoma's sprawling Osage reservation and more recently a cattle ranch. The Nature Conservancy bought the property in 1989. The three hundred buffaloes that were released into the preserve in 1993 are the first of eighteen hundred that will eventually roam the area. Their liberation was accompanied by a black-tie-and-blue-jeans dinner, an Osage ceremony making an honorary chief of Norman H. Schwarzkopf, the retired army general and a member of the Nature Conservancy's board of governors, a western swing band to entertain the pre-release crowd, and speeches paying tribute to the cooperation of environmentalists, cattlemen, Indians who own the mineral rights, and local citizens who hope to see a tourist bonanza in creating the preserve.

The history of the buffalo has many parallels. The growth of the human family has put enormous pressure on the habitats of many plant families, and on the livelihood of many animal families. Some animals have become extinct, some are close to extinction, and many may become extinct during the next century. Fortunately, the image of the Great White Hunter has become tarnished, and animal trophies in houses like Theodore Roosevelt's Sagamore Hill on Long Island seem to look rather reproachfully at their visitors. But the pressure on natural environments continues to increase. The oceans are heavily fished (see chapter 5); deserts are watered and made to bloom; forests are cleared and turned into fields and pastures; and swamps are drained and planted with food crops.

12.4.2 *The Disappearing Wetlands*

"Wetlands" is a term applied to a variety of ecosystems ranging from prairie potholes to vast tidal marshes. Until recently they were treated as disposable assets to be diked, drained, drilled, and turned into cropland, shopping malls, subdivisions, and oil fields. It has taken a long time to recognize their intrinsic value as wetlands, and measures to protect and to restore them are finally being taken. Somewhere between 530 and 860 million hectares (5.3 to 8.6 million km^2) of freshwater wetlands remain in the world (Aselmann and Crutzen 1989). Figure 12.42 summarizes their distribution by latitude. The largest wetland areas lie between 50°N and 70°N, where the peatlands of the former USSR, Canada, Ire-

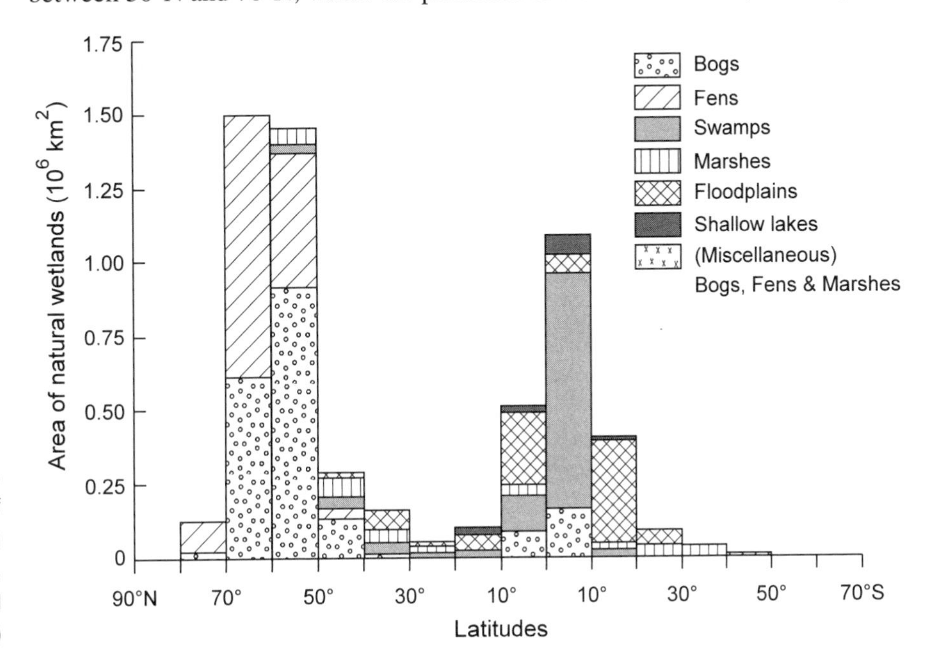

Figure 12.42. The distribution of natural wetlands along 10° latitude bands. (Aselmann and Crutzen 1989)

Table 12.12. Losses of Wetlands in the United States since the 1780s

	Acres	Percentage		Acres	Percentage
Florida	9,286,713	46	Nebraska	1,005,000	35
Texas	8,387,288	52	Colorado	1,000,000	50
Louisiana	7,410,300	46	South Dakota	955,100	35
Arkansas	7,085,000	72	Oregon	868,100	38
Illinois	6,957,500	85	Virginia	774,387	42
Minnesota	6,370,000	42	Wyoming	750,000	38
Mississippi	5,805,000	59	Pennsylvania	627,986	56
Michigan	5,616,600	50	New Jersey	584,040	39
North Carolina	5,400,000	49	Connecticut	497,500	74
Indiana	4,849,367	87	Idaho	491,300	56
California	4,546,000	91	Washington	412,000	31
Ohio	4,517,200	90	Kansas	405,600	48
Wisconsin	4,468,608	46	Arizona	331,000	36
Missouri	4,201,000	87	Montana	306,700	27
Alabama	3,783,800	50	Delaware	256,785	54
Iowa	3,578,100	89	Nevada	251,000	52
North Dakota	2,437,500	49	Utah	244,000	30
Oklahoma	1,892,900	67	New Mexico	238,100	33
South Carolina	1,755,000	27	Massachusetts	229,514	28
Georgia	1,545,000	23	Alaska	200,000	0.1
New York	1,537,000	60	Vermont	121,000	35
Kentucky	1,266,000	81	Rhode Island	37,536	37
Maine	1,260,800	20	West Virginia	32,000	24
Maryland	1,210,000	73	New Hampshire	20,000	9
Tennessee	1,150,000	59	Hawaii	7,000	12

Source: Dahl 1990.

land, and Alaska cover an area of more than 3 million square kilometers. Plate 57 is an aerial photo of typical high-latitude tundra wetland near Kuujuaq in northern Quebec, Canada. The second-largest group of wetlands includes the swamps and floodplains of the southern equatorial regions, particularly in the Amazon basin, in southern Brazil and Argentina, and on the African peneplain. During the past 200 years, more than half of the American wetlands have been lost. Table 12.12 shows that the largest loss of wetlands by area was in Florida (9.3 million acres); the largest percentage loss of wetlands (91%) occurred in California. Today, only 104 million acres of wetlands remain in the lower forty-eight states. These cover 5% of the land area and are being lost at a rate of about 290,000 acres per year (World Resources Institute 1993).

Wetlands have generally had a bad press. Off and on for 400 years, the Everglades region of Florida was described as a series of vast, miasmic swamps, poisonous lagoons, huge dismal marshes without outlet; as a rotting, shallow, inland sea; and as labyrinths of dark trees hung and looped about with snakes and dripping mosses, malignant with tropical fevers and malarias evil to the white man (Douglas 1988). No wonder that the only good wetland was considered to be a drained wetland. We know better now; the uses and values of the major types of wetlands are listed in table 12.13. In the United States many wetlands

Table 12.13. The Uses and Values of the Major Types of Wetlands

	Estuaries (without mangroves)	*Mangroves*	*Open Coasts*	*Flood Plains*	*Fresh-water Marshes*	*Lakes*	*Peat-lands*	*Swamp Forest*
Functions								
1. Groundwater recharge	○	○	○	■	■	■	●	●
2. Groundwater discharge	●	●	●	●	■	●	●	■
3. Flood control	●	■	○	■	■	■	●	■
4. Shoreline stabilization/ erosion control	●	■	●	●	■	○	○	○
5. Sediment/toxicant retention	●	■	●	■	■	■	■	■
6. Nutrient retention	●	■	●	■	■	●	■	■
7. Biomass export	●	■	●	■	●	●	○	●
8. Storm protection/windbreak	●	■	●	○	○	○	○	●
9. Micro-climate stabilization	○	●	○	●	●	●	○	●
10. Water transport	●	●	○	●	○	●	○	○
11. Recreation/tourism	●	●	■	●	●	●	●	●
Products								
1. Forest resources	○	■	○	●	○	○	○	■
2. Wildlife resources	■	●	●	■	■	●	●	●
3. Fisheries	■	■	●	■	■	■	○	●
4. Forage resources	●	●	○	■	■	○	○	○
5. Agricultural resources	○	○	○	■	●	●	●	○
6. Water supply	○	○	○	●	●	■	●	●
Attributes								
1. Biological diversity	■	●	●	■	●	■	●	●
2. Uniqueness to culture/ heritage	●	●	●	●	●	●	●	●

Source: Tolba et al. 1992.
Key: ○ Absent or exceptional; ● present; ■ common and important value of that wetland type.

are valued as magnificent ecosystems and have become major battlegrounds between environmentalists and a variety of drainers and developers. The Clean Water Act of 1972 recognized that wetlands are valuable, virtually irreplaceable biological systems deserving protection. The act therefore requires that anyone seeking to fill a wetland must obtain a permit from the Army Corps of Engineers and submit to review by the EPA. The law has slowed the loss of wetlands, but it has suffered from indifferent enforcement and from the failure of various agencies to agree on what is and is not a wetland. The National Academy of Sciences has since been asked to frame a comprehensive definition of the term. In the meantime, the Clinton administration's wetlands policy keeps most of the important protection for the wetlands that are genuinely worth saving, and includes an executive order pledging "no net loss" of wetlands.

The Everglades region of southern Florida is one of the most important and most fought-over wetlands in America. It includes more than 77,000 square kilometers of uplands, wetlands, rivers, and lakes in three distinct sub-basins: the

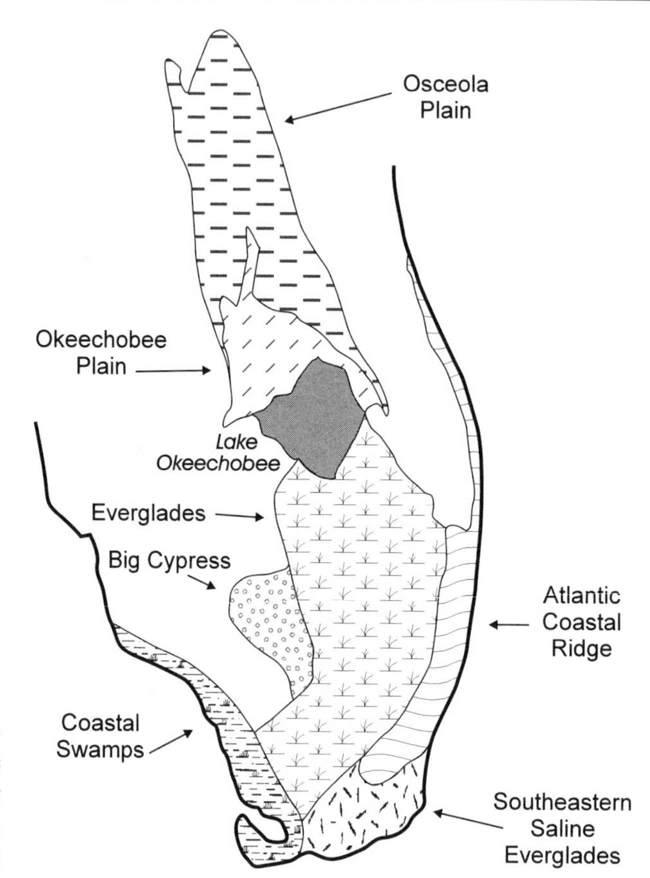

Figure 12.43. The physiographic provinces of the greater Everglades Basin. (Kushlan 1990)

Kissimmee River valley, Lake Okeechobee, and the Everglades (see figs. 12.43 and 12.44). These were formed about 5,000 years ago, when the postglacial rise of sea level slowed sufficiently to allow the establishment of a coastwise levee and the impoundment of an inland freshwater wetland. Since then the Everglades have changed continually in response to peat and marl deposition, plant succession, animal invasions, rainfall cycles, changes in fire frequency and severity, and sea level rise. Before the beginning of the large-scale anthropogenic changes in the late 1800s, the Everglades were an extremely shallow river of pure water and unbroken marsh, Marjorie Stoneman Douglas's "River of Grass" (1988). The river began in Lake Okeechobee and spilled across the lower Florida peninsula in a thin sheet 40 to 60 miles wide (see plate 58). So lazy was the current that water leaving the lake in March would not reach Florida Bay, at the tip of the peninsula 100 miles away, until the following February. An inkling of the magnificence of this natural setting can still be seen at sunset in the Everglades National Park as crimson mist settles over the sawgrass prairie, ospreys alight on the top branches of mangroves, and the natural warfare of life diminishes to an eloquent quiet.

The first dredges began their work in the late 1880s at the Kissimmee lakes, at Lake Okeechobee, and along both coasts. They had only a limited effect on the hydrology of most of the Everglades system. Canals dug in 1907 and 1928 led to saltwater intrusion. The deeper central core of the Everglades was never com-

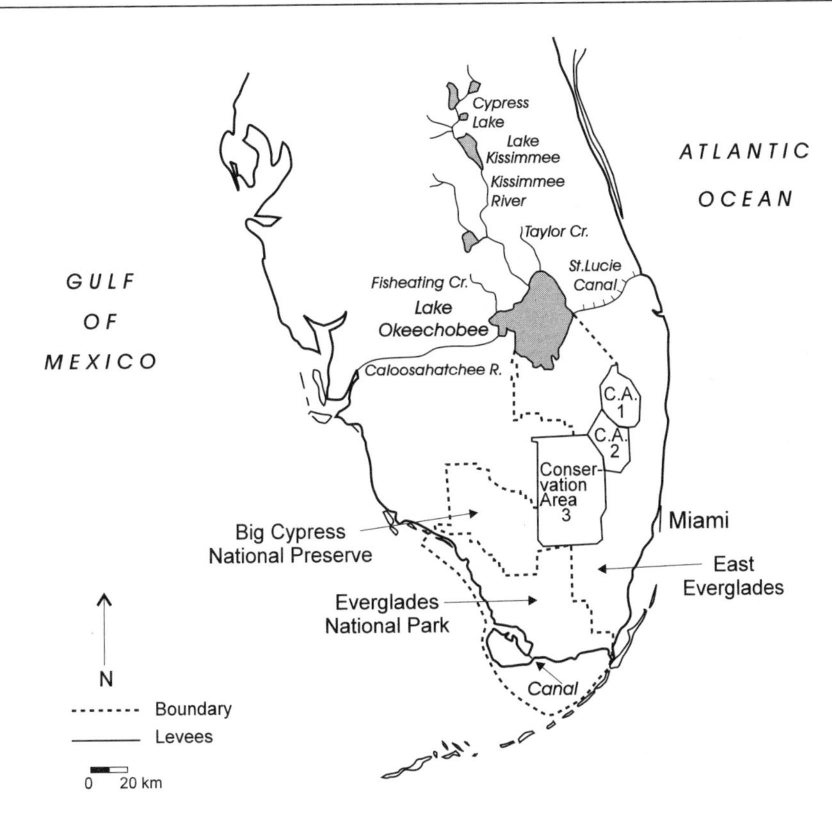

Figure 12.44. Map of south Florida showing the Everglades system. (Kushlan 1990)

pletely drained, and the limited drainage capacity of the canals led to catastrophic floods in the 1940s. As a result, the flood control district was reformed, and the virtually undrainable Everglades were isolated from reclaimable land to the north and east. Between 1961 and 1971, the Kissimmee River was canalized, and its length was reduced by 90 kilometers. Lake Okeechobee had been connected to the Caloosahatchee River in the 1880s to provide more efficient drainage to the Gulf of Mexico. In the 1920s, the lake was surrounded by an earthwork levee, and the St. Lucie canal was constructed to provide an eastern outlet for the lake. Today, all water flow out of Lake Okeechobee is controlled by stage management procedures.

Immediately south of Lake Okeechobee the former wetlands have been drained and farmed since the 1920s. The eastern portion of the Everglades was also drained for farming and subsequently for suburban and even urban development. Currently, 65% of the original Everglades marsh proper has been drained. The biology of the remaining Everglades is seriously affected by changes in the regional hydrology and in water chemistry. Water depth in the Everglades is determined by the interaction of the hydraulic head of upstream flow and in situ rainfall; it varies from a high of 1.5 meters to a low of zero meters, when water retreats below ground in the dry season. Plant and animal populations depend on water level timing and magnitude (Kushlan 1990), and these have been seriously disturbed by current water management practices. Since the Everglades National Park was established in 1947, the number of wading birds has dropped by about 90%. Fourteen animal species are considered endangered, and many others are threatened (see table 12.14).

Table 12.14. Endangered Species, Threatened Species, and Species of Special Concern in the Everglades Planning Area

	Species		*State Designation*	*Federal Designation*
Amphibians and reptiles	American alligator	*Alligator mississippiensis*	SSC	T
	Loggerhead sea turtle	*Caretta caretta*	T	T
	Atlantic green turtle	*Chelonia mydas mydas*	E	E
	American crocodile	*Crocodylus acutus*	E	E
	Leatherback turtle	*Dermochelys coriacea*	E	E
	Indigo snake	*Drymarchon corais*	T	T
	Atlantic hawksbill turtle	*Eretmochelys imbricata imbricata*	E	E
	Gopher tortoise	*Gopherus polyphemus*	SSC	UR
	Atlantic ridley turtle	*Lepidochelys kempii*	E	E
	Florida pine snake	*Pituophis melanoleucus megitus*	SSC	UR
	Gopher frog	*Rana areolata*	SSC	UR
Birds	Roseate spoonbill	*Ajaia ajaja*	SSC	
	Limpkin	*Aramus guarauna*	SSC	
	Burrowing owl	*Athene cunicularia*	SSC	
	Piping plover	*Charadrius nelodus*	T	T
	White-crowned pigeon	*Columba leucocephala*	T	UR
	Kirtland's warbler	*Dendroica kirtlandii*	E	E
	Little blue heron	*Egretta caerulea*	SSC	
	Snowy egret	*Egretta thula*	SSC	
	Reddish egret	*Egretta rufescens*	SSC	UR
	Tricolored heron	*Egretta tricolor*	SSC	
	Swallow-tailed kite	*Elanoides forficatus*		UR
	White ibis	*Eudocimus albus*	SSC	
	Peregrine falcon	*Falco peregrinus*	E	T
	Southeastern kestrel	*Falco sparverius paulus*	T	UR
	Florida sandhill crane	*Grus canadensis pratensis*	T	
	American oystercatcher	*Haematopus palliatus*	SSC	
	Bald eagle	*Haliaeetus leucocephalus*	T	E
	Wood stork	*Mycteria americana*	E	E
	Osprey	*Pandion haliaetus*	SSC	
	Brown pelican	*Pelecanus occidentalis*	SSC	
	Red-cockaded woodpecker	*Picoides borealis*	T	E
	Crested caracara	*Polyborus plancus*	T	T
	Snail kite	*Rostrhamus sociabilis plumbeus*	E	E
	Least tern	*Sterna albifrons*	T	
	Bachman's warbler	*Vermivora bachmanii*	E	E
Mammals	Florida panther	*Felis concolor*	E	E
	Everglades mink	*Mustela vison evergladensis*	T	UR
	Florida mouse	*Peromyscus floridans*	SSC	UR
	Mangrove fox squirrel	*Sciurus niger avicennia*	T	UR
	West Indian manatee	*Trichechus manatus latirostris*	E	E
	Florida black bear	*Ursus americanus floridanus*	T	UR
Invertebrates	Florida tree snail	*Liguus fasciatus*	SSC	
	Bartram's hairstreak butterfly	*Strymon acis bartrami*	UR	

Source: South Florida Water Management District, 1990.
Key: E = endangered; T = threatened; SSC = species of special concern; UR = species recently under review.

This is not the first time that the magnificent plumed birds of the Everglades have suffered from human predation. In the early 1900s, fashion-conscious women wore hats decorated with plumes. It was not only egret plumes (see plate 59) that were in demand; brightly colored birds of every kind were trapped and sent in cages to Havana and New York. Women even wore dead mockingbirds on their overloaded hats. In 1905, the then young Audubon Society hired a young warden, Guy Bradley, to guard the rookeries of American and snowy egrets at the southeast edge of the sawgrass. One day Bradley did not return from the field. The search party found him shot to death in his boat, the victim of one of the

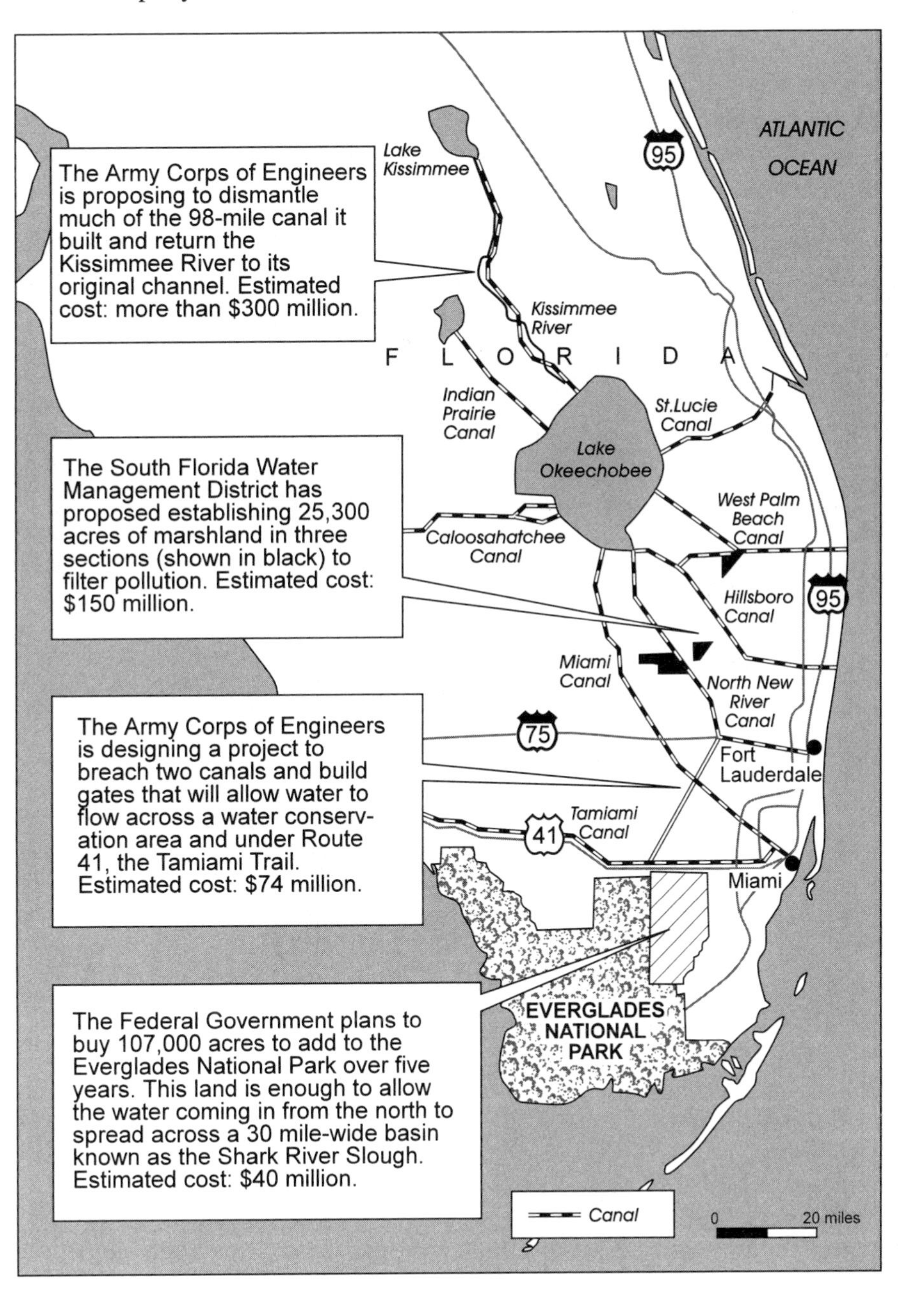

Figure 12.45. Plans to restore the Everglades. (Schneider 1991)

plume hunters (Douglas 1988). Fortunately fashions changed, and the birds of the Everglades recovered.

The fertilizers used in farming the drained Everglade marshes have increased the phosphate content of waters flowing into the remaining Everglades. This has led to the replacement of plant species accustomed to low-nutrient waters by others that thrive on a richer diet. The original flora has also been affected by the introduction of plants that were not native to the Everglades. These include the Brazilian pepper, a tough, fast-growing tree popular among South Florida landscapers, and Melaleuca, a tenacious Australian tree that developers use to drain wet areas (Schneider 1991).

In 1988, the U.S. government sued the state of Florida for failing to enforce water quality laws for agricultural runoff. In 1991, the state of Florida capitulated and joined the federal government and environmentalists in their action against the growers. The suit is part of a tough strategy to restore some of the Everglades (see fig. 12.45). Parts of the gigantic plumbing system that was built during the past century are to be uncoupled, the Kissimmee River is to be reclaimed and restored to its natural channel, excess phosphate is to be removed from water entering the Everglades, and water is to be spread into 107,000 acres of wetland that are being added to the national park. It is unlikely, however, that any close resemblance to the pristine Everglades system will ever be recreated. The economic costs of total restoration to agriculture and to the millions of residents that depend on the Everglades for water supplies and flood control are incalculable. For better or for worse, it is becoming increasingly recognized worldwide that the only way to conserve and to manage wetlands is through their sustainable management for multiple uses (Kushlan 1990).

12.4.3 *The Shrinking Forests*

For many years, the original inhabitants of the Western Hemisphere were thought to have lived in such harmony with nature that they left little, if any, mark on it. Not so. Indians in many parts of the Americas had a major impact on their surroundings. They shaped the ecology by setting fires to clear the land, they modified the size and composition of forests, created and expanded grasslands, and rearranged the land. Soil, local climate, and the movement of water and wildlife were all affected. But the pace of change quickened dramatically with the arrival of the European settlers, so much so that in the last 400 years the vegetation of North America has undergone a radical and continuing transformation unlike anything seen in the preceding several thousand years.

Before the arrival of the white colonists, the Chesapeake Bay region and what is now Maryland were covered with a highly diversified mosaic of trees and shrubs. By the late 1800s forests were reduced to 20% of their presettlement expanse; the rest had been replaced by a patchwork of cultivated and abandoned farms, wetlands, residential lawns, gardens, urban concrete, and asphalt. In some parts of the eastern United States, the forests have reclaimed the land, as farmland has been taken out of production and as agriculture shifted westward. But the presettlement pattern has not returned completely. The original chestnut trees were killed by a canker introduced accidentally from abroad, and the once abundant red cedars are now rare.

Forests have been replaced by agricultural land worldwide. Middle European forests were extensively cleared between 7,000 and 3,000 years ago (Williams 1989, 1991), and were further thinned by the Greek and Roman civilizations.

Table 12.15. Estimates of Preagricultural and Present Areas of Major Ecosystems (in millions of square kilometers)

Ecosystem	*Pre-agricultural*	*Present*	*Reduction*
Tropical closed forest	12.77	12.29	0.48
Other forest	33.51	26.98	6.53
Total forest	46.28	39.27	7.01
Other woodland	15.23	13.10	2.13
Shrubland	12.99	12.12	0.87
Grassland	33.90	27.43	6.47
Tundra	7.34	7.34	—
Desert	15.82	15.57	0.25
Cultivation	0.93	17.56	+16.63

Source: Matthews 1983.

In New Zealand, perhaps 25% of the lowland forests were destroyed in the centuries after the arrival of the Maori settlers in A.D. 950. The spread of European settlements after 1500 led to the destruction of forests in many parts of the globe, not only in North America.

Table 12.15 is an attempt to estimate the changes in the major ecosystems since preagricultural times. The last column shows that about half of the increase in land under cultivation has been at the expense of forests and woodlands. Table 12.16 contains an estimate of forest and woodland distribution in 1980. The figures in the two tables do not quite match; the differences reflect the difficulty of obtaining precise figures for the distribution and history of the several ecosystems. The overall picture is, however, clear: there are still a lot of forests, about half of them in the tropical regions of Africa and America.

Humanity's pressure on many of these forests is great. Agriculture and resettlement, grazing and ranching, the need for fuel wood and charcoal, timber exploitation, and replacement of forests by plantations are all taking their toll. In some regions the forests are decreasing extremely rapidly. In 1940, 67% of the total land area of Costa Rica was covered by primary forest. In 1983, only 17% remained. Figure 12.46 illustrates this dramatic decrease. Population growth in Costa Rica is rapid, and more of the country's primary forests will surely be lost.

Table 12.16. Areas of Forest and Woodlands at the End of 1980 (in millions of square kilometers)

	Boreal Countries	*Temperate Countries*	*Tropical Countries*	*All Countries*
Africa	—	0.081	7.012	7.093
America	2.03	3.108	8.898	14.036
Asia* (excluding former USSR)	—	1.884	3.034	4.918
Pacific	—	0.487	0.426	0.913
Europe (including former USSR)	7.17	2.115	—	9.285
World Total	9.20	7.675	19.37	36.245

Source: Tolba et al. 1992

* Asia includes the Middle and Near East.

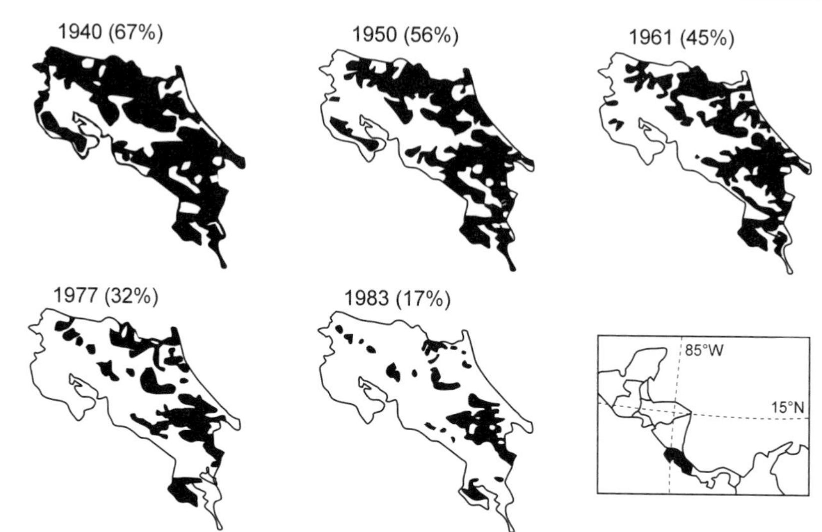

Figure 12.46. Loss of primary forest in Costa Rica, 1940–83. Percentages show forest land as a proportion of total land area. (Whitmore 1990)

During the past decade, a great deal of attention has been focused on deforestation and habitat fragmentation in the Amazon Basin. In part this reflects the great size of the tropical rain forest in this basin: some 4.1 million square kilometers; in part it reflects a deep concern about the potential loss of plant and animal species in this highly biodiverse ecosystem. Until recently, neither the deforestation rate nor the effects of deforestation on habitat degradation were well known. Estimates of the recent deforestation rate have ranged from 20,000 to 80,000 km^2/yr. Fortunately, Skole and Tucker (1993) have now furnished us with rather precise data. They used 1:500,000 scale photographic imagery from Landsat Thematic Mapper data and a geographic information system (GIS) to create a computerized map of deforestation and to evaluate its influence on forest fragmentation and habitat degradation. Plate 60 shows a composite Landsat Thematic Mapper image of southern Rhondônia state in Brazil. Since the areas of forest, of deforestation, and of regrowth are easy to distinguish, the images can be used to construct maps documenting the extent of deforestation of Brazilian Amazonia between 1978 and 1988 (plate 61).

The deforested areas and affected habitats are concentrated along the southern and eastern fringe of Amazonia. The deforested area increased from 78,000 km^2 in 1978 to 230,000 km^2 in 1988. The total area of affected habitat was larger, because habitats are seriously affected for a distance of about 1 km from the edge of deforested areas. The area of affected habitat therefore increased from 208,000 km^2 in 1978 to 588,000 km^2 in 1988. Six percent of all closed-canopy forest had been cleared by 1988, and about 15% of the Amazonian forests had been affected by deforestation-caused habitat destruction, habitat isolation, and edge effects. The distribution of adverse effects on biological diversity is shown in plate 62. While the rate of deforestation between 1978 and 1988 averaged about 15,000 km^2/yr in Brazilian Amazonia, about 38,000 km^2/yr suffered habitat fragmentation and degradation.

At the 1978–88 rate of deforestation, the forests of Brazilian Amazonia would be depleted in about 200 years. The rate has apparently decreased somewhat since 1988, in part due to international pressure on the Brazilian government.

Deforestation between 1978 and 1988 was accelerated by legislation that made deforestation very attractive. Both capital and labor had been induced to move to Amazonia. The precariousness of settlement in Amazonia generated constant new pressures for further frontier expansion. This process can be slowed by cutting official incentives, relieving pressures that generate frontier migration, consolidating existing frontier settlements, and using less predatory forms of new settlement. Since the deforestation rate depends substantially on artificial stimuli, it may well be politically feasible to slow deforestation significantly (Sawyer 1990).

We do not know the effect of deforestation in Amazonia on the extinction of forest species, but we do know that the region shelters an enormous number of species. Tropical rain forests cover only about 7% of the Earth's land surface, but they probably harbor more than half of the 5–30 million species of the entire world biota (Wilson 1988). Studies of biodiversity within clusters of islands have shown that the number of species of birds, reptiles, ants, or other equivalent groups found on each island increases approximately as the fourth root of the island's area. The rule of thumb is that a tenfold increase in area results in a doubling of the number of species. If this rule applies to rain forests, the number of species, N, is given approximately by the equation

$$\frac{N}{N_o} = \left(\frac{A}{A_o}\right)^{1/4}, \tag{12.9}$$

where

N_o = number of species before deforestation
A_o = initial area of rain forest
A = area of remaining rain forest.

When 10% of a rain forest has been cut, $A/A_o = 0.900$ and $N/N_o = 0.974$, i.e., biodiversity has decreased by 2.6%. When 90% of the forest has been eliminated, $A/A_o = 0.100$, $N/N_o = 0.562$, and biodiversity has decreased by 43.8%. Clearly, the effect of cutting the first 10% of Amazonia is much smaller than the effect of cutting the last 10%.

There is reason to doubt that biodiversity in Amazonia will really follow equation (12.9). Forest ecosystems on Caribbean islands have turned out to be more resilient than one would have predicted on the basis of equation (12.9). These islands are densely populated, and their lands have been intensively used and degraded for centuries. In Puerto Rico, human activity has reduced the area of primary forest by 99%, but because of extensive use of coffee shade trees, and because of secondary forest growth, forest cover has never been less than 10%–15%. The massive forest conversion on Puerto Rico did not lead to a correspondingly massive species extinction, certainly nowhere near the 50% predicted by equation (12.9) (Lugo 1988). In the 1980s, there were more bird species (97) on the island than in pre-Columbian times (60). The resiliency of the bird fauna is probably due to its generalist survival strategy and to the location of refuges in secondary forests and coffee plantations on mountaintops along the east-west axis of the island. Experience in the Luquillo Experimental Forest Biosphere Reserve in Puerto Rico (plate 63) has demonstrated that species richness can be partially restored to lands previously used heavily for agriculture, that growing timber does not require the elimination of all natural species richness, and that tropical lands respond to reasonable managed care. A strategy for forest and species conservation in tropical regions should focus on the restoration of forest production on former forest lands where food production is not sustainable. This policy, and the sensible use of secondary forests and tree plantations, will reduce pressure on lands with mature forests or with unique ecological characteristics and may set us on a course that meets the needs of the needy while protecting species diversity (Lugo 1988). In the meantime, species such as the northern spotted owl in the American Northwest are threatened where islands of forest shelter these birds, but where their numbers are too small to sustain their survival.

References

Abelson, P. H. 1993. Pesticides and the Delaney Amendment: Response to Criticisms of His Editorial. *Science* 260:1409–10.

Aselmann, I., and Crutzen, P. J. 1989. Global Distribution of Natural Freshwater Wetlands and Rice Paddies: Their Net Primary Productivity, Seasonality, and Possible Methane Emissions. *Jour. Atm. Chem.* 8:307–58.

Boden, T. A., et al., eds. 1994. *Trends '93: A Compendium of Data on Global Change.* Carbon Dioxide Information Center, Oak Ridge National Laboratory, Oak Ridge, Tenn.

Boyle, E. A., Chapnick, S. D., and Shen, G. T. 1986. Temporal Variability of Lead in the Western North Atlantic. *Jour. Geophys. Res.* 91 (C7):8573–93.

Bradshaw, A. D., Southwood, R., and Warner, F., eds. 1992. *The Treatment and Handling of Wastes.* Chapman and Hall, London.

Brundtland, G. H. 1990. Epilogue. In *Managing Planet Earth*, pp. 137–38. Readings from Scientific American Magazine. W. H. Freeman, New York.

Butler, J. N., Morris, B. F., and Sass, J. 1973. *Pelagic Tar from Bermuda and the Sargasso Sea.* Special Publication 10. Bermuda Biological Station for Research.

Callahan, M. A. et al. 1979. *Water-Related Environmental Fate of 129 Priority Pollutants.* Office of Water Planning and Standards, Report No. EPA-440/4-79-029a, Environmental Protection Agency (EPA). U.S. Government Printing Office, Washington, D.C.

Charlson, R. J. 1988. Have the Concentrations of Tropospheric Aerosol Particles Changed? In *The Changing Atmosphere*, ed. F. S. Rowland and I.S.A. Isaksen, pp. 79–90. John Wiley, New York.

Cicerone, R. J. 1988. How has the Atmospheric Concentration of CO Changed? In *The Changing Atmosphere*, ed. F. S. Rowland and I.S.A. Isaksen, pp. 49–61. John Wiley, New York.

Commission du Rhin. 1988. Consequences de l'incendie du 1 Novembre 1986. In *Rapport d'activité 1986 de la Commission Internationale pour la Protection du Rhin contre la Pollution*, pp. 33–54. B. P. 309, D-5400 Koblenz.

Committee on Alternative Chemical Demilitarization Technologies. 1993. *Alternative Technologies for the Destruction of Chemical Agents and Munitions.* National Research Council. National Academy Press, Washington, D.C.

Committee on Measuring Lead in Critical Populations. 1993. *Measuring Lead Exposure in Infants, Children and Other Sensitive Populations.* National Research Council. National Academy Press, Washington, D.C.

Committee on Pesticides in the Diets of Infants and Children. 1993. *Pesticides in the Diets of Infants and Children.* National Research Council. National Academy Press, Washington, D.C.

Committee on Tropospheric Ozone Formation and Measurement. 1991. *Rethinking the Ozone Problem in Urban and Regional Air Pollution.* National Research Council. National Academy Press, Washington, D.C.

Crutzen, P. J., and Andreae, M. O. 1990. Biomass Burning in the Tropics: Impact on Atmospheric Chemistry and Biogeochemical Cycles. *Science* 250:1669–78.

Custodio, E., Iribar, V., Manzano, B. A., and Galofre, A. 1986. Evolution of Sea Water Intrusion in the Llobregat Delta, Barcelona, Spain. In *Proceedings of the Ninth Salt Water Intrusion Meeting*, ed. R. H. Boekeiman et al. Delft, Netherlands.

Dahl, T. E. 1990. *Wetland Losses in the United States, 1780s to 1980s*. U.S. Department of the Interior.

Delgado, J. P. 1993. Operation Crossroads. *American History* 28 (no. 2):51–59.

Douglas, M. S. 1988. *The Everglades: River of Grass*. Rev. ed. Pineapple Press, Sarasota.

Durousseau, M. 1987. La pollution du Rhin consécutive à l'indêndie intervenu aux établissements Sandoz. In *Les Cahiers de l'Administration de l'Alsace*, pp. 17–26. Suppl. Revue d'Administration, November 1987. Prefecture du Bas-Rhin, Strasbourg.

Eden, H. F., Elero, B. P., and Perkins, J. N. 1993. Nimbus Satellites: Setting the Stage for Mission to Planet Earth. *EOS Transactions* 74:281.

Ehhalt, D. H. 1988. How Has the Atmospheric Concentration of CH_4 Changed? In *The Changing Atmosphere*, ed. F. S. Rowland and I.S.A. Isaksen, pp. 25–32. John Wiley, New York.

European Economic Community (EEC). 1982. Communication from the Council on Dangerous Substances That Could Be Placed on List I of Council Directive 76/464/EEC. *Official Journal of the European Community*, July 14.

Förstner, U., and Wittmann, G.T.W. 1979. *Metal Pollution in the Aquatic Environment*. Springer-Verlag, New York.

Friberg, L., Piscator, M., and Nordberg, G. 1971. *Cadmium in the Environment*. CRC Press, Cleveland, Ohio.

Gleason, J. F., et al. 1993. Record Low Global Ozone in 1992. *Science* 260:523–26.

Gourlay, K. A. 1988. *Poisoners of the Sea*. Zed Books, Atlantic Highlands, N.J.

Graedel, T. E., and Crutzen, P. J. 1990. The Changing Atmosphere. In *Managing Planet Earth*, chap. 2. Readings from Scientific American Magazine. W. H. Freeman, New York.

Grant, J. P. 1995. *The State of the World's Children 1995*. UNICEF. Oxford University Press, Oxford, U.K.

Halpert, M. S., Bell, G. D., Kousky, V. E., and Ropelewski, C. F., eds. 1994. *Fifth Annual Climate Assessment 1993*. Climate Analysis Center, National Oceanic and Atmospheric Administration, Washington, D. C.

Holden, C. 1995. River Blindness Control Pays Off. *Science* 268:503.

Holmes, J. W. 1972. Salinity and the Hydrological Cycle. In *Salinity and Water Use*, pp. 25–40. John Wiley, New York.

Itoh, S. 1988. *The White Egret*. Chronicle Books, San Francisco.

Kellogg, R. L., Maizel, M. S., and Goss, D. W. 1992. *Agricultural Chemical Use and Groundwater Quality: Where Are the Potential Problems?* Soil Conservation Service, Department of Agriculture, Washington, D.C.

Kerr, J. B., and McElroy, C. T. 1993. Evidence for Large Upward Trends of Ultraviolet-B Radiation Linked to Ozone Depletion. *Science* 262:1032–34.

Kerr, R. A. 1993. The Ozone Hole Reaches a New Low. *Science* 262:501.

Kobayashi, I. 1971. Relation between the "Itai-Itai" Disease and the Pollution of River Water by Cadmium from a Mine. Proc. 5th Intern. Conf. San Francisco and Hawaii. *Adv. Water Pollut. Res.* I-25:1–17.

Kushlan, J. A. 1990. The Everglades. In *The Rivers of Florida*, ed. R. J. Livingston, chap. 8. Ecological Studies 83. Springer-Verlag, New York.

la Rivière, J.W.M. 1990. Threats to the World's Water. In *Managing Planet Earth*, chap. 4. Readings from Scientific American Magazine. W. H. Freeman, New York.

Logan, J. A., Prather, M. J., Wofsy, S. C., and McElroy, M. B. 1981. Tropospheric Chemistry: A Global Perspective. *Jour. Geophys. Res.* 86 (C8):7210–7254.

Lugo, A. E. 1988. Estimating Reductions in the Diversity of Tropical Forest Species. In *Biodiversity*, ed. E. O. Wilson and F. M. Peter, chap. 6. National Academy Press, Washington, D.C.

Malle, K. G. 1985. Metallgehalt und Schwebstoffgehalt im Rhein. *Z. Wasser Abwasser Forsch.* 18:207–9.

Marshall, E. 1988. The Sludge Factor. *Science* 242:507–8.

Matthews, E. 1983. Global Vegetation and Land Use: New High Resolution Databases for Climatic Studies. *Jour. Climate and Meteorology*. 22:474–87.

Meybeck, M., Chapman, D. V., and Helmer, R. 1990. *Global Freshwater Quality: A First Assessment.* World Health Organization and United Nations Environmental Programme. Blackwell Reference, Cambridge, Mass.

National Research Council. 1992. *Restoration of Aquatic Ecosystems: Science, Technology and Public Policy.* Committee on Restoration of Aquatic Ecosystems. National Academy Press, Washington, D.C.

Neilson, M. 1983. *Report on the Status of the Open Water of Lake Ontario.* General Directorate of Inland Waters, Environment Canada, Burlington, Ontario.

Nixon, S. W., Hunt, C. D., and Nowicki, B. L. 1986. The Retention of Nutrients (C,N,P), Heavy Metals (Mn,Cd,Pb,Cu) and Petroleum Hydrocarbons in Narragansett Bay, U.S.A. In *Biogeochemical Processes at the Land and Sea Boundary*, ed. P. Lasserre and J. M. Martin, pp. 99–122. Elsevier, New York.

Nriagu, J. O. 1989. The History of Leaded Gasoline. In *Heavy Metals in the Environment*, ed. J.-P. Vernet, pp. 361–66. Page Bros.

Oak Ridge National Laboratory. 1991. *Trends '91: A Compendium of Data on Global Change Highlights.* Carbon Dioxide Information Analysis Center, Oak Ridge National Laboratory, Tennessee.

Oeschger, H., and Siegenthaler, U. 1988. How Has the Atmospheric Concentration of CO_2 Changed? In *The Changing Atmosphere*, ed. F. S. Rowland and I.S.A. Isaksen, pp. 5–23. John Wiley, New York.

Office of Air Quality Planning and Standards. 1992. *National Air Quality and Emissions Trends Report 1991.* Publication No. 450-R-92–001. U.S. Environmental Protection Agency, Washington, D.C.

Office of Health and Environmental Assessment. 1986. *Air Quality Criteria for Lead.* Publication No. EPA-600/8–83/028aF-dF. U.S. Environmental Protection Agency, Washington, D.C.

Pirkle, J. L., et al. 1994. The Decline in Blood Lead Levels in the United States. *Jour. Amer. Med. Assoc.* 272 (4):284–91.

Prospero, J. M. 1978. *The Tropospheric Transport of Pollutants and Other Substances to the Oceans.* National Research Council. National Academy Press, Washington, D.C.

Roberts, G., and Marsh, T. 1987. The Effects of Agricultural Practices on the Nitrate Concentrations in the Surface Water Domestic Supply Sources of Western Europe. In *Water for the Future: Hydrology in Perspective*, pp. 365–80. IAHS Publ. No. 164.

Russell, C. M. 1929. *Good Medicine: The Illustrated Letters of Charles M. Russell.* Doubleday, Garden City, N.Y.

Sawyer, D. 1990. The Future of Deforestation in Amazonia: A Socioeconomic and Political Analysis. In *Alternatives to Deforestation: Steps toward Sustainable Use of the Amazon Rain Forest*, ed. A. B. Anderson, chap. 15. Columbia University Press, New York.

Sawyer, T. K., Nerad, T. A., Daggett, P.-M., and Bodammer, S. M. 1987. Potentially Pathogenic Protozoa in Sediments from Oceanic Sewage-Disposal Sites. In *Biological Processes and Wastes in the Ocean*, chap. 16, vol. 1 of *Oceanic Processes in Marine Pollution*, ed. J. M. Capuzzo and D. R. Kester. Krieger Publishing, Malabar, Fla.

Schneider, K. 1991. Returning Part of Everglades to Nature for $700 million. *New York Times*, March 11, 1991.

Seiler, W., and Conrad, R. 1987. Contribution of Tropical Ecosystems to the Global Budget of Trace Gases, Especially CH_4, H_2, CO, and N_2O. In *The Geophysiology of Amazonia*, ed. R. E. Dickinson. John Wiley, New York.

Shiklomanov, I. A. 1991. The World's Water Resources. In *International Symposium to Commemorate the 25 years of IHD/IMP*. UNESCO, Paris.

Skole, D., and Tucker, C. 1993. Tropical Deforestation and Habitat Fragmentation in the Amazon: Satellite Data from 1978 to 1988. *Science* 260:1905–10.

Smith, R. A. 1994. Water Quality and Health, a Global Perspective. *Geotimes* 39 (Jan.):19–21.

South Florida Water Management District. 1990. Final Draft, Surface Water Improvement and Management Plan for the Everglades, vol. 3, Technical Report.

Stern, P. C., Young, O. R., and Druckman, D., eds. 1992. *Global Environmental Change: Understanding the Human Dimensions*. Committee on Human Dimensions and Global Change. National Academy Press, Washington, D.C.

Stevens, W. K. 1993. Home on the Range (or What's Left of It). *New York Times*, October 19, 1993.

Stolarski, R. S. 1988. Changes in Ozone over the Antarctic. In *The Changing Atmosphere*, ed. F. S. Rowland and I.S.A. Isaksen, pp. 105–19. John Wiley, New York.

Takeuchi, T. 1972. Distribution of Mercury in the Environment of Minamata Bay and the Inland Ariake Sea. In *Environmental Mercury Contamination, International Conference*, ed. R. Hartug and B. D. Dinman, pp. 79–81. Ann Arbor Science Publishers, Ann Arbor, Mich.

Takeuchi, T., et al. 1959. Pathological Observations on the Minamata Diseases. *Acta Pathol. Japan* (suppl.) 9:769–83.

Tarr, J. A., and Ayres, R. U. 1990. The Hudson-Raritan Basin. In *The Earth Transformed by Human Action*, ed. B. L. Turner et al., chap. 38. Cambridge University Press and Clark University, Cambridge, U.K.

Terry, S. 1993. Drinking Water Comes to a Boil. *New York Times*, September 26, 1993.

Thornton, I. 1993. Environmental Geochemistry and Health in the 1990s: A Global Perspective. *Applied Geochemistry* (supp. no. 2): 203–10.

Tolba, M. K., El-Kholy, O. A., El-Hinnawi, E., Holdgate, M. W., McMichael, D. F., and Munn, R. E. 1992. *The World Environment, 1972–1992*. UNEP. Chapman and Hall, London.

United Nations. 1991. *Water and Health*. IHP Humid Tropics Programme Series No. 3. UNESCO Press.

Water Science and Technology Board. 1990. *Ground Water and Soil Contamination Remediation: Toward Compatible Science, Policy, and Public Perception*. National Research Council. National Academy Press, Washington, D.C.

Water Science and Technology Board. 1993. *Ground Water Vulnerability Assessment: Contamination Potential under Conditions of Uncertainty*. National Research Council. National Academy Press, Washington, D.C.

Water Science and Technology Board. 1994. *Alternative for Ground Water Cleanup*. National Research Council. National Academy Press, Washington, D.C.

Watson, R. T., Rohde, H., Oeschger, H., and Siegenthaler, U. 1990. Greenhouse Gases and Aerosols. In *Climate Change: The IPCC Assessment*, ed. J. T. Houghton, G. J. Jenkins, and J. H. Ephraums, pp. 1–40. Cambridge University Press, New York.

White, G. F., Bradley, D. J., and White, A. U. 1972. *Drawers of Water*. University of Chicago Press, Chicago.

Whitmore, T. C. 1990. *An Introduction to Tropical Forests*. Clarendon Press, Oxford.

Williams, M. 1989. Deforestation Past and Present. In *Progress in Human Geography*, ed. E. Arnold, pp. 176–208.

Williams, M. 1991. Forests. In *The Earth as Transformed by Human Action*, ed. B. I. Turner. Cambridge University Press, New York.

Wilson, E. O. 1988. The Current State of Biological Diversity. In *Biodiversity*, ed. E. O. Wilson and F. M. Peter, chap. 1. National Academy Press, Washington, D.C.

World Health Organization (WHO). 1992. World Health Organization and Merck Reach Millions with Free Treatment for River Blindness. WHO Press Release WHO/56, September 23, 1992.

World Health Organization (WHO). 1992. *The International Drinking Water Supply and Sanitation Decade*. CWS Unit, Division of Environmental Health. World Health Organization, Geneva, Switzerland.

World Health Organization Expert Committee on Onchocerciasis. 1987. *Third Report*. World Health Organization Technical Report Series 752. WHO, Geneva.

World Resources Institute. 1993. *The 1993 Information Please Environmental Almanac*. Houghton Mifflin, Boston.

13 Living Dangerously

13.1 Introduction: How Promising Is the Future?

Will we run out of natural resources? Will we drown in our own garbage? This chapter assembles fragments of answers from the preceding chapters. The present century has witnessed great changes in the human condition. Their magnitude and diversity underscore the difficulty of predicting the availability of resources and the quality of the environment during the next century, but we believe that some useful forecasts can be made. Current estimates of the human condition and its future span a very wide range. Many people are very worried. Vice-President Albert Gore (1992, p. 14) has written:

> This life change has caused me to be increasingly impatient with the status quo, with conventional wisdom, with the lazy assumption that we can always muddle through. Such complacency has allowed many kinds of difficult problems to breed and grow, but now, facing a rapidly deteriorating global environment, it threatens absolute disaster. Now no one can afford to assume that the world will somehow solve its problems. We must all become partners in a bold effort to change the very foundation of our civilization.

And somewhat further on (p. 37):

> But research in lieu of action is unconscionable. Those who argue that we should do nothing until we have completed a lot more research are trying to shift the burden of proof even as the crisis deepens. This point is crucial: *a choice to "do nothing" in response to the mounting evidence is actually a choice to continue and even accelerate the reckless environmental destruction that is creating the catastrophe at hand.*

Gore's view of the present state of the Earth and of the "catastrophe at hand" is shared by the World Watch Institute (Brown 1994), but not by Aaron Wildavsky (1995) nor by Dixie Lee Ray and Lou Guzzo (1993), who wrote:

> As the months and years roll by without the appearance of the catastrophes that have been projected by the Greens—and if scientists continue to improve both our knowledge and understanding of natural processes and the degree to which mankind may influence them—the public may get fed up with emotionalism. Then it will become evident that all the predictions of ecological disaster are based upon hypotheses, on imperfect models, and on computer simulations, not on a solid body of truth.

Ray and Guzzo are obviously much more optimistic than Gore and Brown about the quality of our future, and they quote with approval Lord Thomas B. Macaulay's 1830 query: "On what principle is it that, when we see nothing but improvement behind us, we are to expect nothing but deterioration before us?"

During the past several years many who have written and spoken about the future of our planet have taken an intermediate position and have championed the notion of a "sustainable path." Agenda 21 of the 1992 Rio de Janeiro summit is an action plan for "sustainable development." Donella and Dennis Meadows together with Jørgen Randers hope that the world will choose "sustainability"

(1992). Jacques-Yves Cousteau argues for "sustainable social development" (*New York Times*, January 30, 1994). Sandra Postel (1994) pleads for "the formulation of national population and economic policies that will put the world on a sustainable development path, one that will restore a broad-based improvement in the human condition." In this chapter we gather together our data and use them to assess whether it is reasonable to hope for sustained development and even, perhaps, for sustained progress during the next century.

13.2 The Human Population

Since the end of World War II, the fate of the human family has been in doubt. High birthrates and decreasing death rates have produced a dramatic increase in population (see chapter 5). At the same time, the overhanging threat of nuclear war cast a pall over the future of the human race, and the appearance of AIDS has cast doubt on the future of the populations of sub-Saharan Africa and possibly of Asia.

In 1947 the *Bulletin of the Atomic Scientists* introduced a "Doomsday Clock." The position of its hands measured the time to midnight, the outbreak of nuclear war. In 1947 the clock stood at seven minutes to midnight (see fig. 13.1). Between 1947 and 1981 the clock hands oscillated toward and away from midnight

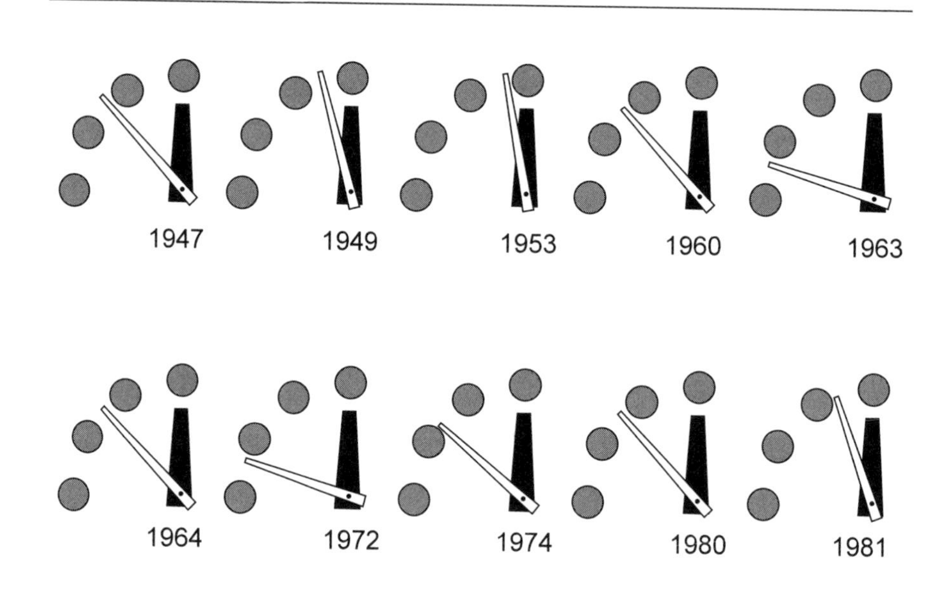

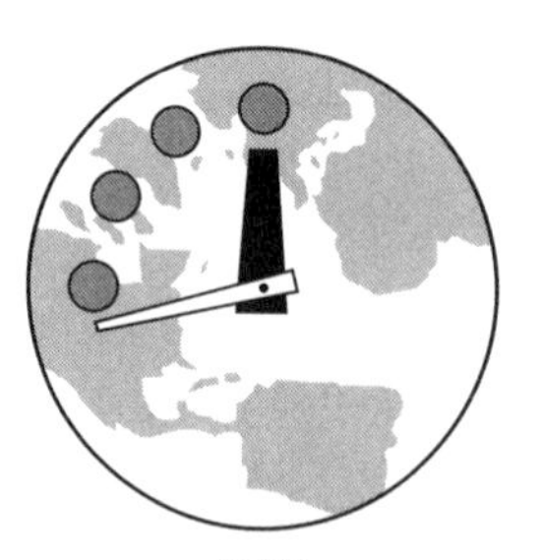

Figure 13.1. The "Doomsday Clock" of the Bulletin of Atomic Scientists. How many minutes to the midnight of nuclear holocaust?

in response to the changing intensity of the cold war. In 1953 the clock stood at 2 minutes to midnight. In 1984 it was once again at 3 minutes to midnight. Finally, the December 1991 issue of the *Bulletin* reset the clock from 10 to 17 minutes to midnight. The Cold War had ended. In 1947 a 15-minute scale was all anyone thought would be needed in their lifetime. The move off-scale to 17 minutes before the hour reflected the conviction that the world had changed in fundamental and positive ways, and that Presidents Bush and Gorbachev had guided their respective nations to a historic intersection of mutual interests.

It is probably too much to hope that major nuclear confrontations are completely a thing of the past. China will almost certainly emerge as a superpower in the twenty-first century. The country is well endowed with natural resources and has a long and distinguished, albeit interrupted, record of scientific and industrial achievements. The future of the former USSR and of the Middle Eastern countries is clouded. The 1995 extension of the Treaty on the Nonproliferation of Nuclear Weapons is surely encouraging, but it would be somewhat of a miracle if humanity were never again threatened by mutual extermination during the course of the next century. If war, AIDS, and other diseases are contained, the world's population in 2100 will probably be between 10 and 16 billion, i.e., between two and three times the present number. To keep the population at the lower figure requires the annual increase in world population to shrink dramatically during the next century. World population will be 16 billion if the mean increase in world population during the next century remains at its present level of about 95 million per year (see chapter 5). Will a doubling or trebling of world population necessarily reduce our quality of life?

13.3 The Availability of Nonrenewable Resources

We are using our nonrenewable resources very rapidly. Since the 1950s this has created fears of major shortages and resource exhaustion during the coming century; but the data in chapters 9 and 11 do not support these fears. The oil crisis of the 1970s was a creature of politics, not of geology. The 1990s are dominated by fears of overproduction and price collapse, not fears of shortages. World reserves of oil and natural gas will probably be largely depleted by 2100, but coal—which can be converted into liquid fuels at a relatively low cost—should still be abundant, as should the nuclear fuels. Whether world population rises to 10 or to 16 billion, there should be no real energy shortages worldwide, especially if energy from nuclear fusion becomes a reality, and if the price of solar energy becomes competitive with that of other energy sources (see chapter 11).

The same, rather cheerful outlook extends to non-fuel commodities (see chapter 10). Iron, aluminum, and copper account for most of the world's bill for metals. The reserves and resources of iron and aluminum at near-present grades are enormous compared to the annual production of these metals, and it is unlikely that we will run out of the ores of these abundant elements. The grade of mined copper ores has decreased by about a factor of ten during the last 150 years; yet the cost of the metal in real terms has not increased significantly. The grade of mined copper ore will surely continue to decrease during the next century, and the copper price may well increase, but we will not "run out." Increased recycling and substitution of copper by aluminum and by nonmetallic materials will probably reduce the need for new copper and will moderate the increase in its price.

The scarce metals and non-metals will probably become scarcer during the twenty-first century, but we will not run out of them either, because the quantity of all these metals in average crustal rocks is several orders of magnitude greater than the quantity in their currently mined ores. The price of the scarce metals in real terms will probably increase, but the beneficiation of ores that contain small amounts of several of these metals will limit the increase, and the small size of the world's bill for the scarce metals will limit the negative effect of increases in their price on the world economy.

13.4 The Availability of Renewable Resources

Of all the renewable resources, food is surely the most important. Unfortunately, the outlook for supplying an adequate diet for 10–16 billion people is not as rosy as the outlook for supplying them with adequate fuel and with other nonrenewable resources. Agricultural productivity has increased rapidly during the twentieth century (see chapter 5), but so has world population. Figure 13.2 shows the effect of the two factors on per capita food production between 1970 and 1992.

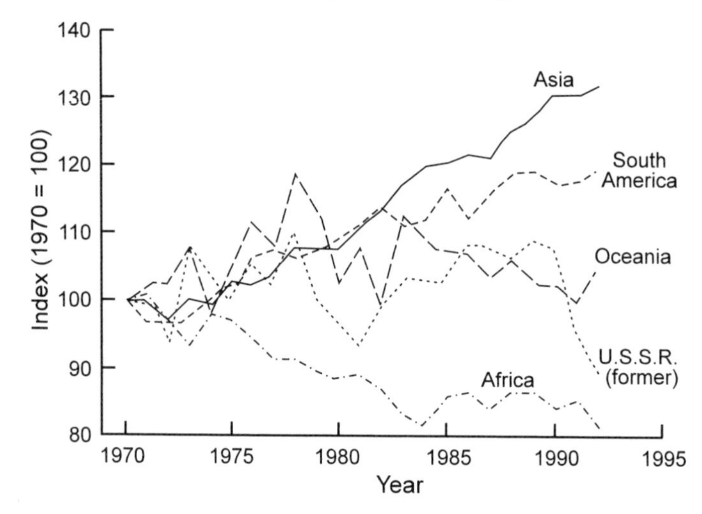

Figure 13.2. Changes in per capita food production by region between 1970 and 1992 (1970 index for all regions = 100). (*World Resources*, 1994–95)

Although world per capita food production increased by about 4%, progress was regionally uneven. In several parts of the world—especially in Asia—per capita food production increased very significantly, but in other areas per capita food production actually decreased between 1970 and 1992. The data in figure 13.3 show that the percentage of malnourished children, defined as children under the age of five who are more than two standard deviations below the desirable weight for their age, declined admirably in a number of countries between 1975 and 1990. In general, the world has been moving slowly in the right direction.

Currently, malnutrition is not due to our inability to produce sufficient food. In fact, agricultural overproduction has been and continues to be a source of national and international friction. Whether and at what cost world food production can be increased by a factor of five by the year 2100 is not yet clear (see, for instance, Ruttan 1993), but the goal does not seem beyond reach if biotechnology

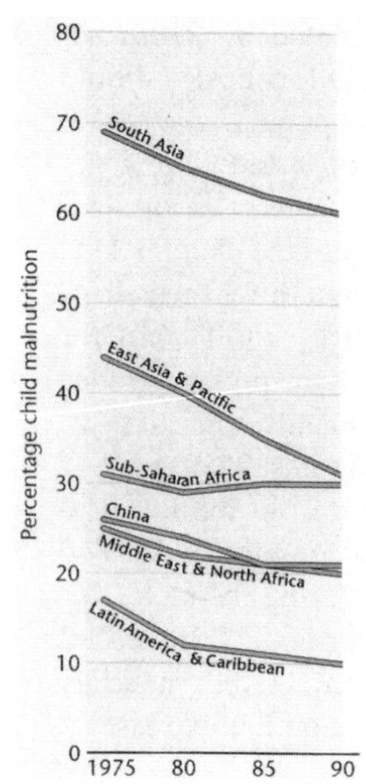

Figure 13.3. The decrease between 1975 and 1990 in the percentage of malnourished children in several countries in Asia and the Americas. (Grant 1992).

and a variety of new ways to produce food contribute as much as can be expected (see chapter 5) with sustained technological progress.

In the process of increasing food production, the need for additional freshwater will become acute. Desalination can supply much of the water needed for human consumption, but the use of desalinated water for agriculture would increase the price of agricultural products very significantly, and it may be cheaper to pipe fresh water from well-watered to dry regions over much greater distances than those to which we have grown accustomed. It remains to be seen whether water transfers of such magnitude will be politically and ecologically palatable.

13.5 Risks, Discomforts, and Disasters

In a particularly suicidal mood, Hamlet bewailed:

> The heartache and the thousand natural shocks
> That flesh is heir to.
> (act 3, scene 1)

Have the natural shocks increased or decreased in number and severity since Shakespeare's time, and are they apt to increase during the coming century? We can venture at least some tentative answers to these questions.

In the United States, as in all of the developed countries, nearly all deaths are due to illnesses (see table 13.1). Only 6.6% of the U.S. deaths in 1992 were due to other causes: 4.0% to accidents and other adverse effects, 1.4% to suicide, and 1.2% to homicide and legal intervention. Motor vehicle accidents accounted for most of the accidental deaths (see table 13.2), while fatalities due to natural disasters were responsible for only a minuscule fraction. Changes in life expectancy

Table 13.1. The Ten Leading Causes of Death in the United States, 1992

Rank	*Cause of Death*	*Number*	*Death Rate*[a]	*Percentage of Total Deaths*
	All causes	2,177,000	853.3	100.0
1.	Heart disease	720,480	282.5	33.1
2.	Cancer	521,090	204.3	23.9
3.	Stroke	143,640	56.3	6.6
4.	Chronic obstructive lung diseases and allied conditions	91,440	35.8	4.2
5.	Accidents and adverse effects	86,310	33.8	4.0
	Motor Vehicle accidents	41,710	16.4	1.9
	All other accidents and adverse effects	44,600	17.5	2.0
6.	Pneumonia and influenza	76,120	29.8	3.5
7.	Diabetes mellitus	50,180	19.7	2.3
8.	Human immunodeficiency virus (HIV) infection[b]	33,590	13.2	1.5
9.	Suicide	29,760	11.7	1.4
10.	Homicide and legal intervention	26,570	10.4	1.2

Source: National Center for Health Statistics, U.S. Dept. of Health and Human Services.

Note: Data are provisional, estimated from a 10 percent sample of deaths. Figures may not add to totals due to rounding. Rates have been recomputed based on revised population estimates.

[a] Per 100,000 population.

[b] HIV is the virus that causes AIDS.

Table 13.2. Principal Types of Accidental Deaths in the United States, 1970–92

Year	*Motor Vehicle*	*Falls*	*Poison (Solid, Liquid)*	*Drowning*	*Fires, Burns*	*Ingestion of Food, Object*	*Firearms*	*Poison (Gases)*
1970	54,633	16,926	3,679	7,860	6,718	2,753	2,406	1,620
1975	45,853	14,896	4,694	8,000	6,071	3,106	2,380	1,577
1980	53,172	13,294	3,089	7,257	5,822	3,249	1,955	1,242
1985	45,901	12,001	4,091	5,316	4,938	3,551	1,649	1,079
1990	46,300	12,400	5,700	5,200	4,300	3,200	1,400	800
1991	43,500	12,200	5,600	4,600	4,200	2,900	1,400	800
1992	40,300	12,400	5,200	4,300	4,000	2,700	1,400	700
			Death Rates per 100,000 Population					
1970	26.8	8.3	1.8	3.9	3.3	1.4	1.2	0.8
1975	21.3	6.9	2.2	3.7	2.8	1.4	1.1	0.7
1980	23.4	5.9	1.4	3.2	2.6	1.4	0.9	0.5
1985	19.2	5.0	1.7	2.2	2.1	1.5	0.7	0.5
1990	18.8	5.0	2.3	2.1	1.7	1.3	0.6	0.3
1991	17.2	4.8	2.2	1.8	1.7	1.1	0.6	0.3
1992	15.8	4.9	2.0	1.7	1.6	1.1	0.5	0.3

Source: National Safety Council.
Note: There were 12,000 other accidental deaths in 1992; the most frequently occurring types were medical complications, machinery, air transport, water transport, mechanical suffocation, and excessive cold.

in America therefore largely reflect the balance between improvements in medical practice, medical care, and nutrition, and countervailing effects of environmental degradation. Figure 13.4 shows that life expectancy at birth in the United States increased from 54.1 years in 1920 to 75.7 years in 1992. The trends of life expectancy have been much the same in all of the developed countries. Clearly, the effects beneficial to health have outweighed detrimental effects during the twentieth century.

Similar trends have prevailed in the developing countries. In many cases, the increase in life expectancy has been even more dramatic than in the developed countries, because life expectancies at the beginning of the twentieth century were so much lower. Even so, life expectancies in the poorest countries, those where the per capita income is less than about $600 per year (see fig. 5.24), are still significantly lower than in the richer countries. This is due in part to the much higher child mortality in the developing world. Fortunately, child mortality rates have been declining rapidly during the past 30 years, as has the number of children per woman (see fig. 13.5). These declines will produce a significant increase in life expectancy in the developing world during the next century.

However, the next century will bring its own risks, discomforts, and disasters. Figure 13.6 is a qualitative but nevertheless useful scheme for classifying natural disasters. Their intensity is plotted as a function of distance from their source. Small disasters, such as minor earthquakes, produce only local damage. Very large earthquakes have severe or very severe effects on a local scale and may

Figure 13.4. Life expectancy at birth in the United States between 1920 and 1992. (National Center for Health Statistics, as reported in the *World Almanac* of 1994)

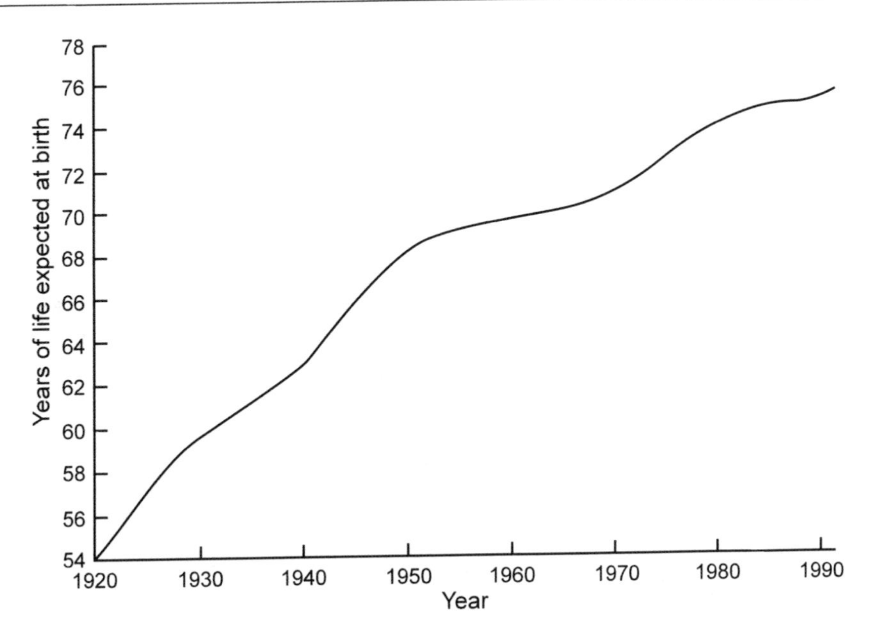

Figure 13.5. Child deaths and child births. Each line on the chart represents, for one developing country, the change in under-five mortality rate (USMR) and total fertility rate (TFR) over the period from 1960 to 1989. The intermediate point on each line represents data for the year 1980. On the right-hand side of the graph is shown the present under-five mortality rate of some of the most populous developing countries today. (Grant 1992); see also Brass and Jolly 1993, and Eubank and Gribble 1993)

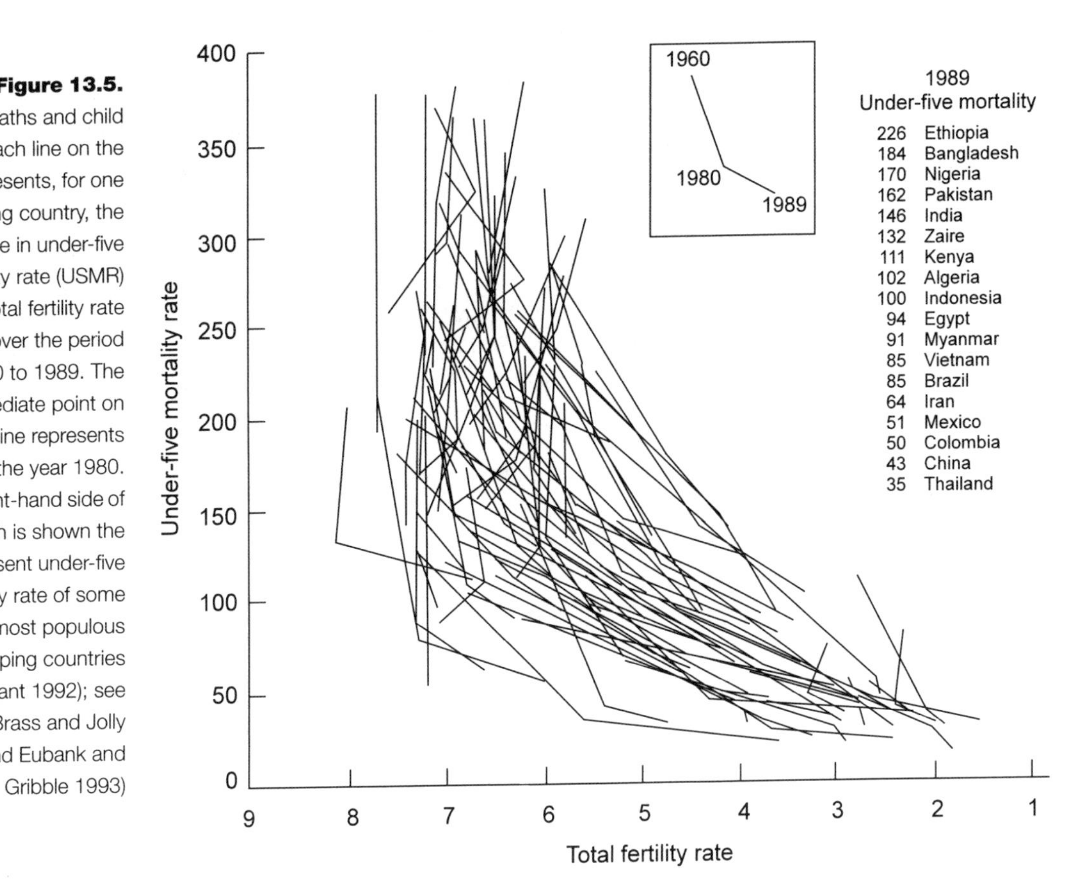

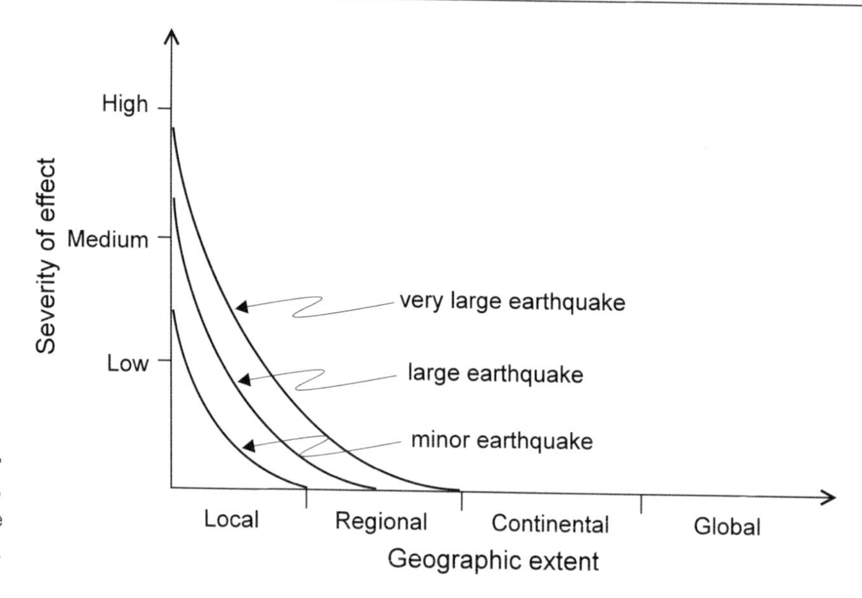

Figure 13.6. The effects of minor, large, and very large earthquakes.

produce moderate damage on a regional scale. Few earthquakes have detrimental effects on a continental or global scale, although individual events can be detected by seismographs the world over (see chapter 8). The severity of earthquake damage can be measured in terms of loss of life and/or property. Table 13.3 lists the six earthquakes that have exacted the highest toll in human lives. "Very severe" on this basis implies a loss in excess of 100,000 lives. The San Francisco earthquake of 1906 was responsible for only 503 deaths; the San Fernando Valley earthquake of 1971 caused 65 deaths, and the 1989 earthquake in the San Francisco Bay Area had a death toll of 62. When a really large earthquake strikes the Los Angeles or the San Francisco Bay Area, the death toll will probably be much larger than one thousand but probably not as large as that of the six largest earthquakes in table 13.3, because many California buildings are now designed to withstand large earthquakes.

In the future, the death toll from large earthquakes will increase due to population increases, but it is apt to be reduced by improvements in building design. Our ability to predict the timing of earthquakes is still quite poor, but there is a good chance that, with sufficient research into the matter, we will be able to do so more effectively during the next century. If so, the death toll exacted by large earthquakes could be reduced significantly.

Many of the comments about earthquakes also apply to volcanic eruptions. The maximum death toll from the largest of these is much smaller than that from the largest earthquakes (see table 13.3). Disasters due to volcanic eruptions plot in a fashion similar to that of earthquakes in area-intensity diagrams. Death tolls from volcanic eruptions will probably decrease somewhat in the twenty-first century as geophysical research leads to the development of better tools for forecasting major volcanic eruptions.

Floods and tidal waves have caused great havoc, particularly in heavily populated flood plains and in low-lying coastal regions (see fig. 13.7). The largest losses of life have been in China. These completely dwarf the loss of life, somewhat in excess of fifty, caused by the major flooding of the Mississippi River in

Table 13.3
The Greatest Disasters Due to the Four Most Destructive Natural Processes (1994 World Almanac)

	Year	*Place*	*Number of Deaths*
Floods and and Tidal Waves	1931	Huang He (Yangtze) River, China	3,700,000
	1887	Huang He (Yangtze) River, China	900,000
	1642	China	300,000
	1939	Northern China	200,000
	1911	Chang Jan River, China	100,000
	1228	Holland	100,000
Earthquakes	1556	Shaanxi, China	830,000
	1737	Calcutta, India	300,000
	526	Antioch, Syria	250,000
	1976	Tangshan, China	242,000
	1923	Yokohama, Japan	200,000
	1927	Nan-shan, China	200,000
Hurricanes, Typhoons, Blizzards, and Other Storms	1970	Cyclone, Bangladesh	300,000
	1991	Cyclone, Bangladesh	70,000
	1942	Hurricane, Bengal, India	40,000
	1965	Windstorm, Bangladesh	30,000
	1963	Windstorm, Bangladesh	22,000
	1965	Windstorm, Bangladesh	17,000
Volcanic Eruptions*	1863	Krakatau, Indonesia	35,000
	1902	Mt. Pelée, Martinique	30,000
	1985	Nevado del Ruiz, Colombia	22,940
	1669	Mt. Etna, Sicily	20,000
	79	Mt. Vesuvius	16,000
	1169	Mt. Etna, Sicily	15,000

* Note the different casualty figures in table 9.3.

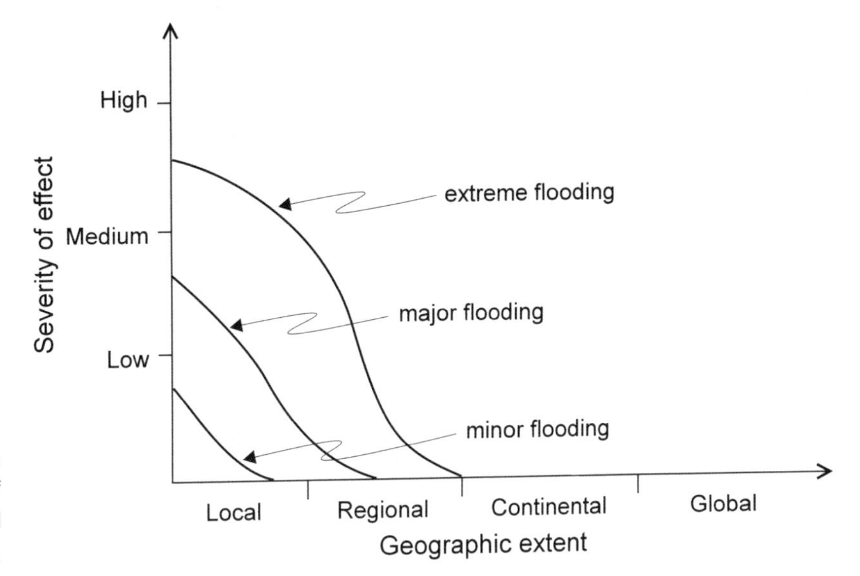

Figure 13.7. The effects of minor, major, and extreme flooding.

1993 (see chapter 4). More extensive flood control measures will probably decrease the death toll due to flooding in the next century. The toll from tidal waves may also decrease somewhat as better warning systems and improved construction techniques are developed, but hurricanes, typhoons, blizzards, and other storms will probably continue to take a heavy toll, particularly in low-lying, heavily populated countries such as Bangladesh.

Even very large natural disasters have played a very minor role in limiting world population growth, and there is reason to believe that their future role will be no greater. A good but not conclusive case can be made for the proposition that advances in geophysics, geohydrology, meteorology, and civil engineering will decrease the overall destructive effects of natural disasters during the next century.

The classification of natural disasters proposed above is also useful for thinking about the effects of anthropogenic disasters. Among industrial accidents, only the 1986 nuclear accident at Chernobyl (see chapter 11) can match the death toll of large natural disasters. Although the number of deaths immediately following this explosion was less than one hundred, the number of premature deaths due to cancer and other radiation effects may ultimately exceed 100,000. The damage produced by the accident was both local and regional. Although radionuclides were spread over much of the globe, the health effects at great distances from the accident are and probably will be very small.

Many human accidents are small but so frequent that their cumulative effect is not negligible. The effects of individual motor vehicle accidents are only local, but in aggregate they accounted for nearly 2% (41,700) of all 1992 deaths in the United States. In assessing the impacts of such anthropogenic accidents on the human race it is probably better to consider their aggregate effect rather than the magnitude of individual accidents. This distinction is important in analyzing the impacts of the several types of environmental pollution. Figure 13.8 shows in a schematic way the evolution of environmental problems related to water pollution. Historically, water pollution by domestic wastes and by related pathogens came first. This type of pollution has largely been brought under control in the industrialized countries (see chapter 12), but it is still the single most murderous type of pollution in the developing countries.

In the developed countries, domestic pollutants were first followed by industrial pollutants, then by nutrients, and—most recently—by microorganic substances (chapter 12). The presence of these pollutants was recognized reasonably early, and steps were and are being taken to decrease their concentrations and to mitigate their effects on public health. In the rapidly developing countries, the timescale on which these pollutants have affected water quality has been telescoped, and these countries are now faced with the formidable task of dealing simultaneously with all four pollutant types. In the slowly developing countries, only domestic and industrial pollutants are currently of importance, but these are the countries where pathogens are exacting a particularly large toll in human lives (see chapter 12).

It is likely, but not certain, that the pattern of increasing water pollution followed by remediation will continue to govern the evolution of water quality both in the developed and developing countries. The reduction of pollutants in the developing countries will be related to their economic growth and political stability. Child mortality in these countries since 1960 has decreased significantly,

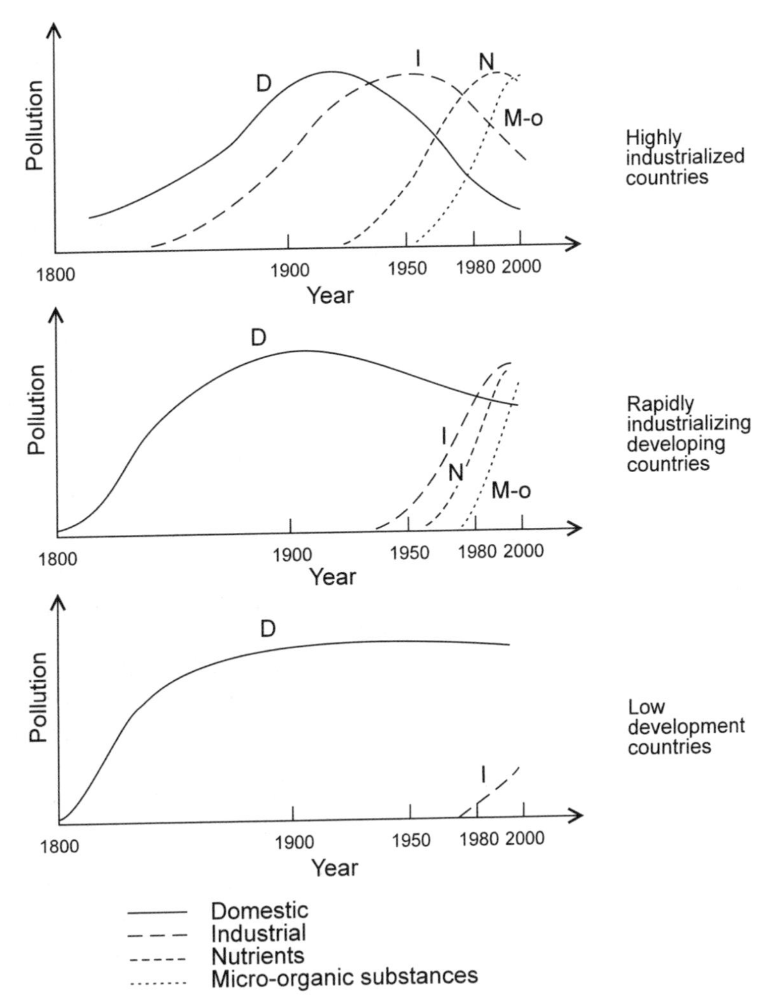

Figure 13.8. The successive development of water pollution problems in countries according to their level and speed of development. (Modified after Meybeck, Chapman, and Helmer 1989)

but it is still many times larger than in the developed world. Its further reduction is surely one of the major tasks for the twenty-first century.

Neither the prospect of global warming nor the reduction of stratospheric ozone fit neatly into the proposed classification scheme of disasters. The increase in CO_2 and in the concentration of the other greenhouse gases has produced no definitely identifiable effects on world climate. If such effects do emerge, they will probably grow gradually, and will be regional and global in scale (see chapter 11). Currently, no serious attempt is being made to forestall a possible global warming. CFC emissions now have regionally important and globally mild effects on ozone levels. The replacement of CFCs by HFCs promises to mitigate and then to solve the problem of ozone depletion during the next century.

Of all of the human environmental impacts, modern warfare is the only one whose death toll exceeds those of the major natural disasters and of water pollution by pathogens. Some 11 million people were killed during World War I between 1914 and 1918. The death toll of World War II between 1939 and 1945 was about 40 million (Cornwell 1969). Casualties during an all-out nuclear war would probably be in excess of 100 million and could exceed one billion. The two world wars, awful as they were, did not materially slow the growth of human populations. However, an all-out war with nuclear and/or biological weapons during the next century could spell the end of human civilization.

13.6 The Future of the Biosphere

Anton Chekhov (1860–1904), the renowned Russian playwright and author of *The Cherry Orchard*, would look out at his own orchard with a twinkle in his eye, and would often say:

> Look, I have planted each tree here and certainly they are dear to me. But this is of no consequence. Before I came here all this was waste land and ravines, all covered with stones and thistles. Then I came and turned this beautiful wilderness into a cultivated, beautiful place. Do you know? Do you know that in three or four hundred years all the earth will become a flourishing garden. And then life will be exceedingly easy and comfortable. [Alexander Kurpin, *To Chekhov's Memory*]

But is that really what we want? Do we really wish to see all wilderness converted into gardens? Most of us would answer "No," because we are stewards of our planet, and because we feel responsible for the welfare of all of its inhabitants, not only for members of the human race. A major challenge for the next century will be to find the proper balance between the needs of humanity and the needs of millions of other species. It has often been said that you have to break some eggs to make an omelette. But what if the only eggs that are available belong to an endangered bird species? Perhaps we will then decide to do without the omelette. But what if the omelette is needed to keep a starving child alive? The omelette then transcends an ordinary meal, and becomes a parable for a major quandary of the twenty-first century. We may do well to remember that extraordinary man, Saint Francis of Assisi, who "called all creatures *brothers* and in a most excellent manner, unknown to others, saw into the hidden nature of things with his discerning heart, like one who had already escaped into the glorious freedom of the sons of God" (Thomas of Celano, as quoted by Tierney, Kagan, and Williams 1976).

13.7 A Summation

The list of dangers facing humanity during the next century is long and formidable; but at the start of which century has the list been shorter or less forbidding? To the authors of this book none of the dangers look insurmountable. But appearances can be deceiving. In 1403 Harry Percy, also known as Hotspur, led a revolt against Henry IV. He was cautioned against the enterprise, but in the famous lines given to him by Shakespeare he replied:

> But I tell you, my lord fool,
> Out of this nettle, danger,
> we pluck this flower, safety.
> (*Henry IV*, part 1, act 2, scene 3)

Unfortunately, Harry Percy was killed not long after in an engagement with forces raised by the sheriff of York in defense of the king.

The course of world history in the twenty-first century is uncertain, but it will surely depend on the sturdiness of human institutions and on the dedication of their members. There will be sufficient natural resources only if there are enough

Figure 13.9.
The Survivors, 1923, charcoal drawing by Käthe Kollwitz. (By permission of the Galerie St. Etienne)

well-trained Earth scientists and engineers to find, extract, and manufacture them. There will be enough food only if farmers are supported by well-trained biologists and agronomists. There will be a pleasant environment only if a whole cadre of environmental scientists is available to forestall future disasters and to remediate those that have befallen us. All these people will be available only if there are adequate educational and research facilities. These must be supported by political institutions that are sufficiently flexible to adapt to the changing needs of the next century, and sufficiently strong to prevent the scars of war (fig. 13.9).

Should we be optimists or pessimists, Cornucopians or Malthusians, as we look forward to the new century? Table 13.4 gives a fairly complete range of options. Only the first and the last of these seem unreasonable: the first, because there are too many charted and uncharted reefs and shoals; the last, because there has been so much improvement during the course of the twentieth century. Espousal of any of the other four options seems entirely defensible. One's choice depends in part on temperament, in part on the relative weight that one attaches to the various indicators of progress and regress. We, the authors, are both cautiously optimistic.

Table 13.4.
A Range of Possible Attitudes as We Look toward the Twenty-first Century

1. Exuberant optimism
2. Optimism
3. Cautious optimism
4. Mild-pessimism
5. Pessimism
6. Despondent pessimism

References

Brass, W., and Jolly, C. L., eds. 1993. *Population Dynamics of Kenya*. National Academy Press, Washington, D.C.

Brown, L. R. 1994. *State of the World 1994*. W. W. Norton, New York.

Cornwell, R. D. 1969. *World History in the Twentieth Century*. Longman Group, London.

Eubank, D. C., and Gribble, J. N., eds. 1993. *Effects of Health Programs on Child Mortality in Sub-Saharan Africa*. National Academy Press, Washington, D.C.

Gore, A. 1992. *Earth in Balance: Ecology and the Human Spirit*. Houghton Mifflin, Boston.

Grant, J. P. 1992. *The State of the World's Children, 1992*. United Nations Children's Fund (UNICEF). Oxford University Press, Oxford.

Hammond, A. L., ed. 1994. *World Resources, 1994–95*. World Resources Institute. Oxford University Press, New York.

Meadows, D. H., Meadows, D. L., and Randers, J. 1992. *Beyond the Limits: Confronting Global Collapse, Envisioning a Sustainable Future*. Chelsea Green, Post Mills, Vermont.

Meybeck, M., Chapman, D. V., and Helmer, R. 1990. *Global Freshwater Quality: A First Assessment*. World Health Organization and United Nations Environmental Programme. Blackwell Reference, Cambridge, Mass.

Postel, S. 1994. Carrying Capacity: Earth's Bottom Line. In *State of the World*, chap. 1. L. R. Brown, Proj. Dir. W. W. Norton, New York.

Ray, D. L., and Guzzo, L. 1993. *Environmental Overkill: Whatever Happened to Common Sense?* Regnery Gateway, Washington, D.C.

Rutten, V. W. 1993. Population Growth, Environmental Change, and Innovation: Implications for Sustainable Growth in Agriculture. In *Populations and Land Use in Developing Countries*, ed. C. L. Jolly and B. B. Torrey, chap. 9. National Academy Press, Washington, D.C.

Tierney, B., Kagan, D., and Williams, L. P. 1976. Thomas of Celano. In *Great Issues of Western Civilization*, 3d ed., pp. 421–22. Random House, New York.

Wildavsky, A. 1995. *But Is It True? A Citizen's Guide to Environmental Health and Safety Issues*. Harvard University Press, Cambridge, Mass.

Appendixes

Appendix A
Unit Prefixes

Prefix	*Symbol*	*Multiple*	*Prefix*	*Symbol*	*Multiple*
tera-	T	10^{12}	deci-	d	10^{-1}
giga-	G	10^{9}	centi-	c	10^{-2}
mega-	M	10^{6}	milli-	m	10^{-3}
kilo-	k	10^{3}	micro-	μ	10^{-6}
hecto-	h	10^{2}	nano-	n	10^{-9}
deka-	da	10	pico-	p	10^{-12}
			femto-	f	10^{-15}
			atto-	a	10^{-18}

Appendix B
Units and Conversion Factors

Unit	*Conversion*
Length	
Ångstrom	1 Å = 0.1 nm = 10^{-8} cm = 10^{-10} m
nanometer	1 nm = 10 Å = 10^{-7} cm = 10^{-9} m
micron = micrometer	1 μm = 10^{-6} m
millimeter	1 mm = 1,000 μm = 0.1 cm = 0.0394 in
centimeter	1 cm = 10,000 μm = 10 mm = 0.394 in
meter	1 m = 100 cm = 39.37 in = 3.2808 ft = 1.094 yd
kilometer	1 km = 3,280.8 ft = 1,000 m = 0.6214 mi
inch	1 in = 2.54 cm = 0.083 ft
foot	1 ft = 30.48 cm = 12 in = 0.3048 m
yard	1 yd = 91.44 cm = 36 in = 3 ft = 0.9144 m
statute mile	1 mi = 5,280 ft = 1,760 yd = 1,609 m = 1.609 km
nautical mile	1 n.m. = 1,852 m = 1.852 km = 1.151 mi
Area	
square centimeter	1 cm^2 = 100 mm^2 = 0.155 in^2
square meter	1 m^2 = 10,000 cm^2 = 10.7636 ft^2 = 1.1962 yd^2
hectare	1 ha = 10^8 cm^2 = 10,000 m^2 = 2.471 a. = 0.01 km^2 = 0.00386 mi^2
square kilometer	1 km^2 = 10^6 m^2 = 100 ha = 0.386 mi^2
square inch	1 in^2 = 6.4516 cm^2
square foot	1 ft^2 = 929.03 cm^2 = 144 in^2 = 0.0929 m^2
square yard	1 yd^2 = 1,296 in^2 = 9 ft^2 = 0.836 m^2
acre	1 a. = 43,560 ft^2 = 4,840 yd^2 = 4,047 m^2 = 0.4047 ha
square mile	1 mi^2 = 640 a. = 2.589 km^2

Appendix B *cont.*

Unit	*Conversion*
Volume	
cubic centimeter	1 cm^3 = 0.0610 in^3
cubic meter	1 m^3 = 35.714 ft^3 = 1.3072 yd^3 = 1,000 l = 264.2 gal
cubic kilometer	1 km^3 = 10^{15} cm^3 = 10^9 m^3 = 0.24 mi^3
cubic inch	1 in^3 = 16.387 cm^3
cubic foot	1 ft^3 = 1,728 in^3 = 0.02832 m^3
cubic yard	1 yd^3 = 27 ft^3 = 0.765 m^3
trillion cubic feet	1 TCF = 28.32 × 10^9 m^3
liter	1 l = 10 dl = 1,000 cm^3 = 61 in^3 = 1.06 qt = 0.2642 gal
deciliter	1 dl = 0.1 l = 100 cm^3 = 6.1 in^3
pint	1 pt = 28.875 in^3 = 0.473 l
quart	1 qt = 57.75 in^3 = 2 pt = 0.473 l
U.S. gallon	1 gal = 231 in^3 = 8 pt = 4 qt = 3.785 l
U.S. barrel of oil	1 bbl = 42 gal = 159 l
acre-foot	1 a.-ft = 43,560 ft^3 = 1,233.6 m^3
Weight	
kilogram	1 kg = 1,000 gm = 2.2046 lb
metric ton	1 m.t = 10^6 gm = 2,204.6 lb = 1,000 kg = 1.1025 st = 0.984 lt
teragram	1 Tg = 10^{12} gm = 10^6 m.t
gigaton	1 Gt = 10^{15} gm = 10^9 m.t
ounce avoirdupois	1 oz = 28.350 gm = 0.9115 oz.t = 0.0625 lb
ounce troy	1 oz.t = 31.103 gm = 1.09714 oz = 0.08333 lb.t
pound troy	1 lb.t = 12 oz.t = 373.24 gm
pound	1 lb = 16 oz = 453.59 gm = 0.4536 kg
short ton	1 st = 2,000 lb = 0.907 m.t.
long ton	1 lt = 2,240 lb = 1.12 st = 1.016 m.t
Concentration	
part per trillion	1 ppt = 10^{-6} ppm, gpt, mg/kg
part per billion	1 ppb = 10^{-3} ppm, gpt, mg/kg
part per million	1 ppm = 1 gpt = 1 mg/kg = 8.345 lb/million gal
gram per metric ton	1 gpt = 1 ppm = 1 mg/kg = 8.345 lb/million gal
milligram per kilogram	1 mg/kg = 1 ppm = 1 gpt = 8.345 lb/million gal
weight percent	1 wt% = 10,000 ppm = 20 lb/st = 10 kg/m.t
pound per short ton	1 lb/st = 0.05 wt% = 500 ppm = 0.5 kg/m.t
kilogram per metric ton	1 kg/m.t = 0.1 wt% = 1,000 ppm = 2 lb/st
microgram per cubic meter	1 $\mu g/m^3$ = 0.001 mg/m^3 = 0.001 μg/l
microgram per liter	1 μg/l = 1 mg/m^3 = 0.001 mg/l = 0.1 μg/dl = 1,000 $\mu g/m^3$
microgram per deciliter	1 μg/dl = 0.01 mg/l = 10 μg/l = 10 mg/m^3 = 10,000 $\mu g/m^3$
milligram per cubic meter	1 mg/m^3 = 1 μg/l = 0.001 mg/l
milligram per liter	1 mg/l = 100 μg/dl = 1,000 μg/l = 1,000 mg/m^3
part per billion by volume	1 ppbv = 0.001 ppmv
part per million by volume	1 ppmv = 1,000 ppbv
volume percent	1 vol.% = 10,000 ppmv

Appendix B *cont.*

Unit	*Conversion*
Concentration, *cont.*	
nanomole per kilogram	1 nM/kg = 1 nmol/kg = 0.001 μM/kg
micromole per kilogram	1 μM/kg = 1 μmol/kg = 0.001 mM/kg = 1,000 nM/kg
millimole per kilogram	1 mM/kg = 1 mmol/kg = 1,000 μM/kg
Dobson unit	1 DU = 1 μm thickness of O_3 upon compression to 1 atm
Density	
gram per cubic centimeter	1 gm/cm^3 = 1 kg/l = 1 m.t/m^3 = 1,000 kg/m^3 = 62.43 lb/ft^3
kilogram per liter	1 kg/l = 1 gm/cm^3 = 1 m.t/m^3 = 1,000 kg/m^3 = 0.03613 lb/ft^3
metric ton per cubic meter	1 m.t/m^3 = 1 gm/cm^3 = 1 kg/l = 1,000 kg/m^3 = 62.43 lb/ft^3
pound per cubic inch	1 lb/in^3 = 27.68 gm/cm^3 = 27,678 kg/m^3
pound per cubic foot	1 lb/ft^3 = 16.02 kg/m^3 = 0.01602 gm/cm^3
Energy—Heat—Work	
electron-volt	1 eV = 1.602×10^{-19} J
erg	1 erg = 10^{-7} J
joule	1 J = 10^7 erg
calorie	1 cal = 4.186 J
kilogram-meter	1 kg-m = 2.344 cal
kilocalorie	1 kcal = 1,000 cal = 3.968 Btu
British thermal unit	1 Btu = 1,054.8 J = 252 cal = 107.5 kg-m = $2.928\text{x}10^{-4}$ kwh
kilowatt-hour	1 kwh = 3,415 Btu = 860.5 kcal
billion kilowatt hours	1 Gkwh = 0.00341 quad
quadrillion Btu	1 quad = 10^{15} Btu = $1.055\text{x}10^{18}$ J = $2.93\text{x}10^{11}$ kwh
Force	
dyne	1 dyn = 10^{-5} N
newton	1 N = 1 kg·m/s^2 = 10^5 dyn
Frequency	
hertz	1 Hz = 1 cps
cycle per second	1 cps = 1 Hz
Power	
watt	1 w = 1 J/s = 0.0569 Btu/min = 0.01434 kcal/min = 0.001341 hp
horsepower	1 hp = 745.7 w
kilowatt	1 kw = 1,000 w = 1.341 hp
megawatt	1 Mw = 10^3 kw = 10^6 w = 1,341 hp
Pressure	
nanobar	1 nbar = 10^{-9} bar
microbar	1 μbar = 10^{-6} bar

Appendix B *cont.*

Unit	*Conversion*
Pressure, *cont.*	
newton per square meter	1 $N/m^2 = 10^{-5}$ bar
centimeter of mercury	1 cm Hg = 0.01316 atm
pound per square inch	1 psi = 0.06804 atm = 0.07031 kg/cm^2
kilogram per square centimeter	1 kg/cm^2 = 0.968 atm
bar	1 bar = 10^5 N/m^2 = 0.9869 atm
atmosphere	1 atm = 1.0133x10^5 N/m^2 = 76.0 cm Hg = 29.92 in Hg
	1 atm = 14.70 psi = 1.033 kg/cm^2 = 1.01325 bar
Radioactivity	
1 Ci	radioactivity of 1 gm radium = 3.7×10^{10} disintegrations per second
Temperature	
Degree Celsius	
= degree Centigrade	°C = (°F – 32) × (5/9) = K – 273.16
Kelvin	K = °C + 273.16 = (°F – 32) × (5/9) + 273.16
Degree Fahrenheit	°F = °C × (9/5) + 32 = (K – 273.16) × (9/5) + 32
Absolute zero	0 K = -273.16 °C = -460 °F
Freezing point of water at 1 atm	273.16 K = 0 °C = 32 °F
Boiling point of water at 1 atm	373.16 K = 100 °C = 212 °F

Mineral Descriptions

Albite White to grayish triclinic mineral of the plagioclase feldspar group, $NaAlSi_3O_8$. Common in granites.

Amphibole Double-chain-hydrated ferromagnesian silicate mineral having the general formula $(Ca,Mg,Fe)_7Si_8O_{22}(OH)_2$ or $(Ca,Na,K)_{0-1}(Ca,Fe,Li,Mg,Mn)_2(Al,Fe,Mg,Mn,Cr,Ti)_5(Si,Al)_8O_{22}(OH,F,Cl)_2$. *See* hornblende. Common in igneous and metamorphic rocks.

Anatase Brown, green, or gray tetragonal mineral, TiO_2. Widespread as a minor constituent of igneous and metamorphic rocks. Also a common detrital mineral. *See* brookite and rutile.

Anhydrite Vitreous/greasy, white orthorhombic mineral, transparent when in perfect crystals, $CaSO_4$. Common in evaporite sediments and present in some hydrothermal ore deposits.

Anorthite White or gray triclinic mineral of the plagioclase feldspar group, $CaAl_2Si_2O_8$. Common in basic igneous rocks.

Apatite Hexagonal mineral of various colors, $Ca_5(PO_4)_3(OH,F,Cl)$, widely distributed in igneous and sedimentary rocks. By far the most abundant mineral containing phosphorus, for which it is mined from some sedimentary and igneous rocks.

Aquamarine Transparent, light blue to sea-green, gem-quality variety of beryl, $Be_3Al_2Si_6O_{18}$. Found in pegmatites.

Aragonite Colorless, white, yellowish, or gray orthorhombic mineral, $CaCO_3$. Commonly secreted by organisms to form shells. Has the same chemical composition as calcite, but a different atomic arrangement. *See* calcite.

Arsenopyrite Steel gray monoclinic mineral, FeAsS. Most abundant and widespread arsenic mineral, occurring mostly in hydrothermal ore deposits.

Asbestos Generic term for fibrous varieties of a number of minerals of widely differing chemical composition. The most important is chrysotile, a variety of serpentine, $Mg_6Si_4O_{10}(OH)_8$; the others, all members of the amphibole group, are anthophyllite, crocidolite (the carcinogenic "blue asbestos"), tremolite, actinolite, and amosite.

Augite Dark green to black monoclinic mineral of the pyroxene group, $(Ca,Mg,Fe)_2(Si,Al)_2O_6$. Common in many basic igneous rocks.

Barite Colorless or white orthorhombic mineral, $BaSO_4$, with a relatively high density (4.5). The most common mineral containing barium. Principal ore of barium, occurring frequently in hydrothermal ore deposits.

Beryl Vitreous, generally green, blue, or white hexagonal mineral, $Be_3Al_2Si_6O_{18}$. Commonly found in pegmatites. Common gem varieties: emerald (deep green), golden beryl (yellow), aquamarine (light blue to sea-green), and morganite (pink).

Biotite Generally shiny, black monoclinic mica with perfect platy cleavage, $K(Mg,Fe)_3(Si_3Al)O_{10}(OH,F)_2$. Commonly found in igneous and metamorphic rocks.

Bischofite Colorless, white monoclinic mineral, $MgCl_2.6H_2O$. Comparatively rare constituent of marine evaporite deposits.

Borax Vitreous, resinous or earthy, colorless or white monoclinic mineral, $Na_2B_4O_5(OH)_4.8H_2O$. Occurs mostly in evaporite deposits of saline lakes and playas.

Brookite Metallic, adamantine brown to black orthorhombic mineral, TiO_2. Common as an accessory mineral in metamorphic rocks (gneisses) and detrital sediments. *See* anatase and rutile.

Calcite Generally white or pale yellow, translucent hexagonal mineral, $CaCO_3$. Main constituent of limestones and marbles. Common in hydrothermal ore deposits. *See* aragonite.

Carnallite Greasy, generally colorless, white or reddish orthorhombic mineral, $KMgCl_3.6H_2O$. Occurs mainly in marine evaporites.

Carnotite Earthy or silky, canary yellow to greenish yellow, strongly radioactive monoclinic mineral, $K_2(UO_2)_2(VO_4)_2.3H_2O$. Ore mineral of uranium and vanadium.

Chalcopyrite Metallic, yellow tetragonal mineral, $CuFeS_2$. Most widely occurring copper mineral and most important copper ore mineral.

Chert *See* Glossary.

Chlorite Group name for generally greenish monoclinic and triclinic sheet silicates resembling mica, with the general formula $(Mg,Al,Fe)_{4-6}(Si,Al)_4O_{10}(OH,O)_8$. Common as alteration products of pyroxenes, amphiboles, biotite, and garnet. Abundant in some schists.

Chromite Metallic, black to brownish black cubic mineral, $(Fe,Mg)(Cr,Al)_2O_4$. Common accessory mineral in peridotite and related mafic and ultramafic rocks, as well as in some sands. Most important chromium ore mineral.

Chrysotile White, gray, or green monoclinic and orthorhombic member of the serpentine group with the formula $Mg_3Si_2O_5(OH)_4$. Economically most important variety of asbestos.

Clinoamphibole Group name for monoclinic amphiboles, $Ca_2(Mg,Fe)_5Si_8O_{22}(OH)_2$. Common in some igneous rocks.

Clinopyroxene Group name for monoclinic pyroxenes, $Ca(Mg,Fe)Si_2O_6$. Common in some igneous rocks.

Coffinite Black tetragonal mineral, $U[SiO_4,(OH)_4]$. Found in some uranium ore deposits.

Corundum Vitreous, generally colorless or gray hexagonal mineral, Al_2O_3. Common accessory mineral in igneous and metamorphic rocks. Gem varieties are ruby (deep red) and sapphire (blue). Commonly used as an abrasive.

Crocidolite Fibrous amphibole, commonly known as "blue asbestos," $Na_2Fe_5(Si_8O_{22})(OH)_2$. Carcinogenic.

Cryolite Vitreous to greasy, white or brownish red monoclinic mineral, Na_3AlF_6. Used in the manufacture of aluminum.

Diamond Extremely hard, highly refractive, usually colorless cubic form of carbon. Occurs in kimberlite dikes and "pipes" (volcanic necks), and in placers. Used as gemstone and in abrasives. Differs from graphite in atomic arrangement and physical properties. *See* graphite.

Diopside White to green monoclinic pyroxene, $CaMgSi_2O_6$. Common in dolomitic limestone metamorphosed or altered by hydrothermal fluids. *See* skarn.

Dolomite White, gray, or pink hexagonal mineral, $CaMg(CO_3)_2$. Chief component of some sedimentary carbonate rocks; also present in some hydrothermal ore deposits.

Enargite Metallic, grayish black orthorhombic mineral, Cu_3AsS_4. Found in hydrothermal ore deposits, where it can be an important copper ore mineral.

Fayalite Yellowish or greenish orthorhombic mineral of the olivine group, Fe_2SiO_4. Common in some basic igneous rocks.

Feldspar Group name for white to gray rock-forming minerals with the general formula $(K,Na,Ca)(Al,Si)_4O_8$. Orthoclase = $KAlSi_3O_8$; Albite = $NaAlSi_3O_8$; Anorthite = $CaAl_2Si_2O_8$; Plagioclase = $Ca_xNa_{(1-x)}Al_{(1+x)}Si_{(3-x)}O_8$. Common in most rock types.

Fluorite Vitreous, colorless, green, yellow, blue, pink, or purple cubic mineral, CaF_2. Common in hydrothermal ore deposits. Major ore mineral of fluorine.

Forsterite Translucent yellow or green orthorhombic mineral of the olivine group, Mg_2SiO_4. Common in mafic and ultramafic igneous rocks, and in metamorphosed dolomites and limestones.

Galena Metallic, gray cubic mineral, PbS. Occurs predominantly in hydrothermal ore deposits. Principal ore mineral of lead.

Garnet Group name for variously colored cubic minerals with the general formula $(Ca,Fe,Mg,Mn)_3(Al,Fe,Cr)_2(SiO_4)_3$. Main varieties are as follows:

Almandine	$Fe_3Al_2(SiO_4)_3$	metamorphic rocks	red-brown
Andradite	$Ca_3Fe_2(SiO_4)_3$	skarns	brown
Grossular	$Ca_3Al_2(SiO_4)_3$	skarns	green
Pyrope	$Mg_3Al_2(SiO_4)_3$	kimberlite and peridotite	deep red
Spessartine	$Mn_3Al_2(SiO_4)_3$	granitic rocks	rocks red-brown
Uvarovite	$Ca_3Cr_2(SiO_4)_3$	metamorphic rocks	green

Gibbsite White monoclinic mineral, $Al(OH)_3$. Common constituent of bauxite, the main aluminum ore.

Goethite Yellow or earthy brown orthorhombic mineral, α-FeO(OH). Weathering product of iron-bearing minerals and a precipitate from marine or from meteoric waters in bogs and springs. Major constituent of limonite.

Gold Metallic, yellow cubic mineral with a density of 19.3. Occurs mainly in hydrothermal deposits and in placers. Main source of the metal.

Graphite Metallic, black-gray hexagonal form of carbon. Formed mainly by metamorphism of carbonaceous material in sediments. Differs greatly from diamond in atomic arrangement and physical properties. *See* diamond.

Gypsum Colorless, white monoclinic mineral, $CaSO_4.2H_2O$. Most common sulfate mineral. Forms mainly in marine evaporites and in saline lakes.

Halite Colorless or white cubic mineral, NaCl. Formed mainly by the evaporation of seawater and brines in isolated basins. Common name: table salt.

Halloysite Pearly, waxy, or dull white porcelainlike monoclinic clay mineral, $Al_2Si_2O_5(OH)_4$ or $Al_2Si_2O_5(OH)_4.2H_2O$. Commonly formed by the alteration of silicate minerals.

Hematite Metallic or earthy, blackish red to brick red, soft hexagonal mineral, α-Fe_2O_3. Commonly forms from the weathering of iron-bearing silicates and widespread as a precipitate mineral in sediments. Most important iron ore mineral.

Hornblende Vitreous, green or bluish green to black monoclinic amphibole with the general formula $Ca_2Na(Mg,Fe)_4(Al,Fe,Ti)_3Si_8O_{22}(O,OH)_2$. Very common in igneous and metamorphic rocks.

Ilmenite Metallic to submetallic gray to black, opaque, hexagonal mineral, $FeTiO_3$. Common accessory mineral in gabbros, diorites, and anorthosites. Can form economically minable ore bodies in igneous rocks, and is obtained commercially from black beach sands.

Kaolinite Common, pearly to earthy, white to brownish triclinic clay mineral, $Al_2Si_2O_5(OH)_4$. Derived from weathering and from hydrothermal alteration of silicate rocks. Principal constituent of kaolin. Much used in making ceramics, refractories, and paper.

Kunzite Lilac-rose to pink gem variety of the monoclinic mineral spodumene, $LiAlSi_2O_6$. Found in pegmatites.

Magnetite Metallic, black, cubic, strongly magnetic, opaque mineral of the spinel group, $(Fe,Mg)Fe_2O_4$ or Fe_3O_4. One of the most abundant and widespread oxides, particularly as an accessory mineral in igneous, sedimentary, and metamorphic rocks, and in meteorites. Common iron ore mineral formed by magmatic segregation, in skarns and in some sediments (iron formations). Commonly called lodestone.

Mica Group name for generally monoclinic sheet silicates with the general formula $(K,Na,Ca)(Mg,Fe,Li,Al)_{2-3}(Al,Si)_4\,O_{10}(OH,F)_2$. Characterized by perfect cleavage into thin, tough, somewhat elastic laminae or plates. Colorless to silvery white, pale brown or yellow, green or black. *Examples*: muscovite, biotite and phlogopite. Common in igneous, sedimentary and metamorphic rocks.

Molybdenite Metallic gray hexagonal mineral, MoS_2. Its soft, flexible, shiny laminae resemble graphite. Most common molybdenum mineral. Found as an accessory mineral in some granites. Most important molybdenum ore mineral.

Muscovite Translucent, colorless to green monoclinic mica, $(K,Na)Al_2(Si_3Al)O_{10}(OH,F)_2$. Present in many granitic rocks and common in metamorphic rocks.

Nepheline White hexagonal mineral, $(Na,K)AlSiO_4$. Common in silica-deficient igneous rocks, such as nepheline syenite.

Olivine Group name for olive green, grayish green, or brownish orthorhombic minerals with the general formula $(Mg,Fe)_2SiO_4$. Common in basic igneous rocks and in metamorphosed dolomites and limestones. *See* fayalite and forsterite.

Orthoamphibole Group name for orthorhombic amphiboles, $(Mg,Fe)_7Si_8O_{22}(OH)_2$. Common in some igneous and metamorphic rocks.

Orthoclase Colorless, white, and pink monoclinic mineral, $KAlSi_3O_8$. Common in felsic igneous rocks, in gneisses and schists, and occasionally in sedimentary rocks.

Orthopyroxene Group name for orthorhombic pyroxenes, $(Mg,Fe)SiO_3$.

Pentlandite Metallic yellow, cubic mineral, $(Fe,Ni)_9S_8$. Nearly always intimately associated with pyrrhotite in mafic igneous rocks. Most important sulfide ore mineral of nickel.

Periclase Colorless to grayish white cubic mineral, MgO. Formed by the metamorphism of dolomitic limestone.

Peridot Olive green gem variety of olivine, $(Mg,Fe)_2SiO_4$.

Perovskite Adamantine or metallic, black to brown orthorhombic mineral, $CaTiO_3$. Accessory mineral in some igneous rocks and pegmatites.

Phlogopite Colorless, brown, green, or yellowish monoclinic mica, $KMg_3(Si_3Al)O_{10}(F,OH)_2$. Found in metamorphosed dolomites and in kimberlites.

Pitchblende Massive, colloform, brown to black variety of uraninite, UO_2. Occurs mainly in hydrothermal deposits. One of the principal ore minerals of uranium.

Plagioclase Group of vitreous, white to gray triclinic feldspars of general formula $Ca_xNa_{(1-x)}$-$Al_{(1+x)}Si_{(3-x)}O_8$. Common in nearly all igneous rocks.

Pyrite = "fool's gold" Metallic yellow, cubic mineral, FeS_2. Most widespread and most abundant sulfide mineral, formed by diverse geological processes. Used to produce sulfur dioxide for sulfuric acid.

Pyrope *See* garnet.

Pyrophyllite Pearly white, gray, pale green or brown monoclinic and triclinic mineral, $Al_2Si_4O_{10}(OH)_2$. Formed by hydrothermal alteration of igneous rocks.

Pyroxene Group of light to dark green and black orthorhombic and monoclinic silicate minerals, closely related in crystal form and composition and having the general formula $(Mg,Fe)SiO_3$ or $(Ca,Mg,Fe)(Mg,Fe)Si_2O_6$. Silicon is sometimes replaced in part by aluminum. *See* augite. Common in igneous and metamorphic rocks.

Pyrrhotite Metallic, bronze monoclinic and hexagonal mineral, Fe_7S_8 to $Fe_{10}S_{11}$. Occurs in mafic igneous rocks and in many magmatic and hydrothermal deposits, particularly in skarns.

Quartz Vitreous, colorless, white, pink, yellow or violet, generally transparent hexagonal mineral, α-SiO_2. Very widespread mineral formed in igneous rocks, incorporated into clastic sedimentary rocks, preserved and formed in metamorphic rocks, and precipitated in many hydrothermal deposits.

Rutile Metallic, adamantine red, brown or yellowish, commonly prismatic, tetragonal mineral, TiO_2. Much more abundant than anatase or brookite. Common accessory mineral in igneous rocks, in gneisses and schists, in sands and in clastic sediments. *See* anatase and brookite.

Serpentine Group name for monoclinic sheet silicates with the general formula $(Mg,Al,Fe)_{2-3}(Si,Al,Fe)_2O_5(OH)_4$ or $Mg_3Si_2O_5(OH)_4$. Formed largely by the alteration of magnesium-bearing minerals and rocks. The variety chrysotile is one of the minerals designated as asbestos.

Smectite Group name for sheet silicates with the general formula $(Ca,Na,Li)_{0-1}$ $(Mg,Fe,Al,Li)_{2-3}(Si,Al)_4O_{10}.nH_2O$.

Sphalerite = blende or zinc blende Resinous, yellow or brown cubic mineral, α-(Zn,Fe,Cd,Mn)S. Most common zinc mineral. Occurs primarily in hydrothermal deposits. Most important zinc ore mineral.

Spinel Group name for hard vitreous white, orange, red, green, blue, or black cubic minerals with composition $(Mg,Fe,Zn,Mn,Ti)(Al,Fe,Cr,Ti)_2O_4$. *See* chromite, magnetite and ulvöspinel.

Spodumene Colorless, grayish white to pink monoclinic mineral, $LiAlSi_2O_6$. Found mainly in pegmatites. The transparent, gem-quality, lilac-pink variety is known as kunzite.

Sulfosalt Group term for sulfide minerals in which both a metal and a semimetal (e.g., As, Sb, Bi) are present. Sulfosalts may be considered double sulfides (i.e., pyrargyrite Ag_3SbS_3 may be considered to be $3Ag_2S.Sb_2S_3$). Examples are the following:

pearceite-polybasite	$(Ag,Cu)_{16}(As,Sb)_2S_{11}$
proustite-pyrargyrite	$Ag_3(As,Sb)S_3$
tennantite-tetrahedrite	$(Cu,Ag)_{10}(Fe,Zn)_2(As,Sb)_4S_{13}$

Sylvite Colorless, white, or gray cubic mineral, KCl. Formed mainly in marine evaporite deposits.

Talc Translucent, colorless, white triclinic mineral, $Mg_3Si_4O_{10}(OH)_2$. Forms scaly crystals. Usually an alteration product of nonaluminous magnesium silicates, such as pyroxenes, amphiboles, and olivine. Commonly found in metamorphic rocks.

Tetrahedrite Metallic, gray, antimony-rich member of the cubic tetrahedrite-tennantite series, $(Cu,Ag)_{10}(Fe,Zn)_2$ $(Sb,As)_4S_{13}$. Widespread as a hydrothermal ore mineral of silver.

Tourmaline Group name for hexagonal ring silicates with the general formula (Na,K,Ca) $(Mg,Fe,Mn,Li)_3(Al,Fe,Cr)_6Si_6O_{18}(BO_3)_3(OH,F)_4$. Generally brown or black, typically found in pegmatites. Some yellow, pink, and emerald-green varieties are used as gems.

Ulvöspinel Metallic, black cubic mineral of the spinel group, Fe_2TiO_4. Usually occurs as fine exsolution lamellae intergrown with magnetite. Commonly associated with gabbro and anorthosite. *See* spinel.

Uraninite Submetallic, pitchy, black cubic mineral, $\approx (U,Th)O_2$. Occurs in pegmatites, in some hydrothermal ore deposits, and in ancient (> 2.2 Ga) placers.

Vermiculite Group of grayish, brownish, platy or micaceous, hydrated monoclinic silicate minerals, closely related to chlorite and to smectite, of the general formula $(Mg,Fe,Al)_3(Al,Si)_4O_{10}(OH)_2.4H_2O$.

Wollastonite Translucent, white, colorless triclinic mineral, $CaSiO_3$. Commonly found in metamorphosed limestone and skarn.

Wurtzite Resinous, brownish black hexagonal mineral, β-ZnS. Found in some hydrothermal ore deposits. Rare compared to sphalerite, α-ZnS.

Hurlbut, Jr., C. S. 1968. *Minerals and Man*. Random House, New York.

Hurlbut, Jr., C. S. 1971. *Dana's Manual of Mineralogy*. 18th ed. John Wiley and Sons, New York.

Klein, C., and Hurlbut, Jr., C. S. 1993. *Manual of Mineralogy*. John Wiley and Sons, New York.

Kraus, E. H., Hunt, W. F., and Ramsdell, L. S. 1936. *Mineralogy*. McGraw-Hill, New York.

Nickel, E. H., and Nichols, M. C. 1991. *Mineral Reference Manual*. Van Nostrand Reinholt, New York.

Palache, C., Berman, H., and Frondel, C. 1944. *The System of Mineralogy*, vol. 1. John Wiley and Sons, New York.

Palache, C., Berman, H., and Frondel, C. 1951. *The System of Mineralogy*, vol. 2. John Wiley and Sons, New York.

Glossary

Acid/acidic igneous rock Igneous rocks that contain more than 60% SiO_2; generally lighter colored and poorer in iron, magnesium, and/or calcium than basic (or "mafic") igneous rocks; also referred to as felsic igneous rocks. *Examples*: granite, rhyolite, granodiorite, dacite.

Acid rain Rainwater is acid if its pH is less than 7, but the term "acid rain" is usually reserved for rainwater with a pH less than 5.5.

Albedo The ratio of the amount of electromagnetic energy reflected by a surface to the amount of energy incident upon it.

Alkalic igneous rock Igneous rock that contains significantly more sodium and/or potassium than is considered average for its silica content.

Alkaline-earth metals The chemical elements beryllium, magnesium, calcium, strontium, barium, and radium.

Alkalinity The total quantity of base per unit volume or weight of a solution, usually determined by titration with a strong acid.

Alluvial plain A level or gently sloping, slightly undulating land surface produced by extensive deposition of alluvium, usually adjacent to a river that periodically overflows its banks.

Alluvium Clay, silt, sand, gravel, or similar unconsolidated material deposited during comparatively recent geologic time by a stream or other body of running water as sorted or semisorted sediment in the bed of a stream or on its flood plain or delta, or as a cone or fan at the base of a mountain slope.

Alpha particle A positively charged particle emitted by some radioactive nuclides; it consists of two protons and two neutrons and is transformed into a helium atom by the acquisition of two electrons.

Andesite A dark-colored, fine-grained extrusive rock that often contains crystals of plagioclase and one or more mafic minerals (e.g., biotite, hornblende, pyroxene), with a groundmass generally composed of very small crystals of the same minerals plus some quartz; the extrusive equivalent of diorite.

Anode furnace A furnace designed to produce a metal by using an electric current.

Anorthosite Igneous rock composed almost entirely of plagioclase feldspar and a small percentage of dark-colored minerals.

Anoxic Devoid of oxygen.

Anthracite Hard coal with a high carbon content and little volatile matter; burns with a clean flame (see fig. 7.13).

Anthropogenic Created, produced, or made by humans.

Anticline A fold, generally convex upward, whose core contains stratigraphically older rocks.

Aquifer A body of rock or sediment that can yield water to a well or spring in significant quantities.

Arithmetic ratio or progression A sequence of terms each of which, after the first, is derived by adding successively a constant quantity; the sequence 6, 10, 14, 18 is an arithmetic progression. *See also* geometric progression.

Asthenosphere A somewhat plastic layer or shell of the Earth below the lithosphere, equivalent to the upper mantle.

Atomic number The number of protons in an atomic nucleus; atoms with the same atomic number belong to the same chemical element.

Basalt A dark, hard, compact, fine-grained volcanic rock composed chiefly of plagioclase, augite, and magnetite (see table 6.1b).

Basement 1. The undifferentiated complex of rocks that underlies the rocks of interest in an area. 2. The crust of the Earth below sedimentary deposits.

Basic igneous rock Igneous rocks that contain less than 60% SiO_2; they are generally darker in color and richer in iron, magnesium, and/or calcium (i.e., "mafic") than acid igneous rocks. *Examples*: gabbro, basalt.

Batholith Generally discordant, irregular, and coarse-grained mass of igneous rock that has more than 100 km^2 of surface exposure and no known floor.

Bauxite Off-white, grayish, brown, yellow, or reddish brown rock composed of a mixture of various amorphous or crystalline hydrous aluminum oxides and aluminum hydroxides (principally gibbsite), together with clay minerals, silt, iron hydroxides and silica; the main ore of aluminum.

Bed load Material moved along a stream bed by flowing water.

Beta particle A negatively charged particle emitted by some radioactive nuclides; physically identical with an electron.

B.I.F., banded iron formation A sedimentary deposit of alternating bands of chert (microcrystalline silica) and iron minerals, usually oxides; commonly used as an iron ore. *See also* taconite.

Biodiversity The diversity of living organisms, often referring to a particular geographic area.

Biofuel or biomass fuel A renewable fuel derived from living organisms or the by-product of living organisms. Biomass fuels include wood, dung, methane gas, and grain alcohol.

Biogenic sediment or rock Sediment or sedimentary rock produced by organic processes or composed of organic remains.

Bioremediation The use of microorganisms to convert toxic organic chemicals into innocuous compounds.

Biota All living organisms of an area; the flora and fauna considered as a unit.

Bitumen A natural flammable substance of variable color, hardness, and volatility, composed principally of a mixture of hydrocarbons.

Bituminous coal A soft coal containing a mixture of carbon compounds that burn with a smoky yellow flame; softer than anthracite but harder than lignite.

Blackbody A theoretically perfect absorber of all incident radiation. It also radiates energy at the maximum possible rate per unit area at each wavelength for any given temperature. No actual substance behaves in this way, although some materials, such as lampblack, approach it. *See also* blackbody radiation.

Blackbody radiation The characteristic thermal radiation emitted by a blackbody at a specific temperature. *See also* blackbody.

Blister copper A mass of almost pure copper produced during an intermediate stage of copper refining.

Bomb *See* volcanic bomb.

Breccia A coarse-grained clastic rock, composed of angular, broken rock fragments held together by a mineral cement or embedded in a fine-grained matrix.

Buffer A substance that minimizes changes in the properties of a chemical system due to the addition of another component; examples include pH and redox buffers.

Calcine To heat a substance to a high temperature but below its melting point, causing loss of moisture, CO_2, and other volatiles.

Calcite compensation depth = C.C.D. The depth in the oceans, about 4 km, below which calcite dissolves more rapidly than it settles out.

Calc-silicate rock A rock consisting mainly of calcium-bearing silicates, such as garnet, diopside, and wollastonite, formed by the metamorphism of impure limestone or dolomite. *See also* skarn.

Capillary action, capillarity The action by which a fluid, such as water, is drawn up (or depressed) in small interstices or tubes as a result of surface tension.

Carbon 14 (^{14}C) A radioisotope of carbon. Half-life = 5,730 years.

Carbonate ooze Fine-grained, unconsolidated sediment composed of carbonate particles, usually calcite and/or aragonite.

Carcinogenicity Capacity or tendency for a substance to produce cancer.

Chain reaction Chemical or nuclear reaction in which the reaction products activate reactant atoms or molecules, thus speeding the reaction and potentially causing an explosion (see fig. 11.29).

Chert Cryptocrystalline (e.g., extremely fine-grained) sedimentary rock consisting largely of quartz and opal, often found as nodules in limestone and as layered deposits; synonymous with flint.

Chlorofluorocarbons (CFCs) A group of compounds, such as freon, which contain carbon, chlorine, and fluorine. These compounds are broken down in the atmosphere by solar ultraviolet radiation, releasing chlorine atoms that can destroy ozone molecules.

Chlorophyll Generally a mixture of two waxy pigments ($C_{55}H_{72}O_5N_4Mg$ and $C_{55}H_{70}O_6N_4Mg$) that occur in plants and serve as a catalyst in photosynthesis. Other forms of chlorophyll occur in diatoms, algae, etc.

Chromitite A rock composed chiefly of the mineral chromite.

Cirque A deep, steep-walled, semicircular basin eroded out by a mountain glacier.

Cladding Metal coating bonded onto another metal.

Clastic Pertaining to a rock or sediment composed principally of broken fragments that are derived from preexisting rocks or minerals and that have been transported some distance from their places of origin; also describes the texture of such a rock.

Clinker A rough, jagged fragment of solidified lava.

Coccolith A general term applied to various microscopic calcareous structural elements or buttonlike plates having many different shapes and averaging about 3 microns in diameter, constructed of calcite or aragonite crystals, and constituting the outer skeletal remains of a coccolithophore.

Coke A combustible material produced by heating bituminous coal and driving off its volatile matter; coke derives its strength from fusing together mineral matter and fixed carbon.

Concretion A rounded mass of mineral matter formed by precipitation from a water solution around a nucleus in sedimentary or pyroclastic rock.

Conglomerate A coarse-grained, clastic sedimentary rock composed of rounded to subangular fragments larger than 2 mm in diameter, often set in a fine-grained matrix of sand or silt, and commonly cemented by calcium carbonate, iron oxide, silica, or hardened clay; the consolidated equivalent of gravel.

Constant dollars Cost or value expressed in dollars corrected for inflation using a specific reference year. *See* current dollars.

Coordination, coordination number The number of nearest-neighbor atoms or ions that surround a given atom or ion in a crystal structure.

Core The central part of the Earth below a depth of 2,900 km, consisting of a fluid *outer core* and a solid *inner core*; both parts are thought to consist largely of iron and nickel.

Corona A faintly colored, luminous ring around a celestial body visible through a haze or thin cloud, especially such a ring around the moon or sun, caused by diffraction of light from suspended matter in the intervening medium.

Cottrell precipitator Electrostatic precipitator of dust, smoke, and fumes from air and gases, particularly in smelters; used to reduce air pollution.

Craton A stable part of the Earth's continental crust that has been little deformed for a prolonged period of geologic time, typically since the Precambrian.

Critical point The temperature and pressure at which the liquid and vapor phases of a pure substance have the same density. *See* supercritical.

Crust The outermost layer or shell of the Earth; the continental crust is about 40 km thick and consists largely of granitic, granodioritic, and metamorphic rocks; the oceanic crust is about 10 km thick and consists largely of basaltic rocks.

Cubic, isometric Crystal structure and symmetry that can be expressed in terms of three equal and mutually perpendicular axes.

Cumulate An igneous rock formed by the accumulation of crystals that settled out from a magma by the action of gravity.

Cupola An upward extension of an igneous intrusion into its roof.

Current dollars Cost or value expressed in dollars at the time quoted. *See* constant dollars.

Cutoff grade The lowest grade of mineralized material that qualifies as ore in a given deposit.

Dacite A fine-grained extrusive rock of the same general composition as andesite, but with a less calcic plagioclase and more quartz; the extrusive equivalent of granodiorite.

Daughter element An element that is the product of a nuclear reaction.

Delta A low, nearly flat alluvial tract of land at or near the mouth of a river, commonly forming a triangular or fan-shaped plain of considerable extent, crossed by many distributaries of the main river, sometimes extending beyond the main trend of the coast, due to the accumulation of sediment supplied by the river in such quantities that it is not removed by tides, waves, and currents.

Demographic transition The transition from a state in which the birthrate greatly exceeds the death rate to one where the two are essentially equal and where the population has stabilized.

Detritus, detrital material Loose fragments, particles, or grains formed by the physical disintegration of rocks.

Deuterated water Water in which much of the hydrogen (^1H) has been replaced by deuterium ($^2H = D$). Also known as *heavy water*.

Diabase A fine-grained intrusive rock that is chemically equivalent to gabbro and basalt; its main components are calcic plagioclase and pyroxene. *See* gabbro.

Diagenesis The chemical, physical, and biologic changes undergone by a sediment during its conversion to sedimentary rock.

Diatom A microscopic, single-celled plant that grows in both marine and fresh water. Diatoms secrete siliceous shells in a great variety of forms, and these may accumulate in sediments to form diatomite.

Dielectric A nonconductor of electricity, especially a substance with electrical conductivity less than a millionth of a mho (mho is a unit of conductance, reciprocal of the ohm, backward spelling of ohm).

Dike A long, often planar mass of igneous rock that cuts across the structure of adjacent rocks.

Dip The acute angle that a stratum or any planar geological feature makes with a horizontal plane.

Discharge (Q) The volume of a fluid that passes through an area or past a station within a given period of time; usually expressed in cubic feet or cubic meters per second.

Dolomite A sedimentary rock containing more than 50% by weight of the mineral dolomite.

Dolomitization The process by which calcium carbonate in limestone is partly or wholly replaced by the mineral dolomite, forming a dolomitic limestone or a dolomite rock.

Dump leaching The process by which tailings accumulated in a pile ("dump") are leached in order to recover soluble elements easily.

Ecosystem A biological community and its environment.

Electrolytic process, electrolysis Chemical change, especially decomposition, produced in an electrolyte by an electric current.

Electrolyte A substance that dissociates into ions in solution or when fused, thereby becoming electrically conducting.

Electron A stable elementary particle of electrical charge −1; its mass is 1/1,850 that of a proton.

Eutrophication The process whereby a body of water becomes highly productive of aquatic plants, such as algae, due to the input of large quantities of nutrients.

Evaporite A sedimentary rock composed primarily of minerals produced by the evaporation of saline solutions as a result of extensive or total evaporation of the solvent; examples of minerals include gypsum, anhydrite, halite (rock salt), and various nitrates and borates.

Extrusive Igneous rock that was erupted onto the surface of the Earth, e.g., lava flows and pyroclastic material such as volcanic ash.

Facies A distinctive group of characteristics of a rock unit, usually reflecting the conditions or environment of its origin; these characteristics differentiate the unit from adjacent or associated units, which may or may not have formed at the same time.

Fast breeder reactor A nuclear reactor that can generate new fissionable material (such as ^{239}Pu) from nonfissionable isotopes (such as ^{238}U) at a faster rate than it consumes fissile material (^{235}U).

Feedback system A system in which part of the output is fed back into the input; feedback can be either positive or negative, and can either destabilize or stabilize a system.

Felsic Adjective applied to igneous rocks containing an abundance of light-colored minerals, mainly feldspars and quartz.

Felsite A general term for any light-colored, fine-grained igneous rock composed chiefly of quartz and feldspar.

Ferromagnesian Containing iron and magnesium; applied to mafic minerals, especially amphiboles, pyroxenes, biotite, and olivine.

Fission A nuclear reaction in which an atomic nucleus splits into fragments (fission products, usually of comparable mass) and neutrons, spontaneously or by absorption of a neutron.

Fissionable material Material that can undergo nuclear fission.

Floodplain, flood plain, floodland A level plain bordering a stream, covered by water during floods.

Flow sheet A diagram illustrating the flow of materials during an industrial process, such as mineral separation and concentration.

Fluid inclusion A cavity in a mineral containing liquid, sometimes also gas and solids.

Flux 1. A flow of matter or energy. 2. Flux material: A mineral or rock added to a furnace charge to promote melting.

Foraminifera Unicellular protozoans characteristically having a calcareous shell with perforations.

Foreshock A small tremor that precedes a larger earthquake or main shock by an interval from seconds to weeks and that originates at or near the focus of the larger earthquake.

Fractional crystallization Crystallization of a liquid during which early-formed crystals do not react with later liquids, with which they are not in equilibrium.

Fractional melting Melting of a mineral aggregate, such as a rock, in which some minerals melt before others, resulting in a melt whose composition changes during progressive melting.

Gabbro A coarse-grained, dark, compact plutonic igneous rock composed mainly of calcic plagioclase and pyroxene (also called "norite"), chemically equivalent to basalt.

Gamma ray High energy photon, such as those released during many nuclear reactions.

Gangue Economically worthless minerals or rock associated with ore.

Geodetic survey A precise survey of a large area in which account is taken of the figure and size of the Earth and corrections are made for Earth curvature.

Geometric ratio, progression A sequence of terms in which the ratio of each term to the preceding one is the same throughout the sequence: 1, 2, 4, 8, 16 are a geometric progression. *See* arithmetic progression.

Geothermal Pertaining to heat in the interior of the Earth.

Gneiss A layered metamorphic rock that shows compositional banding and parallel alignment of minerals.

Granite A coarse-grained igneous or metamorphic rock containing abundant quartz, orthoclase, sodium-rich plagioclase, and micas.

Granodiorite A coarse-grained plutonic igneous rock resembling granite, in which plagioclase is more abundant than orthoclase; biotite, hornblende, and, more rarely, pyroxene are minor components.

Granophyre A porphyritic, fine-grained granitic rock.

Graywacke A dark gray, poorly sorted sandstone containing abundant feldspar and rock fragments often embedded in a clayey matrix.

Greenhouse effect Warming of the lower atmosphere due to the absorbtion of infrared energy emitted by the Earth; water vapor and carbon dioxide are the dominant greenhouse gases.

Greenhouse gas A constituent of the Earth's atmosphere that contributes to the greenhouse effect, e.g., H_2O and CO_2, but also CH_4, N_2O, and CFCs.

Green Revolution The large increase in agricultural productivity due to the development of new varieties of food plants and improved agricultural practices.

Guyot A flat-topped submarine mountain or seamount.

Half-life The time required for half of a given number of atoms of a radioactive nuclide to decay.

Halocarbon A compound of carbon and a halogen (i.e., fluorine, chlorine, bromine, or iodine).

Heavy water Water composed of oxygen and deuterium, the isotope of hydrogen with atomic weight 2; D_2O.

Herbicide Any substance used to destroy plants, especially weeds, or to slow down their growth.

Hexagonal Crystal structure and symmetry that can be expressed in terms of three equal axes intersecting at 60° in one plane, and one axis of different length that is at right angles to the others. *See* cubic/isometric, monoclinic, orthorhombic, tetragonal, and trigonal/triclinic.

Holocene The time period or epoch from about 8,000 or 10,000 years ago to the present.

Hot spot A volcanic center 100–200 km across and persistent for at least a few tens of millions of years, that is thought to be the surface expression of a long-lived rising plume of hot mantle material.

Hydrargyrism Mercury poisoning.

Hydrocarbon Any of numerous organic compounds, such as benzene and methane, that contain only carbon and hydrogen.

Hydrograph A graph showing the stage (height), discharge, velocity, or other characteristics of a body of water over time.

Hydrosphere The waters of the Earth, as distinguished from the rocks (lithosphere), living organisms (biosphere), and the air (atmosphere); includes the waters of the ocean, rivers, lakes, and other bodies of surface water in liquid form on the continents, snow, ice, and glaciers, and liquid water, ice, and water vapor in both the unsaturated and saturated zones below the land surface; included by some, but excluded by others, is water in the atmosphere, which includes water vapor, clouds, and all forms of precipitation while still in the atmosphere.

Hydrothermal deposit A mineral deposit formed by precipitation of ore and gangue minerals in fractures, faults, breccia openings, or other spaces, by replacement or open-space filling, from watery fluids ranging in temperature from ca. 50° to ca. 700°C and ranging in pressure from ca. 10 to ca. 3,000 bars. *See* hydrothermal fluid. Alteration of host rocks is common, but not universal.

Hydrothermal fluid Hot subsurface fluid, consisting predominantly of water, usually containing dissolved volatiles and salts, but generally excluding petroleum and natural gas.

Igneous rock A rock formed by the cooling of magma (molten rock).

Intrusive igneous rock Igneous rock resulting from the subsurface cooling of magma.

Irradiance The radiant energy per unit time per unit area incident upon a surface.

Island arc Chain of volcanic islands formed at a convergent plate boundary parallel to a subduction zone on the overriding plate.

Isometric *See* cubic.

Kaolin A fine, white to yellowish or grayish clay, consisting mostly of kaolinite, used in ceramics and refractories and as a filler or coating for paper and textiles.

Kerogen A mixture of insoluble organic compounds found in many fine-grained sedimentary rocks; a major constituent of oil shales.

Kimberlite Volcanic rock that contains many peridotite fragments from the Earth's mantle, abundant olivine, and garnet; often a source of diamonds.

Komatiite Basaltic rocks distinguished by the presence of ultramafic (highly magnesian) lavas.

K/T boundary The transition from the Cretaceous (K) to the Tertiary (T) period of Earth history, during which a large number of living species became extinct (see fig. 3.9).

Lacustrine Pertaining to, produced by, or formed in a lake or lakes.

Laterite Red soil, the product of intense chemical weathering, rich in oxides of iron and/or aluminum, nearly devoid of bases and primary silicates; usually developed in wet tropical climates.

Layered intrusion An intrusive igneous body in which there are layers of different mineralogical composition.

Levee 1. A long, broad, low ridge or embankment of sand and coarse silt, built by a stream along its banks in times of flood, when water overflowing the normal banks deposits the coarsest part of its load. 2. An artificial embankment built along the bank of a water course or an arm of the sea to protect land from inundation or to confine a stream to its channel.

Lignite A low-grade, brownish black coal, also called "brown coal." The softest of the coals (see fig. 7.13).

Limestone A sedimentary rock consisting mainly of calcium carbonate, usually as the mineral calcite. Limestones are formed by both organic or inorganic processes.

Limiting nutrient An essential biological nutrient which is in short supply and therefore limits the size or health of an ecologic community (e.g., phosphate in lakes and oceans).

Liquefaction Loss of coherence in soils, conversion into quicksand, often due to ground motions associated with earthquakes.

Liquid immiscibility or unmixing Separation of a single liquid into two liquids; an important magmatic process.

Lithosphere 1. The solid part of the Earth, as distinguished from the hydrosphere and atmosphere. 2. The rocky crust of the Earth.

Lognormal distribution A frequency distribution of a parameter x, where the distribution of $log\ x$ is normal.

Luminosity The quantity of radiant energy emitted by an object, e.g., the Sun.

Mafic Adjective applied to igneous rocks containing an abundance of dark-colored ferromagnesian minerals, mainly amphiboles, pyroxenes, biotite, and olivine.

Magma Naturally occurring melt generated in the Earth from which igneous rocks form by cooling.

Magma chamber A reservoir of magma, usually at a depth of a few to tens of kilometers, from which volcanic materials are derived.

Magnetic reversal A change in the Earth's magnetic field during which the magnetic North pole becomes the magnetic South pole, and vice versa.

Manganese nodule A small, typically 3–5 cm, irregular, black to brown, friable, concentrically banded concretionary mass consisting primarily of alternating manganese and iron oxide minerals (see fig. 7.30).

Mantle The zone of the Earth between the base of the crust at a depth of about 40 km and the core at a depth of 2,900 km (see fig. 9.3).

Marl A mixture of 35%–65% clay and 65%–35% calcium carbonate, formed under marine or freshwater conditions.

Matte A mixture of a metal with its oxides and sulfides, produced by smelting certain sulfide ores.

Meander A sinuous curve, bend, loop, turn, or winding in the course of a stream that develops as a stream erodes the outer bank of a curve and deposits sediment against the inner bank.

Metallogenic province An area characterized by a particular assemblage of mineral deposits.

Metamorphic rock A rock whose original mineralogy, texture, and/or composition have been changed during metamorphism, largely due to the effects of temperature and pressure.

Metamorphism Process during which rocks are subjected to changes in temperature and pressure; these can lead to alterations in mineralogy, texture, and composition.

Meteorite A stony or metallic object from interplanetary space that penetrates the atmosphere and impacts the Earth's surface.

Mixing ratio The fraction of a particular component in a mixture.

Mixing time The time required to homogenize an initially inhomogeneous medium.

Moderator, moderating material A substance (such as water, heavy water, or graphite) used in a nuclear reactor to slow down neutrons.

Monoclinic Crystal structure and symmetry that is expressed in terms of three unequal axes, two of which intersect obliquely and are perpendicular to the third.

MOR = Midocean ridge See fig. 7.1.

Mutagenicity The capacity to change the genetic material of living organisms and cause biological mutation.

MVT = Mississippi Valley Type deposit A strata-bound deposit of lead and/or zinc minerals in carbonate rocks, frequently with associated fluorite and barite.

Natural reactor A mass of high-grade, in situ uranium ore that sustains significant nuclear fission over a considerable time without human intervention. The natural reactors at Oklo (Gabon) operated about 2 billion years ago.

Nepheline syenite A plutonic rock composed essentially of alkali feldspar and nepheline. It may also contain an amphibole or a pyroxene.

Neutrino An electrically neutral particle with a mass that is either zero or very small compared to the mass of an electron.

Neutron An electrically neutral particle which can be considered to consist of a proton and an electron, and having a mass slightly greater than that of a proton.

Norite A coarse-grained plutonic rock containing calcic plagioclase as the chief constituent and orthopyroxene as the dominant dark mineral. *See* gabbro.

Normal magnetic polarity A natural remanent magnetization in a mineral or rock that is closely parallel to the present geomagnetic field direction.

Nuclear fusion The combination of two light nuclei to form a heavier nucleus; a process that is frequently accompanied by the release of a large amount of energy.

Nuclear winter Winter conditions caused by increased opacity of the atmosphere following the detonation of a large number of nuclear bombs.

Nuclide A species of atom, characterized by its number of neutrons and protons.

Nuée ardente (ash flow) A "glowing cloud" of hot volcanic gases and pyroclastic material that travels rapidly downslope as a density current in the atmosphere.

Obduction The overriding or overthrusting of oceanic crust onto the leading edge of a continental lithospheric plate.

Oil shale A finely laminated, brown or black sedimentary rock rich in organic matter that yields liquid or gaseous hydrocarbons on heating or distillation.

Onchocerciasis = River blindness A disease caused by a worm (*Onchocerca volvulus*), transmitted by the black fly (*Simulium damnosum*), which breeds in fast-flowing tropical rivers; fibrous nodular tumors grow around adult worms settled in connective tissues and in deep organs; when they invade the eyes, they can cause partial or total blindness.

Ophiolite complex A complex igneous unit in an oceanic plate, consisting chiefly of basalt and gabbro underlain by stratified accumulations of olivine crystals (see fig. 9.2).

Ore A natural deposit of a mineral or aggregate of minerals from which a constituent can be profitably mined or extracted.

Orebody, ore body 1. An accumulation of ore. 2. A body of ore, either massive or disseminated within a rock.

Orthoquartzite A clastic sedimentary rock composed almost exclusively of quartz sand cemented by silica.

Orthorhombic Crystal structure and symmetry that is expressed in terms of three mutually perpendicular axes of different length.

Overburden Barren or very low grade material overlying a mineral deposit, either loose or consolidated, which must be removed prior to mining.

Oxidant A substance capable of oxidizing another substance.

Oxidation The process of the combining of a substance with oxygen.

Ozone hole Very low ozone content of the atmosphere above Antarctica during spring.

Pangea = Pangaea A supercontinent that existed from about 300 to about 200 million years ago and included most of the continental crust of the Earth. The present continents were derived by the fragmentation of Pangea.

Partial pressure The portion of the total pressure of a gas attributable to the presence of a particular gas component.

Pathogen A microorganism or virus that can cause disease.

Pay, pay streak, or zone A structure or stratum that contains very profitable ore.

Peat Dark brown to black, partially carbonized plant material; found in bogs or marshes and used as fertilizer and fuel.

Pegmatite An exceptionally coarse-grained igneous rock, usually found as irregular dikes, lenses, or veins, frequently at the margins of batholiths. Most grains are 1 cm or more in diameter. Their mineralogy is generally, though not always, that of granite and may include rare minerals of lithium, boron, fluorine, niobium, tantalum, uranium, and rare earths.

Pelagic sediment Deep-sea sediment composed of fine-grained material that has settled from surface waters; common constituents are clays, radiolaria, and foraminifera.

Pelite A sediment or sedimentary rock composed of clay or mud-size particles.

Peneplain A nearly flat land surface produced by long-term subaerial erosion.

Peridotite A coarse-grained igneous rock composed largely of olivine, with small amounts of pyroxene and amphibole.

Permeability A measure of the ability of a substance to transmit fluids.

PGE = platinum group element One or more of the following elements: ruthenium, rhodium, palladium, rhenium, osmium, iridium, and platinum.

pH A measure of the acidity or alkalinity of a solution, numerically equal to 7 for neutral solutions at 25°C and 1 atm pressure, increasing with increasing alkalinity and decreasing with increasing acidity.

Phosphorite A sedimentary rock with a high enough content of phosphate minerals to be of economic interest.

Photochemistry Chemical reactions produced by the effect of light or other radiant energy.

Photodissociation The dissociation of atoms and molecules by radiant energy.

Photosynthesis The process by which chlorophyll-containing cells in green plants convert incident light into chemical energy and synthesize organic compounds.

Phytoplankton Floating aquatic plants.

"Pig iron" Iron as it comes from a blast furnace or smelter; usually cast into oblong masses called pigs.

Pillow lava Pillow-shaped masses of lava formed in a subaqueous environment, usually basaltic or andesitic in composition.

Placer Mineral deposit formed by mechanical concentration of dense mineral particles, such as gold, cassiterite, or rutile along the course of rivers or on beaches.

Plagiogranite Igneous rock consisting predominantly of plagioclase, smaller amounts of quartz, and some biotite and amphibole.

Plankton Floating and drifting aquatic organisms.

Plate tectonics The observation that the lithosphere is divided into roughly a dozen plates that move laterally over the plastic asthenosphere; earthquakes, volcanism, and structural deformation occur along the plate boundaries. *See* seafloor spreading.

Pluton A body of igneous rock formed beneath the surface of the Earth by consolidation of magma.

Plutonic Of deep igneous or magmatic origin.

Pneumoconiosis = black lung Any of various diseased conditions of the lungs characterized by fibrous hardening as a result of chronic inhalation of irritating dust particles.

Porosity The percentage of the bulk volume of a rock or soil that is occupied by pore space, whether isolated or connected.

Porphyritic Texture of igneous rocks in which larger crystals (phenocrysts) are set in a finer-grained groundmass.

Porphyry An igneous rock with a porphyritic texture.

Porphyry copper or molybdenum deposit A large body of rock, typically a felsic porphyry, that contains disseminated pyrite, chalcopyrite, and/or molybdenite, and that is minable in bulk on a large scale (generally in open pits; see fig. 9.9).

Potash 1. Potassium carbonate, K_2CO_3. 2. A term loosely used for potassium oxide, potassium hydroxide, or even potassium in such informal expressions as potash feldspar or potash spar.

PV = photovoltaic = photoelectric Of or having to do with the electric effects produced by light, especially as in the emission of electrons by certain substances when subjected to light or radiation of suitable wavelength.

Pyroclastic 1. Rock formed by volcanic explosion. 2. Rock texture of explosive origin.

Quartzite Moderately to intensely metamorphosed sandstone, consisting chiefly of quartz grains solidly cemented with silica.

Radiance Radiant energy emitted per unit time in a specified direction by a unit area of an emitting surface.

Radiolaria One-celled marine animals with siliceous skeletons living in a marine pelagic environment.

Radionuclide A radioactive nuclide.

Railroad ballast Coarse gravel or crushed rock used as a bed for railroads.

Rank [coal] Degree of metamorphism undergone by coal. It is the basis for classifying coal into a natural series from lignite to anthracite.

Rare earth, rare earth element Any of the fifteen elements from lanthanum (atomic number 57) to lutetium (71).

Rare gas, noble gas, inert gas Any of the unreactive gases helium, neon, argon, krypton, xenon, or radon.

Reactor core The part of a nuclear reactor in which nuclear fission reactions take place.

Real terms Informal expression for constant dollars.

Reduce 1. To decrease the valence of a cation by adding electrons. 2. The addition of hydrogen to. 3. To change to a metallic state by removing nonmetallic constituents (generally by smelting).

Reductant, reducing agent A substance that reduces another substance, or brings about reduction, and is itself oxidized in the process.

Reef 1. A mound or ridge-shaped structure built by calcareous organisms, especially corals, and consisting mostly of their remains. 2. A metalliferous mineral deposit, e.g., gold-bearing quartz or sandstone.

Refractory 1. Any of various materials (such as alumina, silica, and magnesia) that do not deform significantly or change chemically at high temperatures. 2. *Plural*. Bricks of such materials and of various shapes used to line furnaces. 3. Said of an ore from which it is difficult or expensive to recover its valuable constituents.

Reserve base The sum total of the demonstrated (i.e., measured and indicated) reserves, marginal reserves, and part of the subeconomic resources of a given element or mineral.

Reserves That portion of the demonstrated resources of an element or mineral that can be recovered economically (see fig. 8.41).

Reservoir rock A permeable, porous geologic formation that yields water, oil, and/or natural gas.

Reverberatory furnace A furnace for roasting ore in which a flame enters a hearth holding the charge of ore plus fluxes; the flame strikes the roof, is deflected (reverberated) downward, and heats the charge.

Reversed magnetic polarity A natural remanent magnetization of a mineral or rock with a direction opposite to the present geomagnetic field direction.

Rhyolite Light-colored, glassy, or finely crystalline extrusive igneous rock, often containing some larger crystals and commonly exhibiting flow texture; the larger crystals are generally quartz or alkali feldspar; the extrusive equivalent of granite.

Richter scale A logarithmic scale of earthquake magnitude, based on the amplitudes of seismic waves generated by earthquakes.

Rock burst Sudden and often violent breaking of a mass of rock from the walls of a tunnel, mine, or deep quarry, caused by failure of highly stressed rock and the rapid or instantaneous release of accumulated strain energy. It may result in closure of a mine opening, or projection of broken rock into it, accompanied by ground tremors, rockfalls, and air concussions.

Runaway greenhouse A process by which the greenhouse effect is amplified due to positive feedbacks leading to extremely high atmospheric temperatures.

Runoff Rainwater that leaves an area as surface drainage.

Sandstone A medium-grained clastic sedimentary rock composed of abundant rounded or angular fragments of sand size, more or less firmly united by a fine-grained cementing material (commonly silica, iron oxide, or calcium carbonate); the consolidated equivalent of sand. The sand particles usually consist of quartz.

Saprolite Soft, earthy, typically clay-rich, thoroughly decomposed rock, formed in place by chemical weathering of igneous, sedimentary, and metamorphic rocks. Characterized by preservation of structures that were present in the unweathered rock.

Schist Strongly foliated metamorphic rock, usually containing mica-rich layers.

Seafloor spreading Creation of new seafloor by the separation of adjacent plates, usually at midocean ridges, and the filling of the openings by lava and igneous intrusives.

Seam A thin bed, layer, or stratum, as of coal or rock.

Seamount Isolated mountain on the seafloor. Its height may exceed 1 km. Flat-topped seamounts are called *guyots*.

Sediment Detrital material or chemical precipitate that settles to the bottom of a river, lake, or ocean, or settles from air onto the Earth's surface.

Sedimentary rock Rock formed by the accumulation and cementation of mineral grains transported by wind, water, or ice to the site of deposition or formed by chemical precipitation at the depositional site.

Sedimentation rate The amount of sediment accumulated during a given period of time, usually expressed as thickness or mass of accumulation per unit time.

Semibituminous coal Coal of rank between bituminous coal and anthracite; harder and more brittle than bituminous coal.

Shale A fine-grained detrital sedimentary rock, formed by the consolidation of clay, silt, or mud; has a finely laminated structure and breaks readily into thin layers approximately parallel to the bedding.

Shear zone A tabular zone of rock that has been crushed and brecciated along many closely spaced parallel fractures.

Shield A large region of stable, ancient basement rocks, usually part of a continent.

Shield volcano A volcano in the shape of a broad dome with gentle slopes, usually built by very fluid basaltic lava flows.

Shock metamorphism The changes produced in rocks and minerals by the passage of high-pressure shock waves, usually created by the impact of large meteorites.

Silicosis A disease of the lungs caused by long-term inhalation of dust containing quartz or silicates.

Sill A relatively thin, tabular sheet of igneous rock intruded between sedimentary beds.

Silt 1. A rock fragment or detrital particle smaller than a fine sand grain and larger than clay, having a diameter in the range 4–62 microns. 2. A loose aggregate of silt size particles.

Siltstone An indurated silt.

Skarn Rock consisting largely of calcium and magnesium silicates (e.g., garnet, diopside, wollastonite) formed in limestone or dolomite by the action of hydrothermal fluids.

Slag Vitreous (glassy) mass left as a residue from the smelting of metallic ore.

Sluice An artificial channel or passage for water, having a gate or valve to regulate the flow.

Sluice gate Gate to control the flow of water in a sluice. *See also* sluice.

Source rock Sedimentary rock in which organic material is transformed to liquid and/or gaseous hydrocarbons; any rock that is the source of metals in ore deposits.

Spectrum The distribution of energy emitted by a radiant source; usually arranged in order of wavelengths.

Spontaneous fission Radioactive decay in which a nuclide decomposes spontaneously into two nuclides and usually several neutrons.

Strain Change in the shape or volume of a body as a result of stress. *See also* stress.

Stratigraphic section Sequence of rock units found in a given region.

Stratum, *pl.* strata A bed or layer of rock that can be readily distinguished from beds above and below.

Stress The force per unit area acting on a part of a solid body. *See also* strain.

Strike The compass direction of a horizontal line in the plane of an inclined structural feature such as a rock bed, dike, sill, fault, or vein. *See also* dip.

Stripping ratio The ratio of the amount of overburden that must be removed to the amount of the ore that can be mined.

Subbituminous coal Coal of rank and with properties intermediate between those of lignite and bituminous coal (see fig. 7.13).

Subduction The descent of one lithospheric plate beneath another at a convergent plate boundary.

Subduction zone A long, narrow belt along which subduction takes place.

Supercritical Said of a system that is at a temperature and pressure higher than its critical temperature and critical pressure. *See also* critical point.

Syenite Plutonic rock containing alkali feldspar (usually orthoclase), a small amount of plagioclase, one or more mafic minerals (commonly hornblende), and little or no quartz.

Symmetry The ordered internal arrangement of atoms in a crystalline solid that renders its appearance, structure, and directional properties invariant to certain operations, such as rotation, reflection, inversion, and their combinations. *See also* cubic, isometric, hexagonal, monoclinic, orthorhombic, tetragonal, and trigonal.

Syncline A fold in which the core contains the stratigraphically younger rocks.

Taconite A low-grade banded iron formation ore suitable for concentration of magnetite and hematite by fine grinding and magnetic separation, from which pellets containing 62%–65% iron can be produced. *See also* B.I.F.

Tails, tailings Those portions of washed or milled ore that are too poor to be treated further, as distinguished from the concentrate, or material of value.

Tarn A small mountain lake within a cirque.

Tar sand A sandstone that originally contained petroleum and from which the lighter hydrocarbon fractions have escaped, leaving a residual asphalt filling the interstices.

Tectonic Pertaining to, causing, or resulting from structural deformation in the Earth's crust.

Tectonite A rock whose fabric reflects a history of deformation, especially solid flow.

Tetragonal Crystal structure and symmetry that is expressed in terms of three mutually perpendicular axes, two of which are of equal length.

Tidal wave, tsunami A great sea wave produced by a large, short-duration disturbance of the ocean floor, principally a shallow submarine earthquake, but also submarine earth movements, subsidence, or volcanic eruptions.

Tomography Any of several techniques for determining the internal structure of a solid object; applied in the expression "mantle tomography," a technique for using seismic waves to map the three-dimensional structure of the Earth's interior.

Topsoil The fertile, dark-colored upper portion of a soil.

Transform fault A strike-slip fault connecting the ends of an offset in a midocean ridge.

Trap 1. Any barrier to the movement of oil and gas, allowing either or both to accumulate in the subsurface. 2. A dark-colored, fine-grained extrusive rock, such as basalt or diabase.

Triclinic Crystal structure and symmetry that is expressed in terms of three unequal axes intersecting at angles differing from 90°.

Tritium A radioactive isotope of hydrogen having two neutrons and one proton in the nucleus: ^3H.

Troposphere The lowest part of the atmosphere, in which temperature generally decreases rapidly with altitude, clouds form, and convection is active. In middle latitudes the troposphere generally includes the first 10–12 km above the Earth's surface (see fig. 3.1).

Turbidite A sediment deposited from a turbidity current. Graded bedding and moderate sorting are characteristic features.

Turbidity current A current in water, air, or other fluid, caused by different amounts of matter in suspension, such as a dry-snow avalanche or a descending cloud of volcanic dust; specifically a current laden with suspended sediment, moving swiftly (under the influence of gravity) down a subaqueous slope and spreading horizontally on the floor of the body of water, having been set and/or maintained in motion by locally churned or stirred-up sediment that gives the current a density greater than that of the surrounding or overlying water.

Turbine An engine or motor driven by the pressure of steam, water, or air against the curved vanes of a wheel or set of wheels.

Tuyère Tube, nozzle, or pipe through which air is forced into a blast furnace or forge to facilitate combustion.

Ultramafic rock Composed mostly of olivine, with small amounts of pyroxenes and amphiboles. *See also* peridotite.

Unconformity A surface that separates two strata that were not deposited continuously, i.e., where a significant interval of time elapsed between the deposition of the two strata.

Unconformity type uranium deposit A hydrothermal uranium deposit formed at an unconformity, typically where oxidized and reduced rocks are in contact (see fig. 9.21).

Vein A fracture in the Earth's crust containing ore minerals. *See also* ore body.

Viscosity A measure of resistance to flow in a liquid.

Volcanic ash Particles under 2 mm in size ejected during a volcanic eruption.

Volcanic bomb A rounded mass of viscous magma, normally larger than fist-size, ejected during a volcanic eruption and cooled in flight; of variable shape; may be vesicular or hollow.

Xenolith A rock fragment foreign to the igneous mass in which it occurs.

Zooplankton The animal population in plankton.

Bates, R. L., and Jackson, J. A., eds. 1980. *Glossary of Geology*. 2d ed. American Geological Institute, Falls Church, Virginia.

Guralnick, D. B., and Friend, J. H., et al., eds. 1966. *Webster's New World Dictionary of the American Language*. World Publishing, Cleveland and New York.

Morris, W., ed. 1973. *The American Heritage Dictionary of the English Language*. American Heritage Publishing and Houghton Mifflin, New York.

Press, F., and Siever, R. 1986. *Earth*. 4th ed. W. H. Freeman, New York.

Skinner, B. J., and Porter, S. C. 1992. *The Dynamic Earth*. 2d ed. John Wiley and Sons, New York.

Credits for Color Plates

Plate 1 Kelly, R. W., ed. 1988. *The Home Planet.* Addison-Wesley Publishing Co., Inc., Reading, Mass.
Plate 2 National Oceanic and Atmospheric Administration
Plate 3 Courtesy of Climate Prediction Center/NCEP and Office of Hydrology/NOAA, U.S. Dept. Commerce
Plate 4 Lands imagery courtesy of Earth Observation Satellite Company, Lanham, Maryland
Plate 5 National Oceanic and Atmospheric Administration
Plate 6 Heinrich D. Holland
Plate 7 Photo courtesy of Hans J. Holland
Plate 8 Kaeppler, A. L. 1980. *Kapa, Hawaiian Bark Cloth.* Boom Books, Hilo Bay, H.I.
Plate 9 Wendell E. Wilson
Plate 10 Courtesy of Carl Francis, Harvard University
Plate 11 Heinrich D. Holland
Plate 12 Heinrich D. Holland
Plate 13 Diagram by Heinrich D. Holland
Plate 14 Ulrich Petersen
Plate 15 Heinrich D. Holland
Plate 16 Photo courtesy of Alice P. Holland
Plate 17 Heinrich D. Holland
Plate 18 Woods Hole Oceanographic Institution, Woods Hole, Mass.
Plate 19 Courtesy of Ocean Drilling Program
Plate 20 Woods Hole Oceanographic Institution, Woods Hole, Mass.
Plate 21 Heinrich D. Holland
Plate 22 Courtesy of Hamersley Iron Pty., Ltd., Perth, Australia.
Plate 23 Courtesy of the U.S. Geological Survey
Plate 24 Reproduced courtesy of *Public Roads*, Federal Highway Administration
Plate 25 Heinrich D. Holland
Plate 26 Copyright © 1994, The Geological Society of America, Inc. (GSA); photo courtesy of Paul Hoffman
Plate 27 Courtesy of Shell Offshore, Inc.
Plate 28 Woods Hole Oceanographic Institution, Woods Hole, Mass.
Plate 29 Courtesy of W. J. Su, Harvard University
Plate 30 Wenkam, R. 1987. *The Edge of Fire: Volcano and Earthquake Country in Western North America and Hawaii.* Sierra Club Books, San Francisco, Calif.
Plate 31 © David Muench, 1995
Plate 32 © 1995 Gary Braasch
Plate 33 © Jeff Gnass Photography, Inc.
Plate 34 Photo courtesy of Patrick Le Fort
Plate 35 Heinrich D. Holland
Plate 36A Courtesy of Carl Francis, Harvard University
Plate 36B Wendell E. Wilson
Plate 37 Heinrich D. Holland
Plate 38 Photo courtesy of Christoph Hohenemser
Plate 39 Ulrich Petersen
Plate 40 Ulrich Petersen

Plate 41	Copyright 1992, The American Geophysical Union
Plate 42	Courtesy of Yankee Atomic Electric Company, Boston, Mass.
Plate 43	© Sigma Photo News
Plate 44	Courtesy of the U.S. Geological Survey
Plate 45	William Campbell/*Time* Magazine
Plate 46	William Campbell/*Time* Magazine
Plate 47	Photo Researchers
Plates 48 and 49	Reprinted with permission from D. J. Stanley and A. G. Warne, Nile Delta: Recent Geological Evolution and Human Impact, *Science* 260:628–634. Copyright 1993, American Association for the Advancement of Science. Photo courtesy of Daniel Jean Stanley
Plate 50	Photo by Alex Schriener; courtesy of UNOCAL Corporation
Plate 51	Photograph by George Turner, courtesy of the Sacramento Municipal Utility District
Plate 52	Courtesy of Jon Brenneis
Plate 53	National Oceanic and Atmospheric Administration
Plates 54 and 55	Reprinted with permission from J. B. Kerr and C. T. McElroy, Evidence for Large Upward Trends of Ultraviolet-B Radiation Linked to Ozone Depletion, *Science* 262:1032–1034. Copyright 1993 American Association for the Advancement of Science. Photos courtesy of Environment Canada
Plate 56	Kellogg, "Updated to 1995," Robert Kellogg, Natural Resources Conservation Service, USDA, personal communication
Plate 57	Heinrich D. Holland
Plate 58	Photograph by Stanley Osolinski
Plate 59	Itoh, S. 1988. *The White Egret*. Chronicle Books, San Francisco
Plates 60, 61A, 61B, and 62	Reprinted with permission from D. Skole and C. Tucker, Tropical Deforestation and Habitat Fragmentation in the Amazon: Satellite Data from 1978 to 1988, *Science* 260: 1905–1910. Copyright 1993, American Association for the Advancement of Science
Plate 63	Heinrich D. Holland

Author Index

Subject Index